CAMBRIDGE

Brighter Thinking

A Level Mathematics for AQA

Student Book 1 (AS/Year 1)

Stephen Ward, Paul Fannon, Vesna Kadelburg and Ben Woolley

CAMBRIDGE
UNIVERSITY PRESS

Shaftesbury Road, Cambridge CB2 8EA, United Kingdom

One Liberty Plaza, 20th Floor, New York, NY 10006, USA

477 Williamstown Road, Port Melbourne, VIC 3207, Australia

314–321, 3rd Floor, Plot 3, Splendor Forum, Jasola District Centre, New Delhi – 110025, India

103 Penang Road, #05–06/07, Visioncrest Commercial, Singapore 238467

Cambridge University Press is part of the University of Cambridge.

It furthers the University's mission by disseminating knowledge in the pursuit of
education, learning and research at the highest international levels of excellence.

www.cambridge.org
Information on this title: www.cambridge.org
www.cambridge.org/978-1-316-64422-5 (Paperback)
www.cambridge.org/978-1-316-64468-3 (Paperback with Cambridge Digital Access edition)

First published 2017

20 19 18 17 16 15 14 13 12

Printed in Great Britain by Ashford Colour Press Ltd.

A catalogue record for this publication is available from the British Library

ISBN 978-1-316-64422-5 Paperback
ISBN 978-1-316-64468-3 Paperback with Cambridge Digital Access edition

Additional resources for this publication at www.cambridge.org/education

Message from AQA

This textbook has been approved by AQA for use with our qualification. This means that we have checked that it broadly covers the specification and we are satisfied with the overall quality. Full details of our approval process can be found on our website.

We approve textbooks because we know how important it is for teachers and students to have the right resources to support their teaching and learning. However, the publisher is ultimately responsible for the editorial control and quality of this book.

Please note that when teaching the A/AS Level Mathematics (7356, 7357) course, you must refer to AQA's specification as your definitive source of information. While this book has been written to match the specification, it cannot provide complete coverage of every aspect of the course.

A wide range of other useful resources can be found on the relevant subject pages of our website: www.aqa.org.uk

IMPORTANT NOTE
AQA has not approved any Cambridge Elevate content

Contents

Introduction

You have probably been told that mathematics is very useful, yet it can often seem like a lot of techniques that just have to be learnt to answer examination questions. You are now getting to the point where you will start to see where some of these techniques can be applied in solving real problems. However as well as seeing how maths can be useful we hope that anyone working through this book will realise that it can also be incredibly frustrating, surprising and ultimately beautiful.

The book is woven around three key themes from the new curriculum.

Proof

Maths is valued because it trains you to think logically and communicate precisely. At a high level maths is far less concerned about answers and more about the clear communication of ideas. It is not about being neat – although that might help! It is about creating a coherent argument which other people can easily follow but find difficult to refute. Have you ever tried looking at your own work? If you cannot follow it yourself it is unlikely anybody else will be able to understand it. In maths we communicate using a variety of means – feel free to use combinations of diagrams, words and algebra to aid your argument. And once you have attempted a proof, try presenting it to your peers. Look critically (but positively) at some other people's attempts. It is only through having your own attempts evaluated and trying to find flaws in other proofs that you will develop sophisticated mathematical thinking. This is why we have included lots of common errors in our Work it out boxes – just in case your friends don't make any mistakes!

Problem solving

Maths is valued because it trains you to look at situations in unusual, creative ways, to persevere and to evaluate solutions along the way. We have been heavily influenced by a great mathematician and maths educator George Polya who believed that students were not just born with problem-solving skills – they were developed by seeing problems being solved and reflecting on their solutions before trying similar problems. You may not realise it but good mathematicians spend most of their time being stuck. You need to spend some time on problems you can't do, trying out different possibilities. If after a while you have not cracked it then look at the solution and try a similar problem. Don't be disheartened if you cannot get it immediately – in fact, the longer you spend puzzling over a problem the more you will learn from the solution. You may never need to integrate a rational function in the future, but we firmly believe that the problem-solving skills you will develop by trying it can be applied to many other situations.

Modelling

Maths is valued because it helps us solve real-world problems. However, maths describes ideal situations and the real world is messy! Modelling is about deciding on the important features needed to describe the essence of a situation and turning that into a mathematical form, then using it to make predictions, compare to reality and possibly improve the model. In many situations the technical maths is actually the easy part – especially with modern technology. Deciding which features of reality to include or ignore and anticipating the consequences of these decisions is the hard part. Yet it is amazing how some fairly drastic assumptions – such as pretending a car is a single point or that people's votes are independent – can result in models which are surprisingly accurate.

More than anything else this book is about making links – links between the different chapters, the topics covered and the themes mentioned here, links to other subjects and links to the real world. We hope that you will grow to see maths as one great complex but beautiful web of interlinking ideas.

Maths is about so much more than examinations, but we hope that if you take on board these ideas (and do plenty of practice!) you will find maths examinations a much more approachable and possibly even enjoyable experience. However always remember that the results of what you write down in a few hours by yourself in silence under exam conditions are not the only measure you should consider when judging your mathematical ability – it is only one variable in a much more complicated mathematical model!

How to use this book

Throughout this book you will notice particular features that are designed to aid your learning. This section provides a brief overview of these features.

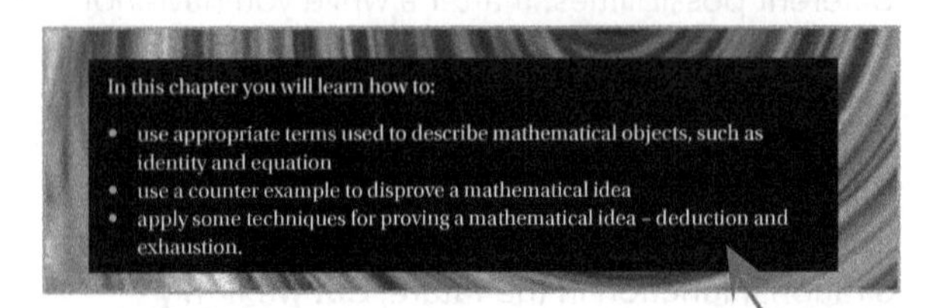

In this chapter you will learn how to:
use appropriate terms used to describe mathematical objects, such as identity and equation
use a counter example to disprove a mathematical idea
apply some techniques for proving a mathematical idea – deduction and exhaustion.

Learning objectives
A short summary of the content that you will learn in each chapter.

Before you start...		
GCSE	You should know the definition of the square root function.	1 Write down the value of $\sqrt{9}$.
GCSE	You should be able to manipulate algebraic expressions.	2 Factorise $4x^2 - 1$.
GCSE	You should know basic angle facts.	3 a What is the sum of the angles in a triangle? b What is the sum of the exterior angles of any polygon?

Before you start
A list of things that you should already know from your previous learning and questions to check that you're ready to start the chapter.

WORKED EXAMPLE

The left-hand side shows you how to set out your working. The right-hand side explains the more difficult steps and helps you understand why a particular method was chosen.

Key point

A summary of the most important methods, facts and formulae.

PROOF

Step-by-step walkthroughs of standard proofs and methods of proof.

Common error

Specific mistakes that are often made. These typically appear next to the point in the Worked example where the error could occur.

WORK IT OUT

Can you identify the correct solution and find the mistakes in the two incorrect solutions?

Tip

Useful guidance, including ways of calculating or checking and using technology.

Colour coding of exercises

The questions in the exercises are designed to provide careful progression, ranging from basic fluency to practice questions. They are uniquely colour-coded, as shown.

1 Disprove the statement $\sqrt{x^2+9} \equiv x+3$.

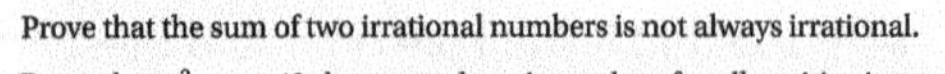

2 Prove that the sum of two irrational numbers is not always irrational.

3 Prove that $n^2 + n + 41$ does not take prime values for all positive integers.

4 Are two lines that never meet necessarily parallel?

5 Prove that the number of factors of a number is not always even.

Black – drill questions. These come in several parts, each with subparts **i** and **ii**. You only need attempt subpart **i** at first; subpart **ii** is essentially the same question, which you can use for further practice if you got part **i** wrong, for homework, or when you revisit the exercise during revision.

Yellow – designed to encourage reflection and discussion

Green – practice questions at a basic level

Blue – practice questions at an intermediate level

Red – practice questions at an advanced level

 indicates content that is for A level students only

Each chapter ends with a **Checklist of learning and understanding** and a **Mixed practice exercise**, which includes **past paper questions** marked with the icon

In between chapters, you will find extra sections that bring together topics in a more synoptic way.

Focus on...

Unique sections relating to the preceding chapters that develop your skills in proof, problem-solving and modelling.

Cross-topic review exercise

Questions covering topics from across the preceding chapters, testing your ability to apply what you have learned.

You will find **practice papers** towards the end of the book, as well as a glossary of key terms (picked out in colour within the chapters), and **answers** to all questions. Full **worked solutions** can be found on the Cambridge Elevate digital platform, along with other essential resources such as a **digital version** of this Student Book.

Maths is all about making links, which is why throughout this book you will find signposts emphasising connections between different topics, applications and suggestions for further research.

Rewind

Reminders of where to find useful information from earlier in your study.

Focus on...

Links to problem-solving, modelling or proof exercises that relate to the topic currently being studied.

Fast forward

Links to topics that you may cover in greater detail later in your study.

Did you know?

Interesting or historical information and links with other subjects to improve your awareness about how mathematics contributes to society.

Some of the links point to the material available only through the **Cambridge Elevate** digital platform.

Elevate

A support sheet for each chapter contains further worked examples and exercises on the most common question types. Extension sheets provide further challenge for the most ambitious.

Gateway to A Level

GCSE transition material which provides a summary of facts and methods you need to know before you start a new topic, with worked examples and practice questions.

Working with the large data set

As part of your course you will work with a large data set. At the time of this Student Book's publication (2017), this data set is on the purchasing of different types of food in different parts of the country in different years – this context may change in future years, but the techniques for working with large data sets will stay the same. This is an opportunity to explore statistics in real life. As well as using the ideas from Chapters 20 and 22 you will use these data to look at four key themes.

Practical difficulties with data

The real world is messy. Often there are difficulties with being overwhelmed by too much data, or there are errors, missing items or ambiguous labels. For example, how do you deal with the fact that milk drinks and milk substitutes are combined together in some years if you want to compare regions over time? When grouping data for a histogram, how big a difference does where you choose to put the class boundaries make?

Using technology

Modern statistics is heavily based on using technology. You will use spreadsheets and graphing packages, looking at the common tools which help simplify calculations and present data effectively. One important technique you can employ with modern technology is simulation. You will look more closely at hypothesis testing by using the data set to simulate the effect of sampling on making inferences about the population.

Thinking critically about statistics

Why might you want to use a pie chart rather than a histogram? Whenever you calculate statistics or represent data sets graphically, some information is lost and some is highlighted. In modern statistics it is important to ask critical questions about how evidence provided by statistics is used to support arguments. One important part of this is the idea of validating statistics. For example, what does it mean when it says that there has been a 100% decrease in the amount of sterilised milk purchased? You will look at ways to interrogate the data to try to understand it more.

Statistical problem solving

Technology can often do calculations for you. The art of modern statistics is deciding what calculations to do on what data. You rarely have exactly the data you want, so you have to make indirect inferences from the data you do have. For example, you probably will not see a newspaper headline saying 'the correlation coefficient between amount of bread purchased and amount of confectionary purchased is –0.52', but you might see one saying 'Filling up on carbs reduces snacking!'. Deciding on an appropriate statistical technique to determine whether bread purchases influence confectionary purchases and then interpreting results is a skill which is hard to test but very valuable in real world statistics.

There are lots of decisions to make. Should you use the mass of butter bought? Or the mass compared to 2001? Or the mass as a percentage of all fats? The answer to your main question depends on such decisions.

You will explore all of these themes with examples and questions in the Elevate Section. You need to get used to working with the variables and contexts presented in the large data set.

Proof and mathematical communication

In this chapter you will learn how to:

- use appropriate terms to describe mathematical objects, such as identity and equation
- use a counter example to disprove a mathematical idea
- apply some techniques for proving a mathematical idea - deduction and exhaustion.

Before you start...

GCSE	You should know the definition of the square root function.	1 Write down the value of $\sqrt{9}$.
GCSE	You should be able to manipulate algebraic expressions.	2 Factorise $4x^2 - 1$.
GCSE	You should know basic angle facts.	3 a What is the sum of the angles in a triangle? b What is the sum of the exterior angles of any polygon?
GCSE	You should know the definition of rational and irrational numbers.	4 Which of these numbers are irrational? π, $0.\dot{3}$, 0.5, $\sqrt{2}$
GCSE	You should be able to work with function notation.	5 If $f(x) = 2x^2 - 3$ find $f(3)$.

Beyond any doubt

Mathematicians are particularly valued for their ability to communicate their ideas precisely and to support their assertions with formal arguments, called **proofs**. In this chapter you will look at the language used by mathematicians and some ways in which they prove their ideas.

Did you know?

The first known use of the equals (=) sign occurs in Robert Recorde's 1557 book, *The Whetstone of Witte*. He explains that he used two parallel lines, 'because no two things can be more equal'.

Section 1: Mathematical structures and arguments

Mathematical ideas may be represented in many different ways, such as tables, diagrams, graphs or words. One of the most fundamental representations is an **equation**. For example:

$$x^2 - 1 = 8$$

An equation is only true for some values of x, in this case, $x = \pm 3$.

Another similar mathematical structure is called an **identity**.

There are some rules that only apply to identities. For example, if two polynomials are identically equal the coefficients of corresponding variables must be the same.

Key point 1.1

An identity is a relation that is true *for all values* of the unknown. It is indicated by the ≡ symbol. For example:

$$x^2 - 1 \equiv (x-1)(x+1)$$

Two statements connected by an identity symbol are **congruent** expressions.

WORKED EXAMPLE 1.1

$2x^2 + 12x - 3 \equiv a(x+p)^2 + q$

Find the values of a, p and q.

$2x^2 + 12x - 3 \equiv a(x+p)^2 + q$
$= a(x^2 + 2xp + p^2) + q$
$= ax^2 + 2apx + ap^2 + q$

Multiply out the brackets to allow coefficients to be compared.

Coefficient of x^2: $2 = a$
Coefficient of x: $12 = 2ap$
Constant term: $-3 = ap^2 + q$

Compare coefficients.

$4p = 12$
$p = 3$

Substitute $a = 2$ into the second equation.

$2 \times 3^2 + q = -3$
$q = -21$

Substitute $a = 2$, $p = 3$ into the third equation.

Tip

Whenever you are simplifying an expression, technically you should use an identity symbol. However, it is common in mathematics to use an equals sign instead.

Tip

A **coefficient** of a variable is the constant in front of that variable. For example, in the quadratic $2x^2 - 3x$, 2 is the coefficient of x^2 and -3 is the coefficient of x.

You can manipulate both equations and inequalities by doing the same thing to both sides. Generally, you will do this by writing lines of working one under another. In more formal work, you can emphasise the logic of the argument by using special symbols.

Key point 1.2

- The symbol ⇒ means that a subsequent statement follows from the previous statement.
  $P \Rightarrow Q$ means 'P implies Q' or 'if P is true then Q is true' or 'P is sufficient for Q.'
- The symbol ⇐ means that the previous statement follows from the subsequent statement.
  $P \Leftarrow Q$ means 'P is implied by Q' or 'if Q is true then P is true' or 'P is necessary for Q.'
- The symbol ⇔ means that a subsequent statement is equivalent to the previous one.
  $P \Leftrightarrow Q$ means 'P is equivalent to Q' or 'Q is true if and only if P is true.' This can also be written as 'Q iff P.'

Gateway to A Level

See Gateway to A Level Section A for a reminder of expanding brackets.

Fast forward

In Chaper 3 you will see that you often have to write quadratics in the form $a(x+p)^2 + q$. The method of comparing coefficients provides an alternative to the procedure suggested in Chapter 3. You can choose to use whichever one you prefer.

WORKED EXAMPLE 1.2

Insert a $\Rightarrow$, $\Leftarrow$ or a $\Leftrightarrow$ symbol in each position marked by $*$.

a $2x+1=9$

$*2x=8$

$*x=4$

b $x=4$

$*x^2=16$

a $2x+1=9$

$\Leftrightarrow 2x=8$

These statements are equivalent: the logic flows both ways.

$\Leftrightarrow x=4$

Again $2x=8$ and $x=4$ are equivalent.

b $x=4$

$\Rightarrow x^2=16$

$x=4$ implies that $x^2=16$ but the reverse is not true as $x^2=16$ implies $x=\pm4$ (not just $x=4$).

When you are solving equations, each line of your working needs to be equivalent to the last if the 'solutions' you get at the end are to be valid.

WORKED EXAMPLE 1.3

A student is attempting to solve the equation $\sqrt{x+6}=x$.

a Find the error with this working.

$$\sqrt{x+6}=x$$
$$x+6=x^2$$
$$x^2-x-6=0$$
$$(x-3)(x+2)=0$$
$$x=3 \text{ or } x=-2$$

b Solve the equation correctly.

a $\sqrt{x+6}=x$

Look at each line in turn to see whether a $\Leftrightarrow$ symbol is valid.

$\Rightarrow x+6=x^2$

They are not equivalent since:

$x+6=x^2$

$\Rightarrow \pm\sqrt{x+6}=x$

The first line implies the second, but the second does not imply the first, so they are not equivalent.

This leads to one incorrect solution, coming from $-\sqrt{x+6}=x$

All subsequent lines are equivalent, so one of the solutions is correct.

b Therefore the negative solution is false.

The correct solution is $x=3$.

Since the LHS of the original equation is positive, the RHS must be positive too.

Gateway to A Level

See Gateway to A Level Sections B and C for a reminder of solving quadratic equations by factorising.

Tip

Squaring an equation is a common way of introducing incorrect solutions, since it prevents lines of working being equivalent.

Tip

LHS and RHS are standard abbreviations for 'left-hand side' and 'right-hand side'.

In practice it is often easier not to worry about whether every line is equivalent, but to be aware that any 'solutions' need to be checked by substituting them back into the original equation. Any that are not correct can then just be deleted.

Dividing by zero can remove solutions, in the same way that squaring can introduce them.

Fast forward

The problem of false solutions will also arise when you solve equations involving logarithms, in Chapter 7.

WORKED EXAMPLE 1.4

Insert an appropriate $\Rightarrow$, $\Leftarrow$ or a $\Leftrightarrow$ symbol in the position marked by $*$. Hence explain why the solution is incomplete.

$x^2 = 6x$

$* \; x = 6$ Divide by x.

The missing symbol is a $\Leftarrow$ — If $x = 6$ then $x^2 = 6x$, but the reverse is not always true.

So $x = 6$ is only one solution – there is also the possibility that $x = 0$.

EXERCISE 1A

1. Where appropriate insert a $\Rightarrow$, $\Leftarrow$ or a $\Leftrightarrow$ symbol in each space.

   a i Shape P is a rectangle __ shape P is a square.
     ii Shape Q is a quadrilateral __ shape Q is a rhombus.

   b i n is even __ n is a whole number.
     ii n is a prime number __ n is a whole number.

   c i A triangle has two equal sides __ a triangle has two equal angles.
     ii Two circles have the same area __ two circles have the same radius.

   d i $x^2 - 2x - 3 = 0$ __ $x = 3$ ii $x^2 - 2x + 1 = 0$ __ $x = 1$

   e i Sam can do 10 press-ups __ Sam can do 100 press-ups.
     ii Niamh is over 21 __ Niamh is over 18.

   f i Neither A nor B is true __ A is false and B is false.
     ii A and B are not both true __ A and B are both not true.

   g i Chris is a boy __ Chris is a footballer.
     ii Shape X is a right-angled triangle __ shape X is an isosceles triangle.

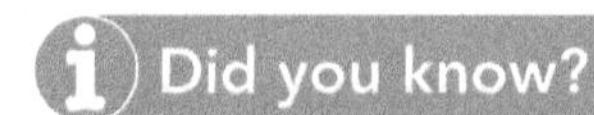

Did you know?

It is assumed here that A is either true or false. The study of this type of logic is called Boolean Algebra.

2. $x^3 - 4x^2 - 3x + 18 = (x + a)(x - b)^2$ for all x.

   Find the values of a and b.

3. $x^4 + 8x^3 + 2x + 16 = (x^3 + a)(x + b)$ for all x.

   Find the values of a and b.

4. What is the flaw in this working?

   Question: For $x = 4$, find the value of $2x + 2$.

   Working: $2 \times 4 = 8 + 2 = 10$

5 Where is the flaw in this argument?

Suppose $1 = 3$

$-1 = 1$ Subtract 2 from both sides.

$1 = 1$ Square both sides.

Therefore the first line is true.

6 **a** Consider the equation:

$\sqrt{x^2+9} = 3x-7$

Add appropriate symbols ($\Leftrightarrow$, $\Leftarrow$ or $\Rightarrow$) in each position marked by $*$ in the solution.

$* \; x^2+9 = 9x^2-42x+49$ Square both sides.

$* \; 0 = 8x^2-42x+40$ Subtract x^2+9 from both sides.

$* \; 0 = 4x^2-21x+20$ Divide both sides by 2.

$* \; 0 = (4x-5)(x-4)$ Factorise.

$* \; x = \frac{5}{4}$ or $x = 4$

b Hence explain the flaw in the solution given.

7 **a** Insert the appropriate symbol ($\Rightarrow$, $\Leftarrow$ or $\Leftrightarrow$) in each position marked by $*$.

$\frac{1}{x^2} = \frac{2}{x}$

$* \; x = 2x^2$

$* \; 0 = 2x^2 - x$

$* \; 0 = x(2x-1)$

$* \; x = 0$ or $x = \frac{1}{2}$

b Hence explain the error in the working.

8 **a** Insert the appropriate symbol ($\Rightarrow$, $\Leftarrow$ or $\Leftrightarrow$) in each position marked by $*$.

$x^2+3x = 4x+12$

$* \; x(x+3) = 4(x+3)$

$* \; x = 4$

b Hence explain the error in the working.

9 Where is the flaw in the following argument?

Suppose two numbers a and b are equal.

$a = b$

$a^2 = ab$ Multiply by a.

$a^2 - b^2 = ab - b^2$ Subtract b^2.

$(a-b)(a+b) = b(a-b)$ Factorise.

$a + b = b$ Cancel $(a-b)$.

$2b = b$ Use the fact that $a = b$.

$2 = 1$ Divide by b.

10 Do you agree with this statement?

Either A or B is true $\Leftrightarrow$ A and B are not both true.

Section 2: Inequality notation

From previous studies, you know that solving a linear inequality is just like solving an equation, as long as you don't multiply or divide by a negative number. The answer is written as an inequality.

For example:

$$2x-5\geqslant 9$$
$$\Leftrightarrow 2x\geqslant 14$$
$$\Leftrightarrow x\geqslant 7$$

This solution can be written in **set notation** as: $\{x : x\geqslant 7\}$.

This is read as 'x such that x is bigger than or equal to 7.'

It can also be written in **interval notation** as: $[7, \infty)$.

This means that the solution lies in the interval from 7 (included) to infinity (not included).

Key point 1.3

- $x \in (a, b)$ means $a < x < b$
- $x \in [a, b]$ means $a \leqslant x \leqslant b$
- $x \in [a, b)$ means $a \leqslant x < b$
- $x \in (a, b]$ means $a < x \leqslant b$

Tip

The $\in$ symbol in set notation means 'is in the set...' or 'belongs to the set...'.

Two different intervals can be combined, using set theory notation.

Key point 1.4

- $A \cup B$ is the **union** of A and B. It means the solution is in A or B or both.
- $A \cap B$ is the **intersection** of A and B. It means the solution is in both A and B.

WORKED EXAMPLE 1.5

Write each inequality in set notation.

a $1\leqslant x<7$
b $x>1$ or $x<-2$

a $\{x : x\geqslant 1\}\cap\{x : x<7\}$ — x is **both** greater than or equal to 1, **and** less than 7 so it is in the intersection.

b $\{x : x>1\}\cup\{x : x<-2\}$ — x is greater than 1 **or** less than -2 so it is in the union.

If there is no solution to the inequality then x is in the empty set, written as $x \in \varnothing$.

EXERCISE 1B

1. Write each inequality in set notation and interval notation.

   a i $x>7$ ii $x<6$ **b** i $x\leqslant 10$ ii $x\geqslant 5$

   c i $0<x\leqslant 1$ ii $5<x<7$ **d** i $x>5$ or $x\leqslant 0$ ii $x\geqslant 10$ or $x<2$

2. Write each interval as an inequality in x.

   a i $[1,4)$ ii $(2,8]$

   b i $[1,3]$ ii $(2,4)$

   c i $(-\infty,5)$ ii $[12,\infty)$

   d i $\{x:0<x<10\}\cap\{x:x\geqslant 8\}$ ii $\{x:1<x<4\}\cap\{x:x\geqslant 3\}$

Section 3: Disproof by counter example

It is not usually possible to prove that something is always true by looking at examples. However, it is possible to use examples to prove that something is not always true. This is called a **counter example**.

WORKED EXAMPLE 1.6

Disprove by counter example that $(x+1)^2\equiv x^2+1$ for all x.

Take $x=2$.

LHS: $(2+1)^2=9$

RHS: $2^2+1=5$

So $x=2$ is a counter example.

When searching for a counter example, try different types of number. Here, any non-zero number will work.

EXERCISE 1C

1. Disprove the statement $\sqrt{x^2+9}\equiv x+3$.
2. Use a counter example to prove that $\sin 2x\not\equiv 2\sin x$.
3. Use a counter example to prove that $\sqrt{x^2}$ is not always x.
4. Prove that the product of two prime numbers is not always odd.
5. Prove that the number of factors of a number is not always even.
6. Prove that the sum of two irrational numbers is not always irrational.
7. Use a counter example to disprove this statement.

   $x<3\Rightarrow x^2<9$
8. Prove that n^2+n+41 does not take prime values for all positive integers.
9. Are any two straight lines that never meet necessarily parallel?

Gateway to A Level

See Gateway to A Level Section D for a reminder of rational and irrational numbers.

Section 4: Proof by deduction

To prove that a result is true, you need to start with what is given in the question and form a series of logical steps to reach the required conclusion.

Algebra is a useful tool which allows you to express ideas in general terms. You will often need to use algebraic expressions for even and odd numbers.

Key point 1.5

It is common to express:

- an even number as $2n$, for any integer n
- an odd number as $2n+1$ (or $2n-1$), for any integer n.

Tip

The word **integer** just means 'whole number'.

Tip

Remember – one example (or even several examples) does not make a proof.

Elevate

See Support Sheet 1 for a further example on algebraic proof and more practice questions.

WORKED EXAMPLE 1.7

Prove that the product of an even and an odd number is always even.

Let the even number be $2n$, for some integer n. — Define a general even number...

Let the odd number be $2m+1$, for some integer m. — ...and a general odd number.

Do not use $2n+1$ as this would be the next integer up from $2n$, which is too specific.

$$2n(2m+1) = 2(2mn+n)$$
$$= 2k \quad \text{for some integer } k$$

Aim to write the product in the form $2k$ to show that it is even.

$\therefore$ the product of an even and an odd number is even. — Make a conclusion.

WORKED EXAMPLE 1.8

Prove that the difference between the squares of consecutive odd numbers is always a multiple of eight.

Let the smaller odd number be $2n-1$.

Let the larger odd number be $2n+1$.

Define two consecutive odd numbers. This time you **do** want n in both.

$$(2n+1)^2 - (2n-1)^2 = (4n^2+4n+1) - (4n^2-4n+1)$$
$$= 4n+4n$$
$$= 8n$$

Square each, and subtract the smaller from the larger.

$\therefore$ the difference between the squares of consecutive odd numbers is always a multiple of 8. — Make a conclusion.

EXERCISE 1D

1. Prove that if n is odd then n^2 is also odd.

2. Prove that the sum of an even number and an odd number is always odd.

3. Prove that the sum of any three consecutive integers is always a multiple of 3.

4. Prove that:

   a the sum of two consecutive multiples of 5 is always odd

   b the product of two consecutive multiples of 5 is always even.

5. Prove that the height, h, in the diagram is given by:

   $$h = \frac{ab}{\sqrt{a^2 + b^2}}$$

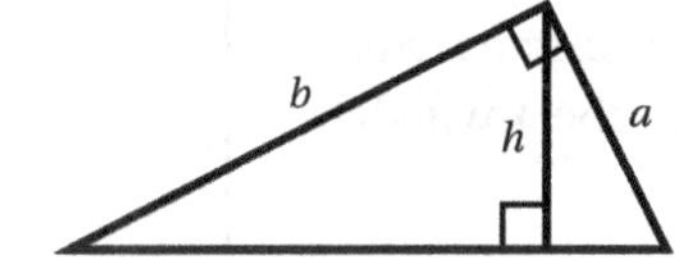

6. Prove that the sum of the interior angles of a hexagon is 720°.

> **Gateway to A Level**
>
> See Gateway to A Level Section E for a reminder of angle facts.

7. Prove that if a number leaves a remainder of 2 when it is divided by 3 then its square leaves a remainder of 1 when divided by 3.

8. a Expand $(x+2)^2$.

   b Prove that $y = x^2 + 4x + 10 \Rightarrow y > 0$.

9. Prove that the exterior angle in a triangle is the sum of the two opposite angles.

10. Prove that $n^2 + 3n + 2$ is never prime if n is a positive integer.

11. a Let n be a four-digit whole number with digits '*abcd*'. Explain why $n = 1000a + 100b + 10c + d$.

    b Prove that n is divisible by 9 if and only if $a + b + c + d$ is a multiple of 9.

    c Prove that n is divisible by 11 if and only if $a - b + c - d$ is divisible by 11.

12. By considering $\left(\sqrt{2}^{\sqrt{2}}\right)^{\sqrt{2}}$ prove that an irrational number raised to an irrational power can be rational.

Section 5: Proof by exhaustion

You should be aware that simply considering some examples does not constitute a mathematical proof. However, in some situations it is possible to check **all** possibilities, which can lead to a valid proof. This is called **proof by exhaustion**.

Did you know?

In traditional mathematics proof by exhaustion was not very common as there are usually too many possibilities to check. The use of computers has made it a much more viable method, but some mathematicians question whether just checking large numbers of possibilities by computer is a valid method of proof. One famous result that has been proved in this way is the **four colour theorem**.

WORKED EXAMPLE 1.9

Prove that 13 is a prime number.

13 is not divisible by 2, 3, 5, 7 or 11.

Since you can write any number as a product of its prime factors, you only need to check whether it has prime factors.

Therefore it must be a prime number.

Tip

You can be more efficient by only checking primes up to the square root of 13, since any factor above this would have to be paired with a factor that is less than the square root.

WORKED EXAMPLE 1.10

A whole number is squared and divided by 3. Prove that the remainder can only be 0 or 1.

You cannot check all whole numbers, but you can split them into three groups when considering division by 3: those that give no remainder, those that give a remainder 1 and those that give a remainder 2. You can then check what squaring does to numbers from each group.

Let n be a whole number.

Now use algebra to write each type of number.

Then n must be:

- a multiple of 3 ($n = 3k$), or
- one more than a multiple of 3 ($n = 3k + 1$), or
- two more than a multiple of 3 ($n = 3k + 2$).

If $n = 3k$ then:

Now check what happens when you square each type of number.

$$\begin{aligned} n^2 &= (3k)^2 \\ &= 9k^2 \\ &= 3(3k^2) \\ &= 3m \end{aligned}$$

which is a multiple of 3.

Continues on next page

If $n = 3k + 1$ then:

$$\begin{aligned} n^2 &= (3k+1)^2 \\ &= 9k^2 + 6k + 1 \\ &= 3(3k^2 + 2k) + 1 \\ &= 3m + 1 \end{aligned}$$

which is one more than a multiple of 3.

If $n = 3k + 2$ then:

$$\begin{aligned} n^2 &= (3k+2)^2 \\ &= 9k^2 + 12k + 4 \\ &= 3(3k^2 + 4k + 1) + 1 \\ &= 3m + 1 \end{aligned}$$

which is one more than a multiple of 3.

So either there is no remainder or the remainder is 1.

Having checked each possible whole number, you can write the conclusion.

EXERCISE 1E

1. Prove that 11 is a prime number.
2. Prove that 83 is a prime number.
3. Prove that all regular polygons with fewer than seven sides have angles with a whole number of degrees.
4. Prove that no square number less than 100 ends in a 7.

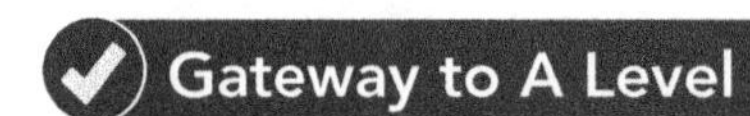

See Gateway to A Level Section F for a reminder of function notation.

5. Let $f(x)$ be the function that gives the number of factors of x. For example $f(10) = 4$ because it has factors 1, 2, 5 and 10. Prove that for any single-digit positive number $f(n) \leqslant n$.
6. Prove that $n^2 + 2$ is not divisible by 4 for the integers 1 to 5.
7. Prove that $n^2 + n$ is always even if $n \in \mathbb{Z}$.
8. Prove that when the square of a whole number is divided by 5, the remainder is always 0, 1 or 4.
9. Prove that $2x^3 + 3x^2 + x$ is always divisible by 6 if x is an integer.

Checklist of learning and understanding

- Mathematical ideas may be expressed through descriptions such as diagrams, equations and identities.
- A mathematical argument can be described by a series of equations or identities put together in a logical order.
  These can be connected using implication symbols: $\Rightarrow$, $\Leftarrow$ or $\Leftrightarrow$.
  - The symbol $\Rightarrow$ means that a subsequent statement follows from the previous one.
  - The symbol $\Leftarrow$ means that the previous statement follows from the subsequent one.
  - The symbol $\Leftrightarrow$ means that a subsequent statement is equivalent to the previous one.
- An identity is a relation that is true **for all values** of the unknown. It is represented by the $\equiv$ symbol.
- Inequalities may be represented by set notation or interval notation.
- One counter example is sufficient to prove that a statement is not always true.
- An algebraic proof is often required to show that a statement is always true.
- Proof by exhaustion involves checking all possibilities; this can only be done if there is a small number of options, or the options can be split up into different cases.

Mixed practice 1

1. Which of these options provides a counter example to the statement: 'n^2+5 is even for every prime number n'?

   A $n=5$ **B** $n=4$ **C** $n=3$ **D** $n=2$

2. Prove that the product of any two odd numbers is always odd.

3. Prove that if n is even then n^2 is divisible by 4.

4. Prove that if $\frac{a}{b} \equiv \frac{c}{d}$ it does not follow that $a \equiv c$ and $b \equiv d$.

5. Prove this statement or disprove it with a counter example.

   'The sum of two numbers is always greater than their difference.'

6. Prove that the product of two rational numbers is always rational.

7. Prove that the sum of the interior angles in an n-sided shape is $(180n-360)°$.

8. Given that $x^3+y^3 \equiv (x+y)(ax^2+bxy+cy^2)$ find the values of a, b and c.

9. Prove this statement: n is odd $\Rightarrow n^2+4n+3$ is a multiple of 4.

10. Prove that the angle, x, from a chord to the centre of a circle is twice the angle, y, to a point on the circumference in the major sector.

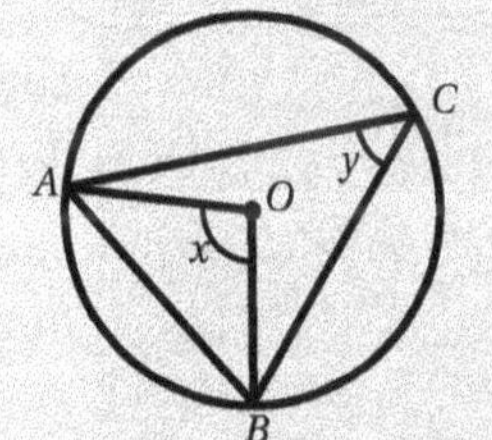

11. Prove that all cube numbers are either multiples of 9 or within one of a multiple of 9.

12. Prove each statement, or disprove it with a counter example.

    a ab is an integer $\Leftrightarrow$ a is an integer and b is an integer.

    b a is irrational and b is irrational $\Leftrightarrow$ ab is irrational.

13. Prove that the product of any three consecutive positive integers is a multiple of 6.

14. Prove that the difference between the squares of any two odd numbers is a multiple of 8.

15. **a** Prove that $n^2-79n+1601$ is not always prime when n is a positive whole number.

    b Prove that n^2-1 is never prime when n is a whole number greater than 2.

16. $x=a^2-b^2$ where a and b are both whole numbers. Prove that x is either odd or a multiple of 4.

Elevate

See Extension Sheet 1 for questions on another principle used in proof.

2 Indices and surds

In this chapter you will learn how to:

- use laws of indices
- work with expressions involving square roots (called surds).

Before you start…

GCSE	You should be able to evaluate expressions involving powers, including working with the order of operations.	1 Evaluate 3×2^3.
GCSE	You should be able to evaluate expressions involving roots.	2 Evaluate $\sqrt[3]{27}$.
GCSE	You should be able to work with the laws of indices.	3 a Evaluate $(x^2)^3$. b Simplify $x^2\times x^5$. c Simplify $\frac{x^{10}}{x^5}$.
GCSE	You should be able to work with negative, fractional and zero indices.	4 Write each expression in the form ax^n. a $\frac{3}{x^2}$ b $\frac{1}{\sqrt[3]{x}}$ c $(4x)^0$
GCSE	You should be able to multiply out two brackets.	5 Expand $(1+x)(2-y)$.
GCSE	You should be able to recognise and use the 'difference of two squares' factorisation.	6 Expand and simplify $(2a+b)(2a-b)$.

Working with powers and roots

Manipulating indices is a very important skill. You will need to use it in several different topics throughout the course. **Surds**, such as $\sqrt{2}$, $\sqrt{3}$ and $\sqrt{5}$, are irrational numbers that occur in the solution of many quadratic equations.

Section 1: Using the laws of indices

From your previous studies you should know these laws of indices.

Key point 2.1

a $a^m \times a^n = a^{m+n}$

b $a^m \div a^n = a^{m-n}$

c $(a^m)^n = a^{m \times n}$

d $a^0 = 1$

e $a^{-n} = \dfrac{1}{a^n}$

f $a^{\frac{m}{n}} = \left(\sqrt[n]{a}\right)^m = \sqrt[n]{a^m}$

Gateway to A Level

See Gateway to A level Section G for revision of the laws of indices.

Tip

Make sure that you can use these rules in both directions, for example, if you see 2^6 you can rewrite it as $(2^3)^2$ and if you see $(2^3)^2$ you can rewrite it as 2^6.

You must be able to combine the rules of indices accurately with the other rules of algebra you already know.

WORKED EXAMPLE 2.1

Simplify $3xy \times 8xy^{-2}$.

$3xy \times 8xy^{-2} = 3 \times 8 \times x \times x \times y \times y^{-2}$

You can rearrange a multiplication into any convenient order.

$= 24x^2y^{-1}$

Apply $a^m \times a^n = a^{m+n}$ (Key point 2.1a) to the *x*s and *y*s.

Fast forward

To prove these rules formally requires a method called mathematical induction. If you study Further Mathematics, you will learn about this.

WORKED EXAMPLE 2.2

Simplify $\dfrac{12a^2b - 9ab^3}{3ab}$.

$\dfrac{12a^2b - 9ab^3}{3ab} = \dfrac{12a^2b}{3ab} - \dfrac{9ab^3}{3ab}$

You can split a fraction if the numerator is a sum or a difference.

$= 4a - 3b^2$

Apply $a^m \div a^n = a^{m-n}$ (Key point 2.1b) to the *a*s and *b*s.

WORKED EXAMPLE 2.3

Write $\dfrac{\sqrt{x}}{5\sqrt[3]{x}}$ in the form kx^p.

$\dfrac{\sqrt{x}}{5\sqrt[3]{x}} = \dfrac{1}{5} \times x^{\frac{1}{2}} \div x^{\frac{1}{3}}$

Rewrite the roots using $a^{\frac{m}{n}} = \sqrt[n]{a^m}$ (Key point 2.1f).

$= \dfrac{1}{5}x^{\frac{1}{6}}$

Now use $a^m \div a^n = a^{m-n}$ (Key point 2.1b).

You also need to be able to manipulate indices to solve equations.

WORKED EXAMPLE 2.4

Solve $x^{\frac{4}{3}} = \frac{1}{81}$.

$x^{\frac{4}{3}} = \frac{1}{81}$

$\left(x^{\frac{4}{3}}\right)^{\frac{3}{4}} = \left(\frac{1}{81}\right)^{\frac{3}{4}}$

Using $(a^m)^n = a^{m\times n}$ (Key point 2.1c), $\left(x^{\frac{4}{3}}\right)^{\frac{3}{4}} = x^{\frac{4}{3}\times\frac{3}{4}} = x^1 = x$ so raise both sides of the equation to the power $\frac{3}{4}$.

$x = \frac{1}{27}$.

WORKED EXAMPLE 2.5

Solve $2^x \times 8^{x-1} = \frac{1}{4^{2x}}$.

$2^x \times 8^{x-1} = \frac{1}{4^{2x}}$

$2^x \times (2^3)^{x-1} = \frac{1}{(2^2)^{2x}}$

Express each term in the same base (2 is easiest), then apply the laws of indices.

$2^x \times 2^{3x-3} = \frac{1}{2^{4x}}$

Use $(a^m)^n = a^{m\times n}$ (Key point 2.1c)...

$2^{4x-3} = 2^{-4x}$

...and then $a^m \times a^n = a^{m+n}$ on the LHS and $a^{-n} = \frac{1}{a^n}$ on the RHS.

$4x - 3 = -4x$

$8x = 3$

$x = \frac{3}{8}$

Equate the powers and solve.

Fast forward

An equation like this, with the unknown (x) in the power, is called an **exponential equation**. In Chapter 7 you will see how to use logarithms to solve more complicated examples.

Be careful when you are multiplying or dividing to combine expressions with different bases. You cannot apply the 'multiplication means add the exponents together' rule because it is **only true when the bases are the same**. There is another rule that works when the bases are different, but only if the exponents are the same. Consider this example.

$$\begin{aligned}3^2 \times 5^2 &= 3\times3\times5\times5\\ &= 3\times5\times3\times5\\ &= 15\times15\\ &= 15^2\end{aligned}$$

Key point 2.2

a $a^n \times b^n = (ab)^n$

b $a^n \div b^n = \left(\frac{a}{b}\right)^n$

WORKED EXAMPLE 2.6

Simplify $\left(\frac{x^6}{8}\right)^{\frac{1}{3}}$.

$\left(\frac{x^6}{8}\right)^{\frac{1}{3}} = \frac{(x^6)^{\frac{1}{3}}}{8^{\frac{1}{3}}}$ — Use $a^n \div b^n = \left(\frac{a}{b}\right)^n$ (Key point 2.2b) to apply the power to each part of the fraction.

$= \frac{x^2}{2}$ — Use $(a^m)^n = a^{m \times n}$ (Key point 2.1c) on the numerator and recognise the cube root of 8 in the denominator.

WORKED EXAMPLE 2.7

Simplify $\frac{(16a^2b^8)^{\frac{1}{2}}}{ab^3}$.

$\frac{(16a^2b^8)^{\frac{1}{2}}}{ab^3} = \frac{16^{\frac{1}{2}}(a^2)^{\frac{1}{2}}(b^8)^{\frac{1}{2}}}{ab^3}$ — Use $a^n \times b^n = (ab)^n$ (Key point 2.2a).

$= \frac{4ab^4}{ab^3}$

$= 4b$ — Apply $a^m \div a^n = a^{m-n}$ (Key point 2.1b) to the *a*s and *b*s.

Common error

Make sure you apply the power to any constant term properly: $(16x^6)^{\frac{1}{2}} = 16^{\frac{1}{2}}(x^6)^{\frac{1}{2}} = 4x^3$ and not $8x^3$.

WORK IT OUT 2.1

Simplify $(16x^2 + 16y^2)^{\frac{1}{2}}$.

Which is the correct solution? Identify the errors made in the incorrect solutions.

Solution 1	Solution 2	Solution 3
$(16x^2+16y^2)^{\frac{1}{2}} = 16^{\frac{1}{2}}(x^2+y^2)^{\frac{1}{2}}$ $= 4(x^2+y^2)^{\frac{1}{2}}$	$(16x^2+16y^2)^{\frac{1}{2}} = 16^{\frac{1}{2}}(x^2+y^2)^{\frac{1}{2}}$ $= 8\sqrt{x^2+y^2}$	$(16x^2+16y^2)^{\frac{1}{2}} = (16x^2)^{\frac{1}{2}} + (16y^2)^{\frac{1}{2}}$ $= 4x+4y$

EXERCISE 2A

1 Simplify these expressions.

a **i** $(x^6)^{\frac{1}{2}}$ **ii** $(x^9)^{\frac{4}{3}}$

b **i** $(4x^{10})^{0.5}$ **ii** $(8x^{12})^{-\frac{1}{3}}$

c **i** $\left(\frac{27x^9}{64}\right)^{-\frac{1}{3}}$ **ii** $\left(\frac{x^4}{y^8}\right)^{-1.5}$

Fast forward

In Chapters 12, 13 and 14 you will need to use the laws of indices to simplify expressions before you can differentiate or integrate them.

2 Solve these equations.

a **i** $x^{-\frac{5}{2}} = 32$ **ii** $x^{\frac{3}{2}} = \frac{8}{27}$

b **i** $2^{x-1} = \frac{4^x}{8}$ **ii** $27^x = \sqrt{3} \times 9^{x+2}$

3 Simplify $(100x^4)^{-\frac{1}{2}}$.

4 Simplify $\frac{x^{\frac{2}{3}} \times x^{\frac{1}{2}}}{x^{\frac{1}{6}}}$.

5 Simplify $\frac{\sqrt[3]{64p^6q^4}}{2q}$.

6 Express $\frac{1-x^3}{\sqrt{x}}$ in the form $x^a - x^b$.

7 Express $\frac{(2-x)(2+x)}{\sqrt[3]{x^2}}$ in the form $ax^n + bx^m$.

8 Simplify $\frac{12ab^{-2} - 16a}{8ab}$, giving your answer as a single fraction.

9 An elementary computer program is known to be able to sort n input values in $k \times n^{1.5}$ microseconds. Observations show that it sorts one million values in half a second. Find the value of k.

10 The volume and surface area of a family of regular solid shapes are related by the formula $V = kA^{1.5}$, where V is given in cubic units and A in square units.

a For one such shape, $A = 81$ and $V = 243$. Find k.

b Hence determine the surface area of a shape with volume $\frac{64}{3}$ cm^3.

11 A square-ended cuboid has volume xy^2, where x and y are lengths. A cuboid for which $x = 2y$ has volume 128 cm^3. Find x.

12 Given that $\sqrt{3^{2a+b}} = \frac{27^a}{3^b}$, express a in terms of b.

13 Simplify $\left(3x^9 + \frac{3}{8}x^9\right)^{\frac{1}{3}}$.

14 Express $\left(3x^{\frac{3}{4}}-x^{-\frac{3}{4}}\right)^2$ in the form $ax^n+\frac{b}{x^n}+c$ where a, b, c and n are to be found.

15 Constants a and b are such that $4^{ax}=b\times 8^x$, find all possible pairs of values of a and b.

16 Any number raised to the power zero is 1, but zero raised to any power is zero, so what is the value of 0^0?

17 What is the value of $2^{2^{2^2}}$?

Section 2: Working with surds

A surd is any number that can only be expressed in terms of roots. For example:

$$2+3\sqrt{5}, \sqrt{3}-\sqrt{7}, \sqrt[3]{2}+\sqrt{2}$$

When working with surds the most important thing to know is that taking the square root of a number is simply raising it to the power of a half, so all the rules for indices also apply to surds.

Gateway to A Level

See Gateway to A level Section H for a reminder of basic manipulations with surds.

Fast forward

In Student Book 2, Chapter 1, you will see a method for proving that surds cannot be written as rational numbers.

WORKED EXAMPLE 2.8

Write $\sqrt{8}+\sqrt{2}$ in the form $\sqrt{a}$.

$\sqrt{8}=\sqrt{4\times 2}$
$=\sqrt{4}\times\sqrt{2}$
$=2\sqrt{2}$

Apply $a^n\times b^n=(ab)^n$ (Key point 2.2a) to $\sqrt{8}$ to split it into a product.

One of the factors of 8 needs to be a square number.

So: $\sqrt{8}+\sqrt{2}=2\sqrt{2}+\sqrt{2}$
$=3\sqrt{2}$

$2\sqrt{2}$ plus another $\sqrt{2}$ is just 3 lots of $\sqrt{2}$.

$=\sqrt{9}\sqrt{2}$
$=\sqrt{18}$

To express it as the square root of a single number use Key point 2.2a again. Note that you need to write 3 as the square root of another number.

WORKED EXAMPLE 2.9

Simplify $(1+\sqrt{2})^2$.

$(1+\sqrt{2})^2=(1+\sqrt{2})(1+\sqrt{2})$
$=1+\sqrt{2}+\sqrt{2}+\sqrt{2}\times\sqrt{2}$

Expand the brackets as normal.

$=1+2\sqrt{2}+2$
$=3+2\sqrt{2}$

Note that $\sqrt{2}\times\sqrt{2}=(\sqrt{2})^2=2$

One important method used when dealing with surds in fractions is called **rationalising the denominator**. This technique removes the surd from the denominator.

WORKED EXAMPLE 2.10

Rationalise the denominator of $\frac{2}{\sqrt{3}}$.

$\frac{2}{\sqrt{3}} = \frac{2\times\sqrt{3}}{\sqrt{3}\times\sqrt{3}}$ — Multiply top and bottom by $\sqrt{3}$.

$= \frac{2\sqrt{3}}{3}$

You can use the difference of two squares to rationalise the denominator of more complicated expressions.

$$a^2 - b^2 = (a-b)(a+b).$$

So for an expression such as $5+\sqrt{3}$, multiplying it by $5-\sqrt{3}$ gives $5^2 - (\sqrt{3})^2 = 22$. Importantly, this product is rational.

Gateway to A Level

See Gateway to A Level Section B for revision of the difference of two squares factorisation.

Key point 2.3

To rationalise the denominator of a fraction, multiply top and bottom by the appropriate expression to give the difference of two squares.

WORKED EXAMPLE 2.11

Rationalise the denominator of $\frac{3}{8-2\sqrt{3}}$.

$\frac{3}{8-2\sqrt{3}} = \frac{3\times(8+2\sqrt{3})}{(8-2\sqrt{3})\times(8+2\sqrt{3})}$ — By inspection, the appropriate term to multiply top and bottom by is $8+2\sqrt{3}$.

$= \frac{24+6\sqrt{3}}{64-4\times3}$ — You don't need to multiply the bottom out - use the identity for the difference of two squares.

$= \frac{24+6\sqrt{3}}{52}$ — There is a common factor of 2 on top and bottom so cancel it.

$= \frac{12+3\sqrt{3}}{26}$

WORK IT OUT 2.2

Rationalise the denominator of $\frac{1}{\sqrt{5}-\sqrt{2}}$.

Which is the correct solution? Identify the errors made in the incorrect solutions.

Solution 1	Solution 2	Solution 3
$\frac{1}{\sqrt{5}-\sqrt{2}}=\frac{1}{\sqrt{3}}$ $=\frac{\sqrt{3}}{3}$	$\frac{1}{\sqrt{5}-\sqrt{2}}=\frac{1\times(\sqrt{5}+\sqrt{2})}{(\sqrt{5}-\sqrt{2})\times(\sqrt{5}+\sqrt{2})}$ $=\frac{\sqrt{5}+\sqrt{2}}{3}$	$\frac{1}{\sqrt{5}-\sqrt{2}}=\frac{1}{\sqrt{5}}-\frac{1}{\sqrt{2}}$ $=\frac{\sqrt{5}}{5}-\frac{\sqrt{2}}{2}$ $=\frac{2\sqrt{5}-5\sqrt{2}}{10}$

EXERCISE 2B

1 Without using a calculator, write each expression in the form $k\sqrt{5}$.

a i $\sqrt{125}$ ii $\sqrt{20}$

b i $7\sqrt{5}-2\sqrt{5}$ ii $\sqrt{5}+9\sqrt{5}-3\sqrt{5}$

c i $3\sqrt{80}-5\sqrt{20}$ ii $\sqrt{125}+7\sqrt{45}$

2 Without using a calculator, write each expression in the form $\sqrt{a}$.

a i $4\sqrt{2}$ ii $10\sqrt{3}$

b i $\sqrt{7}+2\sqrt{7}$ ii $3\sqrt{5}+\sqrt{5}$

c i $\sqrt{3}+\sqrt{75}$ ii $\sqrt{32}+\sqrt{8}$

3 Without using a calculator, write each expression in the form $a+b\sqrt{3}$.

a i $2(3-\sqrt{3})-3(1-\sqrt{3})$ ii $(1+\sqrt{3})-(1-\sqrt{3})$

b i $(1+2\sqrt{3})(2-\sqrt{3})$ ii $(1+\sqrt{3})(2+\sqrt{3})$

c i $(1+\sqrt{3})^2$ ii $(2-\sqrt{3})^2$

4 Without using a calculator, rationalise the denominator of each expression.

a i $\frac{7}{\sqrt{7}}$ ii $\frac{2}{\sqrt{5}}$

b i $\frac{3-\sqrt{6}}{\sqrt{6}}$ ii $\frac{\sqrt{2}+\sqrt{6}}{\sqrt{3}}$

c i $\frac{1}{\sqrt{2}-1}$ ii $\frac{1+\sqrt{5}}{1+\sqrt{7}}$

5 Simplify $\frac{1}{1+\sqrt{n}}+\frac{1}{1-\sqrt{n}}$.

6 Show that $\frac{4}{\sqrt{20}-\sqrt{12}}$ can be written in the form $\sqrt{a}+\sqrt{b}$, where a and b are whole numbers.

7 Show that $\frac{5+\sqrt{2}}{3-2\sqrt{2}}$ can be written in the form $a+b\sqrt{2}$, where a and b are constants to be found.

8 Without using decimal approximations, explain why $3\sqrt{2}$ is larger than $2\sqrt{3}$.

9 Solve $x\sqrt{27}=5x\sqrt{3}+2\sqrt{48}$.

10 Rationalise the denominator of $\frac{1}{2\sqrt{n}-3}$.

11 Given that n is a positive whole number, write $\left(n\sqrt{15}-\sqrt{5}\right)^2$ in the form $a+b\sqrt{3}$.

12 A rectangle has length $a+b\sqrt{2}$ and width $b-a\sqrt{2}$.

a Find the area of the rectangle and write it in the form $m+n\sqrt{2}$.

b Find and simplify an expression for the length of the diagonal of the rectangle.

13 Given that $\frac{5\sqrt{7}-\sqrt{x}}{\sqrt{7}-\sqrt{x}}=8+\sqrt{y}$, where x any y are positive integers and $\sqrt{y}$ is a surd in its simplest form, find the values of x and y.

14 a Write $\sqrt{27}+\sqrt{3}$ in the form $\sqrt{a}$.

b Without using decimal approximations, explain whether $\sqrt{27}-\sqrt{20}$ is greater or less than $\sqrt{5}-\sqrt{3}$.

15 a Show that $a^3-b^3=(a-b)(a^2+ab+b^2)$.

b Hence rationalise the denominator of $\frac{1}{\sqrt[3]{3}-\sqrt[3]{2}}$.

16 Is it always true that $\sqrt{x^2}$ equals x?

Elevate

See Support Sheet 2 for an example of solving equations involving surds and for more practice questions similar to question 9.

Checklist of learning and understanding

- Learn these laws of indices.
  - $a^m \div a^n = a^{m-n}$
  - $(a^m)^n = a^{m \times n}$
  - $a^0 = 1$
  - $a^{-n} = \frac{1}{a^n}$
  - $a^{\frac{1}{n}} = \sqrt[n]{a}$
  - $a^{\frac{m}{n}} = \left(\sqrt[n]{a}\right)^m = \sqrt[n]{a^m}$
  - $a^n \times b^n = (ab)^n$
  - $a^n \div b^n = \left(\frac{a}{b}\right)^n$
- Surds are numbers that can only be expressed in terms of square roots. To rationalise the denominator of a fraction you can multiply top and bottom by the appropriate expression to create a difference of two squares.

Mixed practice 2

1 Which expression is equivalent to $\left(\frac{x}{27x^{-3}}\right)^{-\frac{1}{3}}$?

Choose from these options.

A $\frac{1}{27x^2}$ B $\frac{3}{x^{\frac{4}{3}}}$ C $3x^{\frac{2}{3}}$ D $9x^{-\frac{1}{3}}$

2 Which expression is equivalent to $\frac{(x+\sqrt{x})^2}{x}$?

Choose from these options.

A $1+\frac{1}{x}$ B $1+\frac{1}{x}+2\sqrt{x}$ C $x+1$ D $x+2\sqrt{x}+1$

3 Express $(n+\sqrt{5})^2$ in the form $a+b\sqrt{5}$.

4 Given that $z = xy^2$ and $y = 3x$, express z in terms of x only.

5 Show that $\frac{10}{\sqrt{28}-\sqrt{8}}$ can be written in the form $\sqrt{a}+\sqrt{b}$.

6 If $y = \frac{2}{\sqrt{x}}$, write y^{-4} in the form kx^a.

7 If $3x\sqrt{8} = x\sqrt{2}+\sqrt{32}$, find x.

8 Write $\frac{1-x^2}{\sqrt[4]{x^3}}$ in the form $x^p - x^q$.

[© AQA 2012]

9 Which expression is equivalent to $(x^a)^{\frac{b}{2}} \times (3^b y^b)^{2a}$?

Choose from these options.

A $(3xy)^{ab}$ B $\left(9\sqrt{x}y^2\right)^{ab}$ C $(3xy)^{2ab}$ D $\left(3\sqrt{x^a y^b}\right)^{2ab}$

10 **a** Write down the values of p, q and r, given that:

i $8 = 2^p$ ii $\frac{1}{8} = 2^q$ iii $\sqrt{2} = 2^r$

b Find the value of x for which $\sqrt{2} \times 2^x = \frac{1}{8}$.

[© AQA 2011]

11 Express $\left(x^4 + 7x^3 \times \frac{x}{9}\right)^{-\frac{1}{2}}$ in the form ax^n, where a is a constant to be found.

12 Rationalise the denominator of $\frac{\sqrt{n}+1}{\sqrt{n}-1}$.

13 Simplify $\frac{6^{x+y} \times 15^{x-y}}{3^{2x} \times 10^{x-y}}$. Choose from these options.

A 4^y B 1 C 9^y D $3^x \times 9^y$

14 **a** Find and simplify an expression for $\left(a+b\sqrt{2}\right)^2$.

b By considering $\left(1-\sqrt{2}\right)^4$ prove that $\sqrt{2} < \frac{17}{12}$.

15 **a** Find and simplify an expression for $\left(a+b\sqrt{5}\right)^2$.

b By considering $\left(2-\sqrt{5}\right)^4$ show that $\sqrt{5} < \frac{161}{72}$.

c By considering $\left(2-\sqrt{5}\right)^3$ show that $\sqrt{5} > \frac{38}{17}$.

d **i** Explain why considering $\left(3-\sqrt{5}\right)^3$ gives a worse upper bound for $\sqrt{5}$ than the one found in part **b**.

ii Explain why considering $\left(4-\sqrt{5}\right)^4$ gives a worse upper bound for $\sqrt{5}$ than the one found in part **b**.

Elevate

See Extension Sheet 2 for a selection of more challenging questions.

3 Quadratic functions

In this chapter you will learn how to:

- apply your knowledge of factorisation and the quadratic formula for solving quadratic equations
- recognise the shape and main features of graphs of quadratic functions
- complete the square
- solve quadratic inequalities
- identify the number of solutions of a quadratic equation
- solve disguised quadratic equations.

Before you start...

GCSE	You should be able to multiply out brackets.	1 Expand this expression. $(3x+1)(2x-3)$
GCSE	You should be able to solve quadratic equations by factorising.	2 Solve these equations. a $x^2+x-20=0$ b $2x^2+15x-8=0$ c $5x^2-3x=0$ d $4x^2-9=0$
GCSE	You should be able to use the formula to solve quadratic equations.	3 Solve these equations. a $x^2-4x+2=0$ b $2x^2-10x-5=0$
GCSE	You should be able to solve linear inequalities.	4 Solve this inequality. $5x-1>2x+5$

Quadratic curves are everywhere

Many problems in applications of mathematics involve maximising or minimising a certain quantity. They are common in economics and business (minimising costs and maximising profits), biology (finding the maximum possible size of a population) and physics (electrons move to the lowest energy state). The quadratic function is the simplest function with maximum or minimum points, so it is often used to model such situations.

It also arises in many natural phenomena, such as the motion of a projectile or the dependence of power on voltage in an electric circuit, and can be seen in the shape of the cables in a suspension bridge or in parabolic reflectors that form the basis of satellite dishes and car headlights.

Section 1: Solving quadratic equations

A quadratic function takes the form $f(x) = ax^2 + bx + c$, where a, b and c are constants and $a \neq 0$.

You should be familiar with two methods for solving quadratic equations:

- factorising
- the quadratic formula.

Although your calculator may have an equation solver feature that allows you to type in the coefficients a, b and c to generate solutions, you must be also able to apply these methods, particularly factorising.

Remember that you may first have to rearrange the equation to get it in the form $ax^2 + bx + c = 0$.

Gateway to A Level

See Gateway to A Level Sections B and C for revision of factorising and using the quadratic formula.

Factorising

WORKED EXAMPLE 3.1

Find the values of x for which $(2x-3)^2 = x+6$.

$(2x-3)^2 = x+6$ — Expand the brackets.

$4x^2 - 12x + 9 = x + 6$ — Remember that $(2x-3)^2 = (2x-3)(2x-3)$.

$4x^2 - 13x + 3 = 0$ — Move everything to one side: $f(x) = 0$.

$(4x-1)(x-3) = 0$ — Factorise and solve.

$x = \frac{1}{4}$ or $x = 3$

WORKED EXAMPLE 3.2

Solve the equation $x = 2 + 8x^{-1}$.

$x = 2 + 8x^{-1}$

$x = 2 + \frac{8}{x}$ — This might not look like a quadratic at first but, whether or not you realise that it is, always start by replacing x^{-a} with $\frac{1}{x^a}$.

$x^2 = 2x + 8$ — Now multiply through by x to remove the denominator.

$x^2 - 2x - 8 = 0$ — Move everything to one side: $f(x) = 0$.

$(x-4)(x+2) = 0$ — Factorise and solve.

$x = 4$ or -2

The quadratic formula

If you can't factorise the equation you can use the quadratic formula to solve it.

Key point 3.1

The solutions of $ax^2+bx+c=0$ are given by the quadratic formula

$$x=\frac{-b\pm\sqrt{b^2-4ac}}{2a}.$$

Fast forward

The other alternative is to solve the equation by completing the square. You will see in Section 3 that the quadratic formula actually comes from solving a general quadratic equation by completing the square.

Tip

If you are told to give your answers to a certain number of decimal places or significant figures, or to give exact answers, it means the quadratic won't factorise.

WORKED EXAMPLE 3.3

Solve the equation $x^2+3x=7x+3$. Give your answers in exact form.

$x^2+3x=7x+3$

$x^2-4x-3=0$

Move everything to one side: $f(x)=0$.

$$x=\frac{-(-4)\pm\sqrt{(-4)^2-4\times1\times(-3)}}{2\times1}$$

$$=\frac{4\pm\sqrt{28}}{2}$$

This doesn't factorise, so use the formula, with $a=1, b=-4, c=-3$.

$$=\frac{4\pm2\sqrt{7}}{2}$$

Simplify the surd: $\sqrt{28}=2\sqrt{7}$.

$$=2\pm\sqrt{7}$$

EXERCISE 3A

1 Solve each equation by factorising.

a **i** $3x^2+2x=x^2+3x+6$ **ii** $2x^2+3=17x-7-x^2$

b **i** $9x^2=24x-16$ **ii** $18x^2=2x^2-40x-25$

c **i** $(x-3)(x+2)=14$ **ii** $(2x+3)(x-1)=12$

d **i** $2x=11+\frac{6}{x}$ **ii** $3x+\frac{4}{x}=7$

2 Use the quadratic formula to find the exact solutions of each equation.

a **i** $2x^2+x=x^2+4x-1$ **ii** $x^2-3x+5=6-2x$

b **i** $3x^2-4x+1=5x^2+2x$ **ii** $9x-2=5x^2+1$

c **i** $(x+1)(x+3)=5$ **ii** $(3x+2)(x-1)=2$

d **i** $2x+\frac{1}{x}=6$ **ii** $x=4+\frac{3}{x}$

3. Solve the equation $8x-9=(3x-1)(x+3)$.

4. Solve the equation $6x=5+4x^{-1}$.

5. Find the exact solutions to the equation $x+x^{-1}=3$.

6. Solve the equation $x^2+8k^2=6kx$, giving your answer in terms of k.

7. Rearrange $y=ax^2+bx+c$ to find x in terms of y.

8. The positive difference between the roots of the quadratic equation $x^2+kx+3=0$ is $\sqrt{69}$. Find the possible values of k.

Section 2: Graphs of quadratic functions

All quadratic graphs are one of two possible shapes.

Key point 3.2

The shape of the graph $y=ax^2+bx+c$ depends on the coefficient a.

- If $a>0$ the graph is a positive quadratic.
- If $a<0$ the graph is a negative quadratic.

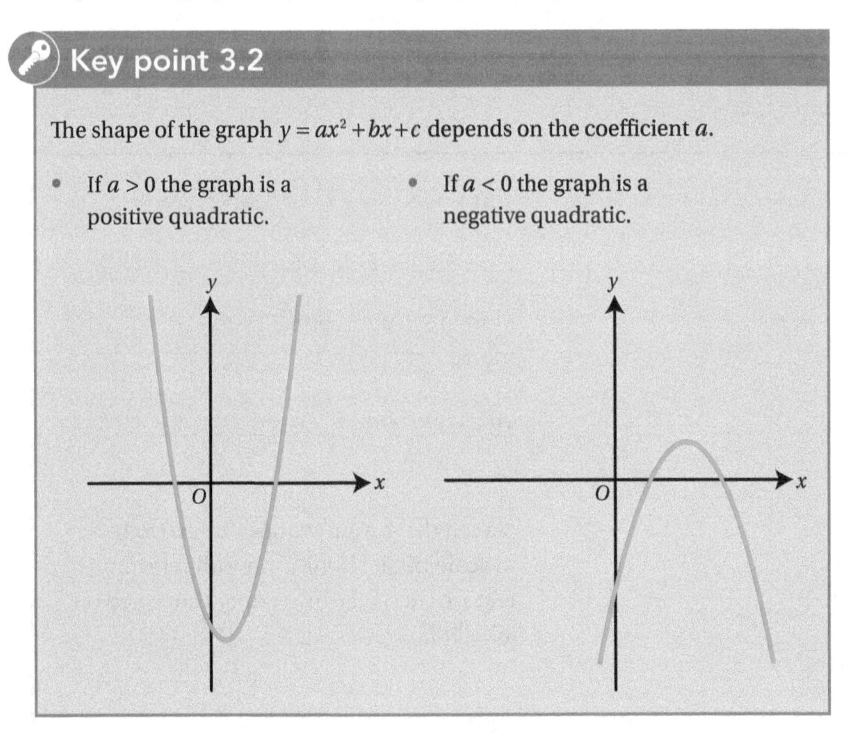

This shape can be located precisely if you know where the graph crosses the axes.

Key point 3.3

The graph $y=ax^2+bx+c$ crosses:

- the y-axis at the point $(0, c)$
- the x-axis at the root(s) of the equation $ax^2+bx+c=0$.

Tip

A root of an equation is another name for a solution.

WORKED EXAMPLE 3.4

Match each equation to the corresponding graph, explaining your reasons.

a $y = 3x^2 - 4x - 1$

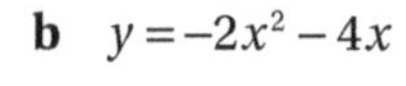

b $y = -2x^2 - 4x$

c $y = -x^2 - 4x + 2$

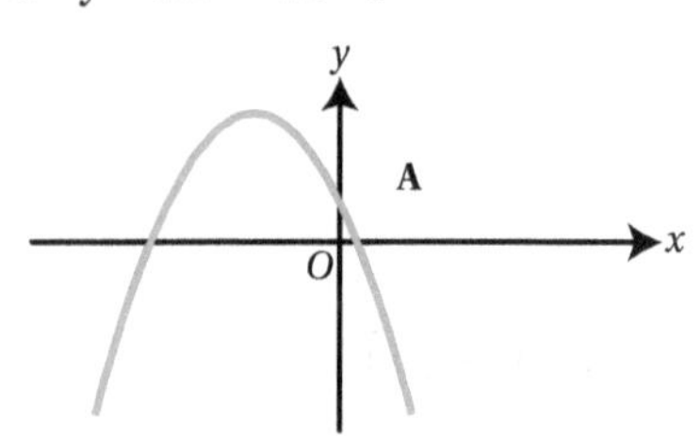

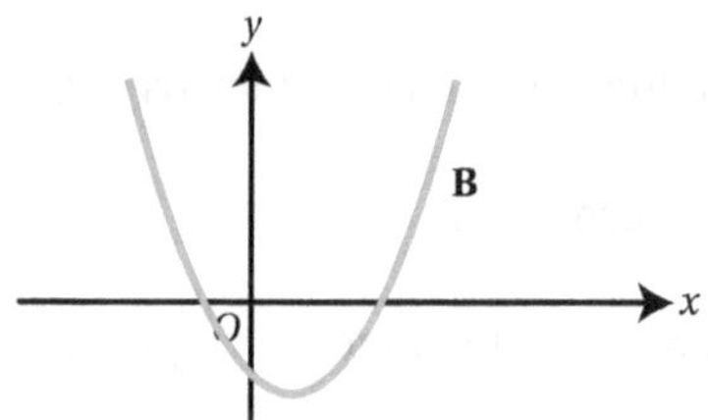

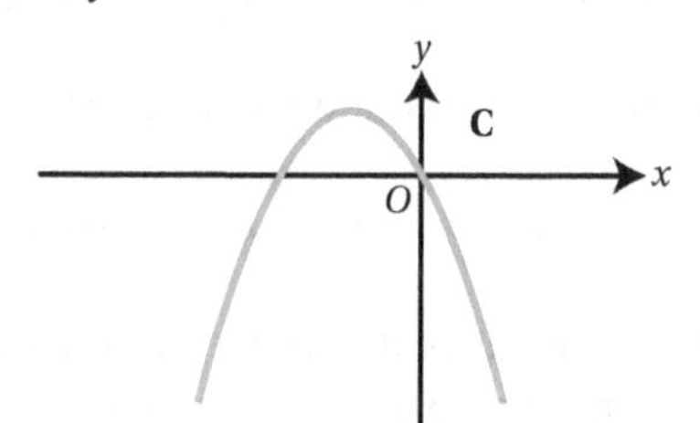

Graph B shows a positive quadratic, so graph B corresponds to equation **a**.

> Graph B is the only positive quadratic.

Graph A has a positive y-intercept, so graph A corresponds to equation **c**.

> You can distinguish between the other two graphs based on their y-intercepts.

Graph C corresponds to equation **b**.

WORKED EXAMPLE 3.5

Sketch the graph of $y = 3x^2 + 2x - 8$.

> This is a positive quadratic as $a > 0$.

When $x = 0$, $y = -8$

> Find the y-intercept.

When $y = 0$:

> Find the x-intercepts. To do this, solve the equation $y = 0$.

$$3x^2 + 2x - 8 = 0$$

$$(3x - 4)(x + 2) = 0$$

> This factorises.

$$x = \frac{4}{3} \text{ or } -2$$

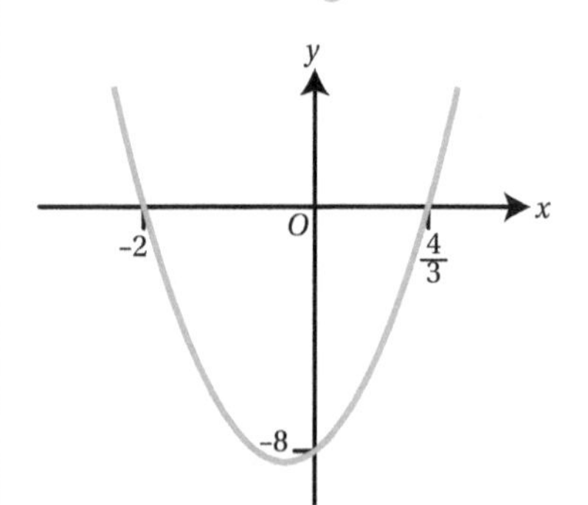

> Sketch the graph. It doesn't have to be to scale but it should show all relevant features and axis intercepts should be labelled.

WORKED EXAMPLE 3.6

Find the equation of the graph, giving your answer in the form $y = ax^2 + bx + c$.

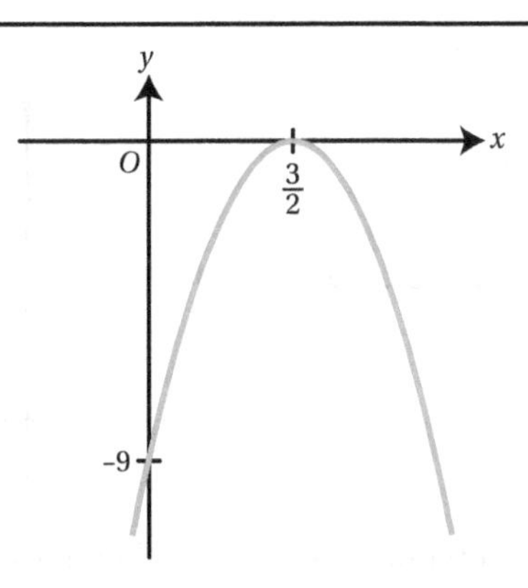

Repeated root at $x = \frac{3}{2}$ $\Rightarrow (2x-3)$ is a repeated factor.	The x-intercepts tell you about the factors.
$y = k(2x-3)^2$	Write the equation in factorised form. The factor $(2x-3)^2$ could be multiplied by any constant so label this unknown constant k.
$-9 = k(0-3)^2$ $-9 = 9k$ $k = -1$	To find the value of k, use the fact that when $x = 0$, $y = -9$.
So the equation is: $y = -(2x-3)^2$ $= -(4x^2-12x+9)$ $= -4x^2+12x-9$	Expand to give the equation in the form required.

EXERCISE 3B

1 Match each equation with its graph.

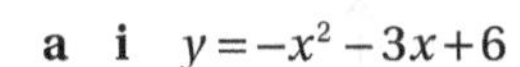

a **i** $y = -x^2 - 3x + 6$

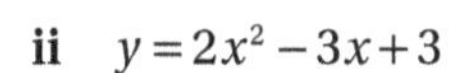

ii $y = 2x^2 - 3x + 3$

iii $y = x^2 - 3x + 6$

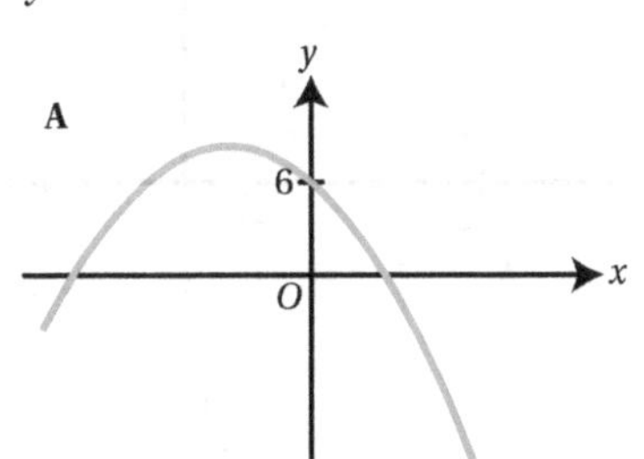

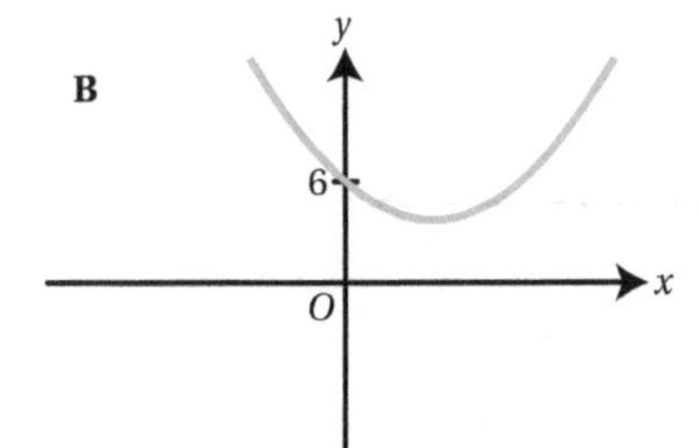

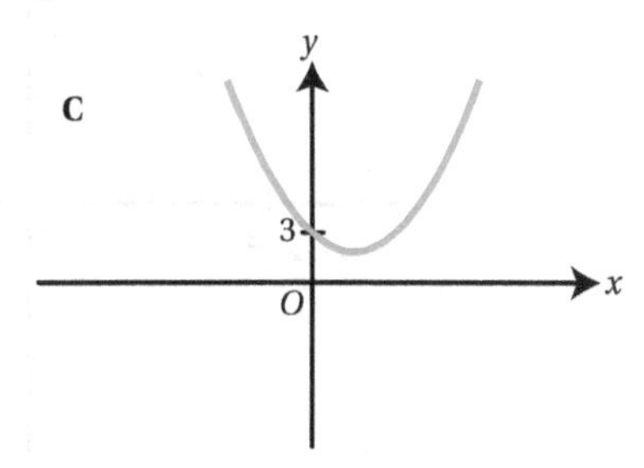

b **i** $y = -x^2 + 2x - 3$ **ii** $y = -x^2 + 2x + 3$ **iii** $y = x^2 + 2x + 3$

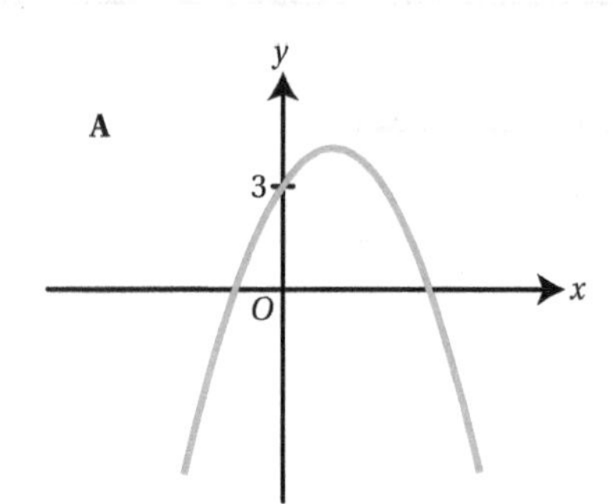

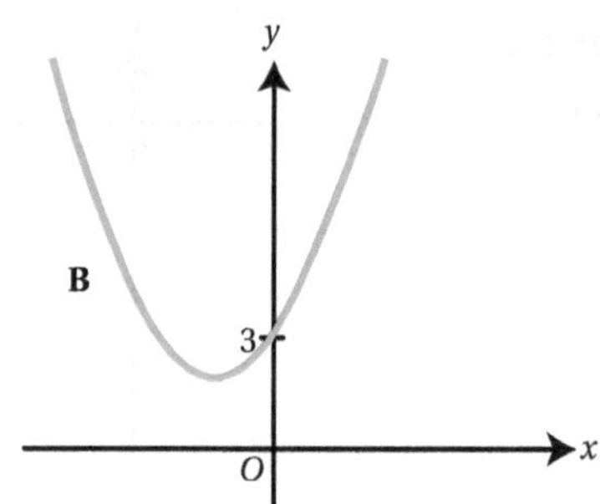

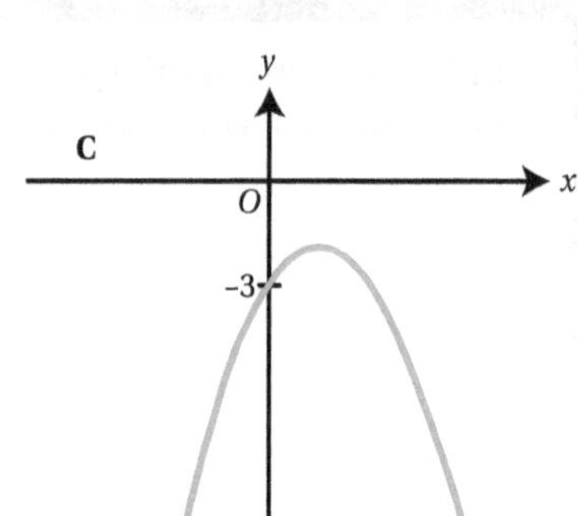

2 Sketch the graph of each quadratic equation, labelling all axis intercepts.

a **i** $y = x^2 - 3x - 10$ **ii** $y = 2x^2 + 11x + 12$

b **i** $y = -3x^2 + 14x - 8$ **ii** $y = 6 - 5x - x^2$

c **i** $y = 3x^2 + 6x$ **ii** $y = 4x - x^2$

d **i** $y = -4x^2 - 20x - 25$ **ii** $y = 4x^2 - 4x + 1$

3 Find the equation of each graph, writing it in the form $y = ax^2 + bx + c$.

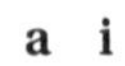
a **i**

ii

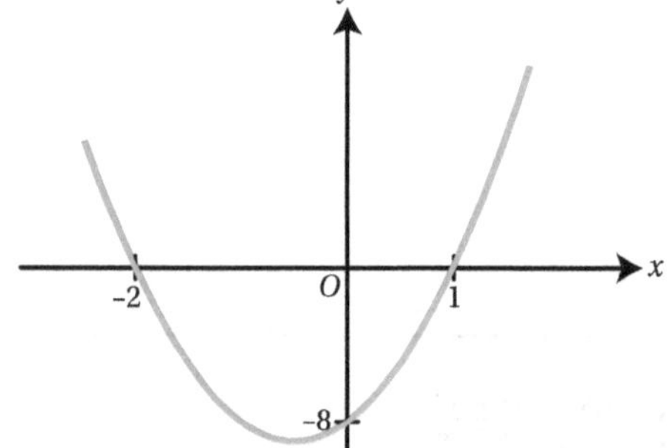

b **i**

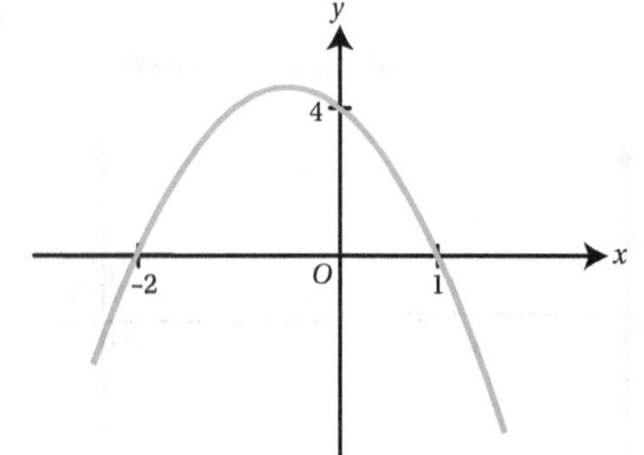

ii

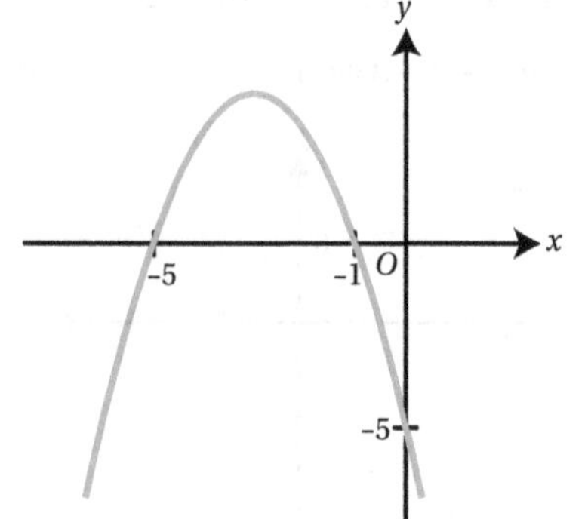

Section 3: Completing the square

It can be useful to rewrite quadratic functions as the square of a term in brackets. Sometimes this just means factorising. For example:

$$x^2+10x+25=(x+5)^2$$

If the function doesn't factorise as a bracket squared (known as a perfect square) then you will need to adjust the constant at the end. For example:

$$x^2+10x=(x+5)^2-25$$

Note that:

- the number inside the brackets is always half the coefficient of x: $\frac{10}{2}=5$
- the constant subtracted at the end is always the square of the number inside the brackets: $5^2=25$

You can use this process, called completing the square, to write any quadratic in the form $a(x+p)^2+q$, as illustrated in the next examples.

Rewind

You are probably already familiar with this method from GCSE, but perhaps not in the more tricky cases shown in Worked examples 3.8 and 3.9. If you don't like this method you can also use comparing coefficients from Worked example 1.1.

WORKED EXAMPLE 3.7

Express x^2-8x+3 in the form $(x+p)^2+q$, stating the values of p and q.

$x^2-8x+3=(x-4)^2-(-4)^2+3$

Halve the coefficient of x.
Subtract $(-4)^2$.
The constant term +3 is still there.

$=(x-4)^2-16+3$

Simplify the constant.

$=(x-4)^2-13$

$\therefore p=-4, q=-13$

WORKED EXAMPLE 3.8

Express x^2+5x+7 in the form $(x+p)^2+q$, stating the values of p and q.

$x^2+5x+7=\left(x+\frac{5}{2}\right)^2-\left(\frac{5}{2}\right)^2+7$

Halve the coefficient of x.
Subtract $\left(\frac{5}{2}\right)^2$.
The constant term +7 is still there.

$=\left(x+\frac{5}{2}\right)^2-\frac{25}{4}+7$

Simplify the constant.

$=\left(x+\frac{5}{2}\right)^2+\frac{3}{4}$

$\therefore p=\frac{5}{2}, q=\frac{3}{4}$

If the coefficient of x^2 isn't 1, you will need to take it out as a factor before completing the square.

Tip

There is no point factorising the coefficient of x^2 from the constant term as that isn't involved in the process of completing the square.

WORKED EXAMPLE 3.9

$2x^2 - 6x + 2 \equiv a(x+p)^2 + q$.

a Find the value of the constants a, p and q.

b Hence solve the equation $2x^2 - 6x + 2 = 0$.

a $2x^2 - 6x + 2 = 2[x^2 - 3x] + 2$ — Take a factor of 2 out of the first two terms.

$= 2\left[\left(x - \frac{3}{2}\right)^2 - \left(\frac{3}{2}\right)^2\right] + 2$ — Now complete the square on the terms inside the brackets.

$= 2\left[\left(x - \frac{3}{2}\right)^2 - \frac{9}{4}\right] + 2$

$= 2\left(x - \frac{3}{2}\right)^2 - \frac{9}{2} + 2$ — Multiply back by the factor 2.

$= 2\left(x - \frac{3}{2}\right)^2 - \frac{5}{2}$ — Simplify the constant.

$\therefore a = 2, p = -\frac{3}{2}, q = -\frac{5}{2}$

b $2x^2 - 6x + 2 = 0$ — 'Hence' means that you must use the result of part **a**.

$2\left(x - \frac{3}{2}\right)^2 - \frac{5}{2} = 0$

$2\left(x - \frac{3}{2}\right)^2 = \frac{5}{2}$ — Now just rearrange to make x the subject.

$\left(x - \frac{3}{2}\right)^2 = \frac{5}{4}$

$x - \frac{3}{2} = \pm\sqrt{\frac{5}{4}}$ — Remember the $\pm$ when taking the square root.

$x - \frac{3}{2} = \pm\frac{\sqrt{5}}{2}$

$x = \frac{3}{2} \pm \frac{\sqrt{5}}{2}$

You might think that this looks exactly like the sort of answer you get by using the quadratic formula – and you'd be right!

You can use exactly the same method as in Worked example 3.9 to solve the general quadratic equation $ax^2 + bx + c = 0$ to establish the quadratic formula.

PROOF 1

Show that if $ax^2+bx+c=0$ then $x=-\frac{b\pm\sqrt{b^2-4ac}}{2a}$.

$$ax^2+bx+c=0$$

First divide by a to make it easier to complete the square on the quadratic expression.

$$x^2+\frac{b}{a}x+\frac{c}{a}=0$$

$$\left(x+\frac{b}{2a}\right)^2-\frac{b^2}{4a^2}+\frac{c}{a}=0$$

Complete the square: halving $\frac{b}{a}$ gives $\frac{b}{2a}$.

$$\left(x+\frac{b}{2a}\right)^2=\frac{b^2}{4a^2}-\frac{c}{a}$$

Now rearrange as before to make x the subject.

$$\left(x+\frac{b}{2a}\right)^2=\frac{b^2-4ac}{4a^2}$$

$$x+\frac{b}{2a}=\pm\sqrt{\frac{b^2-4ac}{4a^2}}$$

Take the square root on both sides, remembering the $\pm$ sign.

$$=\pm\frac{\sqrt{b^2-4ac}}{2a}$$

$$x=-\frac{b}{2a}\pm\frac{\sqrt{b^2-4ac}}{2a}$$

$$=\frac{-b\pm\sqrt{b^2-4ac}}{2a}$$

WORK IT OUT 3.1

Express $-x^2+10x-7$ in the form $a(x+p)^2+q$.

Which is the correct solution? Identify the errors made in the incorrect solutions.

Solution 1	Solution 2	Solution 3
$-x^2+10x-7=-(x-5)^2-25-7$ $=-(x-5)^2-32$	$-x^2+10x-7=-[x^2-10x]-7$ $=-[(x-5)^2-25]-7$ $=-(x-5)^2+25-7$ $=-(x-5)^2+18$	Multiplying by -1: $x^2-10x+7$ Completing the square: $(x-5)^2-25+7\ =(x-5)^2-18$

As well as enabling you to find roots of a quadratic equation, completing the square also allows you to find the coordinates of the maximum or minimum point of a quadratic function.

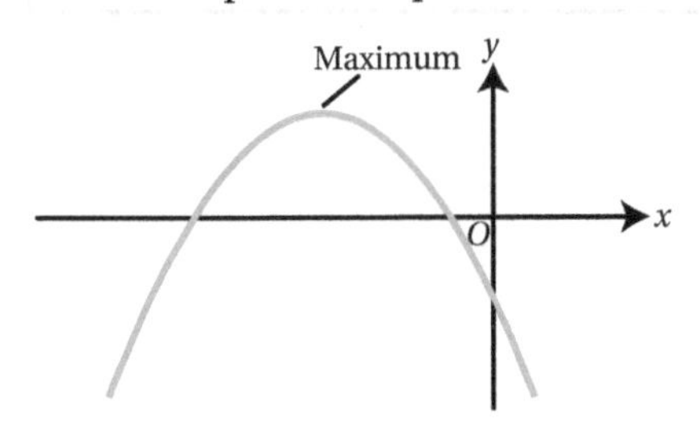

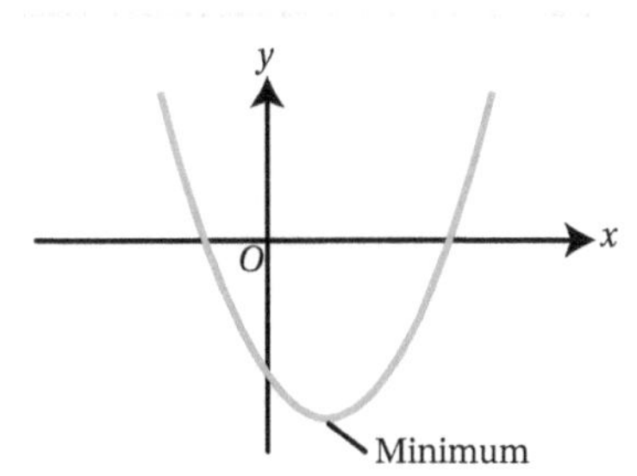

From Worked example 3.7 you know that:

$$x^2 - 8x + 3 = (x-4)^2 - 13$$

Since $(x-4)^2 \geq 0$ for all x, $(x-4)^2 - 13 \geq -13$ for all x.

Then it follows that the smallest value the function can ever take is -13.

This will occur when $(x-4)^2 = 0$, which is when $x = 4$.

So the minimum point of $y = (x-4)^2 - 13$ is $(4, -13)$.

Tip

The maximum or minimum point may also be referred to as the **vertex** of the quadratic, or as the **turning point** – both are general terms to cover either a maximum or minimum.

Key point 3.4

The quadratic $y = a(x+p)^2 + q$ has a turning point at $(-p, q)$.

Fast forward

You will learn about turning points of other functions in Chapter 13.

Quadratic functions have a vertical line of symmetry through their turning point.

If the quadratic has a turning point at (p, q), the equation of the line of symmetry is $x = p$.

WORKED EXAMPLE 3.10

State the coordinates of the turning points of the functions in Worked examples 3.8 and 3.9.

a $y = x^2 + 5x + 7$

b $y = 2x^2 - 6x + 2$

a From Worked example 3.8:

$$x^2 + 5x + 7 = \left(x + \frac{5}{2}\right)^2 + \frac{3}{4}$$

Use the 'completed square' form.

$\therefore$ the coordinates of the turning point are

$\left(-\frac{5}{2}, \frac{3}{4}\right)$.

$y = (x+p)^2 + q$ has a turning point at $(-p, q)$.

b From Worked example 3.9:

$$2x^2 - 6x + 2 = 2\left(x - \frac{3}{2}\right)^2 - \frac{5}{2}$$

Use the 'completed square' form.

$\therefore$ the coordinates of the turning point are

$\left(\frac{3}{2}, -\frac{5}{2}\right)$.

The factor of 2 outside the brackets does not have any effect.

WORKED EXAMPLE 3.11

Find the equation of this quadratic graph.

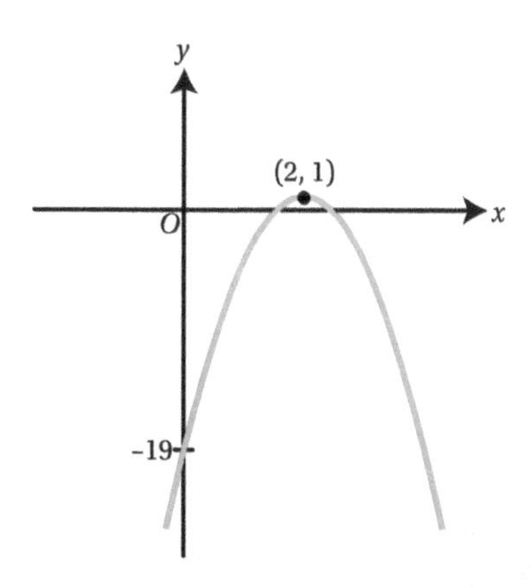

The turning point is at (2, 1) so the function must be of the form $y = a(x-2)^2 + 1$.

Since you are given the coordinates of the turning point, use the 'completed square' form.

When $x = 0$, $y = -19$:
$-19 = a(0-2)^2 + 1$
$-19 = 4a + 1$
$a = -5$

Use the other given point, (0, −19), to find the value of a.

So the equation is $y = -5(x-2)^2 + 1$.

Give the equation. There's no need to express it in the form $y = ax^2 + bx + c$ here.

EXERCISE 3C

1 Write down the coordinates of the vertex of each quadratic function.

a	**i** $y=(x-3)^2+4$	**ii** $y=(x-5)^2+1$
b	**i** $y=2(x-2)^2-1$	**ii** $y=3(x-1)^2-5$
c	**i** $y=(x+1)^2+3$	**ii** $y=(x+7)^2-3$
d	**i** $y=-5(x+2)^2-4$	**ii** $y=-(x+1)^2+5$

2 Write each expression in the form $a(x-k)^2+h$.

a	**i** x^2-6x+4	**ii** $x^2-10x+21$
b	**i** x^2+4x+1	**ii** x^2+6x-3
c	**i** $2x^2-12x+5$	**ii** $3x^2+6x+10$
d	**i** $-x^2+2x-5$	**ii** $-x^2-4x+1$
e	**i** x^2+3x+1	**ii** $x^2-5x+10$
f	**i** $2x^2+6x+15$	**ii** $2x^2-5x-1$

3 Find the equation of each graph, giving it in the form $y = a(x - k)^2 + h$.

a **i**

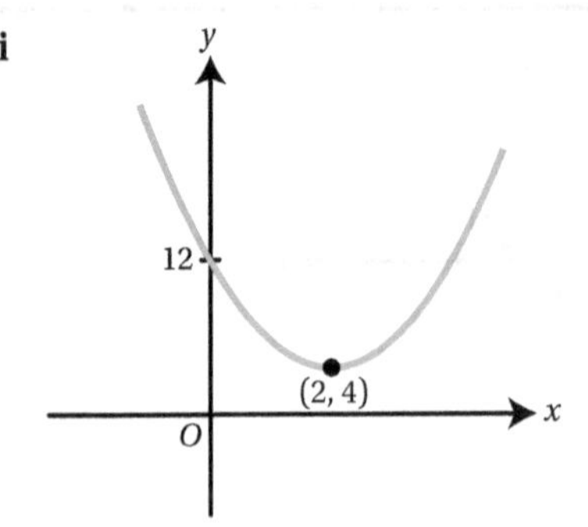

ii

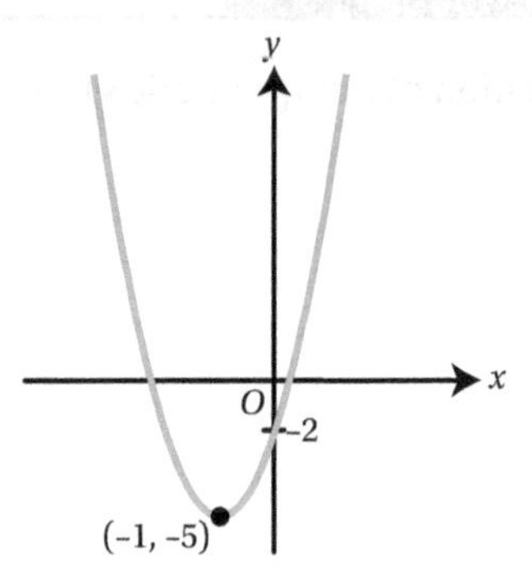

b **i**

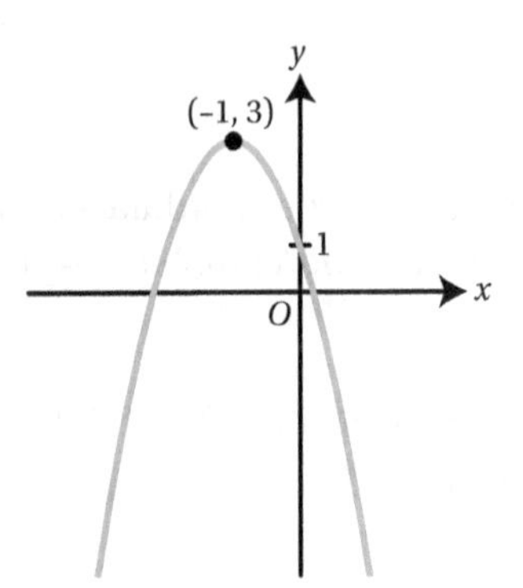

ii

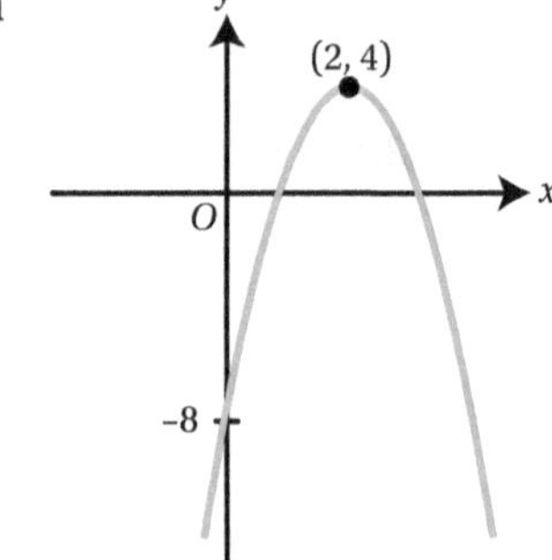

4 $y = x^2 - 6x + 11$

a Write y in the form $(x - a)^2 + b$.

b Find the minimum value of y.

5 The curve $y = a(x + b)^2 + c$ has a minimum point at $(3, 6)$ and passes through the point $(1, 14)$.

a Write down the values of b and c.

b Find the value of a.

6 **a** Express $y = 8x - x^2 - 21$ in the form $a - (x + b)^2$, where a and b are integers to be found.

b Write down the coordinates of the turning point of the graph.

c Hence explain why the equation $8x - x^2 - 21 = 0$ has no real roots.

7 **a** Write $2x^2 + 4x - 1$ in the form $a(x - p)^2 + q$.

b Hence find the exact solutions of the equation $2x^2 + 4x - 1 = 0$.

c Sketch the graph of $y = 2x^2 + 4x - 1$, clearly giving the coordinates of all the axis intercepts and of the minimum point.

d Write down the equation of the line of symmetry of this curve.

8 **a** Write $x^2 + 6x$ in the form $(x + k)^2 + h$.

b Hence find the range of values of p for which $x^2 + 6x = p$ has at least one real solution.

Section 4: Quadratic inequalities

As well as quadratic equations, you will also meet quadratic inequalities, for example: $x^2 < 144$ or $x^2 - 6x - 7 > 0$.

Key point 3.5

To solve quadratic inequalities always sketch the graph.

Gateway to A Level

See Gateway to A Level Section I for a reminder of linear inequalities.

WORKED EXAMPLE 3.12

Solve the inequality $x^2 - 6x - 7 > 0$.

$x^2 - 6x - 7 = 0$

$(x-7)(x+1) = 0$

$x = 7$ or -1 — To sketch the graph you need the roots of the equation.

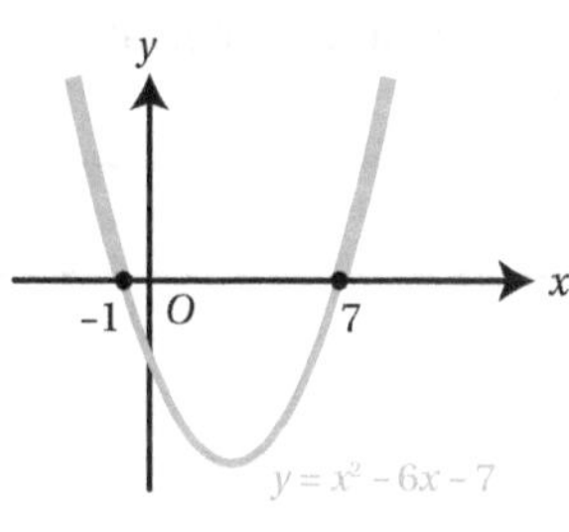

Sketch the graph of $y = x^2 - 6x - 7$.
Since you are solving the inequality $x^2 - 6x - 7 > 0$ you want the part where the graph is positive ($y > 0$).

$\therefore x < -1$ or $x > 7$ — There are two parts of the graph that give the required values of x, so you need to write two inequalities.

As with quadratic equations, you might need to rearrange the inequality first.

WORKED EXAMPLE 3.13

Solve the inequality $5 + 3x - 2x^2 > 1 - 4x$.

$5 + 3x - 2x^2 > 1 - 4x$ — Rearrange: make the coefficient of x^2 positive to make the working easier.

$0 > 2x^2 - 7x - 4$

$2x^2 - 7x - 4 = 0$ — Solve the equation.

$(2x+1)(x-4) = 0$

$x = -\frac{1}{2}$ or 4

Continues on next page

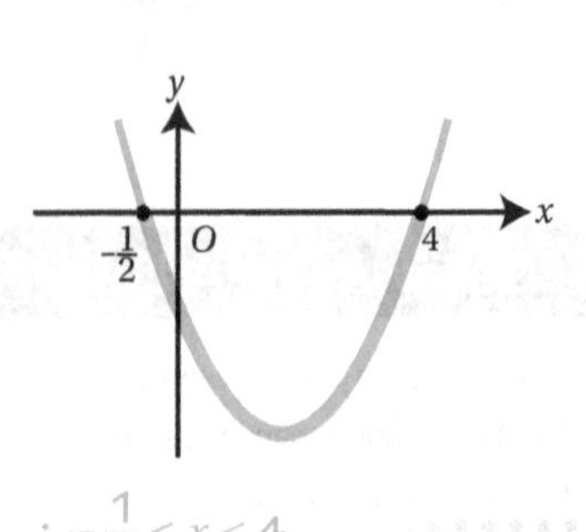

Sketch $y = 2x^2 - 7x - 4$.

Since you are solving the inequality $2x^2 - 7x - 4 < 0$ you want the part where the graph is negative ($y < 0$).

$\therefore -\frac{1}{2} < x < 4$

There is just one part of the graph that gives the required values of x, so write one inequality.

You can illustrate inequalities on a number line. This is particularly useful if there is more than one inequality.

Rewind

Recap Chapter 1 if you need a reminder about interval notation.

WORKED EXAMPLE 3.14

Solve simultaneously $x^2 < 16$ with $x^2 \geqslant 9$.

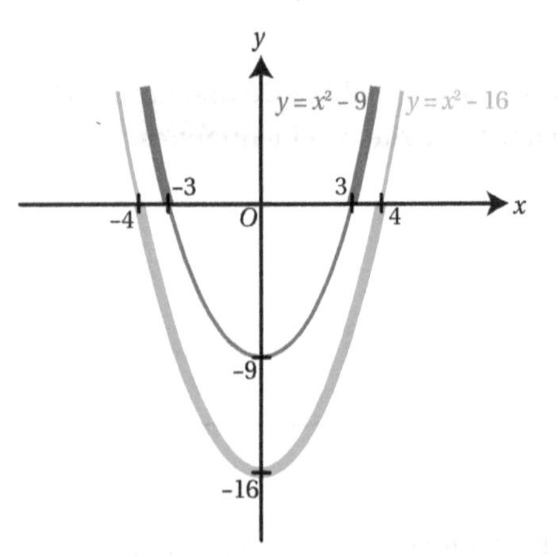

Sketch $y = x^2 - 16$ and $y = x^2 - 9$.

$\therefore -4 < x < 4$ and $x \leqslant -3, x \geqslant 3$

Give the solutions of each.

Illustrate each solution separately on a number line.

The solution to the simultaneous inequality is the region covered by both the red and the green line.

So:

$x \in (-4, -3] \cup [3, 4)$

Remember that you can express inequalities using the interval notation.

EXERCISE 3D

1 Solve these quadratic inequalities.

a **i** $x^2 \leqslant 8$ **ii** $x^2 < 5$

b **i** $x^2 > 6$ **ii** $x^2 \geqslant 12$

c **i** $(x-4)(x+1) > 0$ **ii** $(2x-5)(3x+2) < 0$

d **i** $(3-x)(x+1) < 0$ **ii** $(4-x)(x-2) > 0$

e **i** $(3-x)(12-x) > 0$ **ii** $(2-x)(-2-x) < 0$

2 Solve these inequalities.

a **i** $x^2 - 5x + 6 < 0$ **ii** $x^2 + x - 6 < 0$

b **i** $x^2 - 4x - 12 \geqslant 0$ **ii** $x^2 + 7x + 6 \geqslant 0$

c **i** $12 + x - x^2 > 0$ **ii** $14 - 9x - x^2 < 0$

d **i** $2x^2 + 3x - 5 \leqslant 0$ **ii** $5x^2 + 6x + 1 \leqslant 0$

e **i** $6 + 11x - 2x^2 \leqslant 0$ **ii** $-3x^2 + 16x - 5 \geqslant 0$

3 Solve the inequality $2x^2 > 6 - x$.

4 Find the set of values of x for which $2x^2 + 3x + 1 \leqslant 11 + 4x - x^2$.

5 A ball is thrown upwards and its height h m at time t seconds is given by $h = 7t - 4.9t^2$. How long does the ball spend more than 1.5 m above ground?

6 **a** Solve these inequalities.

i $7x - 5 < 3x + 5$ **ii** $2x^2 - 7x < 4x - 5$

b Hence find the set of values of x for which both $7x - 5 < 3x + 5$ and $2x^2 - 7x < 4x - 5$.

7 Solve simultaneously $x^2 + 6 > 5x$ and $x^2 \geqslant 1$.

8 Find the range of values of x for which both $2x^2 \geqslant 4x$ and $5x^2 - 13x - 6 \leqslant 0$.

9 The cost of producing n items is £$(950 + 63n)$. The items can be sold for £$(280 - 5n)$ **per item**.

How many items can be produced and sold in order to make a profit? Give your answer in the form $M \leqslant n \leqslant N$ where M and N are integers.

Section 5: The discriminant

Fast forward

If you study Further Mathematics you will meet a new type of number, called an imaginary number, which makes it possible to find roots of functions such as this.

Consider trying to apply the quadratic formula to find solutions of $x^2 - 3x + 3 = 0$:

$$x = \frac{3 \pm \sqrt{(-3)^2 - 4 \times 1 \times 3}}{2}$$
$$= \frac{3 \pm \sqrt{-3}}{2}$$

As the square root of a negative number is not a real number, it follows that the expression has no real roots.

This will clearly happen whenever the expression inside the square root, $b^2 - 4ac$, is negative.

Similarly, if $b^2 - 4ac = 0$, then the quadratic formula becomes

$$x = \frac{-b \pm \sqrt{0}}{2a}$$
$$= \frac{-b}{2a}.$$

Then there is just the one root, at $x = -\frac{b}{2a}$.

In all other cases there will be two roots.

The expression $b^2 - 4ac$ is called the **discriminant** of the quadratic (often symbolised by the Greek letter Δ).

These graphs are examples of the three possible situations.

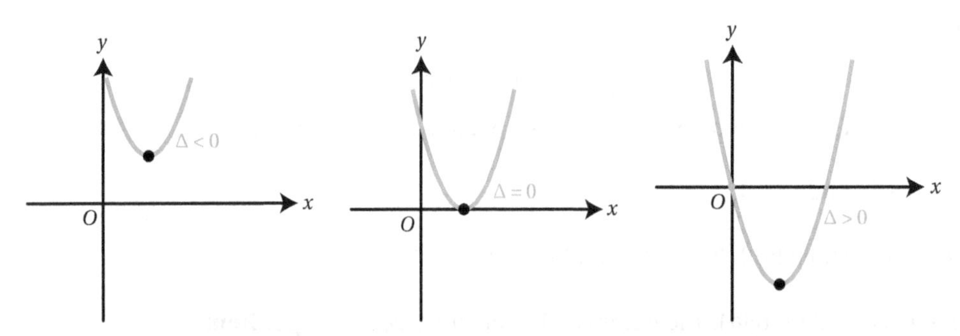

Key point 3.6

For the quadratic equation $ax^2 + bx + c = 0$:

- if $\Delta < 0$ the equation has no real roots
- if $\Delta = 0$ the equation has one (repeated) root
- if $\Delta > 0$ the equation has two distinct real roots

where $\Delta = b^2 - 4ac$ is the discriminant.

Tip

Repeated roots can also be referred to as equal roots.

WORKED EXAMPLE 3.15

Find the exact values of k for which the quadratic equation $kx^2-(k+2)x+3=0$ has a repeated root.

$b^2-4ac=0$ — A repeated root means that $b^2-4ac=0$.

$(k+2)^2-4(k)(3)=0$ — $a=k,\ b=-(k+2),\ c=3$.

$k^2+4k+4-12k=0$ — This is a quadratic equation in k.

$k^2-8k+4=0$

$$k=\frac{8\pm\sqrt{8^2-4\times4}}{2}$$ — It doesn't factorise, so use the quadratic formula.

$$=\frac{8\pm\sqrt{48}}{2}$$

$$=\frac{\left(8\pm4\sqrt{3}\right)}{2}$$

$$=4\pm2\sqrt{3}$$

WORKED EXAMPLE 3.16

Find the set of values of k for which the equation $2x^2-(k+1)x+5-k=0$ has two distinct real solutions.

$b^2-4ac>0$ — Two distinct real roots $\Rightarrow$ $b^2-4ac>0$.

$(k+1)^2-4(2)(5-k)>0$ — $a=2,\ b=-(k+1),\ c=5-k$.

$k^2+2k+1-40+8k>0$

$k^2+10k-39>0$ — This is a quadratic inequality in k.

$k^2+10k-39=0$ — Solve the equation $k^2+10k-39=0$.

$(k+13)(k-3)=0.$

$k=-13, 3$

Continues on next page

Tip

Questions of this type nearly always lead to a quadratic equation or inequality for k.

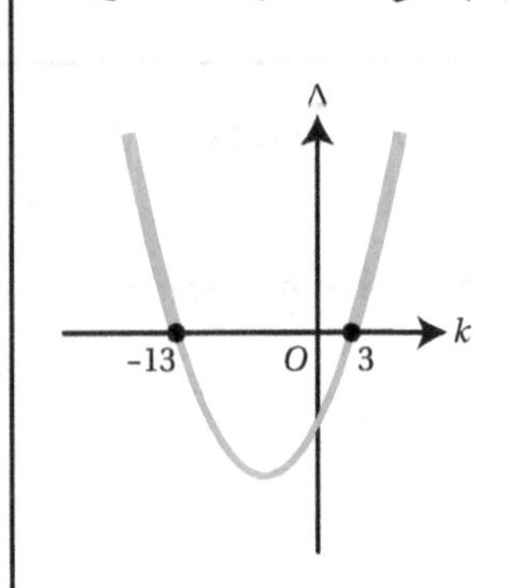

You want the region where $\Delta > 0$.

$\therefore k < -13$ or $k > 3$

State the range of values of k, as required.

Common error

Note that the graph drawn in Worked example 3.16 is not the graph of the original quadratic equation (in the variable x) – it's the graph of Δ against k. The example hasn't solved the original equation, but a quadratic inequality for k that ensures there are two distinct solutions for x in the original equation.

When $\Delta < 0$, the graph does not intersect the x-axis, so it is either entirely above or entirely below it. The two cases are distinguished by the value of a.

Elevate

See Support Sheet 3 for a further example of the type of question in Worked example 3.17 and for more practice questions.

Key point 3.7

For a quadratic function with $\Delta < 0$:

- if $a > 0$ then $y > 0$ for all x.
- if $a < 0$ then $y < 0$ for all x.

WORKED EXAMPLE 3.17

$y = -3x^2 + kx - 12$

Find the values of k for which $y < 0$ for all x.

If $y < 0$ for all x there are no real roots $\Rightarrow \Delta < 0$.

y is a negative quadratic. $y < 0$ means that the graph is entirely below the x-axis. This will happen when $f(x) = 0$ has no real roots.

$b^2 - 4ac < 0$

$k^2 - 4(-3)(-12) < 0$ — $a = -3,\ b = k,\ c = -12$.

$k^2 - 144 < 0$ — This is a quadratic inequality in k.

$k^2 - 144 = 0$ — Solve the equation $k^2 - 144 = 0$.

$(k-12)(k+12) = 0$

$k = -12, 12$

Continues on next page

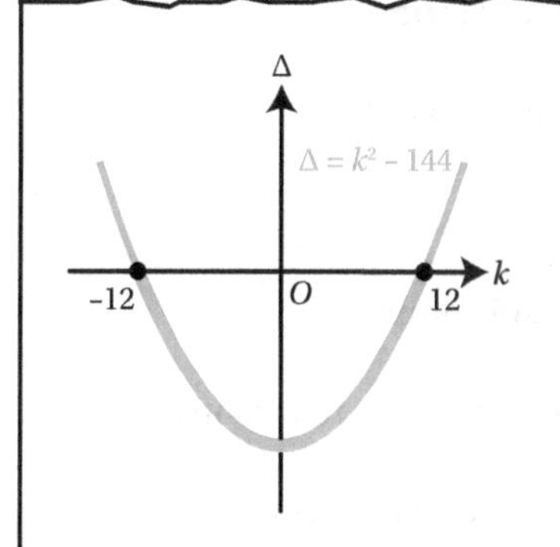

You want the region where $y < 0$.

$\therefore -12 < k < 12$

State the range of values of k as required.

EXERCISE 3E

1 Evaluate the discriminant of each quadratic expression.

a **i** $x^2 + 4x - 5$ **ii** $x^2 - 6x - 8$

b **i** $2x^2 + x + 6$ **ii** $3x^2 - x + 10$

c **i** $3x^2 - 6x + 3$ **ii** $9x^2 - 6x + 1$

d **i** $12 - x - x^2$ **ii** $-x^2 - 3x + 10$

2 State the number of roots for each expression from Question 1.

3 Find the values of k for which:

a **i** the equation $2x^2 - x + 3k = 0$ has two distinct real roots

ii the equation $3x^2 + 5x - k = 0$ has two distinct real roots

b **i** the equation $5x^2 - 2x + (2k - 1) = 0$ has equal roots

ii the equation $2x^2 + 3x - (3k + 1) = 0$ has equal roots

c **i** the equation $-x^2 + 3x + (k - 1) = 0$ has real roots

ii the equation $-2x^2 + 3x - (2k + 1) = 0$ has real roots

d **i** the equation $3kx^2 - 3x + 2 = 0$ has no real solutions

ii the equation $-kx^2 + 5x + 3 = 0$ has no real solutions

e **i** the quadratic expression $(k - 2)x^2 + 3x + 1$ has a repeated root

ii the quadratic expression $-4x^2 + 5x + (2k - 5)$ has a repeated root

f **i** the graph of $y = x^2 - 4x + (3k + 1)$ is tangent to the x-axis

ii the graph of $y = -2kx^2 + x - 4$ is tangent to the x-axis

g **i** the expression $-3x^2 + 5k$ has no real roots

ii the expression $2kx^2 - 3$ has no real roots.

4. Find the values of m for which the quadratic equation $mx^2 - 4x + 2m = 0$ has equal roots.

5. Find the exact values of k such that the equation $3x^2 + (2k+1)x - 4k = 0$ has a repeated root.

6. Find the range of values of c such that $2x^2 - 3x + (2c-1) \geqslant 0$ for all x.

7. Find the set of values of k for which the equation $x^2 - 2kx + 6k = 0$ has no real solutions.

8. Find the range of value of k for which the quadratic equation $kx^2 - (k+3)x - 1 = 0$ has no real roots.

9. Find the range of values of m for which the equation $mx^2 + mx - 2 = 0$ has at least one real root.

10. Find the possible values of m such that $mx^2 + 3x - 4 < 0$ for all x.

Section 6: Disguised quadratics

You will often meet equations that can be turned into quadratics by making a substitution.

WORKED EXAMPLE 3.18

Solve the equation $x^4 - 3x^2 - 4 = 0$.

Let $y = x^2$ — Using the substitution $y = x^2$ turns this into a quadratic equation, where $x^4 = y^2$.

$y^2 - 3y - 4 = 0$ — This is now a standard quadratic equation.

$(y+1)(y-4) = 0$

$y = -1$ or $y = 4$

$x^2 = -1$ (reject) — Now use the substitution to find x.

or

$x^2 = 4$ — Note that some values of y will not lead to a corresponding value of x since square numbers must be postive.

$x = 2$ or -2

Other substitutions may not be so clear. In particular, it is quite common to be given an exponential equation that needs a substitution; look out for an a^x and an a^{2x}.

Fast forward

You will see how to solve more complicated exponential equations in Chapter 7.

WORKED EXAMPLE 3.19

Solve the equation $4^x - 10\times 2^x + 16 = 0$.

$4^x - 10\times 2^x + 16 = 0$
$2^{2x} - 10\times 2^x + 16 = 0$

$4^x = (2^2)^x = 2^{2x}$

Let $y = 2^x$
$y^2 - 10y + 16 = 0$

The substitution $y = 2^x$ turns this into a quadratic equation, since $2^{2x} = (2^x)^2$.

$(y-2)(y-8) = 0$
$y = 2$ or $y = 8$

This is now a standard quadratic equation.

$2^x = 2$
$x = 1$
or
$2^x = 8$
$x = 3$

Use the substitution to find the value of x.

EXERCISE 3F

1. Solve these equations, giving your answers in an exact form.
   - **a** **i** $a^4 - 10a^2 + 21 = 0$ **ii** $x^4 - 7x^2 + 12 = 0$
   - **b** **i** $2x^6 + 7x^3 = 15$ **ii** $a^6 + 7a^3 = 8$
   - **c** **i** $x^2 - 4 = \frac{2}{x^2}$ **ii** $x^2 + \frac{36}{x^2} = 12$
   - **d** **i** $x - 6\sqrt{x} + 8 = 0$ **ii** $x - 10\sqrt{x} + 24 = 0$
   - **e** **i** $3^{2x} - 12\times 3^x + 27 = 0$ **ii** $2^{2x} - 17\times 2^x + 16 = 0$
2. By letting $y = \sqrt{x}$, solve the equation $x - \sqrt{x} - 6 = 0$.
3. Use an appropriate substitution to solve $x^2 + \frac{9}{x^2} = 10$.
4. Use an appropriate substitution to solve $x^3 - 9x^{1.5} + 8 = 0$.
5. **a** Show that the equation $9(1 + 9^{x-1}) = 10\times 3^x$ can be rewritten as $3^{2x} - 10\times 3^x + 9 = 0$.
   b Hence solve the equation $9(1 + 9^{x-1}) = 10\times 3^x$.
6. Solve the equation $5^x = 6 - 5^{1-x}$.
7. Solve the equation $4^{x+0.5} - 17\times 2^x + 8 = 0$.
8. Solve the equation $x = \sqrt{x} + 12$.
9. **a** Find the solution to $x^4 - bx^2 + c = 0$ in terms of b and c.
   b Find a condition on the constants b and c if the equation has two solutions.
   c Find a condition on the constants b and c if the equation has three solutions.

Checklist of learning and understanding

- Quadratic functions have the general form $f(x) = ax^2 + bx + c$. The main features are summarised in this table.

Feature	What to look at	Conclusion
Overall shape	The sign of a	$a > 0$ $a < 0$
y-intercept	The value of c	y-intercept $(0, c)$
Turning point	'Completed square' form: $y = a(x+p)^2 + q$	Turning point $(-p, q)$
x-intercepts	Factorise $f(x) = 0$ directly or use the quadratic formula: $x = \frac{-b \pm \sqrt{b^2 - 4ac}}{2a}$	Roots p and q are given by the x-intercepts $(p, 0)$ and $(q, 0)$
The number of real roots	Discriminant: $\Delta = b^2 - 4ac$	• $\Delta > 0$: two distinct roots • $\Delta > 0$: one repeated root (equal roots, repeated root) • $\Delta < 0$: no real roots

- To solve quadratic inequalities, rearrange to make one side zero and sketch the graph.
- A substitution can transform certain types of equation into a quadratic equation.

Mixed practice 3

1. $2x^2 - 6x + 5 \equiv 2(x+a)^2 + b$. Find the values of a and b.

   Choose from these options.

   A $a = -\frac{3}{2}, b = \frac{1}{2}$ **B** $a = 3, b = -4$ **C** $a = -\frac{3}{2}, b = \frac{11}{4}$ **D** $a = 3, b = \frac{1}{2}$

2. A quadratic function passes through the points $(k, 0)$ and $(k+4, 0)$. Find, in terms of k, the x-coordinate of the turning point.

   Choose from these options.

   A $x = -k - 4$ **B** $x = -k$ **C** $x = k + 1$ **D** $x = k + 2$

3. Solve this equation algebraically.

   $(2x-3)(x-5) = (x-3)^2$

4. By using an appropriate substitution or otherwise, solve this equation algebraically.

   $x^4 - 5x^2 + 4 = 0$

5. The quadratic function $y = (x-a)^2 + b$ has a turning point at $(3, 7)$.

   a State whether this turning point is a maximum or a minimum point.

   b State the values of a and b.

6. The quadratic function $y = a(x-b)^2 + c$ passes through the points $(-2, 0)$ and $(6, 0)$. Its maximum y value is 48. Find the values of a, b and c.

7. The diagram represents the graph of the function $f(x) = (x+p)(x-q)$.

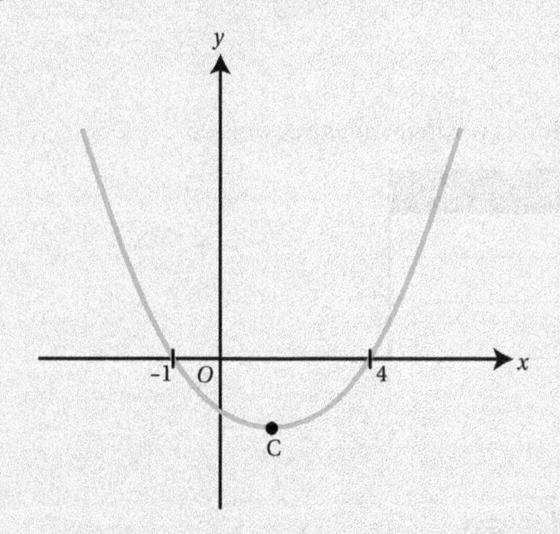

   a Write down the values of p and q if they are both positive.

   b The function has a minimum value at the point C. Find the x coordinate of C.

8 **a** **i** Express $16 - 6x - x^2$ in the form $p - (x + q)^2$, where p and q are integers.

ii Hence write down the maximum value of $16 - 6x - x^2$.

b **i** Factorise $16 - 6x - x^2$.

ii Sketch the curve with equation $y = 16 - 6x - x^2$, stating the values of x where the curve crosses the x-axis and the value of the y-intercept.

[©AQA 2014]

9 For what values of k ($k \neq -1$) does the quadratic equation $(k+1)x^2 + (3k-2)x + k + 1 = 0$ have distinct real roots?

Choose from these options.

A $k > 4, k < 0$ **B** $0 < k < 4$ **C** $k < 4$ **D** $k > 0$

10 Solve simultaneously $x^2 - 2x > 0$ and $x^2 - 4x + 3 \geqslant 0$.

11 The diagram shows the graph of the function $y = ax^2 + bx + c$.

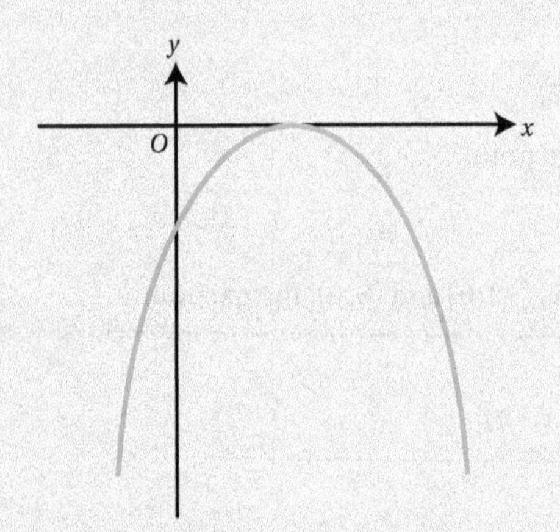

Copy and complete the table to show whether each expression is positive, negative or zero.

Expression	Positive	Negative	Zero
a			
c			
$b^2 - 4ac$			
b			

12 **a** Write $x^2 - 10x + 35$ in the form $(x - p)^2 + q$.

b Hence, or otherwise, find the maximum value of $\frac{1}{(x^2 - 10x + 35)^3}$.

13 Find the set of values of k for which the equation $2kx^2 + (k+1)x + 1 = 0$ has no real roots.

14 Solve algebraically the equation $x^{\frac{1}{4}} + 2x^{-\frac{1}{4}} = 3$.

15 a i Express $2x^2-20x+53$ in the form $2(x-p)^2+q$ where p and q are integers.

ii Use your result from part **a i** to explain why the equation $2x^2-20x+53=0$ has no real roots.

b The quadratic equation $(2k-1)x^2+(k+1)x+k=0$ has distinct real roots.

i Show that $7k^2-6k-1<0$.

ii Hence find the possible values of k.

[©AQA 2010]

16 Alexia and Michaela were both trying to solve a quadratic equation of the form $x^2+bx+c=0$.

Unfortunately Alexia misread the value of b and found that the solutions were 6 and 1.

Michaela misread the value of c and found that the solutions were 4 and 1.

What were the correct solutions?

17 Find the values of k for which the line $y=2x-k$ is tangent to the curve with equation $x^2+y^2=5$.

18 Let α and β denote the roots of the quadratic equation $x^2-kx+(k-1)=0$.

a Express α and β in terms of the real parameter k.

b Given that $\alpha^2+\beta^2=17$, find the possible values of k.

19 Let $q(x)=kx^2+(k-2)x-2$. Show that the equation $q(x)=0$ has real roots for all values of k.

20 Two cars are travelling along two straight roads that are perpendicular to each other and meet at the point O, as shown in the diagram. The first car starts 50 km west of O and travels east at the constant speed of 20 km/h. At the same time, the second car starts 30 km south of O and travels north at the constant speed of 15 km/h.

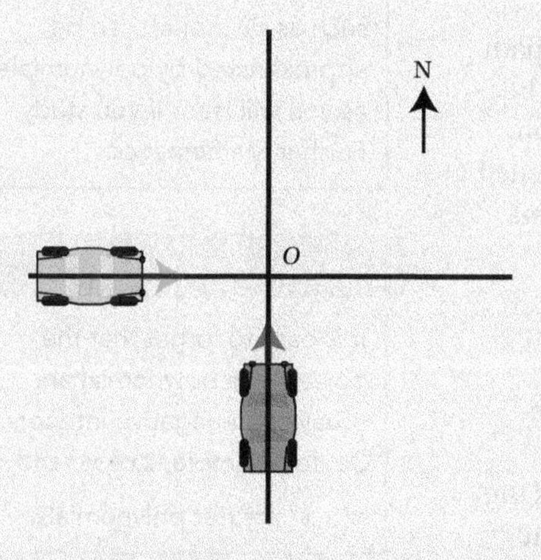

a Show that at time t (hours), the distance d (km) between the two cars satisfies $d^2=625t^2-2900t+3400$.

b Hence find the least distance between the two cars.

Elevate

See Extension Sheet 3 for a selection of more challenging problems.

4 Polynomials

In this chapter you will learn how to:

- define a polynomial
- find the product of two polynomials
- find the quotient of two polynomials
- quickly find factors of a polynomial
- sketch polynomials.

Before you start...

GCSE	You should be able to multiply out brackets.	1 Expand $(2x+1)(x-3)$.
GCSE	You should be able to factorise quadratic expressions.	2 Factorise $x^2-8x+15$.
GCSE	You should be able to use the quadratic formula to solve quadratic equations.	3 Solve $x^2+4x+2=0$.
Chapter 2	You should be able to work with indices.	4 Simplify $x^2 \times x^4$.

Why study polynomials?

In Chapter 3 you were introduced to various properties of quadratic functions. As well as being mathematically interesting, quadratic functions are used to model many real-world situations, such as the path of a projectile. To include more real-world situations, quadratics can be extended to produce expressions that include terms in x^3, x^4 and so on. For example, the relationship between height and mass may be modelled by a cubic equation. This group of functions, called **polynomials**, turns out to be remarkably powerful.

Fast forward

Many other functions, such as $\sin x$ or e^x, can be approximated by polynomials, as you will learn if you study Further Mathematics.

Section 1: Working with polynomials

A polynomial is a function made up of a sum of terms containing non-negative (positive or zero) integer powers of an unknown, such as x. Polynomial functions are classified according to the highest power of the unknown (x) occurring in the function. This is called the **degree** of the polynomial.

Common error

It is easy to forget that the powers in a polynomial are always non-negative integers. So, for example, $2x+x^{-1}$ and $x^2+x^{\frac{1}{3}}$ are not polynomials.

General form of the polynomial	Degree	Name	Example
$f(x)=a$	0	constant function	$y=5$
$f(x)=ax+b$	1	linear function	$y=x+7$
$f(x)=ax^2+bx+c$	2	quadratic function	$y=-3x^2+4x-1$
$f(x)=ax^3+bx^2+cx+d$	3	cubic function	$y=2x^3+7x$
$f(x)=ax^4+bx^3+cx^2+dx+e$	4	quartic function	$y=x^4-x^3+2x+\frac{1}{2}$

Rewind

You met the term 'coefficient' in Chapter 1. Coefficients are the constants $a, b, c\ldots$ in front of the powers of x.

Did you know?

The Greeks knew how to solve quadratic equations – and general cubics and quartics were solved in 14th-century Italy. For over three hundred years nobody was able to produce a general solution to the quintic equation until, in 1821, Niels Abel used a branch of mathematics called group theory to prove that there could never be a 'quintic formula'.

Gateway to A Level

See Gateway to A Level Section A for revision of expanding brackets such as these.

You already know how to expand expressions with two and three brackets, for example, $(2x-1)(x+3)$ or $(3x-2)(x+1)(x-5)$, but if the brackets contain more than two terms each, it is important to take care and organise the multiplication systematically to avoid missing any terms.

WORKED EXAMPLE 4.1

Expand $(x^3+3x^2-2)(x^2-5x+4)$.

$(x^3+3x^2-2)(x^2-5x+4)$

$=x^3(x^2-5x+4)+3x^2(x^2-5x+4)-2(x^2-5x+4)$

$=x^5-5x^4+4x^3$

$\quad +3x^4-15x^3+12x^2$

$\quad -2x^2+10x-8$

Multiply each term inside the first bracket, in turn, by the whole of the expression in the second bracket.

$=x^5-2x^4-11x^3+10x^2+10x-8$

Then collect like terms.

EXERCISE 4A

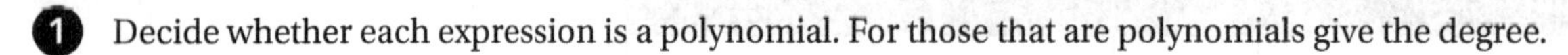

1 Decide whether each expression is a polynomial. For those that are polynomials give the degree.

a $3x^3 - 3x^2 + 2x$ **b** $1 - 3x - x^5$

c $5x^2 - x^{-3}$ **d** $9x^4 - \frac{5}{x}$

e $4e^x + 3e^{2x}$ **f** $x^4 + 5x^2 - 3\sqrt{x}$

g $4x^5 - 3x^3 + 2x^7 - 4$ **h** 1

2 Expand and simplify the brackets in each expression.

a i $(3x-2)(2x^2+4x-7)$ **ii** $(3x+1)(x^2+5x+6)$

b i $(2x+1)(x^3-8x^2+6x-1)$ **ii** $(2x+5)(x^3-6x^2+3)$

c i $(b^2+3b-1)(b^2-2b+4)$ **ii** $(r^2-3r+7)(r^2-8r+2)$

d i $(5-x^2)(x^4-2x^3+1)$ **ii** $(x-x^3)(x^3-x-1)$

3 In what circumstances might you want to expand brackets? In what circumstances is the factorised form more useful?

4 **a** Is it always true that the sum of a polynomial of degree n and a polynomial of degree $n-1$ has degree n?

b Is it always true that the sum of a polynomial of degree n and a polynomial of degree n has degree n?

Section 2: Polynomial division

From Worked example 4.1 you know that:

$$x^5 - 2x^4 - 11x^3 + 10x^2 + 10x - 8 = (x^2 - 5x + 4)(x^3 + 3x^2 - 2)$$

However, if you had just been given $x^5 - 2x^4 - 11x^3 + 10x^2 + 10x - 8$ and told that $x^2 - 5x + 4$ was a factor, it would have been difficult to see that $x^3 + 3x^2 - 2$ was the other factor.

You would have needed to divide $x^5 - 2x^4 - 11x^3 + 10x^2 + 10x - 8$ by $x^2 - 5x + 4$.

$$\frac{x^5 - 2x^4 - 11x^3 + 10x^2 + 10x - 8}{x^2 - 5x + 4} = x^3 + 3x^2 - 2$$

One way to do this is by polynomial long division.

- Divide the leading-order term in the numerator by the leading-order term in the denominator. This is the leading-order term of the answer.
- Multiply this 'answer' term by the whole denominator. Subtract the resulting expression from the numerator.
- Repeat this process until all terms have been accounted for.

There are several ways to set this process out, but the traditional way is given in Worked example 4.2.

Rewind

Another common method is to compare coefficients – this is very similar to the process shown in Chapter 1.

Tip

The resultant polynomial, when one polynomial is divided by another, is called the quotient. In this example, $x^3 + 3x^2 - 2$ is the quotient.

WORKED EXAMPLE 4.2

Calculate $\frac{x^3 - x^2 - 11x - 4}{x - 4}$.

$$\begin{array}{r} x^2 + 3x \quad + 1 \\ (x-4)\overline{)x^3 \quad - x^2 - 11x - 4} \\ \underline{x^3 - 4x^2 \qquad\qquad} \\ 3x^2 - 11x - 4 \\ \underline{3x^2 - 12x \quad} \\ x - 4 \\ \underline{x - 4} \\ 0 \end{array}$$

Divide x^3 by x.

Multiply x^2 by $(x - 4)$.

Subtract to get the remainder. Then divide $3x^2$ by x.

Multiply $3x$ by $(x - 4)$.

Subtract to get the remainder. Then divide x by x.

Multiply 1 by $(x - 4)$.

Subtract – this time there is no remainder.

So: $\frac{x^3 - x^2 - 11x - 4}{x - 4} = x^2 + 3x + 1$

Tip

If you don't like this method, you can also try writing the answer as $ax^2 + bx + c$ and compare coefficients (as illustrated in Worked example 1.1).

WORKED EXAMPLE 4.3

Given that $x - 3$ is a factor of $x^3 + x^2 - 36$, find the other quadratic factor.

$$\begin{array}{r} x^2 + \ 4x + 12 \\ x-3\overline{)x^3 \quad + x^2 + \ 0x - 36} \\ \underline{x^3 - 3x^2 \qquad\qquad\quad} \\ 4x^2 + \ 0x - 36 \\ \underline{4x^2 - 12x \quad\ \ } \\ 12x - 36 \\ \underline{12x - 36} \\ 0 \end{array}$$

Divide x^3 by x. Note that as there is no x term, $0x$ is added to keep everything aligned.

Multiply x^2 by $(x - 3)$.

Subtract to get the remainder. Then divide $4x^2$ by x.

Multiply $4x$ by $(x - 3)$.

Subtract to get the remainder. Then divide $12x$ by x.

Multiply 12 by $(x - 3)$.

Subtract – now there is no remainder.

So, $x^3 + x^2 - 36 = (x - 3)(x^2 + 4x + 12)$

EXERCISE 4B

1. Use polynomial division to simplify each expression.

   a i $\frac{x^3 - x^2 - 7x - 20}{x - 4}$ ii $\frac{x^3 + x^2 + 3x + 27}{x + 3}$

   b i $\frac{x^3 + 4x^2 - 5}{x - 1}$ ii $\frac{x^3 - 4x^2 + 8}{x - 2}$

   c i $\frac{x^3 + 5x + 18}{x + 2}$ ii $\frac{x^3 - 4x - 15}{x - 3}$

2. Use polynomial division to find the other factor when:

   a i $x + 4$ is a factor of $x^3 - 4x^2 - 35x - 12$ ii $x + 3$ is a factor of $x^3 + x^2 - 3x + 9$

   b i $x - 2$ is a factor of $x^3 - 2x - 4$ ii $x - 3$ is a factor of $x^3 - 5x - 12$

   c i $x + 1$ is a factor of $x^3 + x^2 - 2x - 2$ ii $x + 4$ is a factor of $x^3 + 4x^2 + 3x + 12$

Section 3: The factor theorem

Algebraic division allows you to find the other factors of a polynomial, if you know the first, but it can be difficult to find the first factor. A useful tool to help with this is the **factor theorem**.

Focus on...

See Focus on... Proof 1 for a proof of the factor theorem.

Key point 4.1

The factor theorem

If $f(a) = 0$ then $(x - a)$ is a factor of $f(x)$.

Common error

Remember that the sign of the value being substituted and the sign in the linear factor are different: in particular, if $f(-a) = 0$, then $(x + a)$ is a factor.

WORKED EXAMPLE 4.4

Show that $x + 3$ is a factor of $f(x) = x^3 - 6x^2 - 9x + 54$.

$$\begin{aligned} f(-3) &= (-3)^3 - 6(-3)^2 - 9(-3) + 54 \\ &= -27 - 54 + 27 + 54 \\ &= 0 \end{aligned}$$

By the factor theorem, if $f(-3) = 0$ then $x + 3$ is a factor.

Therefore $x + 3$ is a factor of $f(x)$

WORKED EXAMPLE 4.5

Given that $x+3$ is a factor of $f(x) = x^3 - 6x^2 - 9x + 54$, solve the equation $f(x) = 0$.

$$
\begin{array}{r}
x^2 - 9x + 18 \\
x+3\overline{)\,x^3 - 6x^2 - 9x + 54} \\
x^3 + 3x^2 \qquad\qquad \\
\hline
-9x^2 - 9x + 54 \\
-9x^2 - 27x \qquad \\
\hline
18x + 54 \\
18x + 54 \\
\hline
0
\end{array}
$$

Polynomial division will give the other factor (which will be a quadratic).

So: $x^3 - 6x^2 - 9x + 54 = (x+3)(x^2 - 9x + 18)$
$= (x+3)(x-3)(x-6)$

Factorise the quadratic.

$(x+3)(x-3)(x-6) = 0$
$x = -3, 3, 6$

Solve as normal.

If you aren't given the first factor, you will need to use the factor theorem to find it. The only way to do this is to try small integer values until you find one that works.

Tip

If the expression does factorise, you only need to try numbers that are factors of the constant term.

WORKED EXAMPLE 4.6

Solve the equation $p(x) = x^3 + x^2 - 13x + 14 = 0$.

$p(0) = 14$
$p(1) = 3$
$p(-1) = 27$

Substitute into $p(x)$ small integer values that are factors of the constant term 14. Don't forget to try negatives!

$p(2) = 0$
$\therefore (x-2)$ is a factor of $p(x)$.

$p(2) = 0$ so the factor theorem gives the first factor.

$$
\begin{array}{r}
x^2 + 3x - 7 \\
x-2\overline{)\,x^3 + x^2 - 13x + 14} \\
x^3 - 2x^2 \qquad\qquad \\
\hline
3x^2 - 13x + 14 \\
13x^2 - 6x \qquad \\
\hline
-7x + 14 \\
-7x + 14 \\
\hline
0
\end{array}
$$

Polynomial division will give the other factor.

Continues on next page

So: $x^3 + x^2 - 13x + 14 = (x - 2)(x^2 + 3x - 7)$

The quadratic doesn't factorise, so use the formula (or equation solver on your calculator).

$(x-2)(x^2+3x-7)=0$

$x=2$ or $x^2+3x-7=0$

$x=2$ or $x=\frac{-3\pm\sqrt{9+28}}{2}$

So the solutions are:

$x=2, \frac{-3+\sqrt{37}}{2}, \frac{-3-\sqrt{37}}{2}$

WORKED EXAMPLE 4.7

Two of the factors of $q(x) = x^3 + 4x^2 + ax + b$ are $(x-1)$ and $(x+1)$. Find the values of the constants a and b.

$q(1)=0$

$1+4+a+b=0$

$a+b=-5$ (1)

Apply the factor theorem with factor of $(x-1)$.

$q(-1)=0$

$-1+4-a+b=0$

$-a+b=-3$ (2)

Apply the factor theorem with factor of $(x+1)$.

$(1)+(2)$:

$2b=-8$

$b=-4$

$\therefore a=-1$

Solve the two equations simultaneously. Add them to find the value of b.

Tip

If there are unknown coefficients, use the factor theorem rather than polynomial division.

Elevate

See Support Sheet 4 for a further example of this type and for more practice questions.

EXERCISE 4C

1 Decide whether each expression is a factor of $x^3 + 2x^2 - 5x - 6$.

a **i** $x-1$ **ii** $x+1$ **b** **i** $x+6$ **ii** $x-6$

c **i** $x-2$ **ii** $x+2$ **d** **i** $x+3$ **ii** $x-3$

2 Given that $(x-2)$ is a factor of each of these polynomials, factorise them fully.

a **i** x^3+2x^2-5x-6 **ii** x^3-2x^2-x+2 **b** **i** x^3-7x+6 **ii** x^3-3x-2

c **i** x^3-2x^2-4x+8 **ii** $x^3-6x^2+12x-8$ **d** **i** $2x^3-9x^2-2x+24$ **ii** $3x^3+11x^2-40x+12$

3 Factorise each expression fully.

a i x^3+2x^2-x-2 ii x^3+x^2-4x-4

b i $x^3-7x^2+16x-12$ ii $x^3+6x^2+12x+8$

c i $x^3-3x^2+12x-10$ ii $x^3-2x^2+2x-15$

d i $6x^3-11x^2+6x-1$ ii $12x^3+13x^2-37x-30$

4 Without using a calculator, find the roots of each equation.

a i $x^3-2x^2-11x+12=0$ ii $x^3-x^2-17x-15=0$

b i $x^3-5x^2+7x-2=0$ ii $x^3-6x^2+7x-2=0$

c i $x^3+5x^2+11x+10=0$ ii $x^3-4x^2+6x-4=0$

5 a Show that $(x-2)$ is a factor of $\mathrm{p}(x)=x^3-3x^2-10x+24$.

b Hence express $\mathrm{p}(x)$ as the product of three linear factors and solve $\mathrm{p}(x)=0$.

6 a Show that $(x-3)$ is a factor of $\mathrm{p}(x)=x^3-x^2-2x-12$.

b Hence show that $\mathrm{p}(x)=0$ only has one real root.

7 x^3+7x^2+cx+d has factors $(x+1)$ and $(x+2)$.
Find the values of c and d.

8 $\mathrm{f}(x)=x^3+ax^2+bx+60$ has factors $(x-5)$ and $(x-3)$.

a Find the values of a and b.

b Find the remaining factor of $\mathrm{f}(x)$.

9 a Use the factor theorem to find a factor of $\mathrm{p}(x)=x^3-7x^2+6x+14$.

b Hence solve $\mathrm{p}(x)=0$, giving exact answers.

10 The polynomial $x^2+kx-8k$ has a factor $(x-k)$.
Find the possible values of k.

11 The polynomial $x^2-(k+1)x-3$ has a factor $(x-k+1)$.
Find the value of k.

12 The polynomial x^2-5x+6 is a factor of $2x^3-15x^2+ax+b$.
Find the values of a and b.

Section 4: Sketching polynomial functions

The graphs of polynomial functions are smooth curves. You need to know the typical shapes of these functions.

n	$y = x^n$	Positive degree n polynomial	Negative degree n polynomial	x-intercepts	Turning points
2				0, 1 or 2	1
3				1, 2 or 3	0 or 2
4				0, 1, 2, 3, or 4	1 or 3

If you need to sketch a polynomial function, the most useful form is the factorised form. This form shows the x-intercepts.

- If a polynomial has a factor $(x - a)$ then the curve passes through the x-axis at a.
- If a polynomial has a double factor $(x - a)^2$ then the curve touches the x-axis at a.
- If a polynomial has a triple factor $(x - a)^3$ then the curve passes through the x-axis at a, flattening as it does so.

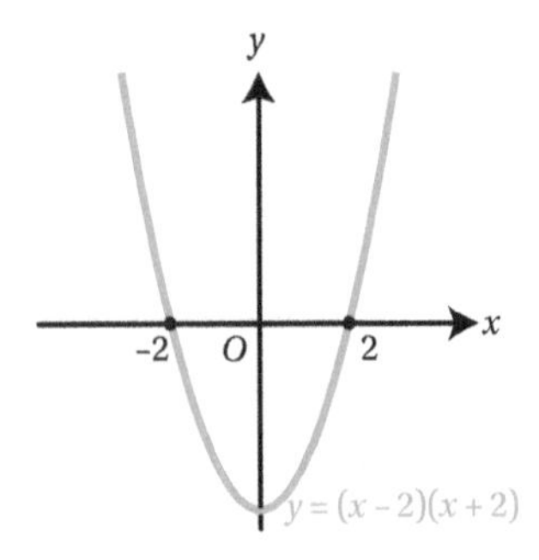

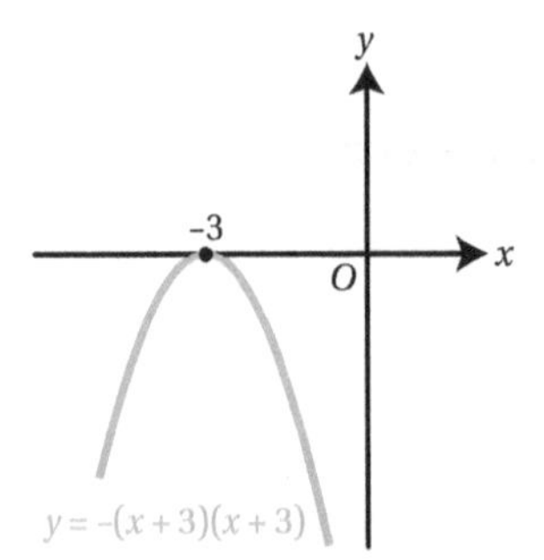

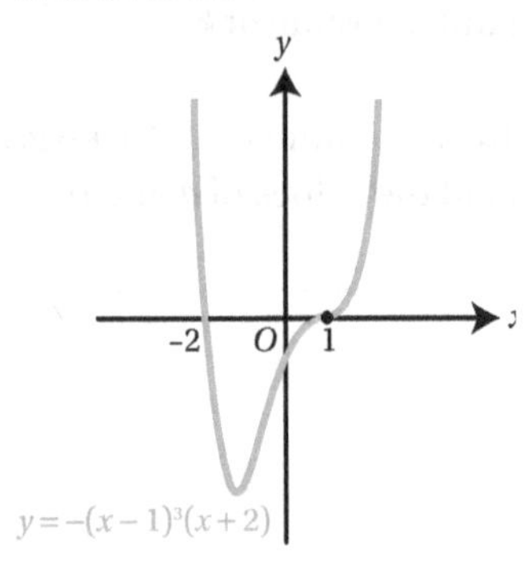

Key point 4.2

To sketch the graph of a polynomial function:

- Classify the order of the polynomial and whether it is positive or negative to deduce the basic shape.
- Set $x = 0$ to find the y-intercept.
- Write the expression in factorised form.
- Identify the x-intercepts.
- Decide how the curve meets the x-axis at each intercept.
- Use all this information to draw a smooth curve.

Fast forward

You don't need to find the coordinates of turning points when sketching these graphs. You will see how to find turning points in Chapter 13.

Common error

It is easy to overlook repeated roots (double or triple factors). These are vital for determining the shape of the graph.

WORKED EXAMPLE 4.8

Sketch the graph of $y = (2 - x)(x - 3)^2$.

Classify the basic shape: this is a negative cubic.

When $x = 0$: $y = 2 \times (-3)^2 = 18$

Find the y-intercept.

When $y = 0$: $x = 2$ or $x = 3$

Find the x intercepts.

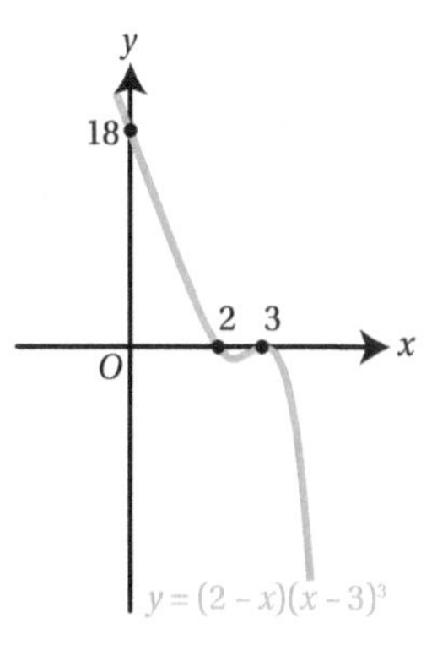

At $x = 2$ the curve passes through the axis.

At $x = 3$ the curve just touches the axis ($x = 3$ is a double root).

Sometimes you need to deduce possible equations for a given curve:

Key point 4.3

To find the equation of a polynomial from its graph:

- Use the shape of the curve and position of the x-intercepts to write down the factors of the polynomial.
- Use any other point to find the constant factor.

Tip

A sketch does not have to be accurate or to scale. It must be approximately the correct shape and all important points – such as axis intercepts – must be clearly labelled.

WORKED EXAMPLE 4.9

Find the equation of this quartic graph.

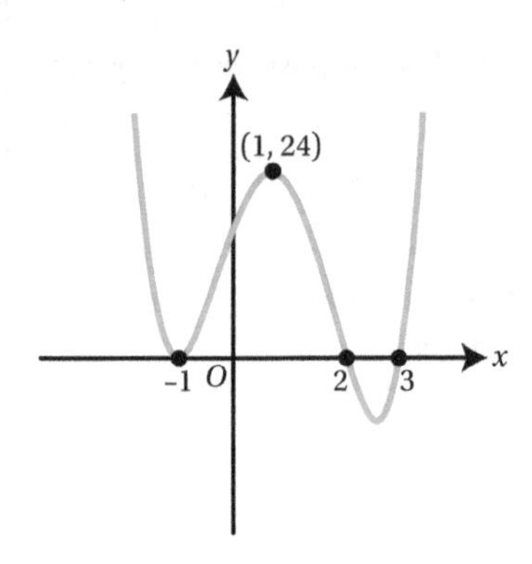

There are single roots at $x = 2$ and $x = 3 \Rightarrow (x - 2)$ and $(x - 3)$ are factors.

The x-intercepts tell you about the factors.

There is a double root at $x = -1 \Rightarrow (x + 1)^2$ is a factor.

$y = k(x+1)^2(x-2)(x-3)$

Write in factorised form. There can still be a constant factor, so call this k.

$24 = k \times (2)^2 \times (-1) \times (-2)$

$24 = 8k$

$k = 3$

Use the fact that when $x = 1$, $y = 24$.

So the equation is $y = 3(x+1)^2(x-2)(x-3)$.

EXERCISE 4D

1 Sketch the graph of each function, labelling all axis intercepts.

a **i** $y = 2(x-2)(x-3)(x-4)$ **ii** $y = 7(x-5)(x+1)(x-3)$

b **i** $y = 4(5-x)(x-3)(x-3)$ **ii** $y = 2(x-1)(2-x)(x-3)$

c **i** $y = -x(x-4)^2$ **ii** $y = (x-2)^2(x+2)$

d **i** $y = x(x^2+4)$ **ii** $y = (x+1)(x^2-3x+7)$

e **i** $y = (1-x)^2(1+x)$ **ii** $y = (2-x)(3-x)^2$

2 Sketch the graph of each function, labelling all axis intercepts.

a **i** $y = x(x-1)(x-2)(2x-3)$ **ii** $y = (x+2)(x+3)(x-2)(x-3)$

b **i** $y = -4(x-3)(x-2)(x+1)(x+3)$ **ii** $y = -5x(x+2)(x-3)(x-4)$

c **i** $y = (x-3)^2(x-2)(x-4)$ **ii** $y = -x^2(x-1)(x+2)$

d **i** $y = 2(x+1)^3(x-3)$ **ii** $y = -x^3(x-4)$

e **i** $y = (x^2+3x+12)(x+1)(3x-1)$ **ii** $y = (x+2)^2(x^2+4)$

3 Find the lowest-order polynomial equation for each graph.

a **i**

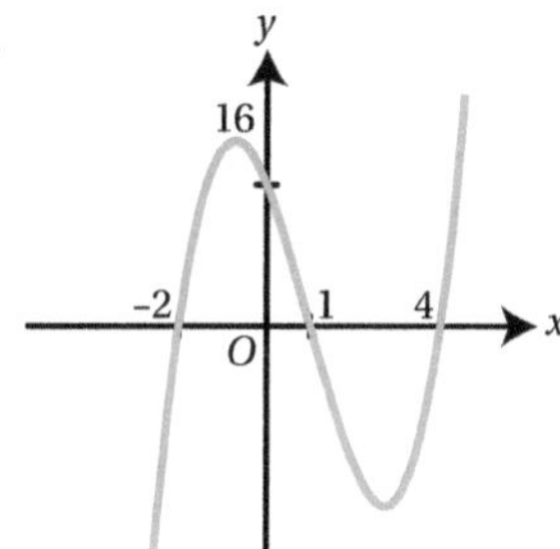

ii

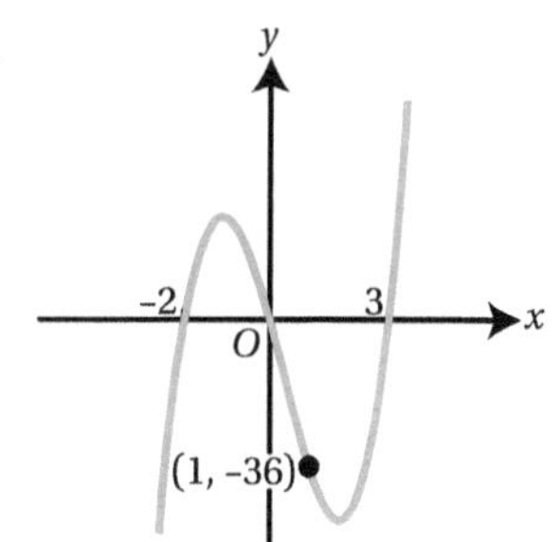

b **i**

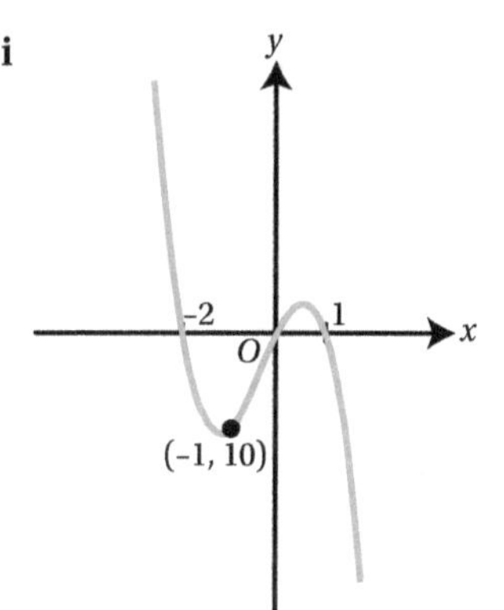

ii

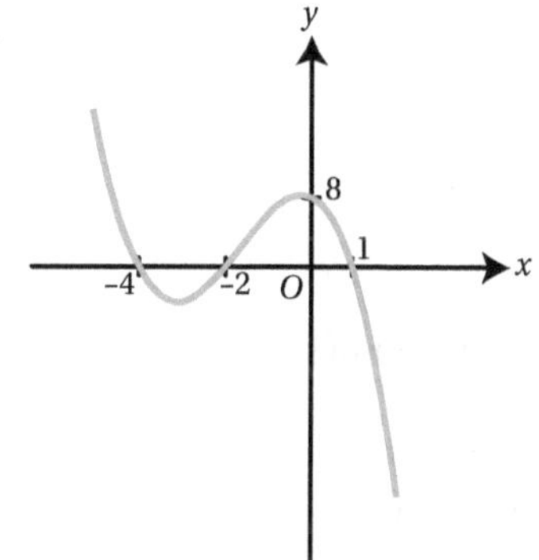

c **i**

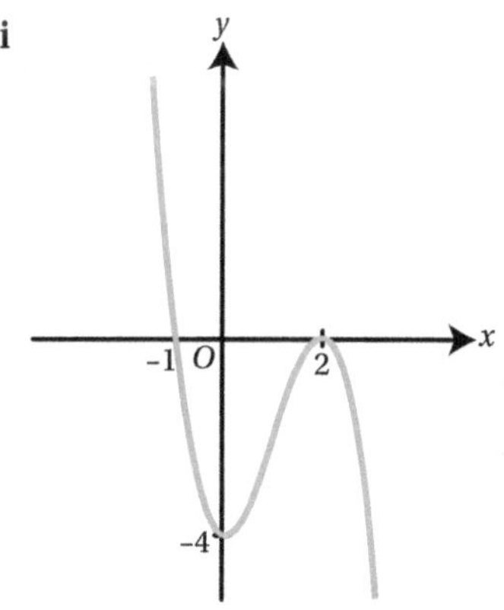

ii

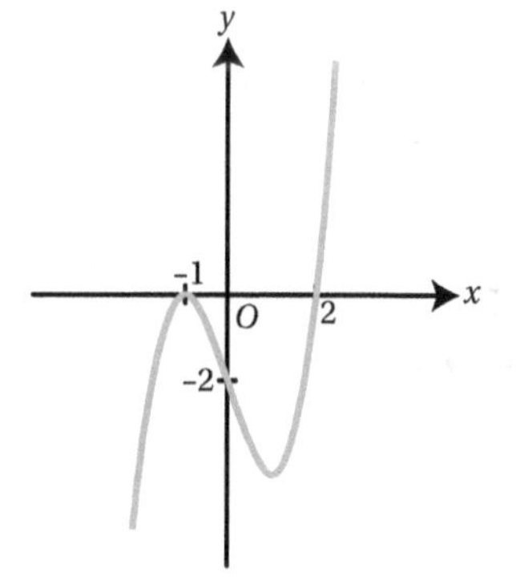

d **i**

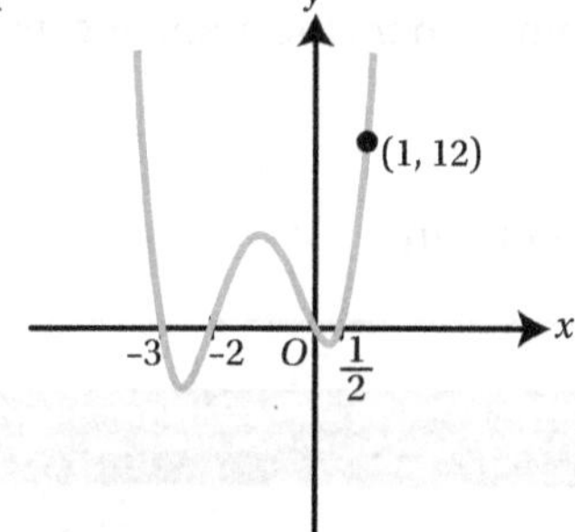

ii

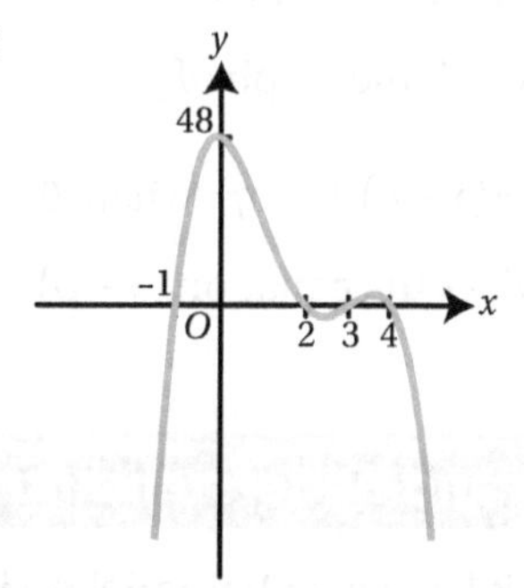

e **i**

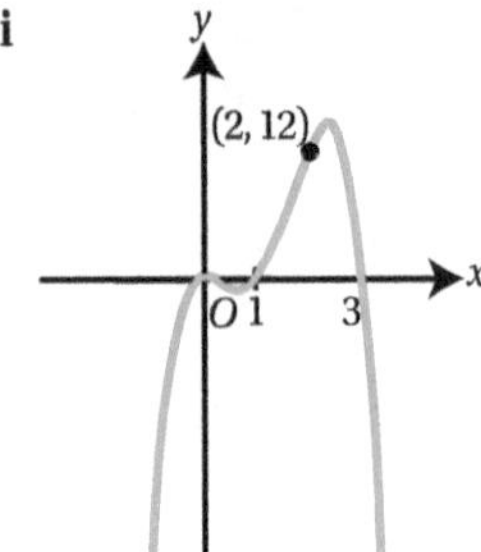

ii

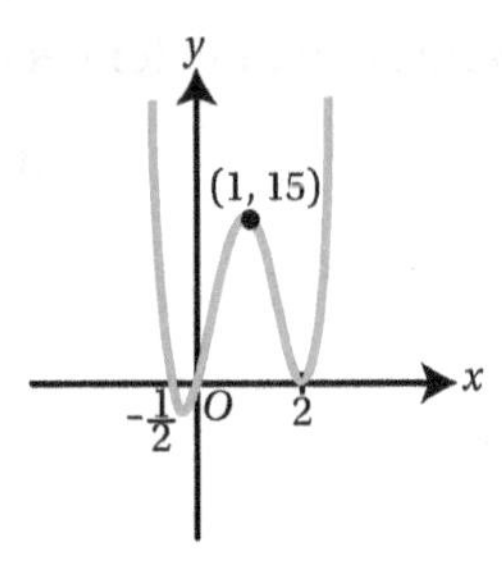

f **i**

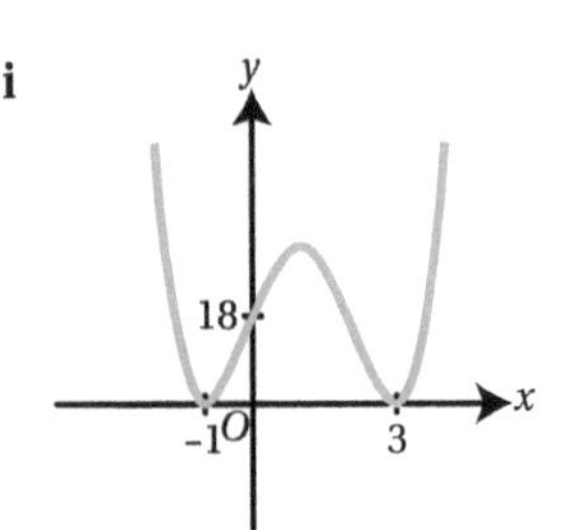

ii

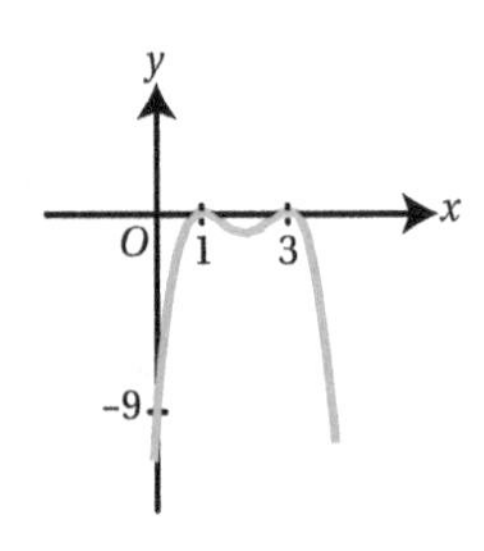

4 **a** Show that $(x-2)$ is a factor of $\text{f}(x)=2x^3-5x^2+x+2$.

b Factorise $\text{f}(x)$.

c Sketch the graph of $y=\text{f}(x)$.

5 Sketch the graph of $y=2(x+2)^2(3-x)$, labelling any axes intercepts.

6 These graphs have equations of the form $y=px^3+qx^2+rx+s$. Find the values of p, q, r and s for each graph.

a

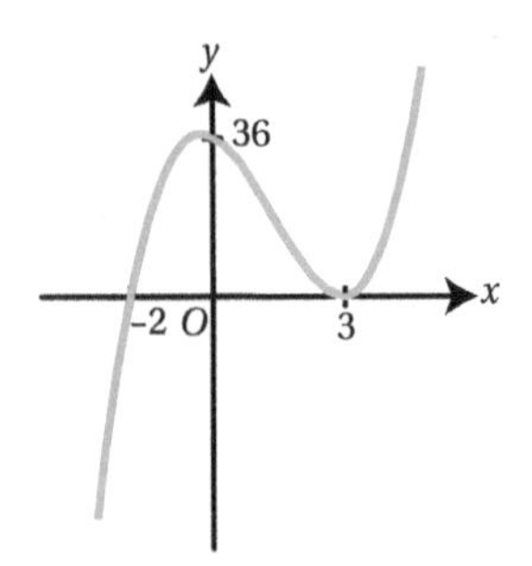

b

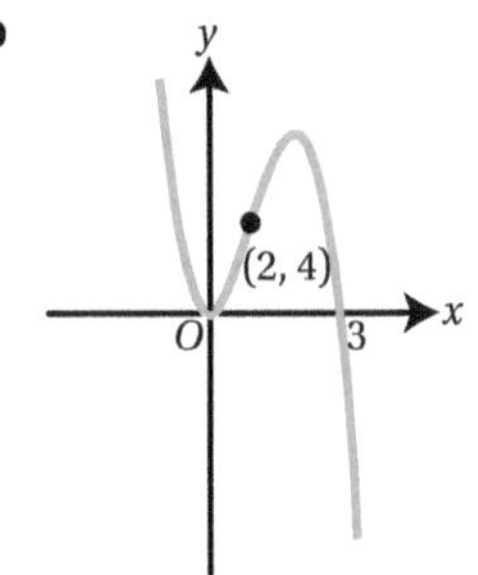

7 **a** Factorise fully x^4-q^4 where q is a positive constant.

b Hence or otherwise sketch the graph of $y=x^4-q^4$, labelling any points where the graph meets an axis.

8 **a** Sketch the graph of $y=(x-p)^2(x-q)$, where $0<p<q$.

b How many solutions does the equation $(x-p)^2(x-q)=k$ have when $k>0$?

Checklist of learning and understanding

- If a linear factor of a cubic is known, **polynomial division** can be used to find the remaining quadratic factor.
- The **factor theorem** provides a quick method for finding factors, so that polynomial division is not always necessary.
  - If $\text{f}(a)=0$ then $(x-a)$ is a factor of $\text{f}(x)$.
- The graphs of polynomial functions are best sketched from their factorised form. **Repeated factors** are particularly important for sketching the graph accurately.

Mixed practice 4

1. A cubic polynomial $f(x)$ satisfies the conditions:

   $f(-2)=0$ $\quad x=3$ is a repeated root

   What is a possible form of $f(x)$?

   Choose from these options.

   A $f(x)=(x-2)(x+3)^2g(x)$ **B** $f(x)=(x-2)(x+3)(x-3)g(x)$

   C $f(x)=(x+2)(x+3)(x-3)g(x)$ **D** $f(x)=(x+2)(x-3)^2g(x)$

2. Determine the number of x-intercepts of the graph of the quartic polynomial $y=(x-a)^2(2x^2-5x+4)$, where a is a real number.

   Choose from these options.

   A 1 **B** 2 **C** 3 **D** 4

3. The diagram shows the graph with equation $y=ax^4+bx^3+cx^2+dx+e$.
   Find the values of a, b, c, d and e.

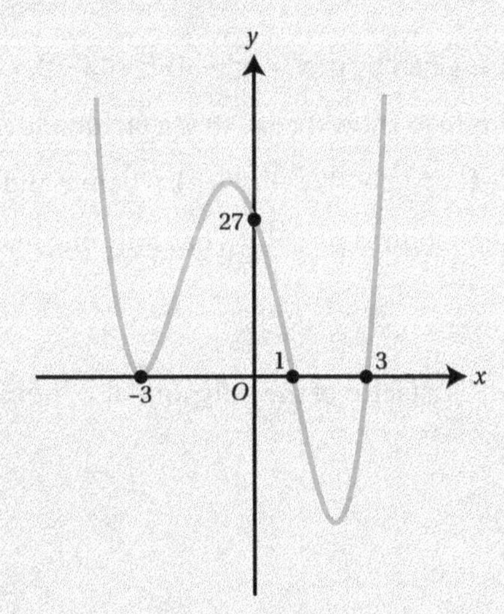

4. Show that $\dfrac{x^3+2x^2-3x-6}{x+2}=x^2+bx+c$ where b and c are integers to be found.

5. **a** Show that $(x-2)$ is a factor of $f(x)=x^3-4x^2+x+6$.

   b Factorise $f(x)$.

   c Sketch the graph of $y=f(x)$.

6. The polynomial $p(x)$ is given by

   $p(x)=x^3+7x^2+7x-15$

   a **i** Use the factor theorem to show that $x+3$ is a factor of $p(x)$.

   ii Express $p(x)$ as the product of three linear factors.

[© AQA 2010]

7 a Given that $(x-3)$ and $(x+2)$ are factors of x^3+ax^2-8x+b, find the values of a and b.

b Hence sketch the graph of $y=x^3+ax^2-8x+b$.

8 Sketch the graph of $y=(x-a)^2(x-b)(x-c)$ where $b<0<a<c$.

9 The cubic polynomial $\mathrm{f}(x)$ is defined by $\mathrm{f}(x)=2x^3+8x^2+5x-6$.

a Use the factor theorem to find a factor of $\mathrm{f}(x)$.

b Hence solve the equation $\mathrm{f}(x)=0$, giving each root in an exact form.

10 a i Show that $(x+2)$ is a factor of $\mathrm{p}(x)=2x^3-x^2-8x+4$.

ii Hence factorise $\mathrm{p}(x)$ completely into linear factors.

b Sketch the graph of $y=2x^3-x^2-8x+4$, and hence solve the inequality $2x^3-x^2-8x+4>0$.

11 a i Sketch the curve with equation $y=x(x-2)^2$.

ii Show that the equation $x(x-2)^2=3$ can be expressed as

$$x^3-4x^2+4x-3=0$$

b The polynomial $\mathrm{p}(x)$ is given by $\mathrm{p}(x)=x^3-4x^2+4x-3$.

i Use the factor theorem to show that $x-3$ is a factor of $\mathrm{p}(x)$.

ii Express $\mathrm{p}(x)$ in the form $(x-3)(x^2+bx+c)$, where b and c are integers.

c Hence show that the equation $x(x-2)^2=3$ has only one real root and state the value of this root.

[© AQA 2011]

12 The polynomial x^2-4x+3 is a factor of the polynomial $x^3+ax^2+27x+b$. Find the values of a and b.

Elevate

See Extension Sheet 4 for a selection of more challenging questions.

5 Using graphs

In this chapter you will learn how to:

- link solving simultaneous equations and the intersection of graphs
- determine the number of intersections between a line and a curve
- use transformations of graphs
- use direct and inverse proportion
- illustrate two-variable inequalities on a graph.

Before you start…

GCSE	You should be able to solve linear inequalities.	1 Solve $3x+1>13$.
GCSE	You should be able to solve simple linear simultaneous equations by elimination.	2 Solve these simultaneous equations. $x+2y=5$ $3x+4y=11$
GCSE	You should be able to solve quadratic equations by factorising or by using the quadratic formula.	3 Solve $x^2+x-1=0$.
Chapter 3	You should be able to use the discriminant to determine the number of solutions of a quadratic equation.	4 How many solutions are there to the equation $x^2+4x+4=0$?
Chapter 2	You should be able to solve equations involving indices.	5 Solve $2^x=8$.
GCSE	You should be able to establish simple direct and inversely proportional relationships.	6 $m=10$ when $n=2$. Find an equation relating m and n if: a m is directly proportional to the square of n b m is inversely proportional to n.

Why use graphs?

Graphs are an alternative way of expressing a relationship between two variables. Understanding the connection between graphs and equations (or inequalities), and being able to switch between the two representations, gives you a wide variety of tools to solve mathematical problems.

Section 1: Intersections of graphs

You already know how to solve linear simultaneous equations, and how to use them to find the point of intersection of two straight lines. You can apply similar ideas to find intersections between curves whose equations involve quadratic functions.

Gateway to A Level

See Gateway to A Level Section J for revision of solving linear simultaneous equations by elimination. This also contains a further example and practice at solving the type of simultaneous equations shown in Worked example 5.1, where one equation is non-linear.

You can generally use technology, such as graphing software or a graphical calculator, to find the intersections of two graphs. However, this only gives approximate solutions. If you need exact solutions you need to use an algebraic method. In many cases the best method is **substitution**, where you replace every occurrence of one variable in one equation by its expression from the other equation.

Key point 5.1

You can find the intersection point(s) of a line and a curve algebraically by solving the two equations simultaneously.

- Substitute from the equation of the line into the equation of the curve to eliminate one of the variables.
- Solve the resulting quadratic equation.

WORKED EXAMPLE 5.1

Find the coordinates of the points of intersection of the line $y = 2x - 1$ and the parabola $y = x^2 - 3x + 5$.

$x^2 - 3x + 5 = 2x - 1$

At the intersection points, the y-coordinates for the two curves are the same, so you can replace y in the first equation by the expression for y from the second equation.

$x^2 - 5x + 6 = 0$

$(x-2)(x-3) = 0$

$x = 2$ or $x = 3$

Rearrange and solve by factorising the quadratic.

$y = 2x - 1$

$x = 2$: $y = 2 \times 2 - 1 = 3$

$x = 3$: $y = 2 \times 3 - 1 = 5$

The coordinates of the points of intersection are (2, 3) and (3, 5).

You also need to find the y-coordinates, by substituting back into one of the equations for y (both should give the same answer). Pick the first equation, as it is easier.

If you use technology to plot the graphs of $y = 2x - 1$ and $y = x^2 - 3x + 5$ you can see how it makes sense that they intersect in two places.

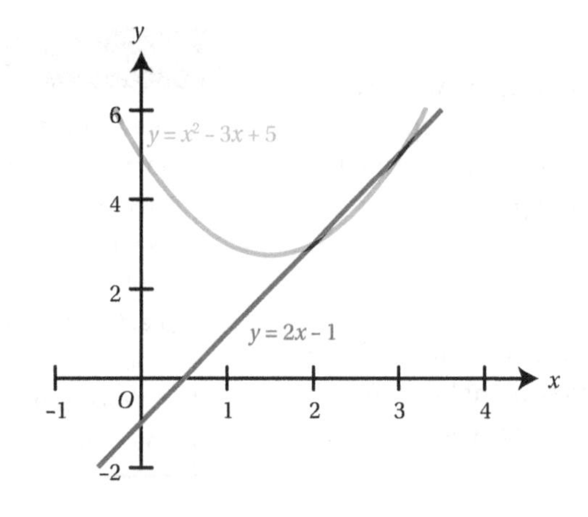

WORK IT OUT 5.1

Solve $x + y = 3$ and $y^2 + 2x^2 = 9$.

Which is the correct solution? Identify the errors made in the incorrect solutions.

Solution 1	Solution 2	Solution 3
$x + y = 3$ (1) $y^2 + 2x^2 = 9$ (2) Squaring (1): $x^2 + y^2 = 9$ (3) (3) – (2): $x^2 = 0$ $x = 0$ Substituting into (1): $y = 3$ Checking this in the second equation: $3^2 + 2 \times 0^2 = 9$ So the solution is $x = 0$, $y = 3$	If $x + y = 3$ then $y = 3 - x$ Substituting into the second equation: $y^2 + 2x^2 = 9$ $(3 - x)^2 + 2x^2 = 9$ $9 - 6x + x^2 + 2x^2 = 9$ $3x^2 - 6x = 0$ Dividing by $3x$: $x - 2 = 0$ $x = 2$ Substituting into $y = 3 - x$: $x = 2$, $y = 1$	Rearranging the first equation: $y = 3 - x$ Substituting into the second equation: $y^2 + 2x^2 = 9$ $(3 - x)^2 + 2x^2 = 9$ $9 - 6x + x^2 + 2x^2 = 9$ $3x^2 - 6x = 0$ $3x(x - 2) = 0$ $x = 0$ or $x = 2$ Substituting into the first equation: $x = 0$, $y = 3$ $x = 2$, $y = 1$

EXERCISE 5A

1 Find the coordinates of the points of intersection of the given curve and the given straight line.

a i $y = x^2 + 2x - 3$ and $y = x - 1$ ii $y = x^2 - 4x + 3$ and $y = 2x - 6$

b i $y = -x^2 + 3x + 9$ and $2x - y = 3$ ii $y = x^2 - 2x + 8$ and $x - y = 6$

2 Solve the simultaneous equations.

a i $x - 2y = 1, 3xy - y^2 = 8$ ii $x + 2y = 3, y^2 + 2xy + 9 = 0$

b i $xy = 3, x + y = 4$ ii $x + y + 8 = 0, xy = 15$

c i $x + y = 5, y = x^2 - 2x + 3$ ii $x - y = 4, y = x^2 + x - 5$

3 Find the coordinates of the points of intersection of $y = \frac{1}{x}$ and $y = 2x$.

4 Solve these equations simultaneously.

$3^x + 2^x = 13$

$3^x - 2^x = 5$

5 Solve these equations simultaneously.

$y = 2^x$

$4^x + y = 72$

6 The sum of two numbers is 8 and their product is 9.75.

a Show that this information can be written as a quadratic equation.

b Hence find the two numbers.

7 Solve the equations $xy + x = 0$, $x^2 + y^2 = 4$.

8 The equations $y = (x-2)(x-3)^2$ and $y = k$ have one solution for all $k < m$. Find the largest value of m.

Section 2: The discriminant revisited

Fast forward

The equation $x^2 - 4x + y^2 + 6y = 12$ is actually a circle. You will study these in Chapter 6.

Sometimes you simply want to know how many intersection points there are, rather than their actual coordinates. You can use the discriminant to determine the number of intersections.

WORKED EXAMPLE 5.2

Find the set of values of k for which the line with equation $x + y = k$ intersects the curve with equation $x^2 - 4x + y^2 + 6y = 12$ at two distinct points.

Equation of the line is:
$y = k - x$

Make y (or x) the subject of the first equation and substitute into the second.

Substitute into the equation of the curve:

$$x^2 - 4x + (k-x)^2 + 6(k-x) = 12$$
$$x^2 - 4x + k^2 - 2kx + x^2 + 6k - 6x = 12$$
$$2x^2 - (10+2k)x + k^2 + 6k = 12$$
$$2x^2 - (10+2k)x + (k^2 + 6k - 12) = 0$$

Rearrange to form a quadratic equation in x.

Two solutions $\Rightarrow \Delta > 0$:

$$b^2 - 4ac > 0$$
$$(10+2k)^2 - 8(k^2 + 6k - 12) > 0$$
$$100 + 40k + 4k^2 - 8k^2 - 48k + 96 > 0$$
$$-4k^2 - 8k + 196 > 0$$
$$k^2 + 2k - 49 < 0$$

There are two intersections so there are two roots of this quadratic.

$k^2 + 2k - 49 = 0$

$$k = \frac{-2 \pm \sqrt{4 + 4 \times 49}}{2}$$
$$= \frac{-2 \pm 2\sqrt{1+49}}{2}$$
$$= -1 \pm \sqrt{50}$$
$$= -1 \pm 5\sqrt{2}$$

This is a quadratic inequality. Solve it as usual, by sketching the graph.

Continues on next page

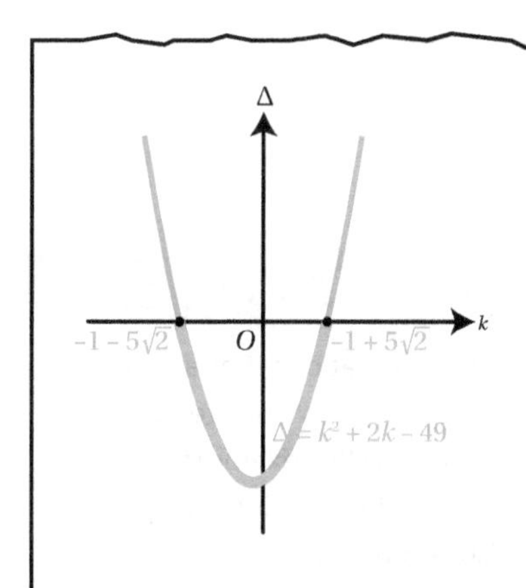

$\therefore -1-5\sqrt{2} < k < -1+5\sqrt{2}$

The graph shows that the required interval is between the roots.

You can see graphically how it is possible for the line to meet the curve once, twice or not at all, depending upon the value of k, which is the y-intercept.

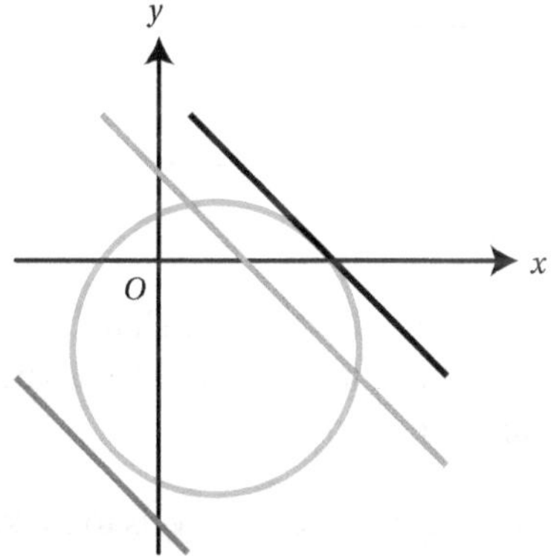

Tip

Questions which talk about the number of intersections (rather than the actual coordinates of the intersections) are often solved using the discriminant.

EXERCISE 5B

1. Show that the line with equation $x - y = 6$ is tangent to the curve with equation $x^2 - 6x + y^2 - 2y + 2 = 0$.

Tip

A **tangent** touches the curve but does not cross at that point. With quadratic equations this means that there are repeated roots so the discriminant is zero.

2. Find the exact values of m for which the line $y = mx + 3$ is tangent to the curve with equation $y = 3x^2 - x + 5$.

3. Let C be the curve with equation $4x^2 + 9y^2 = 36$. Find the exact values of k for which the line $2x + 3y = k$ is tangent to C.

4. Find the values of a for which the curve $y = x^2$ never touches the curve $y = a - (x - a)^2$.

5. Show algebraically that the line $y = kx + 5$ intersects the parabola $y = x^2 + 2$ twice for all values of k.

Section 3: Transforming graphs

From previous study you should know how changing the function changes its graph.

Transformation of $y = \mathrm{f}(x)$	Transformation of graph
$y = \mathrm{f}(x) + c$	Translation by the vector $\begin{pmatrix} 0 \\ c \end{pmatrix}$.
$y = \mathrm{f}(x + d)$	Translation by the vector $\begin{pmatrix} -d \\ 0 \end{pmatrix}$.
$y = p\mathrm{f}(x)$	Stretch parallel to the y-axis, scale factor p.
$y = \mathrm{f}(qx)$	Stretch parallel to the x-axis, scale factor $\frac{1}{q}$.
$y = -\mathrm{f}(x)$	Reflection in the x-axis.
$y = \mathrm{f}(-x)$	Reflection in the y-axis.

Common error

Transformations that affect y-coordinates behave as expected (for example, $\mathrm{f}(x)+c$ moves the function up by c) but transformations that affect x-coordinates do the opposite (for example, $\mathrm{f}(x+c)$ moves the function left by c).

PROOF 2

Prove that $\mathrm{f}(x + d)$ translates the function $\mathrm{f}(x)$ to the left d units.

Let:

$y_1 = \mathrm{f}(x_1)$

$y_2 = \mathrm{f}(x_2 + d)$

Define the variables. You have to take care to avoid assuming that the xs are all the same or the ys are all the same.

$x_1 = x_2 + d \Rightarrow y_1 = y_2$

One way to make $y_1 = y_2$ is to make $x_1 = x_2 + d$.

So when $x_2 = x_1 - d$ the graph of $y_2 = \mathrm{f}(x_2 + d)$ has the same height as the graph of $y_1 = \mathrm{f}(x_1)$.

Interpret this implication geometrically.

This occurs when x_2 is d units to the left of the equivalent point on $y_1 = \mathrm{f}(x_1)$.

WORKED EXAMPLE 5.3

The graph of $y = x^2 + 2x$ is translated 5 units to the left.

Find the equation of the resulting graph in the form $y = ax^2 + bx + c$.

Let $\mathrm{f}(x) = x^2 + 2x$

Relate the transformation to function notation.

Then the new graph is:

$y = \mathrm{f}(x + 5)$

$= (x + 5)^2 + 2(x + 5)$

Replace all xs by $(x + 5)$ in the equation for the function.

$= x^2 + 12x + 35$

WORKED EXAMPLE 5.4

Describe a transformation that transforms the graph of $y = x^2 + 3x$ to the graph of $y = 4x^2 + 6x$.

Let $f(x) = x^2 + 3x$.
Then:
$y = 4x^2 + 6x$
$= (2x)^2 + 3(2x)$
$= f(2x)$

Try to relate the two equations by writing the second function in a similar way to the first.

It is a horizontal stretch with scale factor $\frac{1}{2}$.

Relate the function notation to the transformation.

WORKED EXAMPLE 5.5

The graph of $y = f(x)$ has a single maximum point with coordinates $(4, -3)$. Find the coordinates of the maximum point on the graph of $y = f(-x)$.

The transformation taking $y = f(x)$ to $y = f(-x)$ is the reflection in the y-axis.

Relate function notation to the transformation.

The maximum point is $(-4, -3)$.

Reflection in the y-axis leaves the y-coordinates unchanged and changes x to $-x$.

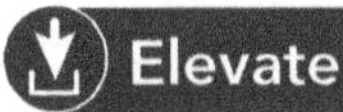

Elevate

See Support Sheet 5 for a further example of this type and for more practice questions.

EXERCISE 5C

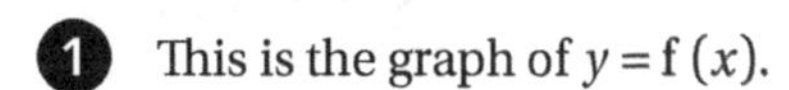

1 This is the graph of $y = f(x)$.

Sketch the graph of each transformed function, giving the position of the minimum and maximum points.

a i $y = f(x) + 3$ ii $y = f(x) + 5$

b i $y = f(x) - 7$ ii $y = f(x) - 0.5$

c i $y = f(x + 2)$ ii $y = f(x + 4)$

d i $y = f(x - 1.5)$ ii $y = f(x - 2)$

e i $y = 3f(x)$ ii $y = 5f(x)$

f i $y = \frac{f(x)}{4}$ ii $y = \frac{f(x)}{2}$

g i $y = f(2x)$ ii $y = f(6x)$

h i $y = f\left(\frac{2x}{3}\right)$ ii $y = f\left(\frac{5x}{6}\right)$

i i $y = -f(x)$ ii $y = f(-x)$

2 Find the new equation of the graph after the transformation is applied to the given graph.

a i $y=3x^2$ after a translation of $\begin{pmatrix}0\\3\end{pmatrix}$.

ii $y=9x^3$ after a translation of $\begin{pmatrix}0\\-7\end{pmatrix}$.

b i $y=7x^3-3x+6$ after a translation of $\begin{pmatrix}0\\-2\end{pmatrix}$.

ii $y=8x^2-7x+1$ after a translation of $\begin{pmatrix}0\\5\end{pmatrix}$.

c i $y=4x^2$ after a translation of $\begin{pmatrix}5\\0\end{pmatrix}$.

ii $y=7x^2$ after a translation of $\begin{pmatrix}-3\\0\end{pmatrix}$.

d i $y=3x^3-5x^2+4$ after a translation of $\begin{pmatrix}-4\\0\end{pmatrix}$.

ii $y=x^3+6x+2$ after a translation of $\begin{pmatrix}3\\0\end{pmatrix}$.

3 Find the required translation to transform:

a i the graph $y=x^2+3x+7$ to the graph $y=x^2+3x+2$

ii the graph $y=x^3-5x$ to the graph $y=x^3-5x-4$

b i the graph $y=x^2+2x+7$ to the graph $y=(x+1)^2+2(x+1)+7$

ii the graph $y=x^2+5x-2$ to the graph $y=(x+5)^2+5(x+5)-2$

c i the graph $y=\sqrt{2x}$ to the graph $y=\sqrt{2x+6}$

ii the graph $y=\sqrt{2x+1}$ to the graph $y=\sqrt{2x-3}$.

4 Find the equation of the graph after the given transformation is applied.

a i $y=3x^2$ after a vertical stretch, factor 7, relative to the x-axis.

ii $y=9x^3$ after a vertical stretch, factor 2, relative to the x-axis.

b i $y=7x^3-3x+6$ after a vertical stretch, factor $\frac{1}{3}$, relative to the x-axis.

ii $y=8x^2-7x+1$ after a vertical stretch, factor $\frac{4}{5}$, relative to the x-axis.

c i $y=4x^2$ after a horizontal stretch, factor 2, relative to the y-axis.

ii $y=7x^2$ after a horizontal stretch, factor 5, relative to the y-axis.

d **i** $y = 3x^3 - 5x^2 + 4$ after a horizontal stretch, factor $\frac{1}{2}$, relative to the y-axis.

ii $y = x^3 + 6x + 2$ after a horizontal stretch, factor $\frac{2}{3}$, relative to the y-axis.

5 Describe the stretch transforming:

a **i** the graph $y = x^2 + 3x + 7$ to the graph $y = 4x^2 + 12x + 28$

ii the graph $y = x^3 - 5x$ to the graph $y = 6x^3 - 30x$

b **i** the graph $y = x^2 + 2x + 7$ to the graph $y = (3x)^2 + 2(3x) + 7$

ii the graph $y = x^2 + 5x - 2$ to the graph $y = (4x)^2 + 5(4x) - 2$

c **i** the graph $y = \sqrt{4x}$ to the graph $y = \sqrt{12x}$

ii the graph $y = \sqrt{2x+1}$ to the graph $y = \sqrt{x+1}$.

6 Find the equation of the graph after the given transformation is applied.

a **i** $y = 3x^2$ after reflection in the x-axis.

ii $y = 9x^3$ after reflection in the x-axis.

b **i** $y = 7x^3 - 3x + 6$ after reflection in the x-axis.

ii $y = 8x^2 - 7x + 1$ after reflection in the x-axis.

c **i** $y = 4x^2$ after reflection in the y-axis.

ii $y = 7x^3$ after reflection in the y-axis.

d **i** $y = 3x^3 - 5x^2 + 4$ after reflection in the y-axis.

ii $y = x^3 + 6x + 2$ after reflection in the y-axis.

7 Describe the transformation of:

a **i** the graph $y = x^2 + 3x + 7$ to the graph $y = -x^2 - 3x - 7$

ii the graph $y = x^3 - 5x$ to the graph $y = 5x - x^3$

b **i** the graph $y = x^2 + 2x + 7$ to the graph $y = x^2 - 2x + 7$

ii the graph $y = x^2 - 5x - 2$ to the graph $y = x^2 + 5x - 2$

c **i** the graph $y = \sqrt{4x}$ to the graph $y = \sqrt{-4x}$

ii the graph $y = \sqrt{2x-1}$ to the graph $y = \sqrt{-1-2x}$.

8 The diagram shows $y = f(x)$.

Sketch on the same graph the original function and

a $y = f\left(\frac{x}{2}\right)$ **b** $y = f(x+1)$.

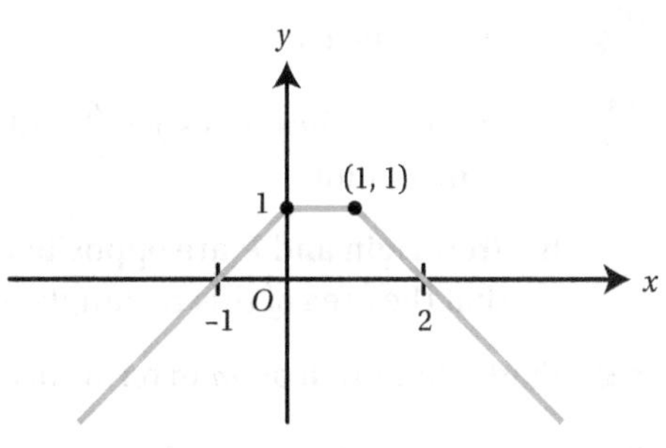

9 Describe the transformation which transforms $y = x^2 + 4x + 6$ into $y = 4x^2 + 8x + 6$.

10 Describe two different transformations which transform $y = 2^x$ into $y = 2^{x+1}$.

11 Describe the stretch factor which transforms $y = x^2$ into $y = kx^2$ if the stretch is:

a vertical **b** horizontal.

12 A function f is defined by $f(x) = x^2 - 2x$. The minimum of $y = kf(x)$ occurs at the same points as the minimum of $y = f(x) - c$. Find the relationship between k and c.

13 The graph $y = x^2 - 4x$ is translated vertically. The graph $y = x^2 - 10x + 28$ is translated horizontally. The result is the same graph. Find the equation of this resulting graph.

Section 4: Graphs of $\frac{a}{x}$ and $\frac{a}{x^2}$

You need to be able to sketch the graphs of $y = \frac{1}{x}$ and $y = \frac{1}{x^2}$.

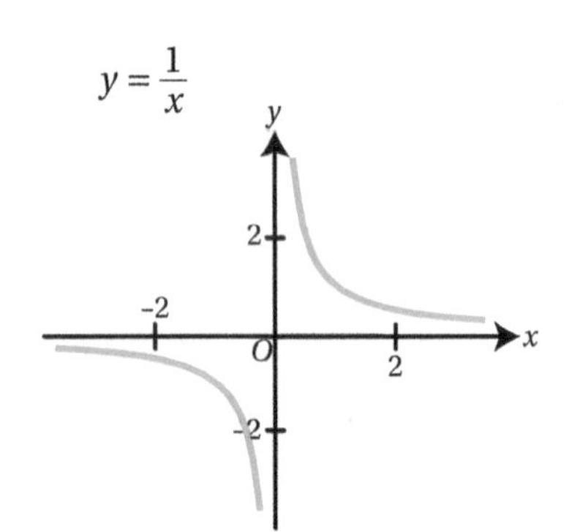

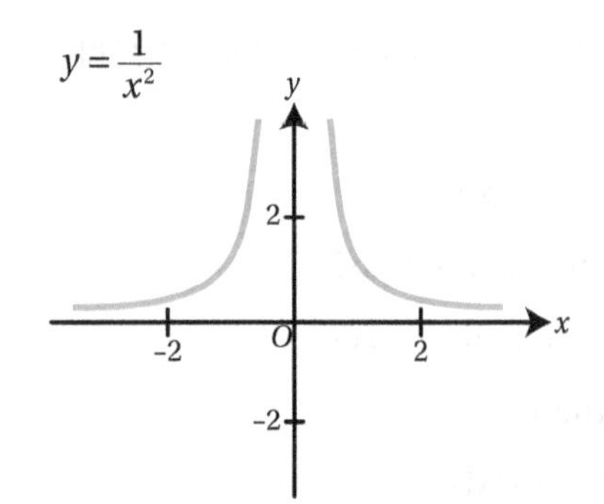

The graphs of $y = \frac{a}{x}$ and $y = \frac{a}{x^2}$ are very similar to the graphs shown here. Each is vertically stretched by a factor of a.

Both graphs have two **asymptotes**. An asymptote is a line to which the curve gets closer and closer without reaching it. The curves shown have asymptotes at $x = 0$ and $y = 0$.

EXERCISE 5D

1 **a** Write down the transformation that changes the graph of $y = \frac{a}{x}$ into the graph of $y = \frac{a}{x-d}$.

b Hence write down the equations of the asymptotes of the graph $y = \frac{a}{x-d}$.

2 Show that the curve $y = \frac{a}{x^2}$ is a horizontal stretch of the curve $y = \frac{1}{x^2}$ and find the stretch factor.

3 **a** Show that the curves $y = \frac{a}{x}$ and $y = \frac{b}{x^2}$ always intersect at exactly one point, P, and find the coordinates of that point.

b The origin and P are opposite vertices of a rectangle with sides parallel to the coordinate axes. Show that the area of this rectangle is independent of b.

4 Find a condition on m in terms of a and c so that the curve $y = \frac{a}{x}$ is a tangent to $y = mx + c$.

5 The function $f(x)$ is a cubic polynomial. Show graphically that the curve $y = \frac{1}{x}$ can intersect this curve in 0, 1, 2, 3 or 4 places.

Section 5: Direct and inverse proportion

When two quantities are proportional they are related in one of two ways, through a constant, k.

Proportionality	Equation
y is proportional to x^n	$y = kx^n$
y is inversely proportional to x^n	$y = \frac{k}{x^n}$

Linear functions are closely related to direct proportion.

Key point 5.2

If $y = mx + c$ then $(y - c)$ is directly proportional to x.

Gateway to A Level

See Gateway to A Level Section K for a reminder of calculations involving direct and inverse proportion.

Tip

The $\propto$ symbol may also be used to denote that two quantities are proportional, so y is proportional to x^n can also be written $y \propto x^n$.

Fast forward

A common example of a straight line being used to make predictions is a line of best fit, used in statistics. You will meet this in Chapter 20.

Straight-line graphs may be used to represent or model a variety of real-life situations. In some cases a linear model is only approximate: when making predictions, you should consider the accuracy and limitations of the model used.

WORKED EXAMPLE 5.6

It takes Ben 12 minutes to drive from his house to the motorway. On the motorway, he drives at an average speed of 65 miles per hour.

a Approximately how long does it take him to drive to York, which is 154 miles away?
b Write down an equation modelling the time, t hours, it takes Ben to drive to a city d miles away.
c Explain why this model only gives approximate times.

a Calculate the time, in hours:

$0.2 + \frac{154}{65} = 2.57$ hours

(about 2 hours 34 minutes)

Time spent on motorway $= \frac{\text{distance}}{\text{speed}}$.

Previous 12 minutes = 0.2 hours.

b $t = 0.2 + \frac{d}{65}$

- The speed on the motorway is not constant.
- It doesn't take into account the time from getting off the motorway in York.
- The 154 miles distance is probably not exact; it doesn't specify where in York Ben is going.

You are modelling the speed as constant, although in reality this is not the case.

The speeds and distances quoted are probably only correct to the nearest integer.

All these considerations mean that the model does not give an exact answer, but it is probably good enough for practical purposes.

EXERCISE 5E

1 a Given that y is proportional to x^2 and $y = 12$ when $x = 2$, find the value of y when $x = 4$.

b Sketch the graph of y against x.

2 a Given that y is proportional to $x - 4$ and $y = 1$ when $x = 6$, find the value of y when $x = 8$.

b Sketch the graph of y against x.

3 a Given that y is inversely proportional to x^2 and $y = 20$ when $x = 1$, find the value of y when $x = 4$.

b Sketch the graph of y against x.

4 a If y is inversely proportional to $x + 1$ and $y = 9$ when $x = 3$ find y when $x = 5$.

b Sketch the graph of y against x.

5 Economists use supply-and-demand curves to model the number of items produced and sold at a particular price. Let £p be the price of one item. Demand (D) is the number of items that can be sold at this price. Supply (S) is the number of items that the producer will make. The graph shows supply and demand in the simplest model, where both vary linearly with price.

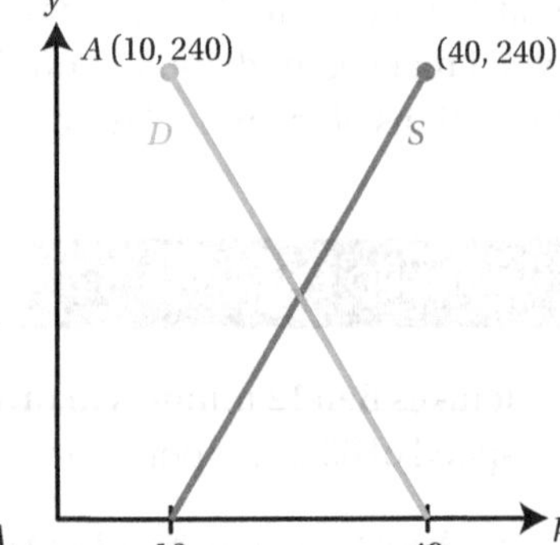

a Show that the equation of line D is $y_D = 320 - 8p$ and find the equation of line S.

b What does the value 320 in the equation of D represent? Suggest why it may not be reasonable to extend the straight line for D beyond point A.

c What is the maximum price that can be charged before there is no more demand?

The market is said to be in equilibrium when supply equals demand.

d Find the equilibrium price of one item.

6 A provider offers two different mobile phone contracts.

A: The a set-up cost is £65, calls cost 3p per minute.

B: There is no set-up cost, calls cost 5p per minute.

a Write down an equation for the total cost, £C, of making m minutes of calls for each contract.

b Hence find out after how many minutes of calls contract A becomes better value.

7 y is inversely proportional to x^2 and z is inversely proportional to y. Sketch a graph of z against x.

8 The strength of the Earth's gravitational field is inversely proportional to the square of the distance from the centre of the Earth. If a satellite is put into orbit, the distance to the centre of the Earth is increased by 10%. Find the percentage decrease in the gravitational field strength.

Section 6: Sketching inequalities in two variables

You have already used a number line to represent inequalities in one variable. If there are two variables, you need to represent the inequality on a graph. For example, this graph shows the inequalities $-1 \leqslant x \leqslant 4$ and $1 < y \leqslant 2$.

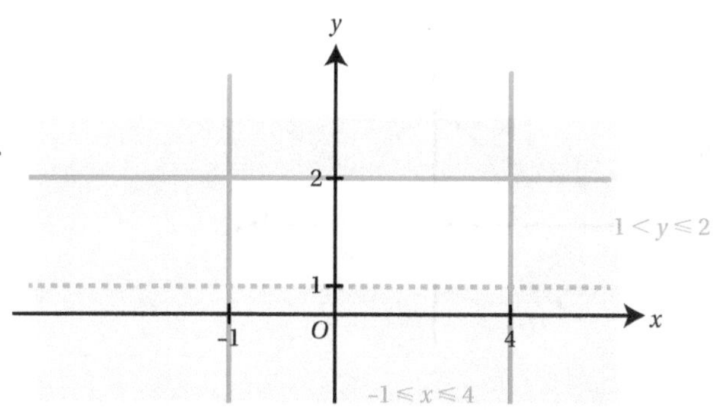

This graph follows the convention that the part satisfying the inequality is left unshaded. This is so that when there are several inequalities on one graph, the region that satisfies all the inequalities is clear.

> **Tip**
>
> A broken line means that the points on that line are not included in the solution.

Even if the inequality involves both variables the solution can still be represented by careful shading. For example, this graph shows $y > x + 1$.

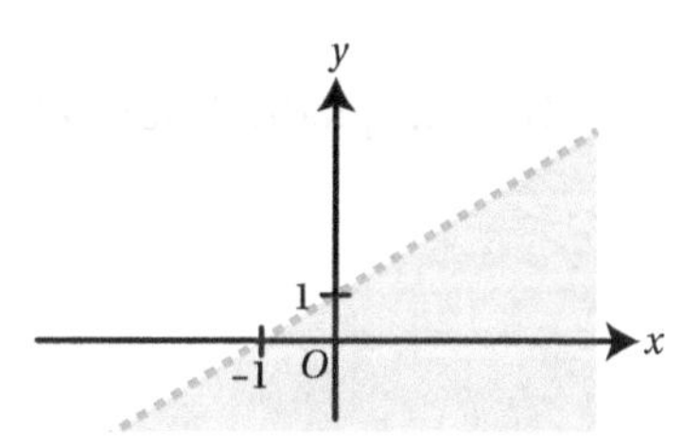

The general process for illustrating inequalities

- Draw the associated line or curve on the graph, using a dashed line if it is not included in the solution.
- Test a convenient point on one side of the curve.
- Shade the side that does not satisfy the inequality.

WORKED EXAMPLE 5.7

a Draw the inequalities $y \geqslant x$ and $y \leqslant 1 - x^2$ on a graph.
b Find the largest value of x that satisfies these inequalities.

a

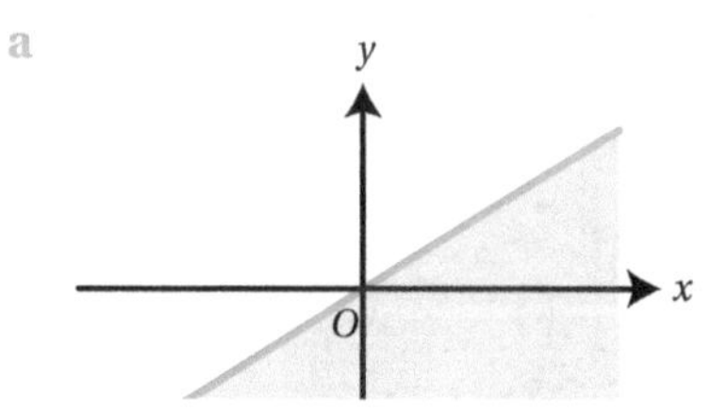

First sketch $y = x$. Use a solid line since it is included in the inequality. Then try the point $(1, 0)$. It does not satisfy the inequality so shade that side of the line.

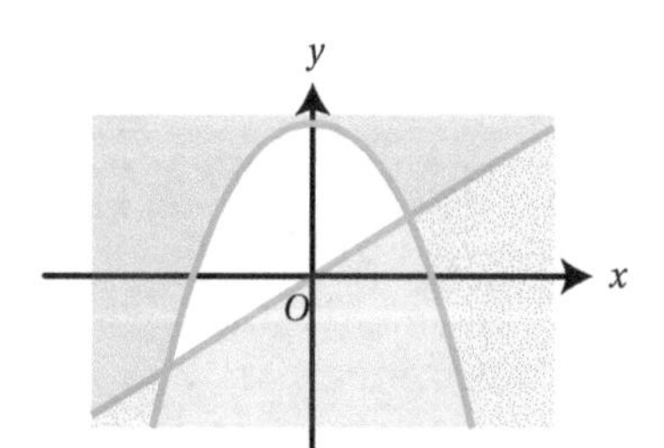

Then sketch $y = 1 - x^2$. Again, use a solid line since it is included in the inequality. Try the point $(0, 0)$. It does satisfy the inequality so shade the other side of the curve.

Continues on next page

b The largest x-value corresponds to the point labelled P.

Now you need to find the point in the unshaded region with the largest x-value.

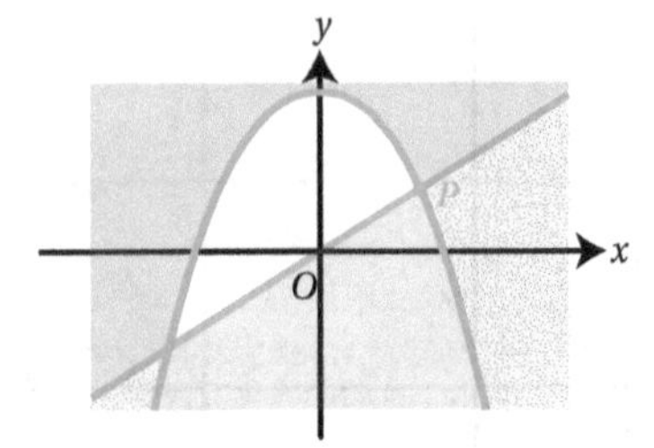

P occurs where:

Solve the simultaneous equations to find this point.

$$x = 1 - x^2$$

$$x^2 + x - 1 = 0$$

$$x = \frac{-1 \pm \sqrt{1^2 - 4 \times 1 \times (-1)}}{2 \times 1}$$

Use the quadratic formula.

$$= \frac{-1 \pm \sqrt{5}}{2}$$

So the largest value of x is $\frac{-1+\sqrt{5}}{2}$.

EXERCISE 5F

1 Illustrate each inequality on a graph.

a **i** $y > 1 + 2x$ **ii** $y < 2 + x$

b **i** $y + x \geqslant 1$ **ii** $y + 2x \geqslant 4$

c **i** $y > x^2$ **ii** $y > -x^2$

d **i** $y > x^2 + 3x + 2$ **ii** $y > x^2 - 7x + 6$

e **i** $y \leqslant x^2 + 2x + 1$ **ii** $y \leqslant x^2 - 7x + 10$

2 Illustrate the region $x > 0, y > 0, x + y < 4$ on a graph.

3 Illustrate the region $y \geqslant x^2$, $y < 4$ on a graph.

4 Illustrate the region $y > x^2 - 4x$ and $y < 2x - x^2$ on a graph.

5 Use inequalities to describe the unshaded region in this graph.

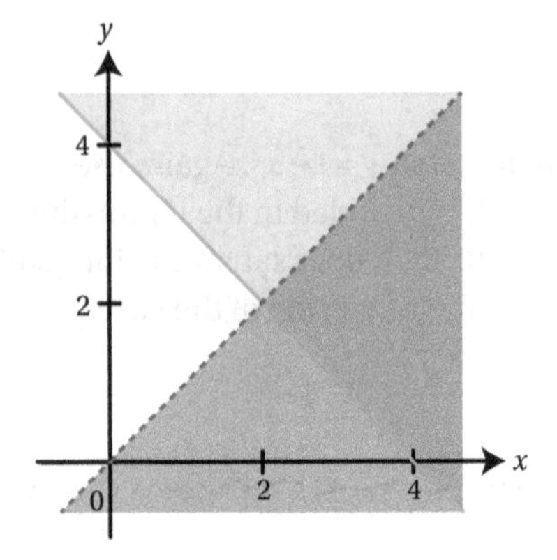

6. The region in this graph is bounded by a parabola and a straight line. Use inequalities to describe the unshaded region.

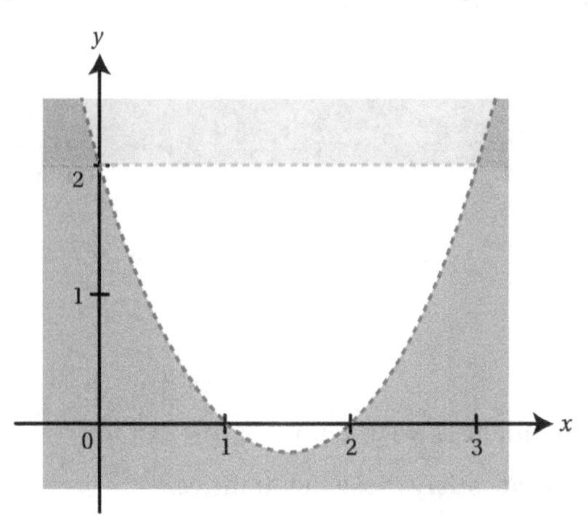

7. Find the largest integer value of x that satisfies $y < 120x - 2x^2$ and $y > 11x$.

8. Sketch a graph of $y > xy$.

Checklist of learning and understanding

- The intersection point(s) of a line and a curve can be found algebraically by solving the two equations simultaneously.
  - Substitute from the equation of the line into the equation of the curve to eliminate one of the variables.
  - Solve the resulting quadratic equation.
- The number of intersections of a quadratic curve and a straight line can be determined by using the discriminant.
- Transforming a function results in a transformation of the graph of the function.

Transformation of $y = \mathrm{f}(x)$	Transformation of graph
$y = \mathrm{f}(x) + c$	Translation by the vector $\begin{pmatrix} 0 \\ c \end{pmatrix}$.
$y = \mathrm{f}(x + d)$	Translation by the vector $\begin{pmatrix} -d \\ 0 \end{pmatrix}$.
$y = p\,\mathrm{f}(x)$	Stretch parallel to the y-axis, scale factor p.
$y = \mathrm{f}(qx)$	Stretch parallel to the x-axis, scale factor $\frac{1}{q}$.
$y = -\mathrm{f}(x)$	Reflection in the x-axis.
$y = \mathrm{f}(-x)$	Reflection in the y-axis.

- You should be able to sketch the graphs of $y = \frac{a}{x}$ and $\frac{a}{x^2}$.
- You should be able to interpret descriptions of the proportionality of two variables and to sketch the associated graph. You can also use a linear model in a variety of contexts and understand that models sometime only give approximate predictions.
- You can represent inequalities in two variables graphically by shading.

Mixed practice 5

1. y is inversely proportional to the cube of x. When $x = 8$, $y = 10$.

   Find the value of y when $x = 64$.

   Choose from these options.

   A $y = \frac{5}{256}$ **B** $y = 5$ **C** $y = 20$ **D** $y = 512$

2. Find the intersection of these graphs.

   $x^2 + y^2 = 25$

   $x + y = 7$

3. **a** Illustrate the region represented by the inequalities $x + y < 3$, $y \geqslant 0$, $y < 2x$.

   b Find the upper bound for the values of y that satisfy these inequalities.

4. Find the transformation that transforms the graph of $y = (x-1)^2$ to the graph of $y = (x+2)^2$.

5. Given that z is proportional to x^2, sketch the graph of z against x.

6. These are the price structures of two taxi companies.

   Company A charges £1.60 per kilometre.

   Company B charges £1.20 per kilometre plus £1.50 call-out charge.

   Find the length of the journey for which the two companies charge the same amount.

7. The graph of $y = \text{f}(x)$ is shown here.

   a On a copy of the diagram, sketch the graph of $y = \text{f}(x-1) + 2$.

   b State the coordinates of the maximum point of the new graph.

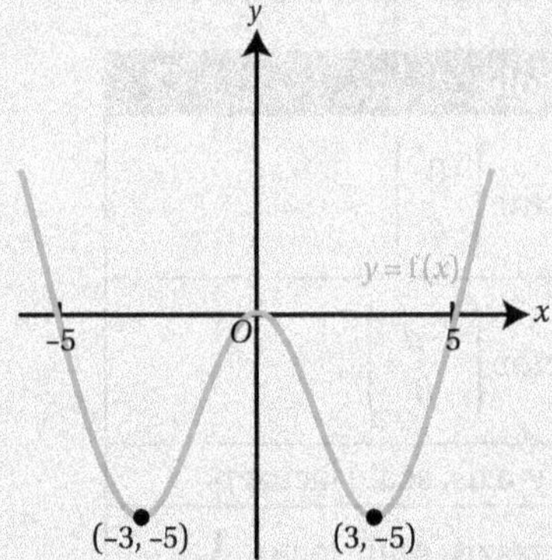

8. The diagram shows part of the graph of $y = \text{f}(x)$.

   Sketch the graph of $y = \text{f}(3x)$.

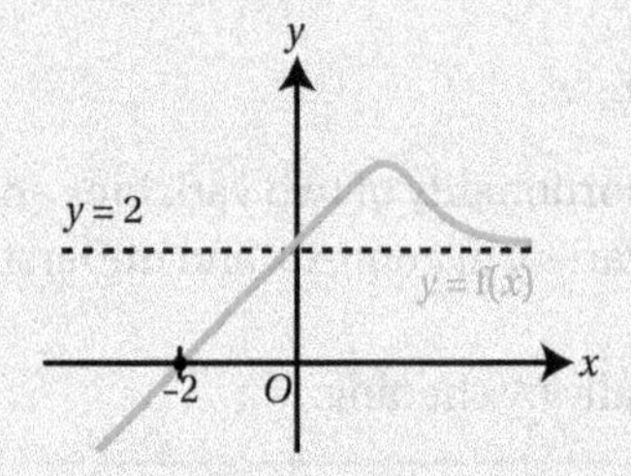

9 A doctor thinks that the weight of a baby can be modelled as a linear function of age. A particular baby weighed 4.1 kg aged 2 weeks, and 4.8 kg aged 5 weeks.

a Taking M as the weight of the baby aged n weeks, show that the straight-line model results in the equation $M = 0.233n + 3.63$, where the coefficients have been rounded to three significant figures.

b Give an interpretation of the values 0.233 and 3.63 in the given equation.

c The normal weight of a healthy one-year-old baby is approximately between 10 and 12 kg. Is the linear model appropriate for babies as old as one year?

10 State which pair of inequalities defines the shaded region.

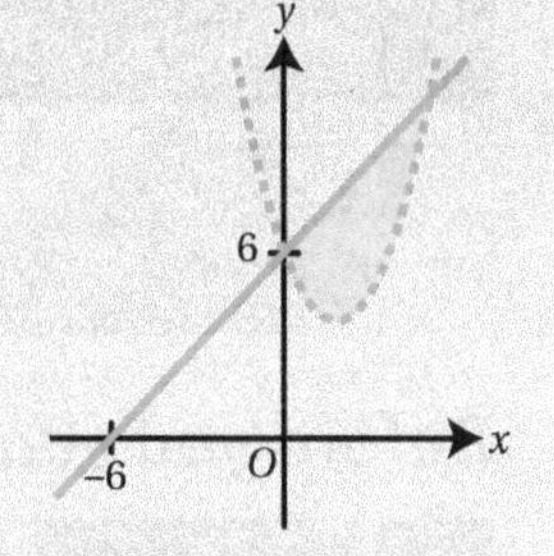

Choose from these options.

A $y < x+6$ and $y < x^2 - 3x + 6$ **B** $y \leqslant x+6$ or $y > x^2 - 3x + 6$

C $y \leqslant x+6$ and $y > x^2 - 3x + 6$ **D** $y \geqslant x+6$ or $y < x^2 - 3x + 6$

11 If x is inversely proportional to y and z is proportional to x^2, sketch the graph of z against y.

12 **a** Describe the single transformation that maps the graph of $y = \sqrt{8x^3 + 1}$ onto the graph of $y = \sqrt{x^3 + 1}$.

b The curve with equation $y = \sqrt{x^3 + 1}$ is translated by $\begin{pmatrix} 2 \\ -0.7 \end{pmatrix}$ to give the curve with equation $y = \text{g}(x)$. Find the value of $\text{g}(4)$.

[© AQA 2013]

13 The curve C has equation $y = 4 - 10x - x^2$ and the line L has equation $y = k(4x - 13)$, where k is a constant.

a Show that the x-coordinates of any points of intersection of the curve C with the line L satisfy the equation

$$x^2 + 2(2k+5)x - (13k+4) = 0$$

b Given that the curve C and the line L intersect in two distinct points, show that

$$4k^2 + 33k + 29 > 0$$

c Solve the inequality $4k^2 + 33k + 29 > 0$.

[© AQA 2011]

14 **a** By using an appropriate substitution, find the exact solutions to the equation $x^4 + 36 = 13x^2$.

b Hence solve the inequality $x^4 + 36 \leqslant 13x^2$.

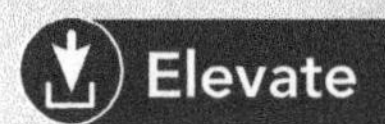

Elevate

See Extension Sheet 5 for questions which look at a different aspect of transformations and symmetry.

6 Coordinate geometry

In this chapter you will learn how to:

- find the distance between two points and the midpoint of two points
- find the equation of a straight line using $y - y_1 = m(x - x_1)$
- determine whether two straight lines are parallel or perpendicular
- find the equation of a circle with a given centre and radius
- solve problems involving intersections of lines and circles.

Before you start…

GCSE	You should be able to find the equation of a straight line in the form $y = mx + c$.	1 Find the equation of a straight line: a with gradient 2 and y-intercept $(0, -1)$ b with gradient -2 and passing through the point with coordinates $(2, 5)$ c passing through the points with coordinates $(1, 3)$ and $(3, 9)$.
GCSE	You should be able to use the fact that parallel lines have the same gradient.	2 A straight lines passes through the points $(0, 1)$ and $(4, p)$ and is parallel to the line with equation $y = 3x + 4$. Find the value of p.
GCSE	You should be able to solve two linear simultaneous equations.	3 Solve these simultaneous equations. $3x - 2y = 13$ $x + 3y = -3$
GCSE	You should be able to use properties of tangents and chords of circles: • the angle in a semi-circle is a right angle • a tangent to the circle is perpendicular to the radius at the point of contact • the radius perpendicular to the chord bisects the chord.	4 Find the values of the angles and lengths marked with letters, giving reasons for your answers. a B y C 5 O z A w x 20° D

Continues on next page

		b c
Chapter 3	You should be able to complete the square.	5 Write $x^2 - 4x - 3$ in the form $(x - p)^2 + q$.
Chapter 5	You should be able to solve linear and quadratic simultaneous equations, and interpret the solution as the intersection points between a line and a curve.	6 a Solve the simultaneous equations. $y = x^2$ $x + y = 6$ b Show that the line $y = 2x + 1$ is a tangent to the parabola $y = -x^2$.

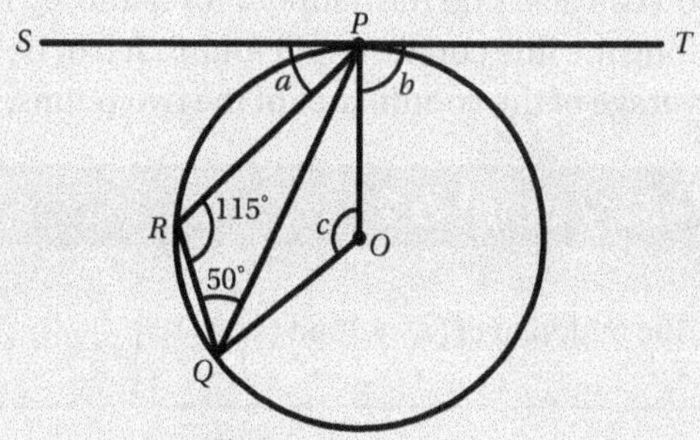

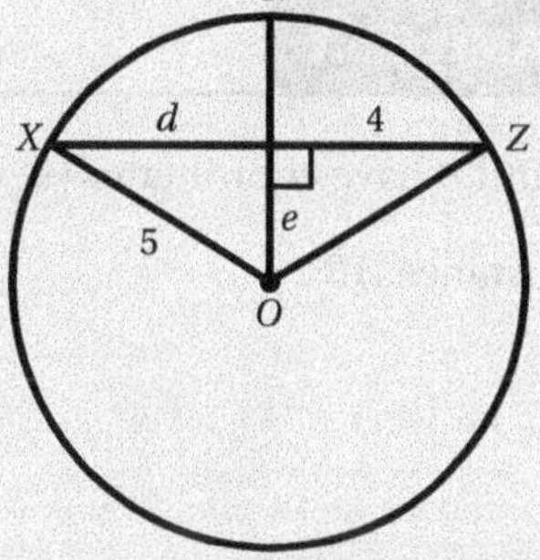

Lines and circles

Straight lines and circles are fundamental shapes in geometry. They can be used to model many real-life objects. You already know several properties of lines and circles, as well as those of other geometrical figures made from them, such as triangles and cones.

In this chapter you will be using coordinates to represent lines and circles. You will use equations to represent those shapes and to find their intersections.

> **Did you know?**
>
> Using equations to represent geometrical shapes is a relatively recent idea in mathematics: it was developed in the 17th century by the French philosopher and mathematician René Descartes. The Cartesian coordinate system is named after him.

Section 1: Distance between two points and midpoint

You may already have met the idea that if you have two points with coordinates (x_1, y_1) and (x_2, y_2) you can find the distance between these two points using Pythagoras' theorem.

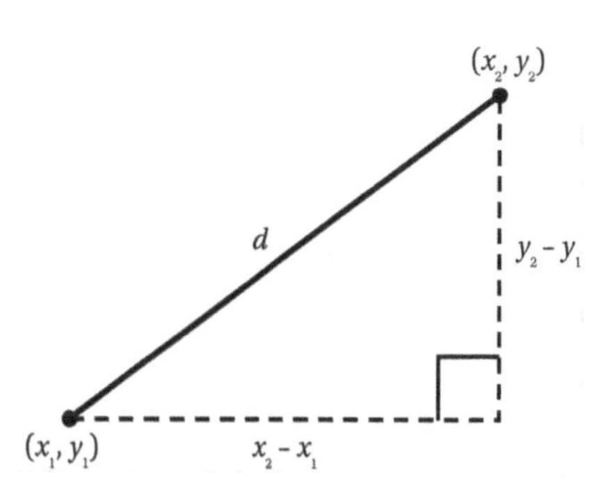

> **Key point 6.1**
>
> The distance between the points (x_1, y_1) and (x_2, y_2) is $\sqrt{(x_2 - x_1)^2 + (y_2 - y_1)^2}$.

If the two points are called A and B you use the notation AB to mean the distance between the two points.

You can also find the midpoint of A and B. This is the point halfway along the line connecting A and B. It can be found by thinking of it as the average of the coordinates of the two points.

Key point 6.2

The midpoint of (x_1, y_1) and (x_2, y_2) is $\left(\frac{x_1+x_2}{2}, \frac{y_1+y_2}{2}\right)$.

Fast forward

You will see how to prove this using vectors in Chapter 15.

WORKED EXAMPLE 6.1

The points A and B have coordinates $(-2, 4)$ and $(5, 2)$. Find:

a the exact distance AB

b the midpoint, M, of A and B.

a The distance is

$AB = \sqrt{(x_2-x_1)^2+(y_2-y_1)^2}$

$= \sqrt{(5-(-2))^2+(2-4)^2}$

$= \sqrt{7^2+(-2)^2}$

$= \sqrt{53}$

Use Key point 6.1 with $x_2 = 5$, $x_1 = -2$ $y_2 = 2$, $y_1 = 4$.

b The midpoint is

$M = \left(\frac{x_1+x_2}{2}, \frac{y_1+y_2}{2}\right)$

$= \left(\frac{-2+5}{2}, \frac{4+2}{2}\right)$

$= (1.5, 3)$

Use Key point 6.2.

Questions can also involve unknown points.

WORKED EXAMPLE 6.2

Find all points of the form (a, a) which are a distance of 5 away from the point $(0, 1)$.

The distance between the points is given by

$5 = \sqrt{(0-a)^2+(1-a)^2}$

Use distance $= \sqrt{(x_2-x_1)^2+(y_2-y_1)^2}$ and set this equal to 5.

$= \sqrt{a^2+1-2a+a^2}$

$= \sqrt{2a^2-2a+1}$

Simplify the expression under the square root.

So $25 = 2a^2-2a+1$

Square both sides to remove the square root.

$0 = 2a^2-2a-24$

$0 = a^2-a-12$

$0 = (a-4)(a+3)$

$a = 4$ or -3

So the points are $(4, 4)$ or $(-3, -3)$

Solve the quadratic equation.

EXERCISE 6A

Gateway to A Level

See Gateway to A Level Section L for more practice on basic questions involving distances between points and midpoints.

1 Find the exact distance between the pairs of points.

a i $(0, 0)$ and $(5, 12)$ ii $(3, 4)$ and $(0, 0)$

b i $(1, 4)$ and $(2, 6)$ ii $(2, 2)$ and $(3, 5)$

c i $(-1, 4)$ and $(3, 2)$ ii $(-1, 3)$ and $(-3, 1)$

d i $(-2, -3)$ and $(-3, 0)$ ii $(-1, -5)$ and $(-2, -1)$

2 Find the midpoint for each of the pairs of points in question 1.

3 Find in terms of a the exact distance between the points $(a, 2a)$ and $(-2a, 8a)$ where $a > 0$.

4 The midpoint of points P and Q is $(1, 1)$.

If point P has coordinates (a, b), find the coordinates of Q.

5 a The point A has coordinates $(0, 1)$, the point B has coordinates $(4, 4)$ and the point C has coordinates $(7, 8)$.
Show that the distance AB equals the distance BC.

b Explain why this does not mean that B is the midpoint of AC.

6 The point $(a, 2a)$ is 3 units away from the point $(3, 1)$. Find the possible values of a.

7 The set of points (x, y) are defined by the property that the distance to the point $(0, 1)$ equals y.

Find the equation connecting x and y.

8 Point A has coordinates (x_1, y_1), point B has coordinates (x_2, y_2) and point M has coordinates $\left(\frac{x_1+x_2}{2}, \frac{y_1+y_2}{2}\right)$.

Prove that $AM = \frac{1}{2}AB$.

9 The points A and B have coordinates $(-2, 4)$ and $(4a, 2a)$. M is the midpoint of A and B.

a Find and simplify in terms of a:

i the distance AB ii the midpoint of A and B.

b If O is the origin show that the ratio $AB : OM$ is independent of a.

10 A spider starts at one corner of a cuboidal room with dimensions 5 m by 5 m by 5 m. It can crawl freely across the surface of the wall.

What is the shortest distance it needs to travel to get to the opposite end of the room?

Section 2: The equation of a straight line

You already know that a non-vertical straight line has an equation of the form $y = mx + c$. In this equation, m is the **gradient** of the line and c is the **y-intercept**.

Gateway to A Level

See Gateway to A Level Section S for revision of $y = mx + c$.

You should also remember that a vertical line has an equation such as $x = 3$, which cannot be written in the form $y = mx + c$.

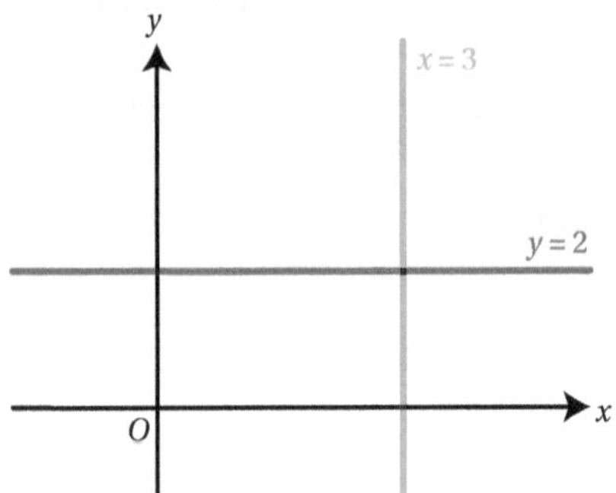

If you know the gradient and y-intercept, you can simply write down the equation but often you don't have this information. It is more common to know two points on the line, or the gradient and one point.

Gateway to A Level

See Gateway to A Level Section L for revision of finding the gradient from two points on the line.

Remember that you can find the gradient from any two points on the line (x_1, y_1) and (x_2, y_2) using $m = \frac{y_2 - y_1}{x_2 - x_1}$.

Equation of a line with given gradient and one point

This diagram shows a straight line passing through the point with coordinates (x_1, y_1). The gradient of the line is m.

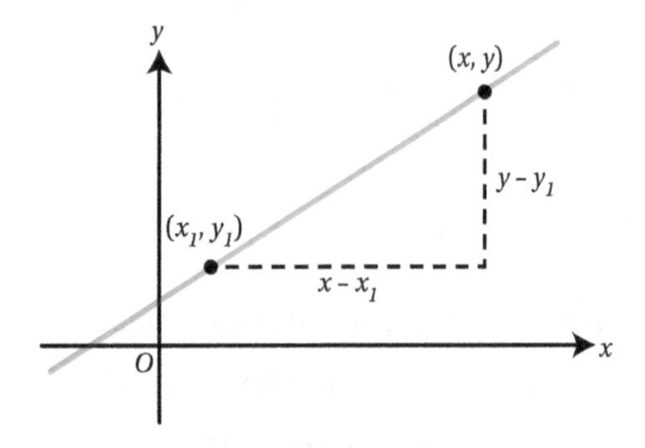

Let (x, y) be any other point on the line. The equation of the line is a rule connecting x and y. Using the dotted sides of the triangle, you can write an equation for the gradient:

$$m = \frac{y - y_1}{x - x_1}$$

Rearranging this equation gives one form of the equation of the line.

Key point 6.3

The line with gradient m through point (x_1, y_1) has equation $y - y_1 = m(x - x_1)$.

This can be rearranged into the form $y = mx + c$ if necessary.

WORKED EXAMPLE 6.3

Find the equation of the line with gradient $\frac{2}{3}$ that passes through the point $(-3, 1)$.
Give your answer in the form $y = mx + c$.

$y - y_1 = m(x - x_1)$

$y - 1 = \frac{2}{3}(x - (-3))$

Use the equation from Key point 6.3.

$y - 1 = \frac{2}{3}x + 2$

$y = \frac{2}{3}x + 3$

Expand the brackets and rearrange.

Equation of a line through two points

If you know two points that the line passes through, you can use their coordinates to find the gradient. You can then use the method from Worked example 6.3 to find the equation of the line.

WORKED EXAMPLE 6.4

Find the y-intercept of the line through the points $(2, -3)$ and $(-1, 4)$.

Gradient:

$m = \frac{y_2 - y_1}{x_2 - x_1}$

$= \frac{4 - (-3)}{-1 - 2}$

$= -\frac{7}{3}$

First find the gradient.

Equation of the line:

$y - y_1 = m(x - x_1)$

$y - (-3) = -\frac{7}{3}(x - 2)$

$y + 3 = -\frac{7}{3}(x - 2)$

Now use the equation from Key point 6.3. You can use either of the two points.

When $x = 0$:

$y + 3 = -\frac{7}{3}(0 - 2)$

$y = \frac{14}{3} - 3$

$y = \frac{5}{3}$

The y-intercept is the point on the line where $x = 0$.

The y-intercept is $\left(0, \frac{5}{3}\right)$.

Worked example 6.4 used the point $(2, -3)$ to find the equation of the line and hence the y-intercept. You should check that using the other point $(-1, 4)$ gives the same y-intercept.

The form $ax+by+c=0$

Sometimes it is convenient to write the equation of a line in a form other than $y=mx+c$; for example, when solving simultaneous equations you may want to write the equations in a form such as $3x+2y=7$.

If you start with the equation of a line in the form $y-y_1=m(x-x_1)$, it is easy to rearrange it into the form $ax+by+c=0$. From this form, it is straightforward to find x- and y-intercepts. If you want to find the gradient of a line given in this form you need to rewrite it as $y=mx+c$.

WORKED EXAMPLE 6.5

Line l passes through the point $(-2, 2)$ and has the same gradient as the line with equation $3x+5y=7$.

a Find the equation of l in the form $ax+by+c=0$, where a, b and c are integers.

b The line l crosses the coordinate axes at points A and B. Find the exact distance AB.

a $3x+5y=7$

$5y=7-3x$

$y=\frac{7}{5}-\frac{3}{5}x$

First find the gradient of l. This requires rewriting $3x+5y=7$ in the form $y=mx+c$.

So $m=-\frac{3}{5}$

The gradient is the coefficient of x.

Equation of l:

$y-2=-\frac{3}{5}(x-(-2))$

Now use $y-y_1=m(x-x_1)$.

$5y-10=-3(x+2)$

$5y-10=-3x-6$

$3x+5y-4=0$

You need all the coefficients to be integers, so multiply through by the common denominator.

b The x-intercept is the point on the line where $y=0$.

Find the coordinates of A and B.

$3x+0-4=0$

$x=\frac{4}{3}$

The y-intercept is the point on the line where $x=0$.

$0+5y-4=0$

$y=\frac{4}{5}$

For the distance between $A\left(\frac{4}{3},0\right)$ and $B\left(0,\frac{4}{5}\right)$:

$AB=\sqrt{\left(0-\frac{4}{3}\right)^2+\left(\frac{4}{5}-0\right)^2}$

Use distance $AB=\sqrt{(x_2-x_1)^2+(y_2-y_1)^2}$.

$=\sqrt{\frac{16}{9}+\frac{16}{25}}$

$=\sqrt{\frac{544}{225}}$

$=\frac{4\sqrt{34}}{15}$

EXERCISE 6B

1 Write down the equation of each line in the form $y - y_1 = m(x - x_1)$.

a i gradient 3, through the point $(4, -1)$
 ii gradient 5, through the point $(-3, 2)$

b i gradient $-\frac{1}{2}$, through the point $(-3, 1)$
 ii gradient $\frac{4}{3}$, through the point $(1, 3)$

c i through points $(3, 7)$ and $(5, 15)$
 ii through points $(4, 1)$ and $(7, 10)$

d i through points $(3, -1)$ and $(-4, 5)$
 ii through points $(-1, -7)$ and $(3, 2)$

2 Find the equation of each line in the form $ax + by + c = 0$, where a, b and c are integers.

a i gradient -2, through the point $(-1, 3)$
 ii gradient -5, through the point $(3, -2)$

b i gradient $\frac{1}{3}$, through the point $(-1, -7)$
 ii gradient $-\frac{3}{2}$, through the point $(2, -5)$

c i through points $(-1, 2)$ and $(2, 7)$
 ii through points $(3, 5)$ and $(5, -4)$

d i through points $(2, 1)$ and $(-3, 7)$
 ii through points $(-3, 2)$ and $(1, -5)$

3 Find the gradient and the x- and y-intercepts of the lines with these equations.

a i $y = 3x + 1$
 ii $y = -2x + 3$

b i $3x - 2y + 5 = 0$
 ii $4x + 5y - 1 = 0$

c i $3y - 3x + 5 = 0$
 ii $y - 4x - 6 = 0$

d i $4x - 3y = 7$
 ii $5x + 2y = 3$

e i $y - 3 = 2(x - 2)$
 ii $y + 1 = 3(x - 5)$

4 Find the coordinates of the point of intersection of each pair of lines.

a i $y = \frac{1}{2}x - 3$ and $y = x + 1$
 ii $y = \frac{2}{3}x + 2$ and $y = 2x - 2$

b i $2x + 3y = 1$ and $x - 3y + 1 = 7$
 ii $3x - y = 3$ and $2x + 5y = 1$

c i $5x + 2y + 3 = 0$ and $3x + 4y - 1 = 0$
 ii $4x - 2y - 1 = 0$ and $3x - 5y + 5 = 0$

d i $ax + by = 3$ and $bx - ay = 7$
 ii $ax + by = 5$ and $bx + ay = -3$

5 a Find the equation of the line through the points with coordinates $(-4, 3)$ and $(5, -1)$. Give your answer in the form $ax + by + c = 0$, where a, b and c are integers.

b A second line has the same gradient as the line in part **a** and passes through the point $(-4, 2)$. Find the equation of this line.

c Find the value of k such that the point $(3k, k)$ lies on the second line.

6 A straight line has gradient $-\frac{2}{3}$ and passes through the point with coordinates $(2, 5)$. It cuts the coordinate axes at points P and Q. Find the area of the triangle OPQ, where O is the origin.

7 Line l_1 passes through the points $(-3, -1)$ and $(10, 12)$. Line l_2 passes through the point $(-16, 12)$ and has gradient $-\frac{3}{2}$.

a Find the equations of l_1 and l_2 in the form $ax + by = c$.

b Find the coordinates of the point of intersection, P, of l_1 and l_2.

l_1 intersects the x-axis at Q and l_2 intersects the x-axis at R.

c Find the area of triangle PQR.

8 A line passes through the points $A(2-2k, k)$ and $B(k-1, 2k+1)$, where k is a constant. The gradient of the line is $\frac{2}{3}$.

a Find the value of k.

b Find the equation of the line.

9 The line with equation $4y-7x+14=0$ crosses the coordinate axes at points A and B. M is the midpoint of AB. Find the distance of M from the origin, giving your answer in the form $\frac{\sqrt{p}}{q}$ where p and q are integers.

10 A line passes through the point $\left(\frac{p}{3}, \frac{2}{p}\right)$ and has gradient $-\frac{6}{p^2}$. It crosses the coordinate axes at points A and B. Show that the area of the triangle OAB is independent of p.

Section 3: Parallel and perpendicular lines

It is useful to be able to tell whether two lines are parallel or perpendicular, without having to draw them accurately. You already know how to decide whether two lines are parallel.

Key point 6.4

Two lines are parallel if they have the same gradient.

WORKED EXAMPLE 6.6

Find the equation of the line that is parallel to $2x+5y-7=0$ and passes through the point $(-1, 3)$. Give your answer in the form $ax+by=c$ where a, b and c are integers.

$2x+5y-7=0$

$5y=-2x+7$

$y=-\frac{2}{5}x+\frac{7}{5}$

Rearrange the equation in order to identify the gradient.

$\therefore m=-\frac{2}{5}$

Parallel lines have the same gradient.

For the line through $(-1, 3)$:

Use $y-y_1=m(x-x_1)$.

$y-3=-\frac{2}{5}(x+1)$

$5y-15=-2x-2$

$2x+5y=13$

Rearrange into the required form.

Tip

The easiest way to identify the gradient is to write the equation of the line in the form $y=mx+c$.

This diagram shows how the gradients of a pair of perpendicular lines are related.

For a line with gradient m you can draw a right-angled triangle with horizontal side length 1 unit and vertical side length m units.

If the line is rotated through 90° then so is the triangle. Now the horizontal side is m units and the vertical side −1.

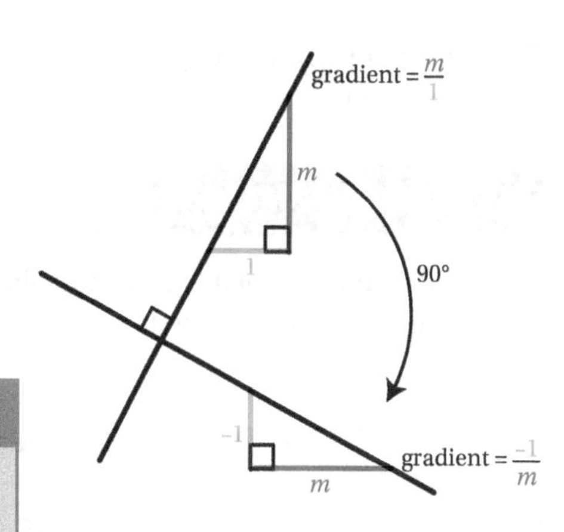

Key point 6.5

- If a line has gradient m, the gradient of any line perpendicular to it is $m_1 = \frac{-1}{m}$.
  Or equivalently:
- Two lines with gradients m_1 and m_2 are perpendicular if $m_1 m_2 = -1$.

For example, lines with gradients $-\frac{2}{3}$ and $\frac{3}{2}$ are perpendicular because $-\frac{2}{3} \times \frac{3}{2} = -1$, and lines with gradients −2 and $\frac{1}{2}$ are perpendicular because $-2 \times \frac{1}{2} = -1$.

WORKED EXAMPLE 6.7

Points A, B and C have coordinates $A(3, -2)$, $B(p, 2)$ and $C(-1, 5)$. Find the possible values of p such that ABC is a right angle.

$m_{AB} = \frac{2-(-2)}{p-3} = \frac{4}{p-3}$ — Find expressions for the gradients of AB and BC.

$m_{BC} = \frac{5-2}{-1-p} = \frac{3}{-1-p}$

$m_{AB} m_{BC} = -1$ — Multiplying the gradients of AB and BC should give −1.

$\frac{4}{p-3} \times \frac{3}{-1-p} = -1$ — Multiply through by the denominators and rearrange.

$12 = -1(p-3)(-1-p)$

$12 = (-p - p^2 + 3 + 3p)$

$p^2 - 2p - 15 = 0$

$(p-5)(p+3) = 0$ — Solve the quadratic equation.

$p = -3$ or 5

One important example of perpendicular lines is the **perpendicular bisector** of a line segment joining two points.

Fast forward

Another important example is the equation of a tangent to a circle, which you will meet in Section 5.

WORKED EXAMPLE 6.8

Find the equation of the perpendicular bisector of the line segment joining points $(4, -3)$ and $(1, 5)$. Give your answer in the form $ax + by = c$.

Gradient of the line segment:

$m = \frac{5-(-3)}{1-4} = -\frac{8}{3}$

You need the gradient of the line segment so you can find the perpendicular gradient.

Perpendicular gradient:

$m_1 = \frac{-1}{\left(-\frac{8}{3}\right)} = \frac{3}{8}$

Use $m_1 = \frac{-1}{m}$

Midpoint of the segment joining $(4, -3)$ to $(1, 5)$:

$y = \frac{-3+5}{2} = 1$

$x = \frac{4+1}{2} = \frac{5}{2}$

The perpendicular bisector passes through the midpoint of the line segment.

Equation of the line:

$y - \frac{1}{2} = \frac{3}{8}(x - 5)$

$16y - 16 = 6x - 15$

$16y - 6x = 1$

EXERCISE 6C

1 Find the gradient of a line perpendicular to each given line.

a **i** $3y = 5x - 2$ **ii** $6y = x - 7$ **b** **i** $4x = 2y - 3$ **ii** $5x + 3 = 2y$

c **i** $y = 4 - 2x$ **ii** $y = 1 - 3x$ **d** **i** $3x - 7y + 1 = 0$ **ii** $x - 5y + 3 = 0$

2 Determine whether each pair of lines is parallel, perpendicular or neither.

a **i** $y = 3 - 4x$ and $y = \frac{1}{4}x - 5$ **ii** $y = 3 - x$ and $y = 5 - x$

b **i** $3x - y + 7 = 0$ and $y - 3x + 5 = 0$ **ii** $5x - 2y + 3 = 0$ and $2y - 5x - 3 = 0$

c **i** $7x + 2y - 3 = 0$ and $2x - 7y + 4 = 0$ **ii** $2x + 4y - 4 = 0$ and $4x + 2y + 1 = 0$

d **i** $5x + 3y - 1 = 0$ and $3y - 5x + 2 = 0$ **ii** $2y - 7x = 3$ and $7x + 2y = 7$

3 Line l_1 has equation $5x - 2y + 3 = 0$.

a Show that the point $P(1, 4)$ lies on l_1.

Line l_2 passes through P and is perpendicular to l_1.

b Find the equation of l_2 in the form $ax + by + c = 0$, where a, b and c are integers.

4 Points A and B have coordinates $A(-2, 3)$ and $B(1, 5)$. O is the origin.

Show that the triangle ABO is right angled, and find its area.

5 a Find the coordinates of the midpoint of the line segment connecting points $A(5, 2)$ and $B(-1, 7)$.

b Hence find the equation of the perpendicular bisector of AB, giving your answer in the form $y = mx + c$.

6 a Find the equation of the line that is parallel to the line with equation $2x + 3y = 6$ and passes through the point $(-4, 1)$.

b The two lines cross the x-axis at points P and Q. Find the distance PQ.

7 Point P has coordinates $(0, 7)$ and point R has coordinates $(12, 4)$. Point Q lies on the x-axis and the angle PQR is a right angle.

Find the possible coordinates of Q, giving your answers in surd form.

8 Points P and Q have coordinates $(-4, 3)$ and $(5, 1)$. Find the equation of the perpendicular bisector of PQ in the form $ax + by + c = 0$, where a, b and c are integers.

9 Point M has coordinates $(3, 5)$. Points A and B lie on the coordinate axes and have coordinates $(0, p)$ and $(q, 0)$, so that $A\hat{M}B$ is a right angle.

a Show that $5p + 3q = 34$.

b Given that $p = 4$, find the value of q and the exact area of the quadrilateral $OAMB$ (where O is the origin).

10 Line l has equation $x - 2y + 3 = 0$. Point P has coordinates $(-1, 6)$.

a Find the equation of the line through P that is perpendicular to l.

b Hence find the shortest distance from P to l.

11 Four points have coordinates $A(k, 2)$, $B(k + 1, k + 2)$, $C(k - 3, k + 4)$ and $D(k - 4, 4)$.

a Show that $ABCD$ is a parallelogram for all values of k.

b Find the value of k for which $ABCD$ is a rectangle.

See Gateway to A level Section E for a reminder of the properties of the parallelogram and rhombus.

Section 4: Equation of a circle

Consider the circle with centre at the point $C(7, 5)$ and radius $r = 4$, as shown in the diagram.

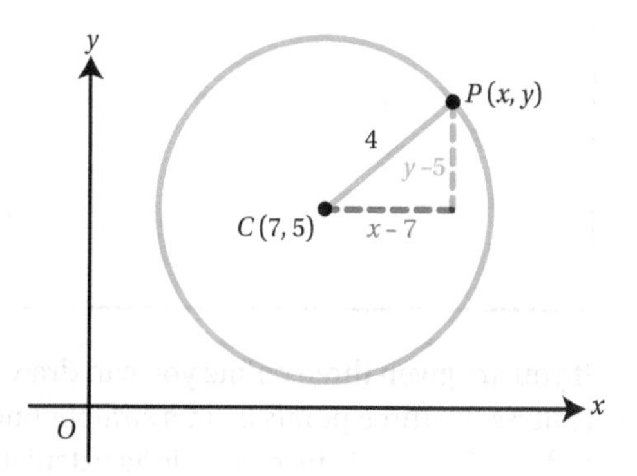

The equation of a circle is a rule satisfied by the coordinates of all the points on the circumference of the circle. Let $P(x, y)$ be a point on the circle. The distance CP is the radius of the circle. Using the formula for the distance between two points (Pythagoras' theorem):

$$(x - 7)^2 + (y - 5)^2 = 4^2$$

This equation is satisfied by the coordinates of any point of the circle.

Key point 6.6

The circle with centre (a, b) and radius r has equation $(x - a)^2 + (y - b)^2 = r^2$.

You can use the equation to check whether a point lies on, inside or outside a given circle.

WORKED EXAMPLE 6.9

A circle has radius 5 and the coordinates of its centre are $(-3, 6)$.

a Write down the equation of the circle.
b Determine whether these points lie on, inside or outside the circle.

i $M(2, 6)$ **ii** $N(3, 1)$

a $(x-(-3))^2+(y-6)^2=5^2$

$(x+3)^2+(y-6)^2=25$

Use $(x-a)^2+(y-b)^2=r^2$
Take care with negative numbers.

b i For point M:

$(2+3)^2+(6-6)^2=5^2+0^2=25$

Point M lies on the circle.

Substitute the coordinates into the equation of the circle.

ii For point N:

$(3+3)^2+(1-6)^2=6^2+(-5)^2=61$

Point N lies outside the circle.

$61 > 25$, so the point is further than 5 units from the centre of the circle.

If you are given an equation of a circle, you can identify the centre and radius. You may need to complete the square first.

WORKED EXAMPLE 6.10

Find the radius and the coordinates of the centre of the circle with equation $x^2-3x+y^2+4y=12$

$$x^2-3x+y^2+4y=12$$

Complete the square for x and y separately.

$$\left(x-\frac{3}{2}\right)^2-\left(\frac{3}{2}\right)^2+(y+2)^2-2^2=12$$

$$\left(x-\frac{3}{2}\right)^2+(y+2)^2=12+\frac{9}{4}+4$$

Rearrange into the form $(x-a)^2+(y-b)^2=r^2$

$$\left(x-\frac{3}{2}\right)^2+(y+2)^2=\frac{73}{4}$$

The centre is $\left(\frac{3}{2}, -2\right)$ and the radius is $\frac{\sqrt{73}}{2}$.

Remember that the number on the right is r^2.

If you are given three points you can draw a circle passing through them (unless the three points lie in a straight line). Finding the centre and radius of the circle involves a long calculation. But there is one special case in which you can use a circle theorem to simplify the calculation.

WORKED EXAMPLE 6.11

Points $A(4, -5)$, $B(2, 9)$ and $C(9, 10)$ lie on a circle.

a Show that AC is the diameter of the circle.

b Hence find the equation of the circle.

a $m_{AB} = \frac{9-(-5)}{2-4} = -7$

$m_{BC} = \frac{10-9}{9-2} = \frac{1}{7}$

If AC is the diameter then ABC is a right angle. You can check this by finding gradients of AB and BC.

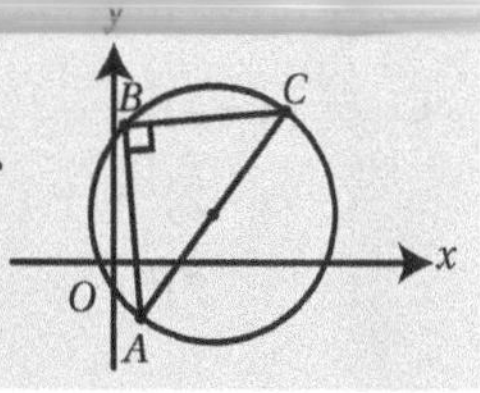

$m_{AB}m_{BC} = -1$ so AB is perpendicular to BC.

For perpendicular lines, $m_1 m_2 = -1$.

Since $ABC = 90°$, AC is the diameter of the circle.

b Find the centre.

Since AC is the diameter, the centre is the midpoint of AC.

$M = \left(\frac{4+9}{2}, \frac{-5+10}{2}\right)$

$= \left(\frac{13}{2}, \frac{5}{2}\right)$

Find the radius.

The radius is half AC.

$AC^2 = (9-4)^2 + (10-(-5))^2 = 250$

$\therefore r = \frac{\sqrt{250}}{2}$

$\left(x - \frac{13}{2}\right)^2 + \left(y - \frac{5}{2}\right)^2 = \frac{250}{4}$

Now use the equation of a circle.

Common error

When finding the radius from two points at either end of a diameter, make sure that you take the square root first to find the length of the diameter and then divide by 2 – not the other way around!

For example, in Worked example 6.11, $AC^2 = 250$ so $AC = \sqrt{250}$ and $r = \frac{\sqrt{250}}{2}$, rather than $r^2 = \frac{250}{2}$ and $r = \sqrt{\frac{250}{2}}$.

Gateway to A Level

See Gateway to A Level Section N for a reminder of circle theorems.

EXERCISE 6D

1. Find the equation of the circle with the given centre and radius.

   a **i** centre $(3, 7)$, radius 4 **ii** centre $(5, 1)$, radius 6

   b **i** centre $(3, -1)$, radius $\sqrt{7}$ **ii** centre $(-4, 2)$, radius $\sqrt{5}$

2. Write down the centre and radius of the circles with the given equations. Hence sketch the graph of each circle.

   a **i** $(x+4)^2 + y^2 = 32$ **ii** $x^2 + (y-7)^2 = 20$

   b **i** $(x-2)^2 + (y-3)^2 = 4$ **ii** $(x+1)^2 + (y-4)^2 = 16$

   c **i** $(x-2)^2 + (y+3)^2 = \frac{9}{4}$ **ii** $(x+1)^2 + (y+5)^2 = \frac{4}{25}$

   d **i** $(x-3)^2 + \left(y - \frac{1}{2}\right)^2 = 6$ **ii** $\left(x + \frac{3}{4}\right)^2 + \left(y - \frac{1}{5}\right)^2 = 3$

3. Find the centre and radius of each circle.

   a **i** $x^2 + 4x + y^2 - 6y + 4 = 0$ **ii** $x^2 - 8x + y^2 + 2y + 8 = 0$

   b **i** $x^2 - 2x + y^2 + 6y + 1 = 0$ **ii** $x^2 - 10x + y^2 + 4y - 1 = 0$

   c **i** $x^2 + 5x + y^2 - y + 2 = 0$ **ii** $x^2 - 3x + y^2 + 7y - 3 = 0$

   d **i** $x^2 + y^2 - 5y = 12$ **ii** $x^2 + y^2 + 3x = 10$

4. Determine whether each point lies on, inside or outside the given circle.

   a **i** point $(1, 7)$, circle centre $(-2, 3)$, radius 5

   ii point $(2, -1)$, circle centre $(-3, 3)$, radius $\sqrt{41}$

   b **i** point $(-1, 1)$, circle centre $(3, 6)$, radius 5

   ii point $(2, 1)$, circle centre $(5, -1)$, radius 7

5. **a** Write down the equation of the circle with centre $(-6, 3)$ and radius $\sqrt{117}$.

   b Find the coordinates of the points where the circle cuts the y-axis.

6. **a** Find the centre and the radius of the circle with equation $x^2 - 5x + y^2 + y = 3$.

   b Determine whether the point $A(-1, 3)$ lies inside or outside the circle.

7. A circle with centre $(3, -5)$ and radius 7 crosses the x-axis at points P and Q. Find the exact distance PQ.

8. Points A, B and C have coordinates $A(-7, 3)$, $B(3, 9)$ and $C(12, -6)$. The angle ABC is a right angle.

   a Find the distance AC.

   b Hence find the equation of the circle passing through the points A, B and C.

9. The circle with equation $(x-p)^2 + (y+3)^2 = 26$, where p is a positive constant, passes through the origin.

   a Find the value of p.

   b Determine whether the point $(3, 2)$ lies inside or outside the circle.

10. A diameter of a circle has endpoints $P(a, b)$ and $Q(c, d)$. Let $Z(x, y)$ be any other point on the circle.

    a Write down the size of the angle PZQ.

    b Hence prove that the equation of the circle can be written as: $(x-a)(x-c) + (y-b)(y-d) = 0$.

Section 5: Solving problems with lines and circles

In this section you will solve a variety of problems involving lines and circles. You will start by looking at intersections, which involves solving simultaneous equations.

Rewind

Before starting this section, you may wish to review the circle theorems listed at the start of the chapter.

For the intersection of a line and a circle, there are three possibilities.

a

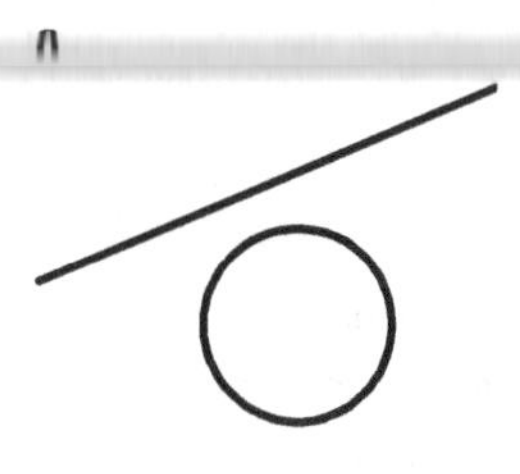

No intersections

b

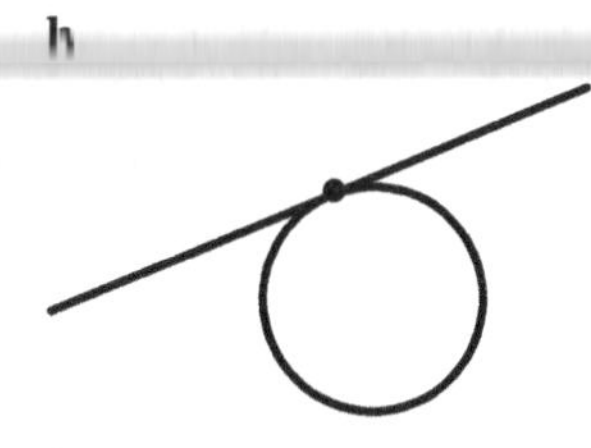

One intersection (tangent line)

c

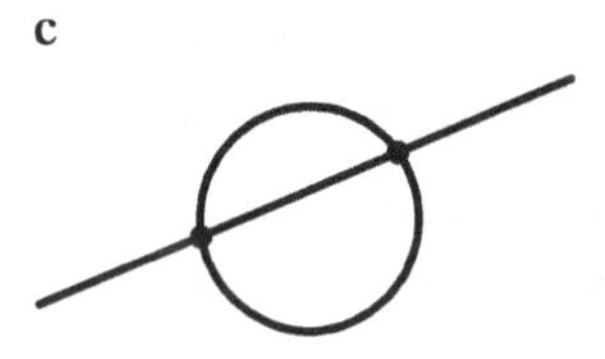

Two intersections

Rewind

See Chapter 5 for a reminder about quadratic simultaneous equations.

Because the equation of the circle is quadratic, you can use the discriminant to determine whether there are two, one or no intersections.

WORKED EXAMPLE 6.12

A circle has centre $(3, 2)$. Find the radius of the circle such that the circle has as a tangent the line with equation $x+5y=20$.

Equation of the circle:

$(x-3)^2+(y-2)^2=r^2$

Equation of the line:

$x+5y=20 \Rightarrow x=20-5y$

Start by writing both equations and trying to find the intersection.

Intersection:

$(17-5y)^2+(y-2)^2=r^2$

$(289-170y+25y^2)+(y^2-4y+4)=r^2$

Substitute x from the second equation into the first.

$26y^2-174y+(293-r^2)=0$

Write in standard quadratic form in order to find the discriminant.

Discriminant must be zero:

$174^2-4(26)(293-r^2)=0$

$104r^2-196=0$

$r^2=1.885$

$\therefore r=1.37$

If the line is a tangent to the circle then the quadratic equation will have only one solution, so $b^2-4ac=0$.

To find the equation of the tangent to a circle at a given point you can use one of the circle theorems: that the tangent is perpendicular to the radius at the point of contact. You can therefore find the gradient of the tangent by using $m_1m_2 = -1$ for perpendicular lines.

Key point 6.7

- The tangent to the circle is perpendicular to the radius at the point of contact.
- The normal is the line through the point of contact and the centre of the circle.

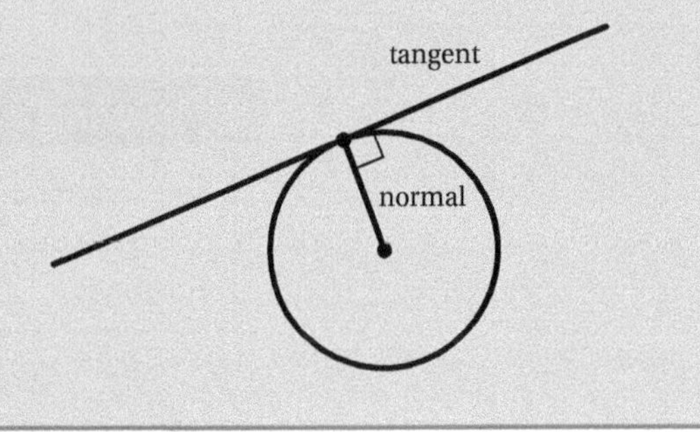

Focus on...

See Focus on ... Problem solving 1 for alternative methods for solving Worked example 6.12, using either equations or geometrical facts.

Fast forward

The line that is perpendicular to the tangent at the point of contact is called a **normal** to the curve. In the case of the circle, the normal is the same line as the radius of the circle.

You will learn about tangents and normals to other curves in Chapter 13.

WORKED EXAMPLE 6.13

Find the equation of the tangent and normal to the circle $(x-3)^2+(y+5)^2=5$ at the point $(2,-7)$.

Coordinates of the centre: $(3,-5)$

The point $(2,-7)$ lies on the circle because $(2-3)^2+(-7+5)^2=5$. Hence it is the point of contact of the tangent.

A diagram is useful here.

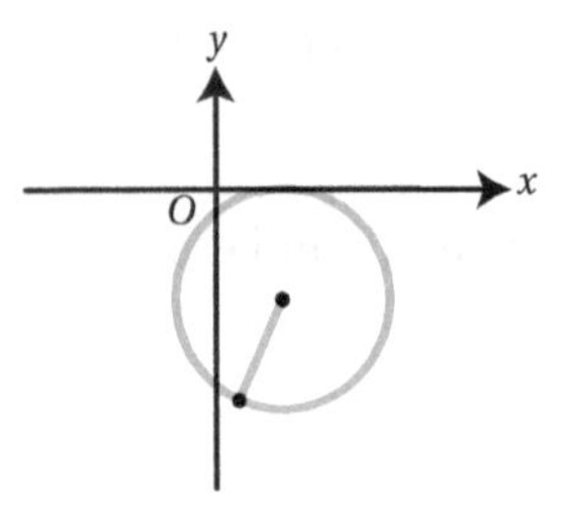

Gradient of the radius:

$$m_1 = \frac{-7-(-5)}{2-3}$$
$$= -\frac{-2}{-1}$$
$$= 2$$

First find the gradient of the radius to the point $(2, -7)$.

Gradient of the tangent:

$$m_2 = -\frac{1}{2}$$

Then use $m_1m_2 = -1$ to find the gradient of the tangent.

Continues on next page

Tangent passes through $(2, -7)$:	Now use the equation of a straight line.
$y-(-7)=-\frac{1}{2}(x-2)$	
$y=-\frac{1}{2}x-6$	
Normal: $y-(-7)=2(x-2)$ $y=2x-11$	The normal is the line connecting $(2, -7)$ to the centre, you already know that its gradient is 2.

Intersection of two circles

Elevate

See Support Sheet 6 for another example of finding tangents to circles and for further practice questions.

There are five possibilities for the relative position of two distinct circles. A clever way to distinguish between them involves comparing the distance, d, between their centres to the radii, r_1 and r_2, of the circles.

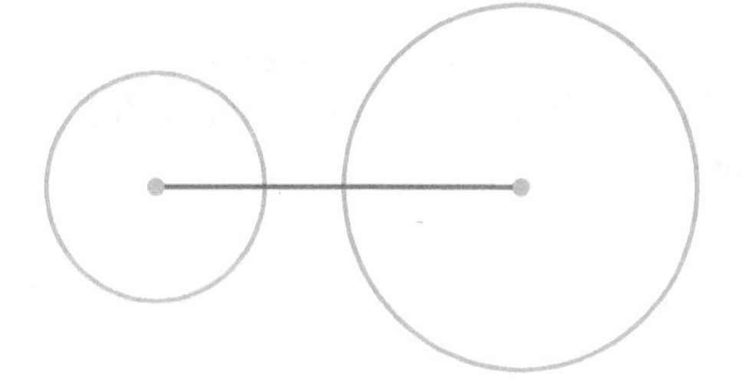

$d > r_1 + r_2$

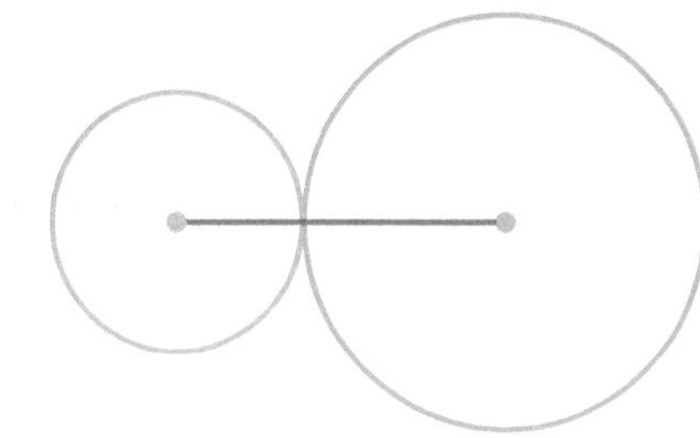

$d = r_1 + r_2$

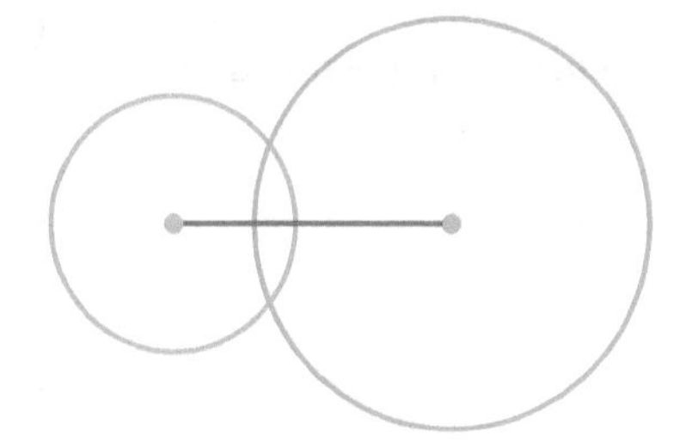

$r_2 - r_1 < d < r_1 + r_2$

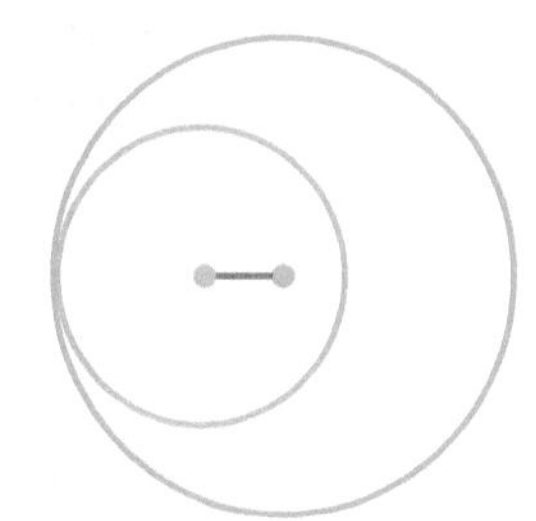

$d = r_2 - r_1$

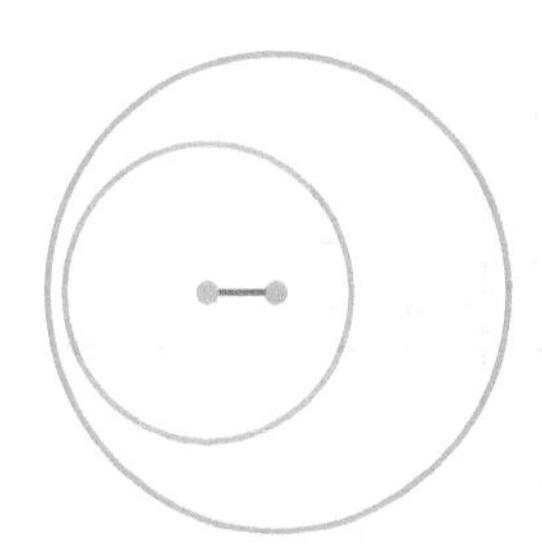

$d < r_1 - r_2$

WORKED EXAMPLE 6.14

Two circles have equations $x^2 - 6x + y^2 - 20y + 45 = 0$ and $(x-15)^2 + (y-5)^2 = r^2$.

a In the case $r = 7$ show that the two circles intersect at two different points.

b Given that the two circles are tangent to each other, find two possible values of r.

a

$$x^2 - 6x + y^2 - 20y + 45 = 0$$
$$(x-3)^2 + (y-10)^2 - 9 - 100 + 45 = 0$$
$$(x-3)^2 + (y-10)^2 = 64$$

First you need to identify the centre and radius of the first circle.

The first circle has centre $(3, 10)$ and radius 8.

The second circle has centre $(15, 5)$ and radius r.

The second circle is already in standard form.

The distance between the centres is:

$d = \sqrt{(3-15)^2 + (10-5)^2} = 13$

You need to compare the distance between the centres to the radii of the circles.

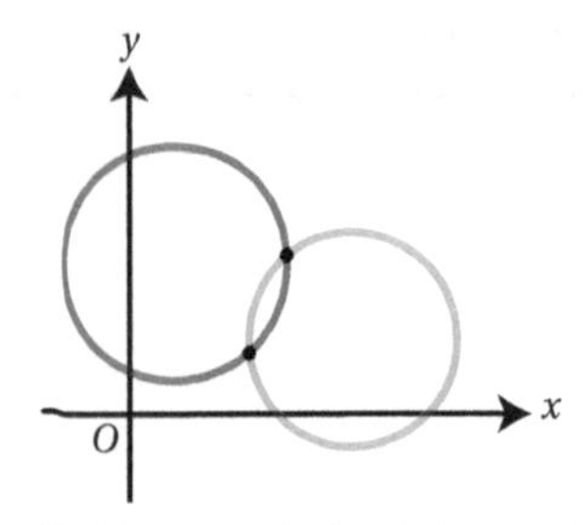

The distance between the centres is less than the sum of the radii but more than their difference.

So the two circles intersect.

b

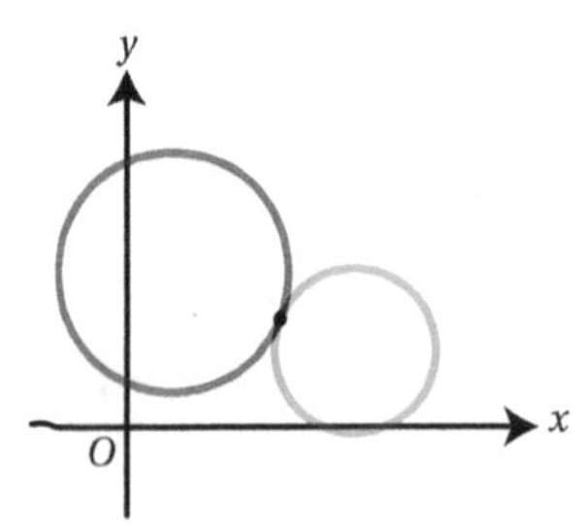

If two circles are tangent to each other then one possibility is that the distance between the centres is equal to the sum of the radii.

$8 + r = 13$

$r = 5$

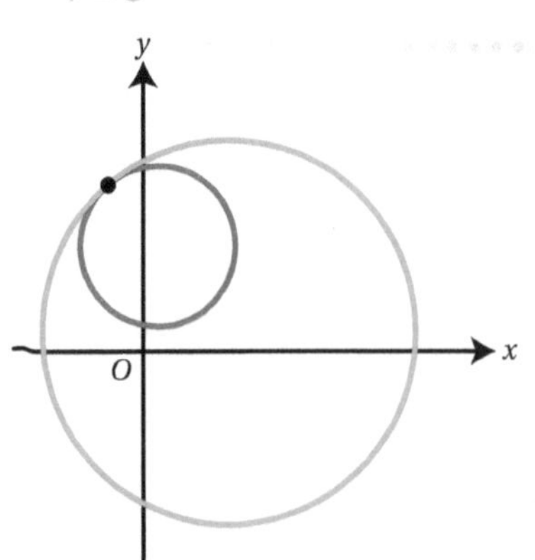

Another possibility is that the circles are tangent to each other at the point on the opposite side of the fixed circle.

$r - 8 = 13$

$r = 21$

Finding points of intersection of two circles can be difficult, as it involves solving two simultaneous quadratic equations. However, there are some special cases where it is possible to use the substitution method as for a straight line intersecting a circle.

WORKED EXAMPLE 6.15

Find the coordinates of the points of intersection of the circles $(x-2)^2+y^2=36$ and $(x-2)^2+(y-10)^2=64$

From the first equation: $(x-2)^2=36-y^2$

Substitute into the second equation:

$$36-y^2+(y-10)^2=64$$
$$36-y^2+y^2-20y+100=64$$
$$20y=72$$
$$y=3.6$$

You need to solve the two simultaneous equations. The term $(x-2)^2$ is common to both equations so you can substitute it from other equation into the other.

$$(x-2)^2+3.6^2=36$$
$$(x-2)^2=23.04$$
$$x-2=\pm4.8$$
$$\therefore x=6.8 \text{ or } -2.8$$

Now substitute back to find the x-coordinates.

The coordinates of the points of intersection are $(6.8, 3.6)$ and $(-2.8, 3.6)$.

EXERCISE 6E

1 Find the equation of:

a i the tangent to the circle $x^2-2x+y^2=15$ at the point $(1, 4)$

ii the tangent to the circle $x^2+y^2+6y=25$ at the point $(-3, 2)$

b i the normal to the circle $x^2+4x+y^2-6y=0$ at the point $(1, 1)$

ii the normal to the circle $x^2-4x+y^2=9$ at the point $(5, 2)$

c i the tangent to the circle with centre $(1, 2)$ and radius $\sqrt{5}$ at the point $(3, 3)$

ii the tangent to the circle with centre $(-3, 1)$ and radius $\sqrt{32}$ at the point $(1, 5)$.

2 Determine whether the line and the circle intersect; where they do, find the coordinates of the point(s) of intersection.

a i $x^2+3x+y^2=24$ and $y=2x+1$

ii $x^2-5x+y^2+y=17$ and $y=2x-1$

b i $x^2-12x+y^2-10y+41=0$ and $2x-y+3=0$

ii $x^2+y^2=32$ and $x+y=8$

c i $x^2+y^2-10y+9=0$ and $x-y=7$

ii $x^2-8x+y^2-6y=25$ and $x+y=20$

3 Determine whether the two circles intersect, are disjoint or tangent to each other.

a **i** $(x-3)^2+(y-5)^2=16$ and $(x+1)^2+(y-2)^2=25$

ii $(x+2)^2+(y-1)^2=8$ and $(x-1)^2+(y+2)^2=8$

b **i** $(x-4)^2+(y+1)^2=20$ and $(x+3)^2+(y-5)^2=17$

ii $x^2+(y-2)^2=10$ and $(x-3)^2+(y+3)^2=12$

c **i** $x^2+y^2=30$ and $(x+1)^2+(y-1)^2=6$

ii $(x+3)^2+(y-2)^2=9$ and $(x+7)^2+(y+1)^2=4$

4 Line l_1 has equation $5x-2y=7$. Line l_2 is perpendicular to l_1 and passes through the point $(3,1)$. Find the coordinates of the point of intersection of the two lines.

5 **a** Show that the point $P(-3,2)$ lies on the circle with equation $(x-1)^2+(y+2)^2=32$.

b Write down the coordinates of the centre of the circle.

c Find the equation of the tangent to the circle at P.

6 Line l_1 has equation $2x+y-10=0$. Line l_2 is perpendicular to l_1 and crosses the x-axis at the point $A(-2,0)$.

a Find the equation of l_2.

b Find the coordinates of M, the point of intersection of l_1 and l_2.

Line l_1 crosses the x-axis at B.

c Find the exact area of the triangle AMB.

7 Find the exact length of the tangent to the circle with equation $x^2+8x+y^2-6y+16=0$ from the point $(4,-2)$.

8 A circle with centre $C(2,5)$ passes through the origin.

a Find the equation of the circle.

b Show that the point $A(0,10)$ lies on the circle.

B is another point on the circle such that the chord AB is perpendicular to the radius OC.

c Find the length of AB, correct to three significant figures.

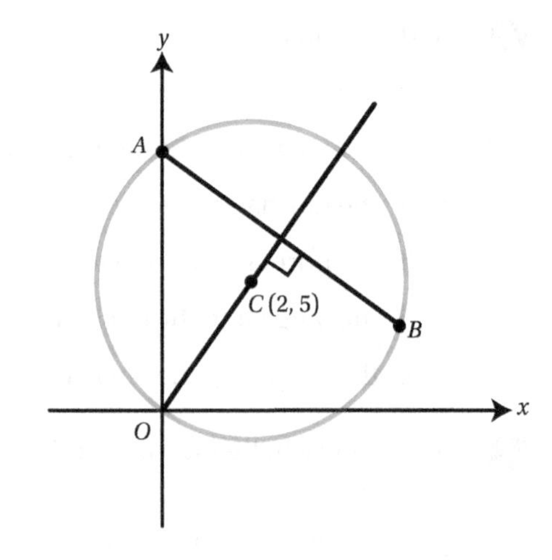

9 Circle C_1 has centre $(-2,5)$ and radius 7. Circle C_2 has centre $(12,5)$.

a Given that the two circles are tangent to each other, find the two possible values for the radius of C_2.

b Given instead that the radius of C_2 is 16, find the coordinates of the intersection points of C_1 and C_2.

10 A circle with the centre at the origin passes through the point $(2, 6)$. The tangent to the circle at $(2, 6)$ cuts the coordinate axes at points P and Q. Find the area of the triangle OPQ.

11 Find the values of k for which the line $y = kx$ is tangent to the circle with centre $(3, 6)$ and radius 2.

12 The line $3x - y = 3$ is tangent to the circle with centre $(5, -1)$ and radius r. Find the value of r.

13 The circle with centre at the origin and radius 5 cuts the negative y-axis at point B. Point $A(4, 3)$ lies on the circle. Let M be the midpoint of the chord AB. The line through O and M cuts the circle at the point P, as shown in the diagram.

a Find the coordinates of M.

b Show that the quadrilateral $OAPB$ is not a rhombus.

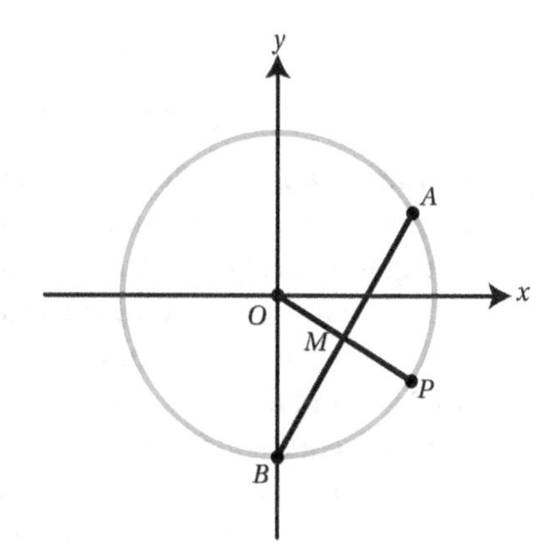

14 A circle has equation $x^2 + y^2 - 10x - 10y + 25 = 0$.

a Show that the circle is tangent to both coordinate axes.

b Show that the point $M(8, 9)$ lies on the circle.

The diagram shows the circle and the tangent at M.

c Find the exact value of the shaded area.

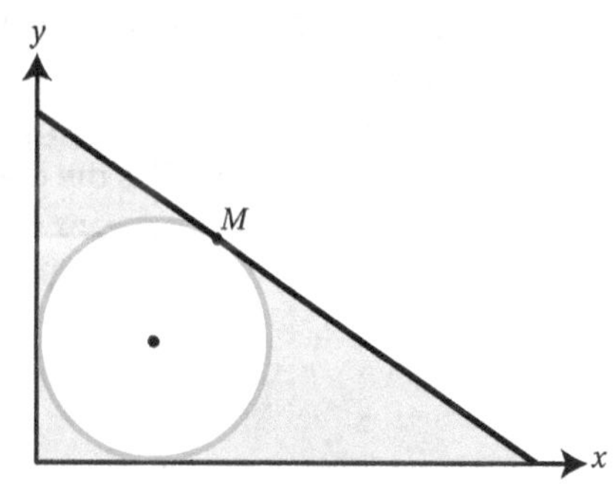

15 Find a condition on a and b such that the curve $x^2 + y^2 = 1$ touches the curve $(x - a)^2 + (y - b)^2 = r^2$ at exactly one point.

16 The circumcircle of a triangle is the circle passing through all three vertices.

Its centre is the point of intersection of the perpendicular bisectors of the sides. A triangle has vertices $A(1, 3)$, $B(5, 7)$ and $C(2, 9)$.

a Find the equations of the perpendicular bisectors of AB and AC.

b Find the coordinates of their intersection, P. This is the centre of the circumcircle.

c Find the equation of the perpendicular bisector of BC and verify that it also passes through P.

d Find the exact value of the radius of the circumcircle of triangle ABC.

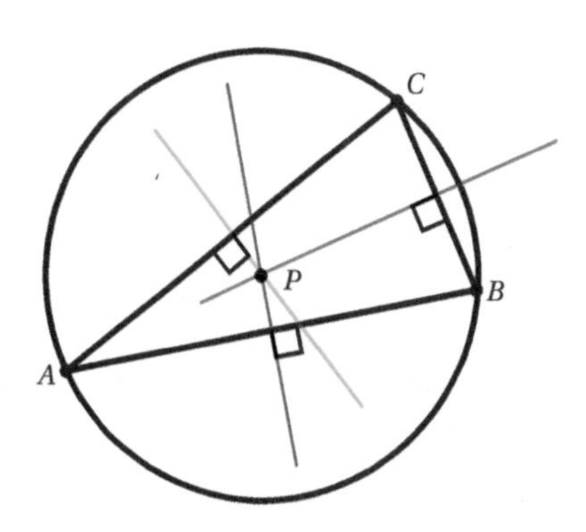

Checklist of learning and understanding

- Given two points $A(x_1, y_1)$ and $B(x_2, y_2)$:
  - the distance $AB = \sqrt{(x_2 - x_1)^2 + (y_2 - y_1)^2}$
  - the midpoint of AB is $\left(\frac{x_1 + x_2}{2}, \frac{y_1 + y_2}{2}\right)$.
- The equation of the straight line through the points (x_1, y_1) and (x_2, y_2) is $y - y_1 = m(x - x_1)$.
  - The gradient is given by $m = \frac{y_2 - y_1}{x_2 - x_1}$.
  - The equation of a line is often written in the form $y = mx + c$ or $ax + by + c = 0$.
- Parallel lines have the same gradient.
- The gradients of perpendicular lines satisfy $m_1 m_2 = -1$.
- A circle with centre (a, b) and radius r has equation $(x - a)^2 + (y - b)^2 = r^2$.
  - You may need to complete the square in order to find the centre and radius.
- You can find the intersection of two lines, a line and a circle, or (sometimes) two circles by solving simultaneous equations.
- You can tell whether two circles intersect, are disjoint or tangent to each other by comparing the sum and the difference of the radii of the circles to the distance between their centres.
- A tangent to a circle is perpendicular to the radius at the point of contact. The normal is in the direction of the radius.
- The angle in a semi-circle is a right angle. It follows that, if A, B and C are three points on a circle and ABC is a right angle, then AC is the diameter of the circle.

Mixed practice 6

1. Find the radius of the circle with equation:

   $x^2 - 8x + y^2 + 6y = 144$

   Choose from these options.

   A $r = 144$ **B** $r = 12$ **C** $r = 13$ **D** $r = 25$

2. Line l_1 has equation $3x - 2y + 7 = 0$.

   a Point $A(2k, k+1)$ lies on l_1. Find the value of k.

   b Point B has coordinates $(-2, p)$. Find the value of p so that AB is perpendicular to l_1.

   c Line l_2 is parallel to l_1 and passes through B. Find the equation of l_2 in the form $ax + by + c = 0$, where a, b and c are integers.

   d l_2 crosses the x-axis at the point C. Find the coordinates of C.

3. Circle C has equation $x^2 - 2x + y^2 - 10y - 19 = 0$.

   a Find the coordinates of the centre, P, of the circle.

   b Show that point $A(7, 2)$ lies on the circle.

   c Point M has coordinates $(1, -1)$. Line l is perpendicular to PA and passes through M. It cuts PA at the point S. Find the coordinates of S.

4. The line AB has equation $3x - 4y + 5 = 0$.

   a The point with coordinates $(p, p+2)$ lies on the line AB. Find the value of the constant p.

   b Find the gradient of AB.

   c The point A has coordinates $(1, 2)$. The point $C(-5, k)$ is such that AC is perpendicular to AB. Find the value of k.

   d The line AB intersects the line with equation $2x - 5y = 6$ at the point D. Find the coordinates of D.

   [© AQA 2013]

5. $y = -3x + 5$ is tangent to the circle C at the point $(4, -7)$. The centre of C is at the point $(k - 4, k + 3)$. Find the value of k.

   Choose from these options.

   A $k = -5$ **B** $k = 17$ **C** $k = 0$ **D** $k = -19$

6. Consider the points $A(4, 3)$, $B(3, -2)$ and $C(9, 2)$.

   a Show that angle BAC is a right angle.

   b Hence find the equation of the circle through A, B and C.

   c Find the equation of the tangent to the circle at B. Give your answer in the form $ax + by + c = 0$, where a, b and c are integers.

7. A circle has centre $(3, 0)$ and radius 5. The line $y = 2x + k$ intersects the circle in two points. Find the set of possible values of k, giving your answers in surd form.

8 A circle has centre $C(7, 12)$ and passes through the point $D(4, 10)$. The tangent to the circle at D cuts the coordinate axes at points A and B. Find the area of the triangle:

a AOB

b ABC.

9 The points $A(-3, 7)$ and $B(5, -1)$ are endpoints of the diameter of a circle. Find the equation of the circle in the form $x^2 + ax + y^2 + by + c = 0$.

10 Find the exact values of k for which the line $y = kx + 3$ is tangent to the circle with centre $(6, 3)$ and radius 2.

11 A circle with centre $C(-3, 2)$ has equation:

$x^2 + y^2 + 6x - 4y = 12$

a Find the y-coordinates of the points where the circle crosses the y-axis.

b Find the radius of the circle.

c The point $P(2, 5)$ lies outside the circle.

i Find the length of CP, giving your answer in the form $\sqrt{n}$, where n is an integer.

ii The point Q lies on the circle so that PQ is a tangent to the circle. Find the length of PQ.

[© AQA 2013]

12 How many values of k are there for which $y = kx - 5$ is tangent to the circle $x^2 + y^2 = 20$?

Choose from these options.

A 0 **B** 1 **C** 2 **D** Infinitely many

13 Find the shortest distance from the point $(-3, 2)$ to the line with equation $3x + 2y = 19$. Give your answer in exact form.

14 Show that each of the circles $x^2 - 6x + y^2 + 10y + 18 = 0$ and $x^2 - 14x + y^2 - 6y + 49 = 0$ lies entirely outside the other one.

15 A circle has centre $(5, 7)$. It crosses the x-axis at points $A(2, 0)$ and $B(p, 0)$, where $p > 2$.

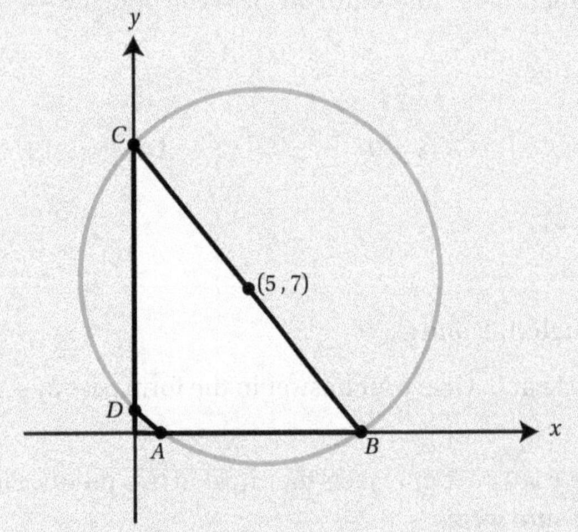

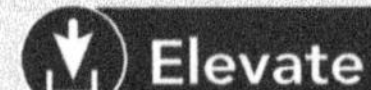

See Extension Sheet 6 for a selection of more challenging problems.

a Find the value of p and hence write down the equation of the circle.

b The circle crosses the y-axis at points C and D. Find the area of the quadrilateral $ABCD$.

7 Logarithms

In this chapter you will learn how to:

- use an operation called a logarithm to undo exponential functions
- use the laws of logarithms
- use logarithms to find exact solutions of some exponential equations
- use the number e.

Before you start...

GCSE	You should be able to work with expressions involving exponents.	1 Answer 'true' or 'false'. a $2\times3^2=36$ b When $x=25$, $4x^{\frac{1}{2}}=10$ c $(2\times3)^7=2^7\times3^7$ d $\frac{1}{2x^3}=2^{-3}x^{-3}$
GCSE	You should be able to evaluate fractional and negative exponents.	2 Evaluate, without using a calculator: a $27^{\frac{4}{3}}$ b $9^{-\frac{3}{2}}$
Chapter 2	You should be able to use laws of indices.	3 Write in the form x^p: a $x^2\times\sqrt{x}$ b $\frac{x\sqrt{x}}{x^2}$
GCSE	You should be able to solve equations involving fractions.	4 Solve this equation. $\frac{x+1}{2x-3}=2$
Chapter 3	You should be able to solve quadratic equations.	5 Solve these equations. a $(x-1)(x+3)=5$ b $2x^2-\frac{1}{x^2}=1$

Discovering logarithms

If you are asked to solve $x^2=3$ for $x\geqslant0$ you could either find a decimal approximation (for example, by using a calculator or trial and improvement) or take the square root.

$$x=\sqrt{3}$$

This statement just says: 'x is the positive value that, when squared, gives 3.'

Similarly, to solve $10^x = 50$, you could use trial and improvement to seek a decimal value.

$$10^1 = 10$$

$$10^2 = 100$$

So x lies between 1 and 2.

$$10^{1.5} \approx 31.6$$

$$10^{1.6} \approx 39.8$$

$$10^{1.7} \approx 50.1$$

So the answer is approximately 1.7.

Just as you can take the square root to answer the question: 'What is the number that, when squared, gives 3?', there is also a function to answer the question: 'What is the number that, when put as the exponent of 10, gives 50?' This function is called a base-10 **logarithm**.

In this chapter you will learn about laws of logarithms and how to use them to solve some problems involving unknown exponents.

Section 1: Introducing logarithms

In the example in the introduction, the solution to the equation $10^x = 50$ can be written as $x = \log_{10} 50$. More generally, the equation $y = 10^x$ can be expressed as $x = \log_{10} y$. In fact the base does not need to be 10, but could be any positive value other than 1.

Tip

Remember that the symbol $\Leftrightarrow$ means the statements are equivalent and you can switch between them.

Key point 7.1

$$a = b^x \Leftrightarrow \log_b a = x$$

WORKED EXAMPLE 7.1

Without using a calculator, evaluate these expressions.

a $\log_2 8$ **b** $\log_{10} 0.01$ **c** $\log_{25} 5$

a $\log_2 8 = 3$ — Since $8 = 2^3$

b $\log_{10} 0.01 = -2$ — Since $0.01 = 10^{-2}$

c $\log_{25} 5 = \frac{1}{2}$ — Since $5 = 25^{\frac{1}{2}}$

You can also use the fact that logarithms are the inverse of raising to an exponent to solve equations and rearrange formulae.

WORKED EXAMPLE 7.2

Make x the subject of each equation.

a $\log_2 x = y+3$ **b** $\log_x (y+1) = 2$

a $\log_2 x = y+3$

$x = 2^{y+3}$

Use $\log_b a = p \Leftrightarrow a = b^p$ (Key point 7.1).

b $\log_x (y+1) = 2$

$x^2 = y+1$

Use $\log_b a = p \Leftrightarrow a = b^p$ (Key point 7.1).

$x = \sqrt{y+1}$

The base of a logarithm has to be positive, so you don't need ±.

Whenever you raise a positive number to a power – positive or negative – the answer is always positive. For example:

$$2^3 = 8 \text{ and } 2^{-3} = \frac{1}{8}$$

So there is no answer to a question such as: 'To what power do you raise 10 to get −3?'

Key point 7.2

The logarithm of a negative number or zero is not a real number.

Did you know?

It is possible to take logarithms of negative numbers if you use complex numbers.

The logarithms of the two most common bases have distinct abbreviations. Since a decimal system is used for counting, base 10 is the default base for a logarithm. Thus, $\log_{10} x$ is usually written simply as $\log x$.

Fast forward

You will study the number e in more detail in Chapter 8.

There is also a special number, called e, which is important when studying rates of change. e is an irrational number, a bit like π, and its value is approximately 2.718 28. The logarithm in base e is called the **natural logarithm** and is denoted by $\ln x$. It follows all the same rules as other logarithms.

Did you know?

Logarithms were introduced by the Scottish mathematician John Napier (1550–1617). He originally studied logarithms in base $\frac{1}{e}$.

Tip

Your calculator has special buttons for log and ln. Most calculators can also evaluate logs in other bases.

Key point 7.3

a $\log_{10} x$ is written as $\log x$

b $\log_e x$ is written as $\ln x$

Remember that taking a logarithm reverses the process of raising to an exponent.

Key point 7.4

a $\log_a(a^x) = x$

b $a^{\log_a x} = x$

As a special case, when $a = \mathrm{e}$:

c $\ln(\mathrm{e}^x) = x$

d $\mathrm{e}^{\ln x} = x$

Common error

You can only use these results when the base of the logarithm and the base of the exponential match and are immediately adjacent in the expression.

WORKED EXAMPLE 7.3

Simplify, without using a calculator:

a $\ln(\mathrm{e}^5)$

b $10^{2+\log 3}$

a $\ln(e^5) = 5$ — Use $\ln(\mathrm{e}^x) = x$.

b $10^{2+\log 3} = 10^2 \times 10^{\log 3}$ — You can't 'cancel' the base 10 and log initially as they aren't adjacent, so first use $x^{a+b} = x^a \times x^b$.

$= 100 \times 3$ — Now use $a^{\log_a x} = x$ with $a = 10$ to 'cancel' 10 and log.

$= 300$

You can solve equations using the method from Worked example 7.3. However, you may prefer to think about 'doing the same thing to both sides.' In Worked example 7.4, after isolating the natural logarithm, you can get rid of it by doing e to the power of both sides and then using Key point 7.4.

WORKED EXAMPLE 7.4

Find the exact solution to the equation $3\ln(2+x) = 6$.

$3\ln(2+x) = 6$

$\ln(2+x) = 2$ — Divide both sides by 3 to get the logarithm by itself.

$e^{\ln(2+x)} = e^2$ — Raise e to the power of both sides.

$2 + x = e^2$ — Then use $\mathrm{e}^{\ln x} = x$ (Key point 7.4) on the LHS to 'cancel' e and ln.

$x = e^2 - 2$ — Now just solve for x. You are asked for an exact answer so leave it in this form – don't try to write it as a decimal.

Remember that $\log x$ is just another value, so it can be treated like any variable. In particular, $\log x \times \log x = (\log x)^2$.

Common error

$(\log x)^2$ is not the same as $\log(x^2)$.

WORKED EXAMPLE 7.5

Simplify $\frac{(\ln x)^2 - 4}{\ln x + 2}$.

$\frac{(\ln x)^2 - 4}{\ln x + 2} = \frac{(\ln x - 2)(\ln x + 2)}{\ln x + 2}$ — Factorise the numerator (difference of two squares).

$= \ln x - 2$ — Cancel common factors in the numerator and denominator.

EXERCISE 7A

1 Evaluate these expressions without using a calculator.

a **i** $\log_3 27$ **ii** $\log_4 16$

b **i** $\log_5 5$ **ii** $\log_3 3$

c **i** $\log_{12} 1$ **ii** $\log_{15} 1$

d **i** $\ln \frac{1}{e}$ **ii** $\log_4 \frac{1}{64}$

e **i** $\log_4 2$ **ii** $\log_{27} 3$

f **i** $\log_8 \sqrt{8}$ **ii** $\ln \sqrt{e}$

g **i** $\log_8 4$ **ii** $\log_{81} 27$

h **i** $\log_{25} 125$ **ii** $\log_{16} 32$

i **i** $\log_4 2\sqrt{2}$ **ii** $\log_9 81\sqrt{3}$

j **i** $\log_{25} 0.2$ **ii** $\log_4 0.5$

Tip

Although you can evaluate logarithms on a calculator, this exercise will help you develop your understanding of the new concept.

2 Use a calculator to evaluate each expression, giving your answer correct to 3 s.f.

a **i** $\log 50$ **ii** $\log\left(\frac{1}{4}\right)$ **b** **i** $\ln 0.1$ **ii** $\ln 10$

c **i** $\log_3 8.5$ **ii** $\log_5 0.6$ **d** **i** $\log_{0.2} 3$ **ii** $\log_{0.8} 0.6$

3 Simplify these expressions.

a **i** $7\ln x - 2\ln x$ **ii** $4\log x + 3\log x$ **b** **i** $\frac{(\log a)^2 + \log a \log b}{\log a \log b}$ **ii** $\frac{(\log a)^2 - 1}{\log a - 1}$

4 Make x the subject of each equation.

a **i** $\log_3 x = y$ **ii** $\log_4 x = 2y$ **b** **i** $\log_a x = 1 + y$ **ii** $\log_a x = y^2$

c **i** $\log_x 3y = 3$ **ii** $\log_x y = 2$ **d** **i** $y = 2 + \ln x$ **ii** $\ln y = \ln x - 2$

5 Find the value of x in each equation.

a **i** $\log_2 x = 5$ **ii** $\log_2 x = 4$

b **i** $\log_5 25 = 5x$ **ii** $\log_{49} 7 = 2x$

c **i** $\log_x 36 = 2$ **ii** $\log_x 10 = \frac{1}{2}$

6 Simplify each expression.

a $e^{2+\ln a}$ b $(e^2)^{\ln a}$

7 Find the value of each expression.

a $\log_{\sqrt{b}} b^3$ b $\log_b b^2 - \log_b \sqrt{b}$

8 Solve the equation $\log_{10}(9x+1)=3$.

9 Solve the equation $\log_8 \sqrt{1-x} = \frac{1}{3}$.

10 Find the exact solution to the equation $\ln(3x-1)=2$.

11 Solve the equation $3(1+\log x) = 6 + \log x$.

12 Find all values of x which satisfy $(\log_3 x)^2 = 4$.

13 Solve the simultaneous equations.

$$\log_3 x + \log_5 y = 6$$
$$\log_3 x - \log_5 y = 2$$

14 Solve $\ln x = 1 + \frac{20}{\ln x}$.

15 Evaluate $\sqrt[6]{(\pi^4 + \pi^5)}$. What do you notice about this result?

Section 2: Laws of logarithms

In just the same way that there are laws of indices, there are also laws of logarithms.

Key point 7.5

The laws of logarithms

a $\log_a xy = \log_a x + \log_a y$

b $\log_a \left(\frac{x}{y}\right) = \log_a x - \log_a y$

c $\log_a x^k = k \log_a x$

d $\log_a 1 = 0$

Tip

A useful particular case of Key point 7.5c is when $k = -1$.

$$\log_a \frac{1}{x} = -\log_a x$$

In fact these laws of logarithms follow from the laws of indices. The first is proved here; the others can be proved similarly.

PROOF 3

$\log_a xy = \log_a x + \log_a y$

Let: $P = \log_a x$ and $Q = \log_a y$. — You will need to manipulate these log terms so it is useful to label them (say P and Q).

So $x = a^P$ and $y = a^Q$. — Although you don't have many algebraic properties of logarithms to work with you do have laws of indices, so use Key point 7.1 to convert to exponential form.

$xy = a^P \times a^Q$
$= a^{P+Q}$ — Use $x^{a+b} = x^a \times x^b$.

Taking logs of both sides:
$\log_a xy = \log_a a^{P+Q}$ — You want $\log_a xy$ so you need to take logs of both sides.

$= P + Q$ — Use $\log_a (a^x) = x$.

$= \log_a x + \log_a y$ — Give the answer in terms of x and y.

You can use these rules to manipulate expressions involving logarithms.

WORKED EXAMPLE 7.6

If $x = \log_{10} a$ and $y = \log_{10} b$, express $\log_{10}\left(\frac{100a^2}{b}\right)$ in terms of x and y.

$\log_{10}\left(\frac{100a^2}{b}\right) = \log_{10} 100a^2 - \log_{10} b$ — First use $\log_a\left(\frac{x}{y}\right) = \log_a x - \log_a y$.

$= \log_{10} 100 + \log_{10} a^2 - \log_{10} b$ — Then use $\log_a xy = \log_a x + \log_a y$.

$= \log_{10} 100 + 2\log_{10} a - \log_{10} b$ — Then use $\log_a x^k = k \log_a x$.

$= 2 + 2\log_{10} a - \log_{10} b$
$= 2 + 2x - y$ — $\log_{10} 100 = 2$

WORK IT OUT 7.1

If $e^{2y} = x+2$, write y in terms of x.

Which is the correct solution? Identify the errors made in the incorrect solutions.

Solution 1	Solution 2	Solution 3
$2y = \ln(x+2)$ $= \ln x + \ln 2$ So $y = \frac{\ln x + \ln 2}{2}$	$2y = \ln(x+2)$ $y = \frac{1}{2}\ln(x+2)$ $y = \ln\sqrt{x+2}$	$2y = \ln(x+2)$ $= \ln x \times \ln 2$ So $y = \frac{1}{2}\ln x \times \frac{1}{2}\ln 2$

The laws of logarithms can also be used to solve equations. The usual tactic is to combine all log terms into one.

WORKED EXAMPLE 7.7

Solve the equation $\log_2 x + \log_2 (x+4) = 5$.

$\log_2 x + \log_2 (x+4) = 5$
$\log_2 (x(x+4)) = 5$

Rewrite one side as a single logarithm.

$2^{\log_2 (x(x+4))} = 2^5$

Raise 2 to the power of both sides...

$x^2 + 4x = 32$

then 'cancel' 2 and $\log_2$.

$x^2 + 4x - 32 = 0$

Solve the quadratic equation.

$(x+8)(x-4) = 0$

$x = -8$ or $x = 4$

When $x = -8$:
LHS $= \log_2(-8) + \log_2(-4)$
is not real so this solution does not work.

Check your solution in the original equation.

When $x = 4$:
LHS $= \log_2 4 + \log_2 8$
$= 2 + 3 = 5 =$ RHS

$\therefore x = 4$

State the final solution.

Tip

As well as knowing the laws of logs it is just as important to know what you can't do!

Common error

One of the most common errors with logs is to say that $\log(x+y) = \log x + \log y$. This is not true!

Elevate

See Support Sheet 7 for a further example of this type and for more practice questions.

Tip

Checking your solutions involves more than just looking for an arithmetic error – it is possible to introduce false solutions through algebraic manipulation.

Rewind

This is exactly the same issue that arose with squaring equations in Chapter 1, where false 'solutions' could be introduced.

EXERCISE 7B

1 If $x = \log a$, $y = \log b$ and $z = \log c$, express these expressions in terms of x, y and z.

a **i** $\log b^7$ **ii** $\log a^2b$

b **i** $\log\left(\frac{ab^2}{c}\right)$ **ii** $\log\left(\frac{a^2}{bc^3}\right)$

c **i** $\log\left(\frac{100}{bc^5}\right)$ **ii** $\log(5b)+\log(2c^2)$

d **i** $\log a^3 - 2\log ab^2$ **ii** $\log(4b)+2\log(5ac)$

2 Solve for x.

a **i** $\log_3(x+2)=2$ **ii** $\log_2(7x+4)=5$

b **i** $\log_3 x - \log_3(x-6)=1$ **ii** $\log_7(x+2)-\log_7 x=1$

c **i** $\log_3(x-8)+\log_3 x=2$ **ii** $\log_2(x-2)-\log_2\left(\frac{1}{x}\right)=3$

3 Find the value of x for which $3\log_b x = \log_b 64$.

4 Solve the equation $\ln x = 2\ln 9 - \ln 3$.

5 Solve the equation $\log(3x+6)=\log(3)+1$.

6 Solve the equation $\log(x+5)-1=\log(x-1)$.

7 If $x = \log a$, $y = \log b$ and $z = \log c$, express these expressions in terms of x, y and z.

a $\log(a^2b)$ **b** $\log\left(\frac{100a}{c}\right)$

8 If $p=\ln x$, $q=\ln y$ and $r=\ln z$, express $3p-2q+r$ as a single logarithm.

9 $a=\ln 2$ and $b=\ln 5$. Find, in terms of a and b:

a $\ln 50$ **b** $\ln 0.16$.

10 Find the exact solution of the equation $2\ln x+\ln 9=3$, giving your answer in the form Ae^B where A and B are rational numbers.

11 Show that there is only one solution to the equation $\log_2(x+2)=3-\log_2 x$.

12 Solve the equation $2\log_3 x+2=\log_3(21x-10)$.

13 Solve the equation $\log_2(5x+18)=3+2\log_2(x+1)$.

14 Solve the equation $\ln(3x-8)+\ln(x+5)=2\ln(x+2)$.

15 Solve the equation $\log_2(x+12)=3+\frac{1}{2}\log_2 x$.

Section 3: Solving exponential equations

One of the main uses of logarithms is to solve equations with the unknown in the exponent.

WORKED EXAMPLE 7.8

Find the exact solution of the equation $3^{x-2} = 5$.

$3^{x-2} = 5$ — Use $a = b^x \Leftrightarrow x = \log_b a$.

$x - 2 = \log_3 5$

$x = \log_3 5 + 2$ — An exact answer is required, so don't calculate as a decimal.

There is often more than one correct way to express your answer. Work it out 7.2 shows some other possibilities.

WORK IT OUT 7.2

Solve the equation $3^{x-2} = 5$.

Which is the correct solution? Identify the errors made in the incorrect solutions.

Solution 1	Solution 2	Solution 3
$3^{x-2} = 5$	$3^{x-2} = 5$	$3^{x-2} = 5$
$\log(3^{x-2}) = \log 5$	$\ln(3^{x-2}) = \ln 5$	$(x-2)\log 3 = \log 5$
$(x-2)\log 3 = \log 5$	$(x-2)\ln 3 = \ln 5$	$x\log 3 - 2\log 3 = \log 5$
$x\log 3 - 2\log 3 = \log 5$	$x - 2 = \ln\left(\frac{5}{3}\right)$	$x\log 3 = 2\log 3 + \log 5$
$x\log 3 = \log 5 + 2\log 3$	$x = \ln\left(\frac{5}{3}\right) + 2$	$x\log 3 = \log 45$
$x = \frac{\log 5 + 2\log 3}{\log 3}$		$x = \frac{\log 45}{\log 3} = \log 42$

When the unknown appears in the power of both sides it is often better to take logs of both sides and work from there. You can choose any base – unless it is specified in the question!

WORKED EXAMPLE 7.9

Solve the equation $2^{x+3} = 5^{x-1}$, giving your answer in the form $\frac{\ln p}{\ln q}$.

$2^{x+3} = 5^{x-1}$ — Take logs of both sides. The question requires you to use ln.

$\ln(2^{x+3}) = \ln(5^{x-1})$

Continues on next page

$(x+3)\ln 2=(x-1)\ln 5$	Use $\log_a x^k = k\log_a x$ on each side.
$x\ln 2+3\ln 2=x\ln 5-\ln 5$ $3\ln 2+\ln 5=x\ln 5-x\ln 2$	Expand the brackets. Group terms with x on one side and terms without x on the other.
$3\ln 2+\ln 5=x(\ln 5-\ln 2)$	Factorise the expression on the right so that x occurs only once in the equation.
$\ln(2^3\times 5)=x\ln\left(\frac{5}{2}\right)$	Use laws of logarithms to reduce to a single log on each side.
$x=\frac{\ln 40}{\ln 2.5}$	Now divide by the coefficient of x to get the answer in the required form.

WORK IT OUT 7.3

Solve the equation $3\times 2^x = 7^{x-2}$.

Which is the correct solution? Identify the errors made in the incorrect solutions.

Solution 1	Solution 2	Solution 3
$\log 3\times\log(2^x)=\log(7^{x-2})$	$\ln 3+x\ln 2=(x-2)\ln 7$	$6^x=7^{x-2}$
$\log 3\times x\log 2=(x-2)\log 7$	$\ln 3+2\ln 7=x\ln 7-x\ln 2$	$x=(x-2)\log_6(7)$
$x\log 3\times\log 2=x\log 7-2\log 7$	$x=\frac{\ln 147}{\ln 3.5}$	$x=x\log_6 7-\log_6 49$
$x\log 6-x\log 7=-2\log 7$		$\log_6 49=x(\log_6 7-1)$
$x\log\left(\frac{7}{6}\right)=2\log 7$		$x=\frac{\log_6 49}{\log_6 7-1}$
$x=\frac{\log(49)}{\log\left(\frac{7}{6}\right)}$		

EXERCISE 7C

1 Solve for x, giving your answers correct to 3 s.f.

a **i** $3^{2x+5}=10$ **ii** $4^{3-x}=7$ **b** **i** $e^{3x-2}=8$ **ii** $e^{5-2x}=11$

c **i** $3\times 4^x=90$ **ii** $1000\times 1.02^x=10\ 000$ **d** **i** $6\times 7^{3x+1}=1.2$ **ii** $5\times 2^{2x-5}=94$

e **i** $3^{2x}=4^{x-1}$ **ii** $5^x=6^{1-x}$ **f** **i** $3\times 2^{3x}=7\times 3^{3x-2}$ **ii** $4\times 8^{x-1}=3\times 5^{2x+1}$

2 Solve these equations, giving your answers in terms of natural logarithms.

a **i** $2\times 3^x=5$ **ii** $5\times 7^x=3$ **b** **i** $4e^x=1$ **ii** $3e^x=1$

c **i** $2^{2x}=5$ **ii** $10^{3x}=7$ **d** **i** $2^{3x+1}=10$ **ii** $5^{2x-3}=4$

e **i** $2^{x+1}=5^x$ **ii** $5^{x-2}=3^x$ **f** **i** $2^{3x}=3e^x$ **ii** $e^{2x}=5\times 2^x$

g **i** $5\times 3^x=8^{x+2}$ **ii** $3^{2x+1}=7\times 2^{x-2}$

3. Solve the equation $5^{4x+3} = 28$, giving your answer correct to 3 s.f.

4. Solve the equation $4 \times 3^{x-5} = 1$, giving your answer correct to 3 s.f.

5. Find, in terms of the natural logarithm, the exact solution of the equation $e^{3-2x} = 10$.

6. Find the exact solution of the equation $10^x = 5 \times 2^{3x}$, giving your answer in the form $x = \frac{\log p}{q + \log r}$, where p, q and r are integers.

7. Solve the equation $2^{3x-1} = 5^{2-x}$, giving your answer in the form $x = \frac{\ln a}{\ln b}$, where a and b are integers.

8. If $3^x \times 4^{2x+1} = 6^{x+2}$, show that $x = \frac{\ln 9}{\ln 8}$.

Disguised quadratics

It is usually impossible to simplify expressions which contain powers with different bases, such as $3^x + 5^x$. This means that you cannot (algebraically) solve the equation $3^x + 5^x = 17$. However, there is a special type of equation that looks very similar to this but which you can solve exactly.

Rewind

You saw these 'disguised quadratics' in Chapter 3 and will meet them again in Chapter 10 on trigonometry.

WORKED EXAMPLE 7.10

Find the exact solution of the equation $4^x + 2^x = 12$.

$4^x + 2^x = 12$

$(2^x)^2 + 2^x = 12$ — $4^x = (2^2)^x = (2^x)^2$

Let $y = 2^x$. — Make a substitution to turn the equation into a quadratic.

Then $y^2 + y = 12$

$y^2 + y - 12 = 0$

$(y+4)(y-3) = 0$ — Solve the quadratic.

$y = -4$ or 3

When $2^x = -4$ there are no solutions. — Return to x. No real power of a positive number can be negative.

When $2^x = 3$, $x = \log_2 3$ — The other solution for y does give a valid solution for x.

EXERCISE 7D

1 Find the exact solution(s) of each equation.

a i $(2^x)^2-6\times2^x+5=0$ ii $(3^x)^2-7\times3^x+12=0$

b i $7^{2x}-9\times7^x+8=0$ ii $5^{2x}-9\times5^x+20=0$

c i $4^x-5\times2^x+6=0$ ii $9^x-6\times3^x+8=0$

d i $9^x-8\times3^x=9$ ii $25^x-5^x=6$

e i $e^{2x}+16e^x=80$ ii $e^{2x}-9e^x+20=0$

f i $25^x-15\times5^x+50=0$ ii $4^x-7\times2^x+12=0$

2 **a** By letting $y=3^x$, show that $3^{2x}-3^{x+2}+20=0$ can be written in the form $y^2-9y+20=0$.

b Hence solve $3^{2x}-3^{x+2}+20=0$, giving your answers correct to 3 s.f.

3 Find the exact solutions of the equation $4^x-10\times2^x+16=0$.

4 Solve the equation $2e^{2x}-9e^x+4=0$, giving your answers in the form $k\ln 2$.

5 Solve the equation $5^{2x+1}-14\times5^x-3=0$.

6 Solve, in exact form, the equation $e^x=8-15e^{-x}$.

Checklist of learning and understanding

- Logarithms allow you to answer the question: 'To what power do I need to raise b to get a?'

$$x=\log_b a \Leftrightarrow b^x=a$$

- You can only take a logarithm of a positive number.
- Logarithms in base 10 are written as $\log x$. Logarithms in base e (natural logarithms) are written as $\ln x$.
- Logarithms and exponentials can be 'cancelled' when they are adjacent and the bases are the same.
  - $\log_a(a^x)=x$
  - $a^{\log_a x}=x$

  When $a=$ e:
  - $\ln(e^x)=x$
  - $e^{\ln x}=x$
- Logarithms obey these laws.
  - $\log_a xy=\log_a x+\log_a y$
  - $\log_a\left(\frac{x}{y}\right)=\log_a x-\log_a y$
  - $\log_a\left(\frac{1}{x}\right)=-\log_a x$
  - $\log_a x^k=k\log_a x$
  - $\log_a 1=0$
- Many exponential equations can be solved by taking a logarithm of both sides and using the laws of logs.
- Some equations can be turned into quadratic equations by using a substitution of the form $y=a^x$.

Mixed practice 7

1. If $8^x = 16$ then what is the value of x?

   Choose from these options.

   A $\log_{16} 8$ B $\sqrt[8]{16}$ C $\frac{16}{8}$ D $\frac{4}{3}$

2. Solve the equation $4\log_a x = \log_a 81$.

3. Solve the equation $\log_5\left(\sqrt{x^2+49}\right) = 2$.

4. Given that $\log_x 4 = 9$, find the value of x correct to 3 s.f.

5. If $\log_a y + \log_a 7 = 4$, express y in terms of a.

6. If $a = \log x$, $b = \log y$ and $c = \log z$, express these expressions in terms of a, b, c.

   a $\log \frac{x^2\sqrt{y}}{z}$ b $\log\sqrt{0.1x}$ c $\log_{100}\left(\frac{y}{z}\right)$

7. Solve the equation $3e^{2x+1} = 17$, giving your answer correct to 3 s.f.

8. The curve $y = 3^{2x-1}$ intersects the line $y = 4$ at the point P. Find the exact value of the x-coordinate of P.

9. If $4\log_b x - \log_b 9 = 2$, then express b in terms of x.

   Choose from these options.

   A $b = \frac{x^4}{9}$ B $b = 2x - 3$ C $b = \frac{x^2}{3}$ D $b = 4x - 9$

10. Solve the simultaneous equations.

    $\ln x + \ln y^2 = 8$

    $\ln x^2 + \ln y = 6$

11. If $y = \ln x - \ln(x+2) + \ln(4-x^2)$, express x in terms of y.

12. Solve, correct to 3 s.f., $3^{2x} - 3^{x+1} - 10 = 0$.

13. Solve the equation $\log(x^2+1) = 1 + 2\log x$.

14. a Given that $\log_a b = c$, express b iterms of a and c.

    b By forming a quadratic equation, show that there is only one value of x which satisfies the equation $2\log_2(x+7) - \log_2(x+5) = 3$.

    [© AQA 2013]

15. a Given that $2\log_k x - \log_k 5 = 1$, express k in terms of x. Give your answer in a form not involving logarithms.

    b Given that $\log_a y = \frac{3}{2}$ and that $\log_4 a = b + 2$, show that $y = 2^p$, where p is an expression in terms of b.

    [© AQA 2011]

16 If $\log_3 y = \log_9 3$, then what is the value of y?

Choose from these options.

A $y = \sqrt{3}$ **B** $y = 9$ **C** $y = 3$ **D** $y = \frac{1}{3}$

17 Find the exact value of x satisfying the equation $2^{3x-2} \times 3^{2x-3} = 36^{x-1}$. Give your answer in simplified form $\frac{\ln p}{\ln q}$, where $p, q \in \mathbb{Z}$.

18 Solve the equation $5 \times 4^{x-1} = \frac{1}{3^{2x}}$, giving your answer in the form $x = \frac{\ln p}{\ln q}$, where p and q are rational numbers.

19 Find the exact solutions to $e^x + e^{-x} = 4$.

20 Find the values of x for which $(\log_3 x)^2 = \log_3 x^3 - 2$.

Elevate

See Extension Sheet 7 for a selection of more challenging problems.

8 Exponential models

In this chapter you will learn how to:

- recognise and use graphs of exponential functions
- use exponential functions in modelling
- use logarithms to transform curved graphs into straight lines.

Before you start...

Chapter 7	You should be able to work with the number e and natural logarithms.	1 Simplify each expression. a $\ln(5e^2)$ b $e^{1+\ln(5)}$
Chapter 7	You should be able to use the laws of logarithms.	2 If $y = 100x^3$ write $\log y$ in the form $n + k \log x$.
Chapter 5	You should be able to transform graphs.	3 Describe the effect of changing $y = f(x)$ into $y = f(2x)$.
Chapter 6	You should be able to work with equations of straight lines.	4 a What is the gradient of $3y + 2x = 5$? b Find the equation of this line in the form $y = mx + c$. 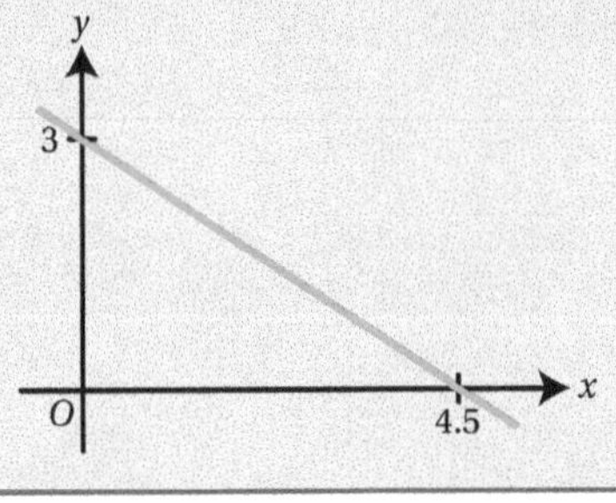

Why use exponential models?

In many situations in the real world the rate of growth of a quantity is approximately proportional to the amount that is there; for example, the more people there are in a country, the more babies will be born. The only functions that have this property are **exponential functions**, of the form $y = Ca^x$.

Section 1: Graphs of exponential functions

The diagram shows the graph of $y = 2^x$.

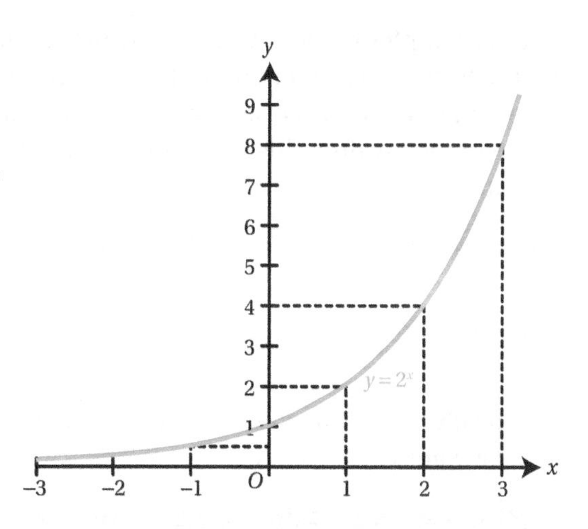

For very large positive values of x the y-value approaches infinity, and for very large negative values of x the y-value approaches (but never reaches) 0. A line to which a function gets increasingly close (but never reaches) is called an **asymptote**. In this case the x-axis is an asymptote to the graph.

Looking at the graphs of exponential functions with different bases, it is possible to make some generalisations.

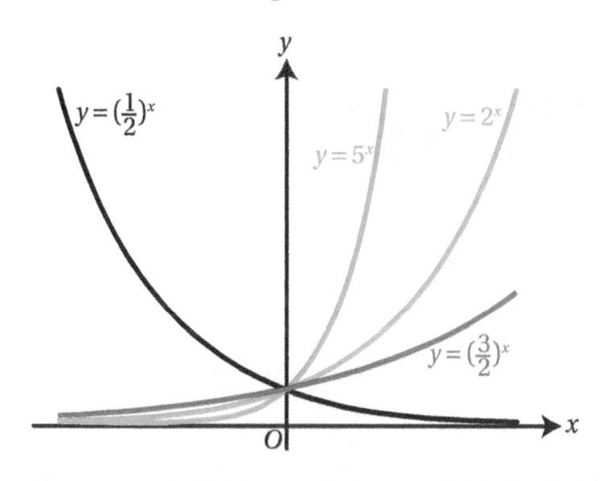

Rewind

Notice that the black line is a reflection of the blue line. This is because $\left(\frac{1}{2}\right)^x = (2^{-1})^x = 2^{-x}$. You know from Chapter 5 that replacing x by $-x$ results in the graph being reflected in the y-axis.

Key point 8.1

For graphs of the form $y = a^x$:

- The y-intercept is always $(0, 1)$ because $a^0 = 1$.
- The graph of the function lies entirely above the x-axis.
- The x-axis is an asymptote.
- If $a > 1$, then as x increases so does y. This is called **exponential growth**.
- If $0 < a < 1$, then as x increases, y decreases. This is called **exponential decay**.

Gradient of an exponential graph

For an exponential growth graph, the gradient also increases with x. In fact, the gradient is exactly proportional to the y-value at every point on the graph.

You can find the gradient of a curved graph by drawing a tangent and calculating its gradient. The diagram shows some tangents to the graph of $y = 2^x$.

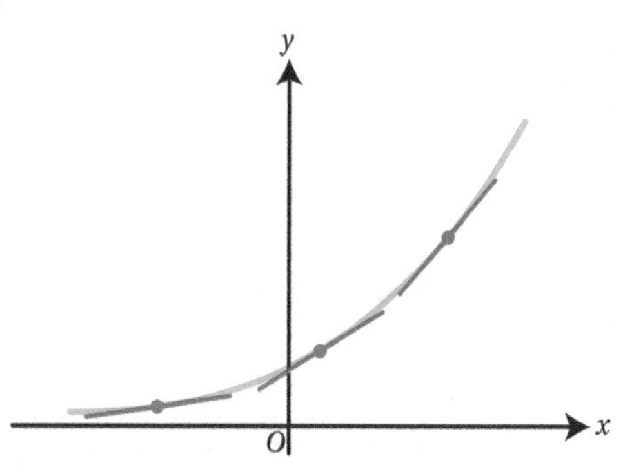

x	−1	0	1	2
y	0.5	1.0	2.0	4.0
gradient	0.35	0.7	1.4	2.8

The table shows the y-values and the gradient for each of the points. You can check that the gradient $\approx 0.69y$.

You can do a similar calculation for the graph of $y = 3.5^x$ and find that the gradient $\approx 1.25y$.

These examples suggest that the constant of proportionality depends on the base of the exponential. There is one special value of the base where this constant is 1, so that the gradient at any point is exactly equal to the y-value. This value is the number e, which you met in Chapter 7.

Tip

Remember that $e \approx 2.7$. Since $e > 1$, e^x is a type of exponential growth, while $e^{-x} = \left(\frac{1}{e}\right)^x$ represents exponential decay.

Key point 8.2

The gradient of e^x is e^x.

This result can be extended to exponential functions of the form e^{kx} for any constant k.

Key point 8.3

The gradient of e^{kx} is ke^{kx}.

PROOF 4

Prove that the gradient of e^{kx} is ke^{kx}.

Consider the graphs of $y = e^x$ and $y = e^{kx}$ with tangents drawn at x_1 and x_2 respectively.

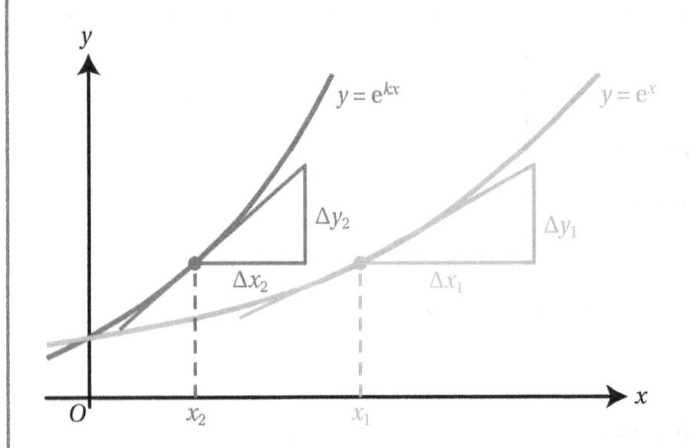

Δx just means a small change in x and Δy a small change in y.

The gradient of $y = e^x$ at the point (x_1, y_1) is $\frac{\Delta y_1}{\Delta x_1} = e^{x_1}$.

You know that the gradient of e^x is e^x.

Since $y = e^{kx}$ is a horizontal stretch by factor $\frac{1}{k}$ of $y = e^x$,

The red graph is a horizontal stretch, factor $\frac{1}{k}$, of everything on the blue graph, including the triangle.

If ... $\Delta y_2 = \Delta y_1$

then ... $\Delta x_2 = \frac{\Delta x_1}{k}$

Continues on next page

So, the gradient of $y = e^{kx}$ at (x_2, y_2) is

$$\frac{\Delta y_2}{\Delta x_2} = \frac{\Delta y_1}{\frac{\Delta x_1}{k}}$$

$$= k\frac{\Delta y_1}{\Delta x_1}$$

$$= ke^{x_1}$$

Since $\frac{\Delta y_1}{\Delta x_1} = e^{x_1}$

But also $x_2 = \frac{x_1}{k} \Rightarrow x_1 = kx_2$.

You now need to write the gradient in terms of the x-coordinate of the point on the red graph.

So the gradient of the red line when $x = x_2$ is ke^{kx_2}.

So, in general the gradient of e^{kx} is ke^{kx}.

WORKED EXAMPLE 8.1

a Find the gradient of $e^{1.2x}$ when $x = 2.6$.
b Find the value of x when the gradient of $e^{1.2x}$ equals 15.
c Find the gradient of the graph of $y = e^{1.2x}$ at the point where $y = 6.5$.

a Gradient $= 1.2e^{1.2x}$
When $x = 2.6$,
gradient $= 1.2e^{1.2 \times 2.6} = 27.2$

The gradient of e^{kx} is ke^{kx}.

b $1.2e^{1.2x} = 15$

The expression for the gradient was found in part **a**.

$e^{1.2x} = 12.5$
$1.2x = \ln 12.5$
$x = 2.1$ (2 s.f.)

This is an exponential equation. Isolate the exponential term and then use logarithms.

c Gradient $= 1.2e^{1.2x}$
$= 1.2y$
$= 1.2 \times 6.5$
$= 7.8$

You know that the gradient is $1.2e^{1.2x}$ and $y = e^{1.2x}$, so the gradient is $1.2y$.

Changing the base of an exponential

The gradient of other exponential functions is more difficult to find. However, any exponential function can be converted into an exponential with base e.

WORKED EXAMPLE 8.2

Given that 3^x can be written in the form e^{kx}, find the value of k.

$3^x = e^{kx}$

$\Rightarrow \ln 3^x = \ln e^{kx}$

This is an exponential equation, so take a logarithm of both sides. Since e is involved, choose ln.

$\Rightarrow x \ln 3 = kx$

$\therefore k = \ln 3$

Use the rule $\log_a x^k = k \log_a x$ on the LHS and 'cancel' the ln and e on the RHS.

Key point 8.4

Any exponential function a^x can be written in the form e^{kx}.

WORKED EXAMPLE 8.3

a Write $\left(\frac{3}{4}\right)^x$ in the form e^{kx}, giving the value of k to 3 s.f.

b Hence find the gradient of $\left(\frac{3}{4}\right)^x$ when $x = 2.5$.

a $\left(\frac{3}{4}\right)^x = e^{kx}$

Take ln of both sides as before and rearrange.

$\Rightarrow \ln\left(\frac{3}{4}\right)^x = \ln e^{kx}$

$\Rightarrow x \ln\left(\frac{3}{4}\right) = kx$

Compare coefficients of x.

$\therefore k = \ln\left(\frac{3}{4}\right) \approx -0.288$

b Gradient of $e^{-0.288x} = -0.288e^{-0.288x}$

The gradient of e^x is ke^{kx}.

When $x = 2.5$,

gradient $= -0.288e^{-0.288 \times 2.5}$

$= -0.14$ (2 s.f.)

EXERCISE 8A

1 Match each exponential graph with its equation.

a

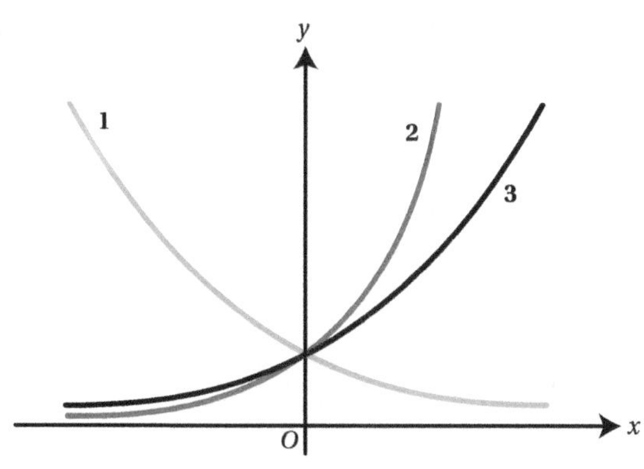

A $y = 1.5^x$ **B** $y = e^x$ **C** $y = 0.7^x$

b

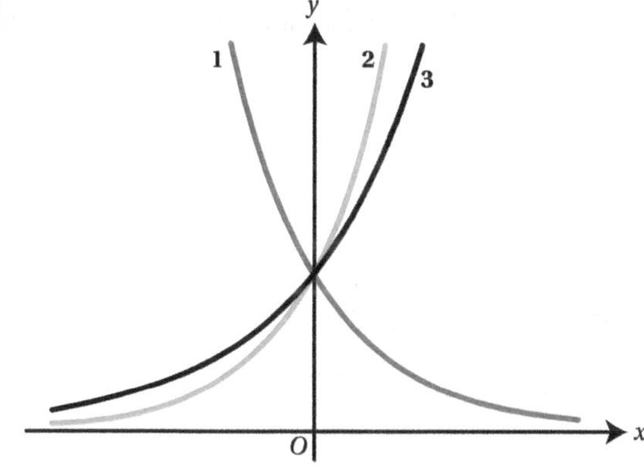

A $y = e^{1.5x}$ **B** $y = e^{2.3x}$ **C** $y = e^{-2x}$

c

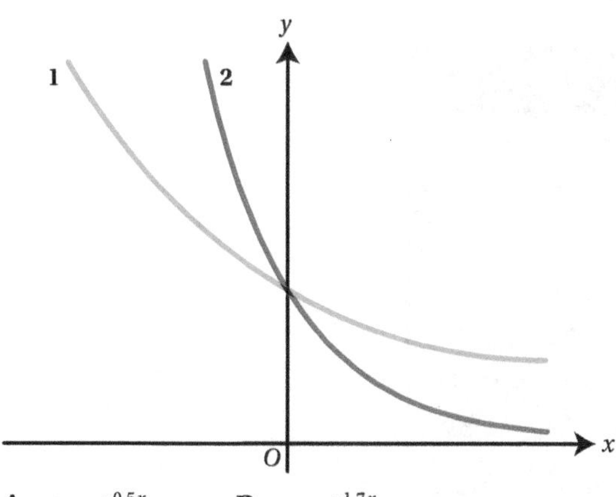

A $y = e^{-0.5x}$ **B** $y = e^{-1.7x}$

2 Find an expression for the gradient of each function.

a **i** $e^{3.2x}$ **ii** $e^{0.6x}$ **b** **i** $e^{-1.3x}$ **ii** e^{-x}

3 Find the gradient of each graph at the given value of x.

a **i** $y = e^{3.5x}$ when $x = 0.8$ **ii** $y = e^{2.9x}$ when $x = -1.2$

b **i** $y = e^{-1.2x}$ when $x = 3.6$ **ii** $y = e^{-0.5x}$ when $x = -0.9$

4 Find the gradient of each graph at the given value of y.

a i $y = e^{1.5x}$ when $y = 17$

ii $y = e^{4x}$ when $y = 0.6$

b i $y = e^{-0.6x}$ when $y = 3.5$

ii $y = e^{-x}$ when $y = 0.5$

5 Find the gradient of the graph of $y = e^{1.5x}$ when:

a $x = -2.1$

b $y = 12$.

6 For the graph of $y = e^{-2.3x}$, find:

a the gradient when $y = 0.5$

b the value of x where the gradient is -2.5.

7 The graph of $y = e^{kx}$ has gradient 26 at the point where $y = 8$.

a Find the value of k.

b Find the gradient of the graph when $x = -1$.

8 The gradient of the graph of $y = e^{ax}$ at the point where $y = 4.6$ is -1.2. Find the value of x at the point where the gradient is -5.

9 a Find the value of k so that $8^x = e^{kx}$.

b Hence find the gradient of the graph of $y = 8^x$ at the point where $x = -0.5$.

10 a Find the value of p such that $0.3^x = e^{px}$.

b Hence find the gradient of the curve $y = 0.3^x$ at the point where $y = 0.065$.

Section 2: Graphs of logarithms

You need to know the graph of the natural logarithm function.

Key point 8.5

The graph of $y = \ln x$:

- passes through the point (0, 1)
- has the y-axis as a vertical asymptote.

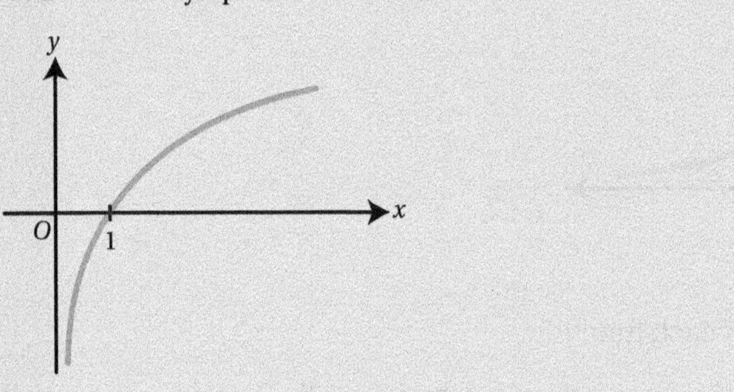

You can combine this fact with your knowledge of other graphs and graph transformations to solve a variety of problems.

EXERCISE 8B

1 a On the same diagram, sketch the graphs of $y = \ln x$ and $y = \frac{2}{x}$.

b Hence state the number of solutions of the equation $x \ln x = 2$.

2 On the same diagram, sketch the graphs of $y = \ln x$ and $y = \ln(x-2)$. Label all intercepts with the coordinate axes.

3 Let k be a positive constant. Use a graphical method to prove that the equation $kx + \ln x = 0$ has exactly one solution.

4 The graph of $y = \ln x$ can be transformed into the graph of $y = \ln(3x)$ either by a translation or by a stretch. Find the translation vector and the scale factor of the stretch.

5 a On the same diagram, sketch the graphs of $y = 2\ln x$ and $y = \ln(x+3)$.

b Find the exact solution of the equation $2\ln x = \ln(x+3)$.

6 $\log_{10} x = k \ln x$

a By raising 10 to the power of both sides, show that $k = \frac{1}{\ln 10}$.

b Describe fully the transformation that transforms the graph of $y = \ln x$ to the graph of $y = \log_{10} x$.

Section 3: Exponential functions and mathematical modelling

You know that the gradient of an exponential function $y = e^{kx}$ is proportional to the y-value. This means that if a quantity grows (or decays) exponentially, then its **rate of change** is proportional to the quantity itself.

This situation, where the rate of change of a quantity is proportional to its size, arises in a number of common examples.

- A population increases at a rate proportional to its size.
- The rate of a chemical reaction is proportional to the amount of the reactant.
- The rate of radioactive decay is proportional to the amount of the substance remaining.
- The value of an investment that is subject to compound interest increases exponentially.

Exponential population growth is an example of a **mathematical model**, in which equations are used to try to capture the important part of a real-world situation. Mathematical models are rarely perfect, so you should always be aware that they might not always work in predicting the real world. For example, there may be other factors that affect the rate of population growth (such as environmental conditions) that are not included in the exponential model.

Focus on...

See Focus on ... Modelling 1 for possible refinements to the basic population growth model to take into account other factors.

WORKED EXAMPLE 8.4

The number of bacteria in a culture medium is modelled by the equation $N = 1000 \times 2^{4t}$, where t is the number of hours elapsed since 08:00.

a What was the size of the population at 08:00?
b At what time will the population first reach one million?
c What does this model predict about the size of the population in the long term? Explain why this is not a realistic prediction.

a $N = 1000 \times 2^0 = 1000$ — This is when $t = 0$.

b $1000 \times 2^{4t} = 1000\,000$ — Solve the equation $N = 1\,000\,000$.

$2^{4t} = 1000$

$\log(2^{4t}) = \log 1000$ — Take logarithms of both sides.

$4t \log 2 = 3$ — Use the rule $\log_a x^k = k \log_a x$.

$t = \frac{3}{4 \log 2} = 2.49$ — Divide both sides by 4 log 2.

The population will first reach 1 million at 10:29. — $0.49 \times 60 = 29$, so this will be 2 hours 29 minutes after 08:00.

c The model predicts that the population will grow indefinitely. — An exponential function grows with increasing gradient.

This is not realistic, as the growth will eventually be limited by lack of food or space.

In Section 1 you only considered exponential functions of the form $y = a^x$ or $y = e^{kx}$. All of these functions have the value 1 when $x = 0$. For an exponential model in which the initial value is different from 1 (such as in Worked example 8.4), the equation must be modified to include another constant.

In many exponential models the quantity varies with time, so in this section t denotes the independent variable.

Key point 8.6

For a function of the form $y = Ae^{kt}$

- The initial value (when $t = 0$) is A.
- The rate of growth is $ky = kAe^{kt}$.

WORKED EXAMPLE 8.5

The mass (m grams) of one of the substances in a chemical reaction is modelled by the equation $m = Ae^{-1.2t}$, where t seconds is the time since the start of the reaction. The initial mass of the substance was 72 g.

a State the value of A.
b Find the rate at which the mass is increasing 5 seconds after the start of the reaction.

a $A = 72$ — A is the value of m when $t = 0$.

b When $t = 5$: — Find m after 5 seconds.

$m = 72e^{-1.2\times5}$
$= 0.178$

The rate of change is: — The rate of change is km.

$-1.2m = -1.2 \times 0.178$
$= -0.214$

The mass is decreasing at the rate of 0.214 grams per second. — The negative rate means that the amount is decreasing.

Sometimes you need to use experimental data to find the parameters in the model.

WORKED EXAMPLE 8.6

A simple model of a population of bacteria states that the number of bacteria (N thousand) grows exponentially, so that:

$N = Ae^{kt}$

where t is time, in minutes, since the start of the experiment.

Initially there were 2000 bacteria and after 5 minutes this number has grown to 7000.

a Find the values of constants A and k.
b According to this model, how many bacteria will there be in the dish after another 5 minutes?
c Give two reasons why this model will not provide a good prediction for the amount of bacteria in the dish 12 hours later.

a When $t = 0$: — Use the equation for $t = 0$ and $t = 5$. Remember that N is in thousands.

$2 = Ae^0$
$A = 2$

When $t = 5$:

$7 = Ae^{5k}$

$7 = 2e^{5k}$ — Use logarithms to solve the equation for k.

$e^{5k} = 3.5$

$k = \frac{1}{5}\ln 3.5 (= 0.251)$

Continues on next page

b $N = 2e^{10k} \approx 24.5$

The model predicts that there will be 24 500 bacteria.

In another 5 minutes, $t = 10$.

You need to use the value of k found in part **a**. Notice that, instead of using the actual value of k, you could use the fact that $e^{10k} = (e^{5k})^2 = 3.5^2$.

c The model predicts that the bacteria population will continue growing indefinitely, but it will eventually slow down as food and space become limiting factors.

The information given in the model is only approximate so in 12 hours errors in this information may cause the prediction to be very different from the correct value.

You are not expected to have any technical biological knowledge, but you may need to apply general experience of the real world to interpret and criticise models.

There are many other possible criticisms of this model, so anything relevant would be acceptable.

WORKED EXAMPLE 8.7

A population of flies grows exponentially, so that its size can be modelled by the equation $N = Ae^{kt}$, where N is the number of flies after t weeks. At the time $t = 0$, the population size is 2400 and it is increasing at the rate of 80 flies per week.

Find the values A and k.

When $t = 0$, $N = A$, so $A = 2400$

A is the initial value.

The rate of increase is kN, so:

$2400k = 80$

$k = \frac{1}{30}$

This is an exponential equation, so the rate of increase of N is kN.

EXERCISE 8C

1 An amount of £C is invested in an account giving p% annual interest.

a Find an expression for the value of the investment after 1 year when:

i the interest is compounded annually

ii $\frac{p}{2}$% interest is added twice a year

iii $\frac{p}{4}$% interest is added four times a year.

b If $\frac{p}{n}$% is added n times a year, explain why the value of the investment after one year is $V = C\left(1 + \frac{p}{100n}\right)^n$.

c Investigate the behaviour of the sequence $\left(1+\frac{1}{n}\right)^n$ as n increases.

d For the case of a (not very realistic!) 100% annual interest rate, find an expression for the value of the investment after x years when the interest is compounded continuously.

2 In a yeast culture, cell numbers are given by $N = 100e^{1.03t}$, where t is measured in hours after the cells are introduced to the culture.

a What is the initial number of cells?

b How many cells will be present after 6 hours?

c How long will it take for the population to reach one thousand?

d At what rate will the population be growing at that point?

3 A technology company is interested in predicting the number of mobile phones being used in the world. The number of mobile phones in billions (N) in t years is predicted to follow the model $N = 2e^{0.1t} + 1$.

a According to the model, how many mobile phones are there currently in the world?

b How many mobile phones does the model predict will exist in 10 years' time? Give your answer to 3 s.f.

4 An algal population on the surface of a pond grows by 10% every day. The area it covers can be modelled by the equation $y = k \times 1.1^t$, where t is measured in days. At 09:00 on Tuesday it covered 10 m^2.

a What area will it cover by 09:00 on Friday?

b Suggest two factors that could be taken into account to make the model more realistic.

5 The mass of a piece of plutonium (M grams) after t seconds is given by $M = ke^{-0.01t}$.

a Sketch the graph of M against t.

b How long will it take to reach 25% of its original mass?

6 A population size is increasing according to an exponential model, $N = N_0e^{at}$, where t is time measured in days. Initially the population size is 450 and is increasing at a rate of 90 per hour.

a Find the values of N_0 and a.

b At what rate is the population increasing when its size is 750?

c How long will the population size take to reach 2000?

7 A radioactive substance decays so that the rate of decay (measured in atoms per day) is numerically equal to 40% of the number of atoms remaining at that time. Initially there were 500 atoms.

Write an equation to model the number of atoms N at time t days.

8 The value of a new car is £6800. One year later the value has decreased to £5440.

a Assuming the value continues to decrease by the same percentage every year, write the model for the price of the car, in the form $V = Pe^{kt}$.

b What does this model predict the value of the car will be in 10 years' time?

A company guarantees to pay at least £300 for any car, no matter how old it is. A refined model is $V = A + Be^{ct}$.

c State the value of A and B.

9 The UK population is currently 70 million and is predicted to grow by 2% each year, due to births and deaths.

a Write down a model for the population of the UK, P, at a time n years from now.

b Give two reasons why this model might not be appropriate when predicting the population of the UK in 2100.

10 A bowl of soup is served at a temperature 35 °C above room temperature. Every 5 minutes, the temperature difference between the soup and the air in the room decreases by 30%. Assuming the room air temperature is constant, the temperature can be modelled by $T = ka^t$, where T is the temperature difference (°C) between the soup and room temperature and t is the time (minutes) since the soup was served.

a What will the temperature of the soup be 7 minutes after serving?

b If the soup were put into a vacuum flask instead of a bowl, how would this affect the value of:

i k ii a?

11 The speed (V metres per second) of a parachutist t seconds after jumping from an aeroplane is modelled by the expression $V = 40(1 - 3^{-0.1t})$.

a Find the initial speed.

b What speed does the model predict that he will eventually reach?

12 The model $I = 100e^{-2x}$ is used to estimate the intensity of light (I) at a distance x metres away from a bulb.

a By what factor has the light intensity dropped between 0 m and 1 m away from the bulb?

b Prove that every 1 m further from the bulb produces the same factor reduction in light intensity.

Section 4: Fitting models to data

In Worked example 8.6, the values of parameters A and k in the model $N = Ae^{kt}$ were found by using the information about the number of bacteria when $t = 0$ and $t = 5$. In many real-life situations models are not exact, or there may be inaccuracies in the measurements. This means that using the number of bacteria when $t = 10$, say, would give slightly different values for N and k.

To get a more reliable result it may be sensible to use data from more than two measurements. Suppose the number of bacteria were measured every 5 minutes for half an hour. Plotting the results might give a graph like this one.

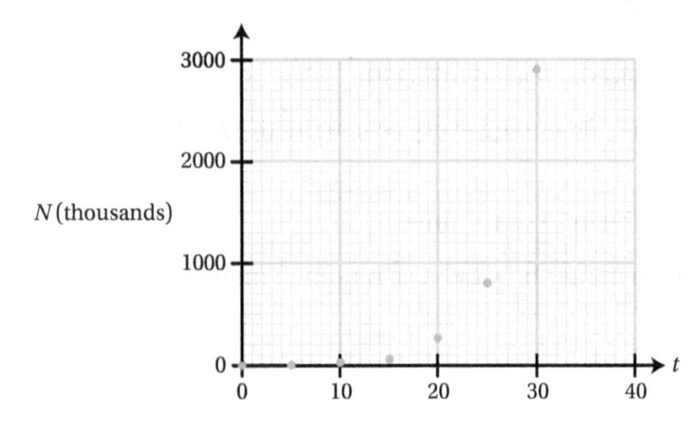

It is difficult to draw an exponential curve that best fits the points but there is a clever trick that turns this curve into a straight line.

The equation for the population size is $N = Ae^{kt}$. Taking a logarithm of both sides:

$$\begin{aligned}\ln N &= \ln(Ae^{kt})\\ &= \ln A + \ln e^{kt}\\ &= \ln A + kt\end{aligned}$$

Writing $y = \ln N$ and remembering that $\ln A$ and k are unknown numbers, this is the equation of a straight line:

$$y = kt + \ln A$$

If the data points are plotted with t on the horizontal axis and $\ln N$ on the vertical axis, they should roughly follow a straight line with gradient k and N-intercept $\ln A$.

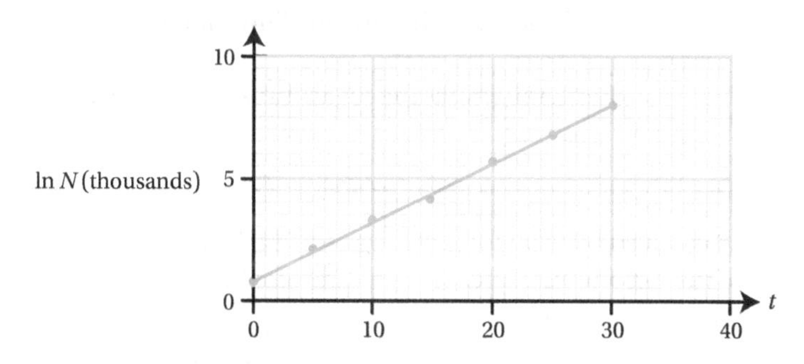

In this example, you can find from the graph that the intercept on the vertical axis is 0.8 and the gradient is $\frac{8-0.8}{30} = 0.24$.

So: $\quad k = 0.24$ and $\ln A = 0.8$

$$\Rightarrow A = e^{0.8} = 2.2$$

Therefore, the experimental data suggest that the model for the bacterial population growth is

$$N = 2.2e^{0.24t}.$$

You can perform a similar calculation when the base of the exponential is unknown. In that case you can take logarithms in any base – although most people would use base 10 or e.

Key point 8.7

If $y = kb^x$ then $\log y = \log k + x \log b$

The graph of $\log y$ against x is a straight line with gradient $\log b$ and y-intercept $\log k$.

Fast forward

You will learn more about lines of best fit in Chapter 20.

Did you know?

Graphs like this are called semi-log graphs and are often used to analyse scientific data.

Rewind

The equation of a straight line was covered in Chapter 6.

Elevate

See Extension Sheet 8 for questions on using logarithmic scales for working with quantities that cover a large range of values.

WORKED EXAMPLE 8.8

The mass of a piece of radioactive material decays exponentially, according to the model $M = Cb^t$, where M is the mass, in grams, t is the time, in seconds, and C and b are constants. A physicist measures the mass several times and plots the points on a graph with t on the horizontal axis and $\log M$ on the vertical axis. The line of best fit has equation $\log M = 1.3 - 1.8t$.

Estimate the values of C and b.

If $M = Cb^t$ then:

$$\log M = \log(Cb^t)$$
$$= \log C + \log b^t$$
$$= \log C + t\log b$$

Take logarithms of both sides to transform to the equation of a straight line.

Use the laws of logarithms.

Comparing this to $\log M = 1.3 - 1.8t$:

$\log C = 1.3$ and $\log b = -1.8$

log C and log b are numbers that should match the coefficients in the straight-line equation.

$C = 10^{1.3} = 20.0$

$b = 10^{-1.8} = 0.0158$

Remember that the logarithms are in base 10.

A variation on the method in Worked example 8.8 can also be used for models of the form $y = ax^n$, where x is the base of the exponential and the power is unknown. In this case you need to take a logarithm of both variables in order to get a straight-line graph.

Did you know?

Many natural and human-made phenomena follow so-called power laws of the form $= ax^n$. Examples include distribution of common words, sizes of cities and corporations, and Kepler's law for planetary orbits.

Key point 8.8

If $y = ax^n$ then $\log y = \log a + n\log x$

The graph of $\log y$ against $\log x$ is a straight line with gradient n and y-intercept $\log a$.

WORKED EXAMPLE 8.9

A scientist thinks that variables x and y are related by an equation of the form $y = ax^n$. She collects the data and plots a scatter graph with $\log x$ on the horizontal axis and $\log y$ on the vertical axis. The points follow a straight line with gradient 2.6 and log y-intercept –0.9.

Find the values of a and n, and hence write an equation for y in terms of x.

If $y = ax^n$ then $\log y = \log a + \log x^n$

$$= \log a + n\log x$$

Take logarithms of both sides to identify the gradient and the log y-intercept.

The gradient is $n = 2.6$.
The y-intercept is $\log a = -0.9$

$\Rightarrow a = 10^{-0.9} = 0.126$

This is a straight line with gradient n and y-intercept $\log a$.

So $y = 0.126x^{2.6}$

Elevate

See Support Sheet 8 for a further example and for more practice questions on this topic.

EXERCISE 8D

1 In each of these examples, variables x and y are related by the equation $y = Ab^x$. You are given the equation of a straight line of $\ln(y)$ as a function of x. Find the values of the constants A and b.

a $\ln y = 1.2x + 0.7$ **b** $\ln y = 3.1 - 0.6x$ **c** $\ln y = 2.3x - 4$

2 In each of these examples, variables x and y are related by the equation $y = Cx^n$. You are given the equation of a straight line of $\ln y$ as a function of $\ln x$. Find the values of the constants C and n.

b $\ln y = 0.7 \ln x + 1.2$ **b** $\ln y = 2.1 \ln x - 4.7$ **c** $\ln y = 2 - 0.9 \ln x$

3 A zoologist is studying the growth of a population of fish in a lake. He thinks that the size of the population can be modelled by the equation $N = Ae^{kt}$, where N is the number of fish and t is the number of months since the fish were first introduced into the lake.

a The zoologist collected some data and wants to plot them on the graph in order to check whether his proposed model is suitable. Assuming his model is correct, state which of these graphs will produce approximately a straight line.

A N against $\log t$ **B** $\log N$ against t **C** $\log N$ against $\log t$

You may now assume that the proposed model is correct.

b Initially, 150 fish were introduced into the lake. Write down the value of A.

c After 10 months there are 780 fish in the lake. Find the value of k.

d Comment on the suitability of this model for predicting the number of fish in the long term.

4 The table shows the population of the five largest cities in a country, to the nearest 100 000.

City rank (n)	1	2	3	4	5
City population (p)	6 100 000	2 900 000	2 200 000	1 500 000	1 200 000

a On suitable scales, plot $\log p$ against $\log n$.

b Hence find the equation linking p and n.

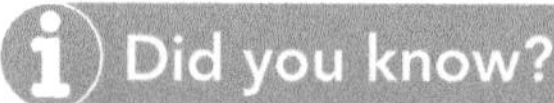

Did you know?

This rule, studied by geographers, is called Zipf's law and is followed remarkably accurately in countries across the world. Similar rules also apply to the frequency with which common words are used.

5 A scientist is modelling exponential decay of the amount of substance in a chemical reaction. She proposes a model of the form $M = Kc^t$ where M is the mass of the substance, in grams, t is the time, in seconds, since the start of the reaction, and K and c are constants. The mass of the substance is recorded for the first six seconds of the reaction. The graph of $\ln M$ against t is shown.

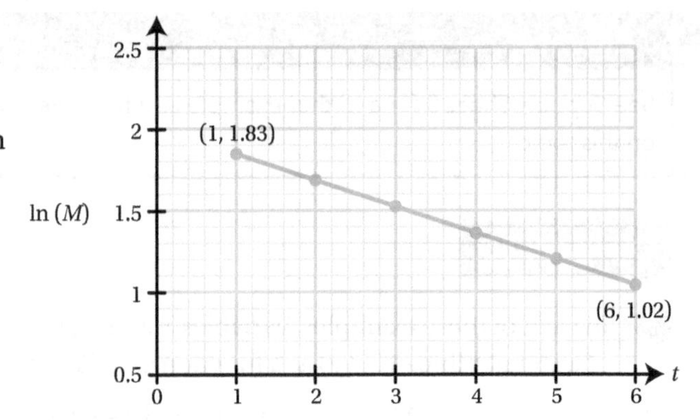

a The points are found to lie on a straight line. Find its equation, giving parameters to 2 s.f.

b Hence find the values of K and c.

c How long, to the nearest second, will it take for the mass of the substance to fall below 1 gram?

6 A model for the size of the population of a city predicts that it will grow according to the equation $P = Ct^n$, where P thousand is the number of people and t is the number of years since the measurements began. The graph shows $y = \ln P$ plotted against $x = \ln t$.

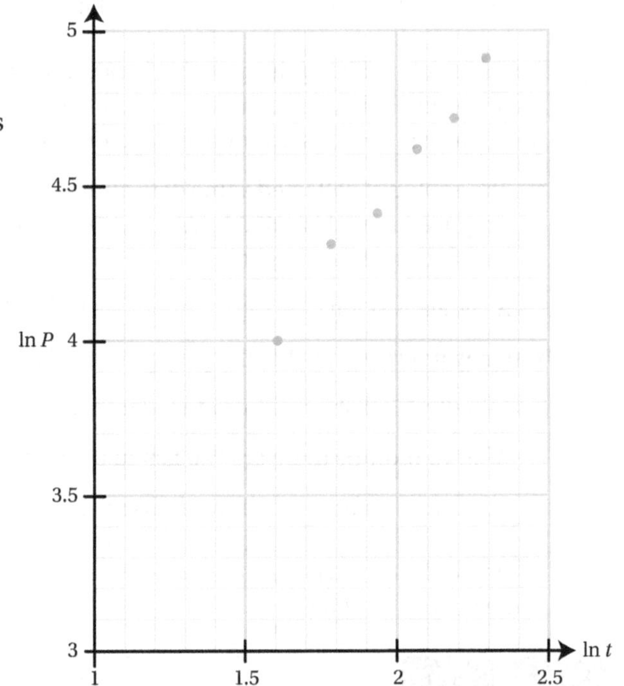

a Find the equation, in the form $y = mx + c$, of a line of best fit for this graph.

b Hence estimate the values of C and n.

c According to this model, after how many years will the population first exceed 200 000?

7 A scientist is investigating the population of mice in a field. She collected some data over a period of time, and recorded them on a graph. Let N denote the number of mice at time t weeks. The graph of $y = \ln N$ against t has equation $y = 5.8 + 0.16t$.

a Find the equation for the size of the population at time t weeks.

b Find the rate at which the population is growing after 8 weeks.

8 The radioactivity level, R mSv, in a patient's body is recorded hourly and a graph of $\ln R$ against t (hours) plotted. The graph has equation $y = \ln 6 - 0.2t$.

a Find the equation for the level of radioactivity at time t.

b How long, to the nearest hour, does it take for this effect to decrease to an average background radiation level of 2.7 mSv?

c Suggest a limitation of this model.

d Create a refined model which overcomes the limitation suggested in part **c**.

9 **a** A common model used in population growth is called the logistic function. It predicts that a population N is related to the time t by the formula

$$N = \frac{1}{1 + Je^{-bt}}$$

Find an expression, in terms of N, which can be plotted against t to form a straight line if a population follows this model. Write down expressions for the gradient and the intercept of the line in terms of b and J.

b By comparing the long term predictions of the logistic function and normal exponential growth explain why the logistic function is a better model for population growth than normal exponential growth.

Checklist of learning and understanding

- For all the graphs $y = a^x$:
  - The y-intercept is always $(0, 1)$, because $a^0 = 1$.
  - The graph of the function lies entirely above the x-axis.
  - The x-axis is an asymptote.
  - If $a > 1$, then as x increases so does y. This is called **exponential growth.**
  - If $0 < a < 1$, then as x increases, y decreases. This is called **exponential decay.**
- Exponential functions are often used to model situations where the rate of growth is proportional to the amount present.
  - The gradient of e^{kx} equals ke^{kx}.
  - In a model of the form $y = Ae^{kt}$ the initial value is A and the rate of change equals ky.
- The graph of $y = \ln x$:
  - passes through $(1, 0)$
  - has the y-axis as an asymptote.

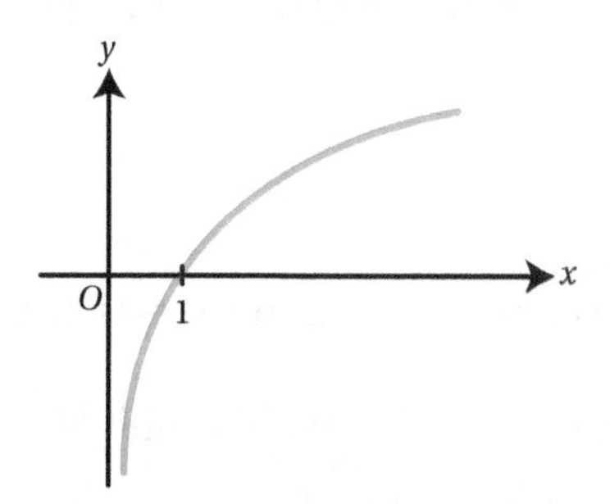

- Logarithms can be used to change some curved graphs into straight lines; this is used to estimate parameters in models.
  - If $y = kb^x$ then $\log y = \log k + x \log b$.
    - The graph of $\log y$ against x is a straight line with gradient $\log b$ and y-intercept $\log k$.
  - If $y = ax^n$ then $\log y = \log a + n \log x$.
    - The graph of $\log y$ against $\log x$ is a straight line with gradient n and y-intercept $\log a$.

Mixed practice 8

1. The value, V, of a new car, in thousands of pounds, follows a model of the form $V = ka^t$, where t is the number of years after it was purchased. Initially it is worth £15 000 but after 3 years it is worth £7680. What are the values of k and a? Choose from these options.

   A $a = 19.7,\ k = 1$ **B** $a = 19.4,\ k = 15\,000$ **C** $a = 0.8,\ k = 15$ **D** $a = 8,\ k = 15$

2. Which of these statements is **not** true for a model of the form $y = a^t$, where t is measured in years from the present?

   Choose from these options.

   A The initial value is 1.

   B The growth rate is proportional to y.

   C The growth rate is proportional to t.

   D There is a constant growth factor each year.

3. **a** Sketch the graph of $y = e^{0.8x}$.

   b Find the gradient of your graph at the point where $x = 3$.

   c Use your graph to determine the number of solutions of the equation $e^{0.8x} = \frac{1}{x}$.

4. The amount of substance in a chemical reaction is decreasing according to the equation $m = 32e^{-0.14t}$, where m grams is the mass of the substance t seconds after the start of the reaction.

   a State the amount of the substance at the start of the reaction.

   b At what rate is the amount of substance decreasing 3 seconds after the start of the reaction?

   c How long will it take for the amount of substance to halve?

5. **a** On the same axes, sketch the graphs of $y = \ln x$ and $y = \frac{3}{x^2}$, labelling any asymptotes and axis intercepts.

   b Hence find the number of solutions of the equation $x^2 \ln x = 3$.

6. The volume, V, of a blob of algae, in cubic centimetres (cm^3), in a jar is modelled by $V = 0.4 \times 2^{0.1t}$, where t is the time, in weeks, after the observation begins.

   a What is the initial volume of the algae?

   b How long does it take for the volume of algae to double?

   c Give two reasons why the model would not be valid for predicting the volume in 10 years' time.

7. A rumour spreads exponentially through a school. When school begins (at 9 a.m.) 18 people know it. By 10 a.m. 42 people know it.

   Let N be the number of people who know the rumour t minutes after school starts.

   a Find constants A and k such that $N = Ae^{kt}$.

   b How many people know the rumour at 10:30 a.m?

   c There are 1200 people in the school. According to the exponential model at what time will everyone know the rumour?

8 The variables y and x are related by the equation $y = ax^n$, where a and n are constants.

A graph of $\ln y$ against $\ln x$ is plotted and found to be a straight line with gradient m and $\ln y$-intercept c.

Express a and n in terms of m and c. Choose from these options.

A $a = c,\ n = m$ **B** $a = \ln c,\ n = \ln m$ **C** $a = c,\ n = e^m$ **D** $a = e^c,\ n = m$

9 A patient is being treated for a condition by having insulin injected. The level of insulin (I) in the blood t minutes after the injection is given by $I = 10e^{-0.05t} + 2$, in units of microunits per millilitre (μ U ml^{-1}).

a What is the level of insulin immediately after the injection?

b There is a danger of coma if insulin levels fall below 1.8 μ U ml^{-1}. According to the model, will this level be reached? Justify your answer.

10 It is thought that the global population of tigers is falling exponentially. Estimates suggest that in 1970 there were 37 000 tigers but by 1980 the number had dropped to 22 000.

a A model of the form $T = ka^n$ is suggested, connecting the number of tigers (T) with the number of years (n) after 1970.

i Show that $22\,000 = ka^{10}$.

ii Write another similar equation and solve the two equations together to find k and a.

b What does the model predict the population will be in 2020?

c When the population reaches 1000 the tiger population will be described as 'near extinction'. In which year will this happen?

11 A zoologist believes that the population of fish in a small lake is growing exponentially. He collects data about the number of fish every 10 days for 50 days. The data are given in this table.

Time (days)	0	10	20	30	40	50
Number of fish	35	42	46	51	62	71

The zoologist proposes a model of the form $N = Ae^{kt}$, where N is the number of fish and t is time, in days. In order to estimate the values of the constants A and k, he plots a graph of t on the horizontal axis and $\ln N$ on the vertical axis.

a Explain why, assuming the zoologist's model is correct, this graph will be approximately a straight line.

b Complete the table of values for the graph.

t	0	10	20	30	40	50
$\ln N$	3.56		3.83	3.93		4.26

c Find the equation of the line of best fit for the table. Hence estimate the values of A and k.

d Use this model to predict the number of fish in the lake when $t = 260$.

e The zoologist finds that the number of fish in the lake after 260 days is actually 720. Suggest one reason why the observed data do not fit the prediction.

12 Quantities m and t are related by an equation of the form $m = at^p$, where a and p are constants. The graph of $\log m$ against $\log t$ is a straight line which passes through the points $(2, 5)$ and $(4, 0)$. Find the values of a and p.

13 The variables y and x are related by an equation of the form $y = ax^n$, where a and n are constants.

Let $Y = \log_{10} y$ and $X = \log_{10} x$

a Show that there is a linear relationship between Y and X.

b The graph of Y against X is shown in the diagram.

Find the value of n and the value of a.

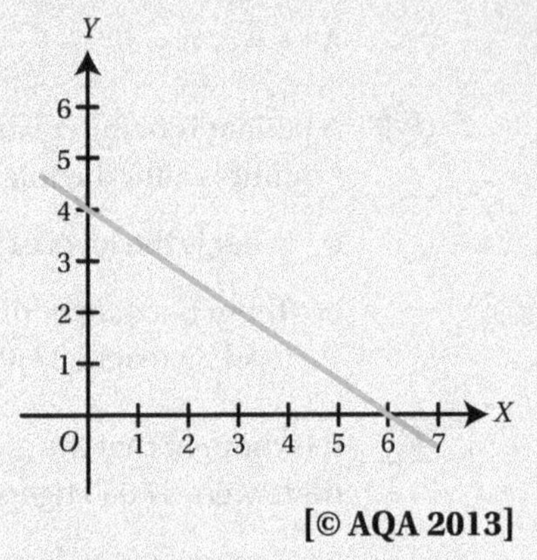

[© AQA 2013]

14 Radioactive decay can be modelled using an equation of the form $m = m_0 e^{-kt}$, where m is the mass of the radioactive substance at time t, m_0 is the initial mass and k is a positive constant. The half-life of a radioactive substance is the length of time it takes for half of the substance to decay. A particular radioactive substance has half-life of 260 years. Find the value of k.

15 The speed, V m s^{-1}, of the parachutist, t seconds after jumping from the aeroplane, is modelled by the equation $V = 42(1 - e^{-0.2t})$.

a What is the initial speed of the parachutist?

b What is the maximum speed that the parachutist could reach?

c When the parachutist reaches 22 m s^{-1} he opens the parachute. How long is he falling before he opens his parachute?

16 When a cup of tea is first made its temperature is 98 °C. After two minutes the temperature has reached 94 °C. The room temperature is 22 °C and the difference between the temperature of the tea and the room temperature decreases exponentially.

a Let T be the temperature of the tea and t be the time, in minutes, since the tea was made. Find the constants C and k so that $T - 22 = Ce^{-kt}$.

b Find the time it takes for the tea to cool to 78 °C.

17 The population of a country, P million, t years from 2010 is modelled by $P = 0.5 \times 2^{0.1t}$

a What is the initial population?

b In what year does the population reach 3 million?

c How long does it take the population to double?

d Show how the model could be refined to take into account separately each of these criticisms.

i The population growth rate in the model is too large and it actually takes 20 years for the population to double.

ii The country has a maximum possible population of 4 million.

18 You are given these facts about measles in the UK:

- There is a background level of about 500 people infected at any time.
- During an epidemic this number can go up to about 20 000.
- It takes about 4 weeks from the peak of an epidemic to return to normal levels.

A suggested model for the number of people infected (I) is given by

$I = 20\,000e^{-0.123t}$

where t is the time in days.

a How long does this model predict it will take for the number of infected people to fall to background levels?

b Suggest three aspects of the real situation that are absent from your model.

c Suggest a possible refinement to the model to take into account one of the limitations suggested in part **b**.

19 A scientist is testing models for the growth and decay of colonies of bacteria.

For a particular colony, which is growing, the model is $P = Ae^{\frac{t}{8}}$, where P is the number of bacteria after a time t minutes and A is a constant.

a This growing colony consists initially of 500 bacteria. Calculate the number of bacteria, according to the model, after one hour. Give your answer to the nearest thousand.

b For a second colony, which is decaying, the model is $Q = 500\,000e^{-\frac{1}{8}t}$, where Q is the number of bacteria after a time t minutes.

Initially, the growing colony has 500 bacteria and, at the same time, the decaying colony has 500 000 bacteria.

i Find the time at which the populations of the two colonies will be equal, giving your answer to the nearest 0.1 of a minute.

ii The population of the growing colony will exceed that of the decaying colony by 45 000 bacteria at time T minutes.

Show that

$$\left(e^{\frac{1}{8}T}\right)^2 - 90e^{\frac{1}{8}T} - 1000 = 0$$

and hence find the value of T, giving your answer to one decimal place.

[© AQA 2012]

9 Binomial expansion

In this chapter you will learn how to:

- expand an expression of the form $(a+b)^n$ for any positive integer n
- find individual terms in the expansion of $(a+b)^n$ for any positive integer n
- use partial expansions of $(a+bx)^n$ to find an approximate value for a number raised to a positive integer power
- understand and use the notations $n!$ and ${}^nC_r$.

Before you start…

GCSE	You should be able to evaluate expressions involving powers, including working with the order of operations.	1 Evaluate: a 2×3^2 b $3-4\times(-2)^3$.
GCSE **Chapter 2**	You should be able to work with the rules of indices.	2 a Evaluate $(2x^3)^4$. b Simplify $x^4\times x^7$. c Simplify $\frac{x^{12}}{x^3}$.
GCSE	You should be able to multiply out two brackets.	3 Expand $(2x+3)^2$.
GCSE **Chapter 3**	You should be able to solve quadratic equations by using the formula or factorising.	4 Solve $x^2+5x+4=0$.

What is the binomial expansion?

A **binomial expression** is one that has two terms, for example, $a+b$.

A power of a binomial expression could be expanded in the usual way; for example, $(a+b)^7$ could be expanded by calculating, at length:

$$(a+b)(a+b)(a+b)(a+b)(a+b)(a+b)(a+b)$$

This is time-consuming and it is easy to make mistakes. Fortunately there is a much quicker approach.

Fast forward

You will see an application of the binomial expansion to the binomial distribution in statistics in Chapter 21.

Section 1: The binomial theorem

Consider these expansions of $(a+b)^n$, produced by the slow method of multiplying out brackets repeatedly.

$$(a+b)^0 = 1 = 1a^0b^0$$
$$(a+b)^1 = a+b = 1a^1b^0 + 1a^0b^1$$
$$(a+b)^2 = a^2+2ab+b^2 = 1a^2b^0 + 2a^1b^1 + 1a^0b^2$$
$$(a+b)^3 = a^3+3a^2b+3ab^2+b^3 = 1a^3b^0 + 3a^2b^1 + 3a^1b^2 + 1a^0b^3$$
$$(a+b)^4 = a^4+4a^3b+6a^2b^2+4ab^3+b^4 = 1a^4b^0 + 4a^3b^1 + 6a^2b^2 + 4a^1b^3 + 1a^0b^4$$

There are several patterns.

- The powers of a and b (coloured blue and green) always add up to n.
- Each power of a and b from 0 up to n is present in one of the terms.
- The pattern of coefficients (coloured in red) in each line is symmetrical.

Tip

Remember that $x^0 = 1$ for all x.

However, you need to be able to find the red coefficients for any integer power n.

Key point 9.1

The **binomial coefficients** can be evaluated by using the ${}^nC_r$ button on your calculator.

${}^nC_r$ can also be written as $\binom{n}{r}$.

For example, you can check that ${}^4C_2 = 6$.

The binomial coefficients also appear as the rows of Pascal's triangle, which is formed by adding two adjacent numbers to produce the number below them in the next row.

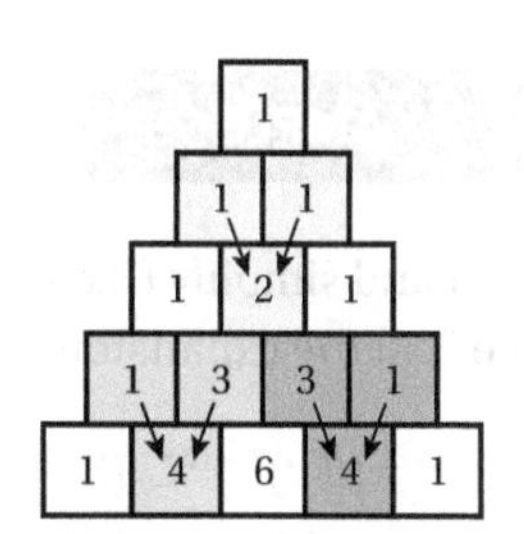

For small values of n, you may remember the relevant row and be able to write the coefficients straight down. For example, to find the binomial coefficient ${}^4C_2$ in Pascal's triangle, look at the row starting 1, 4 ... The coefficients in this row are:

${}^4C_0 = 1$ ${}^4C_1 = 4$ ${}^4C_2 = 6$ ${}^4C_3 = 4$ ${}^4C_4 = 1$

So ${}^4C_2$ is the third number in this row.

You can now form the expansion for any integer value of n:

$$(a+b)^n = {}^nC_0a^nb^0 + {}^nC_1a^{n-1}b^1 + {}^nC_2a^{n-2}b^2 + \ldots + {}^nC_na^0b^n$$

But since the first and last binomial coefficients (${}^nC_0$ and ${}^nC_n$) are always 1, and since $a^0 = b^0 = 1$ for any a and b, this can be simplified.

Key point 9.2

The binomial theorem

$$(a+b)^n = a^n + \binom{n}{1}a^{n-1}b^1 + \binom{n}{2}a^{n-2}b^2 + \ldots + \binom{n}{r}a^{n-r}b^r + \ldots + b^n \; (n \in \mathbb{Z}^+).$$

Tip

Remember that $\binom{n}{r}$ is just another way of writing ${}^nC_r$.

The format given in Key point 9.2 is how the expansion is given in the AQA formula booklet.

In Key point 9.2, a and b can be replaced by any number, letter or expression.

WORKED EXAMPLE 9.1

Expand and simplify $(2+x)^4$.

$(2+x)^4 = 2^4 + \binom{4}{1}(2^3)(x^1) + \binom{4}{2}(2^2)(x^2) + \binom{4}{3} + (2^1)(x^3) + x^4$

Use the formula from Key point 9.2.

$= 1(16) + 4(8)(x) + 6(4)(x^2) + 4(2)(x^3) + 1(x^4)$

$= 16 + 32x + 24x^2 + 8x^3 + x^4$

Find the binomial coefficients (the red numbers) either from your calculator or by looking at Pascal's triangle: you want the row starting 1, 4...

You need to be careful when there is a minus sign or a number in front of x.

WORKED EXAMPLE 9.2

a Expand and simplify $(3x-5)^3$.
b Hence find the expansion of $(3x^4-5)^3$.

a $(3x-5)^3 = (3x)^3 + {}^3C_1(3x)^2(-5)^1 + {}^3C_2(3x)^1(-5)^2 + (-5)^3$

Use the formula from Key point 9.2.

$= 1(27x^3)(1) + 3(9x^2)(-5) + 3(3x)(25) + 1(1)(-125)$

$= 27x^3 - 135x^2 + 225x - 125$

Find the binomial coefficients from your calculator or Pascal's triangle.

Be careful with powers:
$(3x)^3 = 27x^3$
$(-5)^2 = 25$ but $(-5)^3 = -125$

b $(3x^4-5)^3 = 27(x^4)^3 - 135(x^4)^2 + 225(x^4) - 125$

Replace x by x^4 in the above expansion.

$= 27x^{12} - 135x^8 + 225x^4 - 125$

Use the laws of indices to simplify.

Sometimes you will be interested in just a particular term, rather than the whole expansion.

WORKED EXAMPLE 9.3

Find the coefficient of x^5 in the expansion of $(x+2)^8$.

The required term is ${}^8C_3(x)^5(2)^3$.

Write down the required term in the form ${}^nC_r a^{n-r}b^r$ and then substitute $a=x$, $b=2, r=3, n=8$.

${}^8C_3=56$
$(x)^5=x^5$
$(2)^3=8$

Calculate the coefficient and apply the powers to the bracketed terms.

The term is
$56(x^5)(8)=448x^5$

Simplify to identify the coefficient.

$\therefore$ the coefficient is 448.

Tip

You might be asked to give the term or just the coefficient. Make sure that you answer the question you are asked!

WORK IT OUT 9.1

Expand $(1+2x)^4$.

Which is the correct solution? Identify the errors made in the incorrect solutions.

Solution 1	Solution 2	Solution 3
$(1+2x)^4$	$(1+2x)^4$	$(1+2x)^4$
$=1^4+{}^4C_1(2x)+{}^4C_2(2x)^2$ $+{}^4C_3(2x)^3+{}^4C_4(2x)^4$	$=1+4\times2x+6\times2x^2+4\times2x^3+2x^4$	$=1^4(2x)^0+1^3(2x)^1+1^2(2x)^2$ $+1^1(2x)^3+1^0(2x)^4$
$=1+8x+24x^2+32x^3+16x^4$	$=1+8x+12x^2+8x^3+2x^4$	$=1+2x+4x^2+8x^3+16x^2$

EXERCISE 9A

1 Evaluate these binomial coefficients.

a **i** ${}^7C_3$ **ii** ${}^9C_2$ **b** **i** ${}^9C_0$ **ii** ${}^6C_0$

c **i** $\binom{8}{7}$ **ii** $\binom{10}{8}$ **d** **i** $\binom{12}{12}$ **ii** $\binom{20}{20}$

2 Expand and simplify each expression.

a **i** $(1+x)^4$ **ii** $(x+1)^5$ **b** **i** $(x-1)^3$ **ii** $(1-x)^4$

c **i** $(2-x)^5$ **ii** $(3+x)^6$ **d** **i** $(2-3x)^4$ **ii** $(2x-7)^3$

3 a i Find the coefficient of y^3 in the expansion of $(2+y)^4$.

ii Find the coefficient of y^4 in the expansion of $(3+y)^7$.

b i Find the coefficient of b^6 in the expansion of $(2+b)^7$.

ii Find the coefficient of b^3 in the expansion of $(2+b)^8$.

4 a Evaluate the coefficient of y^3 in each expansion.

i $(2+3y)^4$ ii $(5+y)^4$

b Find the term in y^4 in each expansion.

i $(1-2y)^7$ ii $(y-2)^7$

c Find the coefficient of a^2 in each expansion.

i $\left(2a-\frac{1}{2}\right)^5$ ii $(17a+3)^5$

5 Find the expansion of $(3+x)^3$.

6 Which term in the expansion of $(1-2y)^5$ has coefficient:

a 80 b −80?

7 Find the coefficient of x^3 in the expansion of $(1-5x)^9$.

8 Find the term in x^2 in the expansion of $(3-2x)^7$.

9 Find the first four terms in the expansion of $(y+3y^2)^6$ in ascending powers of y.

10 a Find the expansion of $(1+2x)^4$.

b Expand $\left(\frac{x+2}{x}\right)^4$.

11 In the expansion of $\left(1+\frac{x}{k}\right)^8$ the coefficient of x^2 is equal to the coefficient of x^3. Find the value of k.

12 Find the coefficient of x^2 in the expansion of $\left(x+\frac{1}{x}\right)^8$.

13 Find the constant term in the expansion of $(x-2x^{-2})^9$.

Section 2: Binomial coefficients

The easiest way to find binomial coefficients is to use a calculator, but if there are unknowns involved you may need to use a formula. To do so, you need to know a new function, the **factorial** function, $n!$

Key point 9.3

$$n! = n\times(n-1)\times\ldots\times3\times2\times1$$

$0!$ is defined to be 1.

Using this function, you can find a formula for the binomial coefficients.

Key point 9.4

$$^{n}C_{r}=\frac{n!}{r!(n-r)!}$$

This will be given in your formula book.

WORKED EXAMPLE 9.4

Show that $^{6}C_{4}=15$.

$$^{6}C_{4}=\frac{6!}{4!\times(6-4)!}$$ Use the formula with $n=6$ and $r=4$.

$$=\frac{6!}{4!\times 2!}$$

$$=\frac{6\times5\times4\times3\times2\times1}{(4\times3\times2\times1)\times(2\times1)}$$ Use the definition of $n!$ to write out each term in full.

$$=\frac{5\times6}{2}$$ Then cancel common factors from the numerator and denominator.

$$=15$$

Key point 9.5

- $^{n}C_{0}=1$
- $^{n}C_{1}=n$
- $^{n}C_{2}=\frac{n(n-1)}{2}$

Tip

Remember that $^{n}C_{r}$ can also be written as $\binom{n}{r}$.

These expressions are useful when part of an expansion has been given and you need to find the power.

WORKED EXAMPLE 9.5

The first three terms of the expansion of $(1+2x)^{n}$ are given by $1+ax+112x^{2}$. Find the values of n and a.

$$(1+2x)^{n}=\binom{n}{0}(1)^{n}+\binom{n}{1}(1)^{n-1}(2x)+\binom{n}{2}(1)^{n-2}(2x)^{2}+\ldots$$ Write out the expansion of the left-hand side in terms of n.

$$=1+n(2x)+\frac{n(n-1)}{2}4x^{2}+\ldots$$

$$=1+2nx+2n(n-1)x^{2}+\ldots$$ Now use the expressions in Key point 9.5 for $\binom{n}{0}$, $\binom{n}{1}$ and $\binom{n}{2}$.

Comparing coefficients of x^{2}:

$2n(n-1)=112$ You are given that the coefficient of x^{2} is 112 so equate coefficients of x^{2}.

Continues on next page

$2n^2 - 2n - 112 = 0$

$n^2 - n - 56 = 0$

$(n-8)(n+7) = 0$

$n = 8$ or -7

And solve for n

$\therefore n = 8$ as n must be positive.

Comparing coefficients of x:

$2n = a$

$a = 16$

Now that you know n, you can equate coefficients of x to find the value of a.

EXERCISE 9B

1 Evaluate these binomial coefficients without using a calculator.

a i $^7C_2$ ii $^5C_2$

b i $^{10}C_3$ ii $^{12}C_3$

c i $\binom{8}{1}$ ii $\binom{15}{1}$

d i $\binom{15}{0}$ ii $\binom{9}{0}$

e i $^{11}C_{11}$ ii $^8C_8$

f i $\binom{7}{6}$ ii $\binom{12}{11}$

2 Find the value of n in each equation.

a i $^nC_1 = 10$ ii $^nC_1 = 13$

b i $\binom{n}{2} = 45$ ii $\binom{n}{2} = 66$

3 $(1+3x)^n = 1 + 42x + \dots$

Find the value of n.

4 The coefficient of x^2 in the expansion of $(1+2x)^n$ is 264.
Find the value of n.

5 $(2+x)^n = 32 + ax + \dots$

a Find the value of n. **b** Find the value of a.

6 $(1+2x)^n = 1 + 20x + ax^2 + \dots$

a Find the value of n. **b** Find the value of a.

7 $\left(1+\frac{x}{2}\right)^n = 1 + ax + \frac{33}{2}x^2 + \dots$

a Find the value of n.

b Find the value of a.

8 Given that $^nC_k = {^nC_{(k+1)}}$, write n in terms of k.

Elevate

See Extension Sheet 9 for more questions on identities involving binomial coefficients.

Section 3: Applications of the binomial theorem

There are two common applications of the binomial expansion:

- multiplying another expression by a bracket that needs to be expanded with the binomial expansion
- making approximations to numbers raised to a positive integer power.

WORKED EXAMPLE 9.6

Use the binomial theorem to expand and simplify $(5-3x)(2-x)^4$.

$(2-x)^4 = 1(2)^4 + 4(2)^3(-x) + 6(2)^2(-x)^2 + 4(2)^1(-x)^3 + 1(-x)^4$
$= 16 - 32x + 24x^2 - 8x^3 + x^4$

First expand $(2-x)^4$.

So:
$(5-3x)(2-x)^4 = (5-3x)(16-32x+24x^2-8x^3+x^4)$
$= 5[16-32x+24x^2-8x^3+x^4]$
$-3x[16-32x+24x^2-8x^3+x^4]$
$= 80-208x+216x^2-112x^3+29x^4-3x^5$

Then multiply by $(5-3x)$ in the normal way.

Sometimes you will be interested in only one of the terms, rather than the whole expansion.

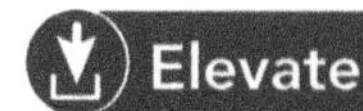

Elevate

See Support Sheet 9 for a further example of this type and for more practice questions.

WORKED EXAMPLE 9.7

a Find the first three terms in the expansion of $(1+2x)^6$.
b Hence find the coefficient of x^2 in the expansion of $(1-x+2x^2)(1+2x)^6$.

a $(1+2x)^6 = 1(1)^6 + 6(1)^5(2x)^1 + 15(1)^4(2x)^2 + \ldots$
$= 1 + 12x + 60x^2 + \ldots$

Expand but stop after the first three terms.

b $(1-x+2x^2)(1+12x+60x^2 \ldots)$

You only need the first three terms of the binomial expansion, as after that all terms have powers of x^3 or higher.

The x^2 term is: $60x^2 - 12x^2 + 2x^2 = 50x^2$

There are three ways to get the x^2 term: $1\times 60x^2$, $-x\times 12x$ and $2x^2\times 1$

So the coefficient of x^2 is 50.

When x is very small, as the powers of x become larger their values become smaller. So the first few terms of a binomial expansion may be used to approximate the entire expansion.

Key point 9.6

If the value of x is close to zero, large powers of x will be extremely small and can be neglected.

WORKED EXAMPLE 9.8

a Find the first three terms in ascending powers of x of the expansion of $(2-x)^5$.

b Use your answer to (a) to find an approximate value of 1.99^5.

a The first three terms of $(2-x)^5$ are:

$1(2)^5+5(2)^4(-x)^1+10(2)^3(-x)^2=32-80x+80x^2$

Expand.

b $2-x=1.99$

$x=0.01$

You need to choose a value of x so that $(2-x)^5=1.99^5$, which means that $2-x=1.99$.

$1.99^5\approx 32-80(0.01)+80(0.01^2)$

$=32-0.8+0.008$

$=31.208$

Now just evaluate the first three terms of the series for $x=0.01$ to give an approximation for 1.99^5.

EXERCISE 9C

1 Find the first three terms, in ascending powers of x of each expression.

a **i** $(2+x)(1+3x)^4$ **ii** $(5+x)(1+2x)^5$

b **i** $(3+4x)(2-3x)^4$ **ii** $(4+5x)(3-2x)^5$

2 **a** Expand and simplify $(1+2x)^5$.

b Hence expand $(3-x)(1+2x)^5$.

3 **a** Find the first four terms, in ascending powers of x, in the expansion of $(2-5x)^7$.

b Hence find the coefficient of x^3 in the expansion of $(1+2x)(2-5x)^7$.

4 **a** Find the first three terms in the expansion of $(3-5x)^4$.

b By choosing a suitable value of x, use your answer to part **a** to find an approximation for 2.995^4 correct to 5 significant figures.

5 **a** Find the first three terms in the expansion of $(8+2x)^5$.

b Use your answer to part **a**, with a suitable value of x, to find an approximate value for 8.02^5.

6 **a** Find the first three terms in the expansion of $(2+3x)^7$.

b Hence find an approximation for:

i 2.3^7 **ii** 2.03^7.

c Which of your answers in part **b** provides a more accurate approximation? Justify your answer.

7 **a** Find the first three terms in the expansion of $(4-x)^7$.

b Hence find the coefficient of x^2 in the expansion of $(2+2x-x^2)(4-x)^7$.

8 **a** Expand $\left(e+\frac{2}{e}\right)^5$.

b Simplify $\left(e+\frac{2}{e}\right)^5+\left(e-\frac{2}{e}\right)^5$.

9 **a** Write the expression $(1+x)^n(1-x)^n$ in the form $(f(x))^n$.

b Find the first three non-zero terms of the expansion of $(1-x)^{10}(1+x)^{10}$ in ascending powers of x.

10 Find the coefficient of x^5 in the expansion of $(1+3x)(1+x)^7$.

11 $(1+x)^3(1+mx)^4=1+nx+93x^2+\ldots$

Find the possible values of m and n.

Checklist of learning and understanding

- The expansion of $(a+b)^n$ can be found by using the formula:

$$(a+b)^n=a^n+\binom{n}{1}a^{n-1}b^1+\binom{n}{2}a^{n-2}b^2+\ldots+\binom{n}{r}a^{n-r}b^r+\ldots+b^n \text{ (for any integer } n)$$

- The binomial coefficients, written ${}^nC_r$ or $\binom{n}{r}$, are given by the formula:

$${}^nC_r=\frac{n!}{r!(n-r)!}$$

where $n!=n\times(n-1)\times\ldots\times3\times2\times1$

- Approximations for numbers raised to a positive integer power can be made using the first few terms of a binomial expansion $(a+bx)^n$. This is valid when bx is small, so that terms with higher powers are negligible.

Mixed practice 9

1. ${}^{n}C_2 = \dfrac{n!}{2!6!}$

   What is the value of n?

   Choose from these options.

   A 2 **B** 4 **C** 6 **D** 8

2. Expand $(2-x)^{12}$ in ascending powers of x up to and including the term in x^3.

3. Fully expand and simplify $(2x+5)^3$.

4. Find the coefficient of y^6 in the expansion of $(3+2y^2)^5$.

5. The constant term in the expansion of $\left(x^2+\dfrac{a}{x^4}\right)^9$ is $-\dfrac{28}{9}$.

   What is the value of a?

   Choose from these options.

   A 3 **B** $-\dfrac{1}{3}$ **C** -6 **D** $-\dfrac{1}{9}$

6. **a** Find the first four terms in the expansion of $\left(1-\dfrac{x}{10}\right)^7$.

   b By choosing an appropriate value of x, find an approximation for 0.99^7.

7. $a = 2-\sqrt{2}$. Using the binomial expansion or otherwise, express a^5 in the form $m+n\sqrt{2}$.

8. Find the constant coefficient in the expansion of $(x^3-2x^{-1})^4$.

9. Fully expand and simplify $\left(x^2-\dfrac{2}{x}\right)^4$.

10. Find the coefficient of c^4 in the expansion of $(2c+5)(c+1)^{14}$.

11. Find the coefficient of x^6 in the expansion of $(1-x^2)(1+x)^5$.

12. In the expansion of $(2+ax)^6$ the coefficient of x^3 is three times the coefficient of x^2.

    Find the value of a.

13. **a** Expand $\left(1+\dfrac{4}{x}\right)^2$.

    b The first four terms of the binomial expansion of $\left(1+\dfrac{x}{4}\right)^8$ in ascending powers of x are $1+ax+bx^2+cx^3$. Find the values of the constants a, b and c.

    c Hence find the coefficient of x in the expansion of $\left(1+\dfrac{4}{x}\right)^2\left(1+\dfrac{x}{4}\right)^8$.

[© AQA 2013]

14 Find the coefficient of x^2 in the expansion of $\left(2x+\frac{1}{\sqrt{x}}\right)^5$.

15 $(1+ax)^n = 1+10x+40x^2+\ldots$

Find the values of a and n.

16 In the expansion of $(1+kx)^4(1+x)^n$ the coefficient of x is 13 and the coefficient of x^2 is 74.

Find the possible values of k and n.

17 **a** Sketch the graph of $y=(x+2)^3$.

b Find the binomial expansion of $(x+2)^3$.

c Find the exact value of 2.001^3.

d Solve the equation $x^3+6x^2+12x+16=0$.

FOCUS ON ... PROOF 1

From general to specific

Key point 4.1 in Chapter 4 states the **factor theorem**:

For any polynomial $f(x)$, if $f(a) = 0$ then $(x - a)$ is a factor of $f(x)$.

This is actually a specific case of a more general result, called the **remainder theorem**:

Let $f(x)$ be a polynomial and suppose you can write $f(x) \equiv (x - a)q(x) + r$ for some polynomial $q(x)$. Then $r = f(a)$.

Note that the statement $f(x) \equiv (x - a)q(x) + r$ is effectively saying that you can write any integer as a product of integers plus a remainder, for example, $13 = 3 \times 4 + 1$.

PROOF 5

Proof of the remainder theorem

$f(x) \equiv (x - a)q(x) + r$ — Start with the given expression.

$\Leftrightarrow f(a) \equiv (a - a)q(a) + r$ — You can substitute any value of x in the identity. $x = a$ is useful because it makes one of the terms equal to zero.

$\Rightarrow f(a) = r$

QUESTIONS

1. Where in the proof is it important that $q(x)$ is a polynomial? Does $f(x)$ have to be a polynomial?

2. Adapt the proof to find the remainder when $f(x)$ is divided by $(bx - a)$.

3. Use the remainder theorem to prove the factor theorem.

4. The factor theorem is stated as: 'If $f(a) = 0$, then $(x - a)$ is a factor of $f(x)$.'

   The full statement of the theorem is, in fact:

   $(x - a)$ is a factor of a polynomial $f(x)$ **if and only if** $f(a) = 0$.

   This statement comprises two implications:

   - $f(a) = 0 \Rightarrow (x - a)$ is a factor of $f(x)$
   - $(x - a)$ is a factor of $f(x) \Rightarrow f(a) = 0$.

   a Prove the second implication:

   If $(x - a)$ is a factor of $f(x)$, then $f(a) = 0$.

   b Did you need to use the remainder theorem in your proof?

FOCUS ON ... PROBLEM SOLVING 1

Alternative approaches

When faced with an unfamiliar problem, most people are happy if they can just reach the solution, by whatever method. However, once you have solved the problem, it is worth thinking about whether there is an alternative strategy. This helps confirm that your solution is correct but, more importantly, it gives you an opportunity to reflect on which approach is best for what type of question.

In Worked example 6.12 (Chapter 6, Section 5), you solved this problem.

A circle has centre (3, 2). Find the radius of the circle such that the circle has as a tangent line with equation $x+5y=20$.

Strategy 1

This is the method used in Worked example 6.12.

- Write the equation of the circle with unknown radius: $(x-3)^2+(y-2)^2=r^2$
- The equation for the intersection of this circle with the line rearranges to: $26y^2-174+(293-r^2)=0$
- If the circle is tangent to the line this equation has only one solution, the discriminant is zero: $174^2-4(26)(293-r^2)=0$
  This gives the solution $r=1.37$.

Strategy 2

This uses the fact that the tangent is perpendicular to the radius at the point of contact. You don't know the coordinates of the point where the line touches the circle, but you do know it lies on the line $x+5y=20$. This means that for every point on the line, $x=20-5y$, so you can write the unknown coordinates as $(20-5y, y)$.

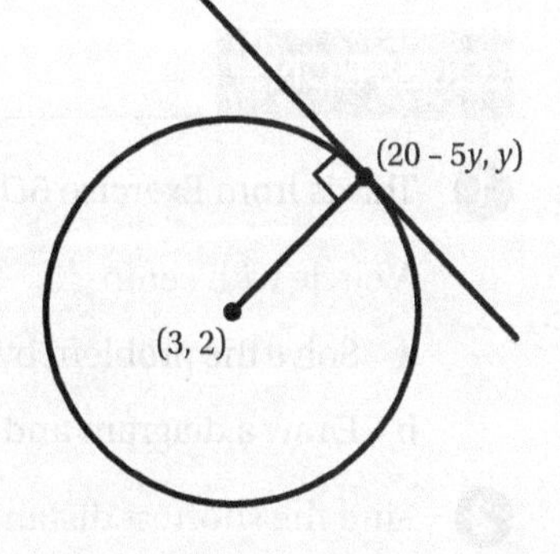

- The line connecting this unknown point to the centre (3, 2) needs to be perpendicular to $x+5y=20$.
- Write down the gradient of the line $x+5y=20$.
- Write down the gradient of a perpendicular line.
- Hence write an equation for the gradient of the line connecting (3, 2) to $(20-5y, y)$.
- You should find that the coordinates of the point of contact are $\left(\frac{85}{26}, \frac{87}{26}\right)$.
- Now you can find the radius, which is the distance of this point from the centre.

Strategy 3

This is based on the same information as Strategy 2, but this time you will find the coordinates of the point of contact by intersecting the line $x+5y=20$ with the line perpendicular to it and passing through the centre (3, 2). See if you can carry out this strategy for yourself.

Rewind

This is the same method used to find the shortest distance from a point to a line in Exercise 6C, Question 10.

Strategy 4

This also uses the fact that the radius is perpendicular to the tangent, but you will find the length of the radius directly, without finding the coordinates of the point of contact first. Instead, you will create a right-angled triangle and find its area in two different ways.

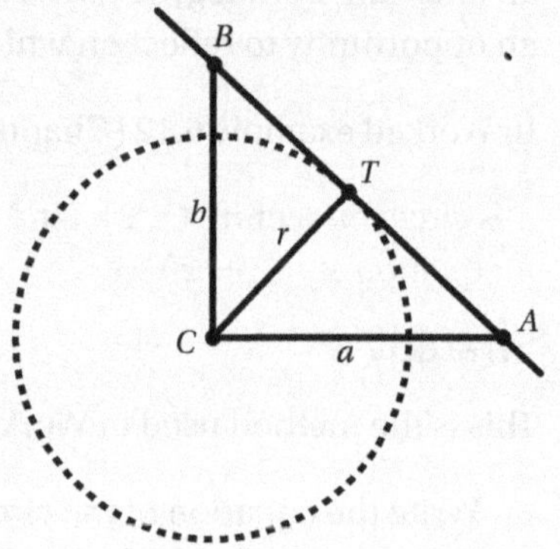

- Draw lines parallel to the coordinate axes to form a right-angled triangle ACB. You can calculate the area of this triangle in two different ways:

$$\text{area} = \frac{1}{2}ab = \frac{1}{2}r(AB)$$

- Points A and B are on the line $x + 5y = 20$, so you can find their coordinates.
A has the same y-coordinate as $C(y = 2)$. Show that $a = 7$.
B has the same x-coordinate as $C(x = 3)$. Show that $b = \frac{7}{5}$.
- Show that the length of AB is $\sqrt{\frac{1274}{25}}$.
- So the area equation gives:

$$A = \frac{1}{2}(7)\left(\frac{7}{5}\right)$$

$$= \frac{1}{2}r\sqrt{\frac{1274}{25}} \Rightarrow r = 1.37$$

QUESTIONS

1 This is from Exercise 6D, Question 7.

A circle with centre $(3, -5)$ and radius 7 crosses the x-axis at points P and Q. Find the exact distance PQ.

a Solve the problem by writing the equation of the circle and finding the coordinates of points P and Q.

b Draw a diagram and label some lengths. Hence use a geometrical method to solve the problem.

2 Find the shortest distance from the origin to the line with equation $5x + 12y = 60$.

How many different ways can you find to solve this problem?

3 This is Question 10 from the Mixed practice exercise in Chapter 6:

Find the exact values of k for which the line $y = kx + 3$ is tangent to the circle with centre $(6, 3)$ and radius 2.

a Solve the problem by finding a quadratic equation for intersections and using the discriminant.

b Draw a diagram and find a different way to solve the problem.

c Would the second method work if the equation of the tangent was $y = kx + 2$ instead of $y = kx + 3$?

Tip

Remember how the gradient of a line is related to the angle it makes with the horizontal.

4 Circle C_1 has centre at the origin and radius 1. Circle C_2 has centre $(4, 3)$ and radius r. Find the value of r such that the two circles touch.

FOCUS ON ... MODELLING 1

Using data to refine the model

In Chapter 8 you learnt that you can use the exponential function to model population growth. This basic model is based on the assumption:

The population growth is proportional to the size of the population.

Fast forward

In Chapter 12 you will learn that this statement can be written as an equation, $\frac{dN}{dt} = kN$.

QUESTIONS

1. This simple population model leads to the equation $N = Ae^{kt}$, where N is the size of the population at time t.

 a Plot the graph of N against t for various values of A and k.

 b What do the constants A and k represent, in the context of this model?

 c Suppose you observe the size of the population at two specific times; for example, $N = 35$ when $t = 5$ and $N = 152$ when $t = 11$. Use technology to find approximate values of A and k.

 d Now use algebra to find the values of A and k.

 e At a later time, you acquire a third observation: $N = 340$ when $t = 18$. What does this tell you about your model?

2. In a refined model, the population is assumed to have a maximum capacity (for example, limited by the amount of food or space available). The rate of growth is proportional both to the size of the population and to the remaining space.

 This is called a **logistic** model, and leads to the equation: $N = \frac{ACe^{kt}}{C - A + Ae^{kt}}$.

 a Which of the parameters $(A, C$ or $k)$ represents the initial population size?

 b Investigate this equation for various values of the parameters. Which parameter represents the maximum capacity?

 c Use technology to find approximate values of the parameters that fit with the three observed data values: $(5, 35)$, $(11, 152)$ and $(18, 340)$.

 Now consider the population model $N = Ae^{-kt}$, where k is a positive constant.

a What is the effect of the negative sign in the equation? How can you interpret this in the context of this problem?

b What does this model predict will happen to the population in the long term?

To counteract population decline, individuals can be added to the population. (For example, this could model controlled immigration in a country where the death rate is larger than a birth rate.)

In a simple immigration model, new individuals are added at a constant rate D. This model leads to the equation: $N = \frac{D}{k} - \frac{1}{k}(D - kA)e^{-kt}$.

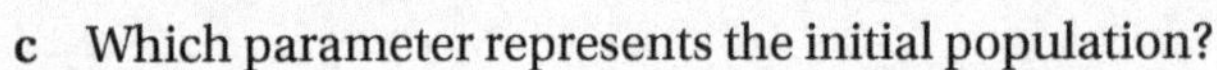

c Which parameter represents the initial population?

d According to this model, what happens to the population in the long term?

e Explain what happens in the case when $D = kA$, and why it happens.

Tip

You should find that the behaviour depends on whether $D > kA$ or $D < kA$.

 A population of a small country was 7 430 000. Ten years later, the population has fallen to 7 026 000.

a For a simple exponential model, $N = Ae^{-kt}$, find the value of the constants A and k.

b By what factor does the population decrease each year? Given that the annual birth rate is 4.7 babies per 1000 people, estimate the annual death rate.

c To counteract the population decline, the government proposes a controlled immigration programme. They want to aim for a stable long-term population of about 7 200 000.

i What annual immigration target should they set?

ii What assumptions about the immigrant population need to be made for the model prediction (a long-term population of 7 200 000) to be valid?

Tip

See Question 3 for a simple immigration model.

CROSS-TOPIC REVIEW EXERCISE 1

1 a i Write $x^2 - 6x + 5$ in the form $(x-h)^2 - k$.

ii Describe a single transformation that transforms the graph of $y = x^2$ into the graph of $y = x^2 - 6x + 5$.

iii Sketch the graph of $y = x^2 - 6x + 5$, marking the coordinates of the axes intercepts and the minimum point.

b i Add the graph of $y = x - 1$ to your sketch.

ii Solve the equation $x^2 - 6x + 5 = x - 1$.

iii Shade the region which satisfies $y \geqslant x^2 - 6x + 5$ and $y \leqslant x - 1$.

2 a Show that $(x-3)$ is a factor of $\mathrm{p}(x) = x^3 - 4x^2 - 3x + 18$.

b Factorise $\mathrm{p}(x)$ completely.

c Hence sketch the graph of $y = \mathrm{p}(x)$.

d Sketch the graph of $y = (x+2)^3 - 4(x+2)^2 - 3(x+2) + 18$.

3 The circle with centre $(12, 9)$ and radius $\sqrt{145}$ intersects the x-axis at points A and B and the y-axis at points C and D. Find the area of the quadrilateral $ABDC$.

4 The line AB has equation $3x + 2y = 7$. The point C has coordinates $(2, -7)$.

a i Find the gradient of AB.

ii The line which passes through C and which is parallel to AB crosses the y-axis at the point D. Find the y-coordinate of D.

b The line with equation $y = 1 - 4x$ intersects the line AB at the point A. Find the coordinates of A.

c The point E has coordinates $(5, k)$. Given that CE has length 5, find the two possible values of the constant k.

5 a $3\log x - \log 4x + \log 4y = \log z$

Find z in terms of x and y, fully simplifying your answer.

b If $\ln K = 2 - \ln c$, show that $K = \frac{A}{c}$, where A is a constant to be found.

6 Use an algebraic method to find the exact solutions of the equation $3\mathrm{e}^{2x} - 7\mathrm{e}^{x} + 2 = 0$.

7 The graph of $y = \ln x$ can be transformed into the graph of $y = \ln kx$ using either a horizontal stretch or a vertical translation. State:

a the stretch factor of the horizontal stretch

b the vertical translation vector.

8 a Expand and simplify $(1+x)^4 + (1-x)^4$.

b Hence show that $(\sqrt{2}+1)^4 + (\sqrt{2}-1)^4$ is an integer and find its value.

9 Given that $\binom{n}{2} = k$, express n in terms of k.

10 The diagram shows the graph with equation $y = C + Ae^{-kt}$. The graph passes through the point $P(2, 3)$.

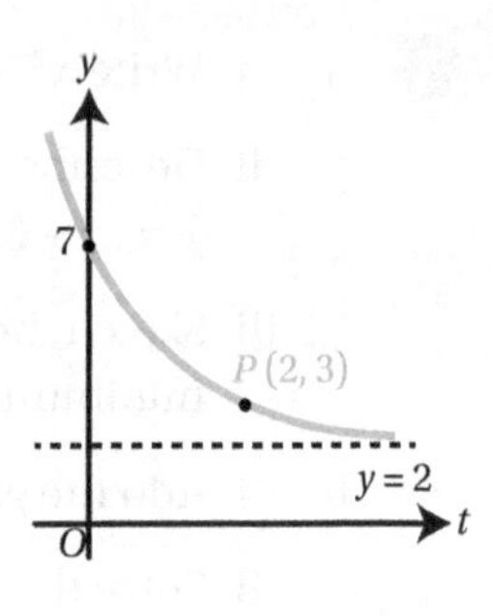

a Write down the value of C and the value of A.

b Find the exact value of k.

11 A model for the radioactive decay of a form of iodine is given by $m = m_0 2^{-\frac{1}{8}t}$

The mass of the iodine after t days is m grams. Its initial mass is m_0 grams.

a Use the given model to find the mass that remains after 10 grams of this form of iodine have decayed for 14 days, giving your answer to the nearest gram.

b A mass of m_0 grams of this form of iodine decays to $\frac{m_0}{16}$ grams in d days. Find the value of d.

c After n days, a mass of this form of iodine has decayed to less than 1% of its initial mass. Find the minimum integer value of n.

[© AQA 2011]

12 The number x satisfies the equation $x^2 = 3x - 1$.

a Show that $x + \frac{1}{x} = 3$.

b i Expand $\left(x+\frac{1}{x}\right)^2$ and $\left(x+\frac{1}{x}\right)^3$.

ii Hence find the values of $x^2 + \frac{1}{x^2}$ and $x^3 + \frac{1}{x^3}$.

13 Show that the graph of $y = x^2 - (m+3)x + (m+1)$ crosses the x-axis for all values of m.

14 a Given that $y = e^x + e^{-x}$ express x in terms of y.

b i Given that x is a real number, find the set of possible values of y.

ii For a fixed y from this set, show that the sum of all the possible values of x is zero.

15 a Use logarithms to solve the equation $2^{3x} = 5$, giving your value of x to three significant figures.

b Given that $\log_a k - \log_a 2 = \frac{2}{3}$, express a in terms of k.

c i By using the binomial expansion, or otherwise, express $(1+2x)^3$ in ascending powers of x.

ii It is given that: $\log_2 [(1+2n)^3 - 8n] = \log_2 (1+2n) + \log_2 [4(1+n^2)]$

By forming and solving a suitable quadratic equation, find the possible values of n.

[© AQA 2015]

16 The temperature T °C of water t minutes after boiling is modelled by

$$T = 100e^{-0.1t}$$

a What temperature is predicted 10 minutes after the water has boiled?

b What temperature does the model predict the water will eventually reach?

c The long term temperature of the water is actually 20 °C. Suggest a refined model which has the same rate of cooling and initial temperature but has a long term temperature of 20 °C.

d How long does it take the refined model to reach the same temperature as found in part **a**?

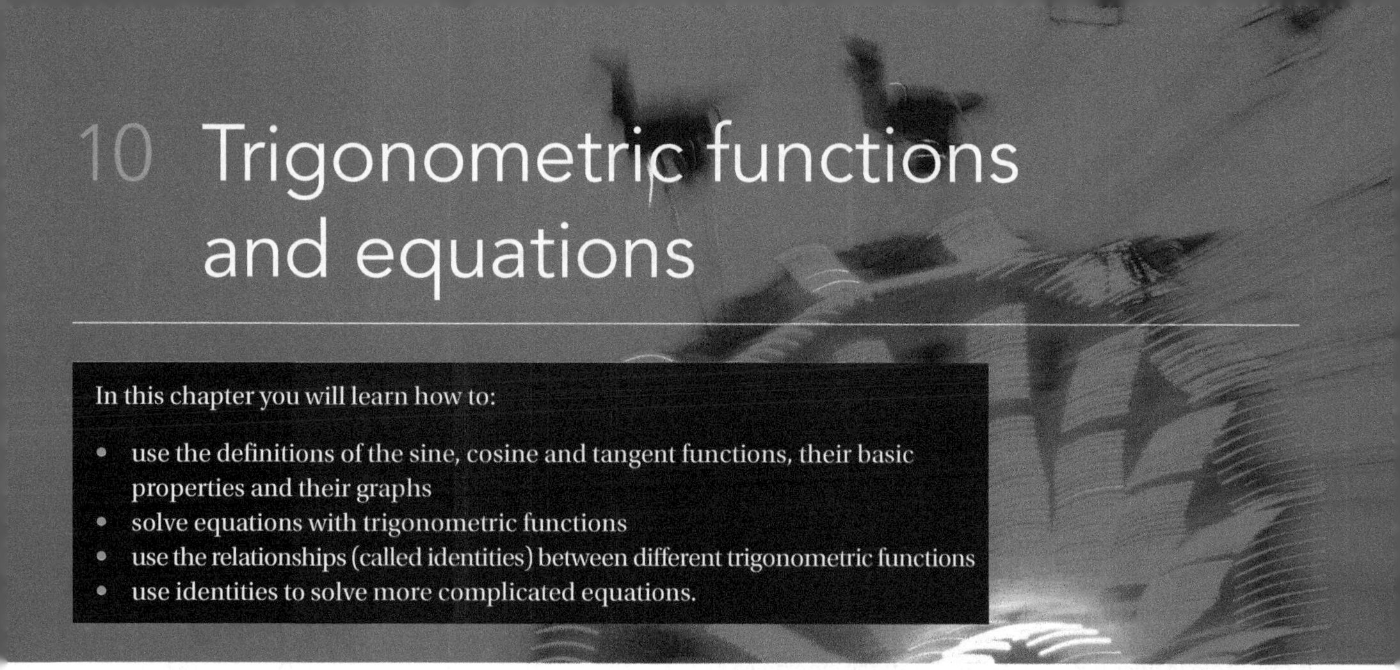

10 Trigonometric functions and equations

In this chapter you will learn how to:

- use the definitions of the sine, cosine and tangent functions, their basic properties and their graphs
- solve equations with trigonometric functions
- use the relationships (called identities) between different trigonometric functions
- use identities to solve more complicated equations.

Before you start…

GCSE	You should be able to use trigonometry in right-angled triangles to find unknown lengths.	1 Find the value of x in the diagram. 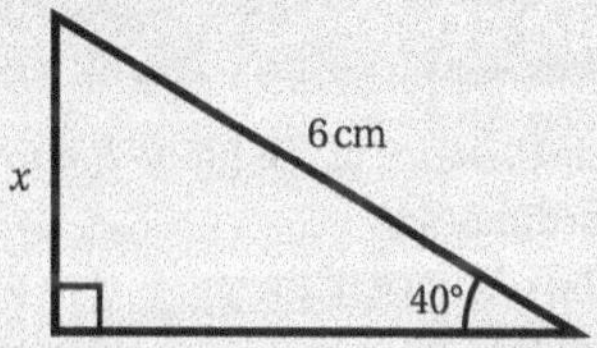
GCSE	You should be able to use trigonometry in right-angled triangles to find unknown angles.	2 Find the value of θ in the diagram. 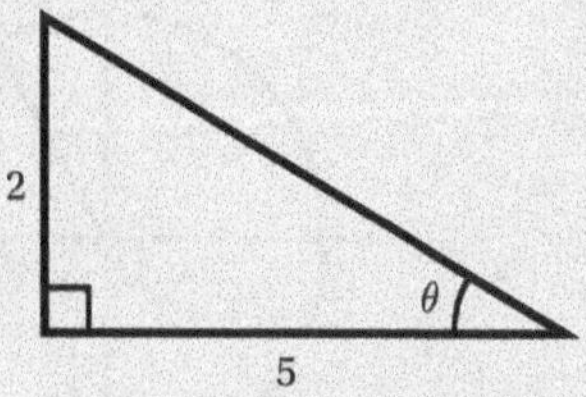
GCSE	You should be able to use Pythagoras' theorem in a right-angled triangle.	3 The lengths of the two shorter sides of a right-angled triangle are 5 cm and 12 cm. Find the length of the hypotenuse.
Chapter 3	You should be able to solve quadratic equations, using the formula or factorising.	4 Solve the equation $x^2 - 2x + 1 = 0$.

What goes around, comes around

There are many real-life situations in which something repeats at regular intervals – such as the height of a fairground ride, the tides of the sea or the vibration of a guitar string. All of these can be modelled using **trigonometric functions**.

You first met sin, cos and tan when working with angles in a right-angled triangle. In this chapter you will find out how they can be used in a variety of other contexts.

Gateway to A Level

See Gateway to A Level Section O for revision of Pythagoras' theorem and trigonometry.

Section 1: Definitions and graphs of the sine and cosine functions

You have already used trigonometric functions in right-angled triangles but in these triangles no angle can exceed 90°. If you want to use trigonometric functions for other purposes it is useful to have a more general definition.

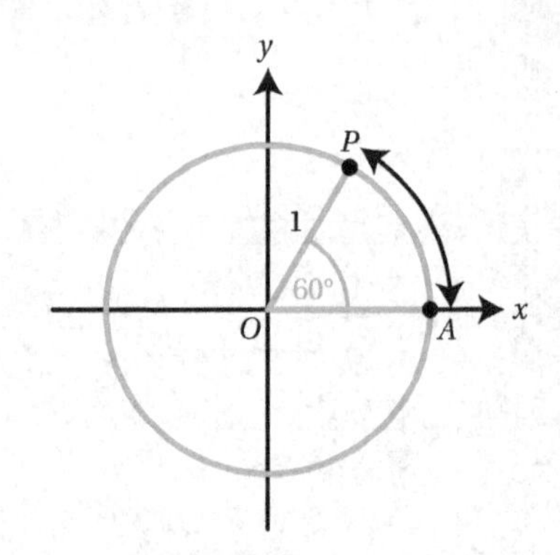

To do this, consider a circle of radius 1 unit centred at the origin (the unit circle). As a point P moves anti-clockwise around the circumference, an angle is formed between OP and the horizontal, for example, angle AOP in the diagram is 60°.

Any angle can be defined in this way. If the angle is greater than 360° the point P rotates more than a full turn. For example, in the diagram point P has rotated one and a quarter turns and represents 450°.

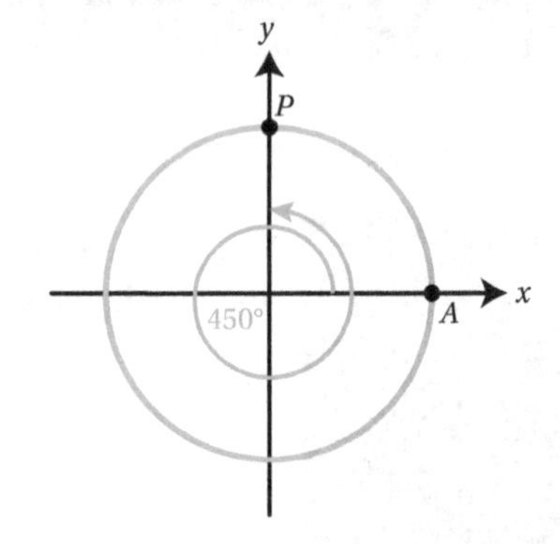

If the angle is negative then point P rotates clockwise.

For an angle θ, the sine and cosine functions are then defined in terms of the distance of the point P to the axes.

Key point 10.1

- $\sin\theta$ is the distance of the point P above the horizontal axis (its y-coordinate).
- $\cos\theta$ is the distance of the point P to the right of the vertical axis (its x-coordinate).

With this definition, you can draw the graphs of $y = \sin x$ and $y = \cos x$ for any value of x.

For $y = \sin x$:

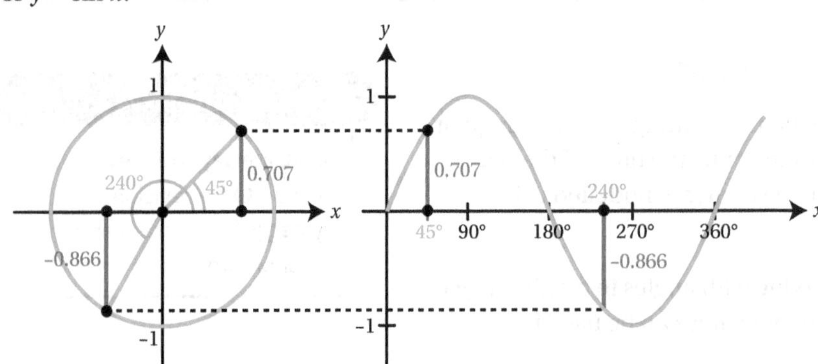

For $y = \cos x$:

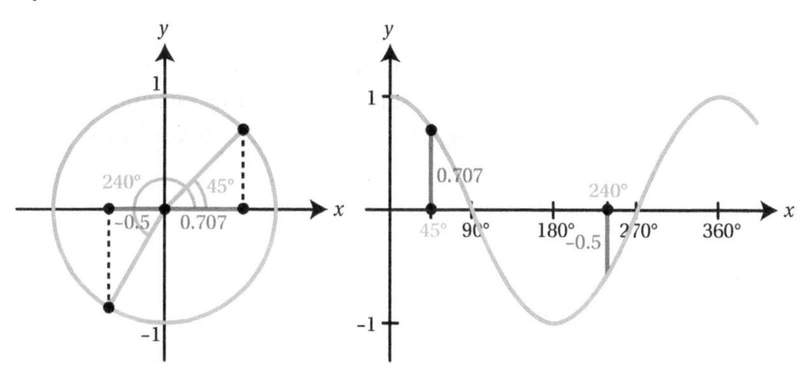

The sine function

Key point 10.2

The graph of $y = \sin x$

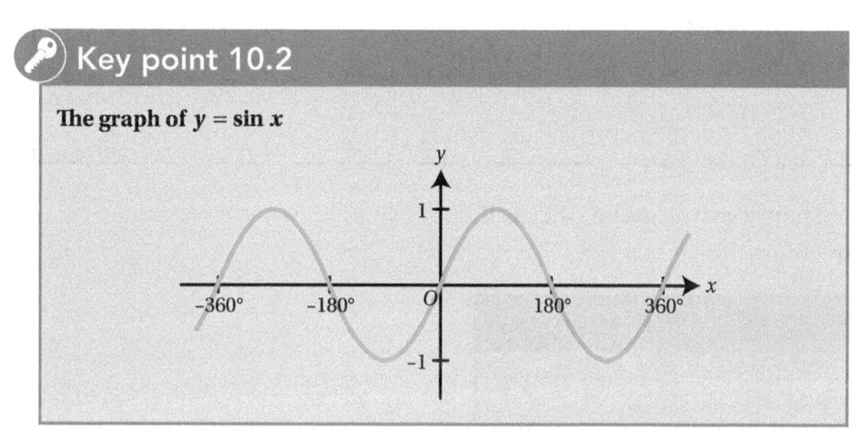

The graph repeats after 360°: the sine function is **periodic** with **period** 360°.

You can also see that the minimum possible value of $\sin x$ is -1 and the maximum value is 1: the sine function has **amplitude** 1.

You can use the symmetries of the sine graph to see how values of $\sin x$ for various angles are related to each other.

WORKED EXAMPLE 10.1

Given that $\sin x = 0.6$ find the value of:

a $\sin(180° - x)$ **b** $\sin(180° + x)$.

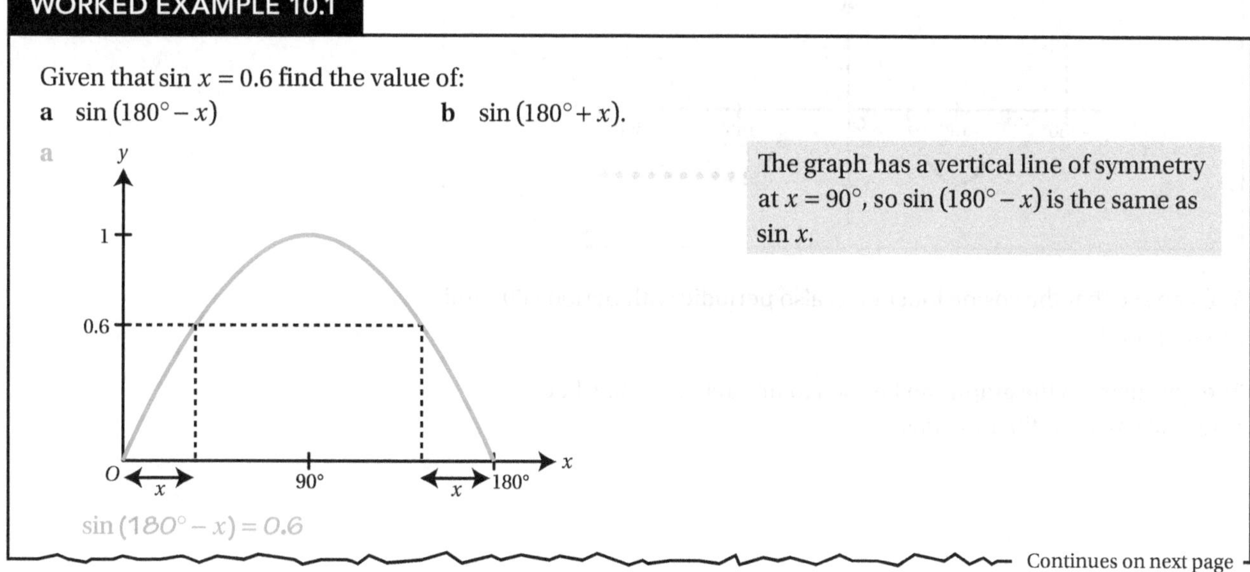

The graph has a vertical line of symmetry at $x = 90°$, so $\sin(180° - x)$ is the same as $\sin x$.

$\sin(180° - x) = 0.6$

Continues on next page

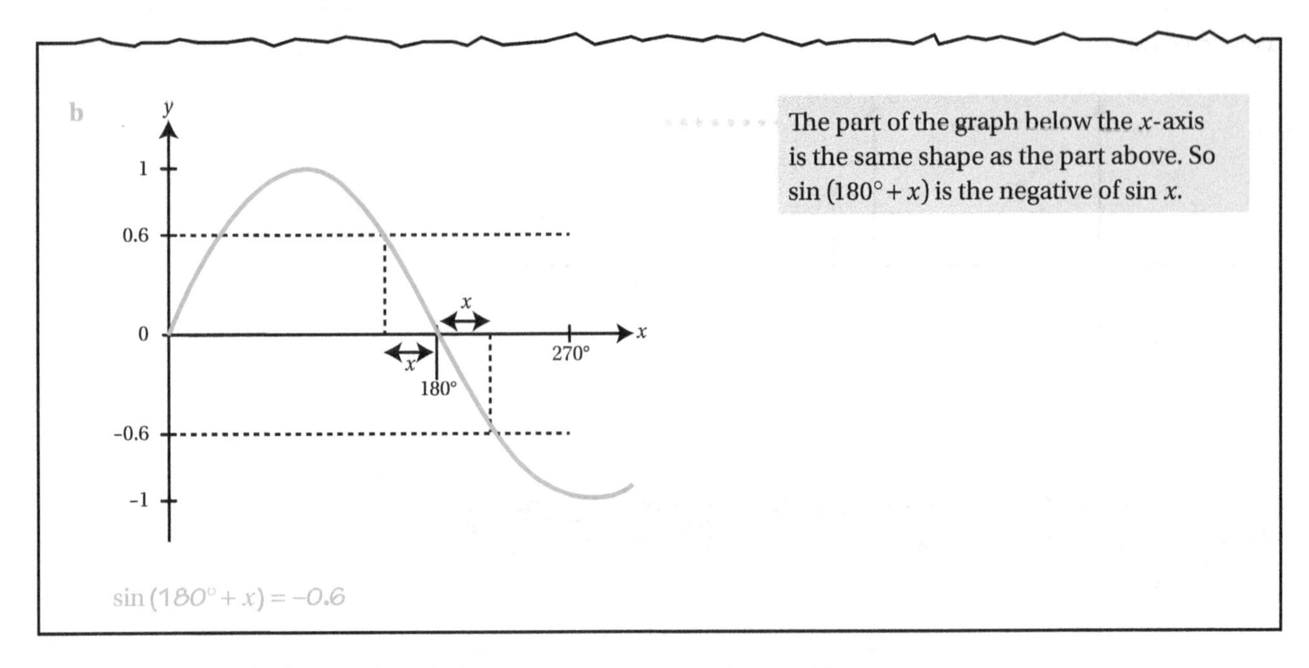

You can find several other similar relationships from the graph. They will be useful when solving trigonometric equations.

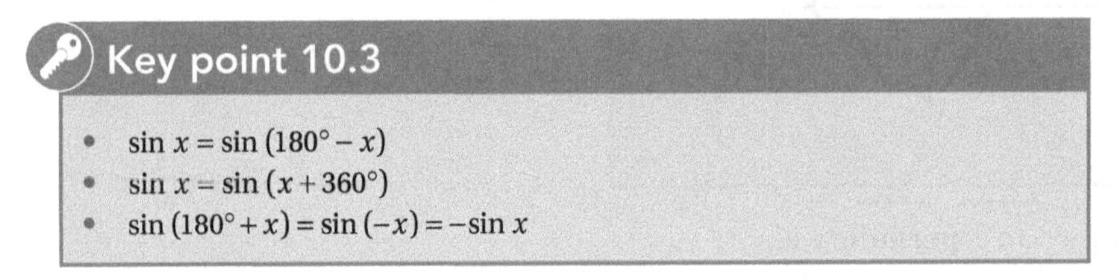

Key point 10.3

- $\sin x = \sin(180° - x)$
- $\sin x = \sin(x + 360°)$
- $\sin(180° + x) = \sin(-x) = -\sin x$

The cosine function

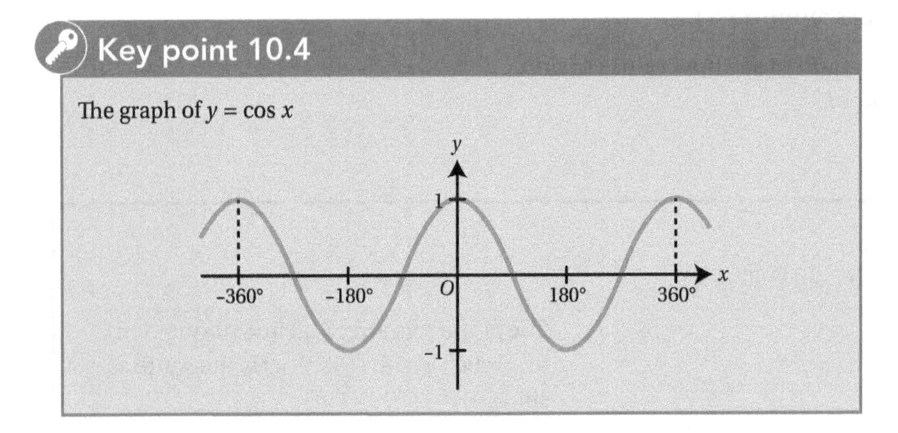

Key point 10.4

The graph of $y = \cos x$

You can see that the cosine function is also periodic with period 360° and has amplitude 1.

The symmetry of the graph can be used to find relationships between values of $\cos x$ for different angles.

WORKED EXAMPLE 10.2

If $\cos 20° = c$, find two values of x between 0° and 360° for which $\cos x = -c$.

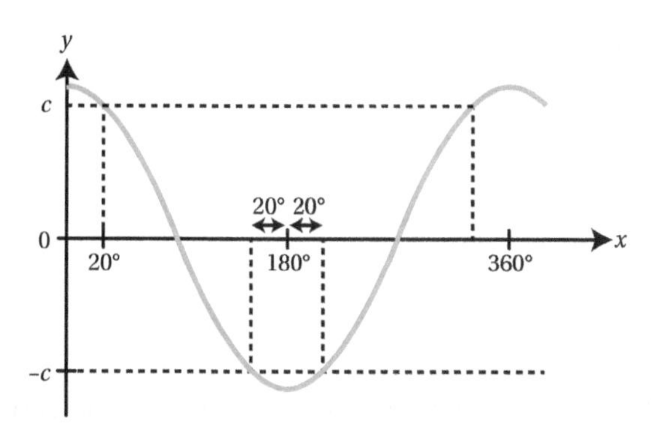

The two values are 20° away from 180° on either side.

$x_1 = 180° - 20° = 160°$

$x_2 = 180° + 20° = 200°$

Key point 10.5

- $\cos x = \cos(-x)$
- $\cos x = \cos(x + 360°)$
- $\cos(180° - x) = \cos(180° + x) = -\cos x$

The unit circle definition can be used to establish a connection between the sine and cosine functions.

WORKED EXAMPLE 10.3

Given that $\sin x = 0.4$, find the value of:

a $\cos(90° - x)$ **b** $\cos(90° + x)$.

a

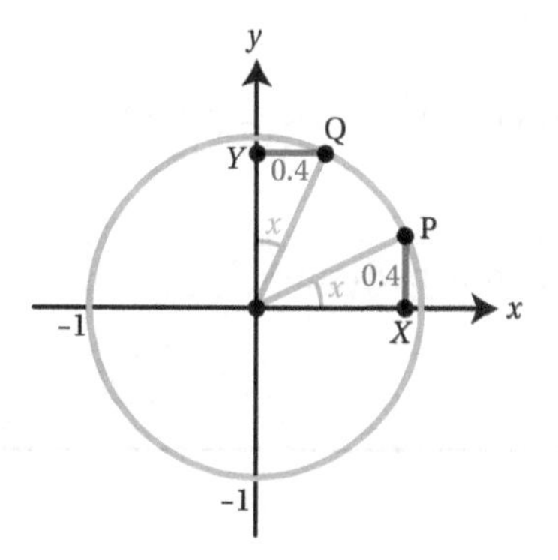

Let P be the point corresponding to x and Q the point corresponding to $90° - x$.

$\cos(90° - x) = 0.4$

$\cos(90° - x) = QY$

Continues on next page

b

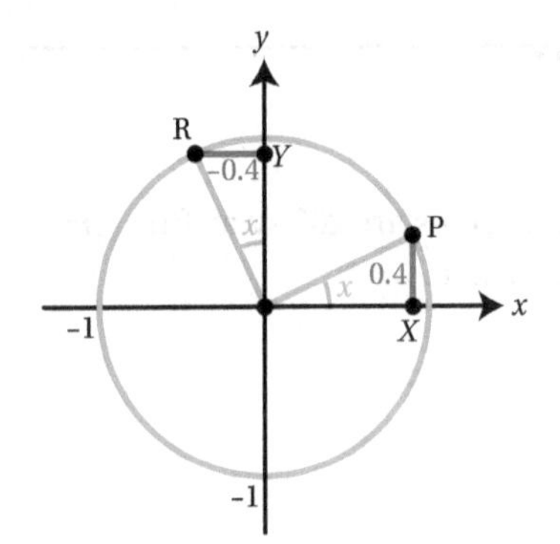

R is the point corresponding to $90° + x$.

$\cos(90° + x) = -0.4$

$\cos(90° + x) = RY$, but this is in the negative direction on the x-axis.

Key point 10.6

It is useful to know these general relationships:

- $\sin x = \cos(x - 90°) = \cos(90° - x)$
- $\cos x = \sin(x + 90°) = \sin(90° - x)$

Tip

You can also see these relationships by noting that the graph of $y = \cos x$ is obtained from the graph of $y = \sin x$ by translating it $90°$ to the left and vice versa.

WORKED EXAMPLE 10.4

Find the two values of x in the interval $0 \leqslant x \leqslant 360$ for which $\sin x = \cos(\sqrt{23})°$.

Give your answers as exact values.

$\sin x = \cos(\sqrt{23})°$

$\cos(90 - x)° = \cos(\sqrt{23})°$.

Use $\sin x = \cos(90 - x)°$ so that both sides are in terms of the same trigonometric function.

Also,

$\cos(90 - x)° = \cos(-\sqrt{23})°$

Since $\cos x = \cos(-x)$, $\cos\left(\sqrt{23}\right)° = \cos\left(-\sqrt{23}\right)°$.

$\therefore 90 - x = \sqrt{23}$ or $-\sqrt{23}$

Remove the cos function from both sides.

$x = 90 - \sqrt{23}$ or $90 + \sqrt{23}$

Rearrange.

EXERCISE 10A

1 Use your calculator to evaluate these ratios, giving your correct answers to 3 s.f.

a **i** $\sin 42°$ **ii** $\cos 168°$ **b** **i** $\sin(-50°)$ **ii** $\cos(-227°)$

2 Use graphs to find the value of each ratio.

a **i** $\sin 90°$ **ii** $\sin 360°$ **b** **i** $\cos 0°$ **ii** $\cos(-180°)$

c **i** $\sin(-90°)$ **ii** $\cos 450°$

3 Given that $\cos 40° = 0.766$ find the value of:

a $\cos 400°$ **b** $\cos 320°$ **c** $\cos(-220°)$ **d** $\cos 140°$

4 Given that $\sin 130° = 0.766$ find the value of:

a $\sin 490°$ **b** $\sin 50°$ **c** $\sin(-130°)$ **d** $\sin 230°$

5 **a** Sketch the graph of $y = \sin x$ for:

i $0° \leqslant x \leqslant 180°$ **ii** $90° \leqslant x \leqslant 360°$

b Sketch the graph of $y = \cos x$ for:

i $-180° \leqslant x \leqslant 180°$ **ii** $0° \leqslant x \leqslant 270°$

> **Elevate**
>
> See Extension Sheet 10 for questions on more complicated periodic functions.

6 Simplify $\sin(x + 360°) + \sin(x + 540°)$.

7 Prove that $\cos(180° + x) + \cos(180° - x) = k \cos x$ where k is a constant to be found.

8 If $\cos x = a$, show that $\sin(x - 90°) = ka$ where k is a constant to be found.

9 Simplify $\sin x + \sin(x + 90°) + \sin(x + 180°) + \sin(x + 270°) + \sin(x + 360°)$

10 Find all values of x that satisfy $\cos x = \sin(\ln 3)$ for $-360° < x < 360°$, giving your answers in exact form.

Section 2: Definition and graph of the tangent function

Another trigonometric function, the tangent function, is defined as the ratio between the sine and the cosine functions.

> **Key point 10.7**
>
> $$\tan x = \frac{\sin x}{\cos x}$$

This is consistent with your previous knowledge of the tangent function. Taking o as the side opposite the angle, a as the side adjacent to it and h as the hypotenuse, then $\sin x = \frac{o}{h}$ and $\cos x = \frac{a}{h}$ and:

$$\frac{\sin x}{\cos x} = \frac{\frac{o}{h}}{\frac{a}{h}} = \frac{o}{a}$$

which is how you previously defined $\tan x$.

You may notice that there is a problem with this definition: when $\cos x$ is zero, you cannot divide by it. Thus the tangent function in undefined for values of x where $\cos x = 0$ (which is when $x = 90°, 270°$ and so on).

You can also see that $\tan x = 0$ whenever $\sin x = 0$, which is when $x = 0°, 180°, 360°$ and so on.

Key point 10.8

The graph of the tangent function

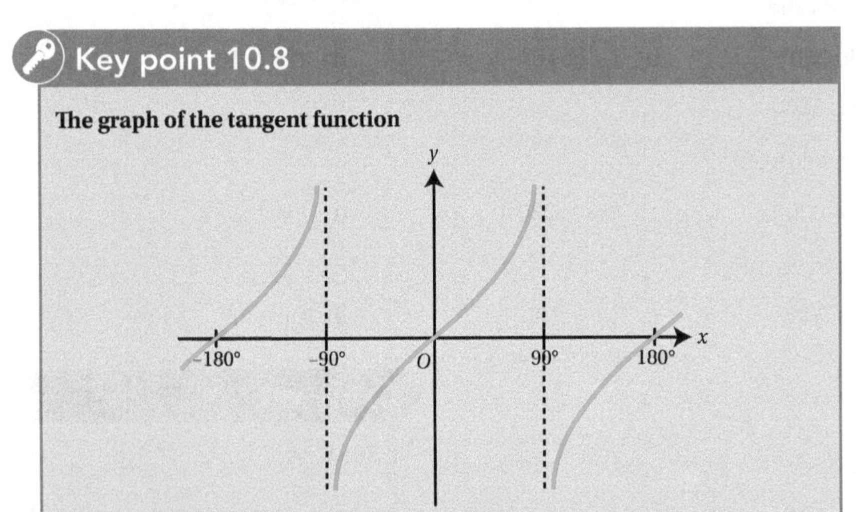

- The tangent function is periodic with period 180°.

$$\tan x = \tan(x + 180°) = \tan(x + 360°) = \ldots$$

- It is undefined for $x = 90°, 270°$ and so on. These lines are vertical **asymptotes**.

Rewind

You met asymptotes in Chapter 5.

EXERCISE 10B

1 Sketch the graph of $y = \tan x$ for:

a $0° \leqslant x \leqslant 360°$ b $-90° \leqslant x \leqslant 270°$

2 Use your calculator to evaluate these ratios, giving your answers correct to 3 s.f.

a i $\tan 32°$ ii $\tan 168°$ b i $\tan(-540°)$ ii $\tan(-128°)$

3 Given that $\tan 20° = 0.364$, use the tan graph to find:

a $\tan 380°$ b $\tan(-160°)$ c $\tan 160°$ d $\tan(-200°)$.

4 Use the properties of sine and cosine to express, in terms of $\tan x$:

a $\tan(90° - x)$ b $\tan(x + 90°)$ c $\tan(-270° - x)$ d $\tan(540° - x)$.

5 Simplify $\tan(x + 360°) - \tan(180° - x)$.

6 Find the two exact values of x in the interval $-180 \leqslant x \leqslant 180$ for which $\tan x = \tan e$.

7 Let $\tan x = t$. Express $\dfrac{\tan(360° - x)}{\tan(270° - x)}$ in terms of t.

Section 3: Trigonometric identities

You have already seen one example of an identity in this chapter: $\frac{\sin x}{\cos x} \equiv \tan x$. There are many other identities involving trigonometric functions. The most important of these is:

Rewind

Recall from Chapter 1 that use of the identity symbol '≡' instead of the equals sign '=' indicates that a mathematical statement is always true.

Key point 10.9

$$\sin^2 x + \cos^2 x \equiv 1$$

Common error

The notation $\sin^2 x$ means $(\sin x)^2$. This is not the same as $\sin(x^2)$.

This result follows immediately from considering the definitions of $\sin x$ and $\cos x$ on the unit circle.

If the point P represents the angle x, then $AP = \sin x$ and $OA = \cos x$.

But triangle OAP is right angled, with hypotenuse 1, so by Pythagoras' theorem $\sin^2 x + \cos^2 x \equiv 1$.

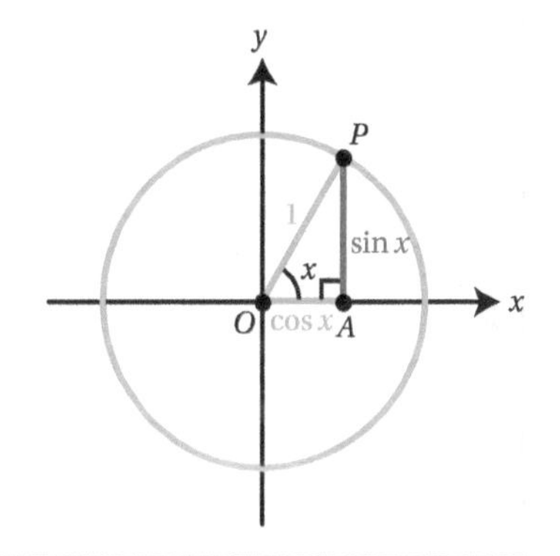

You have already seen examples that used the values of $\sin x$ and $\cos x$, as shown in the diagram, to find the value of $\tan x$. Using the identity $\sin^2 x + \cos^2 x \equiv 1$, you only need to know the value of one of the functions to find the other two.

WORKED EXAMPLE 10.5

Given that $\sin x = \frac{1}{3}$, find the possible values of:

a $\cos x$

b $\tan x$.

a $\sin^2 x + \cos^2 x \equiv 1$

Use the identity $\sin^2 x + \cos^2 x \equiv 1$ to relate cos to sin.

$\left(\frac{1}{3}\right)^2 + \cos^2 x = 1$

$\cos^2 x = 1 - \frac{1}{9}$

$\cos^2 x = \frac{8}{9}$

$\therefore \cos x = \pm\sqrt{\frac{8}{9}}$

Remember to include ± when taking the square root.

$= \pm\frac{2\sqrt{2}}{3}$

b $\tan x \equiv \frac{\sin x}{\cos x}$

Use the identity $\tan x \equiv \frac{\sin x}{\cos x}$ to relate tan to sin and cos...

$= \frac{\frac{1}{3}}{\pm\frac{2\sqrt{2}}{3}}$

...and substitute in the values of sin and cos.

$= \pm\frac{\sqrt{2}}{4}$

Notice that, for a given value of $\sin x$, there are two possible values of $\cos x$ (positive and negative).

You can specify one of the two possible values by restricting x.

WORKED EXAMPLE 10.6

Given that $\tan x = -2$ and $90° < x < 180°$, find the value of $\cos x$.

$\tan x = -2$

$\frac{\sin x}{\cos x} = -2$

$\sin x = -2\cos x$

To introduce cos you need to use $\tan x \equiv \frac{\sin x}{\cos x}$.

$\sin^2 x = 4\cos^2 x$

$1 - \cos^2 x = 4\cos^2 x$

$5\cos^2 x = 1$

$\cos^2 x = \frac{1}{5}$

$\cos x = \pm\frac{1}{\sqrt{5}}$

To remove sin so that you only have cos, use $\sin^2 x + \cos^2 x \equiv 1$. First square the equation.

$90° < x < 180° \Rightarrow \cos x < 0$

$\therefore \cos x = -\frac{1}{\sqrt{5}} = -\frac{\sqrt{5}}{5}$

Since x is between $90°$ and $180°$, $\cos x$ is negative. (You can see this from the graph.)

EXERCISE 10C

1 Find the exact values of $\cos x$ and $\tan x$, given that:

a $\sin x = \frac{1}{3}$ and $0° < x < 90°$ **b** $\sin x = \frac{4}{5}$ and $0° < x < 90°$

2 Find the exact values of $\sin\theta$ and $\tan\theta$ given that:

a $\cos\theta = -\frac{1}{3}$ and $180° < \theta < 270°$ **b** $\cos\theta = -\frac{3}{4}$ and $180° < \theta < 270°$

3 Find the exact value of $\cos x$ if:

a **i** $\sin x = \frac{1}{5}$ and $90° < x < 180°$ **ii** $\sin x = -\frac{1}{2}$ and $270° < x < 360°$

b Find the exact value of $\tan x$ if:

i $\cos x = \frac{3}{5}$ and $-90° < x < 0°$ **ii** $\cos x = -1$ and $90° < x < 270°$

4 **a** Find the possible values of $\cos x$ if $\tan x = \frac{2}{3}$.

b Find the possible values of $\sin x$ if $\tan x = -\frac{1}{2}$.

5 Find the exact value of:

a $3\sin^2 x + 3\cos^2 x$ **b** $\sin^2 5x + \cos^2 5x$ **c** $-2\cos^2 2x - 2\sin^2 2x$ **d** $\frac{3}{2\sin^2 4x} - \frac{3}{2\tan^2 4x}$

6 a Express $3\sin^2 x + 4\cos^2 x$ in terms of $\sin x$ only.

b Express $\cos^2 x - \sin^2 x$ in terms of $\cos x$ only.

7 $\cos\theta = \frac{2}{3}$ and $270° < \theta < 360°$.

Find the exact value of:

a $\sin\theta$ b $\tan\theta$

8 If $\tan\theta = 3$, find, in exact form, the possible values of $\sin\theta$.

9 If $s = \sin x$ and $90° < x < 180°$, express $\cos x$ in terms of s.

10 Express each expression in terms of $\cos x$ only.

a $3 - 2\tan^2 x$ b $\frac{1}{1+\tan^2 x}$

11 Simplify fully

$$\left(\frac{1}{\sin x} - \frac{1}{\tan x}\right)\left(\frac{1}{\sin x} + \frac{1}{\tan x}\right)$$

12 Show that, for all x, $2\tan^2 2x - \frac{2}{\cos^2 2x} = k$, stating the value of the constant k.

13 If $t = \tan x$, express in terms of t:

a $\cos^2 x$ b $\sin^2 x$ c $\cos^2 x - \sin^2 x$ d $\frac{2}{\sin^2 x} + 1$

Section 4: Introducing trigonometric equations

In order to solve equations it is important to be able to 'undo' trigonometric functions. If you are told that the sine of an angle is 0.6, it is not easy to find the actual value. You need to 'undo' the sine function, using its inverse, which is written as $\arcsin x$ or $\sin^{-1} x$.

The inverse cosine function is called $\arccos x$ or $\cos^{-1} x$. The inverse tangent function is called $\arctan x$ or $\tan^{-1} x$.

Rewind

You already know how to use $\sin^{-1}$, $\cos^{-1}$ and $\tan^{-1}$ to calculate angles in a triangle.

Suppose you want to find the x-values that satisfy the equation $\sin x = 0.6$.

Applying the inverse sine to both sides of the equation:

$$\begin{aligned}\sin x &= 0.6\\ x &= \sin^{-1} 0.6\\ &= 36.9° \text{ (3 s.f.)}\end{aligned}$$

Tip

Calculators do not usually have a button labelled arcsin; use the $\sin^{-1}$ button (usually found as SHIFT+sin) instead.

The inverse sine only gives one solution. However, looking at the graph of $y = \sin x$ you can see that there are many points that satisfy this equation.

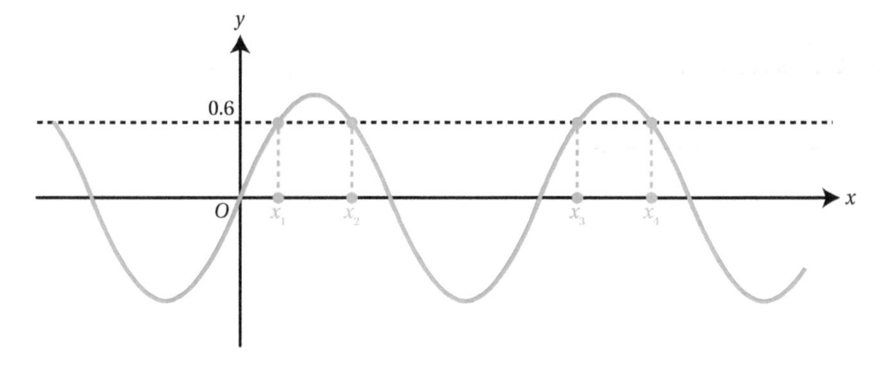

The solutions come in pairs - one in the green section of the graph and one in the blue section. The inverse sine function always gives a solution in the green section that passes through the origin.

Key point 10.10

To find the possible values of x satisfying $\sin x = a$

- Use the calculator to find a solution for $x_1 = \sin^{-1} a$.
- The second solution is given by $x_2 = 180° - x_1$.
- Find other solutions by adding or subtracting 360° to any solution already found.

WORKED EXAMPLE 10.7

Find all possible values of $\theta \in [0, 360°]$ for which $\sin \theta = -0.3$. Give your answers correct to 1 d.p.

$\sin^{-1}(-0.3) = -17.5°$ — Start by taking the inverse sine.

$180° - (-17.5°) = 197.5°$ — The second solution is given by $180° - \theta$.

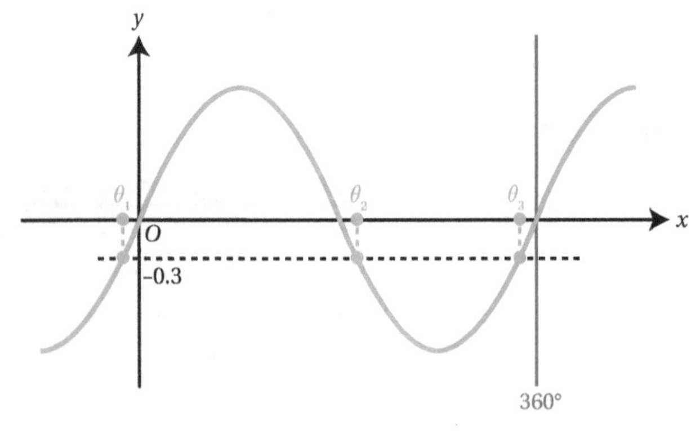

Sketch the graph to see how many solutions there are in the required interval.

There are two solutions.

$-17.5° + 360° = 342.5°$ — The first solution is not in the required interval, so add 360°.

$\theta = 197.5°, 342.5°$ — State the complete list of solutions.

You can apply a similar analysis to the equation $\cos x = k$.

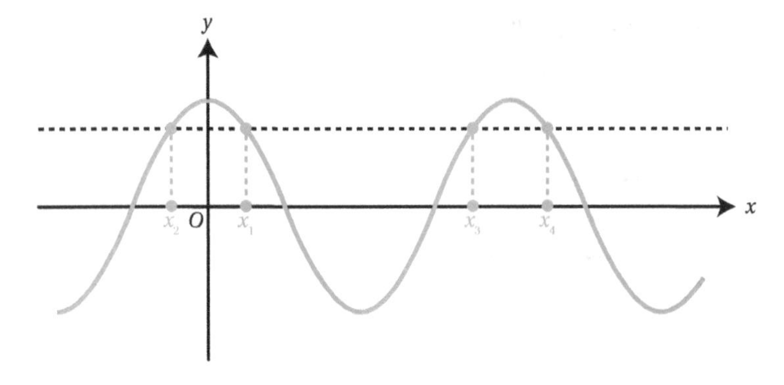

Again, the solutions come in pairs - one in the green section of the graph and one in the blue section. The inverse cosine function always gives a solution in the green section that is between $x = 0°$ and $x = 180°$.

Key point 10.11

To find the possible values of x satisfying $\cos x = a$

- Use the calculator to find a solution for $x_1 = \cos^{-1} a$.
- The second solution is given by $x_2 = -x_1$.
- Find other solutions by adding or subtracting 360° to any solution already found.

Tip

If you are looking for positive solutions, you can find the second solution directly as $x_2 = 360° - x_1$.

It can be difficult to foresee how many times to add or subtract 360°, to make sure that you have found all the solutions in a given interval. Drawing a graph can help: then you can see how many solutions you are looking for and where they are.

WORKED EXAMPLE 10.8

Find the values of $-180° < x < 360°$ for which $\cos x = \frac{\sqrt{2}}{2}$.

$x_1 = \cos^{-1}\left(\frac{\sqrt{2}}{2}\right) = 45°$ — Start by taking the inverse cosine.

$x_2 = -45°$ — The second solution is given by $-x$.

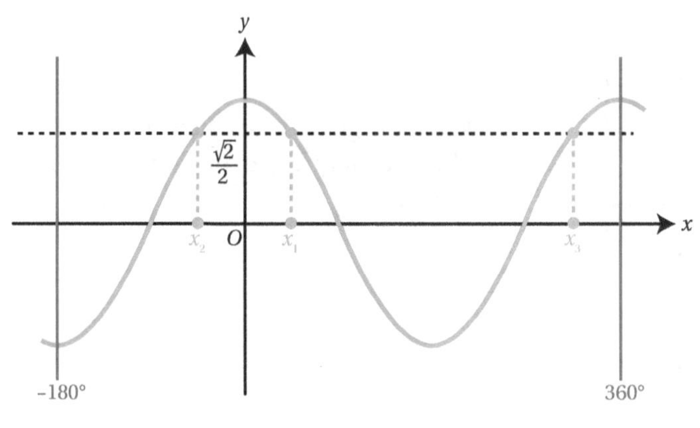

Sketch the graph to see how many solutions there are in the required interval.

There are three solutions.

$x_3 = 360° - 45°$
$= 315°$ — Use the symmetry of the graph to find the other solution.

$\therefore x = -45°, 45°, 315°$ — State the complete list of solutions.

The procedure for solving equations of the type $\tan x = a$ is slightly different, because the tangent function has period 180° rather than 360°.

The inverse tangent function will always give a solution in the green section that passes through the origin.

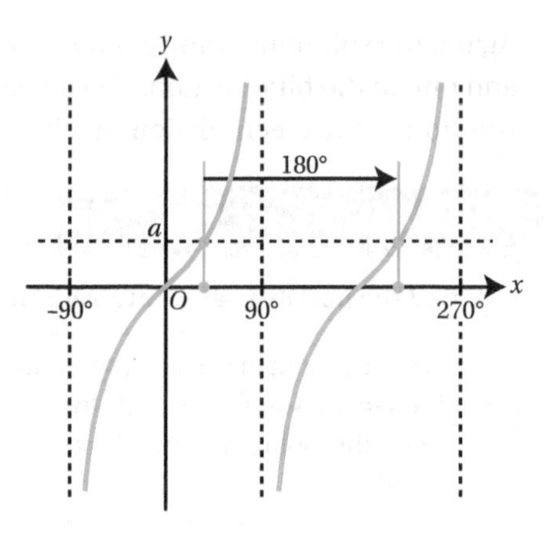

Key point 10.12

To find the possible values of x satisfying $\tan x = a$

- Use the calculator to find a solution for $x_1 = \tan^{-1} a$.
- Find other solutions by adding or subtracting multiples of 180°.

Tip

Always use the ANS button or the stored value rather than the rounded answer when doing subsequent calculations.

WORKED EXAMPLE 10.9

Solve the equation $\tan x = 2.5$ for $-180° < x < 540°$. Give your answers correct to 3 s.f.

$x_1 = \tan^{-1} 2.5 = 68.2°$ — Start by taking the inverse tangent.

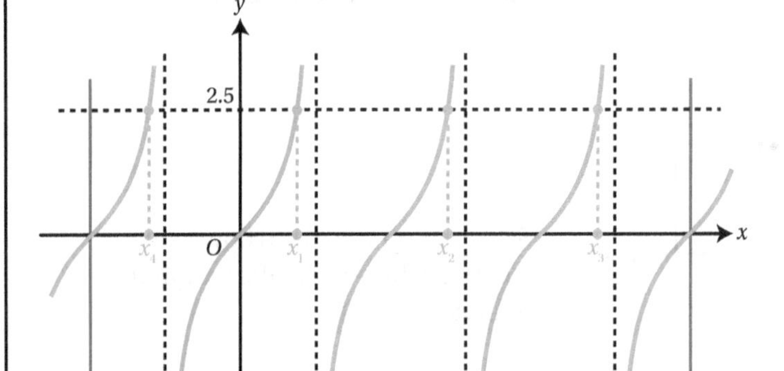

Sketch the graph to see how many solutions there are in the required interval.

There are four solutions.

$x_2 = x_1 + 180° = 248°$

$x_3 = x_2 + 180° = 428°$

$x_4 = x_1 - 180° = -112°$

Find the other solutions by adding or subtracting 180°.

$\therefore x = -112°, 68.2°, 248°, 428°$ — State the complete list of solutions.

EXERCISE 10D

1 Evaluate each expression, giving your answer in degrees correct to 1 d.p.

a **i** $\sin^{-1} 0.7$ **ii** $\sin^{-1} 0.3$ **b** **i** $\cos^{-1}(-0.62)$ **ii** $\cos^{-1}(-0.75)$

c **i** $\tan^{-1} 6.4$ **ii** $\tan^{-1}(-7.1)$

2 Find the values of x between 0° and 360° for which:

a **i** $\sin x = \frac{1}{2}$ **ii** $\sin x = \frac{\sqrt{2}}{2}$ **b** **i** $\cos x = \frac{1}{2}$ **ii** $\cos x = \frac{\sqrt{3}}{2}$

c **i** $\tan x = 1$ **ii** $\tan x = \sqrt{3}$ **d** **i** $\sin x = -\frac{\sqrt{3}}{2}$ **ii** $\sin x = -\frac{1}{2}$

e **i** $\cos x = -\frac{1}{\sqrt{2}}$ **ii** $\cos x = -1$ **f** **i** $\tan x = -\frac{1}{\sqrt{3}}$ **ii** $\tan x = -1$

3 Solve these equations in the given interval. Give your answers correct to 1 d.p.

a **i** $\sin x = 0.45$ for $x \in [0°, 360°]$ **ii** $\sin x = 0.7$ for $x \in [0°, 360°]$

b **i** $\cos x = -0.75$ for $-180° \leqslant x \leqslant 180°$ **ii** $\cos x = -0.2$ for $-180° \leqslant x \leqslant 180°$

c **i** $\tan \theta = \frac{1}{3}$ for $0° \leqslant \theta \leqslant 720°$ **ii** $\tan \theta = \frac{4}{3}$ for $0° \leqslant \theta \leqslant 720°$

d **i** $\sin t = -\frac{2}{3}$ for $t \in [-180°, 360°]$ **ii** $\sin t = -\frac{1}{4}$ for $t \in [-180°, 360°]$

4 Solve these equations.

a **i** $2 \sin \theta + 1 = 1.2$ for $0° < \theta < 360°$ **ii** $4 \sin x + 3 = 2$ for $-90° < x < 270°$

b **i** $3 \cos x - 1 = \frac{1}{3}$ for $0° < x < 360°$ **ii** $5 \cos x + 2 = 4.7$ for $0° < x < 360°$

c **i** $3 \tan t - 1 = 4$ for $-180° < t < 180°$ **ii** $5 \tan t - 3 = 8$ for $0° < t < 360°$

5 Find the values of x between $-180°$ and $180°$ for which $2 \sin x + 1 = 0$.

6 Solve the equation $3 \tan x + 5 = 0$ for $0° \leqslant x \leqslant 360°$.

7 Solve $2 \cos x = \sqrt{3}$ for $x \in [-360°, 360°]$.

8 Find all values of θ in the interval $0° \leqslant \theta \leqslant 360°$ for which $\sin \theta (3 \cos \theta - 2) = 0$.

9 Solve $(\tan x + 1)(5 \sin x - 2) = 0$ for $x \in [-360°, 360°]$.

10 Show by a counter example that $\tan^{-1} x \neq \frac{\sin^{-1} x}{\cos^{-1} x}$.

11 Show by a counter example that $\sin^{-1}(\sin x)$ is not always x.

Section 5: Transformations of trigonometric graphs

Solving the equation $\sin x = 0.6$ for $0° \leqslant x \leqslant 360°$, you can see from the graph that there are two solutions.

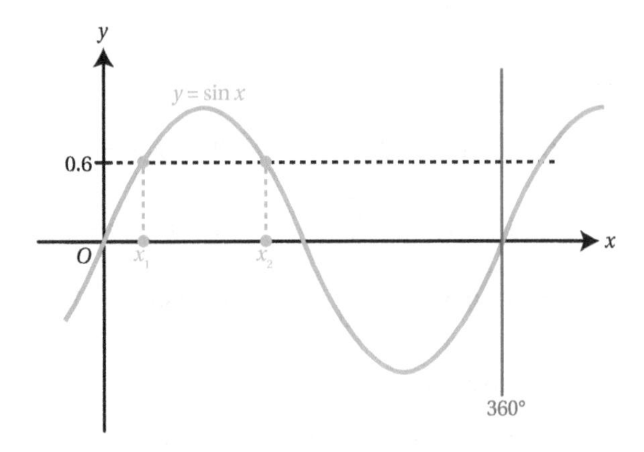

Solving the equation $\sin 2x = 0.6$ for $0° \leqslant x \leqslant 360°$, you can see from the graph that there are four solutions.

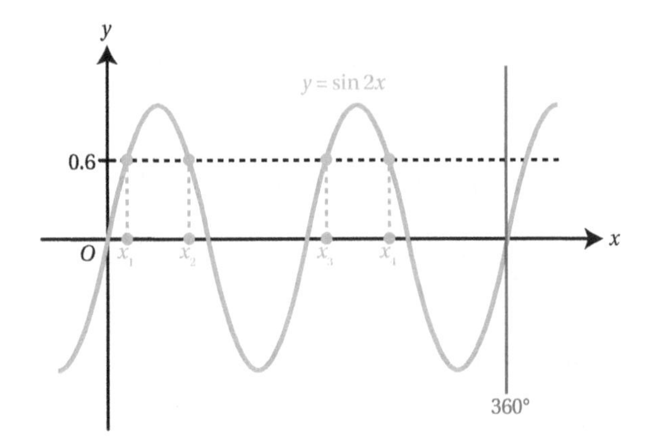

Rewind

Recall from Chapter 5 that replacing x by $2x$ results in the graph being 'squashed' in the horizontal direction.

A method is needed to deal with equations like this. A substitution is a useful way of doing this.

WORKED EXAMPLE 10.10

Solve $3\sin(2x)+1=0$ for $x \in [0°, 360°]$, giving your answers to 3 s.f.

$3\sin(2x)+1=0$ — Rearrange into the form $\sin(2x) = k$.

$\sin(2x) = -\frac{1}{3}$

Let $A = 2x$. — Make a substitution: replace $2x$ by a single letter.

Continues on next page

If $x \in [0°, 360°]$ then $A \in [0°, 720°]$. Rewrite the interval in terms of A.

$\sin A = -\frac{1}{3}$ Solve the equation, as before, for A.

$\sin^{-1}\left(-\frac{1}{3}\right) \approx -19.47°$ Keep an extra decimal place of accuracy at this stage as you will need to use this to calculate the final x-values.

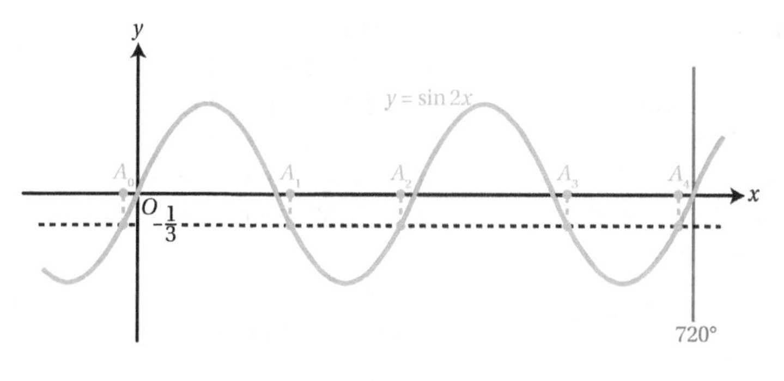

There are four solutions.

$A_1 = 180° - (-19.47°) = 199.47°$
$A_2 = -19.47° + 360° = 340.53°$
$A_3 = 199.47° + 360° = 559.47°$
$A_4 = 340.53° + 360° = 700.53°$

−19.47° is outside the interval, but there are four solutions in the interval.

$x = \frac{A}{2}$
$= 99.7°, 170°, 280°, 350°$ (3 s.f.)

Transform the solutions back into x and only now round to 3 s.f.

Key point 10.13

A four-step summary of this procedure

- Make a substitution (such as $A = 2x$).
- Change the interval for x into the interval for A.
- Solve the equation in the usual way.
- Transform the solutions back into the original variable.

Tip

As highlighted in Worked example 10.10, only round to the required level of accuracy when giving your final answer(s).

WORK IT OUT 10.1

Solve $\tan 3x = 1$ for $-180° < x < 180°$.

Which is the correct solution? Identify the errors made in the incorrect solutions.

Solution A	Solution B	Solution C
Let $A = 3x$ $A_1 = \tan^{-1}(1) = 45°$ $A_2 = 45° - 180 = -135°$ $x = \frac{A}{3}$ $= 15°, -45°$	Let $A = 3x$. $-180° < x < 180°$ $\Rightarrow -540° < A < 540°$ $A_1 = \tan^{-1}(1) = 45°$ $\therefore x_1 = 15°$ $x_2 = 15° + 180° = 195°$ $x_3 = 195° + 180, = 375°$ $x_4 = 15° - 180° = -165°$ $x_5 = -165° - 180° = -345°$ $x_6 = -345° - 180° = -525°$ $x = -525°, -345°, -165°, 15°,$ $195°, 375°$	Let $A = 3x$. $-180° < x < 180°$ $\Rightarrow -540° < A < 540°$ $A_1 = \tan^{-1}(1) = 45°$ $A_2 = 45° + 180° = 225°$ $A_3 = 225° + 180° = 405°$ $A_4 = 45° - 180° = -135°$ $A_5 = -135° - 180° = -315°$ $A_6 = -315° - 180° = -495°$ $x = \frac{A}{3}$ $= -165°, -105°, -45°, 15°,$ $75°, 135°$

Worked example 10.11 shows this method in a more complicated situation.

WORKED EXAMPLE 10.11

Solve the equation $3\tan\left(\frac{1}{2}\theta - 30°\right) = \sqrt{3}$ for $0° \leqslant \theta \leqslant 720°$.

$3\tan\left(\frac{1}{2}\theta - 30°\right) = \sqrt{3}$

$\tan\left(\frac{1}{2}\theta - 30°\right) = \frac{\sqrt{3}}{3}$

Rearrange the equation into the form $\tan\left(\frac{1}{2}\theta - 30°\right) = k$.

Let $A = \frac{1}{2}\theta - 30°$

Make a substitution.

If $0° \leqslant \theta \leqslant 720°$ then $0° \leqslant \frac{1}{2}\theta \leqslant 360°$

Rewrite the interval.

So $-30° \leqslant \frac{1}{2}\theta - 30° \leqslant 330°$

$\tan A = \frac{1}{\sqrt{3}}$

Solve the equation for A.

Continues on next page

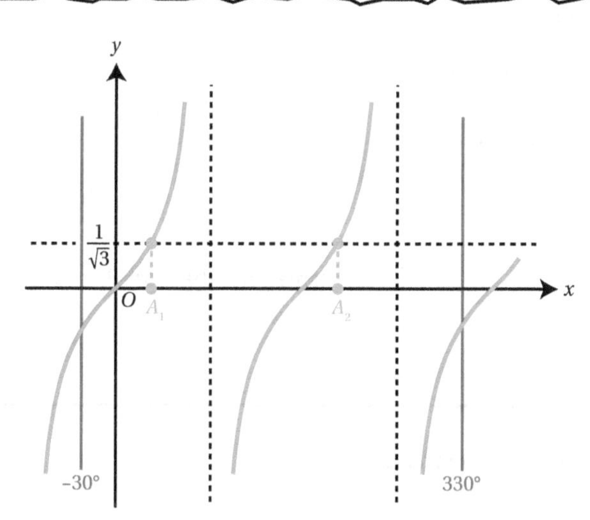

There are two solutions.

$A_1 = \tan^{-1}\left(\frac{1}{\sqrt{3}}\right) = 30°$

$A_2 = 30° + 180° = 210°$

$\theta = 2(A + 30°)$
$= 120°, 480°$

Transform the solutions back into θ.

EXERCISE 10E

1. Solve these equations in the interval $0° \leqslant x \leqslant 360°$, giving your answers correct to 3 s.f.

   a **i** $\sin 2x = 0.7$ **ii** $\sin 3x = -0.2$ **b** **i** $\cos 3x = -0.4$ **ii** $\cos 4x = 1$

   c **i** $\tan 4x = 1.5$ **ii** $\tan 2x = -2$

2. Solve these equations in the interval $-180° \leqslant \theta \leqslant 180°$, giving your answers correct to 3 s.f.

   a **i** $\sin(\theta + 40°) = 0.25$ **ii** $\sin(\theta - 25°) = -0.75$

   b **i** $\cos(\theta - 50°) = -0.9$ **ii** $\cos(\theta + 10°) = 0.3$

   c **i** $\tan(\theta - 45°) = 2$ **ii** $\tan(\theta + 60°) = -3$

3. Find the exact solutions of the equation $\tan 3x = \sqrt{3}$ for $0° < x < 180°$.

4. Solve $2\cos(2x) + 1 = 0$ for $x \in [-180, 180]$.

5. Find the values of θ in the interval $-360° < \theta < 360°$ for which $3\sin\left(\frac{\theta}{2}\right) = -2$.

6. Solve $2\tan\left(\frac{\theta}{3}\right) = 5$ for $0° < \theta < 540°$.

7. Find the values of x in the interval $-360° < x < 360°$ for which $2\sin(2x + 30°) + \sqrt{3} = 0$.

8. Solve $2\cos(3x - 50°) - \sqrt{2} = 0$ for $-90° < x < 90°$.

9. Solve the equation $\frac{\cos 2x + 0.5}{1 - \cos 2x} = 2$ for $x \in (-180°, 180°)$.

10. Find the values of x in the interval $-\sqrt{180}° < x < \sqrt{180}°$ for which $\sin(x^2) = \frac{1}{2}$.

Section 6: Harder trigonometric equations

You have already seen how to solve equations in the form 'trigonometric function = constant'. However, often it is not possible to manipulate the equation into this form, so you need an alternative strategy. Three tactics are often used:

- look for quadratic equations
- take everything over to one side and factorise
- use trigonometric identities.

Tip

Remember that $\cos^2\theta$ means $(\cos\theta)^2$.

WORKED EXAMPLE 10.12

Solve the equation $\cos^2\theta = \frac{4}{9}$ for $\theta \in [0°, 360°]$. Give answers correct to 1 d.p.

$\cos^2\theta = \frac{4}{9}$

$\cos\theta = \pm\frac{2}{3}$

First find possible values of $\cos\theta$.

Remember to include ± when taking the square root.

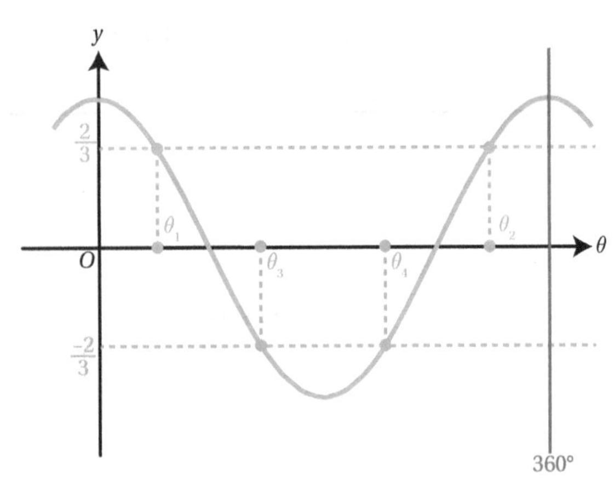

Sketch the graph to see how many solutions there are in the required interval.

There are two solutions in each case.

$\cos^{-1}\left(\frac{2}{3}\right) = 48.2°$

Solve each equation separately.

$\theta_1 = 48.2°$

$\theta_2 = 360° - 48.2° = 311.8°$

$\cos^{-1}\left(-\frac{2}{3}\right) = 131.8°$

$\theta_3 = 131.8°$

$\theta_4 = 360° - 131.8° = 228.2°$

$\therefore \theta = 48.2°, 131.8°, 228.2°, 311.8°$ (1 d.p.)

List all the solutions.

WORKED EXAMPLE 10.13

Solve the equation $3\sin^2 x - 5\sin x + 1 = 0$ for $0° < x < 360°$.

$\sin x = \frac{5 \pm \sqrt{5^2 - 4 \times 3 \times 1}}{2 \times 3}$

$= 1.434$ or 0.2324

This is a quadratic equation in $\sin x$. Since you cannot factorise it, use the quadratic formula.

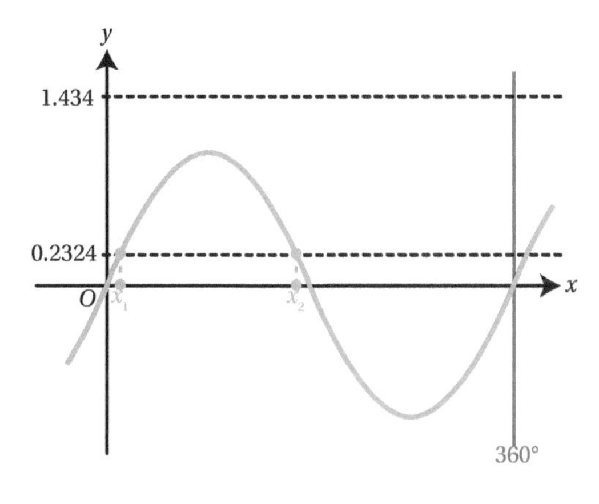

Sketch the graph to see how many solutions there are.

There are two solutions.

$\sin x = 1.434 > 1$ is impossible.

$\sin x$ is always between -1 and 1, so only one of the values is possible.

$\sin x = 0.2324$

$\sin^{-1}(0.2324) = 13.4°$

You can now solve the equation.

$x_1 = 13.4°$

$x_2 = 180° - 13.4° = 167°$

$\therefore x = 13.4°, 167°$ (3 s.f.)

WORKED EXAMPLE 10.14

Solve the equation $3\sin x \cos x = 2\sin x$ for $-180° \leqslant x \leqslant 180°$. Give your answers to 3 s.f.

$3\sin x \cos x = 2\sin x$

$3\sin x \cos x - 2\sin x = 0$

$\sin x(3\cos x - 2) = 0$

This equation contains both sin and cos. However, both sides have a factor of $\sin x$, so you can make one side of the equation zero and factorise.

$\sin x = 0$ or $\cos x = \frac{2}{3}$

You then have two separate equations, each containing only one trigonometric function.

Continues on next page

When $\sin x = 0$:
$\sin^{-1} 0 = 0$

Solve each equation separately. Remember to sketch the graph for each equation to see how many solutions there are.

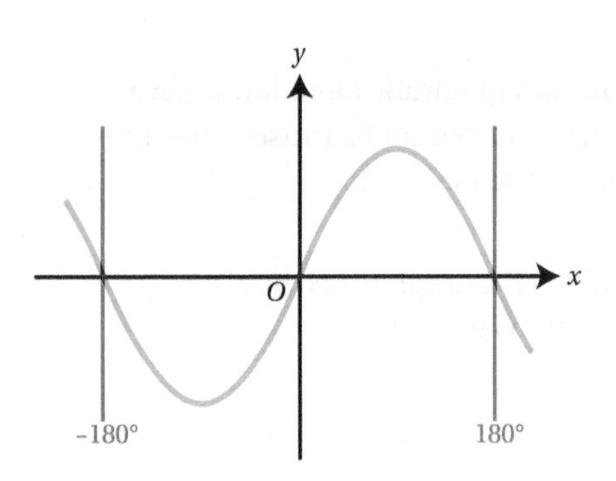

$x_1 = 0°$

$x_2 = 180° - 0° = 180°$

$x_3 = 180° - 360° = -180°$

When $\cos x = \frac{2}{3}$: $\cos^{-1}\left(\frac{2}{3}\right) = 48.2°$

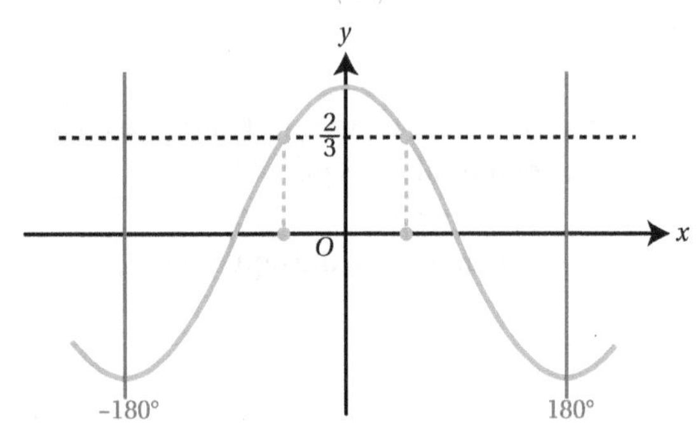

$x_1 = 48.2°$

$x_2 = -48.2°$

$\therefore x = -180°, -48.2°, 0°, 48.2°, 180°$ (3 s.f.)

List all the solutions.

Common error

Do not be tempted to divide both sides of the original equation in Worked example 10.14 by $\sin x$ or you could lose some solutions; $\sin x$ could equal zero.

EXERCISE 10F

1. Solve these equations in the interval $0° \leqslant x \leqslant 360°$, giving your answers correct to 3 s.f.

   a **i** $3\sin^2 x = 2$ **ii** $3\tan^2 x = 5$

   b **i** $\tan^2 x - \tan x - 6 = 0$ **ii** $3\cos^2 x + \cos x - 2 = 0$

   c **i** $4\cos^2 x - 11\cos x + 6 = 0$ **ii** $5\sin^2 x + 6\sin x - 8 = 0$

   d **i** $3\sin^2 x + \sin x = 0$ **ii** $4\tan^2 x + 5\tan x = 0$

2. Solve these equations in the interval $-180° \leqslant \theta \leqslant 180°$, giving your answers correct to 3 s.f.

   a **i** $2\sin\theta - 5\sin\theta\cos\theta = 0$ **ii** $4\cos\theta + 5\sin\theta\cos\theta = 0$

   b **i** $4\sin\theta\cos\theta = \cos\theta$ **ii** $3\sin\theta = 5\sin\theta\cos\theta$

3. Find the values of $x \in (-360°, 360°)$ for which $2\sin x\cos x = \cos x$.

4. **a** Given that $2\sin^2 x - 3\sin x = 2$, find the exact value of $\sin x$.

   b Hence solve the equation $2\sin^2 x - 3\sin x = 2$ for $0 < x < 360°$.

5. Solve the equation $\tan^2 x = \tan x + 12$, $-180° \leqslant x \leqslant 180°$. Give your answers correct to 1 d.p.

6. Find the values of θ in the interval $0° < \theta < 360°$ for which $2\sin 2\theta = \sqrt{3}\sin 2\theta\cos 2\theta$.

7. Solve the equation $3\cos^2 3x + 7\cos 3x + 2 = 0$ for $x \in [0°, 180°]$.

Using identities to solve equations

When there is more than one trigonometric function in an equation, it is often helpful to use an identity to eliminate one of the functions.

Elevate

See Support Sheet 10 for a further example of this type and for more practice questions.

WORKED EXAMPLE 10.15

Find all values of θ in the interval $[-180°, 180°]$ that satisfy the equation $2\sin^2\theta + 3\cos\theta = 1$. Give your answers to three significant figures.

$2\sin^2\theta + 3\cos\theta = 1$

$2(1 - \cos^2\theta) + 3\cos\theta = 1$

The equation contains both sin and cos but you can use $\sin^2\theta + \cos^2\theta \equiv 1$ to replace $\sin^2\theta$ by $1 - \cos^2\theta$.

$2 - 2\cos^2\theta + 3\cos\theta = 1$

$2\cos^2\theta - 3\cos\theta - 1 = 0$

This is a quadratic equation in $\cos\theta$, so write it in the standard form.

$\cos\theta = 1.78$ or -0.281

Solve it, using your calculator or the quadratic formula.

$\cos\theta = 1.78$ is impossible.

cos is always between −1 and 1.

$\cos\theta = -0.281$

Solve the equation as normal.

Continues on next page

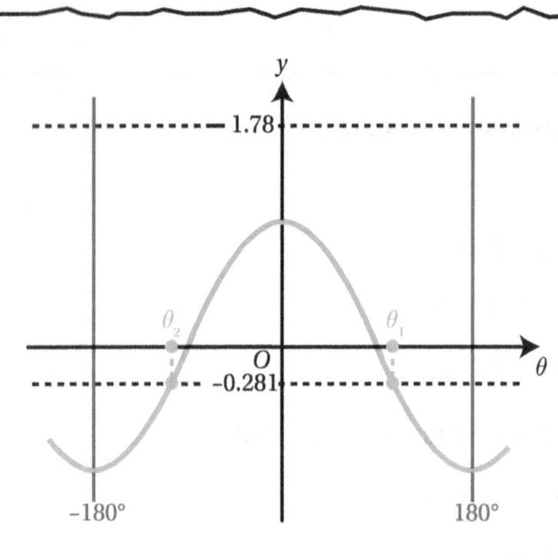

There are two solutions.

$\theta_1 = \cos^{-1}(-0.281) = 106°$

$\theta_2 = -106°$

$\therefore \theta = -106°, 106°$ (3 s.f.)

WORKED EXAMPLE 10.16

Solve the equation $\sin\theta + \sqrt{3}\cos\theta = 0$ for $-360° < \theta < 360°$.

$\sin\theta + \sqrt{3}\cos\theta = 0$

$\sin\theta = -\sqrt{3}\cos\theta$

$\frac{\sin\theta}{\cos\theta} = -\sqrt{3}$

$\tan\theta = -\sqrt{3}$

The equation involves both sin and cos. You can't use $\sin^2 x + \cos^2 x \equiv 1$ to eliminate one of them as neither is squared.

Instead rearrange and use $\frac{\sin\theta}{\cos\theta} \equiv \tan\theta$.

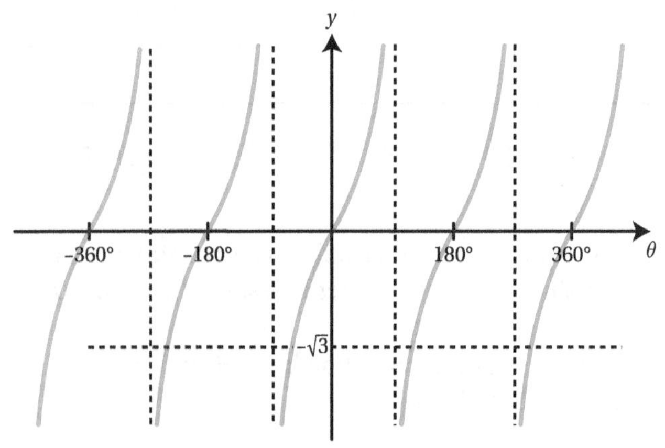

$\theta_1 = \tan^{-1}(-\sqrt{3}) = -60°$

$\theta_2 = -60° + 180° = 120°$

$\theta_3 = 120° + 180° = 300°$

$\theta_4 = -60° - 180° = -240°$

$\therefore \theta = -240°, -60°, 120°, 300°$

Solve this equation as normal.

WORKED EXAMPLE 10.17

Solve the equation $4 \sin 2x = \tan 2x$ in the interval $0° \leqslant x \leqslant 180°$.

$4 \sin 2x = \tan 2x$

$4 \sin 2x = \frac{\sin 2x}{\cos 2x}$

The only identity you can use here is the one for tan.

$4 \sin 2x \cos 2x = \sin 2x$

Multiply both sides by $\cos x$.

$4 \sin 2x \cos 2x - \sin 2x = 0$

$\sin 2x(4 \cos 2x - 1) = 0$

Although you have both sin and cos, just as in Worked example 10.15 both sides contain $\sin x$, so you can make the equation equal to zero and factorise.

$\sin 2x = 0$ or $\cos 2x = \frac{1}{4}$

You now have two equations, each of which only has one trigonometric function.

$0° \leqslant x \leqslant 180° \Rightarrow 0° \leqslant 2x \leqslant 360°$

Now solve each equation separately. Remember to change the interval when finding the value of $2x$.

When $\sin 2x = 0$:

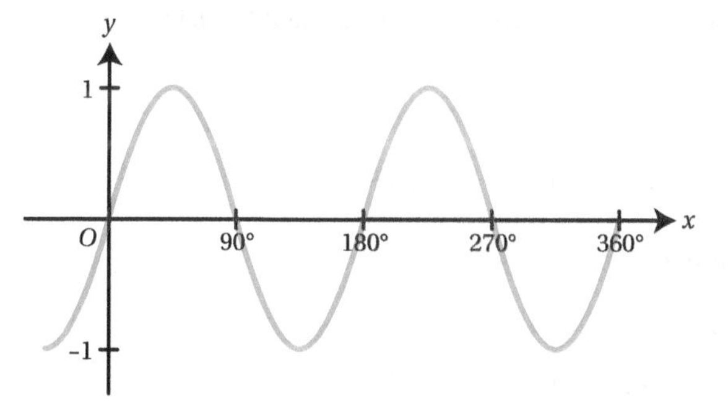

$2x = 0°, 180°, 360°$
$x = 0°, 90°, 180°$

When $\cos 2x = \frac{1}{4}$:

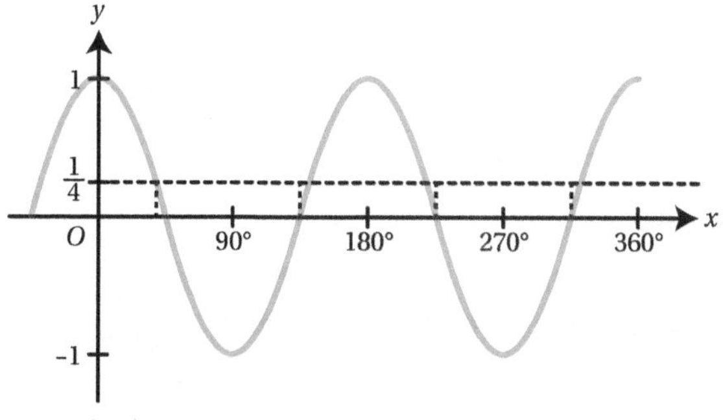

$\cos^{-1}\left(\frac{1}{4}\right) = 75.5°$

$2x = 75.5°$ or $360° - 75.5° = 284°$
$x = 38°, 142°$

$\therefore x = 0°, 38°, 90°, 142°, 180°$

List all the solutions.

EXERCISE 10G

1. By using the identity $\tan x \equiv \frac{\sin x}{\cos x}$, solve these equations for $0° \leqslant x \leqslant 180°$.

   a i $3 \sin x = 2 \cos x$ ii $3 \sin x = 5 \cos x$

   b i $7 \cos x - 3 \sin x = 0$ ii $\sin x - 5 \cos x = 0$

2. Use the identity $\sin^2 x + \cos^2 x \equiv 1$ to solve these equations for x in the interval $[0°, 360°]$.

   a i $7 \sin^2 x + 3 \cos^2 x = 5$ ii $\sin^2 x + 4 \cos^2 x = 2$

   b i $3 \sin^2 x - \cos^2 x = 1$ ii $\cos^2 x - \sin^2 x = 1$

3. Find the values of x in the interval $0° < x < 360°$ for which $\sin 2x + \cos 2x = 0$.

4. Solve the equation $\frac{\cos \theta}{\sin \theta} - 2 = 0$ for $\theta \in [-180°, 180°]$.

5. Solve the equation $\sin x + \frac{\sin^2 x}{\cos x} = 0$ for $0° \leqslant x \leqslant 360°$.

6. Solve the equation $\sin x \tan x = \sin^2 x$ for $-180° \leqslant x \leqslant 180°$.

7. Solve the equation $5 \sin^2 \theta = 4 \cos^2 \theta$ for $-180° \leqslant \theta \leqslant 180°$. Give your answers to the nearest 0.1°.

8. Solve the equation $2 \cos^2 t - \sin t - 1 = 0$ for $0° \leqslant t \leqslant 360°$.

9. Find the values of x in the interval $-180° < x < 180°$ that satisfy $4 \cos^2 x - 5 \sin x - 5 = 0$.

10. a Given that $\cos^2 t + 5 \cos t = 2 \sin^2 t$ find the exact value of $\cos t$.

    b Hence solve the equation $\cos^2 t + 5 \cos t = 2 \sin^2 t$ for $0° \leqslant t \leqslant 360°$.

11. a Given that $6 \sin^2 x + \cos x = 4$, find the exact values of $\cos x$ for $x \in [0°, 360°]$.

    b Hence solve the equation $6 \sin^2 x + \cos x = 4$ for $0° \leqslant x \leqslant 360°$, giving your answers correct to 3 s.f.

12. a Show that the equation $2 \sin^2 x - 3 \sin x \cos x + \cos^2 x = 0$ can be written in the form $2 \tan^2 x - 3 \tan x + 1 = 0$.

    b Hence solve the equation $2 \sin^2 x - 3 \sin x \cos x + \cos^2 x = 0$, giving all solutions in the interval $-180° < x < 180°$.

Checklist of learning and understanding

- The sine and cosine functions are periodic with period 360°.

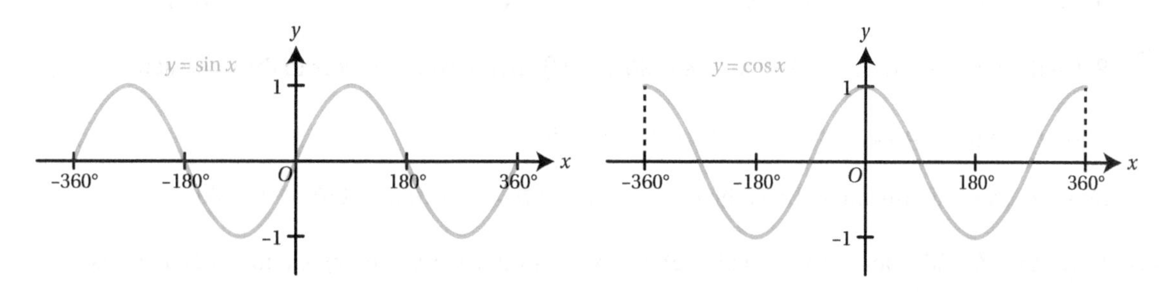

- The sine and cosine functions are related by:
  - $\sin x = -\cos(x+90°) = \cos(90° - x)$
  - $\cos x = \sin(x+90°) = \sin(90° - x)$
- The tangent function is defined by the identity:

$$\tan x \equiv \frac{\sin x}{\cos x}$$

  It is periodic with period 180°.

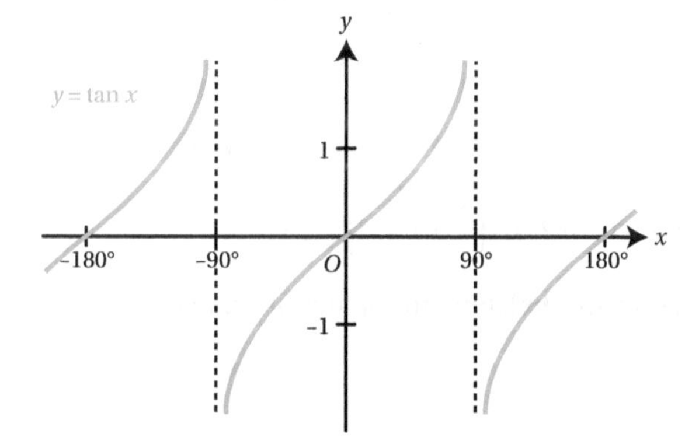

- To solve trigonometric equations:
  - First rearrange into the form $\sin x = k$, $\cos x = k$ or $\tan x = k$
  - Always draw a graph to see how many solutions there are.
  - Find solutions in the interval [0°, 360°]:
    - $\sin x = k$: $x_1 = \sin^{-1} k$, $x_2 = 180° - x_1$
    - $\cos x = k$: $x_1 = \cos^{-1} k$, $x_2 = -x_1$
    - $\tan x = k$: $x_1 = \tan^{-1} k$, $x_2 = x_1 + 180°$
  - Other solutions are found by adding or subtracting multiples of 360°.
- If the angle in the function in the trigonometric equation has been transformed:
  - make a substitution (such as $A = 2x$)
  - change the interval for x into the interval for A
  - solve the equation in the usual way
  - transform the solutions back into the original variable.
- These identities can be used to solve more complicated equations:
  - $\sin^2 x + \cos^2 x \equiv 1$
  - $\tan x \equiv \dfrac{\sin x}{\cos x}$

Mixed practice 10

1. If $\cos(x+180°)=a$, what is the value of $\cos x$? Choose from these options.

   A a **B** $-a$ **C** 0 **D** 1

2. Solve the equation $\tan x=-0.62$ for $x\in(-90°, 270°)$, giving your answers to the nearest 0.1°.

3. Solve the equation $\sqrt{2}\sin\theta+1=0$ for $-360°<\theta<360°$.

4. Solve, to three significant figures, the equation $7\sin^2\theta=9\cos^2\theta$ for $-180°\leqslant\theta\leqslant 180°$.

5. Prove that, for all values of x, the value of the expression $(3\sin x+\cos x)^2+(\sin x-3\cos x)^2$ is an integer and state its value.

   [© AQA 2011]

6. **a** Show that the equation $\cos\theta-2\sin^2\theta+2=0$ can be expressed in the form $2\cos^2\theta+\cos\theta=0$.

   b Hence find all values of $\theta\in[0°, 360°]$ for which $\cos\theta-2\sin^2\theta+2=0$.

7. How many solutions are there to the equation

   $\sin^2 2x=\frac{1}{4}\quad -180°<x<180°$?

   Choose from these options.

   A 2 **B** 4 **C** 6 **D** 8

8. Find the values of x in the interval $0°<x<720°$ for which $2\cos\left(\frac{1}{2}x+45°\right)=\sqrt{3}$.

9. The diagram shows the graph of the function $f(x)=a\sin(bx)$. Find the values of a and b.

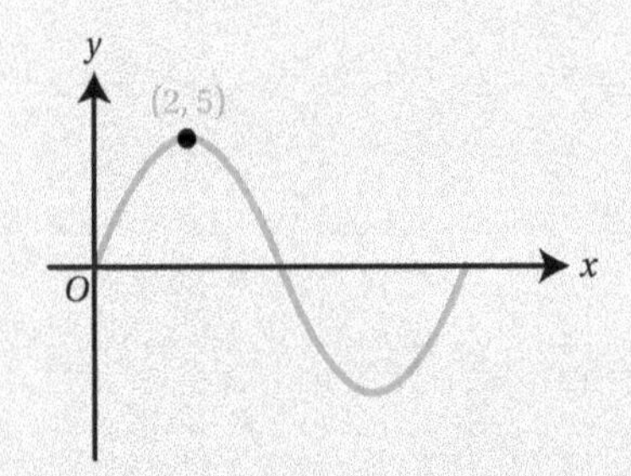

10. Solve the equation $12\sin^2 x+\cos x=6$ for $0°\leqslant x\leqslant 360°$. Give your answers to 3 s.f.

11. **a** Given that $\frac{\cos^2 x+4\sin^2 x}{1-\sin^2 x}=7$, show that $\tan^2 x=\frac{3}{2}$.

    b Hence solve the equation $\frac{\cos^2 2\theta+4\sin^2 2\theta}{1-\sin^2 2\theta}=7$ in the interval $0°<\theta<180°$, giving your values of θ to the nearest degree.

    [© AQA 2014]

12. What is $\tan x+\frac{1}{\tan x}$ equivalent to?

    Choose from these options.

    A $\frac{1}{\sin x\cos x}$ **B** $\tan^2 x$ **C** $\frac{1}{\tan^2 x}$ **D** $\sin x\tan x$

13 Find all values of x in the interval $-90° < x < 90°$ which satisfy $6\cos^2 2x = \sin 2x + 4$.

14 **a** Find the values of k for which the equation $4x^2 - kx + 1 = 0$ has a repeated root.

b Show that the equation $4\sin^2\theta = 5 - k\cos\theta$ can be written as $4\cos^2\theta - k\cos\theta + 1 = 0$.

c Let $f_k(\theta) = 4\cos^2\theta - k\cos\theta + 1$.

i State the number of values of $\cos\theta$ that satisfy the equation $f_4(\theta) = 0$.

ii Find all the values of $\theta \in [-360°, 360°]$ that satisfy the equation $f_4(\theta) = 0$.

iii Find the value of k for which $x = 1$ is a solution of the equation $4x^2 - kx + 1 = 0$.

iv For this value of k, find the number of solutions of the equation $f_k(\theta) = 0$, where $\theta \in [-360°, 360°]$.

11 Triangle geometry

In this chapter you will learn how to:

- use the sine rule to find sides and angles of any triangle
- use the cosine rule to find sides and angles of any triangle
- use a formula for the area of a triangle when you don't know the perpendicular height.

Before you start...

GCSE	You should be able to use trigonometry in right-angled triangles.	1 Find the size of the angle marked x in the diagram. (Right-angled triangle with vertical side 2, hypotenuse 3, angle x at the bottom right.)
GCSE	You should be able to use three-figure bearings.	2 Point A is a on a bearing of $290°$ from B. Find the bearing of B from A.
Chapter 3	You should be able to solve quadratic equations by factorising or using the formula.	3 Solve these equations. a $x^2+x-12=0$ b $x^2-4x-1=0$
Chapter 10	You should be able to solve trigonometric equations.	4 Solve the equation $\sin x = 0.15$ for $0° < x < 180°$.

One of the earliest ideas in mathematics

The first steps in developing trigonometry were made by Babylonian astronomers as early as the second millennium BCE. It is thought that Egyptians used some trigonometric calculations when building the pyramids. Trigonometry was further developed by Greek, Islamic and Indian mathematicians to solve problems in land measurement, building and astronomy.

In this chapter you will use what you already know about trigonometric functions, and revisit results you may know from GCSE that enable you to calculate lengths and angles in triangles.

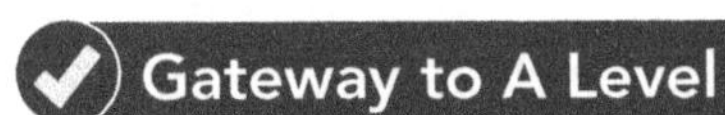

See Gateway to A Level Section O for revision of trigonometry in right-angled triangles.

Section 1: The sine rule

You can use trigonometry to calculate the sizes of sides and angles in triangles without a right angle.

Key point 11.1

The sine rule

$$\frac{a}{\sin A} = \frac{b}{\sin B} = \frac{c}{\sin C}$$

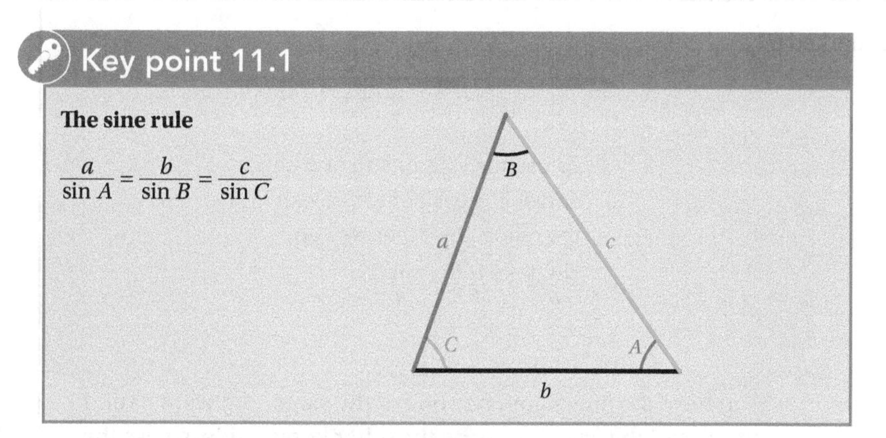

To use the sine rule you need to know the size of an angle and the length of the side opposite it.

Focus on...

See Focus on ... Proof 2 to find out how to prove the sine rule for any triangle.

Tip

Although the sine rule is quoted for all three sides (and their corresponding angles), you only ever need to use any two of these ratios at a time.

Tip

The general convention for labelling triangles is that vertices are given capital letters and their corresponding opposite sides lower case letters. So, for example, side a is opposite vertex A.

WORKED EXAMPLE 11.1

Find the length of side AC.

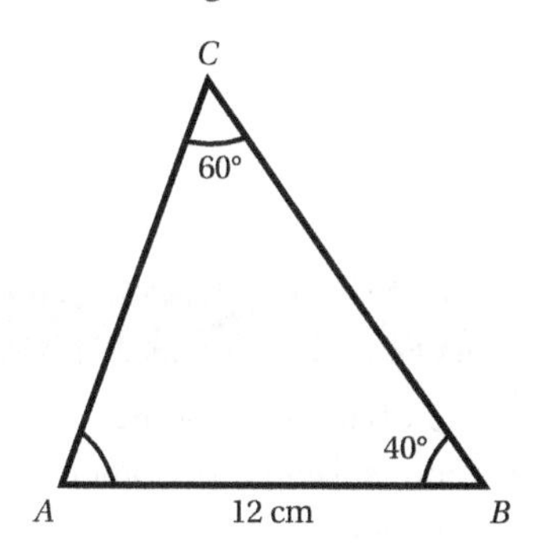

$$\frac{12}{\sin 60°} = \frac{AC}{\sin 40°}$$

You are given the angles opposite sides AB and AC, so use the sine rule with those two sides.

$$AC = \frac{12 \sin 40°}{\sin 60°}$$
$$= 8.91 \text{ cm (3 s.f.)}$$

Rearrange and solve. Although you do not need to include the units in the working, remember to include them in your answer.

You can also use the sine rule to find angles.

WORKED EXAMPLE 11.2

Find, to 1 d.p., the size of the angle marked θ.

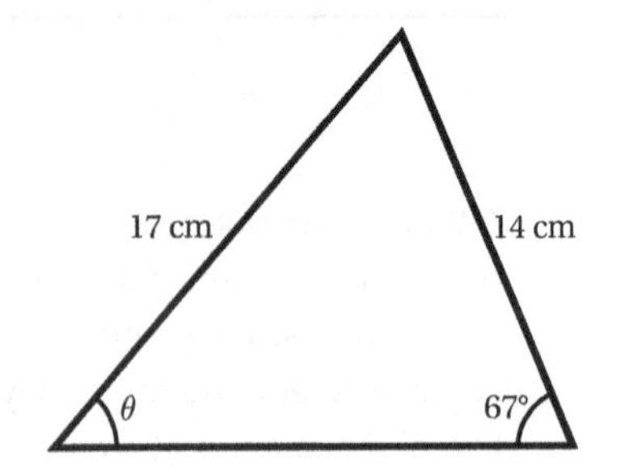

Tip

To use the sine rule to find an angle, it is easier to start with the ratios 'the other way up':

$$\frac{\sin A}{a} = \frac{\sin B}{b} = \frac{\sin C}{c}$$

$$\frac{\sin \theta}{14} = \frac{\sin 67°}{17}$$

You have the angle opposite one of the sides, and want to find the size of the angle opposite the other known side, so use the sine rule with the ratios up the other way: the sines on top.

$$\sin \theta = \frac{14 \sin 67°}{17}$$
$$= 0.758$$
$$\therefore \theta = \sin^{-1} 0.758$$
$$= 49.3°$$

Rearrange and solve.

Notice that you can also find the third angle even though you do not know the length of the side opposite; having used the sine rule to find θ, you can deduce that the final angle must equal $180° - 67° - \theta = 63.7°$.

The ambiguous case of the sine rule

You should remember from your work on trigonometric equations that there is more than one value of θ such that $\sin \theta = 0.758$. Another solution of the equation $\sin \theta = 0.758$ is $180° - 49.3° = 130.7°$.

Rewind

See Chapter 10 for more about solving trigonometric equations.

However, this does not mean that there is another triangle with the given information and an angle of $130.7°$: since one of the other angles is $67°$, this is impossible as it would make the angle sum more than $180°$. All other possible values of θ are outside the interval $(0°, 180°)$, so cannot be angles of a triangle. In this example, there is only one possible value of angle θ.

Worked example 11.3 shows that this is not always the case.

WORKED EXAMPLE 11.3

Find the size of the angle marked θ, giving your answer correct to the nearest degree.

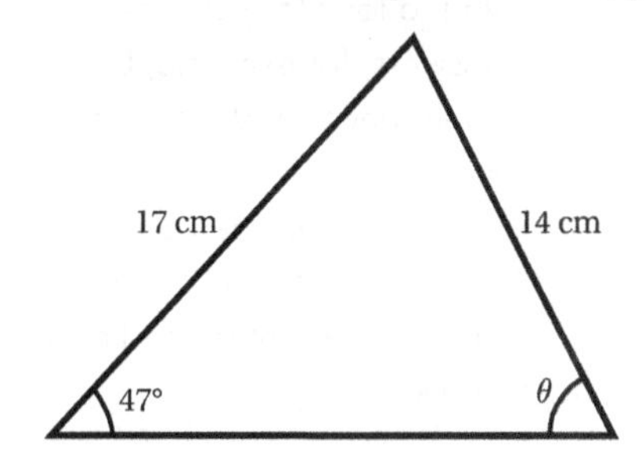

$$\frac{\sin\theta}{17} = \frac{\sin 47^\circ}{14}$$

$$\sin\theta = \frac{17\sin 47^\circ}{14} = 0.888$$

Use the sine rule with the two given sides.

$$\sin^{-1} 0.888 = 62.6^\circ$$

$$\therefore \theta = 62.6^\circ \text{ or } 180^\circ - 62.6^\circ = 117.4^\circ$$

Find the two possible values of θ.

$$62.6^\circ + 47^\circ = 109.6^\circ < 180^\circ$$

$$117.4^\circ + 47^\circ = 164.4^\circ < 180^\circ$$

Check whether both solutions are possible: do the two angles add up to less than 180°?

$$\therefore \theta = 63^\circ \text{ or } 117^\circ$$

Both solutions are possible.

The diagram shows the two possible triangles. In both triangles, the 14 cm side is opposite the angle of 47°, with another side having length 17 cm. As illustrated, if the two triangles are placed adjacent to each other, together they form an isosceles triangle with base angle 47° and matched sides of length 17 cm.

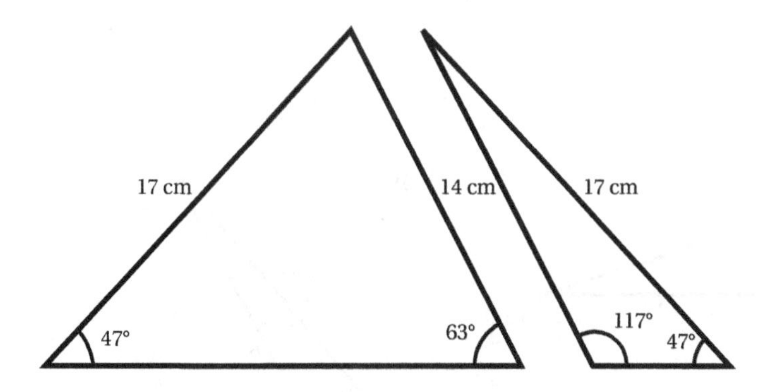

Key point 11.2

When using the sine rule to find an angle, there may be two possible solutions: θ and $180° - \theta$.

Tip

The question will often alert you to look for two possible answers, for example, by specifying that θ is obtuse. However, if it doesn't, you should check whether the second solution is possible by finding the sum of the known angles.

WORK IT OUT 11.1

In triangle ABC, $AB = 10$ cm, $AC = 12$ cm and angle $ABC = 70°$. Find the size of angle ACB.

Which is the correct solution? Identify the errors made in the incorrect solutions.

Solution 1	Solution 2	Solution 3
$\frac{\sin 70°}{12} = \frac{\sin x}{10}$	$\frac{\sin x}{12} = \frac{\sin 70°}{10}$	$\frac{\sin 70°}{12} = \frac{\sin x}{10}$
$\sin x = \frac{10 \sin 70°}{12}$	$\sin x = \frac{12 \sin 70°}{10}$	$\sin x = \frac{10 \sin 70°}{12}$
$= 0.783$	$= 1.13$	$= 0.783$
$x = \sin^{-1} 0.783$	So there are no solutions.	$\sin^{-1} 0.783 = 51.5°$
$= 51.5°$		$\therefore \theta = 51.5°$ or $128°$

EXERCISE 11A

1 Find the lengths of the sides marked with letters.

a **i**

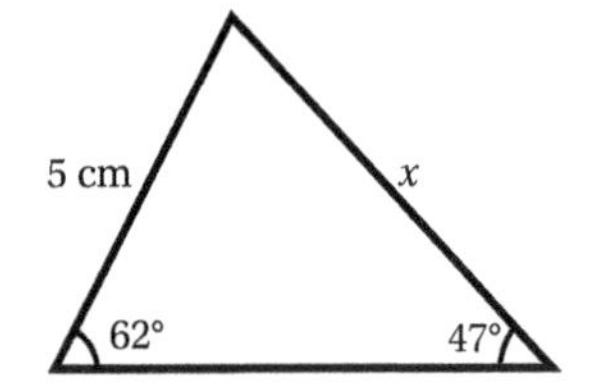

ii

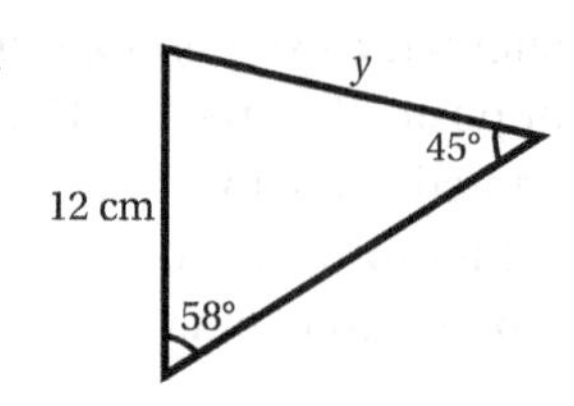

b **i**

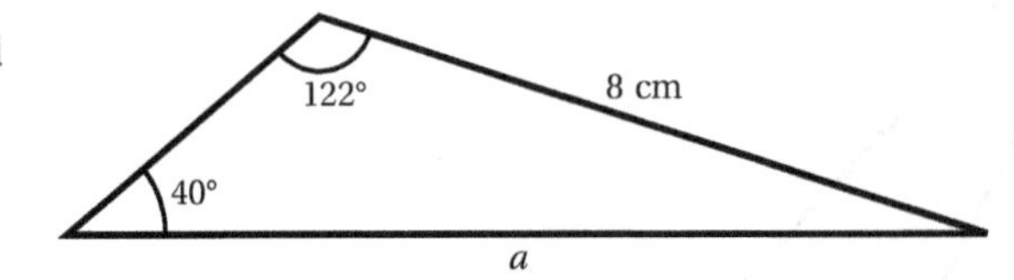

ii

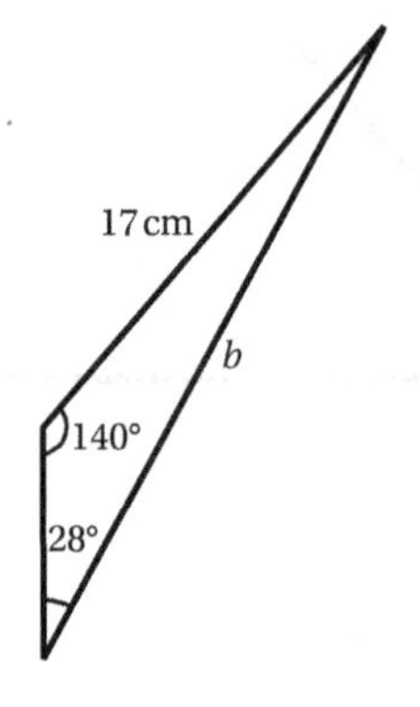

2 Find the sizes of the angles marked with letters, checking whether there is more than one solution in each case.

a **i**

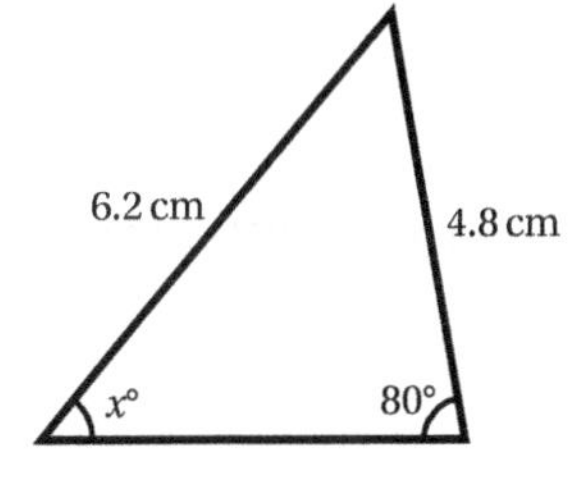

ii

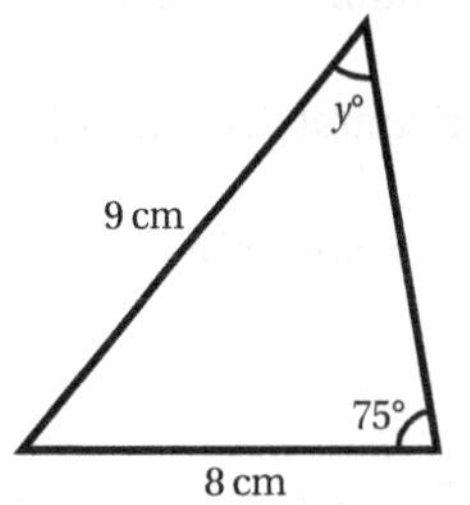

b **i**

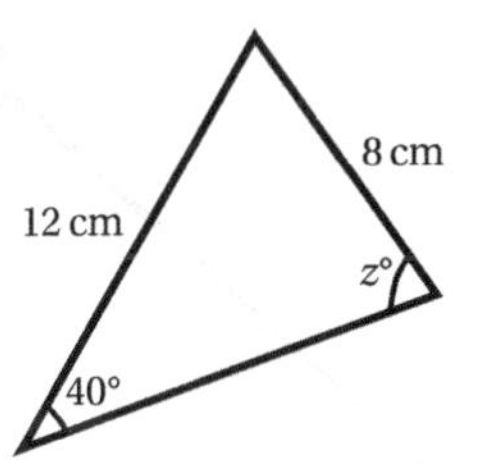

ii

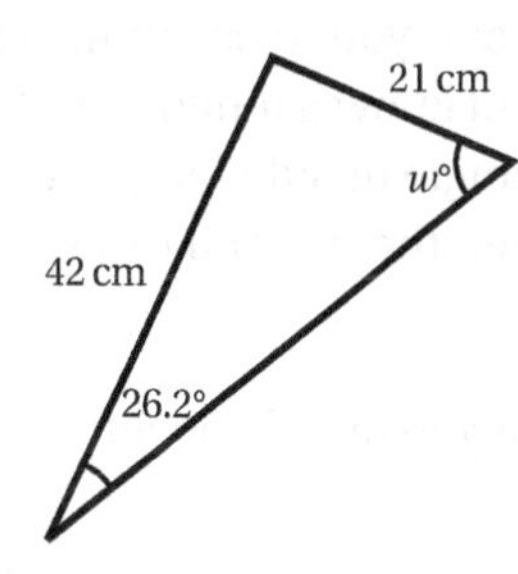

c **i**

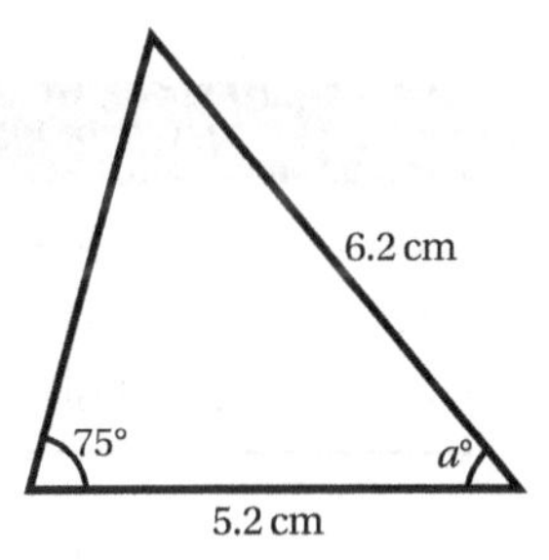

ii

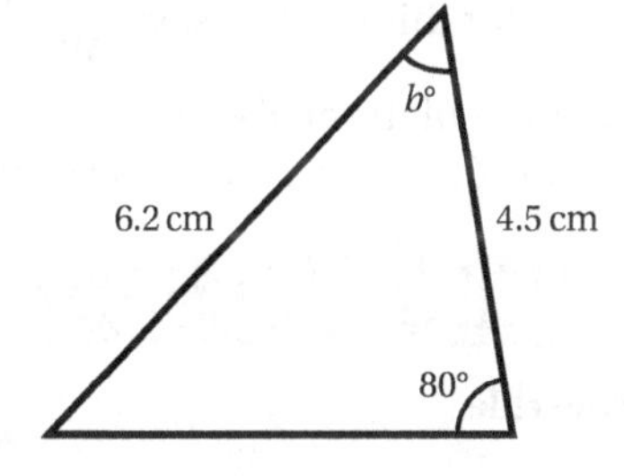

3 Find all the unknown sides and angles of triangle ABC.

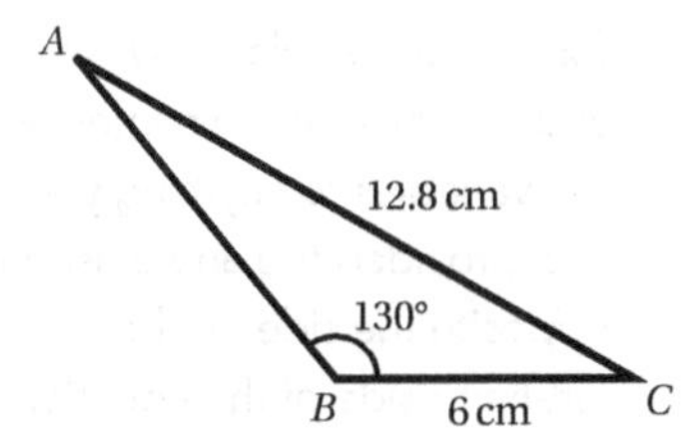

4 In triangle ABC, $AB = 6$ cm, $BC = 8$ cm, $A\hat{C}B = 35°$.

Show that there are two possible triangles with these measurements and find the remaining side and angles for each.

5 In the triangle shown in the diagram, $AB = 6$ cm, $AC = 8$ cm, $AD = 5$ cm and $A\hat{D}B = 75°$. Find the length of the side BC.

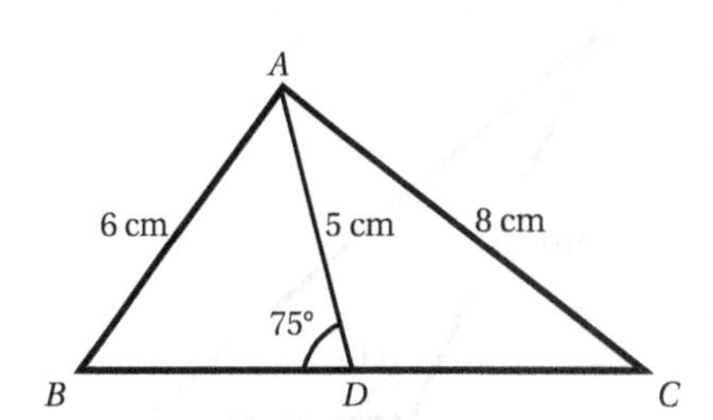

6 A balloon is tethered to a peg in the ground by a string of length 20 m, which makes an angle of 72° to the horizontal. An observer, on the same side of the peg as the balloon, notes that the angle of elevation from him to the balloon is 41° and his angle of depression to the peg is 10°.

Find the horizontal distance of the observer from the peg.

7 Show that it is impossible to draw a triangle ABC with $AB = 12$ cm, $AC = 8$ cm and $A\hat{B}C = 47°$.

Section 2: The cosine rule

The sine rule allows you to calculate angles and sides of a triangle, provided that you know a length and the angle opposite it, together with one other angle or side length. However, if you have two sides and the angle between them, or all three side lengths, you cannot use the sine rule.

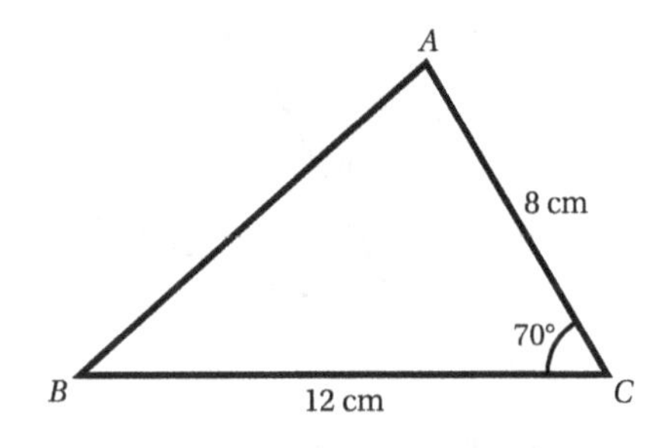

For example, can you find the length of the side AB in the triangle shown in the diagram?

The sine rule for this triangle says: $\frac{AB}{\sin 70°} = \frac{8}{\sin B} = \frac{12}{\sin A}$. But you do not know either of the angles A or B, so it is impossible to find AB from this equation. You need a different strategy.

Key point 11.3

The cosine rule

$a^2 = b^2 + c^2 - 2bc \cos A$

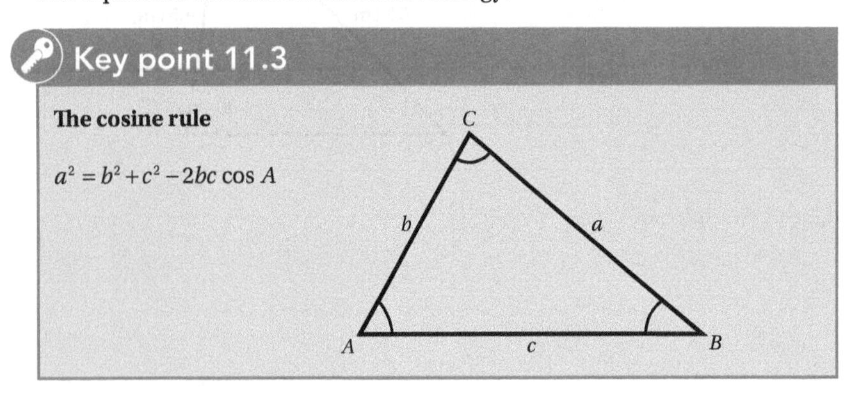

Focus on...

See Focus on ... Proof 2 to find out how to prove the cosine rule for any triangle.

Tip

This is the standard format given for the cosine rule. However, you can change the names of the variables to anything you like, provided the angle used is opposite the side on the left-hand side of the equation.

The cosine rule can be used to find the third side of the triangle when you know the other two sides and the angle between them.

WORKED EXAMPLE 11.4

Find the length of the side PQ, giving your answer to 3 s.f.

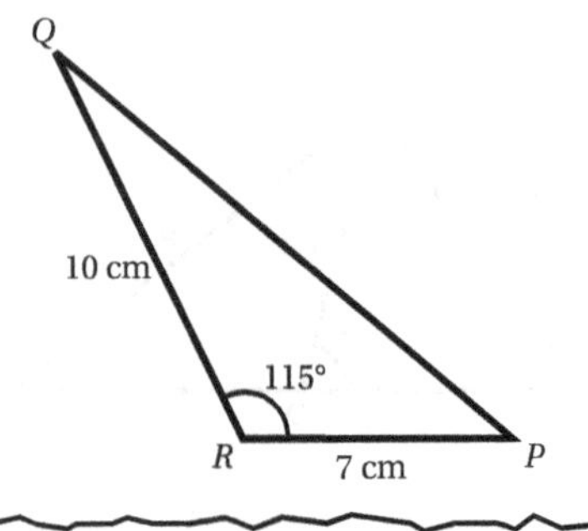

Continues on next page

$$PQ^2 = 7^2 + 10^2 - 2 \times 7 \times 10 \times \cos 115^\circ$$
$$= 208.2$$

As you are given two sides and the angle between them, use the cosine rule.

$$\therefore PQ = \sqrt{208.2}$$
$$= 14.4 \text{ cm}$$

You can also use the cosine rule to find an angle if you know all three sides of a triangle. To help with this, there is a rearrangement of the cosine rule.

Key point 11.4

Rearrangement of the cosine rule

$$\cos A = \frac{b^2 + c^2 - a^2}{2bc}$$

WORKED EXAMPLE 11.5

Find the size of $A\hat{C}B$, correct to the nearest degree.

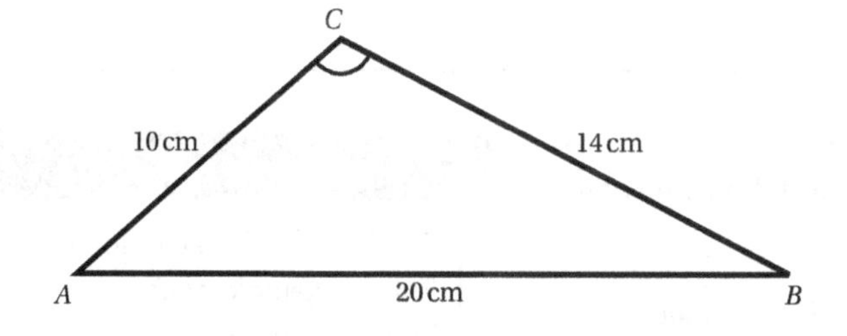

$$\cos C = \frac{a^2 + b^2 - c^2}{2ab}$$
$$= \frac{14^2 + 10^2 - 20^2}{2 \times 14 \times 10}$$
$$= -\frac{104}{280}$$

As you do not know any angles, use the cosine rule. Note that you can change the letters from those given in Key point 11.4 as long as everything is consistent.

$$C = \cos^{-1}\left(-\frac{104}{280}\right)$$
$$= 112^\circ$$

Notice that in the last two examples the angle is obtuse and its cosine is negative. Notice also that when using the cosine rule to find an angle, there is no second solution. Whereas the two solutions for $\sin x = c$ add up to 180°, the two solutions for $\cos x = c$ add up to 360°, so the second solution will always be greater than 180°.

It is possible to use the cosine rule even when the given angle is not opposite the required side. You may need to solve a quadratic equation.

WORKED EXAMPLE 11.6

Find the possible lengths of the side marked a in this triangle.

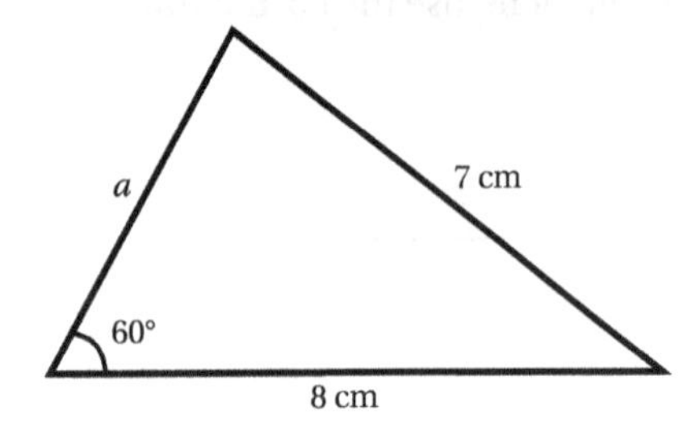

$7^2 = a^2 + 8^2 - 2a \times 8 \cos 60°$ — As all three sides are involved in the question, use the cosine rule. The known angle is opposite the side marked 7.

$49 = a^2 + 64 - 8a$ — $\cos 60° = \frac{1}{2}$.

$a^2 - 8a + 15 = 0$ — Solve the quadratic equation.

$(a-3)(a-5) = 0$

$a = 3$ or $a = 5$

So the length is 3 cm or 5 cm.

It is also possible to answer this question using the sine rule twice, first to find the angle opposite the side marked 8, and then to find the value of a. Try this for yourself.

Elevate

See Support Sheet 11 for a further example of choosing between the sine and cosine rules, and for more practice questions.

Worked example 11.7 illustrates how to select which of the two rules to use. For both the sine and cosine rules, you need to know three measurements in a triangle to find a fourth one.

WORKED EXAMPLE 11.7

In the triangle shown in the diagram, $AB = 6.5$ cm, $AD = 7$ cm, $CD = 5.8$ cm, $\angle ABC = 52°$ and $AC = x$. Find the value of x correct to 1 d.p.

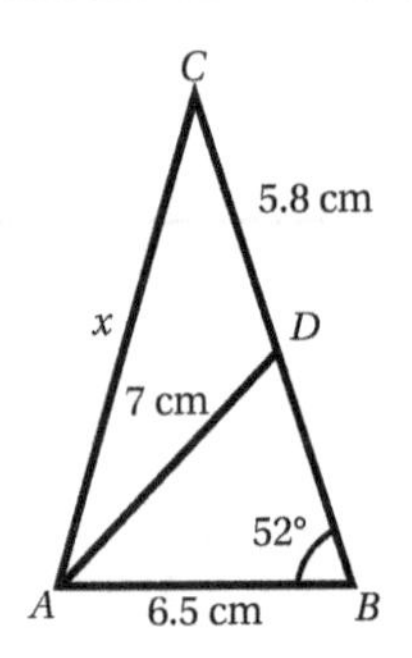

Continues on next page

Using the sine rule in triangle ABD, let angle $ADB = \theta$:

The only triangle in which you know three measurements is ABD. You know two side lengths and an angle opposite one of these, so use the sine rule to find the value of $A\hat{D}B$.

$$\frac{\sin\theta}{6.5} = \frac{\sin 52^\circ}{7}$$

$$\sin\theta = \frac{6.5\sin 52^\circ}{7} = 0.7317$$

$$\theta = \sin^{-1} 0.7317 = 47.03^\circ$$

$180^\circ - 47.03^\circ = 132.97^\circ$

$132.97^\circ + 52^\circ > 180^\circ$

Are there two possible solutions?

$\therefore$ there is only one solution: $\theta = 47.03^\circ$

$A\hat{D}C = 180^\circ - 47.03^\circ = 132.97^\circ$

In triangle ADC, you know two sides and want to find the third. If you knew $A\hat{D}C$, you could use the cosine rule. But you can find this angle more easily.

Using to sine rule in triangle ADC:

Now use the cosine rule.

$$x^2 = 7^2 + 5.8^2 - 2\times 7\times 5.8\cos 132.97^\circ = 137.99$$

$$\therefore x = \sqrt{137.99} = 11.7 \text{ cm}$$

Remember to use the unrounded value of the angle (from the ANS button on your calculator) so as not to lose accuracy in your final answer.

EXERCISE 11B

1 Find the lengths of the sides marked with letters.

a **i**

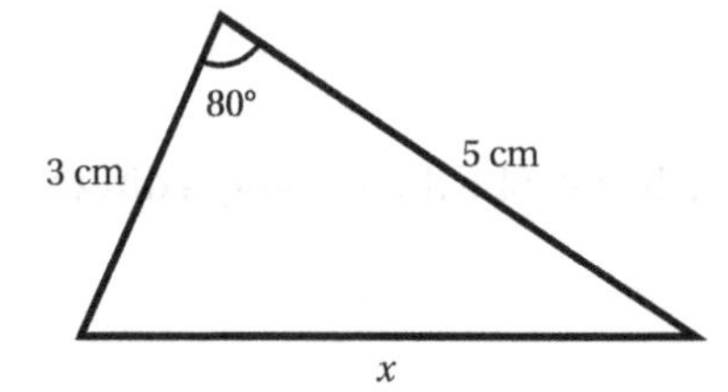

ii

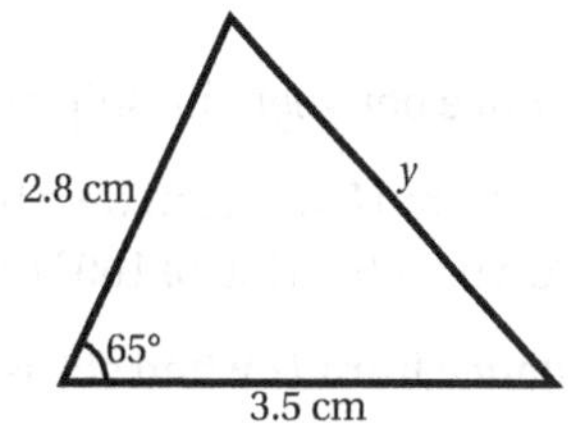

b **i**

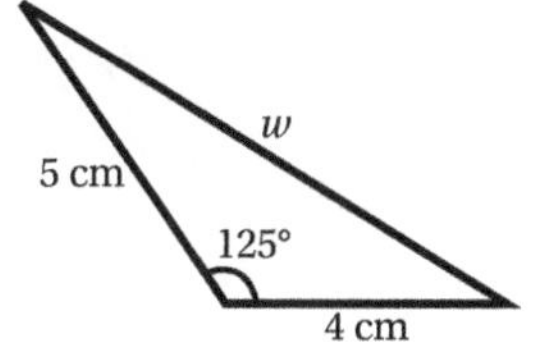

ii

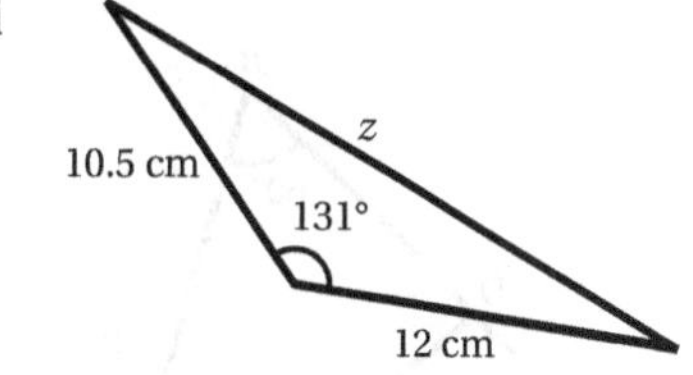

2 Find the sizes of the angles marked with letters.

a **i**

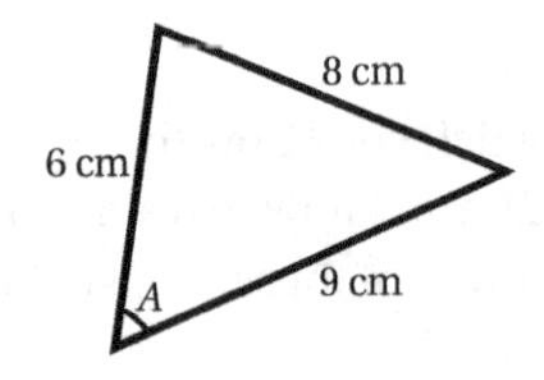

ii

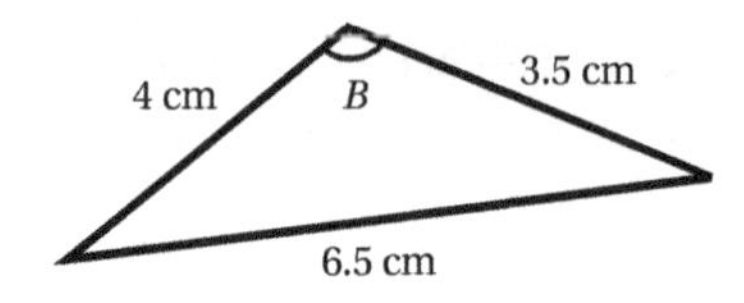

b **i**

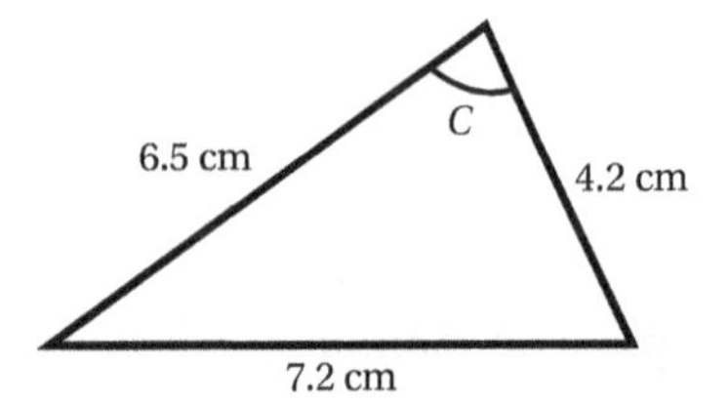

ii

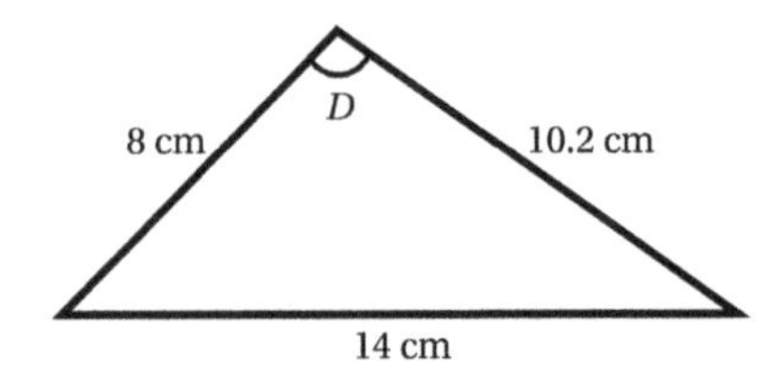

3 **a** Triangle PQR has sides $PQ = 8$ cm, $QR = 12$ cm, $RP = 7$ cm. Find the size of the largest angle.

b Triangle ABC has sides $AB = 4.5$ cm, $BC = 6.2$ cm, $CA = 3.7$ cm. Find the size of the smallest angle.

Tip

The largest angle in a triangle is always opposite the longest side, and the smallest angle opposite the shortest side.

4 Ship S is 2 km from the port on a bearing of $15°$ and boat B is 5 km from the port on a bearing of $130°$, as illustrated in the diagram.

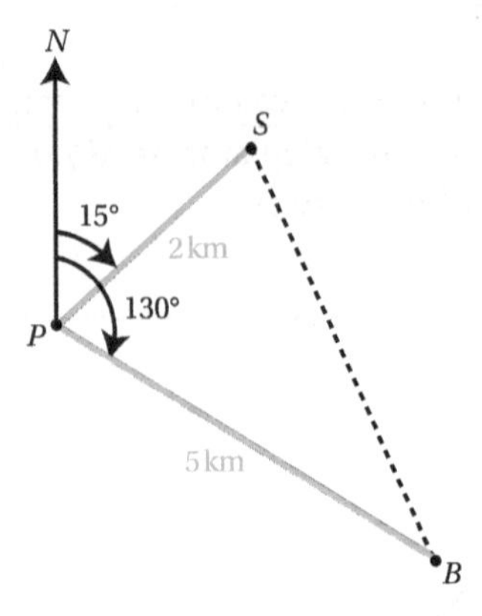

Find the distance between the ship and the boat.

Gateway to A Level

See Gateway to A Level Section E for a reminder of bearings.

5 A cyclist rides from H for 15 km on a bearing of $55°$ until she reaches A. She then changes direction and rides for 25 km on a bearing of $160°$ from A to B.

Find her distance from H when she is at B.

6 Find the value of x in the diagram.

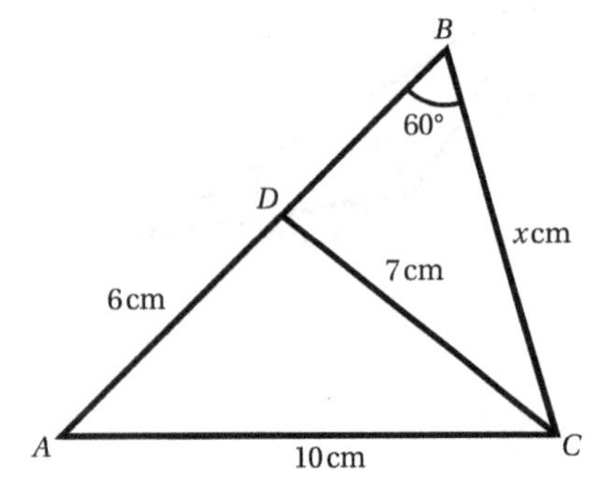

7. In triangle ABC, $AB = (x-3)$ cm, $BC = (x+3)$ cm, $AC = 8$ cm and angle $BAC = 60°$.

   Find the value of x.

8. The longest side of a triangle has length $(3x-2)$ cm. The other two sides have lengths $(2x-3)$ cm and $(x+2)$ cm. The largest angle is $120°$.

   Find the value of x.

9. In triangle KLM, $KL = 4$ cm, $LM = 7$ cm and angle $LKM = 45°$.

   Find the exact length of KM.

Section 3: Area of a triangle

The usual way to calculate the area of a triangle is $A = \frac{1}{2}$ base × height. However, if you don't know the perpendicular height but do know two sides and an angle between them, then there is a more useful formula.

Tip

To use the formula, you need to know the lengths of two sides and the angle between them. If you are given a different set of information, you can use the sine and cosine rules to calculate what you need.

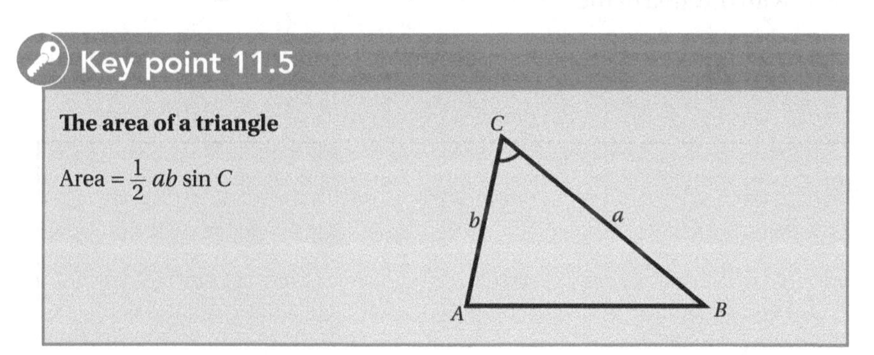

Key point 11.5

The area of a triangle

$$\text{Area} = \frac{1}{2}ab\sin C$$

Did you know

There is also a formula for the area of a triangle that depends on knowing the lengths of all three sides. It is called **Heron's formula**.

$$A = \sqrt{s(s-a)(s-b)(s-c)}$$

where $s = \frac{a+b+c}{2}$

WORKED EXAMPLE 11.8

The area of the triangle shown in the diagram is 52 cm^2. Find the two possible values of angle ABC, correct to 3 s.f.

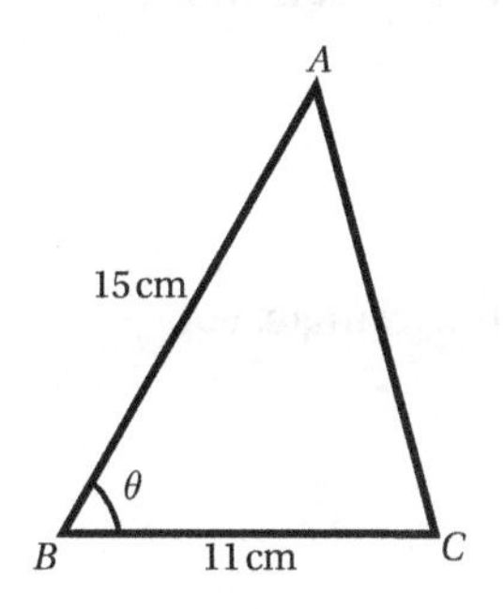

Continues on next page

$A = \frac{1}{2}ab\sin C$ — Use the formula for the area of a triangle.

$52 = \frac{1}{2}(11\times 15)\sin\theta$

$\sin\theta = \frac{2\times 52}{11\times 15}$

$= \frac{104}{165}$

$\theta = \sin^{-1}\left(\frac{104}{165}\right)$ — Remember that with sin there are two possible answers.

$= 39.07°$

$\therefore \theta = 39.1°$ or $180° - 39.1° = 141°$ — The question asks for two possible answers so there is no need to check that they both work.

Worked example 11.9 combines the sine rule with the area of the triangle, working with exact values.

WORKED EXAMPLE 11.9

Triangle PQR is shown in the diagram.

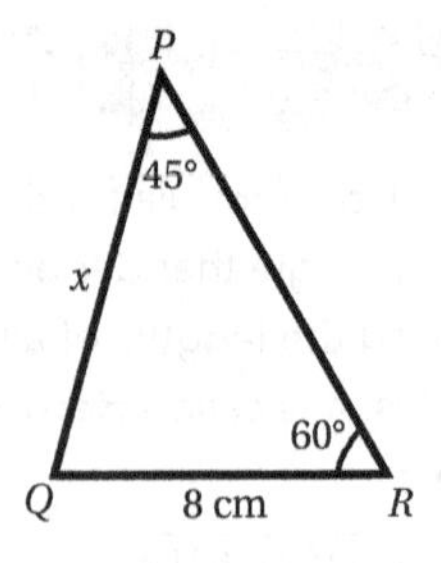

a Calculate the exact value of x.

b Find the area of the triangle, to 3 s.f.

a $\frac{8}{\sin 45°} = \frac{x}{\sin 60°}$ — Since you know two angles and a side, use the sine rule.

$\frac{8}{\frac{\sqrt{2}}{2}} = \frac{x}{\frac{\sqrt{3}}{2}}$ — You need the exact value of x, so use $\sin 45° = \frac{\sqrt{2}}{2}$ and $\sin 60° = \frac{\sqrt{3}}{2}$, which you can get from a calculator.

$\frac{16}{\sqrt{2}} = \frac{2x}{\sqrt{3}}$

$x = \frac{16\sqrt{3}}{2\sqrt{2}} = 4\sqrt{6}$

b $P\hat{Q}R = 180° - 60° - 45° = 75°$ — To use the formula for the area of the triangle, you need $P\hat{Q}R$.

$\text{Area} = \frac{1}{2}(8\times 4\sqrt{6})\sin 75°$ — Now use the area formula.

$= 37.9\text{ cm}^2$ (3 s.f.)

EXERCISE 11C

1 Calculate the area of each triangle.

a **i**

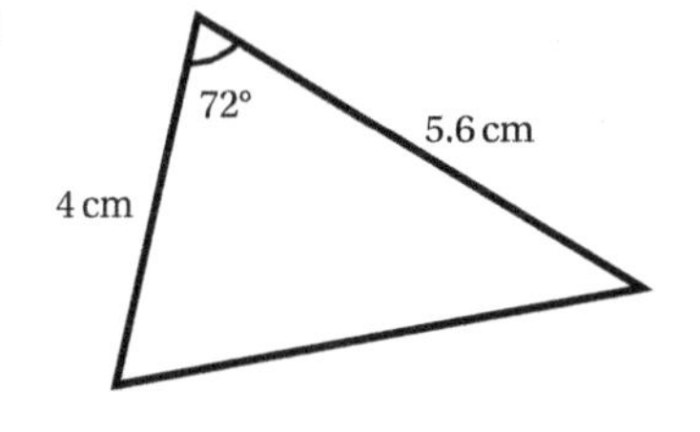

ii

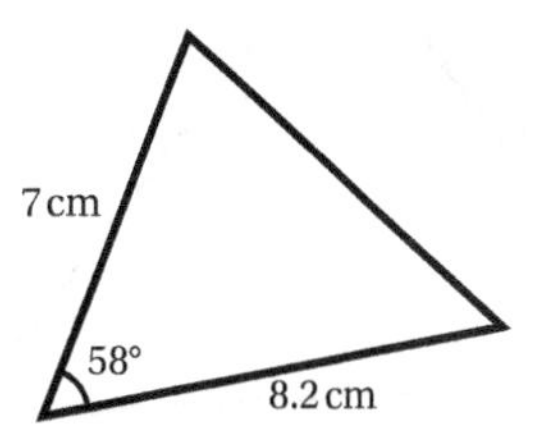

b **i**

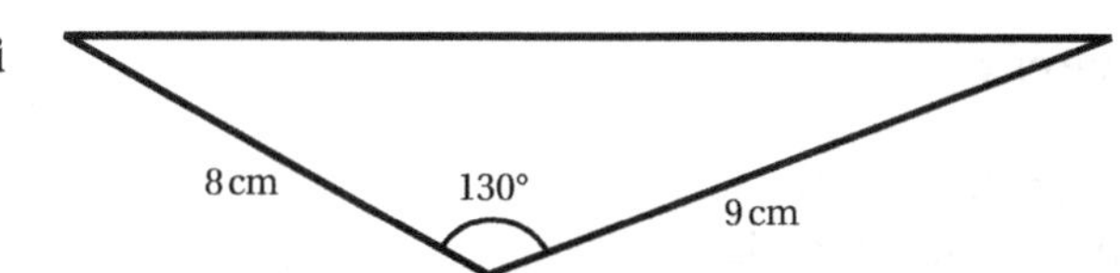

ii

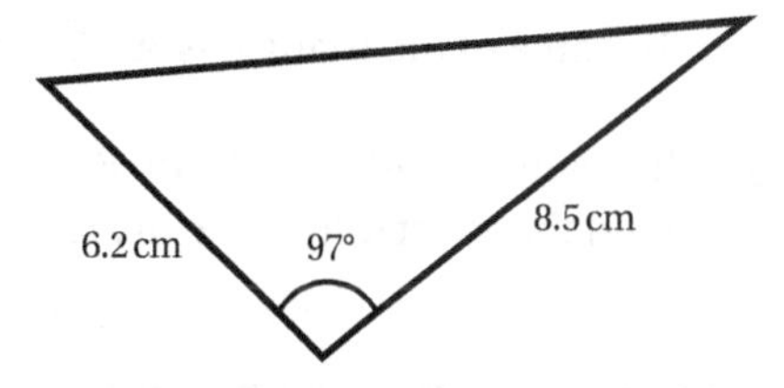

2 The area of each triangle is as shown. Find two possible values of each marked angle.

a

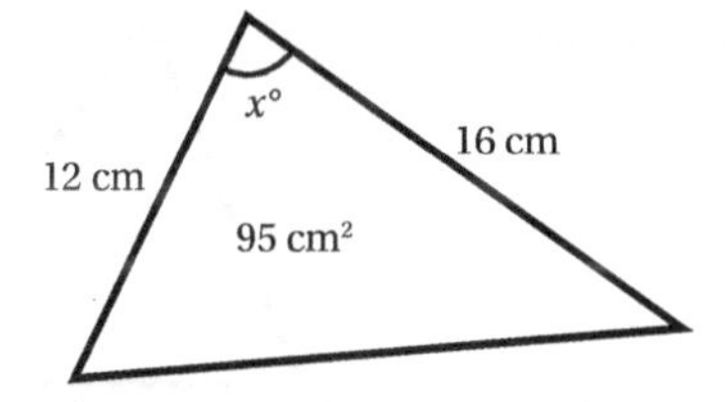

b

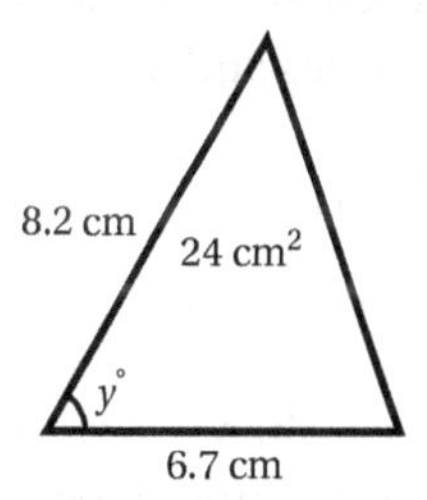

3 In triangle LMN, $LM = 12$ cm, $MN = 7$ cm and angle LMN $= 135°$.
Find the length of the side LN and the area of the triangle.

4 An equilateral triangle has area $25\sqrt{3}$ cm^2. Find the length of each side.

5 In triangle ABC, $AB = x + 3$, $BC = x$ and angle $ABC = 150°$.

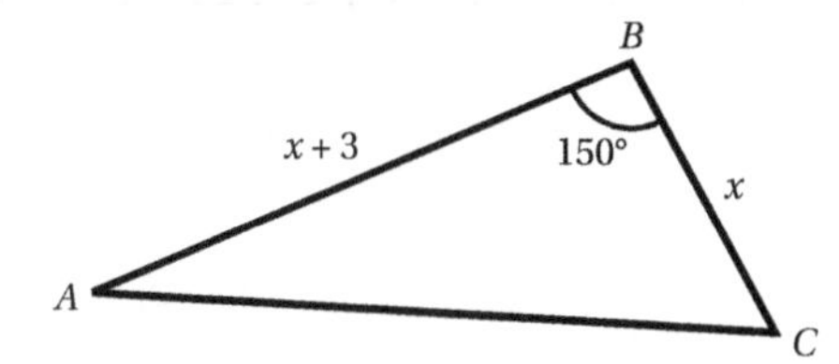

The area of the triangle is 10 square units. Find the value of x.

6 In triangle ABC, $AB = 2x - 3$, $BC = x + 4$ and angle $ABC = 120°$.

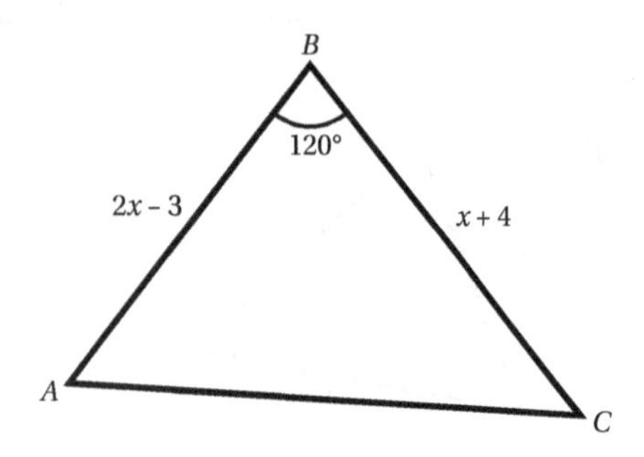

The area of the triangle is $39\sqrt{3}$. Find the value of x.

7 In triangle PQR, $PQ = 8$ cm, $RQ = 7$ cm and angle RPQ $= 60°$.
Find the exact difference in areas between the two possible triangles.

Checklist of learning and understanding

- To calculate the length of a side, when two angles and a side are given, or an angle when two sides and a non-included angle are given, use the **sine rule**:

  $\frac{a}{\sin A} = \frac{b}{\sin B} = \frac{c}{\sin C}$

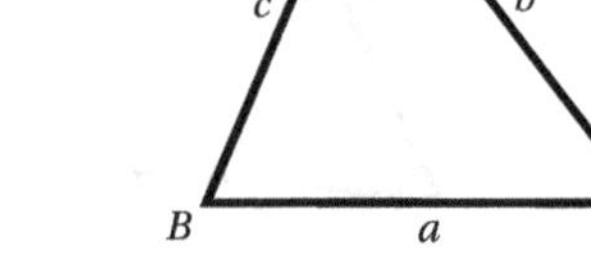

- When the sine rule is used to calculate an angle, there may be two answers: θ and $180° - \theta$.
- To calculate a side when two sides and the angle between them are given, or an angle when all three sides are given, use the **cosine rule**.
  - $a^2 = b^2 + c^2 - 2bc \cos A$
  - $\cos A = \frac{b^2 + c^2 - a^2}{2bc}$
- To find the area of a triangle when two sides and the included angle are known, use the formula:

  area $= \frac{1}{2} ab \sin C$

Mixed practice 11

1. These two triangles are equal in area. Find the value of x.

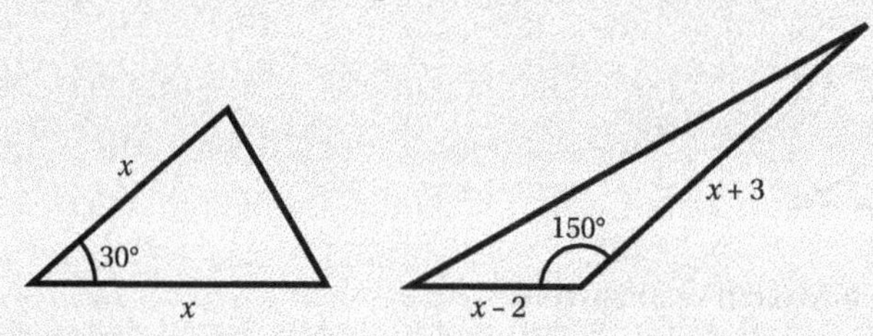

Choose from these options.

A $x = 2$ **B** $x = 4$ **C** $x = 5$ **D** $x = 6$

2. In triangle ABC, $AB = 6.2$ cm, $CA = 8.7$ cm and angle $ACB = 37.5°$. Find the two possible values of angle ABC.

3. The area of this triangle is 12 cm². Find the value of x.

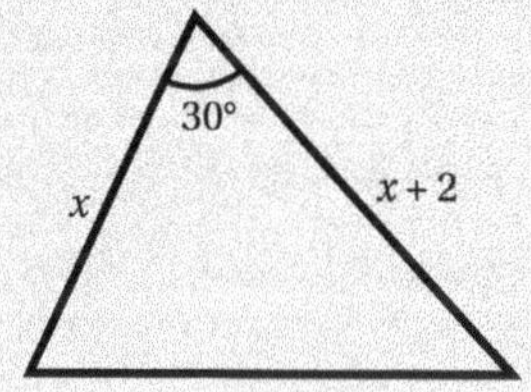

4. A vertical tree of height 12 m stands on horizontal ground. The bottom of the tree is at the point B. Observer A, standing on the ground, sees the top of the tree at an angle of elevation of 56°.

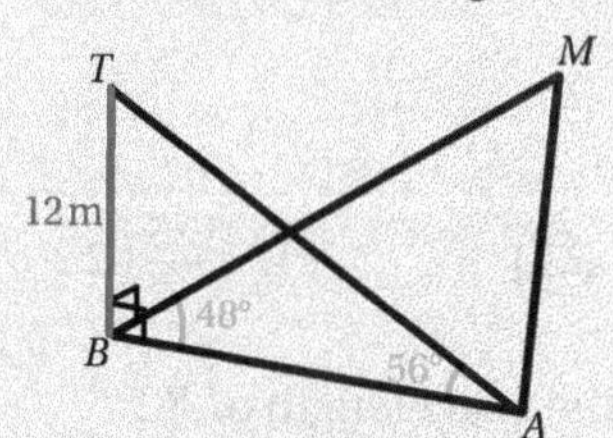

a Find the distance of A from the bottom of the tree.

b Another observer, M, stands the same distance away from the tree, and angle $ABM = 48°$. Find the distance AM.

5. The triangle ABC, shown in the diagram, is such that $AB = 6$ cm, $BC = 15$ cm, angle $BAC = 150°$ and angle $ACB = \theta$.

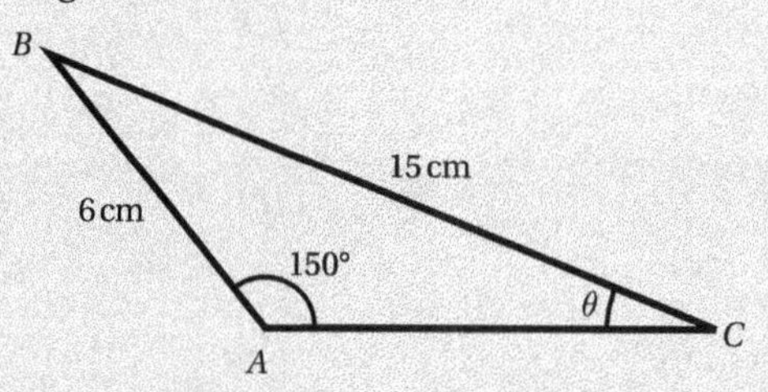

a Show that $\theta = 11.5°$, correct to the nearest 0.1°.

b Calculate the area of triangle ABC, giving your answer in cm² to three significant figures.

[© AQA 2010]

6 The diagram shows a triangle ABC.

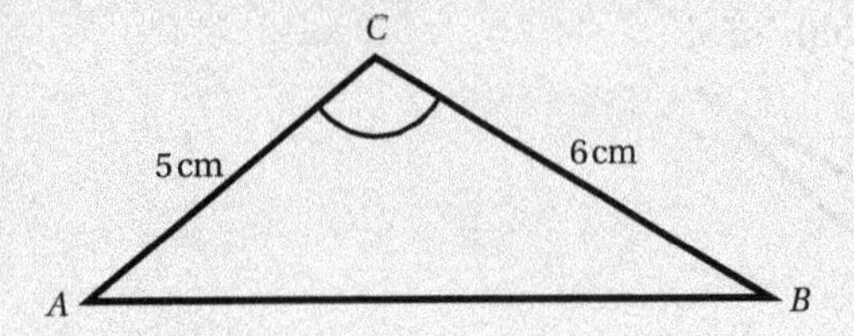

The lengths of AC and BC are 5 cm and 6 cm respectively.

The area of triangle ABC is 12.5 cm^2 and angle ACB is **obtuse**.

a Find the size of angle ACB, giving your answer to the nearest 0.1°.

b Find the length of AB, giving your answer to two significant figures.

[© AQA 2013]

7 Find the **exact** area of this triangle.

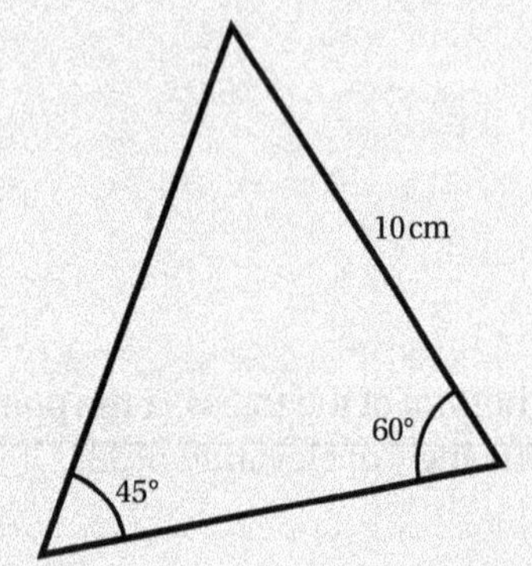

Choose from these options.

A $\frac{25}{2}\sqrt{3}$ **B** $\frac{25}{2}\sqrt{2}$ **C** $\frac{25}{2}\left(3+\sqrt{3}\right)$ **D** $\frac{25}{2}\left(2+\sqrt{2}\right)$

8 In triangle ABC, $AB = 2\sqrt{3}$, $AC = 10$ and angle $BAC = 150°$. Find the exact length of BC.

9 In the acute-angled triangle KLM, $LM = 6.1$ cm, $KM = 4.2$ cm and angle $KLM = 42°$.

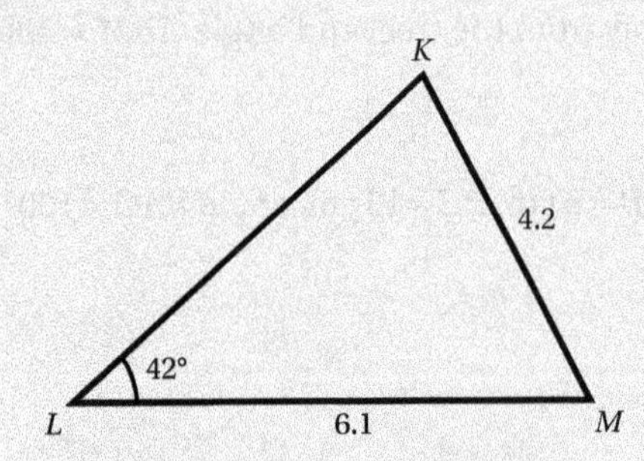

Find the area of the triangle.

10 In triangle ABC, $AB = 10$ cm, $BC = 8$ cm and $CA = 7$ cm.

a Find the exact value of cos ABC.

b Find the exact value of sin ABC.

c Find the exact value of the area of the triangle.

11 Two radar stations, A and B, are 20 km apart. B is due east of A. Station B detects a ship on a bearing of $310°$. The same ship is 15 km from station A.

a Find the two possible bearings of the ship from station A.

b Hence find the distance between the two possible positions of the ship.

12 In triangle ABC, $AB = 5$, $AC = x$ and angle $BAC = \theta$. M is the midpoint of the side AC.

a Use the cosine rule to find an expression for MB^2 in terms of x and θ.

b Given that $BC = MB$, show that $\cos\theta = \frac{3x}{20}$.

c Given that $x = 5$, find the value of the angle θ such that $MB = BC$.

13 A regular pentagon has area 200 cm^2. Find the length of each side.

14 In triangle ABC, $AB = 10$, $BC = 5$, $CA = x$ and angle $BAC = \theta$.

a Show that $x^2 - 20x\cos\theta + 75 = 0$.

b Find the range of values of $\cos\theta$ for which the equation has real roots.

c Hence find the set of values of θ for which it is possible to construct triangle ABC with the given measurements.

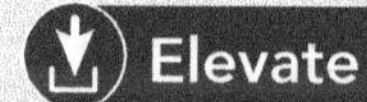

See Extension Sheet 11 for a selection of harder questions with triangles and circles.

12 Differentiation

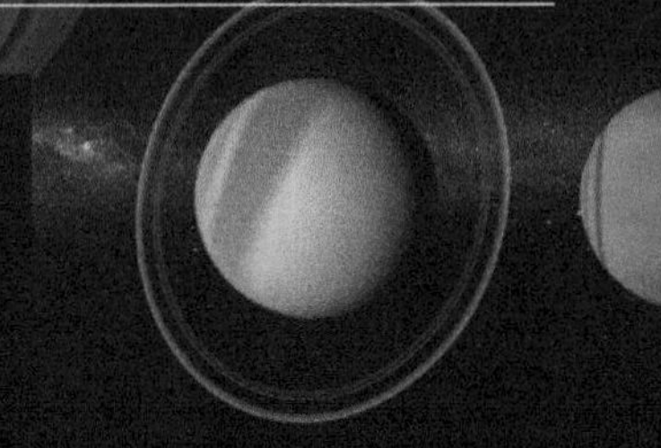

In this chapter you will learn how to:

- sketch the gradient function for a given curve
- find the gradients of curves from first principles
- differentiate x^n
- use differentiation to decide whether a function is increasing or decreasing.

Before you start…

Chapter 2	You should be able to work with indices.	1 Write each expression in the form $nx^a + mx^b$. a $\sqrt[3]{x^2} - \frac{5}{2x}$ b $\frac{2+x}{\sqrt{x}}$
Chapter 3	You should be able to solve linear and quadratic inequalities.	2 Solve these inequalities. a $3x-4 \geqslant 5x+2$ b $x^2-4x-12 \geqslant 0$
Chapter 6	You should know how to find the gradient of a straight line.	3 Find the gradient of the line that passes through the points $(-1, 4)$ and $(7, -2)$.
Chapter 9	You should be familiar with the binomial expansion.	4 Expand $(2+x)^3$.

What is differentiation?

In real life things change: planets move, babies grow and prices rise. Calculus is the study of change, and one of its most important tools is **differentiation** – finding the rate at which the y-coordinate of a curve is changing when the x-coordinate changes. For a straight-line graph, this rate of change is given by the gradient of the line. In this chapter the same idea is applied to curves, where the gradient is different at different points.

Section 1: Sketching derivatives

The first task is to establish exactly what is meant by the gradient of a curve. You already know what this means for a straight line, so this idea will be used to make a more general definition.

Key point 12.1

The **gradient** of a curve at a point P is defined as the gradient of the tangent to the curve at that point.

Rewind

You have already met tangents in Chapters 5 and 6.

Note that when you talk about a tangent just touching a curve and not crossing it, this is meant in a 'local' sense: although the tangent does not cross the curve close to the point P it may intersect the curve at a different point (as shown in the diagram).

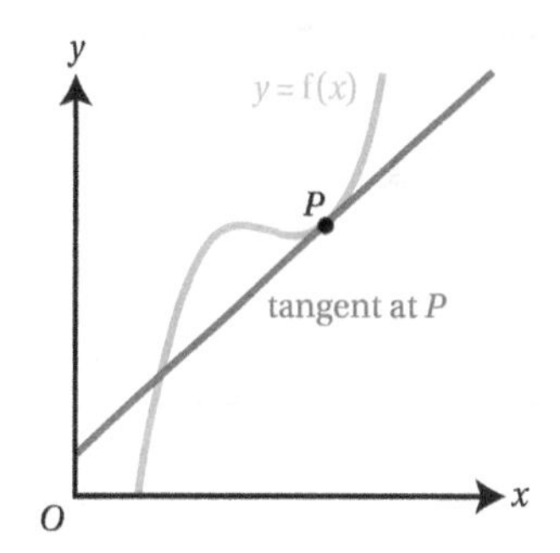

The **derivative** of a function $f(x)$ is another function which gives the gradient of $y = f(x)$ at any point. It is often useful to be able to sketch the derivative (or gradient function).

WORKED EXAMPLE 12.1

Sketch the derivative of this function.

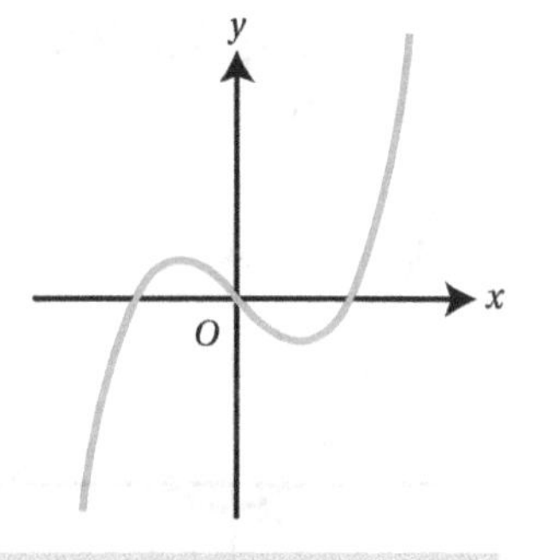

Imagine a point moving along the curve from left to right; you can track the tangent of the curve at the moving point and form the graph of its gradient.

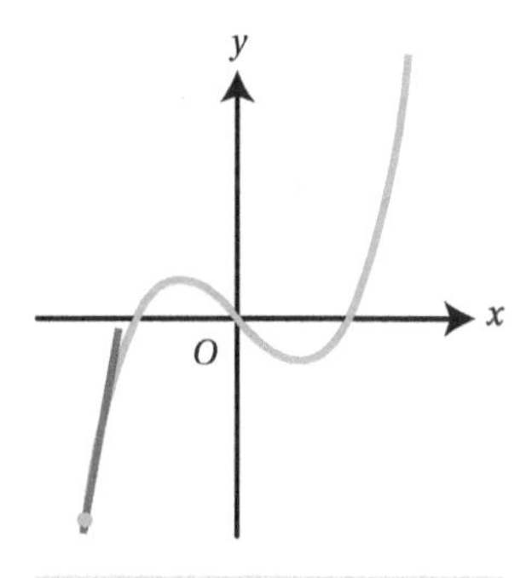

The curve is increasing from left to right, but more and more slowly...

... so the gradient is positive and decreasing.

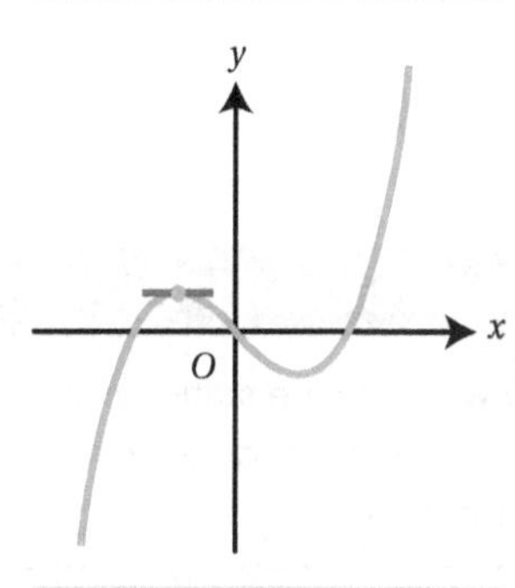

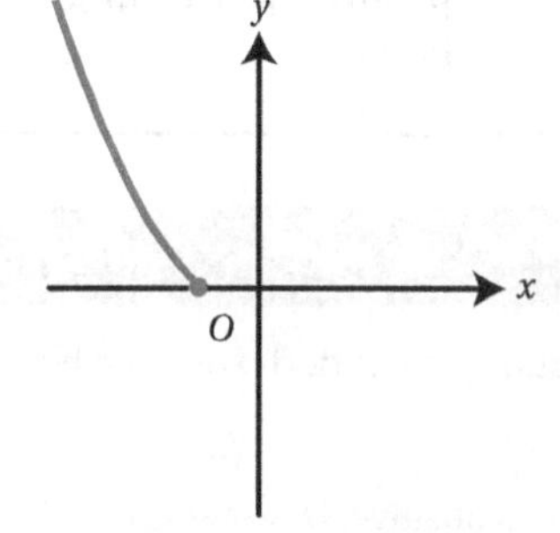

The tangent is horizontal...

... so the gradient is zero.

Continues on next page

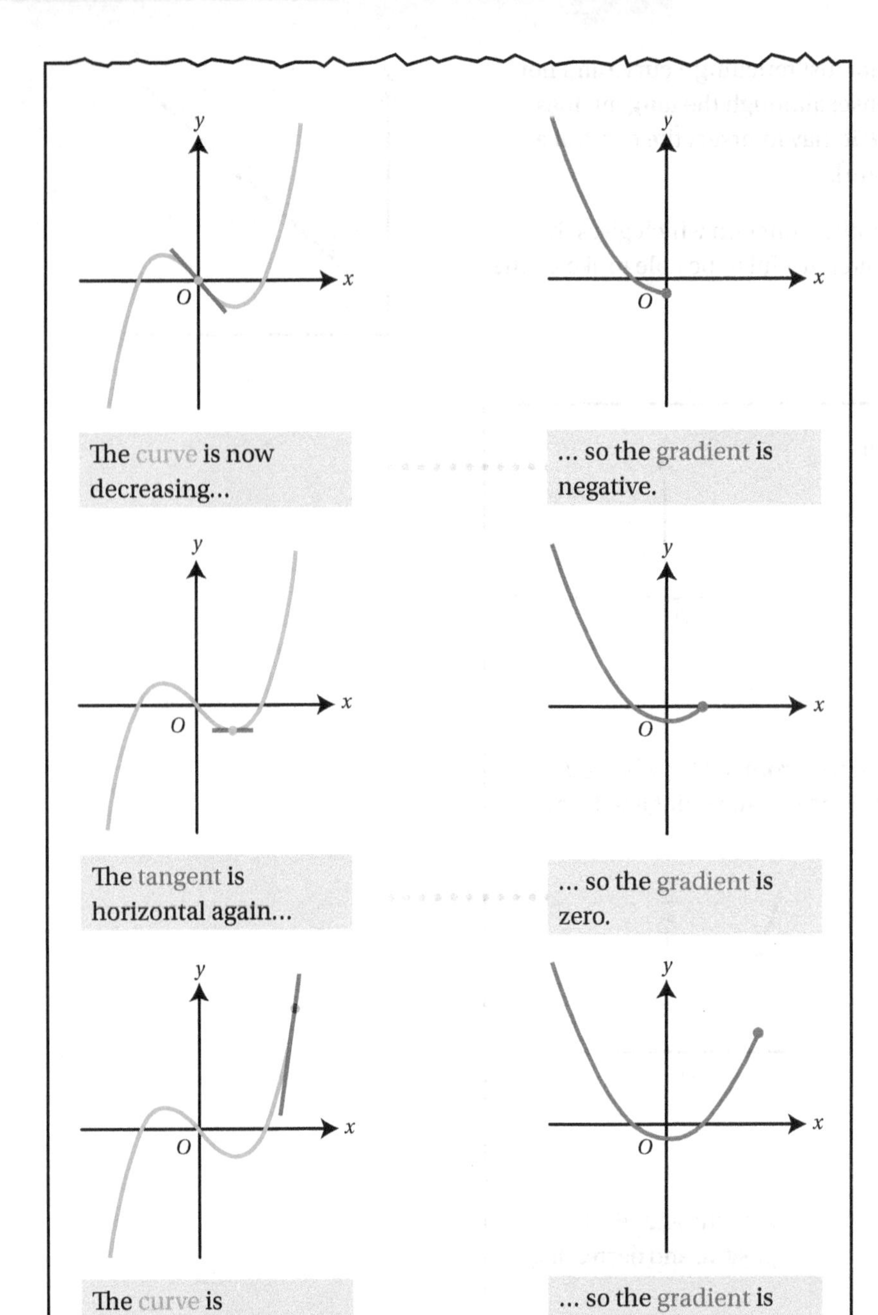

y
x
O
The curve is now decreasing...
... so the gradient is negative.
The tangent is horizontal again...
... so the gradient is zero.
The curve is increasing, and does so faster and faster...
... so the gradient is positive and getting larger.

Key point 12.2

The relationship between the graph of a function and its derivative can be summarised as described.

- When the graph is increasing the gradient is positive.
- When the graph is decreasing the gradient is negative.
- When the tangent is horizontal the gradient is zero. A point on the graph where this happens is called a **stationary point** or **turning point**.

Fast forward

You will examine stationary/turning points in detail in Chapter 13.

EXERCISE 12A

1 Sketch the derivative of each function, showing any intercepts with the x-axis.

a **i**

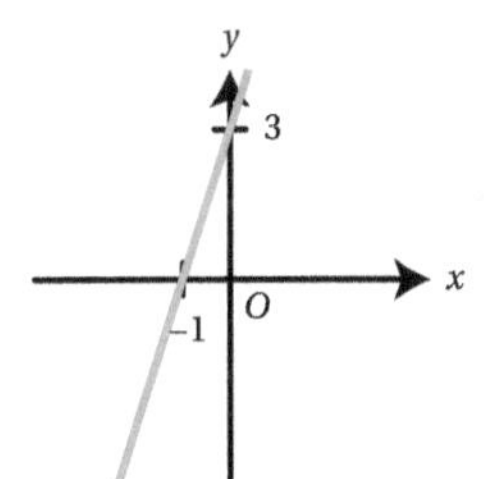

ii

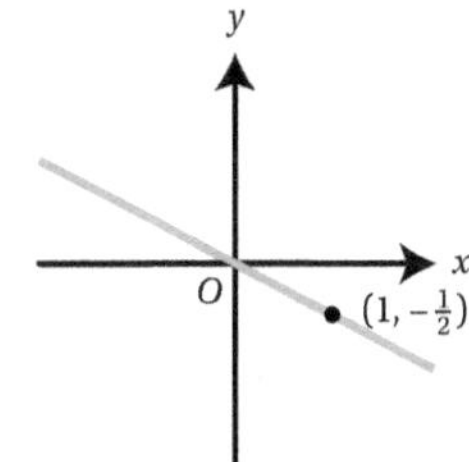

b **i**

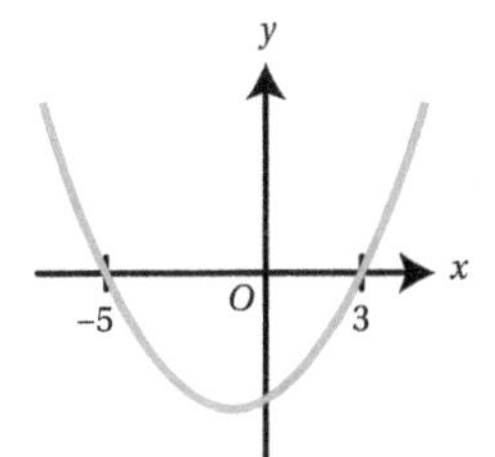

ii

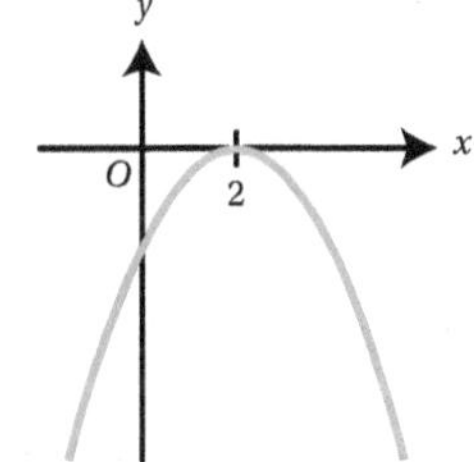

c **i**

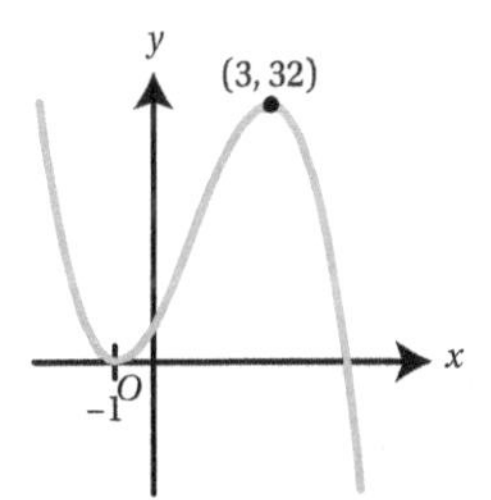

ii

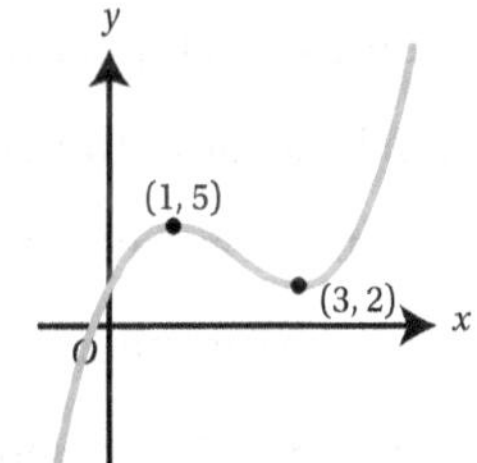

d **i**

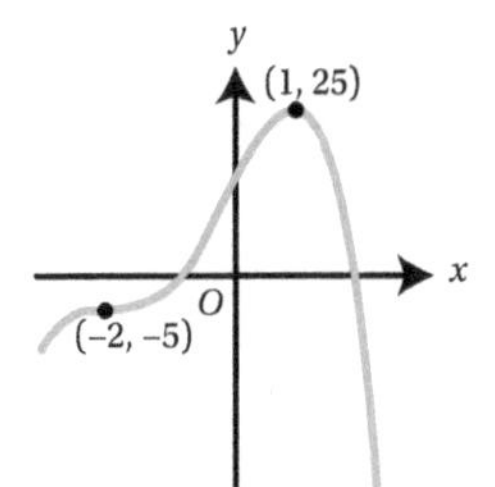

ii

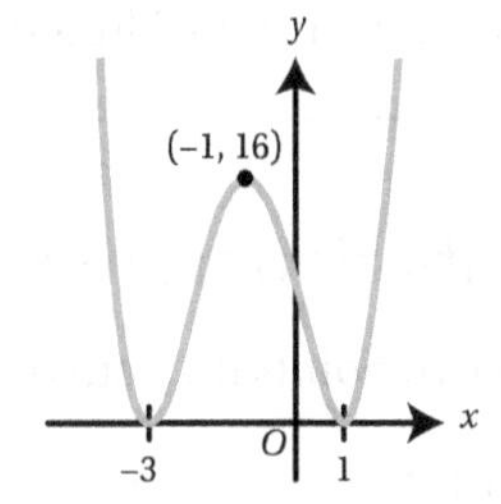

e **i**

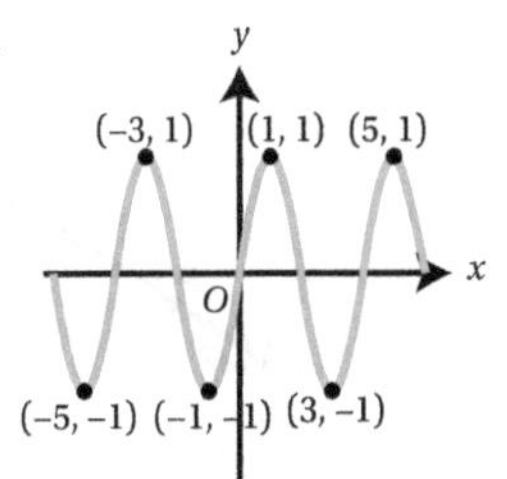

ii

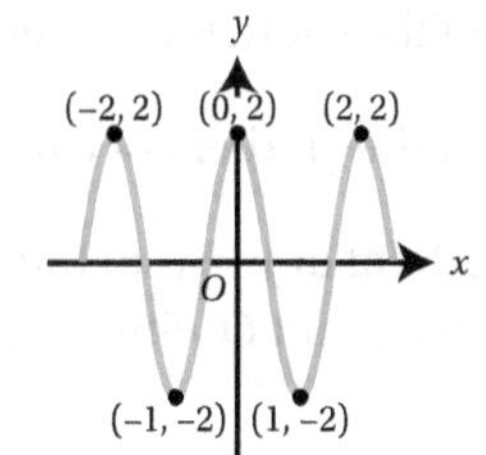

f **i**

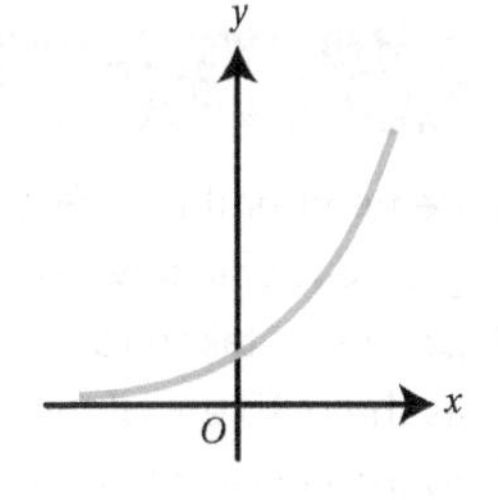

ii

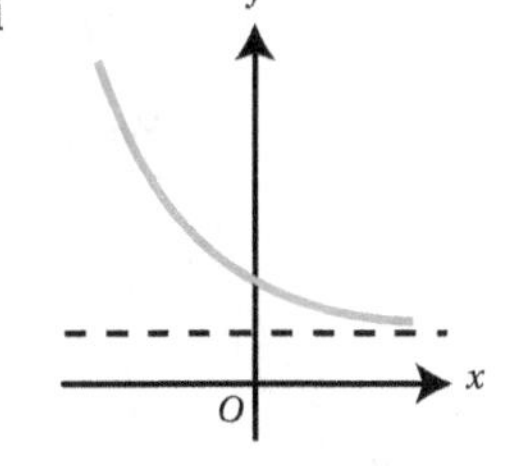

2 Each diagram represents a graph of a function's derivative. Sketch a possible graph for the original function, indicating any stationary points.

a

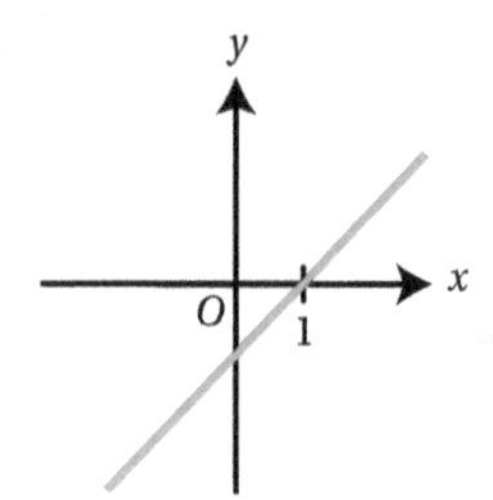

b

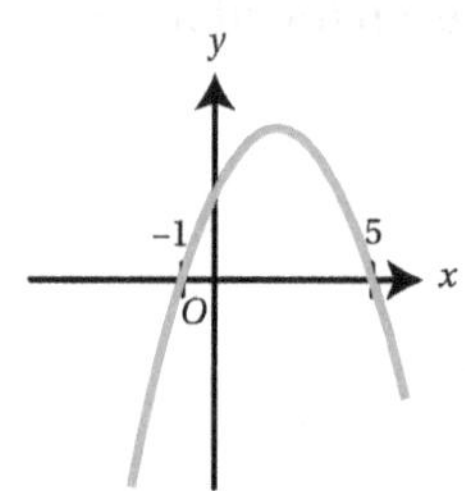

c

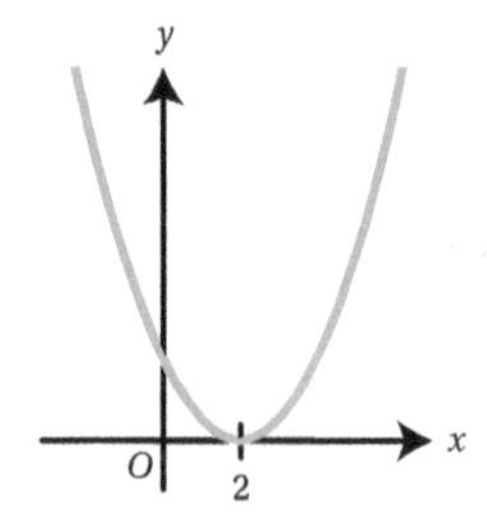

d

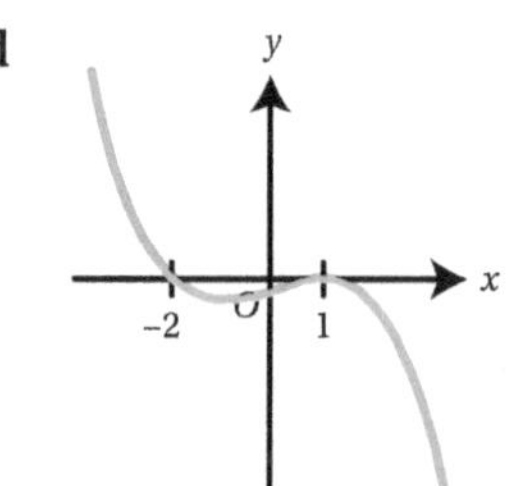

3 Decide whether each statement is true or false. Give a counter example for each statement that is false.

a At a point where the derivative is positive the original function is positive.

b If the original function is negative then the derivative is also negative.

c The derivative crossing the x-axis corresponds to a stationary point on the graph.

d When the derivative is zero the graph is at a local maximum or minimum point.

e If the derivative function is always positive then part of the original function is above the x-axis.

f At the lowest value of the original function the derivative is zero.

Section 2: Differentiation from first principles

The line segment between two points on a curve is called a **chord**.

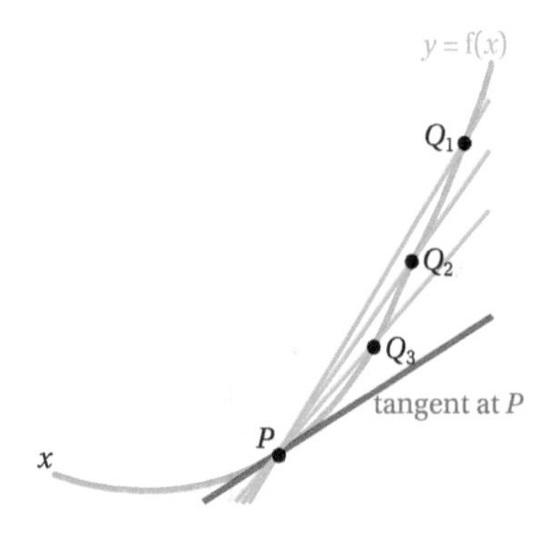

The diagram shows the chord PQ. You can see that the closer the point Q is to P, the closer the gradient of the chord is to the gradient of the tangent at P.

This idea can be used to find the gradient of a function at a given point, P.

For example, to find the gradient of $y = x^2$ at the point P where $x = 3$, consider a chord from P to the point Q with a slightly larger x-coordinate, $x = 3 + h$.

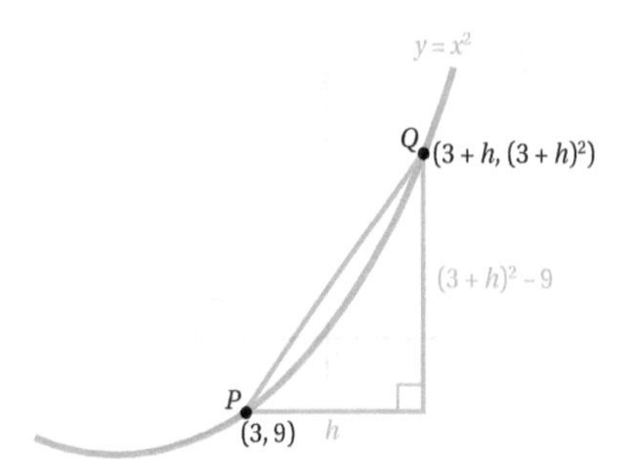

Rewind

Remember that in Section 1 the gradient of a curve at a given point was taken to be the same as the gradient of the tangent to the curve at that point.

The gradient of the chord, m, is:

$$\begin{aligned} m &= \frac{y_2 - y_1}{x_2 - x_1} \\ &= \frac{(3+h)^2 - 9}{(3+h) - 3} \\ &= \frac{9 + 6h + h^2 - 9}{h} \\ &= \frac{6h + h^2}{h} \\ &= 6 + h \end{aligned}$$

Tip

The letter h just denotes a small distance.

As the point Q now approaches P, the distance h tends to zero and so the gradient of the chord, m, tends to 6.

Therefore, the gradient of the tangent at $x = 3$ is 6. But this is the same as saying that the gradient of the curve $y = x^2$ at $x = 3$ is 6.

This idea of the distance h tending to zero is denoted by $\lim\limits_{h \to 0}$, and the process of finding the gradient of the chord is called differentiation from first principles.

The method can be used to find the gradient at a general point on a function $\mathrm{f}(x)$.

Key point 12.3

$$\mathrm{f}'(x) = \lim_{h \to 0} \frac{\mathrm{f}(x+h) - \mathrm{f}(x)}{h}$$

The expression $\mathrm{f}'(x)$ is called the derivative of $\mathrm{f}(x)$. It can also be denoted as f', y' or $\frac{\mathrm{d}y}{\mathrm{d}x}$ where $y = \mathrm{f}(x)$. The process of finding the derivative is called differentiation.

Tip

Differentiation from first principles means using this definition to find the derivative, rather than any of the rules you will meet in the later sections.

You can use this definition to find the derivative of simple polynomial functions.

Did you know?

In the late 17th century there was considerable debate as to whether Gottfried Liebniz or Isaac Newton had invented this idea of considering very small changes in the value of a function, and ultimately the branch of mathematics known as calculus. However, there is evidence that these ideas had been known many centuries before the time of either Liebniz or Newton.

WORKED EXAMPLE 12.2

Prove, using differentiation from first principles, that for $f(x) = x^2 + x$, $f'(x) = 2x + 1$.

$$f'(x) = \lim_{h \to 0} \frac{f(x+h) - f(x)}{h}$$

Use the formula with $f(x) = x^2 + x$, so that $f(x+h) = (x+h)^2 + (x+h)$.

$$= \lim_{h \to 0} \frac{(x+h)^2 + (x+h) - (x^2 + x)}{h}$$

$$= \lim_{h \to 0} \frac{x^2 + 2xh + h^2 + x + h - x^2 - x}{h}$$

Expand $(x+h)^2$ and simplify the expression.

$$= \lim_{h \to 0} \frac{2xh + h^2 + h}{h}$$

$$= \lim_{h \to 0} (2x + h + 1)$$

Then divide top and bottom by h.

$$= 2x + 1$$

Finally, let $h \to 0$.

EXERCISE 12B

1. Prove from first principles that the derivative of $y = 8$ is zero.

2. $f(x) = -4x$

   Use differentiation from first principles to find $f'(x)$.

3. $y = 3x^2$

   Find $\frac{dy}{dx}$ from first principles.

4. Using differentiation from first principles, prove that the derivative of $x^2 + 1$ is $2x$.

5. Differentiate $f(x) = x^2 - 6x$ from first principles.

6. $y = x^2 - 3x + 4$

   Using differentiation from first principles, find y'.

7. **a** Expand $(x+h)^3$.

   b Hence find from first principles the derivative of $y = x^3$.

8. $f(x) = 3x^3 + 2$

   Find $f'(x)$ from first principles.

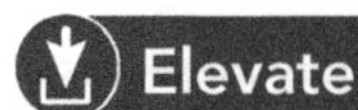

See Extension Sheet 12 for some harder questions on differentiation from first principles.

9. **a** Expand $(x+h)^4$.

   b Hence prove from first principles that if $f(x) = x^4$, then $f'(x) = 4x^3$.

10. **a** Find an expression for the gradient of the chord between the points A and B on the curve $y = f(x)$ with x-coordinates x and $x + h$ respectively.

    b Explain why, in the limit as $h \to 0$, the expression in part **a** becomes $f'(x)$.

11 Given that k is a constant, prove that the derivative of $k\mathrm{f}(x)$ is $k\mathrm{f}'(x)$.

12 If $y = \mathrm{f}(x) + \mathrm{g}(x)$, prove that $y' = \mathrm{f}'(x) + \mathrm{g}'(x)$.

Section 3: Rules of differentiation

By now, you may have noticed a pattern emerging for differentiation.

Key point 12.4

If $y = x^n$ then:

$$\frac{\mathrm{d}y}{\mathrm{d}x} = nx^{n-1}$$

The result in Key point 12.4 is proved here for positive integers, using differentiation from first principles, but the result is true (and you will need to use it) for all rational powers - positive and negative.

PROOF 6

$$\frac{\mathrm{d}y}{\mathrm{d}x} = \lim_{h \to 0} \frac{(x+h)^n - x^n}{h}$$

Use the formula with $\mathrm{f}(x) = x^n$.

$$= \lim_{h \to 0} \frac{x^n + \binom{n}{1}x^{n-1}h + \binom{n}{2}x^{n-2}h^2 + \ldots + h^n - x^n}{h}$$

Expand $(x+h)^n$ using the binomial expansion.

$$= \lim_{h \to 0} \frac{\binom{n}{1}x^{n-1}h + \binom{n}{2}x^{n-2}h^2 + \ldots + h^n}{h}$$

Simplify...

$$= \lim_{h \to 0} \left[\binom{n}{1}x^{n-1} + \binom{n}{2}x^{n-2}h + \ldots + h^{n-1}\right]$$

... and then divide top and bottom by h.

$$= \binom{n}{1}x^{n-1}$$

Let $h \to 0$.

$$= nx^{n-1}$$

$\binom{n}{1} = n$

WORKED EXAMPLE 12.3

$y = x^6$. Find $\frac{\mathrm{d}y}{\mathrm{d}x}$.

$$\frac{\mathrm{d}y}{\mathrm{d}x} = 6x^{6-1}$$

Use $\frac{\mathrm{d}y}{\mathrm{d}x} = nx^{n-1}$.

$$= 6x^5$$

Fast forward

In Student Book 2 you will learn a method called implicit differentiation, which will allow you to extend the proof to all rational powers.

WORKED EXAMPLE 12.4

$y = x^{-5}$. Find $\frac{dy}{dx}$.

$\frac{dy}{dx} = -5x^{-5-1}$ — Use $\frac{dy}{dx} = nx^{n-1}$.

$= -5x^{-6}$

WORKED EXAMPLE 12.5

Differentiate $f(x) = x^{\frac{1}{4}}$

$f'(x) = \frac{1}{4}x^{\frac{1}{4}-1}$ — Use $f'(x) = nx^{n-1}$.

$= \frac{1}{4}x^{-\frac{3}{4}}$

WORKED EXAMPLE 12.6

Find the derivative of $f(x) = x$.

$f(x) = x^1$ — Write x as x^1.

$f'(x) = 1x^{1-1}$ — Now use $f'(x) = nx^{n-1}$.

$= x^0$

$= 1$ — Remember that $x^0 = 1$.

Tip

You do not need to write x^1 every time you differentiate x. You just need to know the derivative will be always be 1.

The results of Exercise 12B suggest some properties of differentiation (in fact you proved these in questions 11 and 12).

Key point 12.5

- If $y = kf(x)$, where k is a constant, then $y' = kf'(x)$.
- If $y = f(x) + g(x)$, then $y' = f'(x) + g'(x)$.

Tip

The second part of Key point 12.5 just says that you can differentiate the terms of a sum separately.

WORKED EXAMPLE 12.7

Find the derivative of $f(x) = 5x^3$.

$f'(x) = 5 \times 3x^2$ — Differentiate x^3 and then multiply by 5.

$= 15x^2$

WORKED EXAMPLE 12.8

Find $\frac{dy}{dx}$ for $y = 8$.

$y = 8x^0$ — Using the fact that $x^0 = 1$, you can write $y = 8$ as $y = 8x^0$.

$\frac{dy}{dx} = 8 \times 0x^{0-1}$ — Now use $\frac{dy}{dx} = nx^{n-1}$.

$= 0$ — As 0 is now a factor, the whole expression is zero.

Tip

You do not need to write out cx^0 every time you differentiate a constant. You just need to know that the derivative will always be zero.

From Worked example 12.8 you can see that the gradient of $y = c$ for any constant c will always be zero. This corresponds to the fact that the gradient of the horizontal line $y = c$ is zero everywhere.

WORKED EXAMPLE 12.9

A curve has equation $y = x^4 - 4x^{\frac{1}{2}} + 5x - 3$. Find $\frac{dy}{dx}$.

$\frac{dy}{dx} = 4x^3 - 4 \times \frac{1}{2}x^{-\frac{1}{2}} + 5$ — Differentiate each term separately. Remember that the derivative of a constant is zero, so the -3 vanishes.

$= 4x^3 - 2x^{-\frac{1}{2}} + 5$

EXERCISE 12C

1 Differentiate each equation.

a **i** $y = x^4$ **ii** $y = x^6$

b **i** $y = 3x^7$ **ii** $y = -4x^5$

c **i** $y = -x$ **ii** $y = 3x$

d **i** $y = 10$ **ii** $y = -3$

e **i** $y = \frac{1}{3}x^6$ **ii** $y = -\frac{3}{4}x^2$

f **i** $y = 4x^3 - 5x^2 + 2x - 8$ **ii** $y = 2x^4 + 3x^3 - x$

g **i** $y = 7x - \frac{1}{2}x^3$ **ii** $y = 2 - 5x^4 + \frac{1}{5}x^5$

2 Find $f'(x)$ for each function.

a **i** $f(x) = x^{\frac{3}{2}}$ **ii** $f(x) = x^{\frac{2}{3}}$ **b** **i** $f(x) = 8x^{\frac{1}{2}}$ **ii** $f(x) = 6x^{\frac{4}{3}}$

c **i** $f(x) = \frac{4}{9}x^{\frac{3}{4}}$ **ii** $f(x) = \frac{3}{5}x^{\frac{5}{6}}$ **d** **i** $f(x) = 3x^4 - 15x^{\frac{2}{5}} - 2$ **ii** $f(x) = x^3 - \frac{3}{5}x^{\frac{5}{3}} + \frac{4}{3}x^{\frac{1}{2}}$

e **i** $f(x)=x^{-1}$ **ii** $f(x)=-x^{-3}$ **f** **i** $f(x)=x^{-\frac{1}{2}}$ **ii** $f(x)=x^{-\frac{2}{3}}$

g **i** $f(x)=-6x^{-\frac{4}{3}}$ **ii** $f(x)=-8x^{-\frac{3}{4}}$ **h** **i** $f(x)=5x-\frac{8}{15}x^{-\frac{5}{2}}$ **ii** $f(x)=-\frac{7}{3}x^{\frac{3}{7}}+\frac{4}{3}x^{-6}$

3 A curve has the equation $y=x^2-3x^{\frac{1}{2}}+5$. Find $\frac{dy}{dx}$.

4 Given $f(x)=4x^3-3x^2+2x^{-\frac{3}{2}}$, find $f'(x)$.

5 Find the derivative of the function $f(x)=12x^{-\frac{1}{3}}+\frac{5x^{\frac{2}{5}}}{6}$.

Section 4: Simplifying into terms of the form ax^n

Notice that there is no rule in Key point 12.5 for differentiating products of functions, $y=f(x)g(x)$, or quotients of functions, $y=\frac{f(x)}{g(x)}$.

Before you can differentiate these, you need to convert them into terms of the form ax^n. This is often done by using the laws of indices.

Fast forward

In Student Book 2 you will learn that there are rules for differentiating products and quotients.

WORKED EXAMPLE 12.10

$y=(x+2)(x-5)$. Find $\frac{dy}{dx}$.

$y=(x+2)(x-5)$
$=x^2-3x-10$

Expand the brackets...

$\frac{dy}{dx}=2x-3$

...then differentiate.

Common error

You can't differentiate products and quotients of functions separately; that is, if:

- $y=f(x)g(x)$, $y'\neq f'(x)g'(x)$
- $y=\frac{f(x)}{g(x)}$, $y'\neq\frac{f'(x)}{g'(x)}$

WORKED EXAMPLE 12.11

A curve has equation $y=\frac{1}{x^3}$. Find y'.

$y=\frac{1}{x^3}$
$=x^{-3}$

First, use the laws of indices to rewrite the function in the form x^n...

$y'=-3x^{-4}$

...then differentiate.

Rewind

If you need to review the laws of indices, see Chapter 2.

WORKED EXAMPLE 12.12

Find the derivative of $f(x) = x^2\sqrt{x}$.

$f(x) = x^2\sqrt{x}$

$= x^2 x^{\frac{1}{2}}$

$= x^{2+\frac{1}{2}}$

$= x^{\frac{5}{2}}$

First, use the laws of indices to rewrite the function in the form x^n...

$f'(x) = \frac{5}{2}x^{\frac{5}{2}-1}$

$= \frac{5}{2}x^{\frac{3}{2}}$

...then differentiate.

WORKED EXAMPLE 12.13

Differentiate $y = \frac{2x-6}{\sqrt{x}}$.

$y = \frac{2x-6}{\sqrt{x}}$

$= \frac{2x-6}{x^{\frac{1}{2}}}$

$= 2x^{1-\frac{1}{2}} - 6x^{-\frac{1}{2}}$

$= 2x^{\frac{1}{2}} - 6x^{-\frac{1}{2}}$

Rewrite the function as a sum of terms in the form x^n.

$y' = 2 \times \frac{1}{2}x^{\frac{1}{2}-1} - 6\left(-\frac{1}{2}\right)x^{-\frac{1}{2}-1}$

$= x^{-\frac{1}{2}} + 3x^{-\frac{3}{2}}$

Now differentiate each term separately.

WORK IT OUT 12.1

Three students' attempts to differentiate $f(x) = \frac{x^2-3x}{x^3}$ are shown.

Which is the correct solution? Identify the errors made in the incorrect solutions.

Solution 1	Solution 2	Solution 3
$f(x) = \frac{x^2-3x}{x^3}$	$f(x) = \frac{x^2-3x}{x^3}$ $= x^{-1} - 3x^{-2}$	$f(x) = \frac{x^2-3x}{x^3}$ $= x^{-3}(x^2-3x)$
$f'(x) = \frac{2x-3}{3x^2}$	$f'(x) = -x^{-2} + 6x^{-3}$	$f'(x) = -3x^{-4}(2x-3)$

EXERCISE 12D

1 Find $\frac{dy}{dx}$ for each equation.

a i $y=\sqrt[3]{x}$ ii $y=\sqrt[5]{x}$ b i $y=8\sqrt[4]{x}$ ii $y=\frac{\sqrt{x}}{3}$

c i $y=-\frac{1}{x}$ ii $y=\frac{1}{x^4}$ d i $y=\frac{3}{x^2}$ ii $y=-\frac{2}{5x^{10}}$

e i $y=\frac{1}{\sqrt{x}}$ ii $y=\frac{1}{\sqrt[3]{x}}$ f i $y=-\frac{10}{\sqrt[5]{x}}$ ii $y=\frac{8}{3\sqrt[4]{x}}$

2 Find $f'(x)$ for each function.

a i $f(x)=(2x-3)(x+4)$ ii $f(x)=3x(x-5)$ b i $f(x)=\sqrt{x}(4x+3)$ ii $f(x)=\sqrt[3]{x}(x-1)$

c i $f(x)=(\sqrt{x}+2x)^2$ ii $f(x)=(\sqrt[4]{x}-4)^2$ d i $f(x)=\left(x+\frac{1}{x}\right)^2$ ii $f(x)=\left(x+\frac{2}{x}\right)\left(x-\frac{2}{x}\right)$

3 Differentiate each function.

a i $f(x)=\frac{3x-2}{x}$ ii $f(x)=\frac{1+4x^2}{2x}$ b i $f(x)=\frac{\sqrt{x}-3}{x^2}$ ii $f(x)=\frac{x^2+4}{\sqrt{x}}$

4 Differentiate $y=x^2(3x-4)$.

5 A curve has the equation $y=2\sqrt{x}(x^3+4)$. Find y'.

6 $y=\sqrt[5]{x^4}$. Find $\frac{dy}{dx}$.

7 Find the derivative of the function $f(x)=\frac{8}{3\sqrt[4]{x^3}}$.

8 A curve has the equation $y=\frac{3x^5-2x}{x^2}$.

a Express $y=\frac{3x^5-2x}{x^2}$ in the form $y=ax^p+bx^q$. b Hence find $\frac{dy}{dx}$.

9 $f(x)=\frac{(x+1)(x+9)}{x}$

Show that $f'(x)=\frac{(x-3)(x+3)}{x^2}$.

10 $f(x)=\frac{9x^2+3}{2\sqrt[3]{x}}$. Find $f'(x)$.

11 Find the derivative of the curve $y=\frac{(2\sqrt{x}-3)^2}{\sqrt{x^3}}$.

Section 5: Interpreting derivatives and second derivatives

The derivative $\frac{dy}{dx}$ has two related interpretations.

- It is the gradient of the graph of y against x.
- It measures how fast y changes when x is changed – this is called the rate of change of y with respect to x.

To calculate the gradient (or the rate of change) at any particular point, simply substitute the value of x into the equation for the derivative.

WORKED EXAMPLE 12.14

Find the gradient of the graph $y = 4x^3$ at the point where $x = 2$.

$\frac{dy}{dx} = 12x^2$ — The gradient is given by the derivative, so find $\frac{dy}{dx}$.

When $x = 2$ — Substitute the given value for x.

$\frac{dy}{dx} = 12 \times 2^2 = 48$

So the gradient is 48. — You can also write this in function notation: if $f(x) = 4x^3$ then $f'(x) = 12x^2$ so $f'(2) = 48$.

If you know the gradient of a graph at a particular point, you can find the value of x at that point. This involves solving an equation.

Tip

Your calculator may be able to give the gradient at a given point, but it cannot find the expression for the derivative.

WORKED EXAMPLE 12.15

Find the values of x for which the graph of $y = x^3 - 7x + 1$ has a gradient of 5.

$\frac{dy}{dx} = 3x^2 - 7$ — The gradient is given by the derivative.

$\frac{dy}{dx} = 5$ — You know that $\frac{dy}{dx} = 5$ so you can form an equation.

$3x^2 - 7 = 5$

$3x^2 = 12$

$x^2 = 4$

$x = 2$ or -2

Increasing and decreasing functions

The sign of the gradient at a point tells you whether the function is increasing or decreasing at that point.

Key point 12.6

- If $\frac{dy}{dx}$ is > 0 the function is increasing – as x gets larger so does y.
- If $\frac{dy}{dx}$ is < 0 the function is decreasing – as x gets larger y gets smaller.

Fast forward

Chapter 13 will show you what happens when $\frac{dy}{dx} = 0$.

A function can increase on some intervals and decrease on others.

WORKED EXAMPLE 12.16

Find the range of values of x for which the function $f(x) = 2x^3 - 6x$ is decreasing.

$f'(x) = 6x^2 - 6$	First find the derivative.
$f'(x) < 0$ $6x^2 - 6 < 0$	A decreasing function has negative gradient.
$6(x^2 - 1) < 0$ $(x-1)(x+1) < 0$	This is a quadratic inequality. To solve it, first sketch the graph.

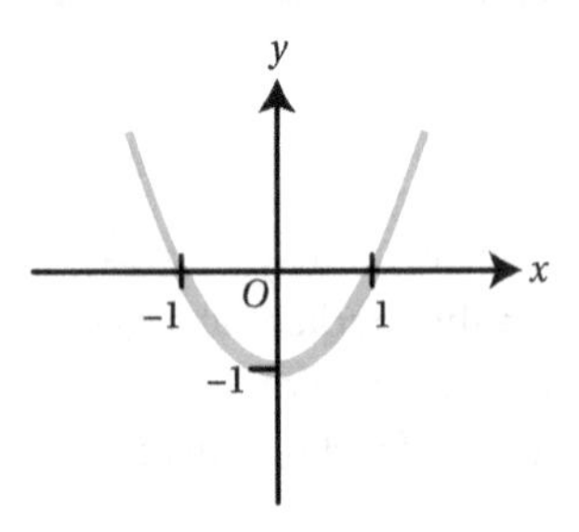

$-1 < x < 1$	The graph is below the x-axis between the two intercepts.

Elevate

See Support Sheet 12 for a further example of increasing/decreasing functions and for more practice questions.

Common error

Make sure that you remember to find the derivative in increasing (or decreasing) function questions: solve the inequality $f'(x) > 0$ (or < 0) and **not** just $f(x) > 0$ (or < 0).

Rewind

See Chapter 3 for a reminder about quadratic inequalities.

Some functions increase (or decrease) for all values of x.

WORKED EXAMPLE 12.17

Show that the function $f(x) = 3x^3 + 5x$ is increasing for all x.

$f'(x) = 9x^2 + 5$	The derivative tells you whether a function is increasing or decreasing.
Since $x^2 \geq 0$, $f'(x) > 0$ for all x.	Squares can never be negative.
Hence $f(x)$ is always increasing.	A positive derivative means the function is increasing.

WORK IT OUT 12.2

Is the function $f(x) = 5x - 3x^2$ increasing or decreasing at $x = -2$?

Which is the correct solution? Identify the errors made in the incorrect solutions.

Solution 1	Solution 2	Solution 3
$f(-2) = 5(-2) - 3(-2)^2$	$f'(x) = 5 - 6x$	$f'(x) = 5 - 6x$
$= -10 - 12$	$f'(-2) = 5 + 12$	$5 - 6x = 0$
$= -22 < 0$	$= 17 > 0$	$x = \frac{5}{6} > 0$
So it is decreasing.	So it is increasing.	So it is increasing.

There is nothing special about the variables y and x. You can just as easily say that $\frac{dB}{dQ}$ is the gradient of the graph of B against Q or that $\frac{d(\text{frequency})}{d(\text{length})}$ measures how fast the frequency of a guitar string changes when you change the length. To emphasise which variables are being used, $\frac{dy}{dx}$ is called the derivative of y with respect to x.

WORKED EXAMPLE 12.18

Given that $a = \sqrt{S}$, find the rate of change of a with respect to S when $S = 9$.

$a = S^{\frac{1}{2}}$

$\frac{da}{dS} = \frac{1}{2}S^{-\frac{1}{2}}$

$= \frac{1}{2\sqrt{S}}$

The rate of change is given by the derivative.

When $S = 9$:

$\frac{da}{dS} = \frac{1}{2\sqrt{9}} = \frac{1}{6}$

Substitute the given value for S.

Higher derivatives

Since the derivative $\frac{dy}{dx}$ is itself a function of x, you can differentiate it with respect to x. The result is called the **second derivative**.

Key point 12.7

The second derivative is the derivative of $\frac{dy}{dx}$ and is given the symbol $\frac{d^2y}{dx^2}$ or $f''(x)$. It measures the rate of change of the gradient.

You can differentiate again to find the third derivative $\left(\frac{d^3y}{dx^3} \text{ or } f'''(x)\right)$, fourth derivative $\left(\frac{d^4y}{dx^4} \text{ or } f^{(4)}(x)\right)$, and so on.

WORKED EXAMPLE 12.19

$f(x) = 5x^3 - 4x$

a Find $f''(x)$.
b Find the rate of change of the gradient of the graph of $y = f(x)$ at the point where $x = -1$.
c Show that, when $x = 0.1$, the graph of $y = f(x)$ is decreasing, but its gradient is increasing.

a $f'(x) = 15x^2 - 4$
$f''(x) = 30x$

Differentiate $f(x)$ and then differentiate the result.

b $f''(-1) = -30$

The rate of change of the gradient means the second derivative.

c $f'(0.1) = 15 \times 0.1^2 - 4$
$= -3.85 < 0$
So the function is decreasing.

A decreasing function has a negative gradient.

$f''(0.1) = 30 \times 0.1 = 3 > 0$
So the gradient is increasing.

The rate of change of gradient is $f''(0.1)$.
If this is positive, the gradient is increasing.

EXERCISE 12E

1 Write each rate of change as a derivative.

- **a** The rate of change of z as t changes.
- **b** The rate of change of Q with respect to P.
- **c** How fast R changes when m is changed.
- **d** How quickly the volume of a balloon (V) changes over time (t).
- **e** The rate of increase of the cost of apples (y) as the mass of the apple (x) increases.
- **f** The rate of change of the rate of change of z as y changes.
- **g** The second derivative of H with respect to m.

2 **a** **i** If $f = 5x^{\frac{1}{3}}$ what is the derivative of f with respect to x?

ii If $p = 3q^5$ what is the derivative of p with respect to q?

b **i** Differentiate $d = 3t + 7t^{-1}$ with respect to t.

ii Differentiate $r = c + \frac{1}{c}$ with respect to c.

c **i** Find the second derivative of $y = 9x^2 + x^3$ with respect to x.

ii Find the second derivative of $z = \frac{3}{t}$ with respect to t.

3 **a** **i** If $y = 5x^2$, find $\frac{dy}{dx}$ when $x = 3$.

ii If $y = x^3 + \frac{1}{x}$, find $\frac{dy}{dx}$ when $x = 1.5$.

b **i** If $A = 7b + 3$, find $\frac{dA}{db}$ when $b = -1$.

ii If $\Phi = \theta^2 + \theta^{-3}$, find $\frac{d\Phi}{d\theta}$ when $\theta = 0.1$.

c **i** Find the gradient of the graph of $A = x^3$ when $x = 2$.

ii Find the gradient of the tangent to the graph of $z = 2a + a^2$ when $a = -6$.

d **i** How quickly does $f = 4T^2$ change as T changes when $T = 3$?

ii How quickly does $g = y^4$ change as y changes when $y = 2$?

e **i** What is the rate of increase of W with respect to p when p is -3 if $W = -p^2$?

ii What is the rate of change of L with respect to c when $c = 6$ if $L = 7\sqrt{c} - 8$?

4 **a** **i** If $y = ax^2 + (1-a)x$ where a is a constant, find $\frac{dy}{dx}$.

ii If $y = x^3 + b^2$ where b is a constant, find $\frac{dy}{dx}$.

b **i** If $Q = \sqrt{ab} + \sqrt{b}$ where b is a constant, find $\frac{dQ}{da}$.

ii If $D = 3(av)^2$ where a is a constant, find $\frac{dD}{dv}$.

5 **a** **i** If $y = x^3 - 5x$, find $\frac{d^2y}{dx^2}$ when $x = 9$.

ii If $y = 8 + 2x^4$, find $\frac{d^2y}{dx^2}$ when $x = 4$.

b **i** If $S = 3A^2 + \frac{1}{A}$, find $\frac{d^2S}{dA^2}$ when $A = 1$.

ii If $J = v - \sqrt{v}$, find $\frac{d^2J}{dv^2}$ when $v = 9$.

c **i** Find the second derivative of B with respect to n if $B = 8n$ and $n = 2$.

ii Find the second derivative of g with respect to r if $g = r^7$ and $r = 1$.

6 **a** **i** If $y = 3x^3$ and $\frac{dy}{dx} = 36$, find x.

ii If $y = x^4 + 2x$ and $\frac{dy}{dx} = 5$, find x.

b **i** If $y = 2x + \frac{8}{x}$ and $\frac{dy}{dx} = -30$, find y.

ii If $y = \sqrt{x} + 3$ and $\frac{dy}{dx} = \frac{1}{6}$, find y.

c **i** If $y = 3x^3$ and $\frac{d^2y}{dx^2} = -54$, find x.

ii If $y = \frac{2}{x}$ and $\frac{d^2y}{dx^2} = \frac{1}{2}$, find x.

7 **a** **i** Find the interval in which $x^2 - x$ is an increasing function.

ii Find the interval in which $x^2 + 2x - 5$ is a decreasing function.

b **i** Find the range of values of x for which $x^3 - 3x$ is an increasing function.

ii Find the range of values of x for which $x^3 + 2x^2 - 5$ is a decreasing function.

c **i** Find the interval in which the gradient of $y = x^3 - 3x$ is decreasing.

ii Find the interval in which the gradient of $y = 2x^3 - x^2$ is increasing.

8 **a** Find the rate of change of $y = \frac{3}{\sqrt{x}}$ at the point where $x = 9$.

b Is the gradient increasing or decreasing at this point?

9 Find the rate of change of the gradient of $f(x) = \sqrt{x} + \frac{1}{x}$ when $x = 4$.

10 Find the range of values of x for which the function $y = 3x^2 - 4x + 1$ is decreasing.

11 Find the values of x for which the graph of $y = \frac{1}{3}x^3 - x^2 - 12x + 1$ has gradient 3.

12 Find the x-coordinates of the points on the curve $y = x^4 - 12x^2 + 3x - 1$ where $\frac{d^2y}{dx^2} = 0$.

13 Show that $y = x^3 + kx + c$ is always increasing if $k > 0$.

14 Find all points of the graph of $y = x^2 - 2x + 1$ where the gradient equals the y-coordinate.

15 Find the range of x-values for which $f(x) = x^3 - 6x^2 + 5$ is an increasing function.

16 The function $f(x) = 12x - 2x^2 - \frac{1}{3}x^3$ is increasing for $a < x < b$. Find the values of the constants a and b.

17 Find the interval in which the gradient of the graph of $y = 7x - x^2 - x^3$ is decreasing.

18 In what interval is the gradient of the graph $y = \frac{1}{4}x^4 + x^3 - \frac{1}{2}x^2 - 3x + 6$ decreasing?

19 Find the range of values of x for which the function $f(x) = x^3 - 6x^2 + 9x + 2$ is decreasing but its gradient is increasing.

20 Find an alternative expression for $\frac{d^n}{dx^n}(x^n)$.

Checklist of learning and understanding

- The gradient (or the derivative) of a function at the point P is the gradient of the tangent to the graph of the function at that point.
- To find the gradient of a function you differentiate.
  - Differentiation from first principles:

$$f'(x) = \lim_{h \to 0} \frac{f(x+h) - f(x)}{h}$$

  - If $f(x) = x^n$, then $f'(x) = nx^{n-1}$.
  - If $y = kf(x)$, where k is a constant, then $y' = kf'(k)$.
  - If $y = f(x) + g(x)$, then $y' = f'(x) + g'(x)$.
- The derivative at a point gives the rate of change of the y-coordinate at that point.
- You can use the derivative to tell whether a function is increasing or decreasing:
  - If $\frac{dy}{dx} > 0$ the function is increasing.
  - If $\frac{dy}{dx} < 0$ the function is decreasing.
- The second derivative, $\frac{d^2y}{dx^2}$, gives the rate of change of gradient.

Mixed practice 12

1. For $y=\dfrac{3x-2}{\sqrt{x}}$, find $\dfrac{dy}{dx}$.

   Choose from these options.

   A $\frac{9}{2}x^{\frac{1}{2}}-x^{-\frac{1}{2}}$ B $3x^{\frac{3}{2}}-2x^{-\frac{1}{2}}$ C $\frac{3}{2}x^{-\frac{1}{2}}+x^{-\frac{3}{2}}$ D $3x^{-\frac{1}{2}}+2x^{-\frac{3}{2}}$

2. A curve has the equation $y=(4x^2-1)(3-x)$. Find $\dfrac{dy}{dx}$.

3. $f(x)=\dfrac{x^2-4}{2x}$. Find $f''(2)$.

4. Given that $f(x)=3\sqrt{x}-\dfrac{2}{\sqrt{x}}$, find:

   a $f'(x)$

   b the gradient of the graph of $y=f(x)$ at the point where $x=4$.

5. $f(x)=x^2+bx+c$. If $f(1)=2$ and $f'(2)=12$ find the values of b and c.

6. **a** Find the gradient of the curve $y=3\sqrt{x}-2$ at the point where it crosses the x-axis.

   b Is the curve increasing or decreasing at this point? Give a reason for your answer.

7. Find the range of values of x for which the function $y=3x^2-4x$ is increasing.

8. Find the rate of change of gradient of the function $y=x^2-2\sqrt{x}$ at the point where $x=9$.

9. A curve C has the equation:

   $$y=\frac{x^3+\sqrt{x}}{x},\quad x>0$$

   a Express $\dfrac{x^3+\sqrt{x}}{x}$ in the form x^p+x^q.

   b **i** Hence find $\dfrac{dy}{dx}$. **ii** Find $\dfrac{d^2y}{dx^2}$.

   [© AQA 2010]

10. A bird flies from a tree. At time t seconds, the bird's height, y metres, above the horizontal ground is given by:

    $$y=\frac{1}{8}t^4-t^2+5,\quad 0\leqslant t\leqslant 4$$

    a Find $\dfrac{dy}{dt}$.

    b **i** Find the rate of change of height of the bird in metres per second when $t=1$.

    ii Determine, with a reason, whether the bird's height above the horizontal ground is increasing or decreasing when $t=1$.

    c Find the value of $\dfrac{d^2y}{dt^2}$ when $t=2$.

    [© AQA 2013]

11. $y=x^2+ax-7$ is increasing for $x>5$. Find a.

    Choose from these options.

    A $a=-10$ B $a=-7$ C $a=5$ D $a=-5$

12 What is the rate of change of the gradient of $y = x^3 + 4x^2 - 2x + 1$ at $x = \frac{1}{2}$?

Choose from these options.

A 6 **B** $\frac{19}{2}$ **C** $\frac{9}{8}$ **D** 11

13 The graph shows the gradient function $y = f'(x)$.

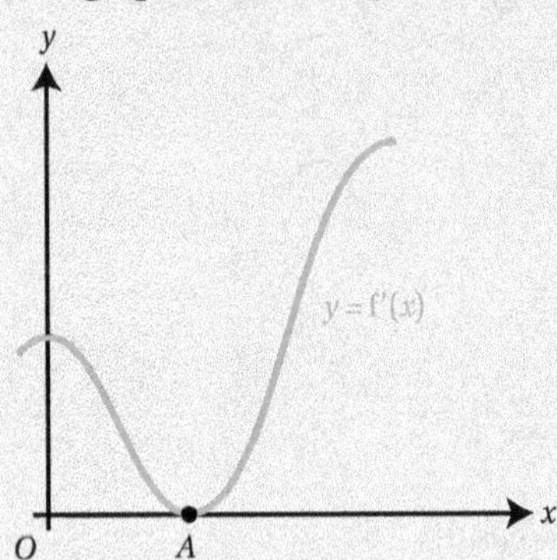

Which of these statements is definitely true at the point A?

Please choose from these options.

A $f(x)$ has a minimum **B** $f(x)$ has a maximum **C** $f(x) = 0$ **D** $f''(x) = 0$

14 Use differentiation from first principles to find $\frac{dy}{dx}$ for $y = x^3 - 5x$.

15 The diagram shows the graph of $y = f'(x)$.

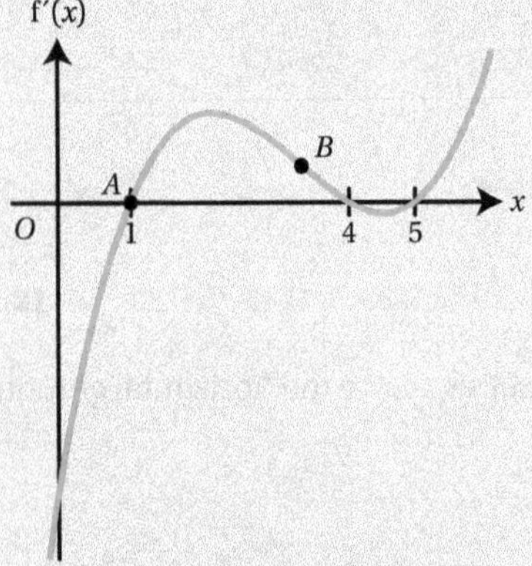

a State the value of the gradient of the graph of $y = f(x)$ at the point marked A.

b Is the function $f(x)$ increasing or decreasing at the point marked B?

c Sketch the function $y = f(x)$.

16 Find the coordinates of the point on the curve $y = \sqrt{x} + 3x$ where the gradient is 5.

17 Find the gradient of the graph of $y = \frac{1}{2\sqrt{x}}$ at the point where the y-coordinate is 3.

18 $f(x) = ax^3 + bx^{-2}$, where a and b are constants. $f'(1) = 18$ and $f''(1) = 18$. Find the values of a and b.

19 $f(x) = \sqrt{x^3} + 15\sqrt{x}$. Find the values of x for which the gradient of f is 9.

20 Find the range of values of x for which the gradient of the graph $y = x^4 - 3x^3 + x$ is decreasing.

13 Applications of differentiation

In this chapter you will learn how to:

- find the equations of tangents and normals to curves at given points
- find maximum and minimum points on curves
- solve problems which involve maximising or minimising quantities.

Before you start...

Chapter 12	You should be able to differentiate functions involving x^n.	1 Differentiate each expression. a $\frac{3x^4-2}{5x^2}$ b $\frac{3}{2\sqrt{x}}$
Chapter 12	You should be able to evaluate second derivatives.	2 Given that $y=\frac{3}{2x^2}$, evaluate $\frac{\mathrm{d}^2y}{\mathrm{d}x^2}$ when $x=-2$.
Chapter 6	You should be able to find the equation of a straight line.	3 Find the equation of the line through the point $(2,1)$ with gradient 3.
Chapter 6	You should be able to find the equation of a perpendicular to a line.	4 Find the equation of the perpendicular to the line with gradient $\frac{3}{4}$ that passes through the point $(2,-3)$.

What can you use differentiation for?

You can apply the techniques from the previous chapter to solve a variety of problems. First, you will look at graphs of functions, learning how to find equations of tangents and normals. These have applications in mechanics when studying collisions, as well in pure mathematics when defining the 'curvature' of a graph. Then you will be finding coordinates of maximum and minimum points and solving practical problems in which you need to minimise or maximise quantities, such as maximising profit or minimising costs in a business.

Section 1: Tangents and normals

The **normal** to a curve at a given point is a straight line that crosses the curve at that point and is perpendicular to the tangent.

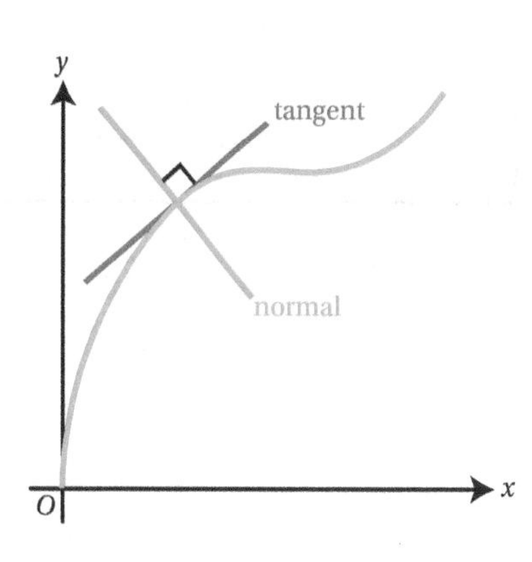

You should know from the previous chapter that the gradient of the tangent at a point is the value of the gradient of the curve at that point; you can find this by substituting the value of x into the equation for the first derivative, $\frac{\mathrm{d}y}{\mathrm{d}x}$.

Then you can find the gradient of the normal by using the fact that if two lines with gradients m_1 and m_2 are perpendicular, then $m_1 m_2 = -1$.

Once you have found the gradient you can use it, together with the coordinates of the point, to find the equation of the tangent and the normal.

Rewind

See Chapter 6 for a reminder of how to find the equation of the line through a given point and with a given gradient.

WORKED EXAMPLE 13.1

A curve has equation $y = x^3 - 5x^2 - x^{\frac{3}{2}} + 22$.

Find the equations of:

a the tangent

b the normal

to the curve at the point $(4, -2)$.

In each case give your answer in the form $ax + by + c = 0$, where a, b and c are integers.

a $y' = 3x^2 - 10x - \frac{3}{2}x^{\frac{1}{2}}$ — First find y'.

At $x = 4$:

$$y' = 3(4)^2 - 10(4) - \frac{3}{2}(4)^{\frac{1}{2}}$$
$$= 48 - 40 - 3$$
$$= 5$$

Then evaluate the derivative at $x = 4$. This will give the gradient of the tangent at $x = 4$.

Equation of tangent:

$$y - (-2) = 5(x - 4)$$
$$y + 2 = 5x - 20$$
$$5x - y - 22 = 0$$

Use $y - y_1 = m(x - x_1)$ to find the equation of the straight line with gradient 5, passing through $(4, -2)$.

b Gradient of normal:

$$m_n = -\frac{1}{5}$$

Find the gradient of the normal, using $m_1 m_2 = -1$.

Equation of normal:

$$y - (-2) = -\frac{1}{5}(x - 4)$$
$$5y + 10 = -x + 4$$
$$x + 5y + 6 = 0$$

Use the equation of the line again. The coordinates of the point are the same, but now use the gradient of the normal.

Key point 13.1

For tangents and normals at the point on the curve $y = f(x)$ where $x = a$

- The gradient of the tangent is $f'(a)$.
- The gradient of the normal is $-\frac{1}{f'(a)}$.
- The coordinates of the point are $x_1 = a, y_1 = f(a)$.

To find the equation of the tangent or the normal use $y - y_1 = m(x - x_1)$ with the appropriate gradient.

Common error

Having found $f'(a)$, don't forget to work out $-\frac{1}{f'(a)}$ for the gradient of the normal.

More difficult questions may give some information about the tangent and require you to find other information.

WORKED EXAMPLE 13.2

The tangent at point P on the curve $y = x^2 + 1$ passes through the origin. Find the possible coordinates of P.

Let P be the point with coordinates (p, q).

You need to find the equation of the tangent at P, but you don't know the coordinates of P.

Then $q = p^2 + 1$

As p lies on the curve, (p, q) must satisfy $y = x^2 + 1$.

$\frac{dy}{dx} = 2x$

The gradient of the tangent is given by $\frac{dy}{dx}$ when $x = p$.

When $x = p$:

$\frac{dy}{dx} = 2p$

$\therefore m = 2p$

Equation of tangent:

Write the equation of the tangent, remembering it passes through (p, q).

$$y - y_1 = m(x - x_1)$$
$$y - q = 2p(x - p)$$
$$y - (p^2 + 1) = 2p(x - p)$$

Since the line passes through $(0, 0)$:

The tangent passes through the origin, so set $x = 0, y = 0$.

$$0 - (p^2 + 1) = 2p(0 - p)$$
$$-p^2 - 1 = -2p^2$$
$$p^2 = 1$$
$$p = 1 \text{ or } -1$$

When $p = 1, q = 2$
When $p = -1, q = 2$
So the coordinates of P are $(1, 2)$ or $(-1, 2)$.

Now use $q = p^2 + 1$ to find the corresponding values of q.

WORKED EXAMPLE 13.3

The point P on the curve $y = \frac{a}{x}$ has x-coordinate 3. The normal to the curve at P is parallel to $x - 3y + 6 = 0$. Find the value of the constant a.

Equation of normal at P:

$x - 3y + 6 = 0$

$$y = \frac{1}{3}x + 2$$

Rearrange the equation of the normal into the form $y = mx + c$ to find its gradient.

So, gradient of curve at P is:

$$m = -\frac{1}{\frac{1}{3}} = -3$$

Find the gradient of the tangent at P, using the relation $m_1 m_2 = -1$.

$y = ax^{-1}$

$y' = -ax^{-2}$

The gradient of the curve is given by y'.

When $x = 3$:

$y' = -a(3)^{-2}$

$$= -\frac{a}{9}$$

Evaluate at $x = 3$.

$$\therefore -\frac{a}{9} = -3$$

$$a = 27$$

The gradient of the tangent and the gradient of the curve at P are the same, so set them equal and solve for a.

EXERCISE 13A

1. Find the equation of the tangent and normal to each curve. In each case, give your answer in the form $ax + by + c = 0$.

   a **i** $y = 2x^2 - 4x$ at $(1, -2)$ **ii** $y = x^3 + 6x + 9$ at $(-1, 2)$

   b **i** $y = \frac{3}{x}$ at $(-3, -1)$ **ii** $y = -\frac{6}{x^2}$ at $\left(2, -\frac{3}{2}\right)$

   c **i** $y = x + \frac{12}{\sqrt{x}}$ at $(4, 10)$ **ii** $y = 4\sqrt[3]{x} - \frac{x}{2}$ at $(8, 4)$

2. Find the equation of the tangent to the curve $y = \frac{x^2 + 4}{\sqrt{x}}$ at the point where $x = 4$.

3. Find the equation of the normal to the curve $y = 2x - \frac{3}{\sqrt{x}}$ at the point where $x = 4$.

4. The normal at the point $x = 2$ to the curve with equation $y = 2x^3 - 4x$ crosses the x-axis at P. Find the coordinates of P.

5. The tangent to the graph of $y = \frac{3}{x^2}$ at $x = 3$ crosses the coordinate axes at points A and B. Find the area of the triangle AOB, where O is the origin.

6. Find the equations of the tangents to the curve $y = 2x^3 - 9x^2 + 12x + 1$ that are parallel to the x-axis.

7. Find the x-coordinates of the points on the curve $y = x^3 - 2x^2$ at which the tangent is parallel to the normal to the curve at the point $(1, -1)$.

8. The normal to the curve $y = ax^3 + bx^2$ at the point $(1, 2)$ is parallel to the line $x - 2y + 14 = 0$. Find the values of a and b.

9. The tangent to the curve $y = p\sqrt{x} + qx$ at the point $\left(1, -\frac{7}{6}\right)$ is parallel to the line $2x + 3y - 12 = 0$. Find the values of p and q.

10. $f(x) = \frac{2}{x}$

    Find the coordinates of the point where the tangent to $y = f(x)$ at $x = 1$ intersects the normal to $y = f(x)$ at $x = -2$.

11. Find the coordinates of the point where the tangent to the curve $y = x^3 - 3x^2$ at $x = 2$ meets the curve again.

12. Find the coordinates of the point on the curve $y = (x-1)^2$ where the normal passes through the point $(4, 0)$.

13. $y = -\frac{1}{2}x + c$ is the equation of a normal to the curve $y = x^2 + 5x + 4$. Find the value of the constant c.

14. A tangent is drawn on the graph $y = \frac{k}{x}$ at the point where $x = a \ (a > 0)$. The tangent intersects the y-axis at P and the x-axis at Q. If O is the origin, show that the area of the triangle OPQ is independent of a.

15. Show that the tangent to the curve $y = x^3 - x$ at the point with x-coordinate a meets the curve again at a point with x-coordinate $-2a$.

Elevate

See Extension Sheet 13 for some harder questions with tangents and normals.

Section 2: Stationary points

In real life you might be interested in maximising profits or minimising the drag on a car. You can use calculus to describe such problems mathematically.

The quantity you wish to maximise or minimise usually depends on another variable. For example, the profit might depend on the selling price of the goods. You can represent this relationship on a graph.

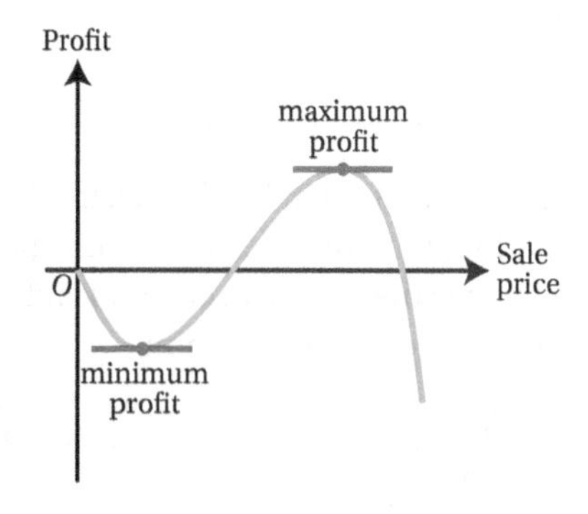

The gradient at both the maximum and the minimum point on the graph is zero.

Key point 13.2

To find local maximum and local minimum points, solve the equation $\frac{dy}{dx} = 0$.

The terms **local maximum** and **local minimum** are used because it is possible that the largest or smallest value of the whole function occurs at the endpoint of the graph, or that there are several points that have zero gradient. These points are just the largest or smallest values of y in their local area.

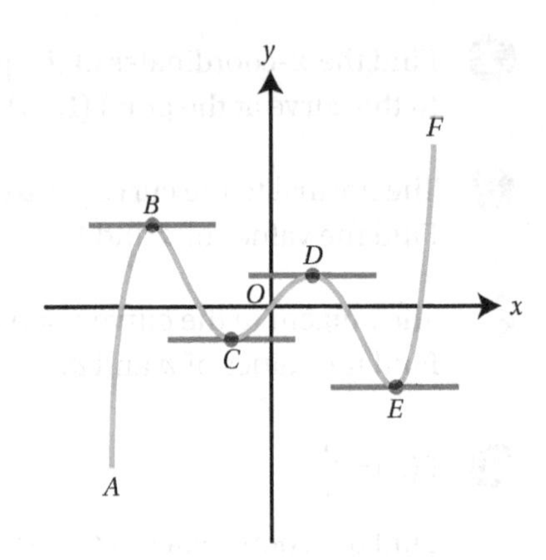

For this function, $y = \mathrm{f}(x)$, defined over the section shown on the graph:

- B and D are local maximum points
- C and E are local minimum points
- A is the point where y is smallest
- F is the point where y is largest.

Points such as B, C, D and E, which have a zero gradient, are called **stationary points.**

WORKED EXAMPLE 13.4

Find the coordinates of the stationary points of $y = 2x^3 - 15x^2 + 24x + 8$.

$\frac{\mathrm{d}y}{\mathrm{d}x} = 6x^2 - 30x + 24$ — First differentiate.

For stationary points $\frac{\mathrm{d}y}{\mathrm{d}x} = 0$: — Then solve the equation $\frac{\mathrm{d}y}{\mathrm{d}x} = 0$.

$6x^2 - 30x + 24 = 0$

$x^2 - 5x + 4 = 0$

$(x-4)(x-1) = 0$

$x = 1$ or $x = 4$

When $x = 1$: — Remember to find the y-coordinate for each point.

$y = 2(1)^3 - 15(1)^2 + 24(1) + 8 = 19$

When $x = 4$:

$y = 2(4)^3 - 15(4)^2 + 24(4) + 8 = -8$

Therefore, the stationary points are: — State the coordinates.

$(1, 19)$ and $(4, -8)$

The calculation in Worked example 13.4 does not identify whether the stationary points found are maximum or minimum points. To determine which is the case, you can look at how the gradient changes on either side of the stationary point.

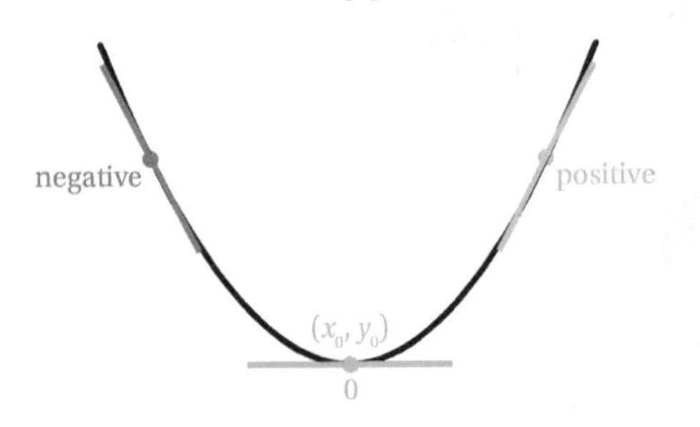

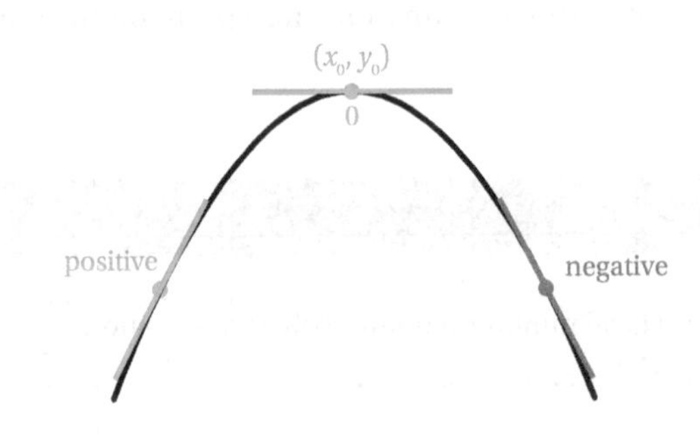

The diagrams show that for a minimum point the gradient goes from negative, through zero, then to positive. For a maximum point the gradient goes from positive, through zero, then to negative.

At a minimum point the gradient $\left(\frac{dy}{dx}\right)$ is changing from negative to positive so the rate of change of the gradient $\left(\frac{d^2y}{dx^2}\right)$ is positive. For a maximum point the gradient is changing from positive to negative so $\frac{d^2y}{dx^2}$ is negative. This leads to a useful test.

Key point 13.3

Given a stationary point (x_0, y_0) of a function $y = f(x)$, if:

- $\frac{d^2y}{dx^2} < 0$ at x_0 then (x_0, y_0) is a *maximum*
- $\frac{d^2y}{dx^2} > 0$ at x_0 then (x_0, y_0) is a *minimum*
- $\frac{d^2y}{dx^2} = 0$ at x_0 then no conclusion can be drawn.

If a question asks you to 'classify' or 'determine the nature' of the stationary points on a curve, you need to decide whether each point is a maximum or minimum.

WORKED EXAMPLE 13.5

Find and classify the stationary points of the curve $y = 2x^3 - 15x^2 + 24x + 8$ (from Worked example 13.4).

The stationary points are $(1, 19)$ and $(4, -8)$.	The stationary points were found in Worked example 13.4.
$\frac{d^2y}{dx^2} = 12x - 30$	Find the second derivative.
At $x = 1$: $\frac{d^2y}{dx^2} = 12(1) - 30$	Evaluate the second derivative at the stationary point with $x = 1$.
$= -18\ (< 0)$ ∴ $(1, 19)$ is a maximum	$\frac{d^2y}{dx^2} < 0$ so the point is a maximum.
At $x = 4$: $\frac{d^2y}{dx^2} = 12(4) - 30$	Evaluate the second derivative at the stationary point with $x = 4$.
$= 18\ (> 0)$ ∴ $(4, -8)$ is a minimum.	$\frac{d^2y}{dx^2} > 0$ so the point is a minimum.

Common error

Make sure you don't attempt to solve the equation $\frac{d^2y}{dx^2} = 0$. Once you have found the x-values of any stationary points, substitute these values into $\frac{d^2y}{dx^2}$ to test whether the points are maximum or minimum.

Fast forward

A You will see in Student Book 2, Chapter 12, that there is another type of stationary point other than a maximum or minimum.

Tip

Remember that the largest or smallest value of a function in a given interval could occur at the endpoint(s) of the interval instead of at the local maximum or minimum.

Elevate

See Support Sheet 13 for a further example of this type and for more practice questions.

One use of stationary points is to help to find the largest and smallest values of a function for a given set of x-values.

WORKED EXAMPLE 13.6

Find the largest and smallest values of $f(x)=2x^3-6x+3$ for $-2\leqslant x\leqslant 3$.

Stationary points:

$$f'(x)=0$$
$$6x^2-6=0$$
$$6(x-1)(x+1)=0$$
$$x=1 \text{ or } x=-1$$

Stationary points give local minimum and maximum values. Solve $f'(x)=0$ to find any.

$$f(1)=2(1)^3-6(1)+3$$
$$=-1$$
$$f(-1)=2(-1)^3-6(-1)+3$$
$$=7$$

Find the value of the function at the stationary points.

End points:

$$f(-2)=2(-2)^3-6(-2)+3$$
$$=-1$$
$$f(3)=2(3)^3-6(3)+3$$
$$=39$$

Check whether the function actually achieves a larger value at either endpoint than it does at its local maximum, and/or a smaller value than at its local minimum.

So, in the specified region, the largest value of $f(x)$ is 39 and the smallest value of $f(x)$ is -1 or $-1\leqslant f(x)\leqslant 39$.

$f(-2)$ is the same as at the local minimum, but $f(3)$ is larger than the value at the local maximum.

EXERCISE 13B

1. Find and classify the stationary points on each curve.

   a **i** $y=x^3-5x^2$ **ii** $y=1+6x^2-x^3$ **b** **i** $y=x^4-8x^2$ **ii** $y=x^4-2x^2+3$

   c **i** $y=4x+\frac{1}{x}$ **ii** $y=4x^2+\frac{27}{x}$ **d** **i** $y=2\sqrt{x}-3x$ **ii** $y=\frac{1}{\sqrt{x}}+4x$

2. Find and classify the stationary points on the curve $y=x^3+3x^2-24x+12$.

3. Prove that the curve $y=x^3-3x^2+4x-1$ has no stationary points.

4. $f(x)=\frac{9x^2+1}{x}$

   a Find the x-coordinates of the stationary points on the curve $y=f(x)$.

   b Determine whether each is a maximum or minimum point.

5. Find the coordinates of the stationary point on the curve $y=x-\sqrt{x}$ and determine its nature.

6. **a** Find the coordinates of the stationary points on the curve with equation $y=x^2\left(6\sqrt[3]{x}-7\right)$.

   b Establish whether each is a maximum or minimum point.

7 The curve $y = x^3 + 10x^2 + kx - 2$ has a stationary point at $x = -8$.

a Find the x-coordinate of the other stationary point.

b Determine the nature of both stationary points.

8 The curve $y = ax^2 + bx - 2$ has a minimum point at $x = -2$ and passes through the point $(1, 13)$. Find the values of a and b.

9 $f(x) = 4 - 9x + 6x^2 - x^3, -2 \leqslant x \leqslant 4$

Find the largest and smallest values of $f(x)$.

10 $f(x) = -\frac{1}{4}x^4 - x^3 + 2x^2 - 10$

Show that $f(x) \leqslant 22$ for all x.

11 a Find the stationary points of the curve $y = 3x^4 - 16x^3 + 18x^2 + 6$.

b Find the set of values of c for which the equation $3x^4 - 16x^3 + 18x^2 + 6 = c$ has four real roots.

12 Find and classify, in terms of k, the stationary points on the curve $y = kx^3 + 6x^2$.

Section 3: Optimisation

You can now apply results from the previous section to solve practical problems that involve maximising or minimising quantities.

WORKED EXAMPLE 13.7

The distance travelled, in metres, by a paper aeroplane of weight w grams, where $w > 1$ is given by $s = \left(1 - \frac{1}{\sqrt{w}}\right)\left(\frac{5}{w}\right)$.

a Find the weight the aeroplane needs to be to travel its maximum distance.

b Find this maximum distance.

a $s = \left(1 - \frac{1}{\sqrt{w}}\right)\left(\frac{5}{w}\right)$

$= \frac{5}{w} - \frac{5}{w\sqrt{w}}$

$= 5w^{-1} - 5w^{-\frac{3}{2}}$

Expand the brackets and use the laws of indices to write each term in the form ax^n.

$\frac{ds}{dw} = -5w^{-2} + \frac{15}{2}w^{-\frac{5}{2}}$

Then find the maximum by solving $\frac{ds}{dw} = 0$.

Stationary points:

$\frac{ds}{dw} = 0$

$-5w^{-2} + \frac{15}{2}w^{-\frac{5}{2}} = 0$

$\frac{15}{2}w^{-\frac{5}{2}} = 5w^{-2}$

Continues on next page

$3w^{-\frac{5}{2}} = 2w^{-2}$ — Multiply through by 2 and divide by 5.

$\frac{3}{w^{\frac{5}{2}}} = \frac{2}{w^2}$ — Rewrite without the negative indices.

$3 = \frac{2w^{\frac{5}{2}}}{w^2}$ — Multiply through by $w^{\frac{5}{2}}$.

$3 = 2w^{\frac{1}{2}}$ — Simplify.

$\sqrt{w} = \frac{3}{2}$ — Solve for w.

$w = \frac{9}{4}$ grams

Nature of stationary point: — You know that there is a stationary point at this value of w but you need to check that this is a maximum.

$\frac{d^2s}{dw^2} = 10w^{-3} - \frac{75}{4}w^{-\frac{7}{2}}$

When $w = \frac{9}{4}$:

$\frac{d^2s}{dw^2} = -0.22 < 0$

$\therefore$ it is a maximum.

b Maximum value of s: — Find the maximum value of the distance by evaluating the original expression for s at $w = \frac{9}{4}$.

$s = \left(1 - \frac{1}{\sqrt{\frac{9}{4}}}\right)\left(\frac{5}{\frac{9}{4}}\right)$

$= \left(1 - \frac{2}{3}\right)\left(\frac{20}{9}\right)$

$= \frac{20}{27}$ m

In more difficult examples a function appears to depend on two different variables. However, these two variables will always be related by a **constraint** that allows one of them to be eliminated. You can then follow the normal procedure for finding maxima or minima.

Two common types of constraint

- A shape has a fixed perimeter, area or volume – this gives an equation relating different variables (height, length, radius...).
- A point lies on a given curve – this gives a relationship between x and y.

See Gateway to A Level Section P for a reminder of formulae for the area and volume of common shapes.

WORKED EXAMPLE 13.8

A wire of length 12 cm is bent to form a rectangle.

a Show that the area, A, is given by $A = 6x - x^2$, where x is the width of the rectangle.

b Find the maximum possible area.

a Let the length of the rectangle be y.
Then: $A = xy$

Introduce a variable, y, for the length in order to write an expression for the area (which is to be maximised).

$P = 2x + 2y$
Since the perimeter $P = 12$:
$12 = 2x + 2y$
$6 = x + y$

Use the information about the perimeter to form a second equation (the constraint).

$y = 6 - x$

Make y the subject.

So:
$A = xy$
$= x(6 - x)$
$= 6x - x^2$

Substitute into the expression for A in order to eliminate y and express A as a function of x only.

b Stationary points:
$\frac{dA}{dx} = 0$
$6 - 2x = 0$
$x = 3$

Find the value of x at which the stationary point of A occurs by solving $\frac{dA}{dx} = 0$.

Nature of stationary point:
$\frac{d^2A}{dx^2} = -2\ (< 0)$

$\therefore$ the point is a maximum.

Check that the stationary point is a maximum.

The maximum value of A is:
$A = x(6 - x)$
$= 3(6 - 3)$
$= 9\ \text{cm}^2$

Find the maximum value of A by substituting $x = 3$ into $A = x(6 - x)$.

Gateway to A Level

See Gateway to A level Section Q for revision of the algebraic manipulation of formulae.

Fast forward

A In Student Book 2 you will need to find stationary points of more complicated functions using A Level differentiation methods.

Common error

Don't forget to substitute the value of x that maximises or minimises the quantity back into the original expression to find the actual maximum or minimum value.

WORKED EXAMPLE 13.9

An open-topped cylindrical can has base radius r and height h.

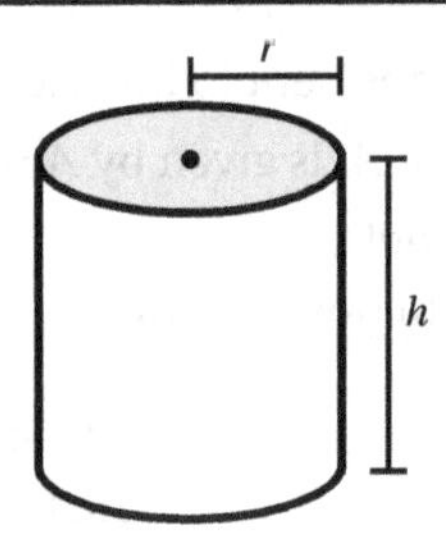

The external surface area $A = 243\pi$ cm^2.

a Show that the volume, V, is given by $V = \frac{\pi}{2}(243r - r^3)$.

b Hence find the maximum capacity of the can.

a $V = \pi r^2 h$ — Write down an expression for the quantity to be maximised.

$A = \pi r^2 + 2\pi rh$ — The surface area is made up of the area of the base (πr^2) and the curved surface area of the cylinder ($2\pi rh$).

Since $A = 243\pi$: — Use the information about the surface area to form a second equation (the constraint).

$$243\pi = \pi r^2 + 2\pi rh$$

$$243 = r^2 + 2rh$$

$$h = \frac{243 - r^2}{2r}$$ — Make h the subject.

So — Substitute into the expression for V in order to eliminate h and express V as a function of r only.

$$V = \pi r^2\left(\frac{243 - r^2}{2r}\right)$$

$$= \frac{\pi r^2}{2r}(243 - r^2)$$

$$= \frac{\pi r}{2}(243 - r^2)$$

$$= \frac{\pi}{2}(243r - r^3)$$

b Stationary points: — Find the value of r at which the stationary point of V occurs by solving $\frac{dV}{dr} = 0$.

$$\frac{dV}{dr} = 0$$

$$\frac{\pi}{2}(243 - 3r^2) = 0$$

$$243 - 3r^2 = 0$$

$$r^2 = 81$$

$$r = 9\ (r > 0)$$

Continues on next page

Nature of stationary point:

$$\frac{d^2V}{dr^2} = \frac{\pi}{2}(-6r)$$
$$= -3\pi r$$
$$= -3\pi(9)$$
$$= -27\pi \ (< 0)$$

$\therefore$ it is a maximum.

Check that the stationary point is a maximum.

Maximum value of V is:

$$V = \frac{\pi}{2}(243r - r^3)$$
$$= \frac{\pi}{2}(243(9) - (9)^3)$$
$$= 729\pi \text{ cm}^3$$

Don't forget to find the maximum value of V by substituting $r = 9$ into $V = \frac{\pi}{2}(243r - r^3)$.

WORKED EXAMPLE 13.10

Let A be a point on the curve $y = x^2$, with x-coordinate a.

Let L be the distance from A to the point $P\left(2, \frac{1}{2}\right)$.

a Write down an expression for L^2 in terms of a.
b Find the minimum possible value of L^2.
c Hence write down the coordinate of the point on the curve $y = x^2$ which is closest to the point $P\left(2, \frac{1}{2}\right)$.

a The coordinates of A are (a, a^2).

For any point of the curve, the y-coordinate equals x^2.

The square of the distance between A and P is:

$$L^2 = (a-2)^2 + \left(a^2 - \frac{1}{2}\right)^2$$

Use $L^2 = (x_2 - x_1)^2 + (y_2 - y_1)^2$

b $L^2 = a^2 - 4a + 4 + a^4 - a^2 + \frac{1}{4}$

$$= a^4 - 4a + \frac{17}{4}$$

Before differentiating to find stationary points, you need to expand the brackets.

$$\frac{d(L^2)}{da} = 4a^3 - 4$$

The minimum point is when the derivative is zero. Here this is the derivative of L^2.

Stationary points:

$$\frac{d(L^2)}{da} = 0$$
$$4a^3 - 4 = 0$$
$$4(a^3 - 1) = 0$$
$$a = 1$$

Continues on next page

Nature of stationary point:

Check that the stationary point is a minimum.

$$\frac{d^2(L^2)}{da^2} = 12a^2$$
$$= 12(1)^2$$
$$= 12\ (>0)$$

$\therefore$ the point is a minumum.

The minimum value of L^2 is:

The question was to find the minimum *value* of L^2.

$$L^2 = 1^4 - 4 + \frac{17}{4}$$
$$= \frac{5}{4}$$

c From part **b**, the minimum value of L occurs when $a = 1$, so the point is $A(a, a^2) = (1, 1)$.

The minimum of L occurs at the same value of a as the minimum of a^2.

Note that the minimum distance from P to the curve is $\frac{\sqrt{5}}{2}$.

EXERCISE 13C

1 **a** **i** Find the maximum value of xy given that $x + 2y = 4$.

ii Find the maximum possible value of xy given that $3x + y = 7$.

b **i** Find the minimum possible value of $a + b$ given that $ab = 3$ and $a, b > 0$.

ii Find the minimum possible value of $2a + b$ given that $ab = 4$ and $a, b > 0$.

c **i** Find the maximum possible value of $4r^2h$ if $2r^2 + rh = 3$ and $r, h > 0$.

ii Find the maximum possible value of rh^2 if $4r^2 + 3h^2 = 12$ and $r, h > 0$.

2 A rectangle has width x metres and length $30 - x$ metres.

a Find the maximum area of the rectangle.

b Show that as x changes the perimeter stays constant and find the value of this perimeter.

3 The sector of a circle with radius r has perimeter 40 cm.

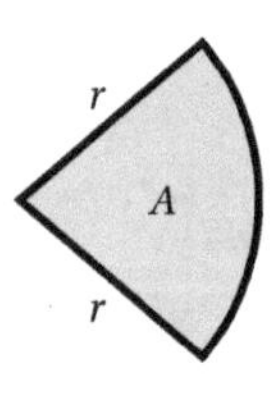

The sector has area A.

a Show that $A = 20r - r^2$.

b **i** Find the value of r for which A is a maximum.

ii Show that this does give a maximum.

4 A square sheet of card, of side, 12 cm has a square of side x cm cut from each of the four corners. The sides are then folded to make a small, open box.

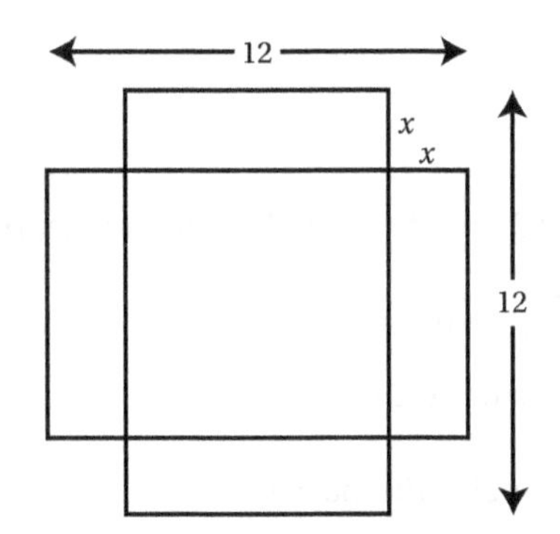

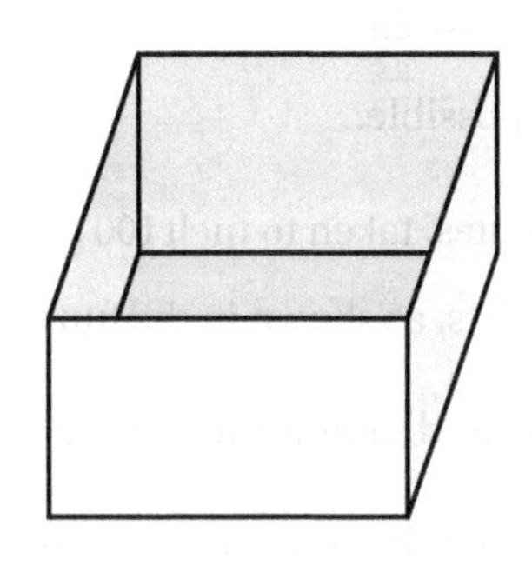

a Show that the volume, V, is given by $V = x(12-2x)^2$.

b Find the value of x for which the volume is the maximum possible, and prove that it is a maximum.

5 Prove that the minimum possible value of the sum of a positive real number and its reciprocal is 2.

6 A solid cylinder has radius r and height h.

The total surface area of the cylinder is 450 cm^2.

a Find an expression for the volume of the cylinder in terms of r only.

b Hence find the maximum possible volume, justifying that the value found is a maximum.

7 A closed carton is in the shape of a cuboid. The base is a square of side x.

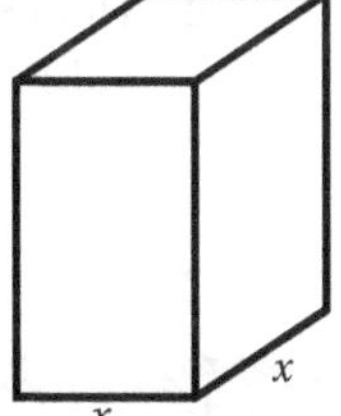

The total surface area is 486 cm^2.

a Find an expression for the volume of the carton in terms of x only.

b Hence find the maximum possible volume, justifying that the value found is a maximum.

8 A certain type of chocolate is sold in boxes that are in the shape of a triangular prism. The cross-section is an equilateral triangle of side x cm. The length is y cm.

The volume of the box needs to be 128 cm^3.

The manufacturer wishes to minimise the surface area.

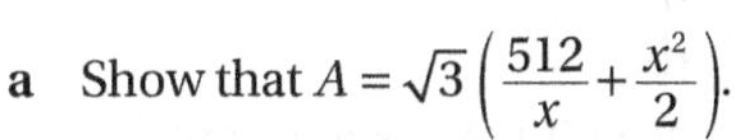

a Show that $A = \sqrt{3}\left(\frac{512}{x} + \frac{x^2}{2}\right)$.

b Find the minimum value of A. Give your answer in exact form.

c Prove that the value found is a minimum.

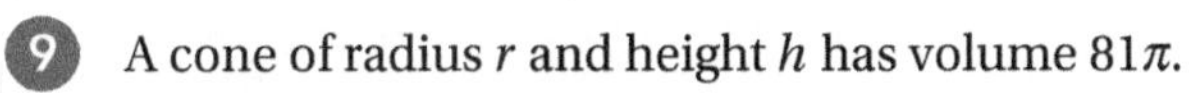

9 A cone of radius r and height h has volume 81π.

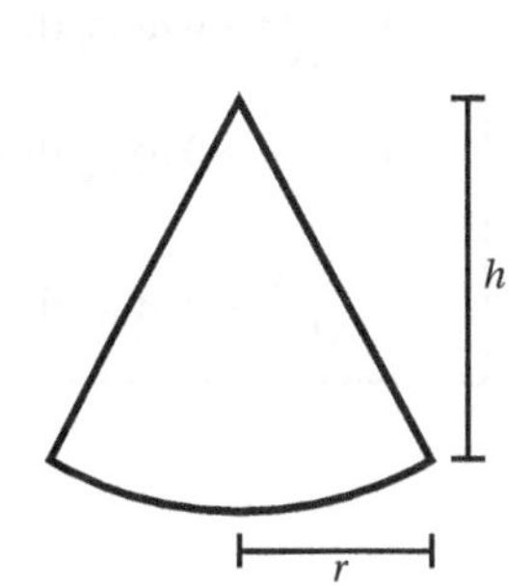

a Show that the curved surface area of the cone is given by $S = \pi\sqrt{r^4 + \frac{243^2}{r^2}}$.

It is required to make the cone so that the curved surface area is the minimum possible.

b By considering stationary points of S^2, or otherwise, find the radius and the height of the cone.

10 The sum of two numbers, x and y, is 6, and $x, y \geqslant 0$. Find the two numbers if the sum of their squares is:

a the minimum possible

b the maximum possible.

11 The time, t, in minutes, taken to melt 100 g of butter depends upon the percentage, p, of the butter that is made of saturated fats, as shown in the function $t = \frac{p^2}{10\,000} + \frac{p}{100} + 2$.

Find the maximum and minimum times to melt 100 g of butter.

12 A 20 cm piece of wire is bent to form an isosceles triangle with base b.

a Show that the area of the triangle is given by $A = \frac{1}{2}\sqrt{100b^2 - 10b^3}$.

b Show that the area of the triangle takes the largest possible value when the triangle is equilateral.

13 The sum of the squares of two positive numbers is a. Prove that their product is the maximum possible when the two numbers are equal.

14 Find the coordinates of the point on the curve $y = x^2$, $x \geqslant 0$, that is closest to the point $(0, 4)$.

15 A cylinder of radius 6 cm and height 6 cm fits perfectly inside a cone, leaving a constant ring of width x around the base of the cylinder.

a Show that the height, h, of the cone is $h = \frac{36}{x} + 6$.

b Find the volume of the cone in terms of x.

c Hence find the minimum value of the volume, justifying that the value you have found is a minimum.

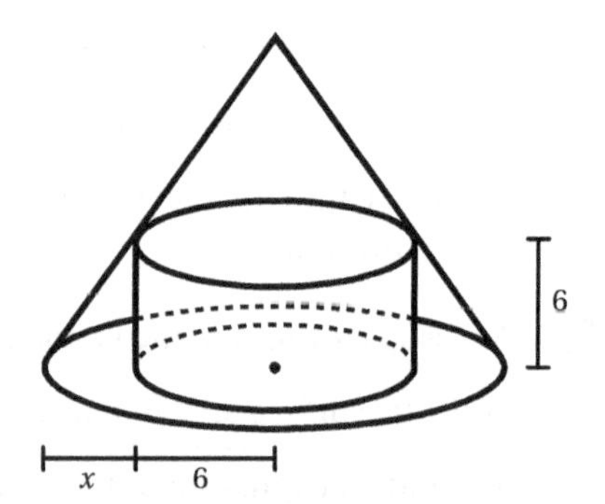

Checklist of learning and understanding

- The **tangent** to a curve at a point (x_0, y_0) has gradient equal to $\frac{dy}{dx}$ evaluated at that point.
- The **normal** to a curve at a point is perpendicular to the tangent at that point.
- **Stationary points** of a function are points where the gradient is zero, so $\frac{dy}{dx} = 0$.
- **The second derivative** can be used to determine the nature of a stationary point. At a stationary point (x_0, y_0), if:
  - $\frac{d^2y}{dx^2} < 0$ at x_0 then (x_0, y_0) is a **maximum**
  - $\frac{d^2y}{dx^2} > 0$ at x_0 then (x_0, y_0) is a **minimum**
  - $\frac{d^2y}{dx^2} = 0$ at x_0 then no conclusion can be drawn.

Mixed practice 13

1. $y = ax^4 + bx^3 + cx^2 + dx + e$. What is the maximum possible number of maximum and minimum points the curve can have?

   Choose from these options.

   A 1 **B** 2 **C** 3 **D** 4

2. The function $y = \mathrm{f}(x)$ has a stationary point, P, at $x = 2$. What is the equation of the tangent at P?

   Choose from these options.

   A $y = \mathrm{f}'(2)$ **B** $x = 2$ **C** $y = \mathrm{f}(2)$ **D** $x = -\frac{1}{2}$

3. Find the equation of the tangent to the curve $y = x^2 - \frac{3}{x}$ at the point where $x = 2$. Give your answer in the form $ax + by = c$, where a, b and c are integers.

4. Find the equation of the normal to the curve $y = (x-2)^3$ when $x = 2$.

5. Find the x-coordinates of the stationary points on the graph of $y = \frac{x^3}{6} - x^2 + x$, and determine their nature.

6. A curve C has equation $y = \frac{1}{4}x^4 - 2x^3 + 4x^2 - 1$.

   a Find the coordinates of the stationary points on C.

   b Determine the nature of each stationary point.

7. A rectangle is drawn inside the region bounded by the curve $y = 4 - x^2$ and the x-axis, so that two of the vertices lie on the axis and the other two on the curve.

   Find the coordinates of vertex A so that the area of the rectangle is the maximum possible.

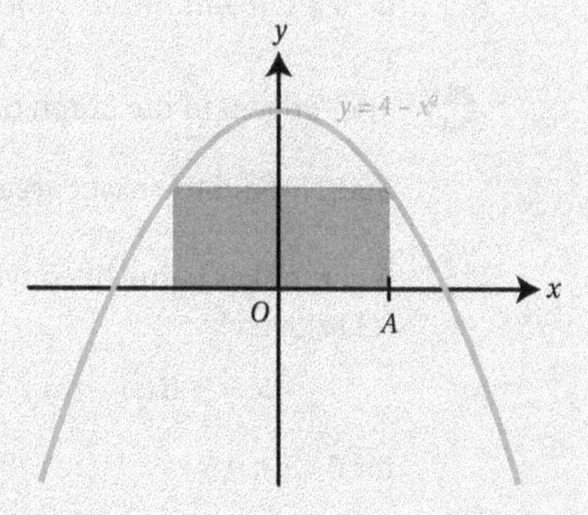

8. A curve has the equation:

   $$y = \frac{12 + x^2\sqrt{x}}{x}, \quad x > 0$$

   a Express $\frac{12 + x^2\sqrt{x}}{x}$ in the form $12x^p + x^q$.

   b **i** Hence find $\frac{\mathrm{d}y}{\mathrm{d}x}$.

   ii Find an equation of the normal to the curve at the point on the curve where $x = 4$.

   iii The curve has a stationary point P. Show that the x-coordinate of P can be written in the form 2^k, where k is a rational number.

[© AQA 2013]

9 The function $y = \frac{1}{3}x^3 - ax^2 + 3ax + 1$, where $a \neq 0$, has one stationary point. What is the value of a?

Choose from these options.

A $a = 2$ **B** $a = 3$ **C** $a = 4$ **D** $a = 5$

10 $\frac{d^2y}{dx^2} > 0$ at the point Q. Which statement is true?

A Q is a local maximum.

B Q is a local minimum.

C Q is either a local maximum or a local minimum.

D More information is required to decide.

11 The curve $y = ax^3 + bx^2 + 8x - 1$ has stationary points at $x = \frac{1}{3}$ and $x = 4$.

Find the values of a and b.

12 $f(x) = x^2 - 4x\sqrt{x} + 4x - 3$.

Show that the curve $y = f(x)$ has two stationary points and determine whether each is a maximum or minimum.

13 A function is defined by $f(x) = x^3 - 9x$ for $-2 \leqslant x \leqslant 5$.

a Find the coordinates of the stationary points on the curve $y = f(x)$.

b Find the minimum and maximum values of $f(x)$.

14 The tangent to the graph of $y = \frac{1}{x^2}$ at the point where $x = 3$ crosses the coordinate axes at points M and N. Find the exact area of the triangle MON.

15 A car tank is being filled with petrol such that the volume in the tank in litres (V) over time in minutes (t) is given by:

$$V = 300(t^2 - t^3) + 4$$

for $0 < t < 0.5$

a How much petrol was initially in the tank?

b After 30 seconds the tank was full. What is the capacity of the tank?

c At what time is petrol flowing in at the greatest rate?

16 A gardener is planting a lawn in the shape of a sector of a circle joined to a rectangle. The sector has radius r and angle 120°.

He needs the area, A, of the lawn to be 200 m².

A fence is to be built around the perimeter of the lawn.

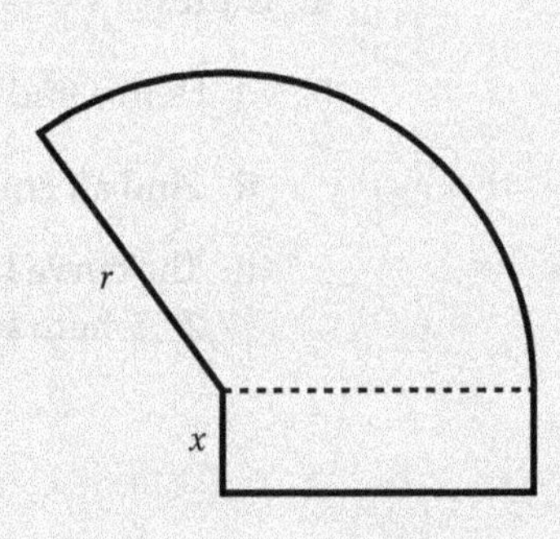

a Show that the length of the fence, P, is given by $P = 2r + \frac{400}{r}$.

b Hence find the minimum length of fence required, justifying that this value is a minimum.

17 The diagram shows a block of wood in the shape of a prism with triangular cross-section. The end faces are right-angled triangles with sides of length $3x$ cm, $4x$ cm and $5x$ cm, and the length of the prism is y cm, as shown in the diagram.

The total surface area of the five faces is 144 cm^2.

a **i** Show that $xy + x^2 = 12$.

ii Hence show that the volume of the block, V cm^3, is given by

$V = 72x - 6x^3$

b **i** Find $\frac{dV}{dx}$.

ii Show that V has a stationary value when $x = 2$.

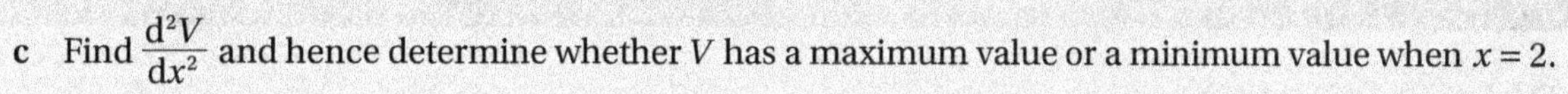

c Find $\frac{d^2V}{dx^2}$ and hence determine whether V has a maximum value or a minimum value when $x = 2$.

[© AQA 2010]

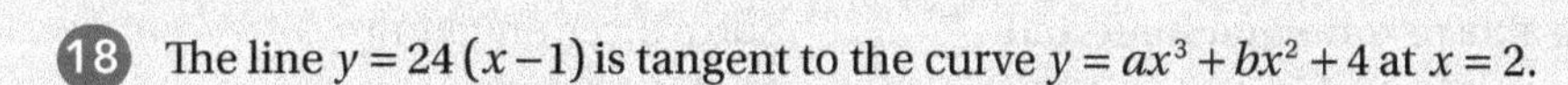

18 The line $y = 24(x - 1)$ is tangent to the curve $y = ax^3 + bx^2 + 4$ at $x = 2$.

a Use the fact that the tangent meets the curve to show that $2a + b = 5$.

b Use the fact that the tangent has the same gradient as the curve to find another relationship between a and b.

c Hence find the values of a and b.

d The line meets the curve again. Find the coordinates of the other point of intersection.

19 On the curve $y = x^3$ a tangent is drawn from the point (a, a^3) and a normal is drawn from the point $(-a, -a^3)$. The tangent and the normal meet on the y-axis. Find the value of a.

20 The curve $y = ax^2 + \frac{24}{x}$ has a stationary point at $y = 18$.

Find the value of a.

14 Integration

In this chapter you will learn how to:

- reverse the process of differentiation (this process is called integration)
- find the equation of a curve, given its derivative and a point on the curve
- find the area between a curve and the x-axis
- find the area between a curve and a straight line.

Before you start...

Chapter 3	You should be able to solve quadratic and cubic equations by factorising.	1 Solve these equations. a $3x^2 - 4x = 0$ b $x^3 - 4x^2 = 0$ c $x^3 - 5x = 0$
Chapter 12	You should be able to differentiate expressions of the form ax^n.	2 Find $\frac{dy}{dx}$ for each function. a $y = 3x^2 - x + 2$ b $y = x^{\frac{1}{2}} - 3x^{-\frac{2}{3}} + 2$
Chapter 12	You should be able to convert an expression to the form ax^n in order to differentiate.	3 Find $\frac{dy}{dx}$ for each function. a $y = \frac{3}{2x^2}$ b $y = 3x\sqrt{x}$

Reversing differentiation

In many areas of mathematics, when you learn a new process you must then learn how to undo it. Reversing differentiation answers the question: 'If I know the equation for the gradient of a curve, can I find the equation of the curve itself?'

This is an important question because, in many applications, the rate of change is easier to measure or model than the quantity itself. For example, you can calculate acceleration, which is the rate of change of velocity, if you know the forces acting on an object. You can then 'undifferentiate' the equation for acceleration to find the equation for velocity.

Undoing the process of differentiation then introduces the possibility of answering another, seemingly unconnected problem: how to find the area under a curve. As you will see in the Mechanics chapters, one application of this is to find the distance travelled from a velocity–time graph.

Section 1: Rules for integration

If you know the function describing the gradient of a curve, you can find the function describing the curve itself by 'undoing' the differentiation. This process of reversing differentiation is called **integration**.

For example, if $\frac{dy}{dx} = 2x$, you know that the equation of the curve, y, must have contained x^2.

However, it could have been:

$$y = x^2 + 2$$

or:

$$y = x^2 - \frac{2}{3}$$

or, in fact:

$$y = x^2 + c, \text{ for any constant } c.$$

Without further information, it is impossible to know what the constant was in the equation of the curve, so:

if $\frac{dy}{dx} = 2x$, then $y = x^2 + c$

or, equivalently, using the integration symbol:

$$\int 2x\,dx = x^2 + c$$

where c is the **constant of integration**.

Fast forward

You will see how to find the constant of integration in Section 4.

The dx in the integral expression simply states that the integration is taking place with respect to the variable x, in the same way that in $\frac{dy}{dx}$ it states that the differentiation is taking place with respect to x. You could equally well write, for example:

$$\int 2t\,dt = t^2 + c$$

Key point 14.1

Since integration is the reverse of differentiation, the formula for differentiation can be reversed to give:

$$\int x^n\,dx = \frac{1}{n+1}x^{n+1} + c$$

for any $n \neq -1$.

Tip

It may be helpful to think of this in words: 'add one to the power and divide by the new power'.

Fast forward

You will learn how to integrate x^{-1} in Student Book 2.

Note the condition $n \neq -1$, which ensures that you are not dividing by zero.

WORKED EXAMPLE 14.1

If $\frac{dy}{dx} = x^3$, find an expression for y.

$y = \int x^3 \, dx$	Integrate to get y.
$= \frac{1}{3+1} x^{3+1} + c$ $\therefore y = \frac{1}{4} x^4 + c$	Use $\int x^n \, dx = \frac{1}{n+1} x^{n+1} + c$.

Common error

Don't forget the $+c$. It is a part of the answer and you must write it every time.

WORKED EXAMPLE 14.2

Find an expression for $\int x^{-4} \, dx$.

$\int x^{-4} \, dx = \frac{1}{-4+1} x^{-4+1}$ $= -\frac{1}{3} x^{-3} + c$	Use $\int x^n \, dx = \frac{1}{n+1} x^{n+1} + c$.

WORKED EXAMPLE 14.3

If $f'(x) = x^{\frac{1}{2}}$, find $f(x)$.

$f(x) = \int x^{\frac{1}{2}} \, dx$	Integrate to find $f(x)$.
$= \frac{1}{\frac{1}{2}+1} x^{\frac{1}{2}+1} + c$	Use $\int x^n \, dx = \frac{1}{n+1} x^{n+1} + c$.
$= \frac{1}{\frac{3}{2}} x^{\frac{3}{2}} + c$ $= \frac{2}{3} x^{\frac{3}{2}} + c$	Simplify.

Tip

The first part of Key point 14.2 just says that you can take a constant out as a factor and integrate the remaining function. The second says that you can integrate terms of a sum separately.

Key point 14.2

Important properties of integration

- $\int k \, f(x) \, dx = k \int f(x) \, dx$, where k is a constant
- $\int f(x) + g(x) \, dx = \int f(x) \, dx + \int g(x) \, dx$

Rewind

These properties arise directly from the properties of differentiation in Chapter 12.

WORKED EXAMPLE 14.4

Find $\int 3x^5 \, dx$.

$\int 3x^5 \, dx = 3 \times \frac{1}{6} x^6 + c$ — Integrate x^5 and then multiply by 3.

$= \frac{1}{2} x^6 + c$

WORKED EXAMPLE 14.5

Find $\int 7 \, dx$.

$\int 7 \, dx = \int 7x^0 \, dx$ — Using the fact that $x^0 = 1$, you can write 7 as $7x^0$.

$= 7 \times \frac{1}{1} x^1 + c$ — Now integrate x^0.

$= 7x + c$

Tip

You don't need to write out $k = kx^0$ every time you integrate a constant. You just need to know that the integral will be $kx + c$.

From Worked example 14.5 you can see that $\int k \, dx = kx + c$ for any constant k.

WORKED EXAMPLE 14.6

A curve has gradient $\frac{dy}{dx} = 4x^{-\frac{1}{2}} - 3x + 8$. Find an expression for y.

$y = \int 4x^{-\frac{1}{2}} - 3x + 8 \, dx$ — Integrate to find y.

$= 4 \times \frac{1}{\frac{1}{2}} x^{\frac{1}{2}} - 3 \times \frac{1}{2} x^2 + 8x + c$ — Integrate each term separately. Remember that the integral of a constant, k, is just kx.

$= 8x^{\frac{1}{2}} - \frac{3}{2} x^2 + 8x + c$

EXERCISE 14A

1 Find an expression for y for each derivative.

a **i** $\frac{dy}{dx} = x^5$ **ii** $\frac{dy}{dx} = x^6$ **b** **i** $\frac{dy}{dx} = 3x^2$ **ii** $\frac{dy}{dx} = -5x^4$

c **i** $\frac{dy}{dx} = 2x^3$ **ii** $\frac{dy}{dx} = -3x^7$ **d** **i** $\frac{dy}{dx} = x$ **ii** $\frac{dy}{dx} = -4x$

e **i** $\frac{dy}{dx} = 0$ **ii** $\frac{dy}{dx} = 1$ **f** **i** $\frac{dy}{dx} = -\frac{1}{6} x^3$ **ii** $\frac{dy}{dx} = \frac{1}{3} x^8$

g **i** $\frac{dy}{dx}=\frac{14x^6}{5}$ **ii** $\frac{dy}{dx}=\frac{10x}{7}$ **h** **i** $\frac{dy}{dx}=3x^5-x+4$ **ii** $\frac{dy}{dx}=2x^4+4x^3-8x-1$

i **i** $\frac{dy}{dx}=\frac{x^2}{2}+2x-\frac{3}{4}$ **ii** $\frac{dy}{dx}=x^2-\frac{3x^5}{8}$

2 Work out these integrals.

a **i** $\int x^{\frac{3}{2}}\,dx$ **ii** $\int x^{\frac{1}{3}}dx$ **b** **i** $\int 5x^{\frac{2}{3}}\,dx$ **ii** $\int 14x^{\frac{3}{4}}\,dx$

c **i** $\int \frac{5}{4}x^{\frac{1}{4}}\,dx$ **ii** $\int -\frac{3}{2}x^{\frac{1}{2}}\,dx$ **d** **i** $\int -\frac{7x^{\frac{4}{3}}}{6}\,dx$ **ii** $\int \frac{14x^{\frac{2}{5}}}{15}\,dx$

e **i** $\int 5x-7x^{\frac{5}{2}}\,dx$ **ii** $\int 6+\frac{x}{3}-12x^{\frac{1}{2}}\,dx$ **f** **i** $\int x^{-2}\,dx$ **ii** $\int x^{-3}\,dx$

g **i** $\int x^{-\frac{1}{3}}\,dx$ **ii** $\int x^{-\frac{1}{4}}\,dx$ **h** **i** $\int -4x^{-\frac{2}{3}}\,dx$ **ii** $\int -5x^{\frac{3}{2}}\,dx$

i **i** $\int x^{-\frac{1}{2}}-\frac{6x^{-4}}{5}\,dx$ **ii** $\int 3x^{-5}+7x^{-\frac{1}{6}}\,dx$

3 Work out an expression for each integral.

a **i** $\int 3\,dt$ **ii** $\int 7\,dz$ **b** **i** $\int q^5\,dq$ **ii** $\int r^{10}\,dr$

c **i** $\int 12g^{\frac{3}{5}}\,dg$ **ii** $\int 6y^{\frac{7}{2}}\,dy$ **d** **i** $\int 5a^{-\frac{3}{4}}\,da$ **ii** $\int 7p^{-\frac{4}{3}}\,dp$

4 If $f'(x)=\frac{x^3}{2}-6x^{-\frac{5}{3}}+2$, find $f(x)$.

5 Find $\int x^{\frac{3}{2}}-x^{-\frac{3}{2}}\,dx$.

6 Find $\int \frac{x^5-3x^{-3}+x}{2}\,dx$.

Section 2: Simplifying into terms of the form ax^n

Just as for differentiation, before you can integrate products or quotients of functions you need to convert them into terms of the form ax^n, often by using the laws of indices.

Fast forward

In Student Book 2 you will learn methods for integrating products and quotients directly.

WORKED EXAMPLE 14.7

Find $\int (2x-1)(x+3)\,dx$.

$\int (2x-1)(x+3)\,dx=\int 2x^2+5x-3\,dx$ Expand the brackets.

$=\frac{2}{3}x^3+\frac{5}{2}x^2-3x+c$ Then integrate.

Common error

You cannot integrate products and quotients of functions separately:

- $\int f(x)g(x)dx \neq \int f(x)dx \times \int g(x)dx$
- $\int \frac{f(x)}{g(x)}dx \neq \frac{\int f(x)dx}{\int g(x)dx}$

WORKED EXAMPLE 14.8

Find $\int \frac{1}{x^2}\,dx$.

$\int \frac{1}{x^2}\,dx = \int x^{-2}\,dx$ — First use the laws of indices to rewrite the function in the form x^n.

$= -x^{-1} + c$ — Then integrate.

WORKED EXAMPLE 14.9

Find $\int 5x^2\sqrt[3]{x}\,dx$.

$\int 5x^2\sqrt[3]{x}\,dx = \int 5x^2x^{\frac{1}{3}}\,dx$ — First use the laws of indices to rewrite the function in the form ax^n.

$= \int 5x^{\frac{7}{3}}\,dx$

$= 5 \times \frac{3}{10}x^{\frac{10}{3}} + c$ — Then integrate: dividing by $\frac{10}{3}$ is the same as multiplying by $\frac{3}{10}$.

$= \frac{3}{2}x^{\frac{10}{3}} + c$

WORKED EXAMPLE 14.10

Find $\int \frac{(x-3)^2}{\sqrt{x}}\,dx$.

$\int \frac{(x-3)^2}{\sqrt{x}}\,dx = \int \frac{x^2-6x+9}{\sqrt{x}}\,dx$ — Expand the brackets first.

$= \int \frac{x^2-6x+9}{x^{\frac{1}{2}}}\,dx$ — Then replace $\sqrt{x}$ with $x^{\frac{1}{2}}$.

$= \int x^{2-\frac{1}{2}} - 6x^{1-\frac{1}{2}} + 9x^{-\frac{1}{2}}\,dx$ — And simplify, using the laws of indices.

$= \int x^{\frac{3}{2}} - 6x^{\frac{1}{2}} + 9x^{-\frac{1}{2}}\,dx$

$= \frac{2}{5}x^{\frac{5}{2}} - 6 \times \frac{2}{3}x^{\frac{3}{2}} + 9 \times 2x^{\frac{1}{2}} + c$ — Then integrate: dividing by $\frac{3}{2}$ is the same as multiplying by $\frac{2}{3}$; dividing by $\frac{1}{2}$ is the same as multiplying by 2.

$= \frac{2}{5}x^{\frac{5}{2}} - 4x^{\frac{3}{2}} + 18x^{\frac{1}{2}} + c$

WORK IT OUT 14.1

Three students are trying to work out the integral $\int 2x(x^2-2)\,dx$.

Which is the correct solution? Identify the errors made in the incorrect solutions and try to explain them.

Solution A	Solution B	Solution C
$\int 2x(x^2-2)\,dx = \int 2x^3-4x\,dx$ $=\frac{2x^4}{4}-\frac{4x^2}{2}+c$ $=\frac{x^4}{2}-2x^2+c$	$\int 2x(x^2-2)\,dx = \frac{2x^2}{2}\left(\frac{x^3}{3}-2x\right)+c$ $=x^2\left(\frac{x^3}{3}-2x\right)+c$	$\int 2x(x^2-2)\,dx = 2x\int x^2-2\,dx$ $=2x\left(\frac{x^3}{3}-2x\right)+c$

EXERCISE 14B

1 Work out these integrals.

a i $\int \sqrt{x}\,dx$ ii $\int \sqrt[4]{x}\,dx$ **b** i $\int 6\sqrt[5]{x}\,dx$ ii $\int \frac{\sqrt[3]{x}}{6}\,dx$

c i $\int \frac{1}{x^3}\,dx$ ii $\int -\frac{1}{x^7}\,dx$ **d** i $\int -\frac{8}{x^5}\,dx$ ii $\int \frac{10}{3x^6}\,dx$

e i $\int \frac{1}{\sqrt{x}}\,dx$ ii $\int \frac{1}{\sqrt[3]{x}}\,dx$ **f** i $\int -\frac{15}{\sqrt[6]{x}}\,dx$ ii $\int \frac{9}{8\sqrt[4]{x}}\,dx$

2 Work out these integrals.

a i $\int (2x+3)(x+1)\,dx$ ii $\int 3x^2(x-2)\,dx$

b i $\int \sqrt{x}(5x+4)\,dx$ ii $\int \sqrt[3]{x}\,(2x^2+3)\,dx$

c i $\int \left(\sqrt[4]{x}+3x\right)^2\,dx$ ii $\int \left(\sqrt{x}-x^2\right)^2\,dx$

d i $\int \left(x-\frac{1}{x}\right)^2\,dx$ ii $\int \left(2x+\frac{1}{x^2}\right)\left(2x-\frac{1}{x^2}\right)dx$

3 Work out these integrals.

a i $\int \frac{7x-2}{x^3}\,dx$ ii $\int \frac{2+5x}{4x^3}\,dx$ **b** i $\int \frac{\sqrt{x}-3}{x^2}\,dx$ ii $\int \frac{x^2-6x}{\sqrt{x}}\,dx$

4 Find $\int 10x\sqrt{x}\,dx$.

5 Find $\int x^2\left(\sqrt[3]{x}-12\right)dx$.

6 Find $\int 9\sqrt[4]{x^5}\,dx$.

7 Find $\int \frac{1}{3\sqrt[3]{x^2}}\,dx$.

8 Find $\int \left(\sqrt{x}-2\right)\left(\frac{1}{\sqrt{x}}-1\right)dx$.

9 Find $\int \frac{1}{3x^3}+\frac{1}{4x^4}\,dx$.

10 $f(x)=\frac{6+x}{\sqrt[3]{x}}$

a Express $f(x)$ in the form ax^n+bx^m.

b Hence find $\int \frac{6+x}{\sqrt[3]{x}}\,dx$.

11 $y=\frac{(x+2)(x-2)}{2\sqrt{x}}$.

a Show that $y=ax^{\frac{3}{2}}+bx^{-\frac{1}{2}}$ where a and b are constants to be found.

b Hence find $\int y\,dx$.

12 Show that $\int 12\sqrt{x}-\frac{4}{\sqrt{x}}\,dx=8\sqrt{x}(x-1)+c$.

13 Find $\int \frac{(x+4)^2}{8x\sqrt{x}}\,dx$.

Section 3: Finding the equation of a curve

Consider again $\frac{dy}{dx}=2x$ which you met at the start of this chapter. You know that the original function has equation $y=x^2+c$ for some constant value c.

If you are also told that the curve passes through the point (1, –1), you can specify which of the family of curves your function must be, and thus find the value of $c=-2$.

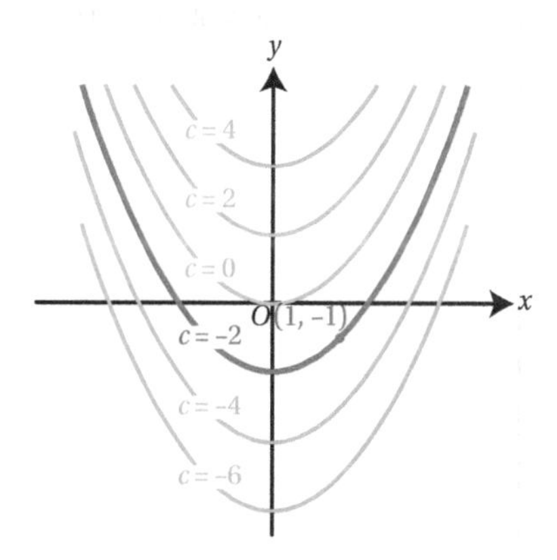

WORKED EXAMPLE 14.11

The gradient of a curve is given by $\frac{dy}{dx}=3x^2-8x+5$ and the curve passes through the point $(1,-4)$. Find the equation of the curve.

$y=\int 3x^2-8x+5\,dx$
$=x^3-4x^2+5x+c$

To find y from $\frac{dy}{dx}$ you need to integrate. Don't forget to include '$+c$'.

When $x=1$, $y=-4$:
$-4=(1)^3-4(1)^2+5(1)+c$
$-4=1-4+5+c$
$c=-6$

Substitute $x=1$, $y=-4$ in order to find c.

$\therefore y=x^3-4x^2+5x-6$

State the final equation.

Worked example 14.11 illustrates the general procedure for finding the equation of a curve, given its gradient function.

Key point 14.3

To find the equation for y given the gradient $\frac{dy}{dx}$ and one point (p, q) on the curve:

- integrate $\frac{dy}{dx}$ to find an equation for y in terms of x, remembering $+ c$
- find the value of c by substituting $x = p$ and $y = q$ into the equation
- rewrite the equation for y, using the value of c that has been found

Fast forward

Equations for second derivatives occur frequently in work on velocity and acceleration – see Chapter 16.

Sometimes you are given the second derivative; in this case you need to integrate twice to find the equation of the original curve. Be careful: you will need a $+c$ each time you integrate.

WORKED EXAMPLE 14.12

A curve has second derivative $\frac{d^2y}{dx^2} = 3 - 6x$. It passes through the point $(1, 2)$ and its gradient at this point is -3. Find the equation of the curve.

$$\frac{dy}{dx} = \int 3 - 6x \, dx$$
$$= 3x - 3x^2 + c$$

Integrating the second derivative once gives the first derivative.

When $x = 1$, $\frac{dy}{dx} = -3$:

You can find c by using the given value of the gradient at $x = 1$.

$$-3 = 3 - 3 + c$$
$$c = -3$$

$$\therefore \frac{dy}{dx} = 3x - 3x^2 - 3$$

State the full equation for the derivative.

$$y = \int 3x - 3x^2 - 3 dx$$
$$= \frac{3x^2}{2} - x^3 - 3x + d$$

Now integrate again to find y. There will be another constant. Use a different letter for this constant, for example d.

When $x = 1$, $y = 2$:

Now use the given x and y values to find d.

$$2 = \frac{3}{2} - 1 - 3 + d$$
$$d = \frac{9}{2}$$

$$\therefore y = \frac{3x^2}{2} - x^3 - 3x + \frac{9}{2}$$

Finally state the equation for y.

WORK IT OUT 14.2

Three students attempt to solve this problem.

The second differential of a curve is $\frac{d^2y}{dx^2}=3x^2$. When $x=2$, $y=3$ and $\frac{dy}{dx}=-1$. Find the equation of the curve.

Which is the correct solution? Identify the errors made in the incorrect solutions.

Solution A	Solution B	Solution C
$\frac{dy}{dx}=\int 3x^2\,dx$	$\frac{dy}{dx}=\int 3x^2\,dx$	$\frac{dy}{dx}=\int 3x^2\,dx$
$=x^3+c$	$=x^3+c$	$=x^3+c$
$y=\int x^3+c\ dx$	$y=\int x^3+c\ dx$	When $x=2$, $\frac{dy}{dx}=-1$:
$=\frac{x^4}{4}+cx+c$	$=\frac{x^4}{4}+cx$	$-1=2^3+c$
		$c=-9$
When $x=2$, $y=3$:	When $x=2$, $y=3$:	$y=\int x^3-9\,dx$
$3=4+2c+c$	$3=4+2c$	$=\frac{x^4}{4}-9x+d$
$3=4+3c$	$c=-\frac{1}{2}$	When $x=2$, $y=3$:
$c=-\frac{1}{3}$	$\therefore y=\frac{x^4}{4}-\frac{1}{2}$	$3=4-18+d$
$\therefore y=\frac{x^4}{4}-\frac{1}{3}x-\frac{1}{3}$		$d=17$
		$\therefore y=\frac{x^4}{4}-9x+17$

EXERCISE 14C

1 Find the equation of the curve when:

a **i** $\frac{dy}{dx}=x$ and the curve passes through $(-2,7)$

ii $\frac{dy}{dx}=6x^2$ and the curve passes through $(0,5)$

b **i** $\frac{dy}{dx}=\frac{1}{x^3}$ and the curve passes through $(1,-1)$

ii $\frac{dy}{dx}=\frac{1}{x^2}$ and the curve passes through $(1,3)$

c **i** $\frac{dy}{dx}=3x-5$ and the curve passes through $(2,6)$

ii $\frac{dy}{dx}=3-2x^3$ and the curve passes through $(1,5)$

d **i** $\frac{dy}{dx}=3\sqrt{x}$ and the curve passes through $(9,-2)$

ii $\frac{dy}{dx}=\frac{1}{\sqrt{x}}$ and the curve passes through $(4,8)$.

2 Find the equation of the curve when:

a i $\frac{d^2y}{dx^2}=6x$ and the curve has gradient 2 at the point (1, 5)

ii $\frac{d^2y}{dx^2}=\frac{6}{x^3}$ and the curve has gradient −1 at the point (1, 3)

b i $f''(x)=1-2x$, $f'(2)=1$ and $f(2)=-1$

ii $f''(x)=3x^2-x$, $f'(1)=3$ and $f(1)=10$.

3 A curve has gradient $\frac{dy}{dx}=x-\frac{1}{x^2}$ and passes through the point (1, 3). Find the equation of the curve.

4 $f'(x)=\frac{2x-1}{\sqrt{x}}$ and $f(4)=2$. Find $f(x)$.

5 $f'(x)=\sqrt{x}(5x-4)$ and $f(1)=\frac{19}{3}$. Find $f(4)$.

6 The gradient of the curve at any point is directly proportional to the x-coordinate of that point. The curve passes through point A with coordinates (3, 2). The gradient of the curve at A is 12. Find the equation of the curve.

7 The gradient of a curve is $\frac{dy}{dx}=x^2-4$.

a Find the x-coordinate of the maximum point, justifying that it is a maximum.

b Given that the curve passes through the point (0, 2) show that the y-coordinate of the maximum point is $7\frac{1}{3}$.

8 A curve has second derivative $\frac{d^2y}{dx^2}=2-x^2$. At the point (1, 3) the gradient of the curve is 5. Find the equation of the curve in the form $y=f(x)$.

9 $f''(x)=12x-\frac{5}{\sqrt[3]{x}}$. $f'(1)=-\frac{9}{2}$ and $f(0)=2$. Find $f(8)$.

10 The gradient of the normal to a curve at any point is equal to the square of the x-coordinate at that point. Given that the curve passes through the point (2, 3) find the equation of the curve in the form $y=f(x)$.

Section 4: Definite integration

So far you have been carrying out a process known as **indefinite integration**: it is indefinite because it gives an unknown constant each time, for example:

$$\int x^2\,dx=\frac{1}{3}x^3+c$$

A similar process, called **definite integration** gives a numerical answer. You evaluate the indefinite integral at two points and take the difference of the two results:

$$\int_2^3 x^2\,dx=\left[\frac{1}{3}x^3+c\right]_2^3$$
$$=\left(\frac{1}{3}\times3^3+c\right)-\left(\frac{1}{3}\times2^3+c\right)$$
$$=6\frac{1}{3}$$

Tip

The constant of integration is cancelled out in the subtraction so there is no need to include it in the calculation at all.

The small numbers 2 and 3 are the **limits of integration**; **2** is the **lower limit** and **3** is the **upper limit**.

Notice the square-bracket notation which means that the integration has taken place but the limits have not yet been applied.

This process of definite integration is summarised in what is known as the fundamental theorem of calculus.

Tip

You may be able to evaluate definite integrals on your calculator. Even when you are asked to find the exact value of an integral, you can use your calculator to check the answer.

Key point 14.4

The fundamental theorem of calculus

$$\int_a^b f(x)\,dx = F(b) - F(a)$$

where $\frac{d}{dx}F(x) = f(x)$

In the given example, $f(x) = x^2$ and $F(x) = \frac{1}{3}x^3$.

WORKED EXAMPLE 14.13

Evaluate $\int_2^4 x^3 - 2x\,dx$.

$\int_2^4 x^3 - 2x\,dx = \left[\frac{1}{4}x^4 - x^2\right]_2^4$ — Integrate, using square brackets to indicate that the limits still need to be applied.

$= \left(\frac{1}{4}\times 4^4 - 4^2\right) - \left(\frac{1}{4}\times 2^4 - 2^2\right)$ — Evaluate the integral at the upper and lower limits and subtract.

$= (64 - 16) - (4 - 4)$

$= 48$

WORKED EXAMPLE 14.14

Find the exact value of $\int_8^9 \frac{3}{\sqrt{x}}\,dx$.

$\int_8^9 \frac{3}{\sqrt{x}}\,dx = \int_8^9 3x^{-\frac{1}{2}}\,dx$ — First rewrite in the form ax^n.

$= \left[3\times 2x^{\frac{1}{2}}\right]_8^9$ — Then integrate: dividing by $\frac{1}{2}$ is the same as multiplying by 2.

$= \left[6x^{\frac{1}{2}}\right]_8^9$

Continues on next page

$=\left(6\times9^{\frac{1}{2}}\right)-\left(6\times8^{\frac{1}{2}}\right)$ Evaluate at the upper and lower limits and subtract.

$=6\sqrt{9}-6\sqrt{8}$ Write the fractional power as root in order to evaluate.

$=18-12\sqrt{2}$ Remember that $\sqrt{8}=\sqrt{4\times2}=2\sqrt{2}$.

EXERCISE 14D

1 Evaluate these definite integrals, giving exact answers.

a i $\int_{2}^{6} x^3\,dx$ ii $\int_{4}^{5} x^4\,dx$ b i $\int_{-2}^{2} 3x^5\,dx$ ii $\int_{-3}^{3} 2x^2\,dx$

c i $\int_{-3}^{-1} 6x^2-3\,dx$ ii $\int_{-4}^{-2} 3x^2-4\,dx$ d i $\int_{1}^{4} x^2+x\,dx$ ii $\int_{-2}^{1} 3x^2-5x\,dx$

e i $\int_{4}^{9} 2\sqrt{x}\,dx$ ii $\int_{8}^{27} 6\sqrt[3]{x}\,dx$ f i $\int_{1}^{16} \frac{6}{\sqrt{x}}\,dx$ ii $\int_{-3}^{-1} \frac{3}{x^2}\,dx$

g i $\int_{-4}^{-2} 1+\frac{2}{x^2}\,dx$ ii $\int_{-3}^{-1} 2x-\frac{1}{x^2}\,dx$

2 Evaluate this integral, giving your answer in the form $a+b\sqrt{2}$.

$\int_{2}^{8} 3\sqrt{x}-2x\,dx$

3 Show that:

$$\int_{1}^{3} 9\sqrt{x}-\frac{4}{\sqrt{x}}\,dx=a+b\sqrt{3}$$

where a and b are integers to be found.

4 Find, in terms of k, $\int_{1}^{k} 2-\frac{1}{x^2}\,dx$.

5 Find, terms of a, $\int_{-a}^{2} x^2(4x-3)\,dx$.

6 Find the value of a such that $\int_{1}^{a} \sqrt{t}\;dt=42$.

7 $\int_{-2}^{p} 1-\frac{10}{3x^2}\,dx=-\frac{2}{3}$

Find all possible values of p.

8 Given that $\int_{3}^{9} f(x)\,dx=7$, evaluate $\int_{3}^{9} 2f(x)+1\,dx$.

9. Given that $\int_{-1}^{4} f(x)\,dx = 10$, evaluate $\int_{-1}^{4} \frac{f(x)}{2} + x - 1\,dx$.

10. **a** **i** Evaluate $\int_{3}^{5} 2x\,dx$. **ii** Evaluate $\int_{5}^{3} 2x\,dx$.

b **i** Find an expression for $\int_{b}^{a} 2x\,dx$. **ii** Find an expression for $\int_{a}^{b} 2x\,dx$.

c Suggest a general relationship between $\int_{b}^{a} f(x)\,dx$ and $\int_{a}^{b} f(x)\,dx$.

11. **a** Find the definite integral $\int_{1}^{a} x^2\,dx$ in terms of a.

b Let $f(a) = \int_{1}^{a} x^2\,dx$. Find $f'(a)$.

c Find $f'(a)$ for:

i $f(a) = \int_{2}^{a} x^2\,dx$ **ii** $f(a) = \int_{-1}^{a} x^2\,dx$.

What do you notice?

d Find $f'(a)$ for each function.

i $f(a) = \int_{0}^{a} 4x^3 - 2x\,dx$ **ii** $f(a) = \int_{1}^{a} \frac{3}{x^2}\,dx$ **iii** $f(a) = \int_{-2}^{a} 1 - 3x^2\,dx$

Section 5: Geometrical significance of definite integration

Now you can find a numerical value for an integral, the natural question to ask is, 'What does this number mean?' The (somewhat surprising) answer is that the definite integral represents the area under a curve.

More precisely, $\int_{a}^{b} y\,dx$ is the area enclosed between the graph of $y = f(x)$, the x-axis and the lines $x = a$ and $x = b$.

Key point 14.5

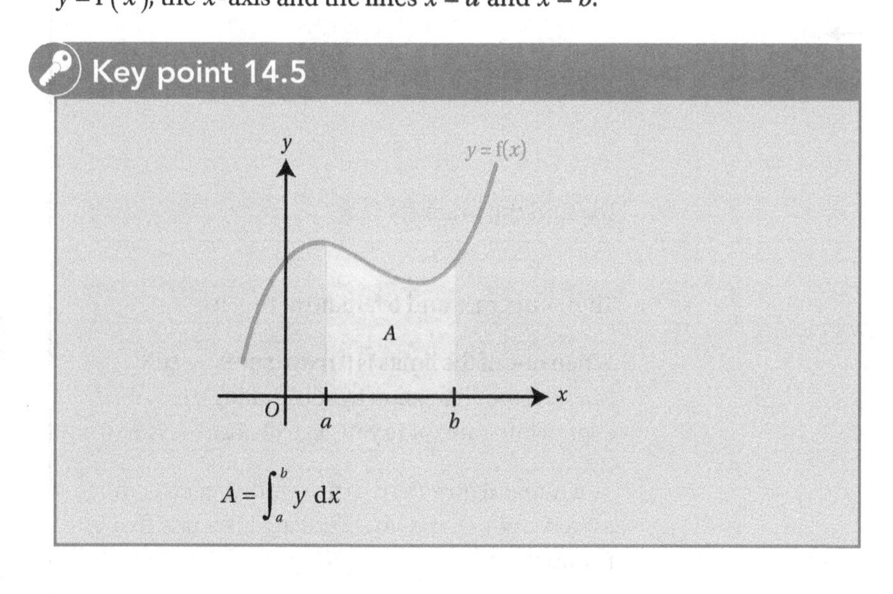

$$A = \int_{a}^{b} y\,dx$$

Did you know?

The ancient Greeks had methods for finding gradients of curves and areas under graphs. However, it took over 2000 years to develop the theory that formally proves that the two are related; that areas can be found by reversing differentiation. This was finally accomplished in the 17th century by Isaac Newton and Gottfried Leibniz.

WORKED EXAMPLE 14.15

Find the exact area enclosed between the graph of $y = 5x - x^2$, the x-axis and the lines $x = 1$ and $x = 3$.

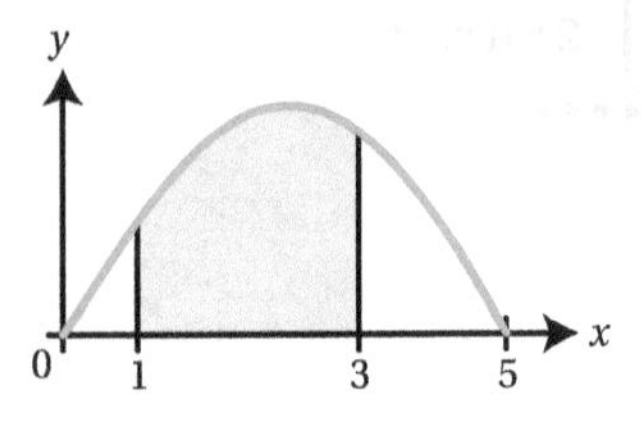

If no diagram is given, make sure you sketch the graph and identify the region required.

$$\text{Area} = \int_1^3 5x - x^2 \, dx$$

Write down the required definite integral. The limits are given by the x-coordinates at the ends of the shaded region.

$$= \left[\frac{5}{2}x^2 - \frac{1}{3}x^3\right]_1^3$$

$$= \left(\frac{5}{2} \times 9 - \frac{1}{3} \times 27\right) - \left(\frac{5}{2} - \frac{1}{3}\right) = \frac{34}{3}$$

Integrate and write the expression inside square brackets, then evaluate the expression at the limits and subtract.

When the curve is entirely below the x-axis the integral will give a negative value. Take the area to be the positive value.

WORKED EXAMPLE 14.16

Find the area of the region A shaded in the diagram:

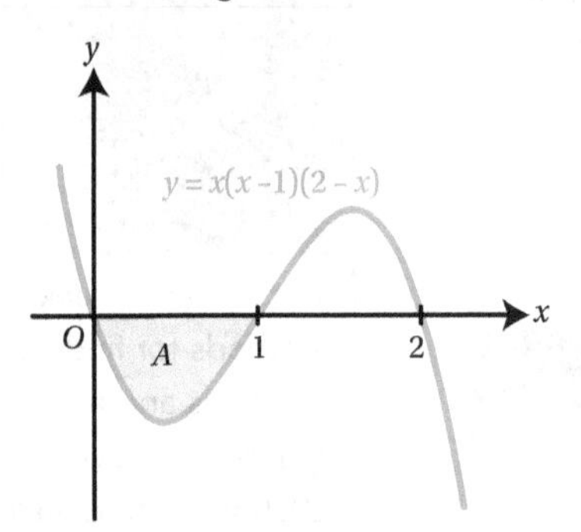

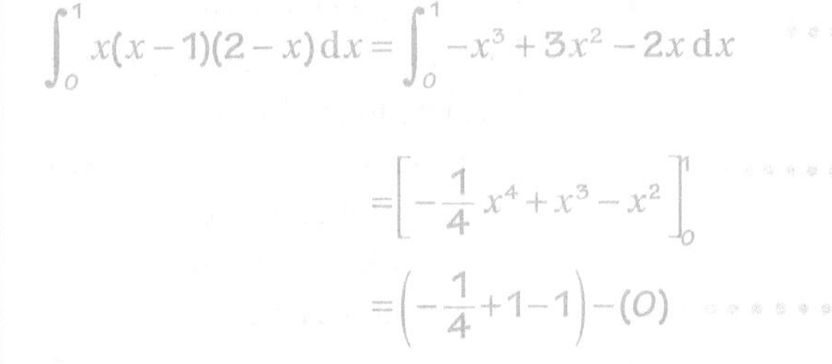

$$\int_0^1 x(x-1)(2-x) \, dx = \int_0^1 -x^3 + 3x^2 - 2x \, dx$$

Expand the brackets.

$$= \left[-\frac{1}{4}x^4 + x^3 - x^2\right]_0^1$$

Then integrate and evaluate at the limits.

$$= \left(-\frac{1}{4} + 1 - 1\right) - (0)$$

When one of the limits is 0 make sure you still evaluate the expression by substituting 0 into the expression – the value you get will not always be 0.

$$= -\frac{1}{4}$$

$$\therefore A = \frac{1}{4}$$

The integral gives a negative value because the area is under the x-axis. The area is a positive number.

The area between a curve and a straight line

You can find the area enclosed between a curve and a straight line by finding the area under each and subtracting the smaller area from the larger area. To find the area under a straight line you can either integrate or find the area of the triangle or trapezium formed.

WORKED EXAMPLE 14.17

The diagram shows the line with equation $y = x + 7$ and the curve with equation $y = x^2 - 6x + 13$.

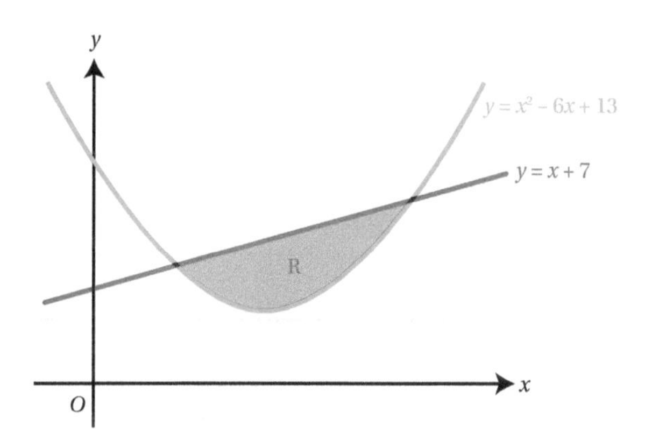

Find the exact area of the region R.

For intersection points:

First find the x-coordinates of the intersections.

$x^2 - 6x + 13 = x + 7$

$x^2 - 7x + 6 = 0$

$(x - 1)(x - 6) = 0$

$x = 1, 6$

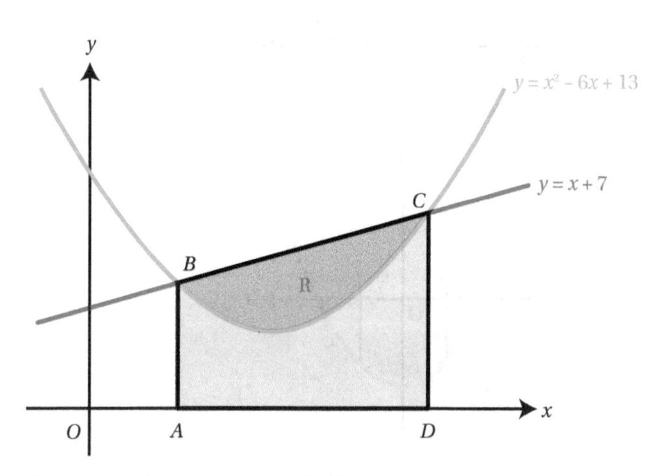

Area of R = **area of trapezium** $ABCD$ - area under curve

Continues on next page

When $x = 1$: $y = 1 + 7 = 8$

When $x = 6$: $y = 6 + 7 = 13$

Find the lengths of the two parallel sides of the trapezium.

Area of trapezium $= \left(\frac{8+13}{2}\right) \times 5 = \frac{105}{2}$

Use $A = \left(\frac{a+b}{2}\right)h$.

$$\int_1^6 x^2 - 6x + 13 \, dx = \left[\frac{1}{3}x^3 - 3x^2 + 13x\right]_1^6$$

Find the area under the curve by integrating.

$$= \left(\frac{1}{3} \times 6^3 - 3 \times 6^2 + 13 \times 6\right) - \left(\frac{1}{3} \times 1^3 - 3 \times 1^2 + 13 \times 1\right)$$

$$= 42 - \frac{31}{3}$$

$$= \frac{95}{3}$$

Area of $R = \frac{105}{2} - \frac{95}{3} = \frac{125}{6}$

EXERCISE 14E

1 Find the area of the shaded region in each graph.

a **i**

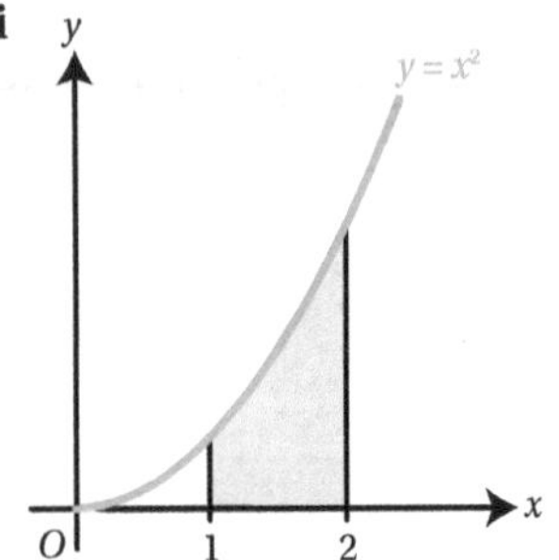

ii

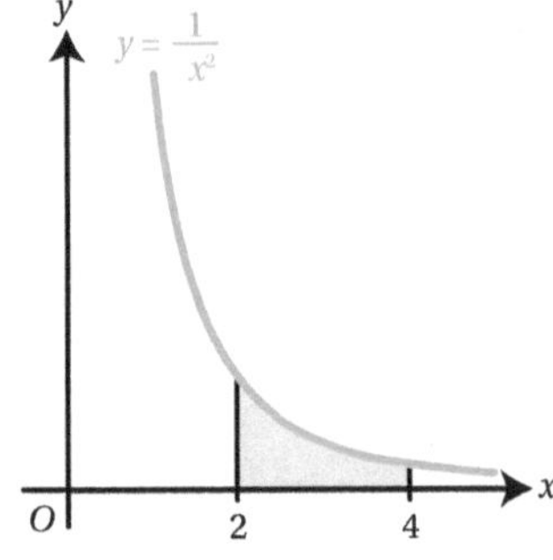

b **i**

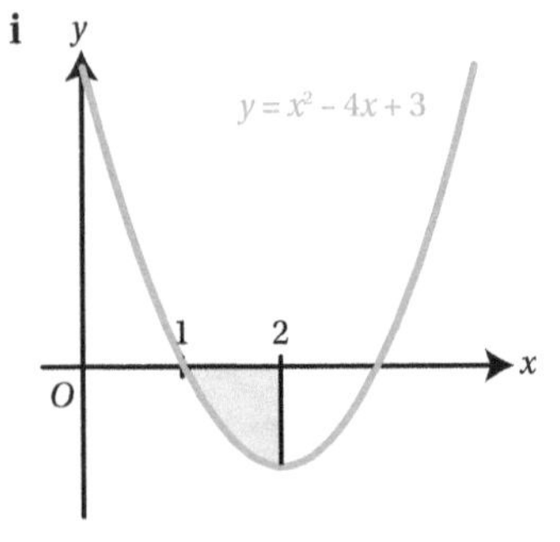

ii

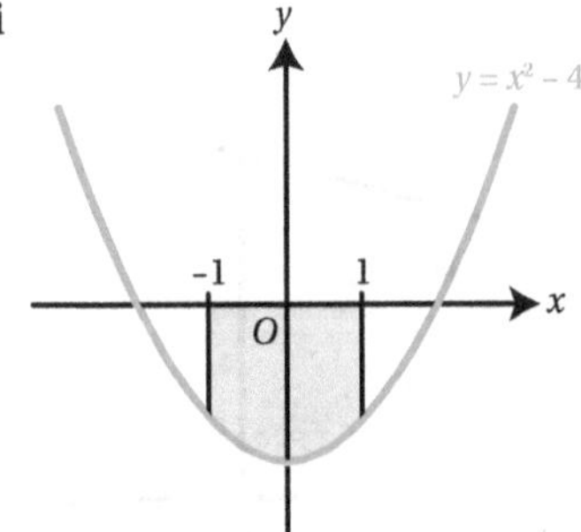

2 Find the area of the shaded region on each graph

a **i**

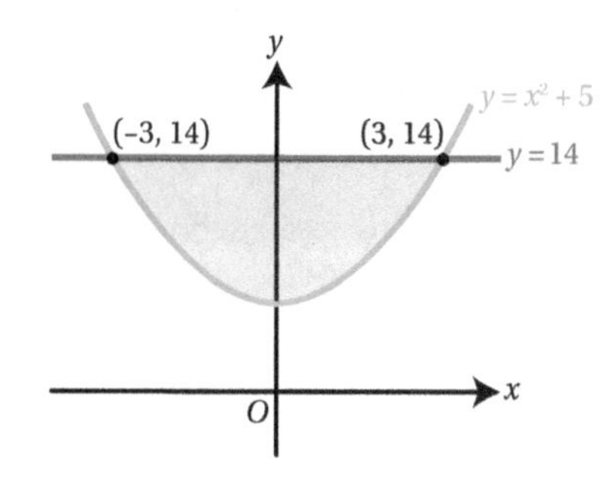

ii

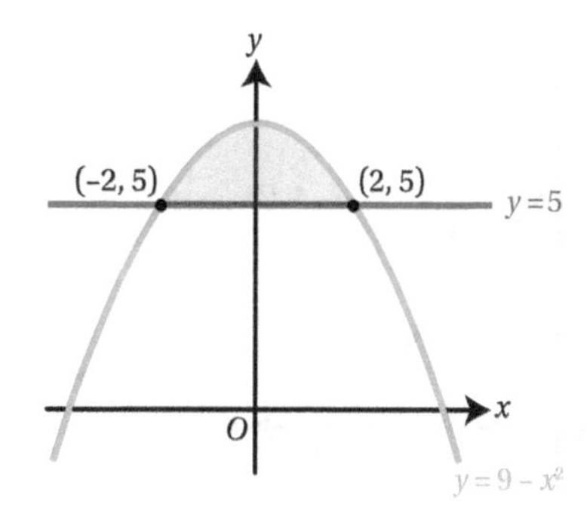

b **i**

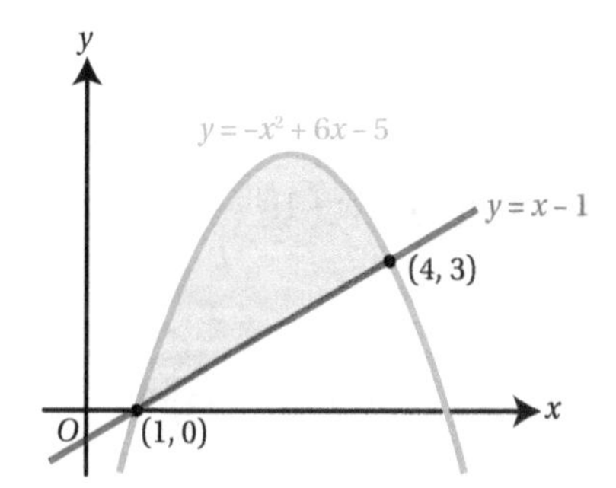

ii

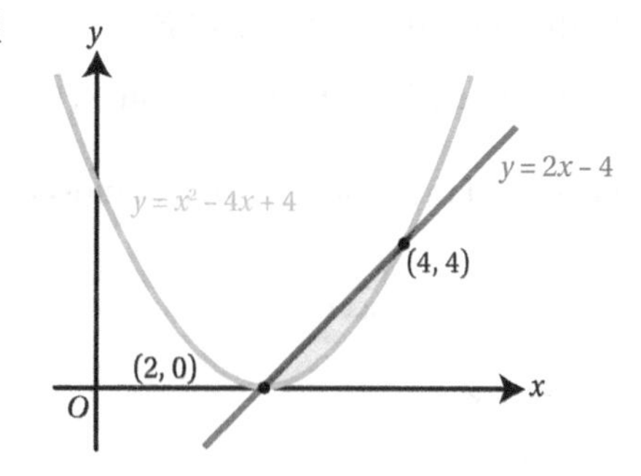

3 Find the area of each region described. You may want to sketch the graph first.

a **i** Between the curve $y = x^2 + 2$, the x-axis and the lines $x = 2$ and $x = 5$.

ii Between the curve $y = 2x^3 + 1$, the x-axis and the lines $x = 1$ and $x = 2$.

b **i** Enclosed between the graph of $y = 4 - x^2$ and the x-axis.

ii Enclosed between the graph of $y = x^2 - 1$ and the x-axis.

4 The curve with equation $y = x^3 - 2x^2 + 2x - 15$ crosses the x-axis at the point P.

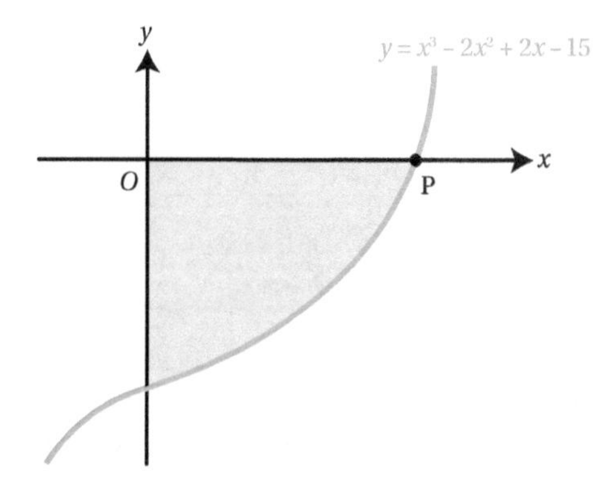

a Find the coordinates of P.

b Find the area of the shaded region bounded by the curve and the coordinate axes.

5 The curve with equation $y=12+4x-x^2$ crosses the y-axis at A and the positive x-axis at B.

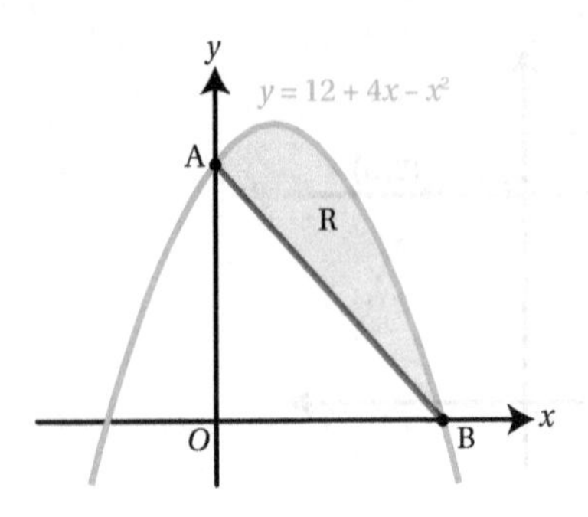

a Find the coordinates of A and B.

b Find the area of the shaded region R.

6 Find the area of the shaded region enclosed between $y=4x-x^2-1$ and $y=x+1$.

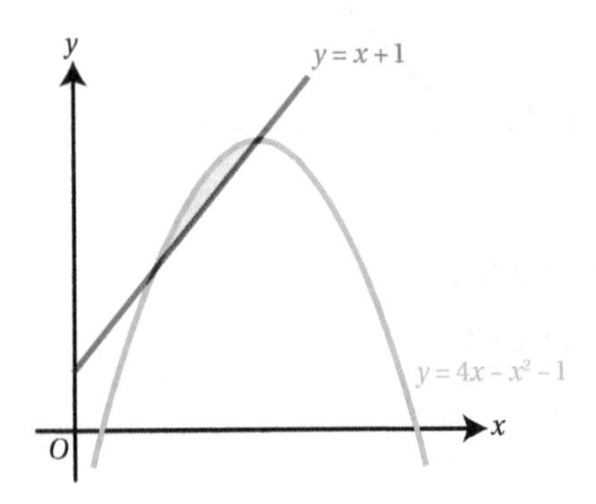

7 The diagram shows the graph of $y=\sqrt{x}$. The area of the shaded region is 18. Find the value of k.

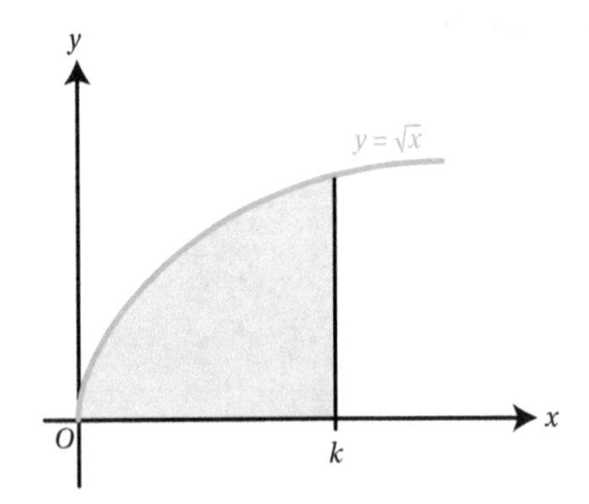

8 **a** Factorise $x^2-7x+10$.

b Find the area enclosed by the curve $y=7x-x^2-10$ and the x-axis.

9 **a** Write down the coordinates of the points where the graph of $y = x^2 - kx$ crosses the x-axis.

b The area shaded in the diagram is bounded by the curve $y = x^2 - kx$, the x-axis and the lines $x = 0$ and $x = 3$.

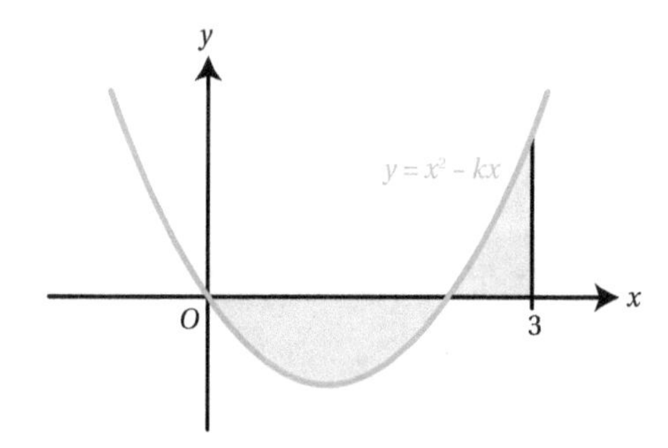

The area below the x-axis is equal to the area above the x-axis. Find the value of k.

10 The diagram shows part of the curve with equation $y = \frac{4}{x^2} + 2x - 2$ and the line with equation $y = -x + 5$.

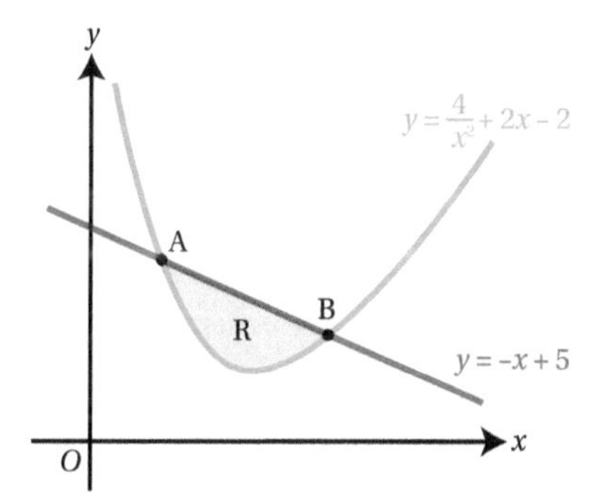

a **i** Show that the x-coordinates of the points of intersection A and B satisfy $3x^3 - 7x^2 + 4 = 0$

ii Hence find the x-coordinates of A and B showing all your working clearly.

b Find the area of the region R.

11 The region enclosed between the curve $y = x^3 - 6x^2 + 10x$, the straight line $y = -2x + 7$ and the x-axis is shown in the diagram.

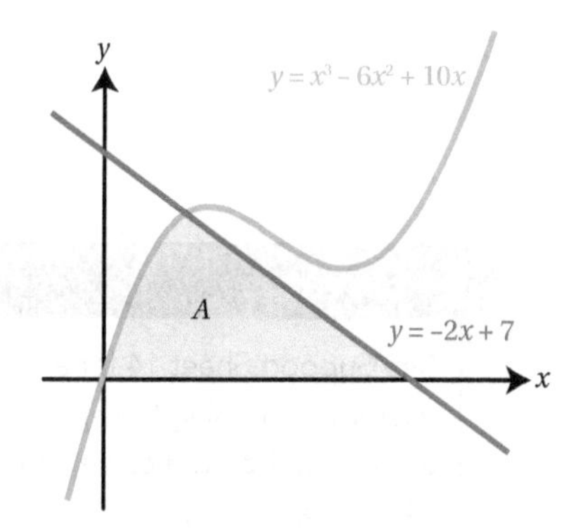

Find the exact value of A.

12 The diagram shows the graphs of $y = 36 - x^2$ and $y = 18x - x^2$. The shaded region is bounded by the two curves and the x-axis.

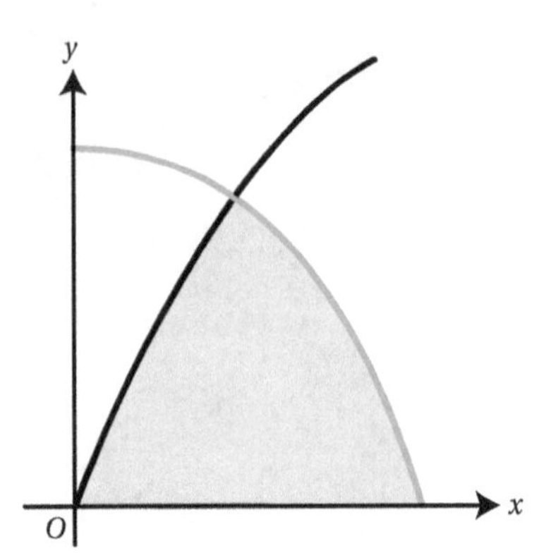

Find the area of the shaded region.

13 The area enclosed between the curve $y = x^2$ and the line $y = mx$ is $10\frac{2}{3}$. Find the value of m if $m > 0$.

14 The equation of the curve in the diagram is $y = \sqrt{x}$.

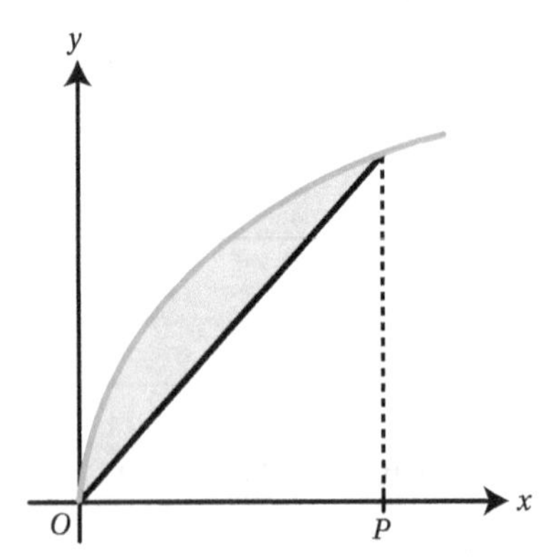

Find the area of the shaded region, in terms of P.

15 The diagram shows a part of the parabola $y = a^2 - x^2$. The x-coordinate of point P is $\frac{a}{2}$.

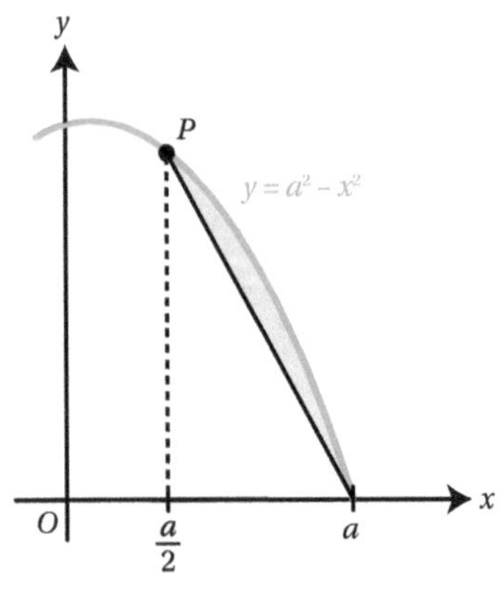

Find the area of the shaded region, in terms of a.

Elevate

See Support Sheet 14 for an example of finding the area above and below a curve, along with more questions.

Checklist of learning and understanding

- **Integration** is the reverse process of differentiation.
- If you know $\frac{dy}{dx}$, the indefinite integral gives an expression for y, with an unknown **constant of integration**. You can find this constant if you know the coordinates of a point on the curve.
- For all rational $n \neq -1$:

$$\int x^n \, dx = \frac{1}{n+1} x^{n+1} + c$$

- The **definite integral** $\int_a^b f(x)dx$ is found by evaluating the integrated expression at the upper limit b and then subtracting the integrated expression evaluated at the lower limit a.
- The **area** between the curve $y = f(x)$, the x-axis and lines $x = a$ and $x = b$ is given by:

$$A = \int_a^b y \, dx$$

provided that the part of the curve between $x = a$ and $x = b$ lies entirely above the x-axis.
- If the curve goes **below the x-axis**, then the integral of the part below the axis will be **negative**.

Mixed practice 14

1. $f(x) = \int 4x+5\,dx$.

   Find $f'(x)$.

   Choose from these options.

   A $4x+5$ **B** $2x^2+5x+c$ **C** 4 **D** $2x^2+5x$

2. A curve has gradient $\frac{dy}{dx} = 3x - \sqrt{x}$ and passes through the point $(4, -1)$. Find the equation of the curve.

3. Find the indefinite integral $\int \frac{1+x\sqrt{x}}{x^2}\,dx$.

4. Given that $f'(x) = (1-x)(\sqrt{x}+2)$, and that $f(1) = 3$, find an expression for $f(x)$.

5. Show that the area of this shaded region is $\frac{9}{2}$.

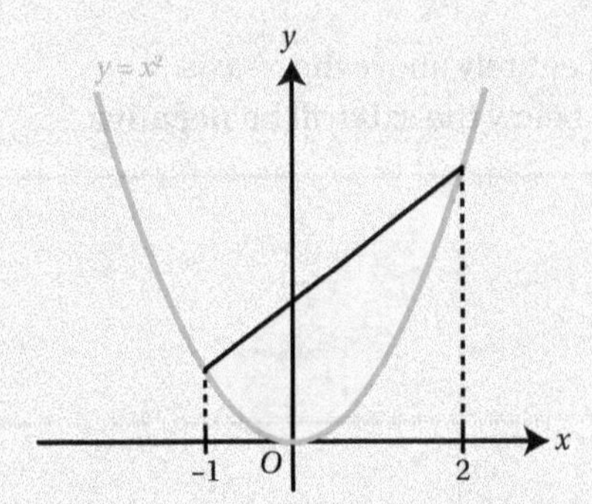

6. **a** Find the exact value of $\int_2^{2\sqrt{3}} 3 - \frac{12}{x^2}\,dx$.

   Give your answer in the form $a + b\sqrt{3}$, where a and b are integers.

   b The diagram shows the curve with equation $y = 3 - \frac{12}{x^2}$. The curve crosses the x-axis at $x = 2$.

   The shaded region is bounded by the curve, the y-axis and the lines $y = 0$ and $y = 2$.

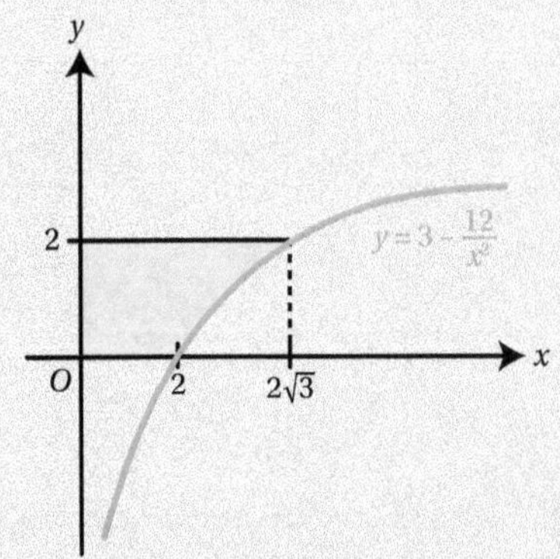

   Find the area of the shaded region.

7. Find the value of $\int_1^4 \left(x^{\frac{3}{2}} - 1\right)^2 dx$.

[© AQA 2012]

8 $f'(x) = \dfrac{4x^2 - 3\sqrt{x}}{x}$ and $f(1) = 2$.

Find the value of $f(4)$.

Choose from these options.

A 6.5 **B** 8 **C** 20 **D** 26

9 $\int_1^a 2x - 3\,dx = 6, a > 0$

What is the value of a?

Choose from these options.

A $a = 3$ **B** $a = 4$ **C** $a = 4.5$ **D** $a = 6$

10 The gradient, $\frac{dy}{dx}$, of a curve at the point (x, y) is given by:

$\frac{dy}{dx} = 10x^4 - 6x^2 + 5$

The curve passes through the point $P(1, 4)$.

Find the equation of the curve.

[© AQA 2013]

11 Find the positive value of a for which $\int_0^a x^3 - x\,dx = 0$.

12 Find the area enclosed between the graph of $y = k^2 - x^2$ and the x-axis, giving your answer in terms of k.

13 Let $f(x) = -x^3 + x^2 + 8x - 12$.

a **i** Show that $(x - 2)$ is a factor of $f(x)$ and hence factorise $f(x)$ completely.

ii Sketch the graph of $y = f(x)$, labelling clearly the points where the curve crosses the coordinate axes.

b Find the exact area enclosed by the x-axis and the graph of $y = f(x)$.

14 The diagram shows the graph of $y = x^n$ for $n > 1$.

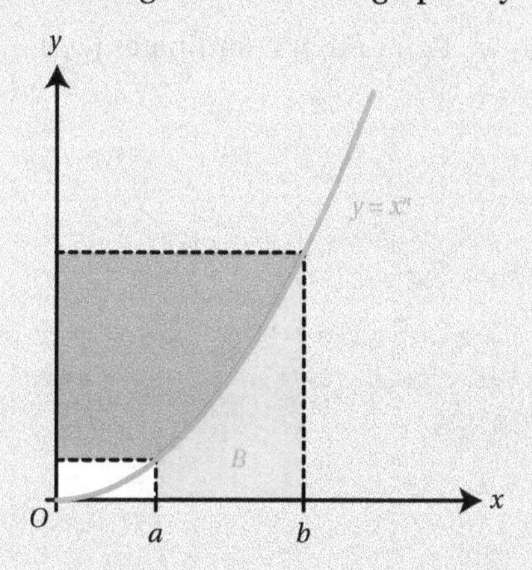

a **i** Write down an expression for the area of the white rectangle.

ii The area of the blue shaded region is B. Find an expression for B in terms of a, b and n.

b The red area is three times as large as the blue area. Find the value of n.

15 The diagram shows a parabola with equation $y = a^2 - x^2\ (a > 0)$. The parabola crosses the x-axis at points A and B, and the y-axis at point C.

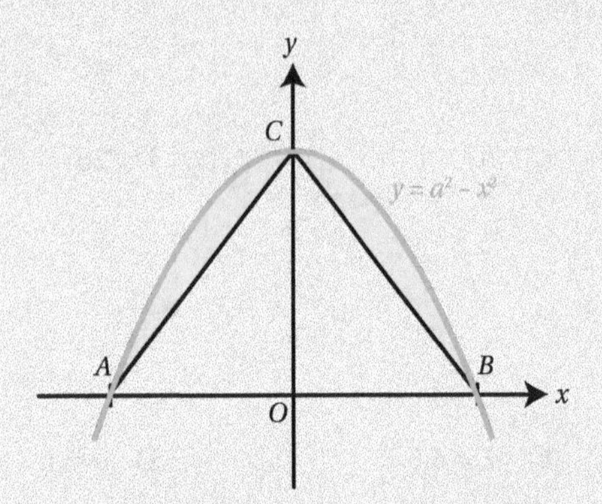

a **i** Write down the coordinates of A, B and C.

ii Find, in terms of a, the area of the shaded region.

b Point P lies on the parabola. The x-coordinate of P is p.

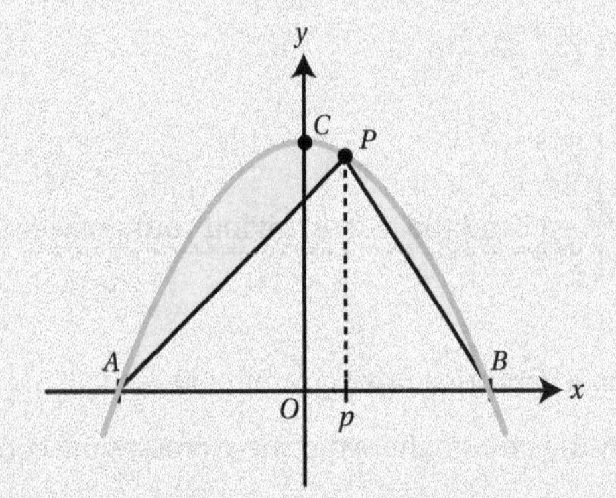

The value of p varies between the x-coordinates of A and B.

Find the minimum value of the shaded area.

16 The diagram shows part of the curve with equation $y = 4 - x^2$. Point P has coordinates $(p, 4 - p^2)$ and point Q has coordinates $(2, 0)$.

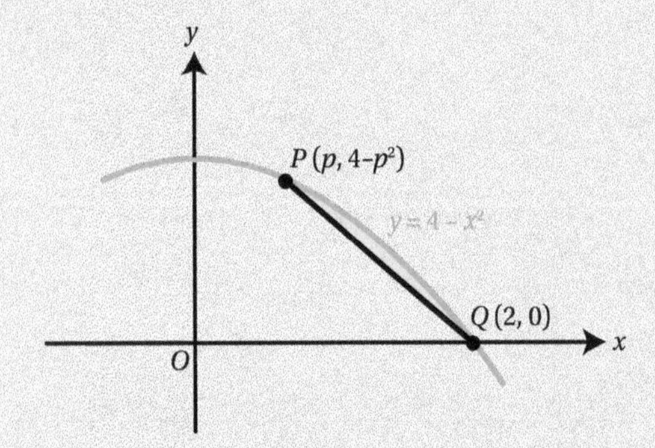

The shaded region is bounded by the curve and the chord PQ. Show that the area of the shaded region is $\frac{1}{6}(2 - p)^3$.

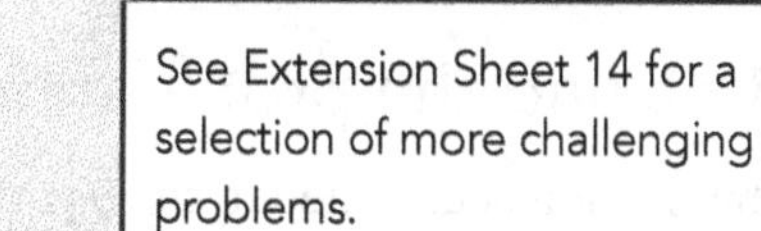

Elevate
See Extension Sheet 14 for a selection of more challenging problems.

17 The function $f(x)$ has a stationary point at $(3, 19)$

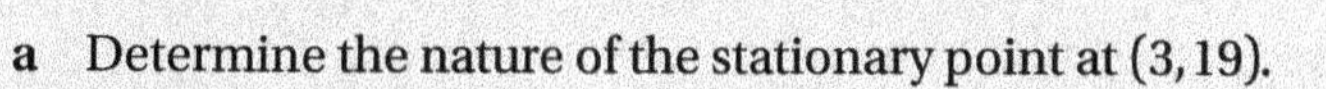

and $f''(x) = 6x + 6$.

a Determine the nature of the stationary point at $(3, 19)$.

b Find an expression for $f(x)$.

18 The diagram shows the graph of $y = 6\sqrt{x}$ and the tangent to the graph at the point $(9, 18)$. The tangent crosses the y-axis at the point B.

Find the area of the shaded region.

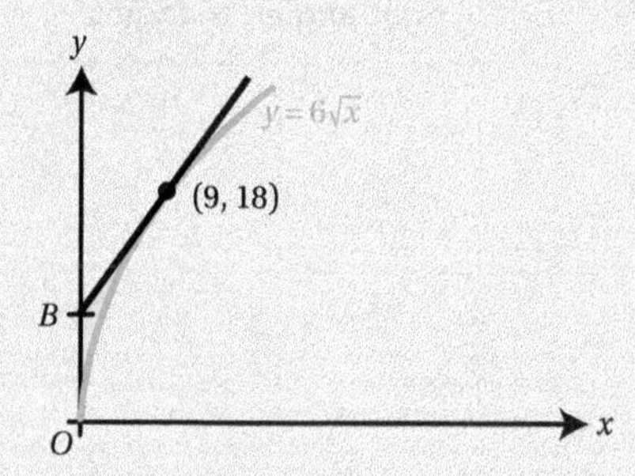

FOCUS ON ... PROOF 2

Breaking the problem down

This section is intended to show you how you can prove new results by breaking them down into previously known ones. It will use as examples the sine and cosine rules.

Tip

When writing a proof, you need to know which results you can assume. Here, assume that you can use trigonometry in right-angled triangles.

Start with a specific example. Consider a triangle with $AB = 7$ units, $\angle BAC = 55°$ and $\angle ACB = 80°$. What is the length of BC?

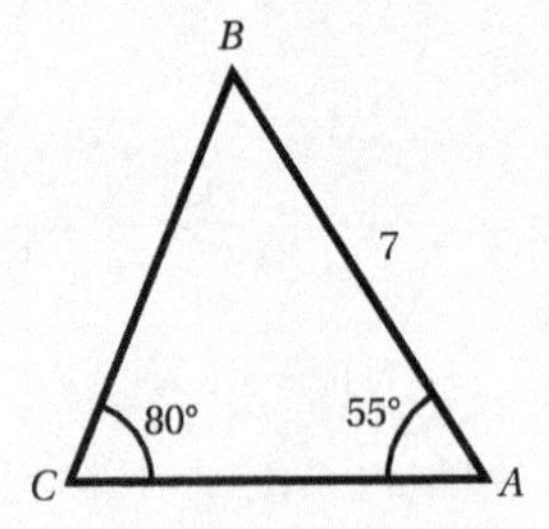

There are no right angles in the diagram, but you can create some by drawing the line BD perpendicular to AC.

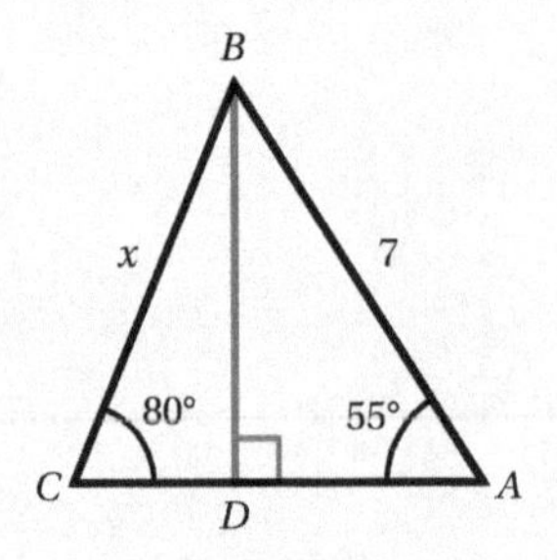

You now have two right-angled triangles, which share the side BD.

Use triangle ABD to write an expression for BD.

Use triangle BCD to write an expression for BD.

Comparing the two expressions for BD:

$$x \sin 80° = 7 \sin 55°$$

Rearranging gives:

$$x = \frac{7 \sin 55°}{\sin 80°} = 5.82$$

QUESTIONS

1. Use the given example to write a general proof of the sine rule: $\frac{a}{\sin A} = \frac{c}{\sin C}$.

2. How is the diagram different if the angle C is obtuse? Does the proof still work?

3. Use this diagram, and your knowledge of right-angled triangles and Pythagoras' theorem, to prove the cosine rule: $c^2 = a^2 + b^2 - 2ab \cos C$.

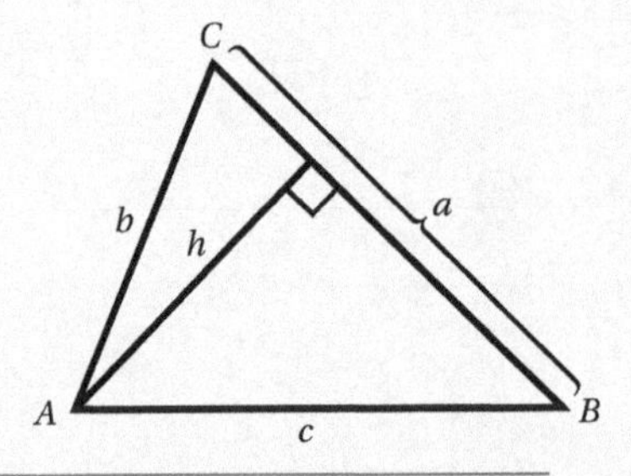

FOCUS ON ... PROBLEM SOLVING 2

Choosing variables

Solving a problem often starts with writing some equations to represent the situation. While in exam questions you are often given instructions such as: 'Express the volume in terms of r,' in real applications you need to decide for yourself which variables are relevant to the question. Sometimes there is more than one possible choice, and some choices need simpler equations than others.

In this section you will focus on selecting variables and writing equations. Some of the resulting equations cannot be solved algebraically, so you will need to use technology.

WORKED EXAMPLE

A closed cylindrical can has a fixed volume of 540π cm^3. Find the minimum possible surface area of the can.

Let r be the radius and h the height of the cylinder. — Define the variables: the surface area depends on the radius and the height of the can.

Surface area:
$S = 2\pi r^2 + 2\pi rh$ — Write an expression for the surface area.

Volume:
$V = \pi r^2 h = 540\pi$ — The surface area depends on two variables. You can eliminate one of them by using the expression for the volume.

$\Rightarrow h = \frac{540}{r^2}$ — You now have a choice: you can express h in terms of r, or r in terms of h. The latter would involve square roots, so choose the former.

$\Rightarrow S = 2\pi r^2 + 2\pi r\left(\frac{540}{r^2}\right)$

$S = 2\pi r^2 + \frac{1080\pi}{r}$

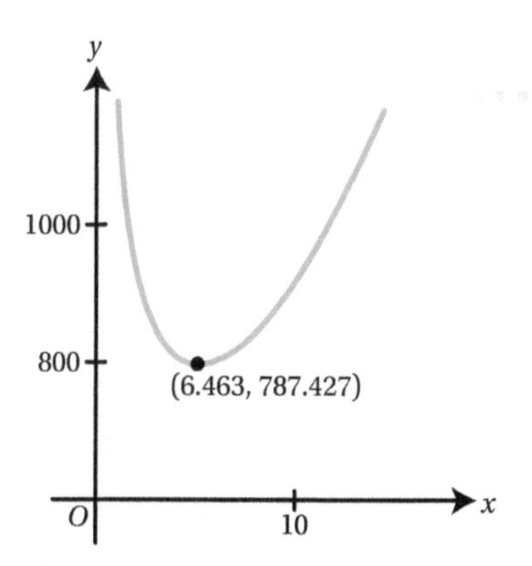

You can use graphing software to sketch the graph of S against r and find the minimum value.

The graph shows that the minimum value of the surface area is 787 cm^2.

QUESTIONS

1. Express the surface area in terms of h instead. How does this change the calculation?

2. An open box is in the shape of a square-based prism. The volume of the box is 4000 cm^3. Find the minimum possible surface area of the box.

3. A closed box in the shape of a square-based prism has surface area 2500 cm^2. Find the maximum possible volume of the box.

4. Find the shortest distance from the point $(4, 1)$ and the curve with equation $y = \sqrt{x}$.

5. A chord connects two points on the parabola $y = x^2$. Find the point on the parabola where the tangent is parallel to this chord.

6. The normal to the graph of $y = x^2$ crosses the axes at points A and B. Given that the area of triangle AOB is 3 units^2, find its perimeter.

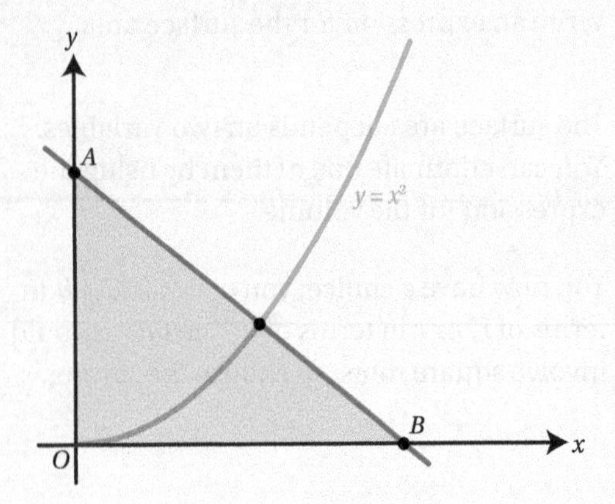

FOCUS ON ... MODELLING 2

The sunrise equation

The 'sunrise equation' can be used to calculate the approximate time the sun rises and sets at a specified location on Earth on a specified day.

$$H = \left| \frac{1}{15} \arccos\left[-\tan(L) \tan\left(23.44 \sin\left(\frac{360(D+284)}{365} \right)^{\circ} \right) \right] \right|$$

In this equation:

- H is the number of hours before local noon when sunrise occurs. 'Local noon' is when the sun is directly overhead at that particular location. (You may need to add an hour for summer time.)
- L is the latitude. It is measured in degrees, between $-90°$ and $90°$; points in the northern hemisphere have positive latitude.
- D is the day of the year, with $D = 1$ being 1 January.
- The modulus sign, $| \; |$, means that you take the positive value even if the value of arccos is negative.
- Sunset occurs H hours after local noon.

As well as calculating the times of sunrise and sunset, the equation can be used to generate a 'Day and night world map', which shows the parts of the world that are in daylight at a particular point in time.

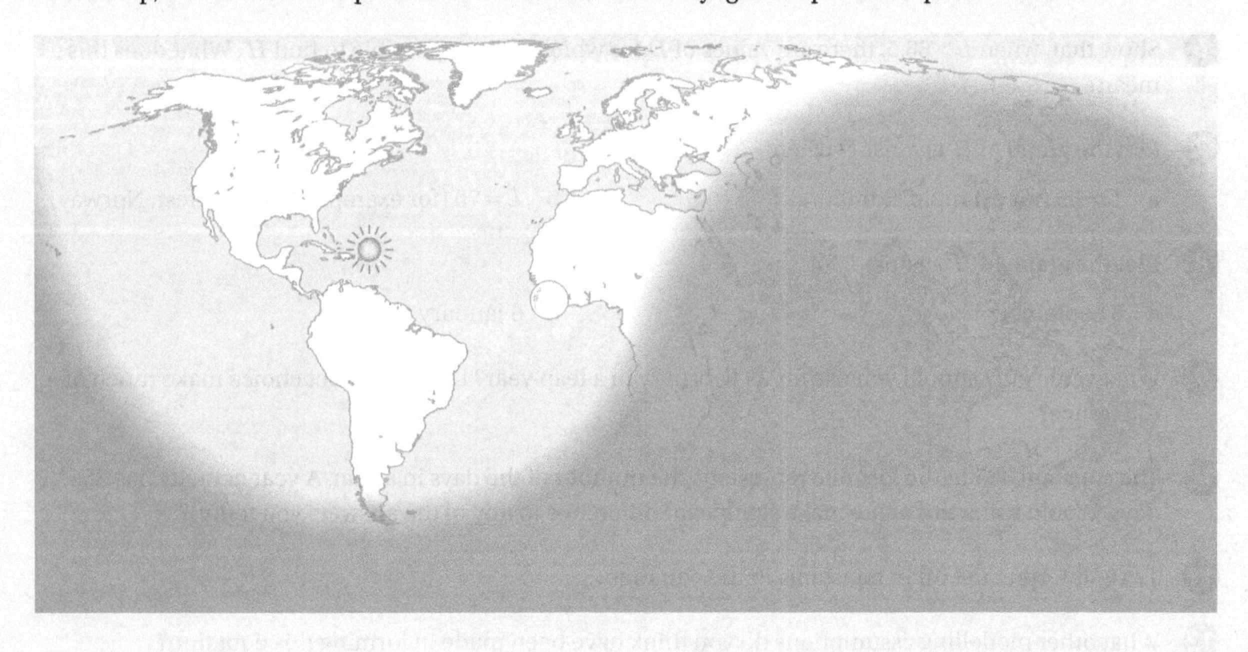

QUESTIONS

1. London is at the latitude of around 53° north. Work out the time that the formula predicts sunrise will occur:

   a on 16 February **b** on 5 July.

2. Predict the time of sunset, relative to local noon, in Rio de Janeiro (about 22° south) on 12 August. What would you need to consider to find the actual time of sunset?

3. Sydney, Australia, is at about 33° south. Find the approximate length of the day (the time between sunrise and sunset) in Sydney on:

   a 8 March **b** 26 August.

4. State the largest possible value of $23.44 \sin x$. Hence find the length of the longest day in Cairo, located at 30° north.

5. Use the sunrise equation to show that, on the days of the vernal and autumnal equinoxes (20 March and 22 September) there are exactly 12 hours of daylight, regardless of the location on Earth.

6. Show that, when $L > 66.5$, there are values of D for which it is not possible to find H. What does this mean?

7. Plot the graph of H against D when:

   a $L = 55$ (for example, Edinburgh) **b** $L = 70$ (for example, Hammerfest, Norway).

8. Plot the graph of H against L for:

   a 6 September **b** 6 January.

9. What value of D should you use for 29 February in a leap year? Does the exact choice make much of a difference?

10. The constant 365 in the formula represents the number of the days in a year. A year actually has 365.25 days. Would using this value make significant difference to any of the answers you found?

11. Try to interpret the other constants in the equation.

12. What other modelling assumptions do you think have been made in forming this equation?

CROSS-TOPIC REVIEW EXERCISE 2

1. Find the exact period of the function $f(x) = \sin 4x + \sin 6x$.

2. Find the exact solutions of the equation $\sin 2\theta = \sqrt{3} \cos 2\theta$ for $0 \leqslant \theta \leqslant 180°$.

3. If $f(x) = ax^3 + bx^2 + 4x - 3$, given that $f'(2) = 0$ and $f''(2) = 10$, find the values of $a, b \in \mathbb{R}$.

4. A polynomial is defined by $f(x) = x^3 - 5x^2 - x + 5$.

   a **i** Show that $(x+1)$ is a factor of $f(x)$.

   ii Hence factorise $f(x)$ completely.

   b The graph of $y = f(x)$ is shown in the diagram.

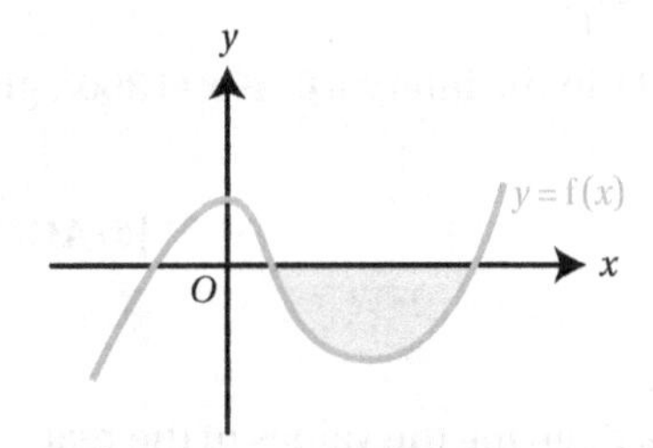

   Find the exact value of the area of the shaded region.

5. **a** The polynomial $f(x)$ is given by $f(x) = x^3 - 4x + 15$.

   i Use the factor theorem to show that $x + 3$ is a factor of $f(x)$.

   ii Express $f(x)$ in the form $(x+3)(x^2 + px + q)$, where p and q are integers.

   b A curve has equation $y = x^4 - 8x^2 + 60x + 7$.

   i Find $\frac{dy}{dx}$.

   ii Show that the x-coordinates of any stationary points of the curve satisfy the equation:

$$x^3 - 4x + 15 = 0$$

   iii Use the results above to show that the only stationary point of the curve occurs when $x = -3$.

   iv Find the value of $\frac{d^2y}{dx^2}$ when $x = -3$.

   v Hence determine, with a reason, whether the curve has a maximum point or a minimum point when $x = -3$.

[© AQA 2013]

6. The triangle in the diagram has sides of length $AB = x + 1$ units, $BC = x + 3$ units, $CA = 2x + 1$ units and angle $A\hat{B}C = 120°$. Find the value of x.

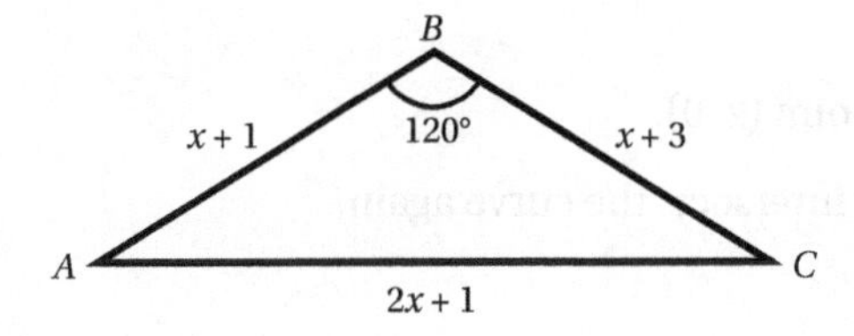

7 a Show that $\frac{1}{\cos x} - \cos x = \sin x \tan x$.

b You are given that $\sin x - \cos x = \frac{1}{\sin x} - \frac{1}{\cos x}$.

i Show that $\tan x = 1$.

ii Hence find the value of $x \in [180°, 360°]$.

8 a Write down the two solutions of the equation $\tan(x+30°) = \tan 79°$ in the interval $0° \leqslant x \leqslant 360°$.

b Describe a single geometrical transformation that maps the graph of $y = \tan x$ onto the graph of $y = \tan(x+30°)$.

c i Given that $5+\sin^2\theta = (5+3\cos\theta)\cos\theta$, show that $\cos\theta = \frac{3}{4}$.

ii Hence solve the equation $5+\sin^2 2x = (5+3\cos 2x)\cos 2x$ in the interval $0° \leqslant x \leqslant 360°$, giving your values of x to three significant figures.

[© AQA 2013]

9 Consider the function $f(x) = \frac{(x-2)(x-6)}{\sqrt{x}}$.

a Show that this can be written in the form $f(x) = x^a - 8x^b + 12x^c$, giving the values of the real numbers a, b and c.

b Find the equation of the normal to $f(x)$ at the point $x = 4$.

c The normal intersects the x-axis at the point P and the y-axis at the point Q.

i State the coordinates of P and Q.

ii Give the exact area of the triangle POQ.

10 The diagram shows the graph of $y = 9 - x^2$ and the tangent to the graph at $x = 1$.

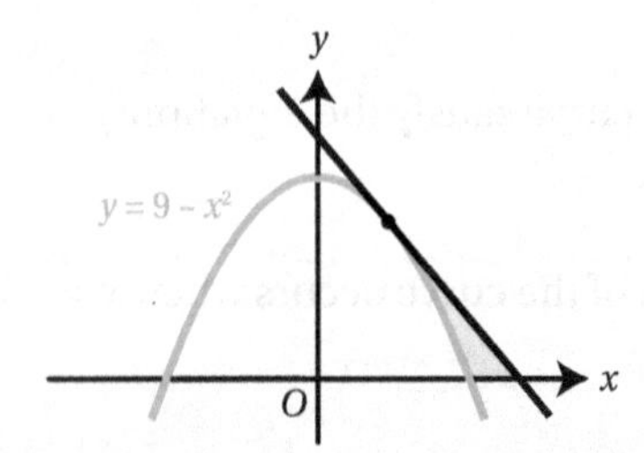

Find the area of the shaded region.

11 a Expand and simplify $(x+h)^4 - x^4$.

b Hence prove from first principles that the derivative of x^4 is $4x^3$.

12 A curve has gradient $f'(x) = 2x - 6$ and passes through the point $(2, 0)$.

a Find the equation of the curve.

b Find the equation of the normal to the curve at the point $(2, 0)$.

c Find the coordinates of the points where the normal intersects the curve again.

CROSS-TOPIC REVIEW EXERCISE 2

13 The diagram shows curves with equations $y=(x-3)^2$ and $y=a(x-p)^2+q$.

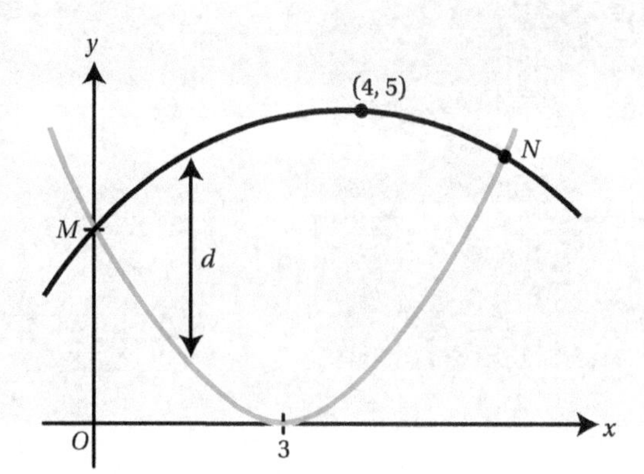

a Find the values of a, p and q.

b Find the coordinates of the intersection points, M and N, of the two curves.

c The vertical distance between the two curves is denoted by d, as shown in the diagram.

i Write an expression for d in terms of x.

ii Hence find the maximum vertical distance between the two curves, on the part of the curves between points M and N.

14 The point $A(4, 3)$ lies on the circle with equation $x^2+y^2=25$.

The point B lies on the parabola $y=10-x^2$.

The tangent to the circle at A is parallel to the tangent to the parabola at B.

Find the coordinates of B.

15 Given that $A=\arcsin\left(\frac{1}{3}\right)$ find the exact value of $\cos A$.

16 The diagram shows a parabola with equation $y=ax^2$ and a circle, with the centre on the y-axis, that passes through the origin. The radius of the circle is r.

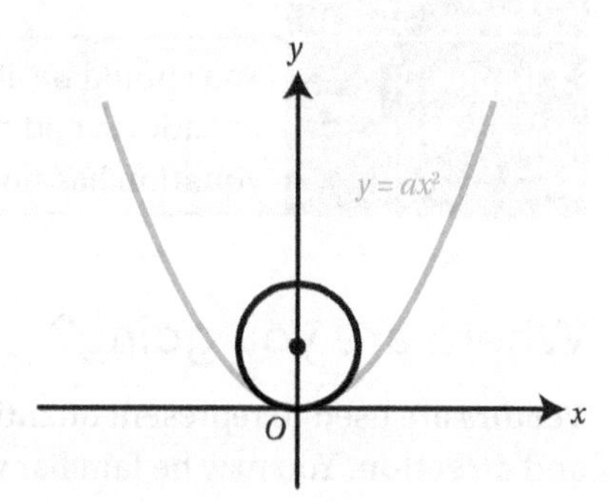

a Show that the y-coordinates of any intersections of the circle and the parabola satisfy the equation $y^2-2yr+\frac{y}{a}=0$.

b Hence find, in terms of a, the largest value of r for which the circle and the parabola have only one common point.

17 a i Simplify $(a^2+b^2)^2-2a^2b^2$.

ii Hence show that $\cos^4\theta+\sin^4\theta=1-2\sin^2\theta\cos^2\theta$.

b Show that $(\cos\theta-\sin\theta)^4=(1-2\sin\theta\cos\theta)^2$.

15 Vectors

In this chapter you will learn how to:

- represent two-dimensional vectors using the base vectors **i** and **j**
- find the magnitude and direction of a vector
- add and subtract vectors, and multiply vectors by a scalar
- recognise when two vectors are parallel
- find unit vectors
- work with positions and displacement of points in the plane
- use vectors to solve problems about geometrical figures.

Before you start...

GCSE	You should be able to represent vectors on a grid and write them as column vectors.	1 Write this as a column vector.
GCSE	You should be able to use Pythagoras' theorem and trigonometry in right-angled triangles.	2 Find the length of the side BC and the size of the angle ABC. C, 2, A, 5, B
Chapter 3	You should be able to solve quadratic equations, and recognise when a quadratic equation has no solutions.	3 State the number of solutions of the equation $6x^2+9x+1=0$.

Where are you going?

Vectors are used to represent quantities that have both **magnitude** (size) and **direction**. You may be familiar with some examples of vectors, such as velocity and force. By contrast, **scalar** quantities (such as mass) can be fully described by a single number. In pure mathematics, vectors are used to represent displacement from one point to another, and thus to describe geometrical figures. They also have many applications in spatial modelling problems, for example, describing flight paths or positions of characters in a computer game.

Fast forward

This chapter will focus on vectors in two dimensions. In Student Book 2, the skills developed here will be used for vectors in three dimensions.

In this chapter you will learn about different ways to represent vectors and use them to solve geometrical problems. In Chapters 18 and 19 you will use vectors to work with forces.

Section 1: Describing vectors

You already know that there are two ways to describe a vector: you can draw an arrow or write it as a column vector.

- The arrow shows the magnitude (length) and the direction of the vector explicitly.
- The numbers in a column vector are called the **components** of the vector. For example, the column vector $\begin{pmatrix} 5 \\ -3 \end{pmatrix}$ has horizontal component 5 and vertical component −3.

See Gateway to A Level Section R for revision of column vectors and vector diagrams.

There is another way to write a vector, using its components. Vectors of length 1 unit in the horizontal direction are denoted as **i** and in the vertical direction by **j**. Then any vector can be expressed in terms of **i** and **j**, as shown in the diagram. **i** and **j** are called **base vectors**.

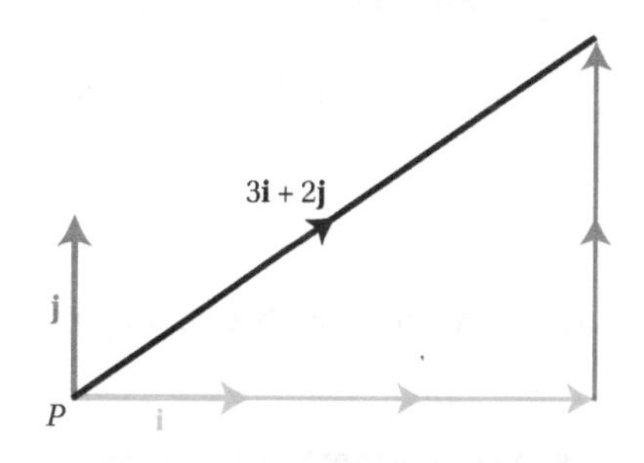

To emphasise that something is a vector, rather than a scalar (number), it is written in **bold type**, for example, $\mathbf{a} = 3\mathbf{i} + 2\mathbf{j}$. When writing by hand, you should underline letters representing vectors.

WORKED EXAMPLE 15.1

a Use **i**, **j** notation to write vectors **a**, **b** and **c**.

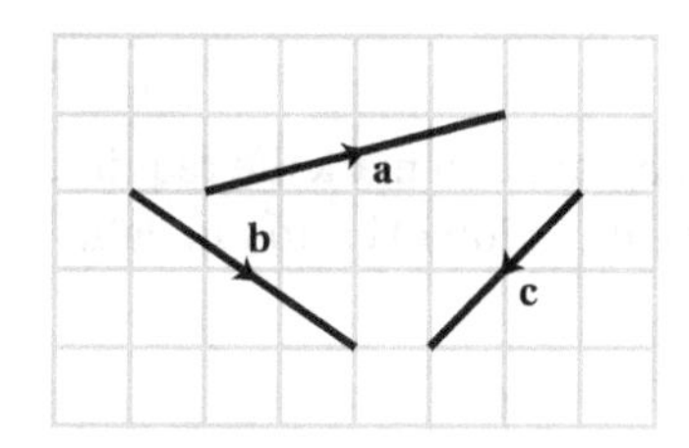

b Write **p**, **q** and **r** as column vectors.

i $\mathbf{p} = 4\mathbf{i} - 4\mathbf{j}$

ii $\mathbf{q} = 2\mathbf{j}$

iii $\mathbf{r} = \mathbf{j} - 4\mathbf{i}$

a $\underline{a} = 4\underline{i} + \underline{j}$

$\underline{b} = 3\underline{i} - 2\underline{j}$

$\underline{c} = -2\underline{i} - 2\underline{j}$

The coefficient of **i** represents the number of horizontal units moved and the coefficient of **j** the number of vertical units moved. Positive directions are to the right and up.

b $\underline{p} = \begin{pmatrix} 4 \\ -4 \end{pmatrix}$

This represents a move of 4 units in the horizontal direction and −4 in the vertical direction.

$\underline{q} = \begin{pmatrix} 0 \\ 2 \end{pmatrix}$

There are zero units in the **i**-direction, so the horizontal component is 0.

$\underline{r} = \begin{pmatrix} -4 \\ 1 \end{pmatrix}$

Be careful: the question gives **j** before **i** this time. The **j** component is 1 and the **i** component is −4.

Magnitude and direction

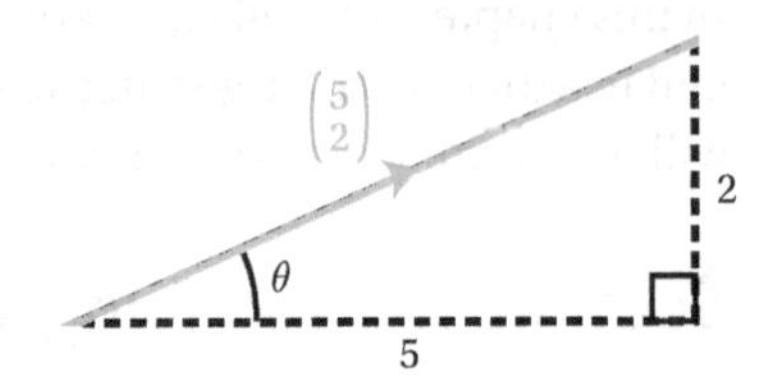

If you are given a vector in component form you can work out its magnitude and direction.

The magnitude of the vector is its length and you can find it by using Pythagoras' theorem. The **modulus** symbol, $|\,\mathbf{a}\,|$, is used to represent the magnitude of a vector. The direction is given by an angle, which you find by using trigonometry.

Key point 15.1

- The **magnitude** of a vector is its length. If $\mathbf{a} = \begin{pmatrix}p\\q\end{pmatrix}$ then $|\,\mathbf{a}\,| = \sqrt{p^2 + q^2}$.
- The direction of a vector is the angle it makes with a specified direction.

Tip

When finding the direction of a column vector it is important to draw a diagram to be sure to calculate the correct angle.

If you aren't told where to measure the angle from you can choose – but it is important to specify clearly, for example, 'above the **i**-direction'.

The only vector with zero magnitude is $\begin{pmatrix}0\\0\end{pmatrix}$, called the **zero vector**.

WORKED EXAMPLE 15.2

a Find the exact magnitude of the vector $\mathbf{a} = 5\mathbf{i} + 3\mathbf{j}$ and the angle it makes with the direction of $\mathbf{i}$.

b Find the angle the vector $\mathbf{b} = \begin{pmatrix}4\\-7\end{pmatrix}$ makes with the vector $-\mathbf{j}$.

a

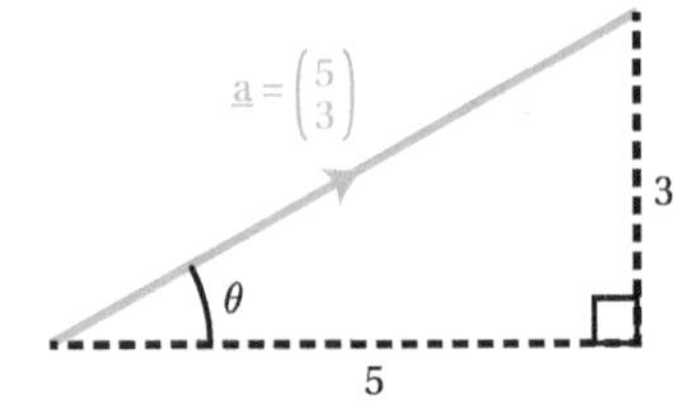

Draw a diagram and mark the required angle. The direction of **i** is to the right.

$|\underline{a}| = \sqrt{25+9}$

$= \sqrt{34}$

Use Pythagoras' theorem to find the length (magnitude).

$\tan\theta = \frac{3}{5}$

$\theta = 31.0°$ (3 s.f.)

Continues on next page

b The direction is 31.0° above the direction of **i**.

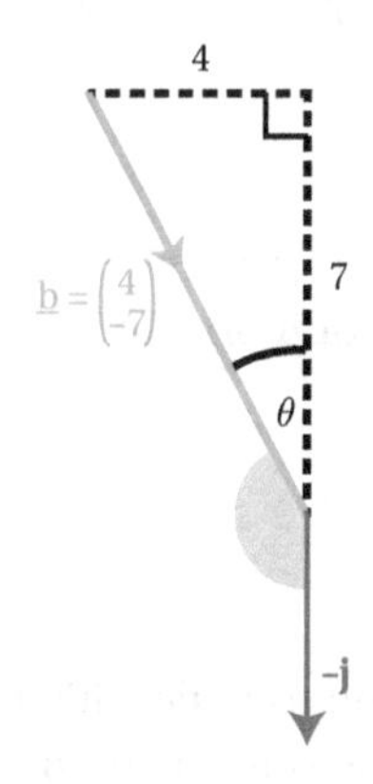

The vector **−j** points straight down. The required angle is marked in blue.

$\tan\theta = \frac{4}{7}$

$\theta = 29.7°$

Use a right-angled triangle to find first.

The required angle is $180 - \theta = 150.3°$

WORK IT OUT 15.1

Find the direction the vector $-5\mathbf{i} + 2\mathbf{j}$ makes with the positive **i** direction.

Which is the correct solution?

Identify the errors made in the incorrect solutions.

Solution 1	Solution 2	Solution 3
$\tan^{-1}\left(\frac{2}{-5}\right) = -21.8°$ So the angle is 21.8° below the horizontal.	$\tan^{-1}\left(\frac{2}{5}\right) = 21.8°$ $180° - 21.8° = 158°$ So the angle is 158°.	$\tan^{-1}\left(\frac{5}{2}\right) = 68.2°$ $180° - 68.2° = 112°$ So the angle is 112°.

You can also use trigonometry to find the components of a vector, if you know its magnitude and direction.

Fast forward

This method will be particularly useful when you learn about resolving forces in Student Book 2.

WORKED EXAMPLE 15.3

Vector **v** has magnitude 7 and makes an angle of 62° above the **i**-direction. Write **v** as a column vector.

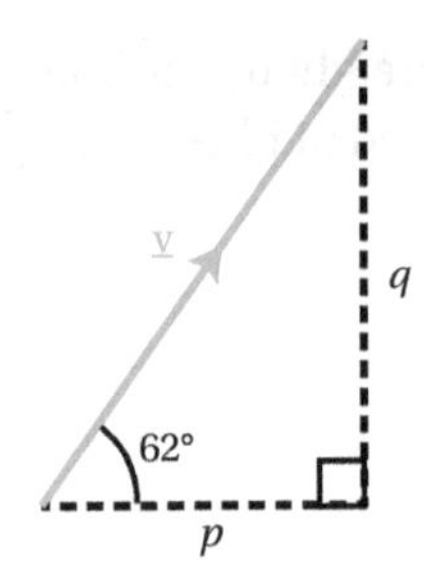

Draw a diagram showing the given information and the components (labelled p and q).

$$\underline{v} = \begin{pmatrix} p \\ q \end{pmatrix}$$

You have a right-angled triangle with a known hypotenuse and one angle, so use trigonometry.

$p = 7\cos 62° = 3.29$

$q = 7\sin 62° = 6.18$

$$\therefore \underline{v} = \begin{pmatrix} 3.29 \\ 6.18 \end{pmatrix}$$

Unit vectors

Vectors of length one are often useful, so are given a special name.

Key point 15.2

A **unit vector** is a vector with magnitude one unit.

You have already met two special examples of unit vectors: the base vectors **i** and **j**.

WORKED EXAMPLE 15.4

a Which of these vectors is a unit vector?

$$\mathbf{a} = \begin{pmatrix} -1 \\ 2 \end{pmatrix}, \mathbf{b} = \begin{pmatrix} \frac{1}{\sqrt{3}} \\ \sqrt{\frac{2}{3}} \end{pmatrix}$$

b Find two possible values of k such that $\mathbf{c} = (12k)\mathbf{i} - (9k)\mathbf{j}$ is a unit vector.

a $|\underline{a}| = \sqrt{1+4}$

$= \sqrt{5} \neq 1$

Use Pythagoras' theorem to find the magnitudes.

$|\underline{b}| = \sqrt{\frac{1}{3} + \frac{2}{3}}$

$= 1$

$\underline{b}$ is a unit vector.

A unit vector has magnitude one.

Continues on next page

b $|\mathbf{c}|=1$ — **c** is a unit vector so it has magnitude 1.

$\sqrt{(12k)^2+(-9k)^2}=1$ — Use Key point 15.1 (or Pythagoras' theorem) to find an expression for the magnitude of **c** in terms of k.

$\sqrt{144k^2+81k^2}=1$

$\sqrt{225k^2}=1$

$225k^2=1$ — Square the equation to remove the square root, and solve for k.

$k^2=\frac{1}{225}$

$k=\pm\frac{1}{15}$ — Remember the + and – when square rooting.

EXERCISE 15A

1 Use **i**, **j** notation to write these vectors.

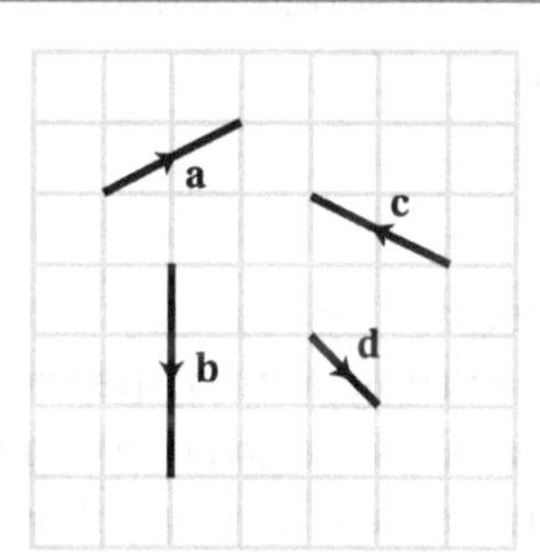

2 Represent each vector on a grid.

a i $\begin{pmatrix}3\\-5\end{pmatrix}$ ii $\begin{pmatrix}-2\\-4\end{pmatrix}$

b i $-4\mathbf{i}+3\mathbf{j}$ ii $2\mathbf{i}-\mathbf{j}$

c i $6\mathbf{j}$ ii $-2\mathbf{i}$

3 For each vector from question 2, find:

A the magnitude

B the angle it makes with the direction of **i**

C the angle it makes with the direction of **j**.

4 Decide which of these vectors are unit vectors.

a i $\frac{1}{2}\mathbf{i}+\frac{1}{3}\mathbf{j}$ ii $-\mathbf{j}$

b i $(\cos\theta)\mathbf{i}+(\sin\theta)\mathbf{j}$ ii $(3\cos\theta)\mathbf{i}+(3\sin\theta)\mathbf{j}$

5. Find the values of c such that $(2c)\mathbf{i}+(5c)\mathbf{j}$ is a unit vector.

6. $\begin{pmatrix} 3k \\ -2k \end{pmatrix}$ is a unit vector. Find the possible values of k.

7. Find the values of t such that the vector $\begin{pmatrix} 4t+10 \\ -2t \end{pmatrix}$ has magnitude $2\sqrt{10}$.

8. Vector $\mathbf{v}$ has magnitude 12 and is on a bearing of 027°. Find the components of $\mathbf{v}$.

9. Vector $\mathbf{b}$ has magnitude $2\sqrt{3}$ and makes an angle of 150° with the positive $\mathbf{i}$-direction. Write $\mathbf{b}$ as a column vector, giving your answers in surd form.

Section 2: Operations with vectors

In order to use vectors to solve problems you need to be able to perform some algebraic operations with them: adding, subtracting and multiplying by a scalar.

Fast forward

If you study Further Mathematics you will learn two different ways of multiplying one vector by another vector.

Adding vectors

If two vectors are represented by arrows, you can perform vector addition by joining the starting point of the second vector to the end point of the first. If the vectors are given in component form, you just add the corresponding components. The sum of two vectors is also called the **resultant vector.**

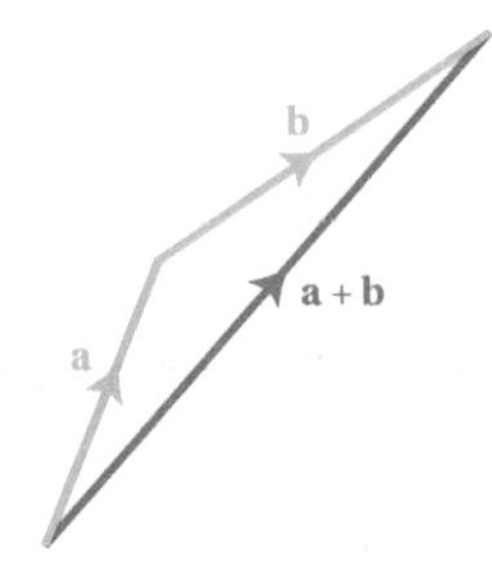

What if the two vectors are not in this position? Remember that vectors represent a length in a given direction, but don't tell you anything about **where** this length actually is. So vectors can be 'moved around', as long as their magnitude and direction remain unchanged.

Common error

Two vectors do not have to start and finish at the same points to be equal. They just need to have the same length and point in the same direction.

WORKED EXAMPLE 15.5

a Represent the vector $\mathbf{a}+\mathbf{b}$ on the grid.

b Write $\mathbf{a}$ and $\mathbf{b}$ as column vectors and hence find $\mathbf{a}+\mathbf{b}$ as a column vector.

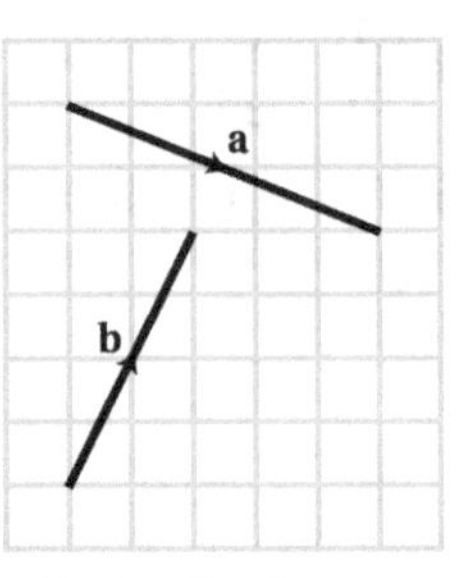

Continues on next page

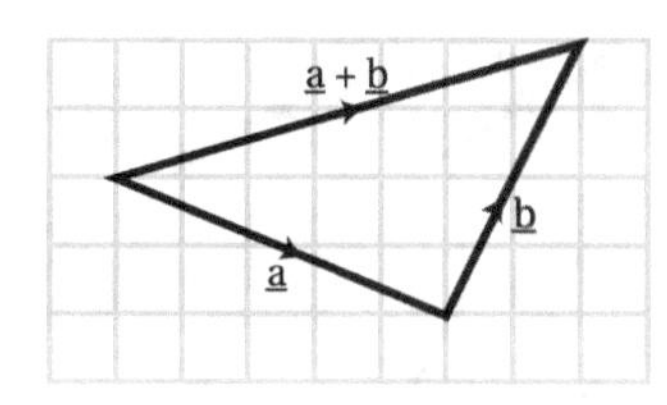

Move vector **b** so that its start point coincides with the end point of vector **a**. Then complete the triangle with **a** + **b**.

b $\underline{a} = \begin{pmatrix} 5 \\ -2 \end{pmatrix}, \underline{b} = \begin{pmatrix} 2 \\ 4 \end{pmatrix}$

Add the components.

$$\underline{a} + \underline{b} = \begin{pmatrix} 5+2 \\ -2+4 \end{pmatrix}$$

$$= \begin{pmatrix} 7 \\ 2 \end{pmatrix}$$

Another way of visualising vector addition is as a diagonal of the parallelogram formed by the two vectors. In this case the vectors are moved so that they have a common starting point.

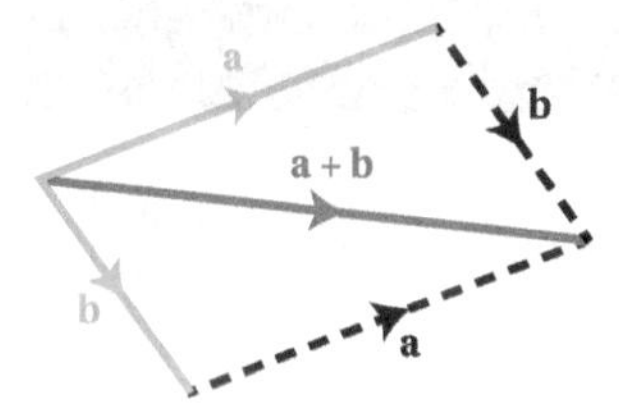

Subtracting vectors

Making a vector negative reverses the direction of the vector. Subtracting a vector is the same as adding its negative. So to subtract two vectors you need to reverse the direction of the second vector and then add it to the first. In component form you subtract corresponding components.

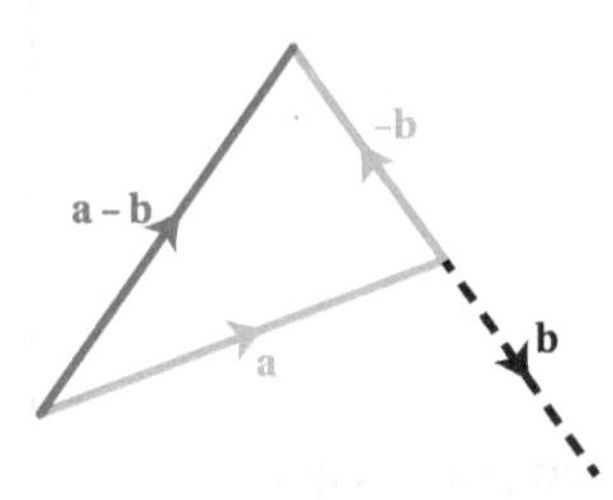

The difference of two vectors can also be represented by the diagonal of the parallelogram formed by the two vectors: the arrow points from the end of **b** towards the end of **a**.

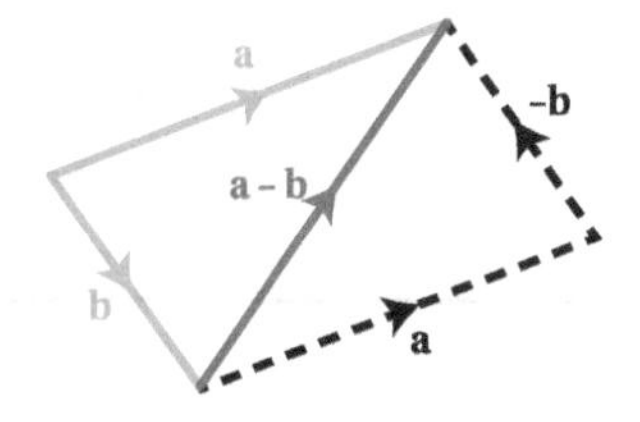

Scalar multiplication (multiplying a vector by a number) changes the magnitude (length) of the vector, leaving the direction the same. So the arrow points in the same direction but has a different length. If the scalar is negative the vector will point in opposite direction (but still along the same line). In component form, each component is multiplied by the scalar.

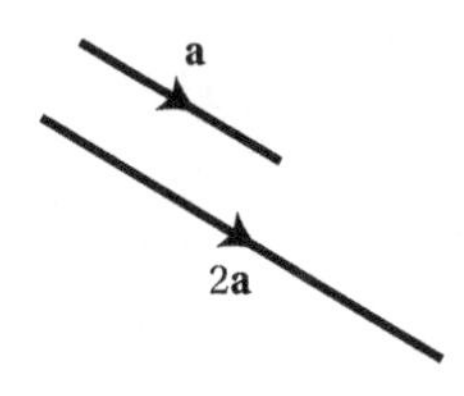

$$2\begin{pmatrix} 5 \\ -3 \end{pmatrix} = \begin{pmatrix} 10 \\ -6 \end{pmatrix}$$

Equal and parallel vectors

Two vectors are **equal** if they have the same magnitude and direction. All their components are equal, although they may have different start and end points.

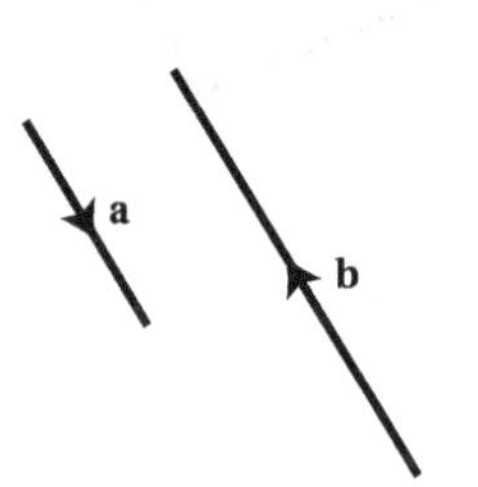

Parallel vectors are scalar multiples of each other. This is because multiplying a vector by a scalar does not change the angle of the vector (if the scalar is negative it will just reverse the direction of the vector).

$\begin{pmatrix} 3 \\ -2 \end{pmatrix}$ is parallel to $\begin{pmatrix} -9 \\ 6 \end{pmatrix}$ since $\begin{pmatrix} -9 \\ 6 \end{pmatrix} = -3\begin{pmatrix} 3 \\ -2 \end{pmatrix}$

Key point 15.3

If vectors **a** and **b** are parallel, you can write $\mathbf{b} = t\mathbf{a}$ for some scalar t.

WORKED EXAMPLE 15.6

Given the vectors $\mathbf{a} = \begin{pmatrix} 4 \\ -3 \end{pmatrix}$, $\mathbf{b} = \begin{pmatrix} -1 \\ 1 \end{pmatrix}$ and $\mathbf{c} = \begin{pmatrix} 11 \\ -9 \end{pmatrix}$, find scalars k and m such that $k\mathbf{a} - m\mathbf{b} = \mathbf{c}$.

$k\underline{a} - m\underline{b} = \begin{pmatrix} 4k \\ -3k \end{pmatrix} - \begin{pmatrix} -m \\ m \end{pmatrix}$

Write $k\mathbf{a} - m\mathbf{b}$ in component form ...

$= \begin{pmatrix} 4k + m \\ -3k - m \end{pmatrix}$

$k\underline{a} - m\underline{b} = \underline{c}$

... and compare it to **c**.

$\begin{pmatrix} 4k + m \\ -3k - m \end{pmatrix} = \begin{pmatrix} 11 \\ -9 \end{pmatrix}$

$\Rightarrow \begin{cases} 4k + m = 11 \\ -3k - m = -9 \end{cases}$

Both components have to be equal.

$\Rightarrow k = 2,\ m = 3$

Solve the simultaneous equations; add them to eliminate m.

WORKED EXAMPLE 15.7

$\mathbf{a} = \begin{pmatrix} 1 \\ 2 \end{pmatrix}$, $\mathbf{b} = \begin{pmatrix} -3 \\ p \end{pmatrix}$ and $\mathbf{c} = \begin{pmatrix} -2 \\ 5 \end{pmatrix}$

a Find the value of p such that **b** is parallel to **a**.

b Find the value of scalar k such that $\mathbf{a} + k\mathbf{c}$ is parallel to vector $\begin{pmatrix} 10 \\ 23 \end{pmatrix}$.

a $\underline{b} = t\underline{a}$ for some scalar t.

If vectors **a** and **b** are parallel you can write $\mathbf{b} = t\mathbf{a}$ for some scalar t.

$$\begin{pmatrix} -3 \\ p \end{pmatrix} = t\begin{pmatrix} 1 \\ 2 \end{pmatrix} = \begin{pmatrix} t \\ 2t \end{pmatrix}$$

$$\Rightarrow \begin{cases} -3 = t \\ p = 2t \end{cases}$$

If two vectors are equal then all their components are equal.

$$\Rightarrow p = -6$$

b $\underline{a} + k\underline{c} = \begin{pmatrix} 1 \\ 2 \end{pmatrix} + \begin{pmatrix} -2k \\ 5k \end{pmatrix}$

You can write the vector $\mathbf{a} + k\mathbf{c}$ in terms of k...

$$= \begin{pmatrix} 1-2k \\ 2+5k \end{pmatrix}$$

$\underline{a} + k\underline{c}$ is parallel to $\begin{pmatrix} 10 \\ 23 \end{pmatrix}$:

...and then use $\mathbf{a} + k\mathbf{c} = t\begin{pmatrix} 10 \\ 23 \end{pmatrix}$.

$$\begin{pmatrix} 1-2k \\ 2+5k \end{pmatrix} = t\begin{pmatrix} 10 \\ 23 \end{pmatrix}$$

$$\Rightarrow \begin{cases} 1-2k = 10t \\ 2+5k = 23t \end{cases}$$

If two vectors are equal then all their components are equal.

$$\Rightarrow \begin{cases} 10t + 2k = 1 \\ 23t - 5k = 2 \end{cases}$$

Solve the simultaneous equations: you only need to find the value of k, so eliminate t.

$$\Rightarrow \begin{cases} 230t + 46k = 23 \\ 230t - 50k = 20 \end{cases}$$

$$\Rightarrow 96k = 3$$

$$\Rightarrow k = \frac{1}{32}$$

You can use scalar multiplication to find a vector with a given magnitude acting in the same direction as a given vector. In particular, you can find a unit vector in a given direction.

WORKED EXAMPLE 15.8

Find two unit vectors parallel to $\begin{pmatrix} 3 \\ 4 \end{pmatrix}$.

Let $\underline{a} = t\begin{pmatrix} 3 \\ 4 \end{pmatrix} = \begin{pmatrix} 3t \\ 4t \end{pmatrix}$.

If a vector parallel to $\begin{pmatrix} 3 \\ 4 \end{pmatrix}$ you can write it as $t\begin{pmatrix} 3 \\ 4 \end{pmatrix}$.

Then $|\underline{a}| = 1$

$$\sqrt{(3t)^2 + (4t)^2} = 1$$

$$9t^2 + 16t^2 = 1^2$$

$$25t^2 = 1$$

$$t = \pm\sqrt{\frac{1}{25}}$$

$$t = \pm\frac{1}{5}$$

A unit vector has magnitude 1. Use Key point 15.1 (or Pythagoras' theorem) to find an expression for the magnitude in terms of t.

Don't forget ± when taking the square root.

$$\therefore \underline{a} = \pm\frac{1}{5}\begin{pmatrix} 3 \\ 4 \end{pmatrix}$$

$$= \begin{pmatrix} 0.6 \\ 0.8 \end{pmatrix} \text{ or } \begin{pmatrix} -0.6 \\ -0.8 \end{pmatrix}$$

EXERCISE 15B

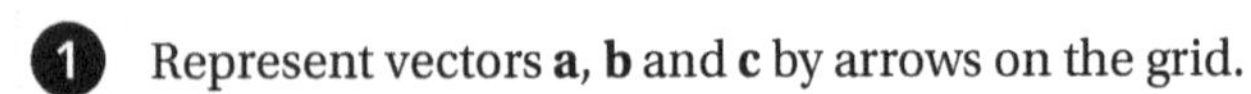

1 Represent vectors **a**, **b** and **c** by arrows on the grid.

a $\mathbf{a} = \mathbf{u} + \mathbf{v}$, $\mathbf{b} = -2\mathbf{u}$, $\mathbf{c} = \mathbf{v} - \mathbf{u}$

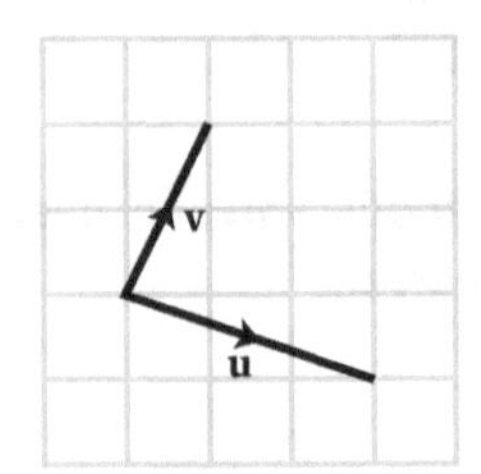

b $\mathbf{a} = 3\mathbf{v}$, $\mathbf{b} = \mathbf{u} - \mathbf{v}$, $\mathbf{c} = \mathbf{u} + \mathbf{v}$

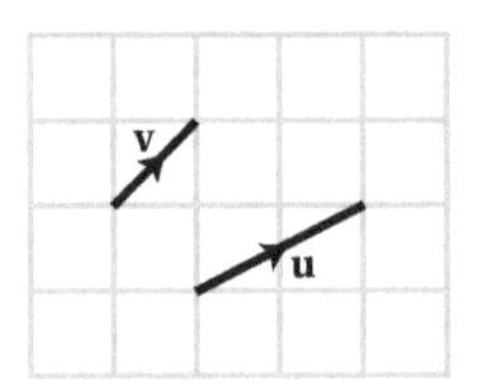

2 Let $\mathbf{a} = \begin{pmatrix} 7 \\ 1 \end{pmatrix}$, $\mathbf{b} = \begin{pmatrix} 5 \\ -2 \end{pmatrix}$ and $\mathbf{c} = \begin{pmatrix} 1 \\ 1 \end{pmatrix}$.

Write down these vectors.

a i $3\mathbf{a}$ ii $4\mathbf{b}$

b i $\mathbf{a} - \mathbf{b}$ ii $\mathbf{b} + \mathbf{c}$

c i $2\mathbf{b} + \mathbf{c}$ ii $\mathbf{a} - 2\mathbf{b}$

d i $\mathbf{a} + \mathbf{b} - 2\mathbf{c}$ ii $3\mathbf{a} - \mathbf{b} + \mathbf{c}$

3 Let $\mathbf{a} = \mathbf{i} + 2\mathbf{j}$, $\mathbf{b} = \mathbf{i}$ and $\mathbf{c} = 2\mathbf{i} - \mathbf{j}$. Write down these vectors.

a i $-5\mathbf{b}$ ii $4\mathbf{a}$ **b** i $\mathbf{c} - \mathbf{a}$ ii $\mathbf{a} - \mathbf{b}$

c i $\mathbf{a} - \mathbf{b} + 2\mathbf{c}$ ii $4\mathbf{c} - 3\mathbf{b}$

4 Given that $\mathbf{a} = 4\mathbf{i} - 2\mathbf{j}$, find the vector $\mathbf{b}$ such that:

a i $\mathbf{a} + \mathbf{b}$ is the zero vector ii $2\mathbf{a} + 3\mathbf{b}$ is the zero vector

b i $\mathbf{a} - \mathbf{b} = \mathbf{j}$ ii $\mathbf{a} + 2\mathbf{b} = 3\mathbf{j}$

5 Decide which pairs of vectors are parallel.

a i $\begin{pmatrix} -4 \\ 1 \end{pmatrix}$ and $\begin{pmatrix} -8 \\ 2 \end{pmatrix}$ ii $\begin{pmatrix} 3 \\ -1 \end{pmatrix}$ and $\begin{pmatrix} -6 \\ 2 \end{pmatrix}$

b i $\begin{pmatrix} 2 \\ 5 \end{pmatrix}$ and $\begin{pmatrix} 4 \\ 9 \end{pmatrix}$ ii $\begin{pmatrix} 3 \\ -1 \end{pmatrix}$ and $\begin{pmatrix} -6 \\ -2 \end{pmatrix}$

c i $3\mathbf{i} - 4\mathbf{j}$ and $1.5\mathbf{i} - 2\mathbf{j}$ ii $9\mathbf{i} - 2\mathbf{j}$ and $4.5\mathbf{i} + \mathbf{j}$

6 Given that $\mathbf{a} = \begin{pmatrix} -1 \\ 1 \end{pmatrix}$ and $\mathbf{b} = \begin{pmatrix} 5 \\ 3 \end{pmatrix}$ find vector $\mathbf{x}$ such that $3\mathbf{a} + 4\mathbf{x} = \mathbf{b}$.

7 Given that $\mathbf{a} = 3\mathbf{i} - 2\mathbf{j}$, $\mathbf{b} = \mathbf{i} - \mathbf{j}$ and $\mathbf{c} = \mathbf{i}$, find the value of the scalar t such that $\mathbf{a} + t\mathbf{b} = \mathbf{c}$.

8 Given that $\mathbf{a} = \begin{pmatrix} 2 \\ 0 \end{pmatrix}$ and $\mathbf{b} = \begin{pmatrix} 3 \\ 1 \end{pmatrix}$, find the value of the scalar p such that $\mathbf{a} + p\mathbf{b}$ is parallel to the vector $\begin{pmatrix} 3 \\ 2 \end{pmatrix}$.

9 Given that $\mathbf{x} = 2\mathbf{i} + 3\mathbf{j}$ and $\mathbf{y} = 4\mathbf{i} + \mathbf{j}$, find the value of the scalar λ such that $\lambda\mathbf{x} + \mathbf{y}$ is parallel to vector $\mathbf{j}$.

10 Given that $\mathbf{a} = \mathbf{i} - \mathbf{j}$, $\mathbf{b} = 2q\mathbf{i} + \mathbf{j}$ and that $p\mathbf{a} + \mathbf{b}$ is parallel to vector $\mathbf{i} + \mathbf{j}$, express q in terms of p.

11 Find the value of k such that the vector $\begin{pmatrix} 3k+1 \\ k-3 \end{pmatrix}$ is parallel to $\begin{pmatrix} -4 \\ 1 \end{pmatrix}$.

12 Find two vectors of magnitude 20 parallel to $\begin{pmatrix} 6 \\ -8 \end{pmatrix}$.

13 Find two vectors of magnitude 3 parallel to $\begin{pmatrix} 4 \\ 2 \end{pmatrix}$.

14 Find two unit vectors parallel to $\begin{pmatrix} -2 \\ 1 \end{pmatrix}$.

Elevate

See Support Sheet 15 for another example of finding parallel vectors of a given magnitude and for further practice questions.

15 **a** Show that vectors $\mathbf{v} = \begin{pmatrix} t-1 \\ 2t+1 \end{pmatrix}$ and $\mathbf{u} = \begin{pmatrix} t^2-1 \\ 2t^2+3t+1 \end{pmatrix}$ are parallel for all values of t.

b Find the value of t for which the two vectors are equal.

Section 3: Position and displacement vectors

Vectors are used to represent **displacements** between points. You can think of a vector as describing how to get from one point to another.

For example, to get from A to B in this diagram, you need to move 5 units to the right and 3 units up. This is represented by the **displacement vector** $\overrightarrow{AB}=\begin{pmatrix}5\\3\end{pmatrix}$.

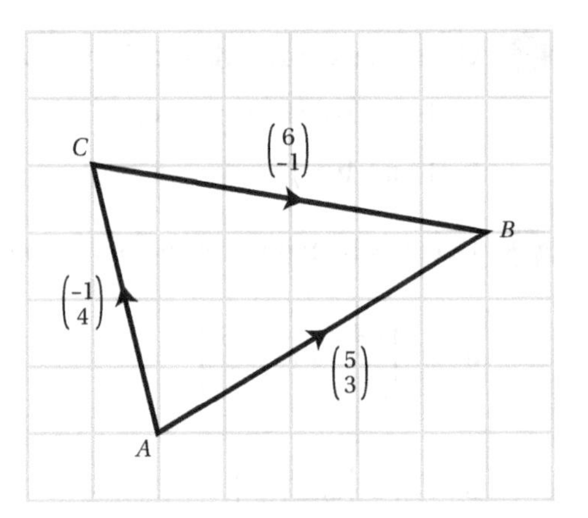

If there is a third point, C, there are two different ways of getting from A to B: either directly or via C. Both of those journeys represent the same total displacement, so you can write $\overrightarrow{AB}=\overrightarrow{AC}+\overrightarrow{CB}$ or, using

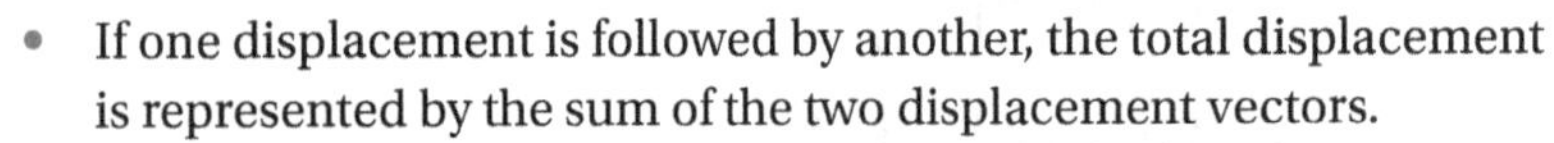

components, $\begin{pmatrix}5\\3\end{pmatrix}=\begin{pmatrix}-1\\4\end{pmatrix}+\begin{pmatrix}6\\-1\end{pmatrix}$.

- If one displacement is followed by another, the total displacement is represented by the sum of the two displacement vectors.
- Multiplying a displacement vector by a scalar represents a displacement in the same direction but with a different magnitude - acting for a different distance.
- Making a displacement negative represents travelling the same distance in the opposite direction.

WORKED EXAMPLE 15.9

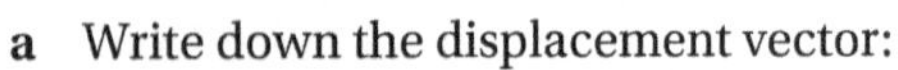

a Write down the displacement vector:

i from A to B **ii** from C to A.

b Plot on the grid points D and E such that:

i $\overrightarrow{AD}=2\overrightarrow{AB}$ **ii** $\overrightarrow{CE}=-\overrightarrow{CA}$.

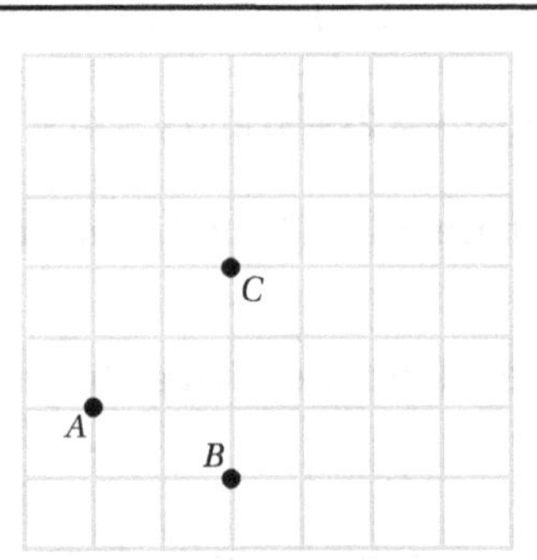

a **i** $\overrightarrow{AB}=\begin{pmatrix}2\\-1\end{pmatrix}$ **ii** $\overrightarrow{CA}=\begin{pmatrix}-2\\-2\end{pmatrix}$

b

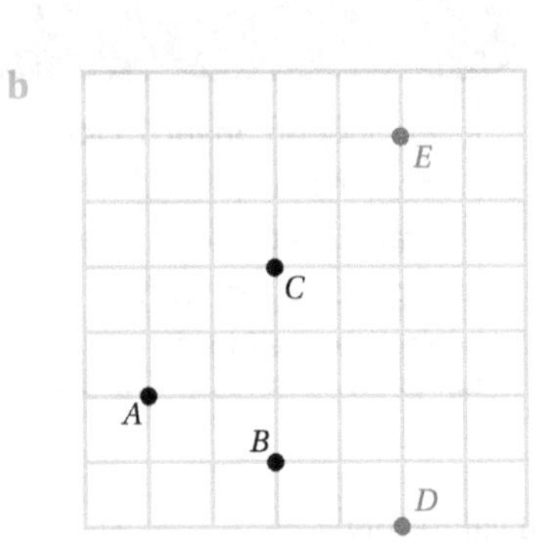

$\overrightarrow{AD}$ is twice as long as $\overrightarrow{AB}$ (in the same direction).

$\overrightarrow{CE}$ is the same length as $\overrightarrow{CA}$ but in the opposite direction.

WORKED EXAMPLE 15.10

The diagram shows points M, N, P, Q such that $\overrightarrow{MN} = 3\mathbf{i} - 2\mathbf{j}$, $\overrightarrow{NP} = \mathbf{i} + \mathbf{j}$ and $\overrightarrow{MQ} = -2\mathbf{j}$.

Write each of these vectors in component form.

a $\overrightarrow{MP}$ **b** $\overrightarrow{PM}$ **c** $\overrightarrow{PQ}$

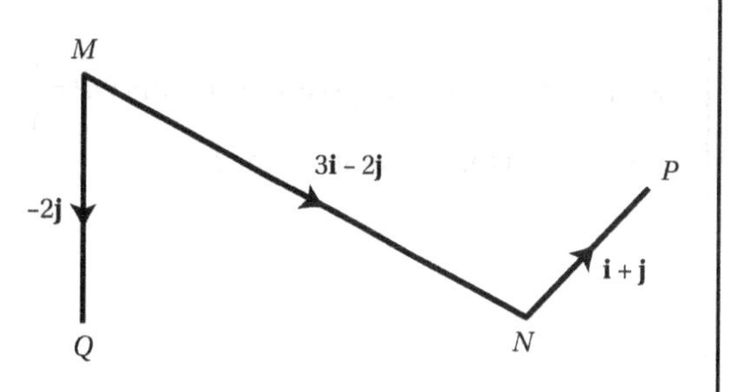

a $\overrightarrow{MP} = \overrightarrow{MN} + \overrightarrow{NP}$

$= (3\mathbf{i} - 2\mathbf{j}) + (\mathbf{i} + \mathbf{j})$

$= 4\mathbf{i} - \mathbf{j}$

You can get from M to P via N.

b $\overrightarrow{PM} = -\overrightarrow{MP}$

$= -4\mathbf{i} + \mathbf{j}$

You have already found MP.

c $\overrightarrow{PQ} = \overrightarrow{PM} + \overrightarrow{MQ}$

$= (-4\mathbf{i} + \mathbf{j}) + (-2\mathbf{j})$

$= -4\mathbf{i} - \mathbf{j}$

You can get from P to Q via M, using the answers from previous parts.

Displacement vectors tell you how to get from one point to another, but say nothing about where the points actually are. To describe the position of a point you need a fixed **origin**, O. Then the **position vector** of a point P is its displacement from the origin, $\overrightarrow{OP}$.

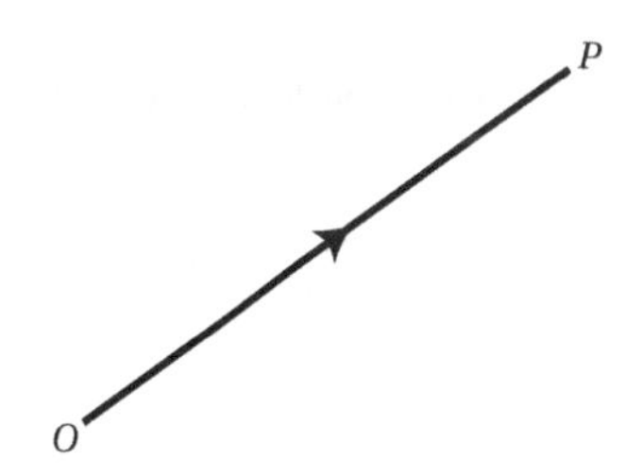

It is common to denote the position vector by the same letter as the point, for example $\overrightarrow{OA} = \mathbf{a}$.

If you know the position vectors of two points A and B you can find the displacement $\overrightarrow{AB}$, as shown in the diagram.

$$\begin{aligned}\overrightarrow{AB} &= \overrightarrow{AO} + \overrightarrow{OB} \\ &= -\overrightarrow{OA} + \overrightarrow{OB} \\ &= \overrightarrow{OB} - \overrightarrow{OA}\end{aligned}$$

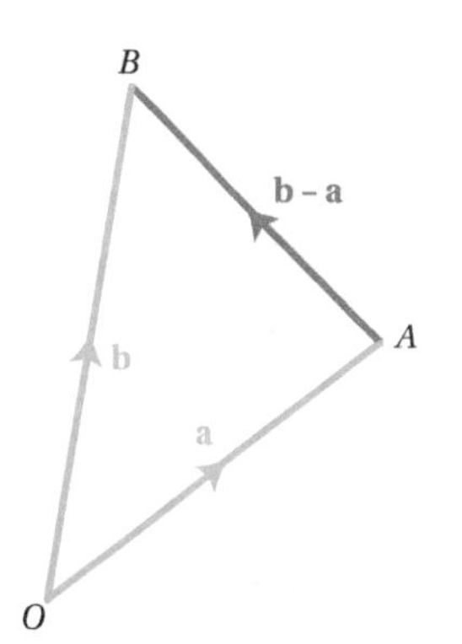

Key point 15.4

If points A and B have position vectors $\mathbf{a}$ and $\mathbf{b}$ then $\overrightarrow{AB} = \mathbf{b} - \mathbf{a}$.

Tip

You can think of this as being the 'journey' $-\mathbf{a}$ followed by the 'journey' $\mathbf{b}$.

WORKED EXAMPLE 15.11

Points P and Q have position vectors $\overrightarrow{OP}=\begin{pmatrix}3\\-5\end{pmatrix}$ and $\overrightarrow{OQ}=\begin{pmatrix}-2\\1\end{pmatrix}$.

Find the displacement vector $\overrightarrow{PQ}$.

$\overrightarrow{PQ}=\overrightarrow{OQ}-\overrightarrow{OP}$ — Use the result from Key point 15.4.

$=\begin{pmatrix}-2\\1\end{pmatrix}-\begin{pmatrix}3\\-5\end{pmatrix}$

$=\begin{pmatrix}-5\\6\end{pmatrix}$

Position vectors are closely related to coordinates. If the base vectors **i** and **j** have directions along the coordinate axes, then the components of the position vector are simply the coordinates of the point.

WORKED EXAMPLE 15.12

Points A and B have coordinates $(3, -1)$ and $(5, 0)$, respectively.

a Write as column vectors:

i the position vectors of A and B

ii the displacement vector $\overrightarrow{AB}$.

b Find the position vector of the point C such that $\overrightarrow{CA}=\begin{pmatrix}-4\\2\end{pmatrix}$

a i $\underline{a}=\begin{pmatrix}3\\-1\end{pmatrix}$ $\underline{b}=\begin{pmatrix}5\\0\end{pmatrix}$ — The components of the position vectors are the coordinates of the points.

ii $\overrightarrow{AB}=\underline{b}-\underline{a}$ — Use $\overrightarrow{AB}=\mathbf{b}-\mathbf{a}$.

$=\begin{pmatrix}5\\0\end{pmatrix}-\begin{pmatrix}3\\-1\end{pmatrix}$

$=\begin{pmatrix}2\\1\end{pmatrix}$

b $\overrightarrow{CA}=\underline{a}-\underline{c}$ — Use $\overrightarrow{CA}=\mathbf{a}-\mathbf{c}$.

$\begin{pmatrix}-4\\2\end{pmatrix}=\begin{pmatrix}3\\-1\end{pmatrix}-\underline{c}$

$\underline{c}=\begin{pmatrix}3\\-1\end{pmatrix}-\begin{pmatrix}-4\\2\end{pmatrix}$ — Rearrange to find the position vector of C.

$=\begin{pmatrix}7\\-3\end{pmatrix}$

The coordinates of C are $(7, -3)$.

Distance between two points

Key point 15.5

If points A and B have position vectors **a** and **b**, then the distance between them, AB, is equal to the magnitude of the vector $\overrightarrow{AB}$.

$$AB = |\overrightarrow{AB}| = |\mathbf{b} - \mathbf{a}|$$

WORKED EXAMPLE 15.13

a Find the distance between points A and B with position vectors $\mathbf{a} = \begin{pmatrix} -3 \\ 2 \end{pmatrix}$ and $\mathbf{b} = \begin{pmatrix} -4 \\ 0 \end{pmatrix}$.

b Point C has position vector $\mathbf{c} = \begin{pmatrix} 2 \\ p \end{pmatrix}$. Find the exact values of p such that $AC = 3AB$.

a $\mathbf{b} - \mathbf{a} = \begin{pmatrix} -4 \\ 0 \end{pmatrix} - \begin{pmatrix} -3 \\ 2 \end{pmatrix}$

First find the vector $\overrightarrow{AB} = \mathbf{b} - \mathbf{a}$.

$= \begin{pmatrix} -1 \\ -2 \end{pmatrix}$

$\therefore AB = \sqrt{1^2 + 2^2}$

$= \sqrt{5}$

Then use Key point 15.1 to find the magnitude.

b $\mathbf{c} - \mathbf{a} = \begin{pmatrix} 2 \\ p \end{pmatrix} - \begin{pmatrix} -3 \\ 2 \end{pmatrix}$

First find an expression for the vector $\overrightarrow{AC} = \mathbf{c} - \mathbf{a}$ in terms of p.

$= \begin{pmatrix} 5 \\ p-2 \end{pmatrix}$

$AC = \sqrt{5^2 + (p-2)^2}$

Then use Key point 15.1 to find the magnitude.

$AC = 3AB$

$\sqrt{25 + (p-2)^2} = 3\sqrt{5}$

Form an equation.

$25 + (p-2)^2 = 45$

Square both sides to remove the root.

$(p-2)^2 = 20$

$p - 2 = \pm\sqrt{20}$

$p = 2 \pm 2\sqrt{5}$

You don't need to expand the brackets. Take the square root, remembering $\pm$.

WORK IT OUT 15.2

Points P and Q have coordinates $P(-1, 3)$ and $Q(7, -2)$. Find the vector $\overrightarrow{QP}$.

Which is the correct solution? Identify the errors made in the incorrect solutions.

Solution 1	Solution 2	Solution 3
$\overrightarrow{QP} = \mathbf{p} - \mathbf{q}$ $= \begin{pmatrix} -1 \\ 3 \end{pmatrix} - \begin{pmatrix} 7 \\ -2 \end{pmatrix}$ $= \begin{pmatrix} -8 \\ 5 \end{pmatrix}$	$\overrightarrow{QP} = \mathbf{p} + \mathbf{q}$ $= \begin{pmatrix} -1 \\ 3 \end{pmatrix} + \begin{pmatrix} 7 \\ -2 \end{pmatrix}$ $= \begin{pmatrix} 6 \\ 1 \end{pmatrix}$	$\overrightarrow{QP} = \mathbf{q} - \mathbf{p}$ $= \begin{pmatrix} 7 \\ -2 \end{pmatrix} - \begin{pmatrix} -1 \\ 3 \end{pmatrix}$ $= \begin{pmatrix} 8 \\ -5 \end{pmatrix}$

EXERCISE 15C

1 a Write down the displacement vectors, in component form:

i from A to C ii from C to B

iii from B to C iv from D to A.

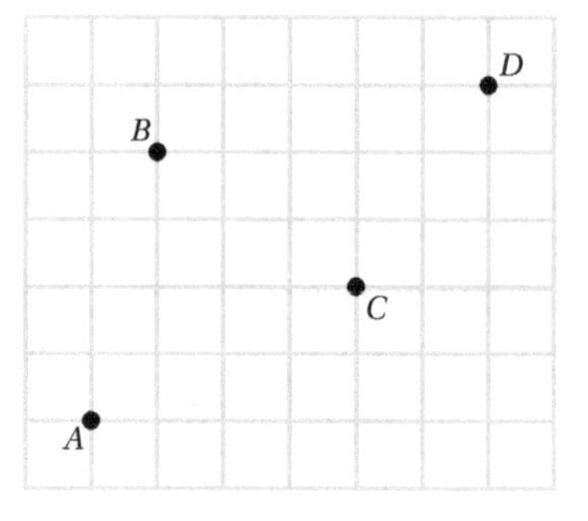

b Mark points P, Q and R on the grid such that:

i $\overrightarrow{AP} = \begin{pmatrix} 3 \\ -2 \end{pmatrix}$ ii $\overrightarrow{QC} = \begin{pmatrix} -1 \\ 3 \end{pmatrix}$

iii $\overrightarrow{RD} = \begin{pmatrix} 0 \\ 4 \end{pmatrix}$.

2 Four points have coordinates $A(-1, 3)$, $B(3, -3)$, $C(0, -5)$ and $D(-3, -2)$. Write, in component form:

a i the position vector of A ii the position vector of B

b i the displacement from B to D ii the displacement from C to B.

c i $\overrightarrow{AB}$ ii $\overrightarrow{DA}$.

3 Points A, B, C and D are as in question 2. Find the coordinates of the points satisfying the condition:

a i point P is such that $\overrightarrow{AP} = \begin{pmatrix} 3 \\ -2 \end{pmatrix}$ ii point Q is such that $\overrightarrow{CQ} = \begin{pmatrix} -4 \\ 1 \end{pmatrix}$

b i point R is such that $\overrightarrow{RB} = \begin{pmatrix} 2 \\ -1 \end{pmatrix}$ ii point S is such that $\overrightarrow{SD} = \begin{pmatrix} -5 \\ -2 \end{pmatrix}$

c i point X is such that $\overrightarrow{AX} = 3\overrightarrow{AB}$ ii point Y is such that $\overrightarrow{BY} = 2\overrightarrow{BD}$

d i point Z is such that $\overrightarrow{CZ} = -2\overrightarrow{AB}$ ii point W is such that $\overrightarrow{DW} = -4\overrightarrow{CA}$.

4 Points A, B, C and D have position vectors $\mathbf{a} = 3\mathbf{i} - 7\mathbf{j}$, $\mathbf{b} = -3\mathbf{i} - 7\mathbf{j}$, $\mathbf{c} = \mathbf{i} - 5\mathbf{j}$, $\mathbf{d} = -4\mathbf{j}$. Work out the distances:

a i AB ii CD b i DA ii BC.

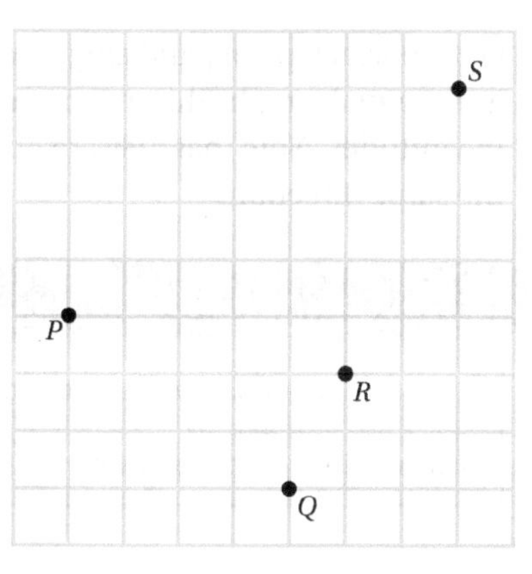

5 The diagram shows four points, P, Q, R and S, such that $\overrightarrow{PQ} = 4\mathbf{i} - 3\mathbf{j}$, $\overrightarrow{QR} = \mathbf{i} + 2\mathbf{j}$ and $\overrightarrow{RS} = 2\mathbf{i} + 5\mathbf{j}$.

Express each of these vectors in component form.

a $\overrightarrow{PR}$ **b** $\overrightarrow{SP}$

6 Points A, B, C and D have coordinates $A(3, -2)$, $B(-4, 1)$, $C(1, -5)$ and $D(0, 7)$.

a Write $\overrightarrow{AC}$ and $\overrightarrow{DB}$ as column vectors.

b Find the coordinates of the point E such that $\overrightarrow{AE} = \begin{pmatrix} -3 \\ 5 \end{pmatrix}$.

7 Points A, B and C have coordinates $A(1, 3)$, $B(5, -1)$ and $C(2, -8)$. Point D is such that:
$\overrightarrow{AD} = \overrightarrow{BC} + 2p\overrightarrow{AB} + 3q\overrightarrow{AC} = \overrightarrow{AB} + 2p\overrightarrow{AC} + 3q\overrightarrow{BC}$

Find the coordinates of D.

8 $\overrightarrow{AB} = \begin{pmatrix} 6 \\ -3 \end{pmatrix}$ and $\overrightarrow{AC} = \frac{4}{3}\overrightarrow{AB}$.

a Write down $\overrightarrow{AC}$.

Point P is such that $AB = 2BP$.

b If $\overrightarrow{AP} = \begin{pmatrix} p \\ q \end{pmatrix}$, write down vectors $\overrightarrow{BP}$ and $\overrightarrow{CP}$ in terms of p and q.

c Show that CP is independent of the values of p and q and evaluate it.

Section 4: Using vectors to solve geometrical problems

In this section you will learn how to describe and prove various properties of geometrical figures.

Parallelograms and rhombuses

Consider four points A, B, C and D such that $\overrightarrow{AB} = \overrightarrow{DC}$, as shown in the diagram.

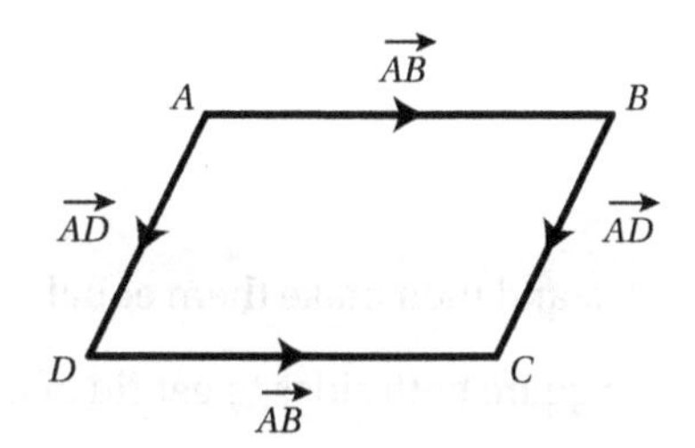

As these vectors are equal, the opposite sides AB and DC are parallel and have equal length, so the shape is a **parallelogram**. It follows that the other two sides are also equal and parallel, so $\overrightarrow{BC} = \overrightarrow{AD}$.

One special type of parallelogram is a **rhombus**, in which all four sides are of equal length. This means that vectors $\overrightarrow{AB}$ and $\overrightarrow{BC}$ have equal magnitudes. Note that they are not equal vectors because they don't have the same direction.

Key point 15.6

If $\overrightarrow{AB} = \overrightarrow{DC}$ then $ABCD$ is a parallelogram.

If in addition $|\overrightarrow{AB}| = |\overrightarrow{BC}|$ then $ABCD$ is a rhombus.

Fast forward

If you study Further Mathematics you will find out how to find an angle between two vectors.

WORKED EXAMPLE 15.14

Four points have coordinates $A(2, -1)$, $B(k, k+1)$, $C(2k-3, 2k+2)$ and $D(k-1, k)$.

a Show that $ABCD$ is a parallelogram for all values of k.

b Find the value of k for which $ABCD$ is a rhombus.

a

$$\overrightarrow{AB} = \begin{pmatrix} k \\ k+1 \end{pmatrix} - \begin{pmatrix} 2 \\ -1 \end{pmatrix}$$

For $ABCD$ to be a parallelogram, you need to show that $\overrightarrow{AB} = \overrightarrow{DC}$.

$$= \begin{pmatrix} k-2 \\ k+2 \end{pmatrix}$$

$$\overrightarrow{DC} = \begin{pmatrix} 2k-3 \\ 2k+2 \end{pmatrix} - \begin{pmatrix} k-1 \\ k \end{pmatrix}$$

$$= \begin{pmatrix} k-2 \\ k+2 \end{pmatrix}$$

$\overrightarrow{AB} = \overrightarrow{DC}$ so $ABCD$ is a parallelogram.

b

$$|\overrightarrow{AB}| = \sqrt{(k-2)^2 + (k+2)^2}$$

For $ABCD$ to be a rhombus, $|\overrightarrow{AB}| = |\overrightarrow{BC}|$ so find expressions for these two distances...

$$\overrightarrow{BC} = \begin{pmatrix} 2k-3 \\ 2k+2 \end{pmatrix} - \begin{pmatrix} k \\ k+1 \end{pmatrix}$$

$$= \begin{pmatrix} k-3 \\ k+1 \end{pmatrix}$$

$$|\overrightarrow{BC}| = \sqrt{(k-3)^2 + (k+1)^2}$$

$$|\overrightarrow{AB}| = |\overrightarrow{BC}|$$

... and then make them equal.

$$\therefore (k-2)^2 + (k+2)^2 = (k-3)^2 + (k+1)^2$$

Square both sides to get rid of the roots.

$$2k^2 + 8 = 2k^2 - 4k + 10$$

$$4k = 2$$

$$k = \frac{1}{2}$$

Straight lines

If points A, B and C lie in a straight line then vectors $\overrightarrow{AB}$ and $\overrightarrow{BC}$ are parallel, so $\overrightarrow{BC} = k\overrightarrow{AB}$ for some scalar k. This scalar also gives the ratio of the lengths BC and AB.

Tip

If points lie on a straight line, they are said to be **collinear**.

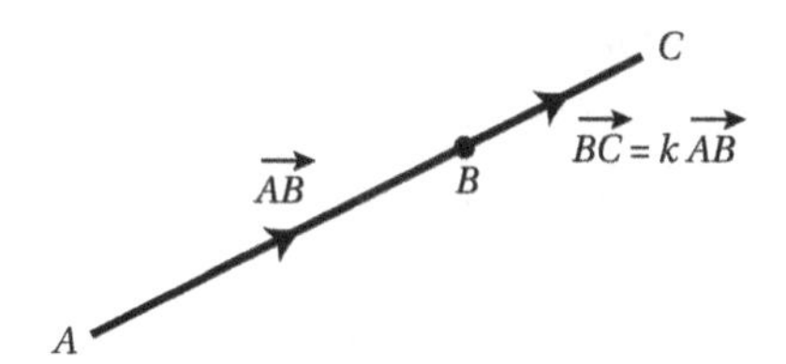

WORKED EXAMPLE 15.15

Three points have the coordinates $A(-3, 2)$, $B(6, 5)$ and $C(12, 7)$.

a Show that A, B and C lie on a straight line.
b Find the ratio $AB : BC$.

a $\overrightarrow{AB} = \begin{pmatrix} 6 \\ 5 \end{pmatrix} - \begin{pmatrix} -3 \\ 2 \end{pmatrix}$

You need to show that $\overrightarrow{AB}$ and $\overrightarrow{BC}$ are parallel. First find these two vectors.

$= \begin{pmatrix} 9 \\ 3 \end{pmatrix}$

$\overrightarrow{BC} = \begin{pmatrix} 12 \\ 7 \end{pmatrix} - \begin{pmatrix} 6 \\ 5 \end{pmatrix}$

$= \begin{pmatrix} 6 \\ 2 \end{pmatrix}$

If $\overrightarrow{BC} = k\overrightarrow{AB}$:

Then try to find a scalar k so that $\overrightarrow{BC} = k\overrightarrow{AB}$. You need to show that k has the same value for both components.

$$\begin{cases} 6 = 9k \Rightarrow k = \frac{2}{3} \\ 2 = 3k \Rightarrow k = \frac{2}{3} \end{cases}$$

$\therefore \overrightarrow{BC} = \frac{2}{3}\overrightarrow{AB}$

Since AB and BC are parallel and contain a common point B, then A, B and C lie on a straight line.

Conclude that A, B and C lie on a straight line.

b $BC = \frac{2}{3} AB$

Since $\overrightarrow{BC} = \frac{2}{3}\overrightarrow{AB}$, the length $BC = \frac{2}{3} AB$.

$\therefore AB : BC = 3 : 2$

Midpoint

Consider points A and B with position vectors **a** and **b** and let M be the midpoint of AB. The point M can be expressed in terms of **a** and **b**.

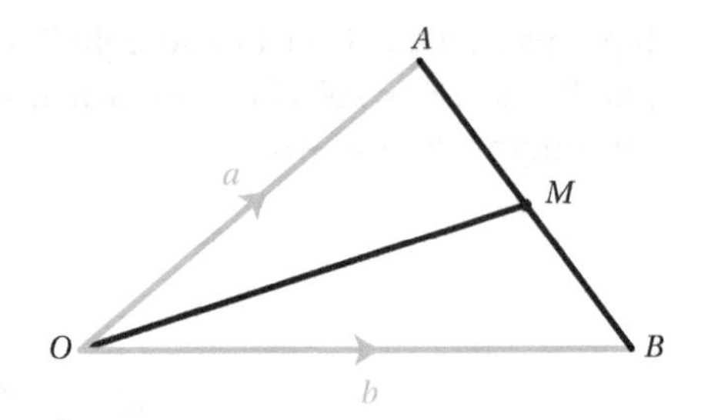

Key point 15.7

The midpoint of the line segment connecting points with position vectors **a** and **b** has position vector $\frac{1}{2}(\mathbf{a}+\mathbf{b})$.

PROOF 7

$\overrightarrow{OM} = \overrightarrow{OA} + \frac{1}{2}\overrightarrow{AB}$ — You can get from O to M by going from O to A and then halfway from A to B.

$= \mathbf{a} + \frac{1}{2}(\mathbf{b}-\mathbf{a})$ — You know that $\overrightarrow{OA} = \mathbf{a}$ and $\overrightarrow{AB} = \mathbf{b}-\mathbf{a}$.

$= \frac{1}{2}\mathbf{a} + \frac{1}{2}\mathbf{b}$

$= \frac{1}{2}(\mathbf{a}+\mathbf{b})$

Worked example 15.16 shows that the diagonals of a particular parallelogram bisect each other. See question 16 in Mixed practice 15 for a general proof of this fact.

Elevate

See Extension Sheet 15 for questions leading to proofs of some geometrical properties of triangles that involve midpoints.

WORKED EXAMPLE 15.16

Points A, B and C have position vectors $\mathbf{a} = \begin{pmatrix} 3 \\ -5 \end{pmatrix}$, $\mathbf{b} = \begin{pmatrix} 4 \\ 2 \end{pmatrix}$ and $\mathbf{c} = \begin{pmatrix} -3 \\ -1 \end{pmatrix}$.

a Find the position vector of the point D such that $ABCD$ is a parallelogram.

M is the midpoint of the diagonal AC.

b Find the position vector of M.

c Show that M is also the midpoint of BD.

a $\overrightarrow{AB} = \overrightarrow{DC}$ — For $ABCD$ to be a parallelogram, $\overrightarrow{AB} = \overrightarrow{DC}$.

$\begin{pmatrix} 4 \\ 2 \end{pmatrix} - \begin{pmatrix} 3 \\ -5 \end{pmatrix} = \begin{pmatrix} -3 \\ -1 \end{pmatrix} - \mathbf{d}$ — Use $AB = \mathbf{b} - \mathbf{a}$ and $DC = \mathbf{c} - \mathbf{d}$.

$\begin{pmatrix} 1 \\ 7 \end{pmatrix} = \begin{pmatrix} -3 \\ -1 \end{pmatrix} - \mathbf{d}$ — Rearrange to find **d**.

$\mathbf{d} = \begin{pmatrix} -3 \\ -1 \end{pmatrix} - \begin{pmatrix} 1 \\ 7 \end{pmatrix}$

$= \begin{pmatrix} -4 \\ -8 \end{pmatrix}$

Continues on next page

b $\mathbf{m} = \frac{1}{2}(\mathbf{a}+\mathbf{c})$

$= \frac{1}{2}\left(\begin{pmatrix} 3 \\ -5 \end{pmatrix} + \begin{pmatrix} -3 \\ -1 \end{pmatrix}\right)$

$= \frac{1}{2}\begin{pmatrix} 0 \\ -6 \end{pmatrix}$

$= \begin{pmatrix} 0 \\ -3 \end{pmatrix}$

Use the result for the midpoint from Key point 15.8.

c $\overrightarrow{BM} = \begin{pmatrix} 0 \\ -3 \end{pmatrix} - \begin{pmatrix} 4 \\ 2 \end{pmatrix}$

$= \begin{pmatrix} -4 \\ -5 \end{pmatrix}$

$\overrightarrow{MD} = \begin{pmatrix} -4 \\ -8 \end{pmatrix} - \begin{pmatrix} 0 \\ -3 \end{pmatrix}$

$= \begin{pmatrix} -4 \\ -5 \end{pmatrix}$

$\overrightarrow{BM} = \overrightarrow{MD}$ so M is the midpoint of BD.

If M is halfway between B and D then $\overrightarrow{BM} = \overrightarrow{MD}$. You could have also used the condition that if M is the midpoint of BD then $\mathbf{m} = \frac{1}{2}(\mathbf{b}+\mathbf{d})$.

Vectors are very useful for proving that two lines are parallel, even if they don't have the same length. The final example combines midpoints with parallel lines.

Tip

Remember that two vectors are parallel if one is a scalar multiple of the other.

WORKED EXAMPLE 15.17

The vertices of triangle ABC have position vectors **a**, **b** and **c**. M and N are the midpoints of sides AB and AC.

a Express the position vectors of M and N in terms of **a**, **b** and **c**.
b Prove that MN is parallel to BC and half its length.

a M is the midpoint of AB so $\mathbf{m} = \frac{1}{2}(\mathbf{a}+\mathbf{b})$.

N is the midpoint of AC so $\mathbf{n} = \frac{1}{2}(\mathbf{a}+\mathbf{c})$.

Find the position vectors of midpoints using the result from Key point 15.8.

b $\overrightarrow{BC} = \mathbf{c} - \mathbf{b}$

You want to compare vectors $\overrightarrow{MN}$ and $\overrightarrow{BC}$.

$\overrightarrow{MN} = \mathbf{n} - \mathbf{m}$

$= \frac{1}{2}(\mathbf{a}+\mathbf{c}) - \frac{1}{2}(\mathbf{a}+\mathbf{b})$

Use the result from part **a** on $\overrightarrow{MN}$.

$= \frac{1}{2}\mathbf{c} - \frac{1}{2}\mathbf{b}$

$\overrightarrow{MN} = \frac{1}{2}\overrightarrow{BC}$

So MN is parallel to BC and half its length.

Since $\overrightarrow{MN} = \frac{1}{2}\overrightarrow{BC}$ the two vectors are parallel.

EXERCISE 15D

1 The diagram shows a parallelogram $ABCD$ with $\overrightarrow{AB} = \mathbf{a}$ and $\overrightarrow{AD} = \mathbf{b}$. M is the midpoint of BC and N is the midpoint of CD. Express each vector in terms of $\mathbf{a}$ and $\mathbf{b}$.

a i $\overrightarrow{BC}$ ii $\overrightarrow{AC}$

b i $\overrightarrow{CD}$ ii $\overrightarrow{ND}$

c i $\overrightarrow{AM}$ ii $\overrightarrow{MN}$

2 In the parallelogram $ABCD$, $\overrightarrow{AB} = \mathbf{a}$ and $\overrightarrow{AD} = \mathbf{b}$. M is the midpoint of BC, Q is the point on the extended line AB such that $BQ = \frac{1}{2}AB$ and P is the point on the extended line BC such that $BC : CP = 3 : 1$, as shown on the diagram.

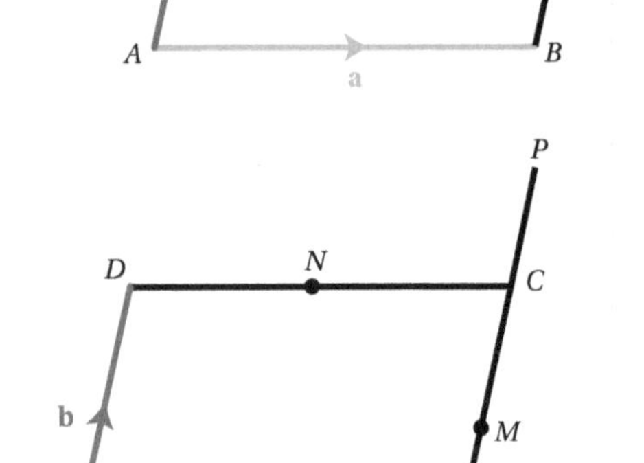

Express each vector in terms of $\mathbf{a}$ and $\mathbf{b}$.

a i $\overrightarrow{AP}$ ii $\overrightarrow{AM}$

b i $\overrightarrow{QD}$ ii $\overrightarrow{MQ}$

c i $\overrightarrow{DQ}$ ii $\overrightarrow{PQ}$

3 For each coordinate set, determine whether the three points A, B and C are collinear. For those that are, find the ratio $AB : BC$.

a i $A(3,8)$, $B(-1,2)$, $C(51,80)$ ii $A(4,-1)$, $B(6,3)$, $C(11,13)$

b i $A(2,1)$, $B(-11,14)$, $C(4,3)$ ii $A(3,5)$, $B(1,7)$, $C(-6,3)$

c i $A(4,2a)$, $B(1,3a+2)$, $C(10,4)$ ii $A(4,2a)$, $B(1,2a+1)$, $C(7,2a-1)$

d i $A(a^2+2a, 3)$, $B(a^2+a-1, 1)$, $C(3a^2+2a-2, 4a-1)$

ii $A(a^2-2a, 5)$, $B(a^2+a+2, 5)$, $C(a^2+2a, -4)$

4 The points A and B have position vectors $\mathbf{a} = 7\mathbf{i}+18\mathbf{j}$ and $\mathbf{b} = -3\mathbf{i}+6\mathbf{j}$. Find the position vector of the midpoint of AB.

5 $\overrightarrow{AB} = \begin{pmatrix} 3 \\ -7 \end{pmatrix}$ and $\overrightarrow{CA} = \begin{pmatrix} -9 \\ 21 \end{pmatrix}$

Show that A, B and C are collinear and find the ratio $AB : BC$.

6 Points A, B and C have position vector $\mathbf{a} = \begin{pmatrix} 3 \\ -5 \end{pmatrix}$, $\mathbf{b} = \begin{pmatrix} -1 \\ 4 \end{pmatrix}$ and $\mathbf{c} = \begin{pmatrix} 4 \\ 5 \end{pmatrix}$.

a Find the position vector of the point D such that $ABCD$ is a parallelogram.

b Determine whether $ABCD$ is a rhombus.

7 Points A, B, C and D have position vectors $\mathbf{a} = \begin{pmatrix} 3 \\ -1 \end{pmatrix}$, $\mathbf{b} = \begin{pmatrix} 5 \\ 0 \end{pmatrix}$, $\mathbf{c} = \begin{pmatrix} 7 \\ 8 \end{pmatrix}$ and $\mathbf{d} = \begin{pmatrix} 4 \\ 3 \end{pmatrix}$.

Point E is the midpoint of BC.

a Find the position vector of E.

b Show that $ABED$ is a parallelogram.

8 The vertices of a quadrilateral $PQRS$ have coordinates $P(-2,1)$, $Q(5,-3)$, $R(6,0)$ and $S(-1,5)$. The midpoints of the sides PQ, QR, RS and SP are A, B, C and D.

Prove that $ABCD$ is a parallelogram.

9 Points A and B have position vectors $\mathbf{a} = \begin{pmatrix} 2 \\ 1 \end{pmatrix}$ and $\mathbf{b} = \begin{pmatrix} -1 \\ 3 \end{pmatrix}$. Point C lies on AB such that $AC : BC = 2 : 3$.

Find the position vector of C.

10 Points A, B, C and D have position vectors $\mathbf{a}$, $\mathbf{b}$, $\mathbf{c}$ and $\mathbf{d}$.

M is the midpoint of AB and N is the midpoint of BC.

P is the midpoint of CD and Q is the midpoint of AD.

By finding the vectors $\overrightarrow{MN}$ and $\overrightarrow{QP}$, prove that $MNPQ$ forms a parallelogram.

11 Points A and B have coordinates $A(10, 1)$ and $B(2, 7)$. Point C lies on the line segment AB such that $AC : BC = x : 1 - x$, where $0 < x < 1$.

a Find the coordinates of C, in terms of x.

Point D has coordinates $D(3, 2)$ and $CD = \sqrt{26}$.

b Find the value of x.

12 $ABCD$ is a parallelogram with $\overrightarrow{AB} = \mathbf{p}$ and $\overrightarrow{BC} = \mathbf{q}$. Let M be the midpoint of the diagonal AC.

a Express $\overrightarrow{AM}$ in terms of $\mathbf{p}$ and $\mathbf{q}$.

b Show that M is also a midpoint of the diagonal BD.

13 Four points have coordinates $A(2, -1)$, $B(k, k+1)$, $C(2k-3, 3k+2)$ and $D(k-1, 2k)$.

a Show that $ABCD$ is a parallelogram for all values of k.

b Show that there is no value of k for which $ABCD$ is a rhombus.

Checklist of learning and understanding

- A **vector** is a quantity that has both **magnitude** and **direction.**
- The magnitude of vector **a** is written $|\mathbf{a}|$.
- There are two ways to represent a vector:
  - by drawing a directed arrow – in this case the magnitude of the vector is represented by the length of the arrow.
  - by stating its **components**, either as a column vector or by using base vectors **i** and **j**.

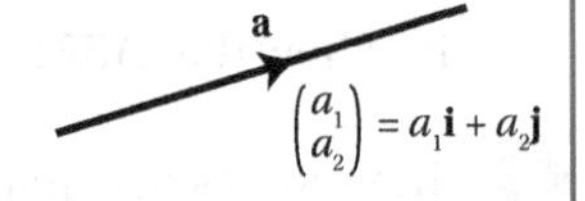

- If a vector is represented by its components then you can calculate its magnitude by using Pythagoras' theorem and its direction by finding the angle it makes with a specified direction (such as the horizontal or vertical).
- You can perform three operations with vectors: addition, subtraction and multiplication by a scalar.
- Two vectors are parallel if they have the same direction but different magnitude. If vectors **a** and **b** are parallel then $\mathbf{b} = t\mathbf{a}$ for some scalar t.
  - A **unit vector** is a vector with magnitude 1.
- Vectors can represent positions of points or displacements between two points.
  - The **position vector** of a point is a vector from the origin to that point.
  - If points A and B have position vectors **a** and **b** then the **displacement** from A to B is $AB = \mathbf{b} - \mathbf{a}$.
  - The **distance** between two points is the magnitude of the displacement between them: $AB = |\overrightarrow{AB}| = |\mathbf{b} - \mathbf{a}|$.

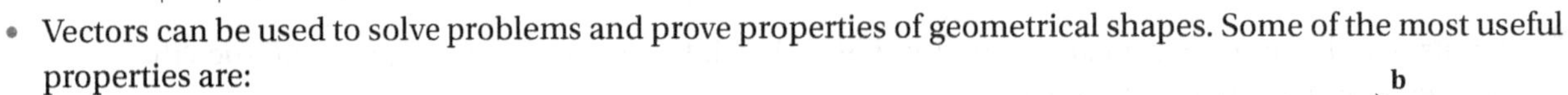

- Vectors can be used to solve problems and prove properties of geometrical shapes. Some of the most useful properties are:
  - If a shape is a parallelogram then the vectors corresponding to the opposite sides are equal.
  - If the shape is a rhombus then the vectors corresponding to the adjacent sides have equal magnitudes.

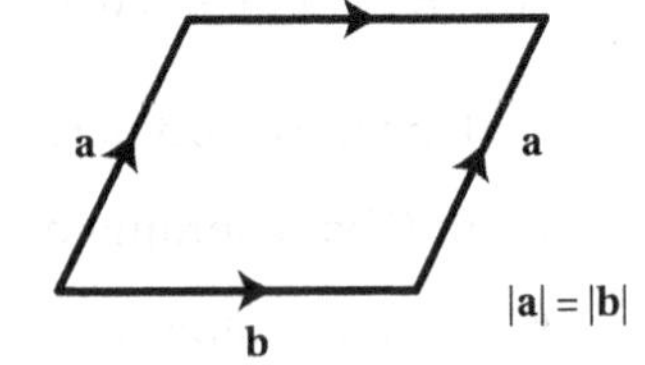

  - The midpoint of the line segment joining points with position vectors **a** and **b** has position vector $\frac{1}{2}(\mathbf{a}+\mathbf{b})$.

  - You can use vectors to show that two lines are parallel.

Mixed practice 15

1. The vectors $\mathbf{F}_1 = \begin{pmatrix} -7 \\ 2 \end{pmatrix}$, $\mathbf{F}_2 = \begin{pmatrix} 3 \\ 4 \end{pmatrix}$ and $\mathbf{F}_3$ are added. Their sum is $\mathbf{R} = \begin{pmatrix} -6 \\ 5 \end{pmatrix}$. What is the vector $\mathbf{F}_3$?

   Choose from these options.

   A $\begin{pmatrix} -10 \\ 11 \end{pmatrix}$ B $\begin{pmatrix} 5 \\ -6 \end{pmatrix}$ C $\begin{pmatrix} -2 \\ -1 \end{pmatrix}$ D $\begin{pmatrix} 4 \\ -6 \end{pmatrix}$

2. Points A and B have position vectors $\mathbf{a} = 3\mathbf{i} - \mathbf{j}$ and $\mathbf{b} = 2\mathbf{j}$. Find the exact distance between A and B.

3. a Given the points $P(-5, 2)$ and $Q(1, -3)$, write vector $\overrightarrow{PQ}$ in the form $a\mathbf{i} + b\mathbf{j}$.

   b Point R is such that $\overrightarrow{RQ} = \mathbf{i} - 4\mathbf{j}$. Find the coordinates of R.

4. Points A and B have position vectors $\mathbf{a} = \begin{pmatrix} 12 \\ -7 \end{pmatrix}$ and $\mathbf{b} = \begin{pmatrix} -3 \\ 5 \end{pmatrix}$.

   M is the midpoint of AB.

   a Find the position vector of M.

   b Find the exact distance BM.

5. Points A, B and C have position vectors $\mathbf{a} = 3\mathbf{i} - \mathbf{j}$, $\mathbf{b} = \mathbf{i} + 2\mathbf{j}$ and $\mathbf{c} = 4\mathbf{i} + \mathbf{j}$. Point D is such that $ABCD$ is a parallelogram. Find the position vector of D.

6. The diagram shows points P, Q and R such that $\overrightarrow{PQ} = \mathbf{a}$ and $\overrightarrow{PR} = 3\mathbf{b}$. Points M and N are on PQ and PR such that $PM = \frac{1}{3}PQ$ and $\overrightarrow{PN} = \mathbf{b}$.

   Express $\overrightarrow{MN}$ in terms of $\mathbf{a}$ and $\mathbf{b}$ and hence prove that MN is parallel to QR.

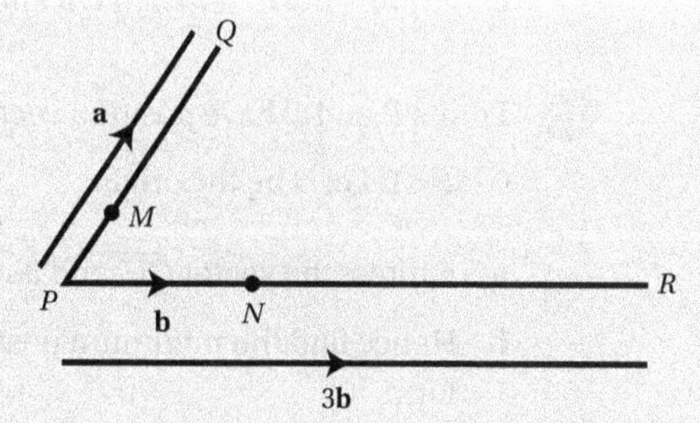

7. Find a vector of magnitude $5\sqrt{10}$ that is parallel to $2\mathbf{i} - 6\mathbf{j}$.

   Choose from these options.

   A $10\mathbf{i} - 30\mathbf{j}$ B $10\sqrt{10}\mathbf{i} - 30\sqrt{10}\mathbf{j}$ C $\frac{2}{5\sqrt{10}}\mathbf{i} - \frac{6}{5\sqrt{10}}\mathbf{j}$ D $5\mathbf{i} - 15\mathbf{j}$

8. OAB is a triangle with $\overrightarrow{OA} = \mathbf{a}$ and $\overrightarrow{OB} = \mathbf{b}$. M is the midpoint of AB and G is a point on OM such that $OG : GM = 2 : 1$. N is the midpoint of OA. Use vectors to prove that the points B, G and N are collinear.

9. Find the magnitude of the vector $(3\sin\theta)\mathbf{i} + (5\cos\theta)\mathbf{j}$ in terms of $\cos\theta$.

10. Points M and N have coordinates $M(-6, 1)$ and $N(3, 5)$. Find a unit vector parallel to $\overrightarrow{MN}$.

11. Points P and Q have coordinates $(1, -8)$ and $(10, -2)$. N is a point on PQ such that $PN : NQ = 1 : 2$.

    a Find the coordinates of N.

    b Calculate the magnitudes of $\overrightarrow{OP}$, $\overrightarrow{ON}$ and $\overrightarrow{PN}$. Hence show that ONP is a right angle.

12 The point A has position vector $\mathbf{a}=\begin{pmatrix}2k\\3k-5\end{pmatrix}$, and the point B has position vector $\mathbf{b}=\begin{pmatrix}2k-3\\5k+2\end{pmatrix}$.

The vector $\overrightarrow{AB}$ is parallel to $\begin{pmatrix}1\\1\end{pmatrix}$. Find the value of the constant k.

Choose from these options.

A $k=-5$ **B** $k=-2$ **C** $k=3$ **D** $k=1$

13 Points A, B, C and D have position vectors $\mathbf{a}$, $\mathbf{b}$, $\mathbf{c}$ and $\mathbf{d}$. M, N, P and Q are midpoints of AB, BC, CD and DA.

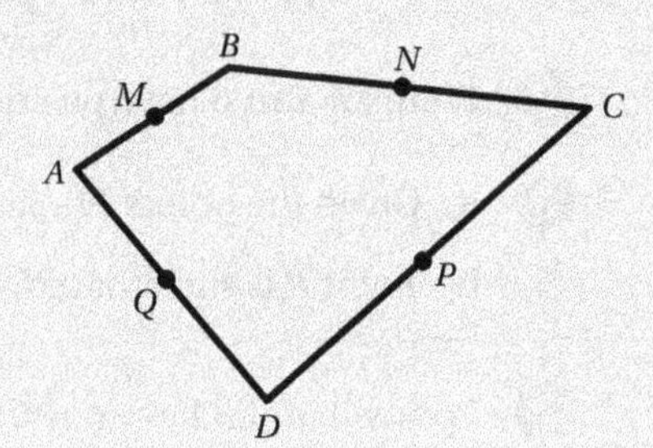

a Express vectors $\overrightarrow{MN}$ and $\overrightarrow{PQ}$ in terms of $\mathbf{a}$, $\mathbf{b}$, $\mathbf{c}$ and $\mathbf{d}$.

b What type of quadrilateral is $MNPQ$?

14 Points A and B have position vectors $\mathbf{a}$ and $\mathbf{b}$. O is the origin and point P is such that $OAPB$ is a parallelogram.

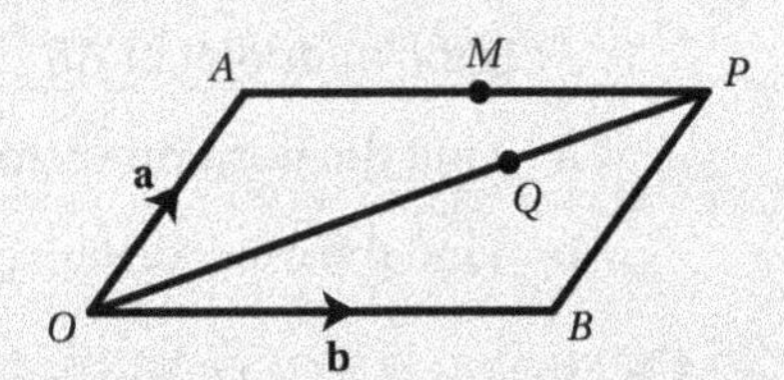

a Write down the position vector of P in terms of $\mathbf{a}$ and $\mathbf{b}$.

b Find the position vector of M, the midpoint of AP.

Point Q lies on OP. Let $\overrightarrow{OQ}=t\overrightarrow{OP}$.

c Express $\overrightarrow{BQ}$ in terms of t, $\mathbf{a}$ and $\mathbf{b}$. Hence find the value of t for which BQM is a straight line.

15 Points P and Q have position vectors $\mathbf{p}=\begin{pmatrix}5\\2\end{pmatrix}$ and $\mathbf{q}=\begin{pmatrix}2\\3\end{pmatrix}$. Point H lies on PQ and $\overrightarrow{PH}=t\overrightarrow{PQ}$, with $0<t<1$. Let O be the origin.

a Express the vector $\overrightarrow{OH}$ and its length in terms of t.

b Hence find the minimum possible distance of H from the origin, giving your answer in exact form.

16 Points A and B have position vectors $\mathbf{a}$ and $\mathbf{b}$. O is the origin and point P is such that $OAPB$ is a parallelogram.

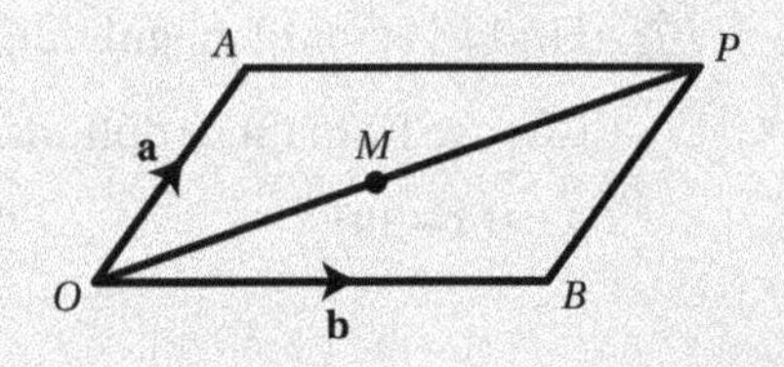

M is the midpoint of OP.

a Show that M lies on AB and determine the ratio $AM : MB$.

b What conclusion can be drawn about the diagonals of a parallelogram?

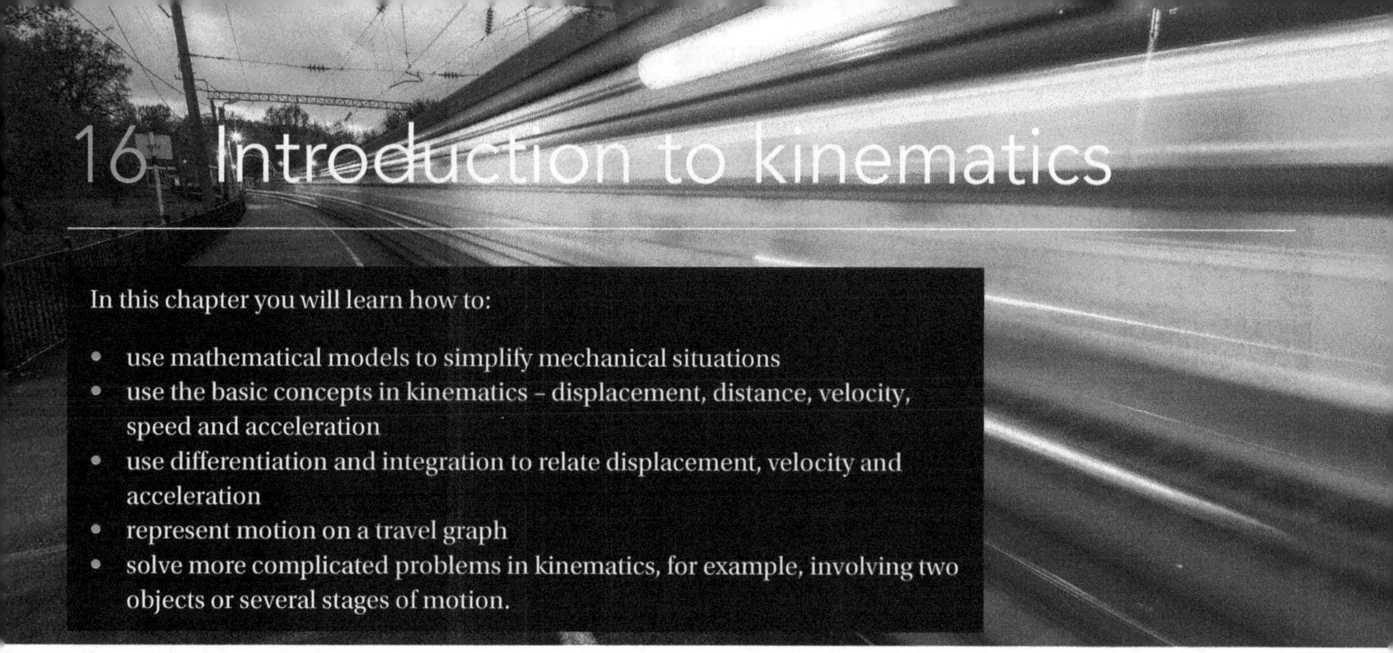

16 Introduction to kinematics

In this chapter you will learn how to:

- use mathematical models to simplify mechanical situations
- use the basic concepts in kinematics – displacement, distance, velocity, speed and acceleration
- use differentiation and integration to relate displacement, velocity and acceleration
- represent motion on a travel graph
- solve more complicated problems in kinematics, for example, involving two objects or several stages of motion.

Before you start…

GCSE	You should be able to find the gradient of a straight line connecting two points.	1 Consider the points $A(2, 5)$, $B(-1, 3)$ and $C(7, -2)$. Find the gradient of the straight line connecting: a A and C b B and A.
GCSE	You should be able to find areas of triangles and trapeziums.	2 Find the areas of the shaded regions marked S and T. 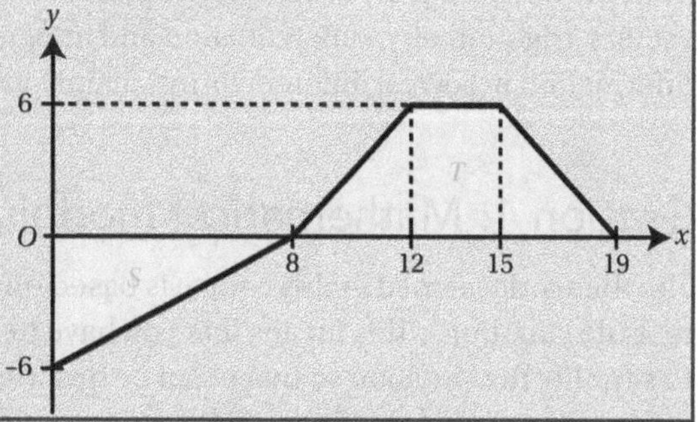
GCSE	You should be able to interpret displacement-time and velocity-time graphs.	3 Use this velocity-time graph to find: a the acceleration of the object during the first 5 seconds b the distance travelled during the whole 8 seconds. 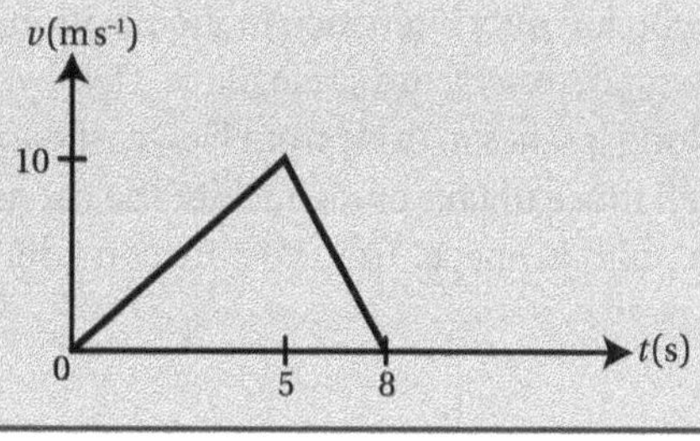

Chapter 12	You should be able to differentiate polynomials.	4 Given that $y=3x^2-4x+\frac{5}{x}$, find: a $\frac{dy}{dx}$ b the gradient of the curve when $x=-1$.
Chapter 13	You should be able to find stationary points.	5 Find the coordinates of the maximum point on the graph of $y=-x^3+12x+5$.
Chapter 14	You should be able to use integration to find an area under a graph.	6 Find the area enclosed by the graph of $y=6x-3x^2$ and the x-axis.
Chapter 14	You should be able to find the constant of integration.	7 A curve has gradient $\frac{dy}{dx}=5-6x^2$ and passes through the point $(1, 2)$. Find the equation of the curve.

What is kinematics?

Kinematics is the study of the motion of objects: how their position, velocity and acceleration depend on time, as well as how they are related to each other. It is not concerned with what causes the motion: that is the purpose of the study of dynamics, which you will meet in Chapter 18. Kinematics and dynamics together form a branch of applied mathematics called **mechanics**.

You may have studied some mechanics before, probably within the field of physics. In this course, the focus is on applying mathematical techniques from previous chapters to analyse problems in mechanics. You will use vectors, trigonometry, differentiation and integration, and represent information in several different forms, mainly graphs and equations.

Section 1: Mathematical models in mechanics

The theory developed in this course is based on **mathematical models** of real-life situations. This means that you have to make some assumptions to simplify the situation so that it can be described mathematically. It is important to consider how realistic these assumptions are, and whether improving the model would lead to substantially different results.

In the case of a car, for example, there are several different types of motion going on. The car might be travelling in a straight line, while the wheels are rotating around the axles and the wipers are moving left and right. Even if you are only interested in the position of the car, the front and the back aren't in exactly the same place. However, if all you need to know is how long it takes to drive from Newcastle to Manchester, then you can ignore all those details and consider the car as a single object, occupying a single point in space. Then you are considering the car to be a **particle**.

Key point 16.1

The **particle model** is a mathematical model that assumes that an object occupies a single point in space.

Fast forward

In Student Book 2, you will encounter situations in which an object cannot be modelled as a particle; for example, you will study moments, which affect how objects rotate.

This does not mean that the object is very small. You could consider an aeroplane as a particle if all you were interested in was its distance from the destination – the length of the aeroplane is negligible compared to the length of its journey. However, if you wanted to look at how different wing flaps move during turbulence, then the particle model would not be appropriate.

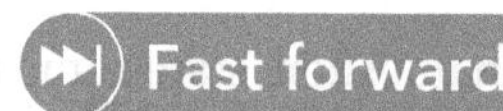

Fast forward

Vector equations of motion are studied in Student Book 2, Chapter 16.

Another assumption you will often make is that the object moves in a **straight line**. This means that its position can be described by a single number, for example, its distance from the starting point. If the object is allowed to move in two or three dimensions then you need **vectors** to describe its position.

WORKED EXAMPLE 16.1

The stretch of the A1(M) motorway between Junction 14 (Alconbury) and Junction 17 (Peterborough) is often cited as the longest straight stretch of UK motorway. The distance along the motorway between the two junctions is 16.8 km, while the straight-line distance is 16.6 km.

A car travels along the motorway at an average speed of 110 km/h.

a Find, to 2 s.f., the percentage difference between the times taken to travel the actual distance and the straight-line distance.

b Hence comment on whether the straight-line model is appropriate for this stretch of motorway.

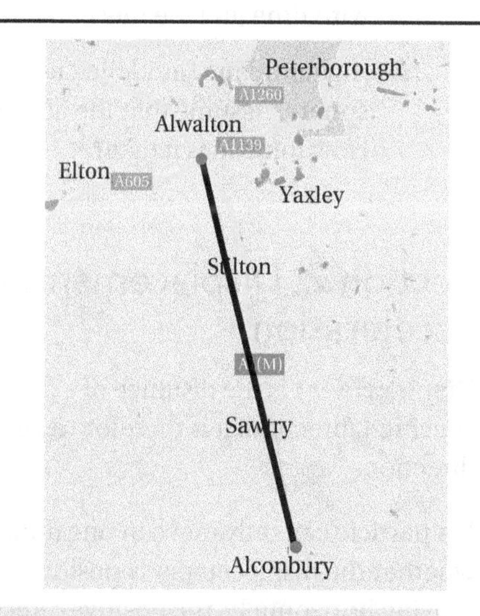

a Actual time:

$t_1 = \frac{16.8}{110}$

$= 0.1527$ hours

time $= \frac{\text{distance}}{\text{average speed}}$.

Straight-line model:

$t_2 = \frac{16.6}{110}$

$= 0.1509$ hours

Percentage difference:

$\frac{0.1527 - 0.1509}{0.1527} \times 100 = 1.2\%$

Make sure you use several decimal places of accuracy for t_1 and t_2 (or the 'ANS' button on your calculator) to avoid affecting the second significant figure of accuracy in your answer.

b This is a small difference, so the straight-line model is appropriate.

1.2% seems an acceptable error compared to the benefits gained from the simplicity of the straight-line model.

EXERCISE 16A

1. Discuss whether the particle model is appropriate in each of these situations.

   a You want to calculate how long it would take a car to complete the journey from Bristol to Birmingham.

   b You are designing a car park.

   c You want to predict the motion of a large box on a smooth floor when a force is applied at one corner.

   d You want to decide how a football should be kicked so that it curves towards the goal.

2. For each question, state some factors that have been ignored in the described model. Discuss how including each factor would affect the answer to the question.

   a A box is modelled as a particle. The box falls from the top of a building. The only force acting on the box is gravity. How long does it take for the box to reach the ground?

   b A snooker ball is hit towards a cushion with the cue making an angle of 30° with the cushion. The ball is modelled as a particle. Will it hit another ball (lying in a specified position)?

   c A bus (of a given mass, modelled as a particle) travels between two cities. You know how its speed varies with time and the fuel consumption at various speeds. How much fuel does it need?

   d An aeroplane (modelled as a particle) flies between London and Tokyo. The Earth is modelled as a sphere. Assume that the aeroplane flies in a straight line at a constant height and at a constant speed. How long does it take?

Section 2: Displacement, velocity and acceleration

Displacement is the distance of a particle from its initial position in a specified direction; it is therefore a **vector** as it has both magnitude and direction.

If a particle can only move in one dimension, its direction is indicated by whether the displacement is positive or negative. In general, for a particle moving horizontally, the positive direction will be to the right.

If the particle doesn't change direction during its motion then the total distance travelled is the same as the magnitude of the displacement from the starting point. However, if it changes direction then the total distance travelled is not necessarily the same as the final displacement.

Fast forward

You will learn more about the uses of vectors in mechanics in Chapters 18 and 19.

Tip

Remember that distance is a scalar whereas displacement is a vector. Therefore if a particle moves, say, 2 m from A to B and then returns from B to A its distance travelled is 4 m but its displacement is 0 m.

WORKED EXAMPLE 16.2

Points A, B and C lie, in that order, in a straight line, with $AB = 150$ m and $BC = 260$ m. The direction of positive displacement is from A towards B. A particle travels from A to C and then from C to B.

Find:

a the displacement from C to B
b the final displacement from A
c the total distance travelled by the particle.

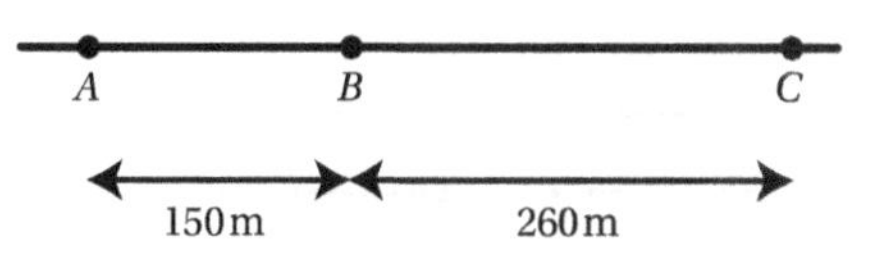

Always draw a diagram to illustrate the situation.

a C to $B = -260$ m

B is to the left of C, so the displacement is negative.

b $AB = 150$ m

The particle ends up at B. This is to the right of A, so the displacement is positive.

c $AC = 150 + 260 = 410$ m
$CB = 260$ m
Total distance $= AC + CB = 670$ m

To find the total distance, you need to add the distance from A to C to the distance from C to B.

If a particle covers the distance of 100 m in 20 seconds, then **on average** it covers 5 metres per second; its **average speed** is 5 m s^{-1}.

If you take into account the direction of motion as well as the speed, you get the **velocity**, which is a vector. If the particle changes direction during the motion, this means that the direction of the velocity vector is changing. The average velocity is equal to overall displacement divided by time.

Tip

Just as distance is the scalar counterpart of the vector displacement, speed is the scalar counterpart of the vector velocity.

Key point 16.2

- Average speed $= \dfrac{\text{total distance}}{\text{time taken}}$
- Average velocity $= \dfrac{\text{final displacement} - \text{initial displacement}}{\text{time taken}}$

The rate of change of velocity is called **acceleration**. Just like velocity, acceleration is a vector.

The velocity is usually denoted by the letter v and acceleration by the letter a. The units of velocity are m s^{-1} because you are dividing a displacement (in metres) by time (in seconds). For acceleration, you are dividing velocity by time, so the units are $\dfrac{\text{m s}^{-1}}{\text{s}} = \text{m s}^{-2}$.

You now know the definitions of the basic quantities needed to describe the motion of a particle.

Key point 16.3

Scalar quantities	Vector quantities
time [seconds]	
distance [m]	displacement [m]
speed [m s^{-1}]	velocity [m s^{-1}]
magnitude of acceleration [m s^{-2}]	acceleration [m s^{-2}]

WORKED EXAMPLE 16.3

Three points, A, B and C, lie in a straight line, as shown in the diagram.

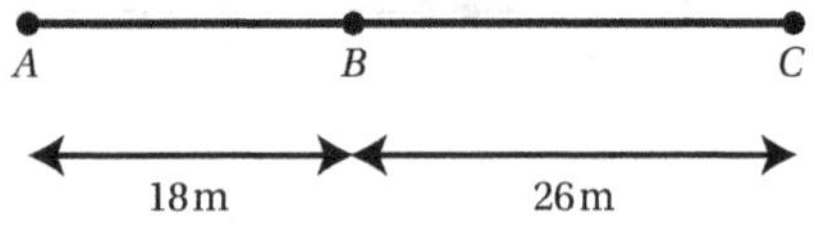

A particle starts at A, moving towards B with a speed of 2.5 m s^{-1}. It passes B 6 seconds later with a speed of 3.1 m s^{-1}. After a further 7.5 seconds it reaches C, where it stops and moves back towards B, which it passes 8.5 seconds later with a speed of 4.1 m s^{-1}. Find:

a the average velocity of the particle as it is moving from A to B

b the average speed of the particle for the whole journey.

a $v = \frac{18-0}{6}$

$= 3\text{ m s}^{-1}$

Average velocity $= \frac{\text{final displacement} - \text{initial displacement}}{\text{time}}$

b The total distance $= AB + BC + CB$

$= 18 + 26 + 26$

$= 70\text{ m}$

Find the total distance. Note that this is not the same as the displacement as the particle has changed direction during the motion.

The total time $= 6 + 7.5 + 8.5$

$= 22\text{ s}$

Find the total time.

The average speed $= \frac{70}{22}$

$= 3.2\text{ m s}^{-1}$ (2 s.f.)

Average speed $= \frac{\text{total distance}}{\text{total time}}$

In all the examples so far you have measured distance in metres and time in seconds as these are the fundamental units of the SI system. However, in real-life applications it is sometimes more convenient to use other units, such as kilometres or hours. You need to be able to convert the derived units for velocity and acceleration.

WORKED EXAMPLE 16.4

Convert:

a 2.6 m s^{-1} into kilometres per hour
b 170 km h^{-2} into m s^{-2}.

a $2.6\text{ m s}^{-1} = \dfrac{0.0026\text{ km}}{\frac{1}{3600}\text{ h}}$
$= 9.36\text{ km h}^{-1}$

2.6 m = 0.0026 km

1 second (s) = $\frac{1}{3600}$ hour (h)

b $170\text{ km h}^{-2} = \dfrac{170\,000\text{ m}}{(3600\text{ s})^2}$
$= 0.0131\text{ m s}^{-2}$

170 km = 170 000 m
1 h = 3600 s
Remember that the hours are squared.

Did you know?

SI stands for *Système International*, the international system of units. It is the most widely used metric system, in which all units can be expressed in terms of seven fundamental (base) units. It was established in the mid-20th century.

EXERCISE 16B

Questions 1 and 2 refer to the four points, A, B, C and D, that lie in a straight line with distances between them as shown in the diagram.

The displacement is measured from left to right.

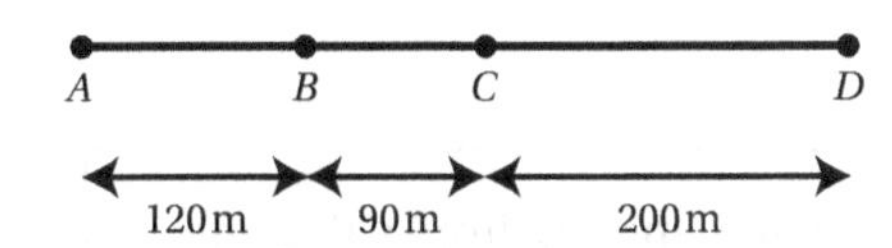

1 Find:

a i the displacement from D to A ii the displacement from D to B

b i the distance from D to B ii the distance from C to A

c i the total displacement when a particle travels from B to C and then to A

ii the total displacement when a particle travels from C to D and then to A.

2 Find the average speed and average velocity for a particle that travels from:

a i A to C in 23 seconds and then from C to B in 18 seconds

ii B to D in 38 seconds and then from D to A in 43 seconds

b i B to D in 16 seconds and then back to B in 22 seconds

ii A to C in 26 seconds and then back to A in 18 seconds.

3 Write these quantities in the specified units, giving your answers to three significant figures (3 s.f.).

a i 3.6 km h^{-1} in m s^{-1} ii 62 km h^{-1} in m s^{-1} b i 5.2 m s^{-1} in km h^{-1} ii 0.26 m s^{-1} in km h^{-1}

c i 120 km h^{-2} in m s^{-2} ii 450 km h^{-2} in m s^{-2} d i 0.82 m s^{-2} in km h^{-2} ii 2.7 m s^{-2} in km h^{-2}

Section 3: Kinematics and calculus

As a particle moves, its displacement, s, velocity, v, and acceleration, a, may be constantly changing. Thus they are all functions of **time**.

If you consider a small change in time from t_1 to t_2 on a displacement-time graph, the average velocity in that time interval is the gradient of the chord between two points on the graph.

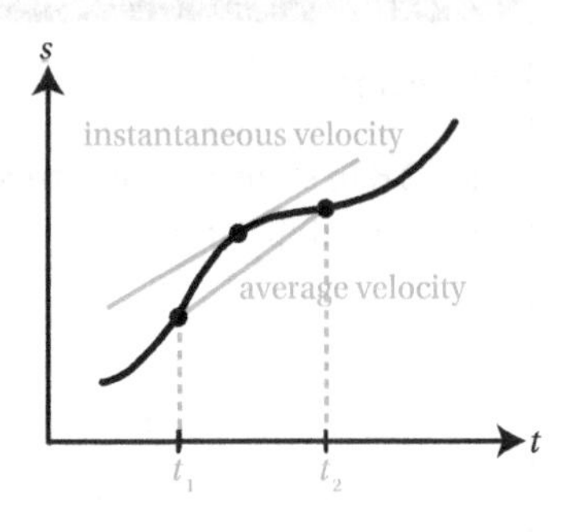

The **instantaneous velocity** at any particular point in time is given by the gradient of a tangent to the graph at that point. This is the same process you went through when finding the gradient of a function, so the instantaneous velocity is the derivative of the displacement.

Similarly, the acceleration is the rate of change of velocity over time. The **instantaneous acceleration** is the derivative of the velocity function (and the gradient of the velocity-time graph).

Rewind

See Chapter 12 if you need a reminder about differentiation from first principles.

Key point 16.4

If the displacement (s) is given as a function of time, t, then:

- the velocity is $v = \frac{ds}{dt}$
- the acceleration is $a = \frac{dv}{dt} = \frac{d^2s}{dt^2}$.

Fast forward

A In Student Book 2 you will need to differentiate more complicated displacement and velocity functions using A Level differentiation methods.

WORKED EXAMPLE 16.5

A toy car moves so that its displacement, s metres, from a flag after t seconds is given by the equation $s = 2.1 + 1.7t^2 - 0.2t^3$. Find:

a the instantaneous velocity and acceleration of the car after 5 seconds
b the initial displacement of the car from the flag
c the average velocity during the first 5 seconds of the motion.

a $v = \frac{ds}{dt}$
$= 3.4t - 0.6t^2$ — To find the velocity, differentiate the displacement equation.

When $t = 5$: — Then substitute in the value of t.
$v = 3.4(5) - 0.6(5)^2$
$= 2 \text{ m s}^{-1}$

$a = \frac{d^2s}{dt^2}$ — To find the acceleration, differentiate again.
$= 3.4 - 1.2t$

When $t = 5$: — Then substitute in the value of t.
$a = 3.4 - 1.2(5)$
$= -2.6 \text{ m s}^{-2}$

Continues on next page

b When $t = 0$:

$s = 2.1 + 1.7(0)^2 - 0.2(0)^3$

$= 2.1\text{ m}$

The initial displacement is when $t = 0$.

c When $t = 5$:

$s = 2.1 + 1.7(5)^2 - 0.2(5)^3$

$= 19.6\text{ m}$

To find the average velocity, you need the final displacement from the starting point. This is the value of s when $t = 5$.

The average velocity is:

$v = \frac{19.6 - 2.1}{5}$

$= 3.5\text{ m s}^{-1}$

The average velocity is $\frac{\text{final displacement} - \text{initial displacement}}{\text{total time}}$.

Common error

You should not assume that the initial displacement is zero.

Sometimes you know the equation for the velocity and want to find how the displacement changes with time. This requires integration: since the velocity is the derivative of the displacement, it follows that the displacement is the integral of the velocity.

Sometimes you know how the acceleration depends on time. You can integrate this equation to find the velocity. Integrating it again gives the equation for the displacement.

Key point 16.5

$$v = \int a\,\mathrm{d}t$$

$$s = \int v\,\mathrm{d}t$$

Fast forward

You will see in Chapter 18 that acceleration is related to force. This explains why, in many models, you can predict the equation for acceleration.

Note that the equations for velocity and displacement are indefinite integrals, which means that the result will involve a constant of integration. You can determine this by using the values of v or s that are given to you (usually, but not always, when $t = 0$).

Fast forward

A In Student Book 2 you will need to integrate more complicated velocity and acceleration functions using A Level integration methods.

WORKED EXAMPLE 16.6

An object moves in a straight line, with the acceleration given by the equation $a = (3t^2 - 6t)\ \text{m s}^{-2}$, where t is measured in seconds. The initial velocity of the object is $1.6\ \text{m s}^{-1}$.

a Find the equation for the velocity of the particle.
b Find the equation for the displacement from the initial position at time t.
c Find the displacement and the velocity of the object after 2 seconds.

a $v = \int 3t^2 - 6t\ \text{d}t$

$= t^3 - 3t^2 + c$

To find the velocity, integrate the acceleration equation. Don't forget the $+c$!

When $t = 0$, $v = 1.6$:

$1.6 = 0 - 0 + c$

$c = 1.6$

$\therefore v = t^3 - 3t^2 + 1.6$

Find the constant of integration by using the initial values of t and v.

b $s = \int t^3 - 3t^2 + 1.6\ \text{d}t$

$= \frac{1}{4}t^4 - t^3 + 1.6t + d$

To find the displacement, integrate the velocity equation. There is another constant $+d$.

When $t = 0$, $s = 0$:

$0 = 0 - 0 + 0 + d$

$d = 0$

$\therefore s = \frac{1}{4}t^4 - t^3 + 1.6t$

When $t = 0$ the object is at the starting point, so its displacement is zero.

c When $t = 2$:

$s = -0.8\ \text{m}$

$v = -2.4\ \text{m s}^{-1}$

To find the values at a specific point in time, substitute in the given value of t.
Note that the negative values of s and v mean that at $t = 2$ the particle is to the left of the starting position and also that it is moving to the left (assuming right is positive).

If you are only interested in the change in displacement between two times (rather than finding the displacement as a function of t) you can use a definite integral.

Key point 16.6

The change in displacement between times t_1 and t_2 is $\int_{t_1}^{t_2} v\ \text{d}t$.

Common error

When finding the total distance travelled, it is easy to forget that the particle might have changed direction. Always check for this before just integrating and evaluating between t_1 and t_2.

WORKED EXAMPLE 16.7

A particle moves in a straight line so that its velocity at time t is given by $v = 3t - t^2$. Find the change in the displacement of the particle between $t = 2$ and $t = 5$.

$$\int_2^5 (3t - t^2)\,dt = \left[\frac{3}{2}t^2 - \frac{1}{3}t^3\right]_2^5$$

You could find an equation for s and then evaluate the difference between s when $t = 2$ and s when $t = 5$. This is the same as evaluating a definite integral.

$$= \left(-\frac{25}{6}\right) - \left(\frac{10}{3}\right)$$
$$= -7.5$$

The particle has moved 7.5 units to the left.

However, as discussed in Section 2, if you want the total distance travelled between times t_1 and t_2 it is important to know whether the particle changes direction. If it does you will have to find the distance travelled for each part of the motion separately.

WORKED EXAMPLE 16.8

A particle moves with velocity $v = t^3 - 7t^2 + 10t$, where t is measured in seconds and v in m s^{-1}.

a Find the displacement from the starting point after 5 seconds.
b **i** Find the times when the particle is instantaneously at rest.
ii Sketch the velocity–time graph
c Find the distance travelled by the particle during the first 5 seconds.
d Hence find the average velocity and the average speed of the particle.

a $$\int_0^5 t^3 - 7t^2 + 10t\,dt = \left[\frac{t^4}{4} - \frac{7t^3}{3} + 5t^2\right]_0^5$$

The displacement from the starting point is the integral between $t = 0$ and $t = 5$.

$$= \left(-\frac{125}{12}\right) - (0)$$
$$= -10.4 \text{ m (3 s.f.)}$$

The negative sign means that the particle is to the left of its starting point.

b i $v = 0$

$t^3 - 7t^2 + 10t = 0$

$t(t-2)(t-5) = 0$

$t = 0$ or $t = 2$ or $t = 5$

The particle is instantaneously at rest when $v = 0$.

ii

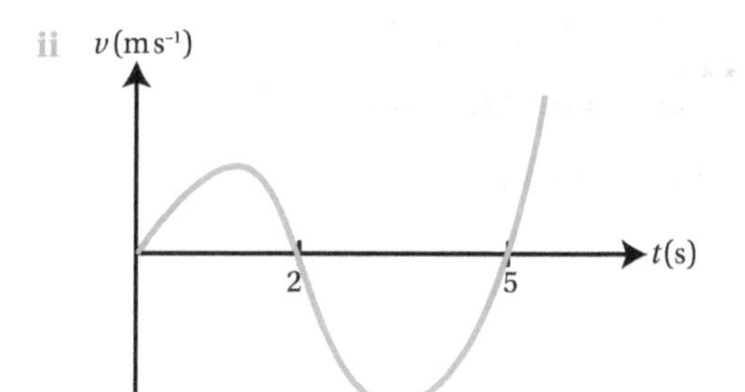

This is a positive cubic with roots at $t = 0$, 2 and 5. Note that from $t = 0$ to $t = 2$, $v > 0$ and so the particle moves to the right. From $t = 2$ to $t = 5$, $v < 0$ so the particle moves to the left.

Continues on next page

c $\int_0^2 t^3 - 7t^2 + 10t \, dt = \left[\frac{t^4}{4} - \frac{7t^3}{3} + 5t^2\right]_0^2$

Integrate from $t = 0$ to $t = 2$ to get the change in displacement to the right.

$= \left(\frac{16}{3}\right) - (0)$

$= \frac{16}{3}$

$\int_2^5 t^3 - 7t^2 + 10t \, dt = \left[\frac{t^4}{4} - \frac{7t^3}{3} + 5t^2\right]_2^5$

Integrate from $t = 2$ to $t = 5$ to get the change in displacement to the left (which will be negative).

$= \left(-\frac{125}{12}\right) - \left(\frac{16}{3}\right)$

$= -15.75$

The distance is:

Add the distance travelled to the right and the distance travelled to the left.

$d = \frac{16}{3} + 15.75$

$= 21.1 \text{ m (3 s.f.)}$

d The average velocity $= -\frac{10.4}{5}$

To find the average velocity, use the final displacement.

$= -2.08 \text{ m s}^{-1} \text{ (3 s.f.)}$

The average speed $= \frac{21.1}{5}$

To find the average speed use the total distance.

$= 4.22 \text{ m s}^{-1} \text{ (3 s.f.)}$

EXERCISE 16C

In questions 1–5, s is measured in metres, t in seconds, v in m s^{-1} and a in m s^{-2}.

1 A particle moves in a straight line. Its displacement from the point A is given by s. Find the equations for the velocity and acceleration in terms of t when:

a **i** $s = 3t^3 - 4t + 2$ **ii** $s = t^4 - 3t^2 + 4t$

b **i** $s = \frac{t^2}{5} - 4.2t$ **ii** $s = -0.5t^3 + \frac{t^2}{3}$.

2 For each part of question 1, find the values of the displacement, velocity and acceleration initially and after 3 seconds.

3 A particle moves in a straight line, with its velocity given by v. The initial displacement from point A is s_0. Find the equation for the displacement from A when:

a **i** $v = 3t - 4,\ s_0 = 2$ **ii** $v = 1 - 2t,\ s_0 = 1$

b **i** $v = 5 - 3t^2,\ s_0 = 0$ **ii** $v = t^2 - 2t,\ s_0 = 0$.

4 For each part of question 3, find the displacement from A after 5 seconds.

5 A particle moves in a straight line with given acceleration a. The initial velocity and displacement from point A are also given. Find the equations for the velocity and displacement, in terms of t, when:

a **i** $a=1-2t,\ v_0=3,\ s_0=0$ **ii** $a=4t+2,\ v_0=-2,\ s_0=0$

b **i** $a=-5,\ v_0=3,\ s_0=5$ **ii** $a=3,\ v_0=-2,\ s_0=7$.

6 A particle moves in a straight line. Its velocity, v m s^{-1}, at time t seconds is given by $v=2.4t-1.5t^2$.

a Find the acceleration of the particle after 2 seconds.

b Find the velocity at the point when the acceleration is zero.

The displacement of the particle from its initial position is s m.

c Find an expression for s in terms of t.

7 An object moves in a straight line. t seconds after it passes point O its displacement, s m, from O is given by $s=0.1t^3-1.2t^2+3.5t$.

a Find the speed and acceleration of the object 6 seconds after passing O.

b What is the speed of the object when it first returns to O?

c Find the first time when the velocity is zero, and the displacement of the object from O at this time.

8 A car starts from rest and moves in a straight line. Its acceleration, a m s^{-2}, is given by $a=0.12t^2-1.44t+4.32$.

a Find the equations for the car's velocity and its displacement from the starting point.

b Find the velocity and the displacement at the point when the acceleration is zero.

9 A particle moves in a straight line. Its displacement from the point P is s metres and its acceleration is $a=(1-0.6t)$ m s^{-2}. The particle is initially 25 m from P and moving away from P with the velocity of 7.5 m s^{-1}.

a Find an expression for the velocity in terms of t.

b Find the particle's displacement from P after 10 seconds.

c Find the particle's displacement from P at the time when its acceleration is –2 m s^{-2}.

10 A particle moves in a straight line, with its velocity given by $v=(0.05t^3-t^2+4.8t)$ m s^{-1}. The particle passes point A when $t=0$.

a Explain how you can tell that the particle is initially at rest, and find two other times when the velocity is zero.

b Find the velocity and the displacement of the particle from A when $t=6$ s.

c Find the total distance travelled during the first 12 seconds.

11 A particle moves in a straight line with acceleration $a=(2-6t)$ m s^{-2}, where the time is measured in seconds. When $t=2$ its velocity is –8 m s^{-1}. Find the average velocity of the particle between $t=5$ and $t=8$.

12 A particle moves in a straight line. Its velocity, v m s^{-1}, at time t seconds is given by $v=(t+3)(t-2)(t-7)$. Find the average speed of the particle during the first 7 seconds.

13 The velocity of a particle is given by:

$$v(t) = \begin{cases} 5t - \frac{1}{2}t^2 & \text{for } 0 \leq t \leq 5 \\ 7.5 - \frac{1}{2}t & \text{for } 5 < t \leq 15 \end{cases}$$

The average speed of the particle during the first T seconds is 4 m s^{-1}.

Find the value of T, where $T < 15$.

14 A particle moves in a straight line. Its displacement, s m, from point O is given by $s = 24t - 3t^2$, where t is measured in seconds. The average velocity of the particle during the first T seconds is 9 m s^{-1}.

Find its average speed during this time.

Section 4: Using travel graphs

The information on how displacement and velocity change with time may be described by equations, as you saw in the previous section, or it may be represented on a graph. Sometimes the graph can be drawn even without finding the equations, and calculations can be done straight from the graph.

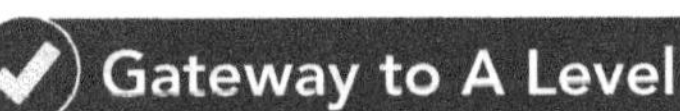

Gateway to A Level

See Gateway to A Level Section S for revision of travel graphs from GCSE.

Fast forward

You will meet constant acceleration equations of motion in Chapter 17.

Displacement–time graphs

On a displacement-time graph (s–t graph), time is shown on the horizontal axis and displacement (measured from some specified reference point) on the vertical axis. Remember that the displacement can be negative - this means that the particle is on the other side of the reference point.

Key point 16.7

On a displacement-time graph:

- average velocity is the gradient of the chord between two points
- instantaneous velocity is the gradient of the tangent to the graph.

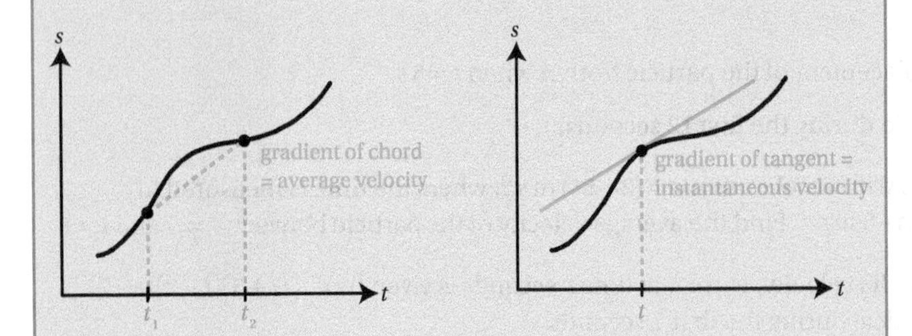

Focus on...

The concept of average velocity has some unexpected applications in pure Mathematics, as explored in Focus on... Proof 3.

WORKED EXAMPLE 16.9

A small boat moves in a straight line. The graph shows its displacement, in metres, from the lighthouse.

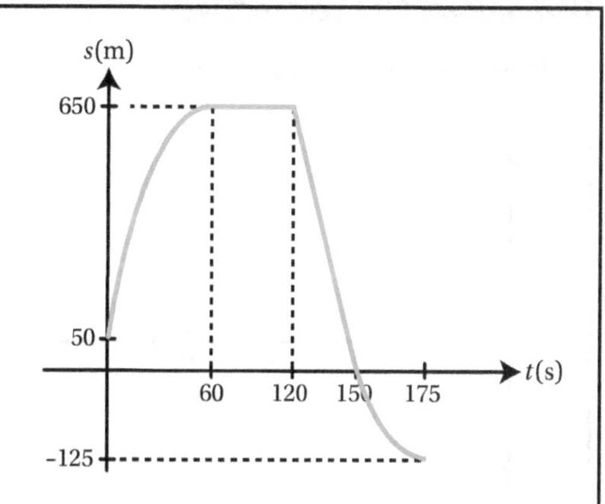

a How far from the lighthouse does the boat start?

b Describe how the displacement and the velocity of the boat are changing.

c Find the velocity of the boat between $t = 120$ and $t = 150$ seconds.

d Find the average velocity for the first 175 seconds.

a The boat starts 50 m from the lighthouse.	This is the displacement when $t = 0$.
b For the first 60 seconds, the boat moves away from the lighthouse, slowing down.	The displacement is increasing. The gradient of the graph is positive but decreasing.
For the next 60 seconds, the boat is stationary. Its velocity is zero.	The displacement is not changing.
From 120 seconds the boat moves back towards the lighthouse with constant velocity.	The displacement is decreasing. The graph is straight, so the velocity is constant.
It passes the lighthouse at 150 seconds and continues to move away from it.	When $t = 150$ the displacement is zero. Negative displacement means that the boat has passed the lighthouse.
c $v = \dfrac{0-650}{30} = -21.7\ \text{m s}^{-1}$ (3 s.f.)	Velocity is the gradient between the points (120, 650) and (150, 0). The answer should be negative.
d Initial displacement = 50 m Final displacement = −125 m The average velocity is: $\dfrac{-125-50}{175} = -1\ \text{m s}^{-1}$	Average velocity = $\dfrac{\text{final displacement} - \text{initial displacement}}{\text{time}}$

Remember that the gradient of the displacement–time graph represents the **velocity**. If you want to know the **speed**, you need to take the magnitude of the velocity. For motion in one dimension, this simply means taking the modulus of the number, so that negative numbers become positive.

WORKED EXAMPLE 16.10

The motion of a particle is represented on this displacement–time graph.

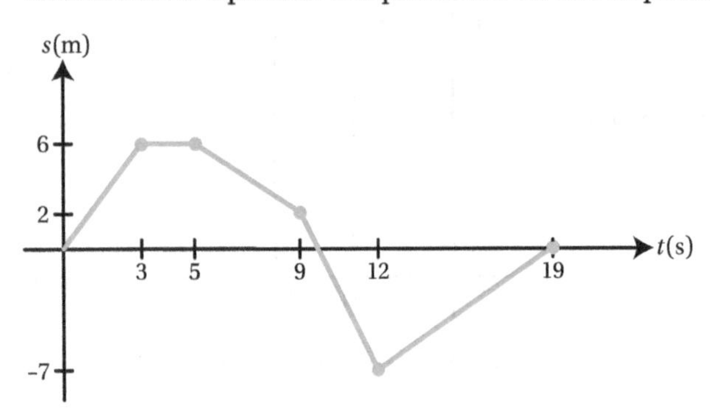

Find:

a the maximum velocity of the particle
b the maximum speed of the particle.

a The velocities are, in order:

$\frac{6}{3} = 2\text{ m s}^{-1}$ $\frac{0}{2} = 0\text{ m s}^{-1}$ $\frac{-4}{4} = -1\text{ m s}^{-1}$

$\frac{-9}{3} = -3\text{ m s}^{-1}$ $\frac{7}{7} = 1\text{ m s}^{-1}$

So the maximum velocity is 2 m s^{-1}.

Find the velocity on each segment, using $\frac{\text{change in displacement}}{\text{time}}$

b The maximum speed is 3 m s^{-1}.

Speed is the magnitude of velocity (negative signs can be disregarded).

Velocity–time graphs

Motion can also be shown on a velocity-time graph (v–t graph), which shows time on the horizontal axis and velocity on the vertical axis.

Key point 16.8

On a velocity-time graph:

- the gradient is the acceleration
- the total area between the graph and the horizontal axis is the distance travelled.

Gateway to A Level

You need to be able to find the area of triangles and trapezia. See Gateway to A Level Section P for a reminder.

WORKED EXAMPLE 16.11

The diagram shows the velocity–time graph for a particle moving in a straight line.

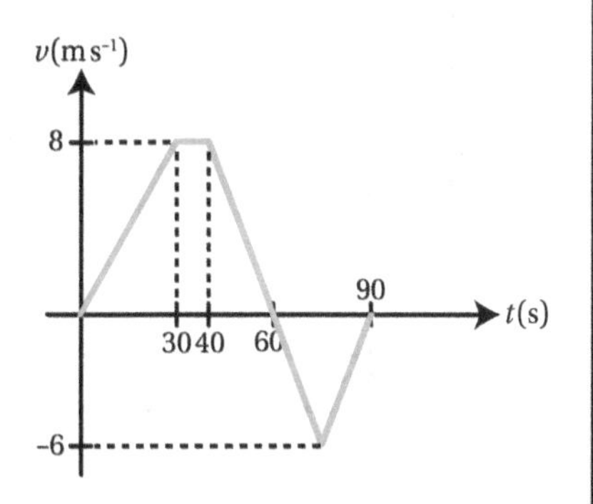

a Find the acceleration of the particle:

  i during the first 30 seconds

  ii between $t = 40$ and $t = 60$.

b What happens when $t = 60$?

c Find the distance travelled by the particle in the first 90 seconds.

a i $a = \frac{8}{30}$

$= 0.267\text{ m s}^{-2}$

The acceleration is the gradient of the graph.

ii $a = -\frac{8}{20}$

$= -0.4\text{ m s}^{-2}$

Between $t = 40$ and $t = 60$ the gradient is negative.

b The particle changes direction.

The velocity changes from positive to negative.

c Total distance $= \left(\frac{60+10}{2} \times 8\right) + \frac{30 \times 6}{2}$

$= 370\text{ m}$

The distance is the total area between the x-axis and the graph, so add the areas of the trapezium and the triangle.

Elevate

See Support Sheet 16 for a similar example of finding the total distance travelled where integration is needed, along with more questions.

You need to be able to draw a velocity–time graph from a given description of motion. You can also write equations to find missing information.

WORKED EXAMPLE 16.12

A car moves in a straight line. When $t = 0$, it passes the point P with the velocity of 12 m s^{-1}. It accelerates for 5 seconds with acceleration of 1.2 m s^{-2} until it reaches the velocity of $V\text{ m s}^{-1}$. It moves with constant velocity for T seconds and then decelerates at 0.8 m s^{-2} until it comes to rest.

a Sketch a velocity–time graph for the car's journey.

b Find the value of V.

c Find the time that the car spends decelerating.

d The total distance travelled by the car is 600 m. Find the value of T.

Continues on next page

a

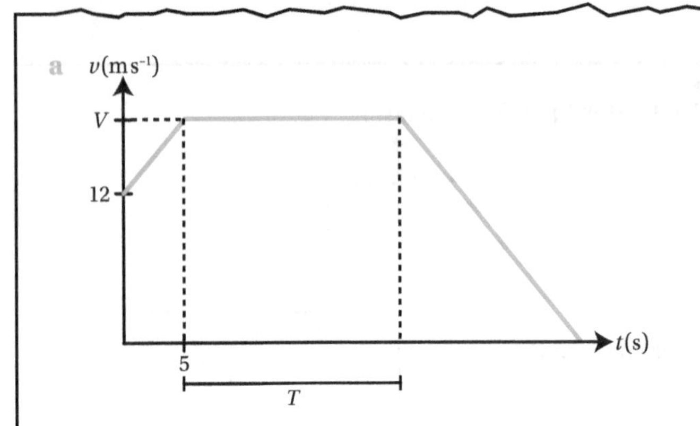

Make sure that the straight line when the car is accelerating has a steeper gradient (1.2) than the when it is decelerating (–0.8).

b $\frac{V-12}{5} = 1.2$

$t = 18$

Acceleration = $\frac{\text{change in velocity}}{\text{time}}$

c $\frac{0-18}{t} = -0.8$

$t = 22.5\text{ s}$

The velocity changes from 18 to 0. The acceleration is negative.

d

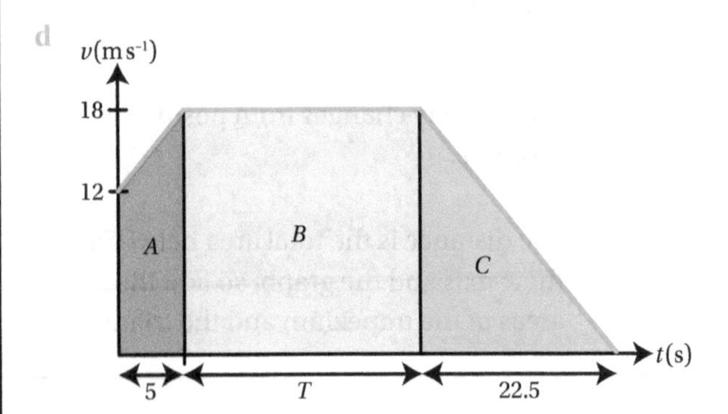

The distance is the area under the graph. You can split it into several parts as shown, for example, on the diagram.

$\frac{18+12}{2} \times 5 + 18T + \frac{22.5 \times 18}{2} = 600$

$18T = 600 - 277.5$

$18T = 322.5$

$T = 17.9\ (3\text{ s.f.})$

A is a trapezium with parallel sides of length 12 and 18 and height 5.

B is a rectangle with base *T* and height 18.

C is a triangle with base 22.5 and height 18.

Since the displacement and the velocity are related to each other, you should be able to draw the velocity-time graph from the displacement-time graph, and vice versa.

 Rewind

This is the same as the relationship between the graph of a function and its derivative, which you met in Chapter 12.

WORKED EXAMPLE 16.13

Match up each velocity-time graph with the corresponding displacement-time graph.

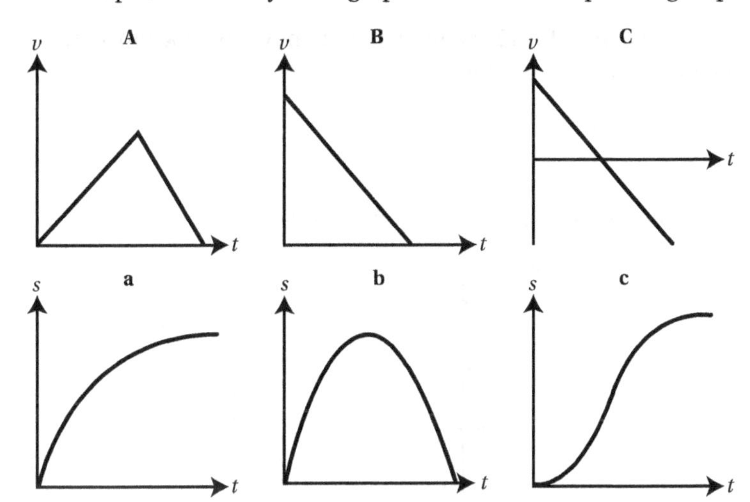

A corresponds to c.

In graph A the velocity increases from zero and then decreases back to zero. This means that the displacement-time graphs starts with a zero gradient and, after a while, s stays constant.

B corresponds to a.

The velocity starts positive and decreases towards zero; so the gradient of the displacement graph decreases towards zero but never becomes negative.

C corresponds to b.

The velocity starts positive but decreases to zero and then becomes negative. This means that the particle stops and turns around, so the displacement starts decreasing and returns to zero.

EXERCISE 16D

1 For each velocity-time graph, find:

i the acceleration from A to B and from C to D

ii the total distance travelled

iii the average speed

iv the average velocity.

a

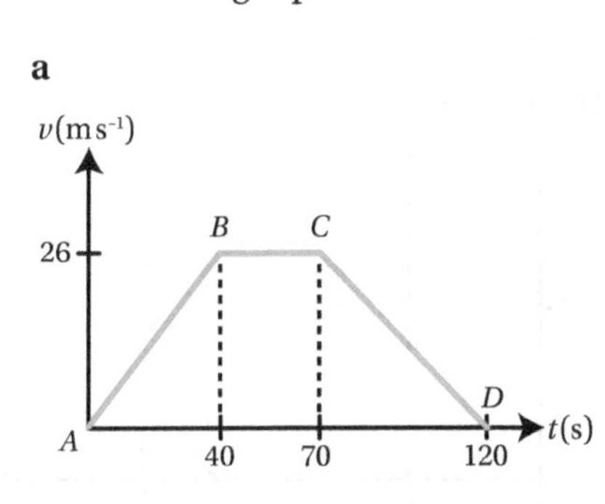

b

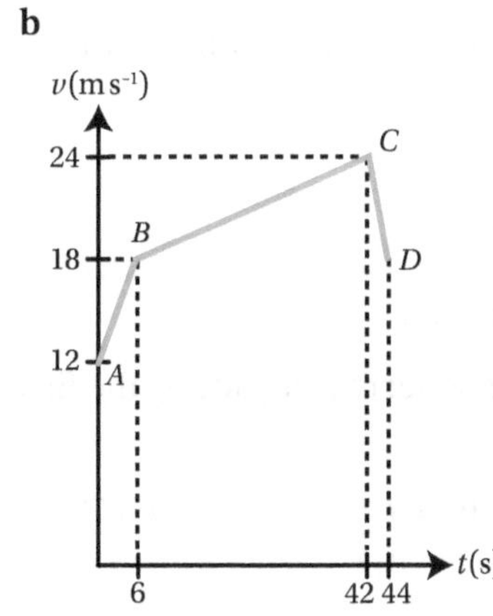

c

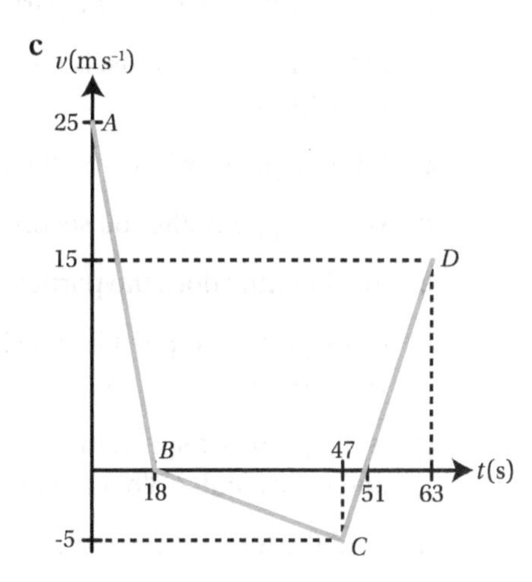

2 For each description of motion, draw the velocity–time graph and find the total distance travelled.

a A particle accelerates uniformly from $20\ \text{m s}^{-1}$ to $32\ \text{m s}^{-1}$ in 15 seconds, then moves with constant speed for 25 seconds and finally decelerates uniformly and comes to rest in another 10 seconds.

b An object starts from rest and accelerates at $2.5\ \text{m s}^{-2}$ for 12 seconds. It then moves with a constant velocity for 8 seconds and finally decelerates at $6\ \text{m s}^{-2}$ until it comes to rest.

c A particle accelerates uniformly from $11\ \text{m s}^{-1}$ to $26\ \text{m s}^{-1}$ with the acceleration of $0.4\ \text{m s}^{-2}$. It then decelerates at $2\ \text{m s}^{-2}$ until it comes to rest.

3 For each displacement–time graph, draw the corresponding velocity–time graph.

a

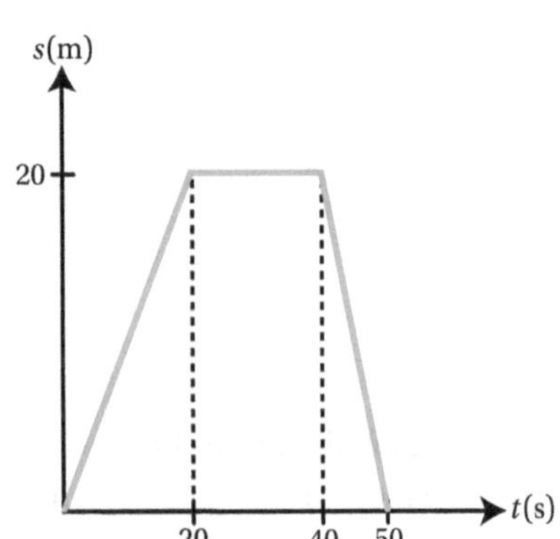

b

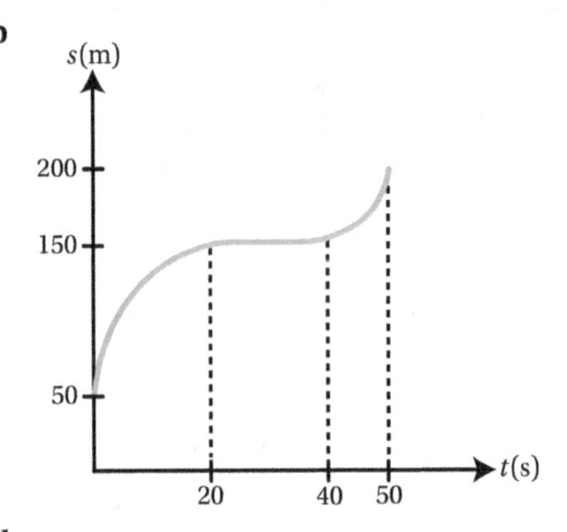

c

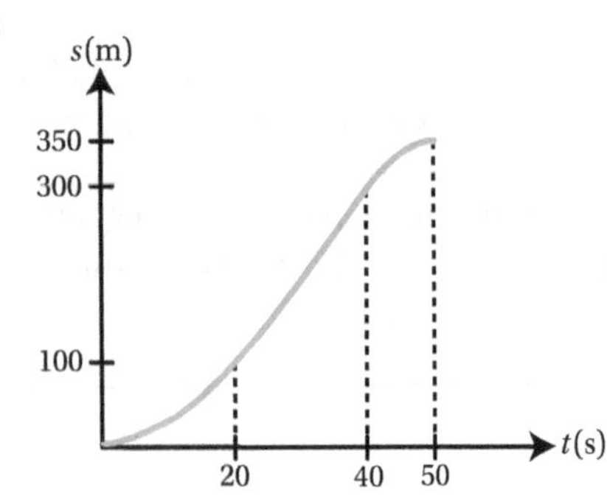

d

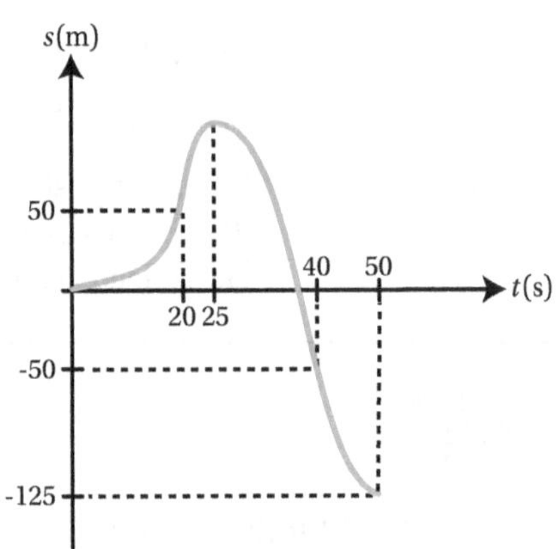

4 A particle moves in a straight line. Its displacement from point P is shown on the displacement–time graph.

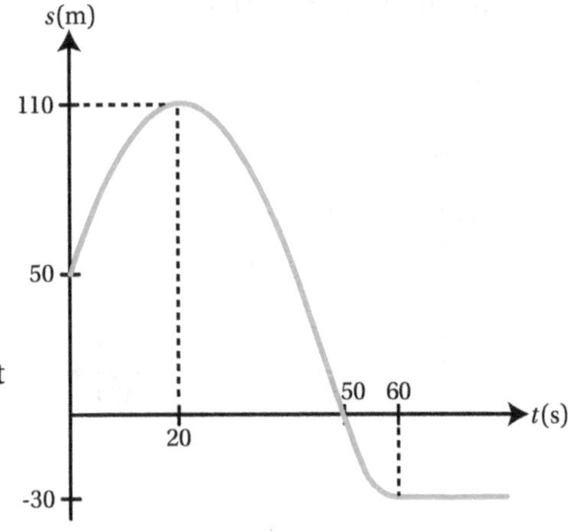

a How far from P does the particle start?

b In the first 20 seconds, is the particle moving towards P or away from it?

c What happens when $t = 20$ seconds?

d What happens after 60 seconds?

e At what time does the particle pass P?

f Is the particle's speed increasing or decreasing during the first 20 seconds?

g Is the particle's speed increasing or decreasing between 50 and 60 seconds? What about its velocity?

h Find the total distance travelled by the particle in the first 60 seconds.

5 A car moves in a straight line. It passes the point P with a velocity of 16 m s^{-1}. It continues to move with constant velocity for 20 seconds and then decelerates at a constant rate of 0.8 m s^{-2} until it comes to rest.

a Represent the car's journey on a velocity–time graph.

b How far from P does the car stop?

6 A particle moves in a straight line. It starts from point A when $t = 0$.

Its motion during 45 seconds is represented on this velocity–time graph.

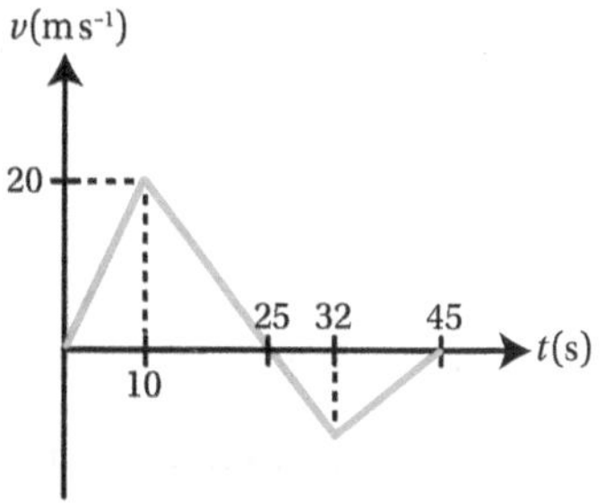

a Find the acceleration of the particle during the first 10 seconds.

b At what time does the particle change direction?

The total distance travelled by the particle is 325 m.

c Find the velocity of the particle when $t = 32$ s.

d How far from P is the particle at the end of the 45 seconds?

7 A particle moves in a straight line. Its motion, for the first 25 seconds, is represented on this velocity-time-graph.

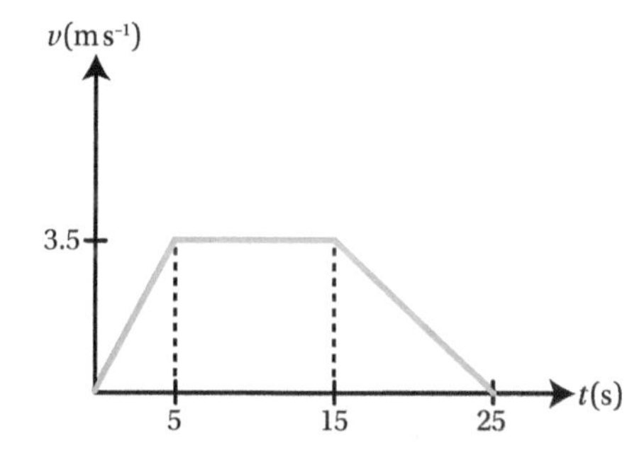

Find the average speed of the particle during the first 25 seconds.

8 The velocity of an object is given by $v = 18t - 6t^2$, where time is measured in seconds and velocity in m s^{-1}.

a Sketch the velocity-time graph for $0 \leqslant t \leqslant 3$.

b Hence find the average speed of the particle during the first 3 seconds.

9 The displacement–time graph represents the motion of a particle moving in a straight line. The particle passes point A when $t = 0$.

The particle is at point B when $t = 12$ and at point C when $t = 30$.

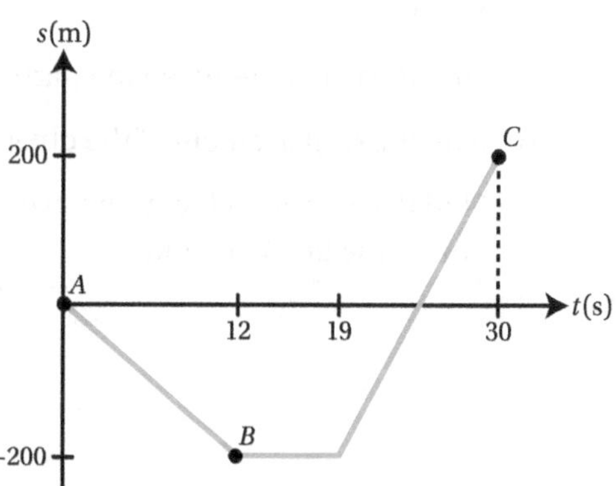

a Describe what happens between $t = 12$ and $t = 19$.

b Write down the displacement of C from A. Hence find the average velocity of the particle during the 30 seconds.

c Find the average speed of the particle during the 30 seconds.

10 Sarah runs in a straight line at a constant speed of 6.2 m s^{-1}. When $t = 0$ she passes Helen who immediately starts running with constant acceleration of 1.3 m s^{-2}. Helen accelerates for 6 seconds and then continues to run at a constant speed until she catches up with Sarah.

a Draw a velocity-time graph to illustrate the motion of both girls.

b How long does it take for Helen to catch up with Sarah?

11 A particle moves in a straight line. The diagram shows its velocity-time graph. The total distance travelled during the 55 seconds is 275 m.

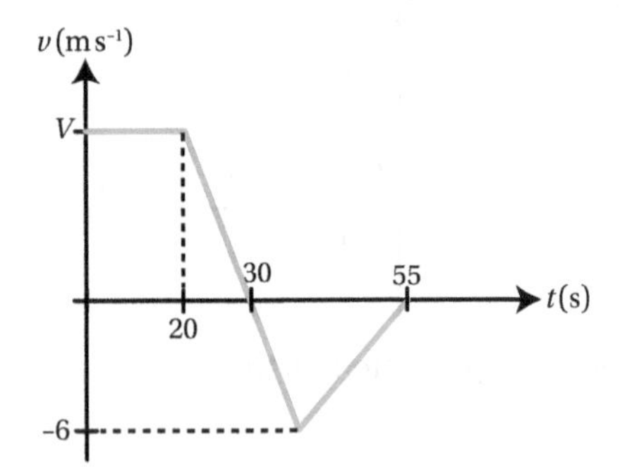

a Find the value of V.

b Find the deceleration of the particle when $t = 30$.

c Which of the graphs show the displacement-time graph for the same particle?

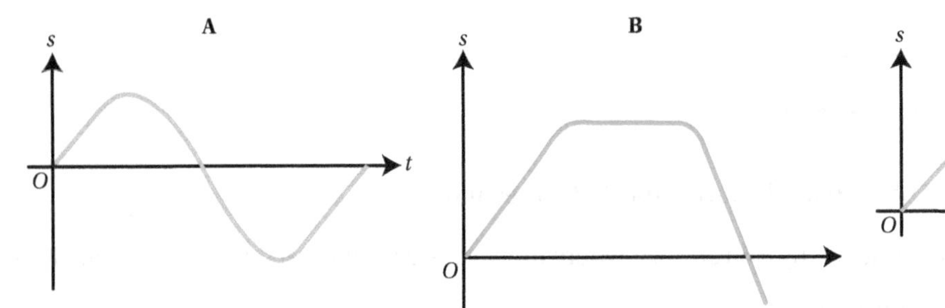

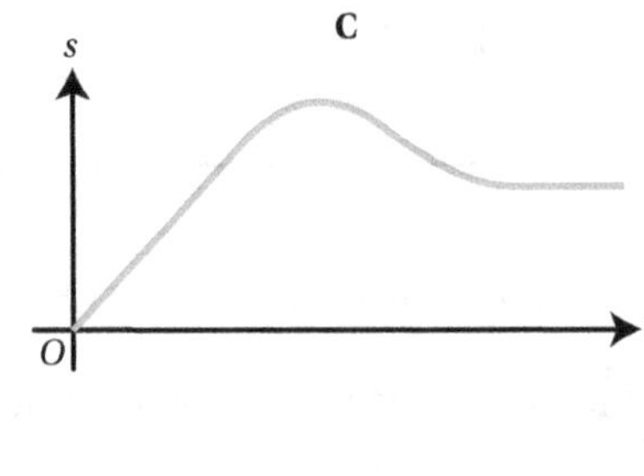

12 The diagram shows the velocity-time graph for an object moving in a straight line. When $t = 0$ the object is at point O. The equation for the velocity is $v = 3t(t - 4)(t - 7)$, where time is measured in seconds and displacement in metres.

a Find the two times when the particle changes direction.

b Find the displacement of the object from O when $t = 7$ s.

c Find the average velocity and average speed of the object during the first 7 seconds.

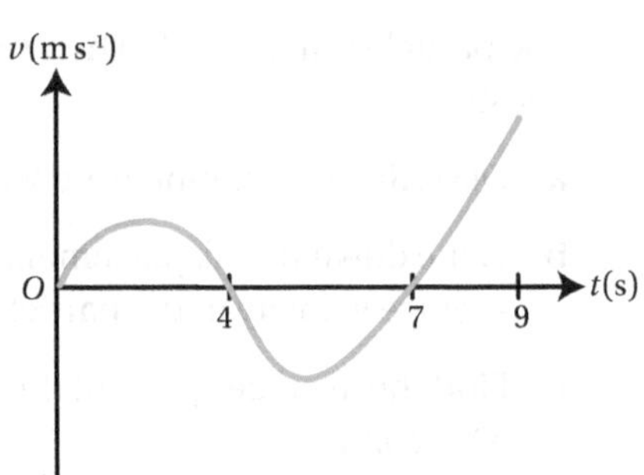

13 A particle moves in a straight line, starting from rest at point P. It accelerates for 5 seconds, until it reaches a speed of 16 m s^{-1}. It maintains this speed for T seconds and then decelerates at 2 m s^{-2} until it comes to rest at point Q.

a Sketch the velocity–time graph to represent the motion of the particle.

b Given that the average speed of the particle on the journey from P to Q is 12 m s^{-1}, find the value of T.

14 Peter and Jack are running in a race. They both start from rest.

Peter accelerates uniformly, then moves at a constant speed, V, for 5 seconds and then decelerates uniformly, coming to rest at the finish line.

Jack accelerates uniformly to the same speed, V, and then decelerates immediately, coming to rest at the finish line. He finishes the race x seconds after Peter.

Find the value of x.

15 The diagram shows the velocity–time graph of a particle moving in a straight line.

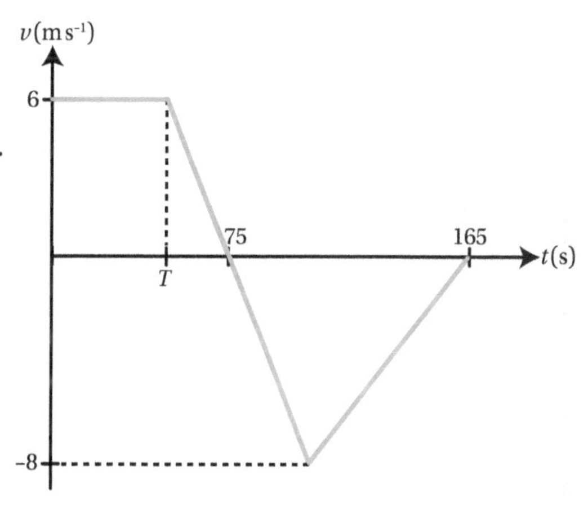

a After 165 seconds the particle returns to the starting point. Find the value of T.

b At what time does the particle have maximum speed?

c Draw the displacement–time graph for this particle.

Section 5: Solving problems in kinematics

You will now look at more complicated problems in which you need to interpret the question and extract relevant information from the given context. You may also need to combine information from graphs and equations.

WORKED EXAMPLE 16.14

A boat moves in a straight line. At $t = 0$ it passes a rock and it is moving with velocity 4 m s^{-1}. Its acceleration, a m s^{-2}, is given by $a = 12 - 6t$.

a Find the time when the boat changes direction.
b Find the maximum velocity of the boat.
c At what time does the boat pass the rock again?

Continues on next page

a $v = \int 12 - 6t \, \mathrm{d}t$

$= 12t - 3t^2 + c$

When $t = 0, v = 4$:

$4 = 0 - 0 + c$

$c = 4$

$\therefore v = 12t - 3t^2 + 4$

First find the velocity.

$12t - 3t^2 + 4 = 0$

The boat changes direction when $v = 0$.

$t = -0.309$ or $t = 4.31$

$\therefore t = 4.31\text{ s}$ (3 s.f.)

Use the quadratic formula or the equation solver on your calculator. Time cannot be negative, so discard $t = -0.309$.

b For maximum velocity $\frac{\mathrm{d}v}{\mathrm{d}t} = 0$:

$\frac{\mathrm{d}v}{\mathrm{d}t} = 12 - 6t$

$0 = 12 - 6t$

$t = 2\text{ s}$

When the velocity is maximum, $\frac{\mathrm{d}v}{\mathrm{d}t} = 0$. Note that this is the same as $a = 0$.

To check this is a maximum:

$\frac{\mathrm{d}^2v}{\mathrm{d}t^2} = -6 \ (< 0)$

So the stationary point is a maximum.

Check that this is a maximum by using the second derivative.

The maximum velocity is:

$v = 12(2) - 3(2)^2 + 4$

$= 16\text{ m s}^{-1}$

Substitute $t = 2$ into the velocity function.

c $s = \int 12t - 3t^2 + 4 \, \mathrm{d}t$

$= 6t^2 - t^3 + 4t + d$

Find the displacement, s, from the rock.

When $t = 0, s = 0$:

$0 = 0 - 0 + 0 + d$

$d = 0$

$\therefore s = 6t^2 - t^3 + 4t$

When $t = 0$, the boat passes the rock, so $s = 0$.

$6t^2 - t^3 + 4t = 0$

$t(-t^2 + 6t + 4) = 0$

$t = 0$ or $-t^2 + 6t + 4 = 0$

$t = 0$ or $t = -0.61$ or $t = 6.61$

The boat passes the rock again when $s = 0$.

$\therefore t = 6.61\text{ s}$ (3 s.f.)

The boat passes the rock again after 6.61 seconds.

You are looking for a positive value of t. (You already know that the boat passes the rock when $t = 0$.)

Worked example 16.14 illustrates some common situations you need to be familiar with.

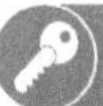

Key point 16.9

- An object is **instantaneously at rest** or changes direction when $v = 0$.
- It reaches maximum/minimum velocity when $a = 0$.
- It returns to the starting point when $s = s_0$.

Common error

Be careful: s_0 is not always 0.

You also need to be able to deal with two-stage problems, in which the equation for velocity changes after a certain time.

WORKED EXAMPLE 16.15

A car accelerates from its parking space and moves in a straight line. Its velocity, v m s^{-1}, at time t seconds, satisfies:

$$v = \begin{cases} 14t - 5t^2 & \text{for } 0 \leqslant t \leqslant 2 \\ 13 - \dfrac{20}{t^2} & \text{for } t > 2 \end{cases}$$

a Find the displacement of the car from the parking space when:

i $t = 2$

ii $t = 5$.

b At what time is the car 35 m from the parking space?

a i $s = \int_0^2 14t - 5t^2 \, dt$

$= \left[7t^2 - \frac{5}{3}t^3\right]_0^2$

$= 14.7$ m (3 s.f.)

From $t = 0$ to $t = 2$, the first velocity equation applies.

Find the displacement by integrating.

ii $\int_2^5 13 - \frac{20}{t^2} \, dt = \left[13t + \frac{20}{t}\right]_2^5$

$= 33$

From $t = 2$ to $t = 5$, use the second velocity equation.

$s = 33 + 14.7$

$= 47.7$ m (3 s.f.)

The car was already 14.7 m from the parking space when $t = 2$.

b

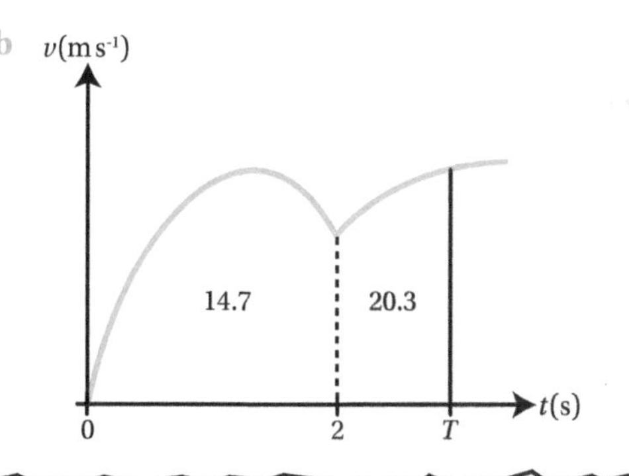

In the first 2 seconds, the car travels 14.7 m from the parking space. Therefore it is 35 m from the space at some time later greater than 2 seconds, so you need to use the second equation.

Continues on next page

$$14.7 + \int_2^T 13 - \frac{20}{t}\,dt = 35$$

The total displacement of 35 m is made up of the 14.7 m travelled in the first 2 seconds, plus the displacement between $t = 2$ and $t = T$.

$$\left[13t + \frac{20}{t}\right]_2^T = 20.3$$

$$\left(13T + \frac{20}{T}\right) - (36) = 20.3$$

$$13T^2 - 56.3T + 20 = 0$$

$$T = 0.390 \text{ or } T = 3.94$$

This can be rearranged into a quadratic equation.

$$\therefore T = 3.94 \text{ s (3 s.f.)}$$

You know that you are looking for $T > 2$ so discard the smaller value.

If the velocity-time graph is made up of straight line segments, it is easy to use areas to answer questions about displacement

WORKED EXAMPLE 16.16

A particle moving in a straight line passes point P when $t = 0$. Its velocity, v m s^{-1}, satisfies:

$$v = \begin{cases} 6.3 + 2.1t & \text{for} \quad 0 \leqslant t \leqslant 6 \\ 29.7 - 1.8t & \text{for} \quad t > 6 \end{cases}$$

where t is measured in seconds.

a Find the displacement of the particle from P when $t = 10$.

b Find the maximum displacement of the particle from P.

c The displacement of point Q from point P is 90 m.

  i Find the first time when the particle passes Q.

  ii How long does it take for the particle to return to Q?

When $t = 0$, $v = 6.3$

When $t = 6$, $v = 18.9$

For the x-intercept:

$v = 29.7 - 1.8t$

$0 = 29.7 - 1.8t$

$t = 16.5$

Since the velocity-time graph is made up of straight line segments, you don't need to use integration to find areas. So start by sketching the graph, labelling all the relevant coordinates.

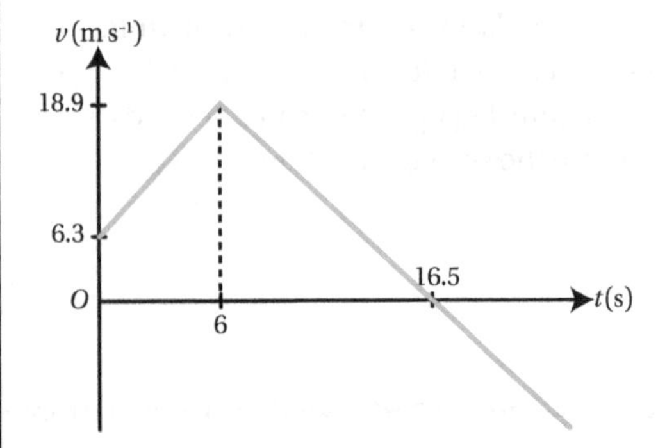

Continues on next page

a When $t = 10$:

$v = 29.7 - 1.8 \times 10 = 11.7$

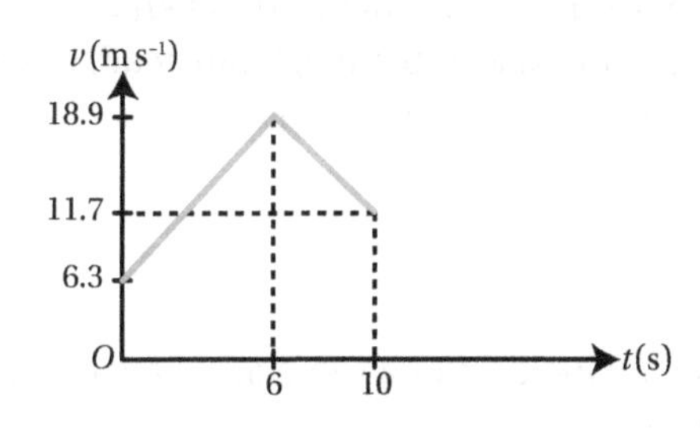

$\frac{6}{2}(6.3 + 18.9) + \frac{4}{2}(18.9 + 11.7) = 136.8\text{ m}$

The displacement is the area under the graph between $t = 0$ and $t = 10$. This is made up of two trapezia.

b

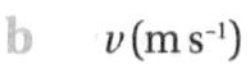

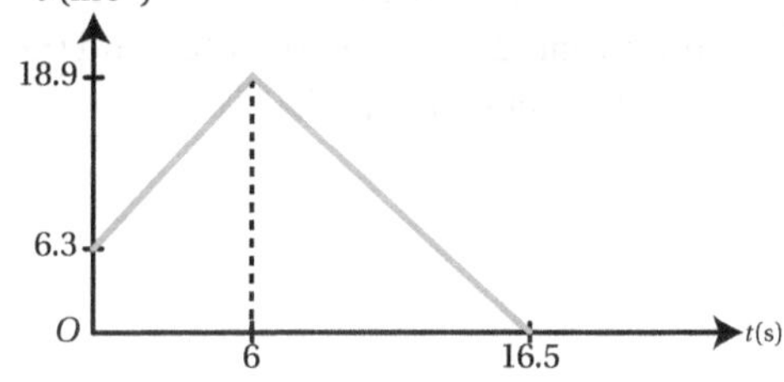

$\frac{6}{2}(6.3 + 18.9) + \frac{10.5 \times 18.9}{2} = 175\text{ m (3 s.f)}$

The particle moves away from P from $t = 0$ to $t = 16.5$, because that's when the velocity is positive. For $t > 16.5$ it moves back towards P.
So the maximum displacement is when $t = 16.5$.
This is equal to the combined areas of the trapezium and the triangle.

c i The displacement in the first 6 seconds:

$\frac{6}{2}(6.3 + 18.9) = 75.6\text{ m}$

The particle is at Q when $t = T_1$.

The displacement from $t = 6$ to $t = T_1$:

$90 - 75.6 = 14.4\text{ m}$

You can see from part **b** that the particle moves 175 m away from P and then comes back. This means that it will pass through Q twice: once in the first 16.5 seconds (when $t = T_1$) and once on the way back (when $t = T_2$).

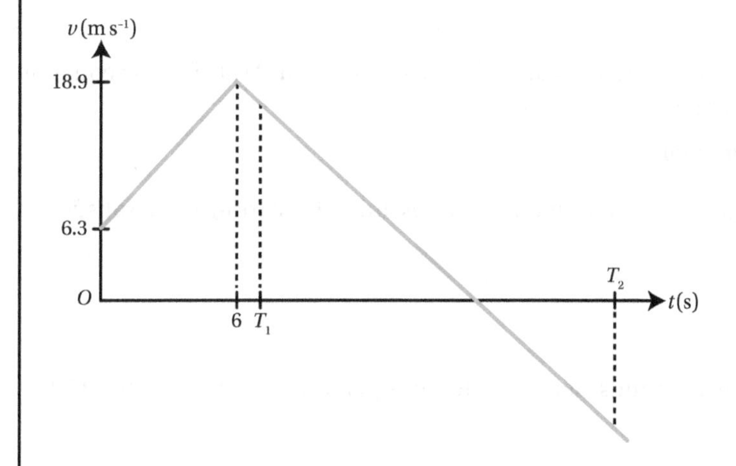

Continues on next page

$$\int_6^{T_1} 29.7 - 1.8t \, dt = 14.4$$

$$\left[29.7t - 0.9t^2\right]_6^{T_1} = 14.4$$

$$(29.7T_1 - 0.9T_1^2) - (145.8) = 14.4$$

$$0.9T_1^2 - 29.9T_1 + 160.2 = 0$$

$$T_1 = 6.79 \text{ or } T_1 = 26.2$$

You need to find T_1 so that the area of the trapezium is 14.4. But you don't know either the base or the height. Instead, you can integrate the equation for v.

$\therefore$ the particle is at Q after 6.79 seconds.

You are looking for a value of T_1 between 6 and 16.5.

ii

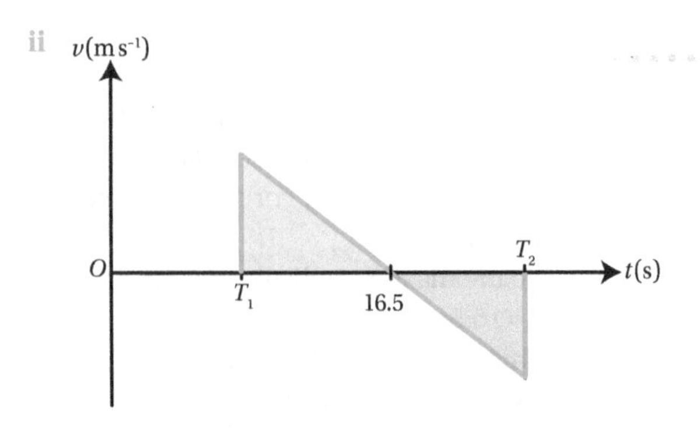

Between T_1 and T_2, the particle stops, changes direction and returns to Q. This means that the areas of the blue and the green triangle are equal.

$$16.5 - T_1 = T_2 - 16.5$$

$$T_2 - 16.5 = 9.71$$

$$T_2 = 26.2$$

The particle returns to Q when $t = 26.2$ seconds.

The solution to part **ii** is the second value from the quadratic equation in part **i**. This is because integrating the velocity equation gives the displacement. If you are in any doubt, check with the graph.

EXERCISE 16E

1. A particle moving in a straight line passes point P when $t = 0$. Its displacement from P satisfies the equation $s = 3.6t - 1.2t^2$, where s is measured in metres and t in seconds.

   Find the time when the particle changes direction.

2. Ellie runs in a straight line. At time t seconds, her displacement, s metres, from the starting point satisfies the equation $s = 1.8t^2 - 0.2t^3$.

   a Show that Ellie starts running from rest.

   b Find her maximum velocity.

3. A toy car's velocity, v m s^{-1}, depends on time, t seconds, according to the equation $v = -0.8t^2 + 1.44t + 9.72$.

   Find:

   a the displacement from the starting point at the moment when the car comes to instantaneous rest

   b the deceleration of the car at this point.

4 A particle moves in a straight line. Its velocity at time t seconds is $v = (11.2t - 3t^2)$ m s^{-1}. The particle is at A when $t = 0$.

How long does it take for it to return to A?

5 The velocity of a car, v m s^{-1}, at time t seconds, satisfies:

$$v = \begin{cases} 4t - 0.5t^2 & \text{for} \quad 0 \leqslant t \leqslant 4 \\ 0.4t^2 - 6.4t + 27.2 & \text{for} \quad t > 4 \end{cases}$$

a Find the acceleration of the car after 5 seconds.

b Find the car's displacement from the starting point when:

i $t = 3$

ii $t = 10$.

6 A dog runs past a tree when $t = 0$ with the speed of 3.7 m s^{-1}. It accelerates for 5 seconds so that its speed satisfies:

$v = (u + 0.4t)$ m s^{-1}

a Write down the value of u.

For the next 5 seconds, the dog decelerates and its speed satisfies $v = \left(\frac{142.5}{t^2}\right)$ m s^{-1}.

b Find the dog's final speed.

c Find the dog's final displacement from the tree.

d At what time is the dog 32 m from the tree?

7 A particle moving in a straight line accelerates from rest. For the first 7 seconds its acceleration at time t seconds satisfies $a = (6.5 - 1.3t)$ m s^{-1}. Subsequently the particle moves with constant acceleration.

Find the equation for the velocity of the particle in terms of t when:

a $0 \leqslant t \leqslant 7$

b $t > 7$.

8 The acceleration of a particle moving in a straight line, a m s^{-2}, satisfies $a = 0.1(t - 5)^2$ for $0 \leqslant t \leqslant 5$ seconds. The particle is initially at rest.

a Explain why the velocity of the particle is positive for $0 \leqslant t < 5$.

b Find the average speed of the particle between $t = 0$ and $t = 5$.

c By sketching the graph of v, show that there is a value of t at which the instantaneous speed equals the average speed.

9 A particle moves in a straight line so that its acceleration at time t seconds is given by $a = 3t^2 - 14t + 10$ (measured in m s^{-2}). The particle starts from rest when $t = 0$.

a Find the equation for the velocity of the particle in terms of t.

b Find:

i the maximum velocity

ii the maximum speed of the particle in the first 5 seconds.

10 A particle is moving in a straight line, starting from point A when $t = 0$. Its velocity, in m s^{-1}, is given by:

$$v = \begin{cases} 0.01(12t^2 - t^3) & \text{for } 0 \leqslant t \leqslant 10 \\ 10 - 0.8t & \text{for } 10 < t < 12.5 \end{cases}$$

where t is measured in seconds.

a Find the distance travelled by the particle in the first 10 seconds.

b After how long is the particle 17 m from A?

11 An object moves in a straight line with velocity $v = 20t^2 - t^3$. The average velocity over the first T seconds equals the instantaneous velocity after $\frac{T}{2}$ seconds.

Find the value of T.

12 Cars can go over speed bumps at 5 km h^{-1}. For an average car, the maximum acceleration is 1.4 m s^{-2} and the maximum deceleration is 3.2 m s^{-2}.

How far apart should the speed bumps be placed to restrict the maximum speed to 30 km h^{-1}?

Checklist of learning and understanding

- The displacement, velocity and acceleration of a particle moving in a straight line are vectors: they can have positive or negative values.
- The instantaneous velocity and acceleration can be found by differentiating the displacement equations:

$$v = \frac{ds}{dt}, \quad a = \frac{dv}{dt}$$

- Integrating the velocity equation gives the displacement equation; integrating the acceleration equation gives the velocity equation:

$$s = \int v \, dt, \quad v = \int a \, dt$$

- The constant of integration can be found from the initial displacement or velocity.
- The change of displacement between time t_1 and t_2 can be found by using the definite integral:

$$s = \int_{t_1}^{t_2} v \, dt$$

  - If the particle does not change direction (so the velocity is always positive) then this integral gives the total distance travelled.
- On a velocity-time graph:
  - the acceleration is the gradient
  - the distance equals the area between the velocity-time graph and the t-axis.
- The average velocity equals change of displacement divided by time.
- The average speed equals total distance divided by time.

Mixed practice 16

1. A particle accelerates from rest with acceleration $a = (3 - 2t)\text{ m s}^{-2}$. Find the maximum velocity in the first 3 seconds of motion.

   Choose from these options.

   A $\frac{3}{2}$ B $\frac{9}{4}\text{ m s}^{-1}$ C 3 m s^{-1} D 2 m s^{-1}

2. An object moves in a straight line so that its velocity, $v\text{ m s}^{-1}$, is given by the equation $v = 3t^2 - 8t$.

   a Find the acceleration of the object after 2 seconds.

   b Find the equation for the displacement from the initial position after t seconds.

3. A particle moves in a straight line with velocity $v = (3 - t^{-2})\text{ m s}^{-1}$ for $t \geqslant 1$.

   a Find the distance travelled between $t = 1$ and $t = 5$.

   b Hence find the average speed of the particle between $t = 1$ and $t = 5$.

4. The velocity–time graph represents the motion of a particle moving in a straight line.

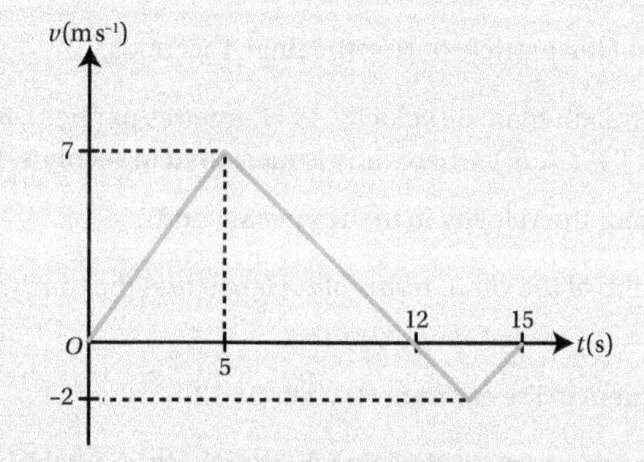

   a Find the acceleration of the particle between $t = 0$ and $t = 5$.

   b State the times when the particle is instantaneously at rest.

   c Find the average speed of the particle for the first 15 seconds.

5. The graph shows how the speed of a cyclist, Hannah, varies as she travels for 21 seconds along a straight horizontal road.

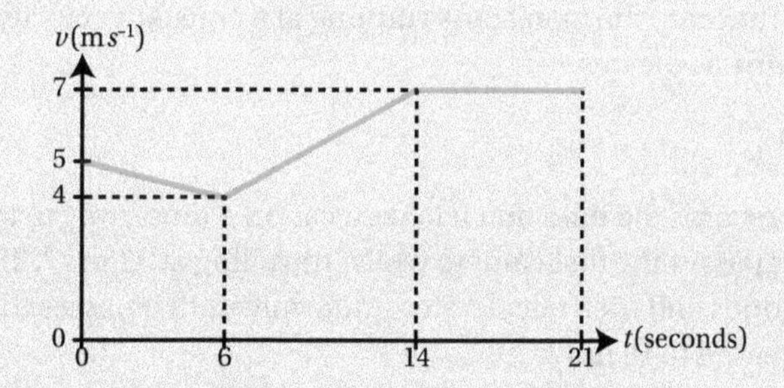

   a Find the distance travelled by Hannah in the 21 seconds.

   b Find Hannah's average speed during the 21 seconds.

[© AQA 2013]

6 A particle moves with velocity $v = (4t^3 - 36t)$ m s^{-1}.

Find the distance travelled in the first 4 seconds.

Choose from these options.

A 12 m **B** 32 m **C** 112 m **D** 130 m

7 The velocity-time graph shows the motion of a particle moving in a straight line. The total distance travelled during the 12 seconds is 360 m.

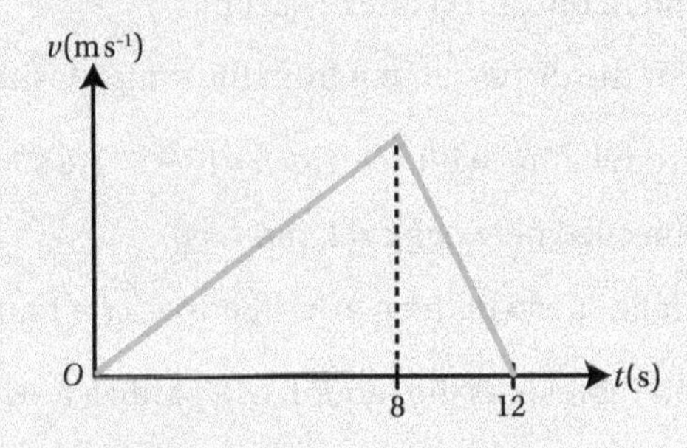

Find the acceleration of the particle during the final 4 seconds.

8 A car travels along a straight road. Its velocity, in kilometres per hour, is given by $v = 40 + 10t - 0.5t^2$ (for $0 \leqslant t \leqslant 20$), where time is measured in seconds. It passes point A when $t = 0$.

a Write an equation for the velocity in metres per second.

b Find the acceleration of the car in terms of t. Hence find the time when the car has maximum velocity.

c Find the displacement of the car from A when $t = 12$.

d The car is modelled as a particle. Explain whether this is a suitable modelling assumption in these situations.

i Finding how long the car takes to overtake a stationary van of length 6.2 m.

ii Finding how long the car take to pass through a tunnel of length 380 m.

9 A bus travels in a straight line. When it passes a man its speed is 8.5 m s^{-1}. It decelerates uniformly until it comes to rest at the bus stop 44.2 m away.

As the bus passes the man, the man starts running at a constant velocity, V m s^{-1}. He arrives at the bus stop at the same time as the bus.

Find the value of V.

10 A pair of cameras records the time that it takes a car on a motorway to travel a distance of 2000 metres. A car passes the first camera whilst travelling at 32 m s^{-1}. The car continues at this speed for 12.5 seconds and then decelerates uniformly until it passes the second camera when its speed has decreased to 18 m s^{-1}.

a Calculate the distance travelled by the car in the first 12.5 seconds.

b Find the time for which the car is decelerating.

c Sketch a speed–time graph for the car on this 2000 metre stretch of motorway.

d Find the average speed of the car on this 2000 metre stretch of motorway.

[© AQA 2011]

11 A particle moves with velocity v m s^{-1}, where:

$$v = \begin{cases} 0.16t^3 - 0.12t + 10.6 & \text{for} \quad 0 \leq t \leq 5 \\ 40 - 2t & \text{for} \quad t > 5 \end{cases}$$

Find the two times when the particle is 200 m away from the starting point.

12 A car is travelling along a road which has a speed limit of 90 km h^{-1}. The speed of cars on the road is monitored via average speed check cameras, which calculate the average speed of a car by measuring how long it takes to travel a specified distance.

The car starts from rest next to one of the cameras. Its velocity is given by $v = \frac{1}{5}t(t-10)^2$, where v is in metres per second and t is in seconds. It comes to rest after 10 seconds.

It stays stationary for T seconds and then starts moving again with a constant acceleration of 3.5 m s^{-2}. The velocity–time graph of the car's motion is shown in the diagram.

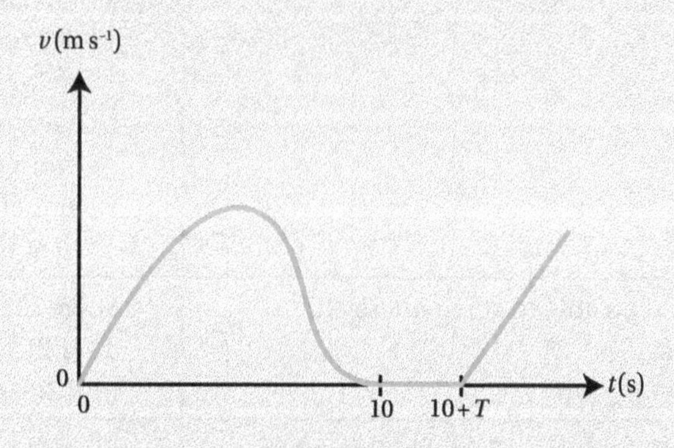

The second camera is positioned 300 m away from the first one.

a Find the time the car takes to reach the second camera after it has started from rest the second time.

b Show that the car's speed exceeded 90 km h^{-1} during both stages of motion.

c The cameras did not detect the car breaking the speed limit. Find the smallest possible value of T.

13 A particle is moving in a straight line so that its displacement from the starting point, s metres, is given by $s = 0.8t^3 - 0.12t^4$. Find the maximum speed of the particle during the first 6 seconds.

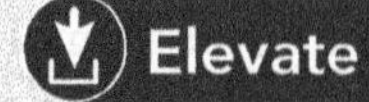

See Extension Sheet 16 for some questions that challenge your visualisation of mechanics problems.

17 Motion with constant acceleration

In this chapter you will learn how to:

- derive equations for motion with constant acceleration
- use constant acceleration equations for horizontal motion
- apply constant acceleration equations to vertical motion under gravity
- solve multi-stage problems.

Before you start...

Chapter 16	You should be able to to use integration to find velocity and displacement from acceleration.	1 A particle moves in a straight line. Its acceleration is given by $a = 2 - 3t^2$. a Given that its velocity at $t = 0$ is $4.2\ \mathrm{m\,s^{-1}}$, find the equation for the velocity at time t. b Given that its initial displacement from point A is 14 m, find the equation for the displacement from A at time t.
Chapter 3	You should be able to solve quadratic equations.	2 Solve these quadratic equations. a $4.2t^2 - 11.5t + 2.6 = 0$ b $12t - 4.9t^2 = 5.2$
Chapter 3	You should be able to find the vertex of a parabola.	3 Find the coordinates of the vertex of the parabola with equation: a $y = 12.2x - 36.1x^2$ b $y = 2.1x^2 - 6.3x + 7$.

When is constant acceleration appropriate?

There are many situations in mechanics in which the acceleration can be modelled as constant, for example, vertical motion under gravity or an object slowing down due to a constant friction force. Techniques from the previous chapter may be used to derive special equations for acceleration, velocity and displacement that apply **only** when the acceleration is constant.

Section 1: Deriving the constant acceleration formulae

If you know that a particle moves with constant acceleration, you can use integration to find equations for its velocity and displacement.

Key point 17.1

For a particle moving with constant acceleration a and initial velocity u:

- the velocity at time t is $v = u + at$
- the displacement from the starting point is $s = ut + \frac{1}{2}at^2$.

These formulae will be given in your formula book.

WORKED EXAMPLE 17.1

Prove that if the acceleration, a, is constant and the initial velocity is u then:

a the velocity at time t is given by $v = u + at$

b the displacement from the starting point is $s = ut + \frac{1}{2}at^2$

a $v = \int a \, dt$ — Link a and v using Key point 16.5

$= at + c$ — a is a constant so $\int a \, dt = a \int 1 \, dt$.

When $t = 0$, $v = u$: — Use the initial condition.

$u = 0 + c$

$c = u$

So $v = at + u$

i.e. $v = u + at$

b $s = \int v \, dt$ — Link s and v using Key point 16.5.

$= \int u + at \, dt$ — $v = u + at$ from part **a**.

$= ut + \frac{1}{2}at^2 + d$

When $t = 0$, $s = 0$ so $d = 0$. — Since s is measured relative to the starting point.

So $s = ut + \frac{1}{2}at^2$

In these equations, the acceleration can be either positive or negative. Negative acceleration is called **deceleration**.

Common error

If an object has a deceleration of 3 m s^{-2} then it has an acceleration of -3 m s^{-2}, so you will need to substitute $a = -3$ into the equations in Key point 17.1.

The two equations in Key point 17.1 can be combined to form another useful equation.

Key point 17.2

$$s = vt - \frac{1}{2}at^2$$

This will be given in your formula book.

This is proved in Exercise 17A, question 3.

You can also derive these formulae by looking at the velocity–time graph. If the acceleration is constant, then the graph is a straight line with gradient a.

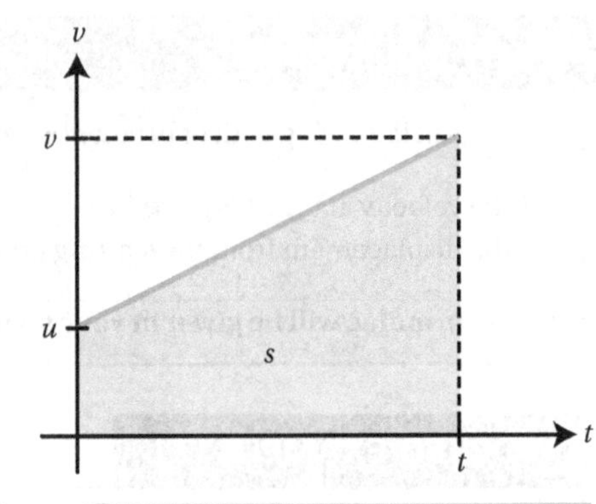

At time t, the velocity is v. Since the gradient of the graph is a:

$$a = \frac{v-u}{t} \Leftrightarrow v = u + at$$

The area under the velocity–time graph is the distance. On this graph, the velocity is always positive so this is the same as the displacement. You can use the formula for the area of a trapezium to find an expression for the displacement.

Key point 17.3

If the initial velocity is u and the velocity at time t is v, then:

$$s = \frac{1}{2}(u+v)t$$

This will be given in your formula book.

Tip

This formula can be interpreted to mean that the displacement is equal to average velocity multiplied by time.

All the formulae derived so far tell you how the velocity and the displacement vary with time. To find an equation not involving time, you need to eliminate t from some of the equations derived.

WORKED EXAMPLE 17.2

Use the equations $v = u + at$ and $s = \frac{1}{2}(u+v)t$ to derive an equation for v^2 in terms of u, a and s.

$v = u + at$

$t = \frac{v-u}{a}$ — Make t the subject of the first equation...

$s = \frac{1}{2}(u+v)t$ — ... and then substitute t into the second to eliminate it.

$= \frac{1}{2}(u+v)\left(\frac{v-u}{a}\right)$

$= \frac{v^2-u^2}{2a}$ — Notice the difference of two squares.

So, $v^2 = u^2 + 2as$ — Rearrange to make v^2 the subject.

Key point 17.4

If the initial velocity is u, when the particle's displacement from the starting point is s, its velocity satisfies:

$$v^2 = u^2 + 2as$$

This will be given in your formula book.

Notice that this equation gives two possible values of v since you can take either the positive or the negative square root. This can be clearly seen on the displacement–time graph.

If the particle moves away from the starting point and then back again, it will pass through the point A twice (at times labelled t_1 and t_2 on the diagram). You can see that the gradient at those two points is numerically the same size but has opposite signs, meaning that the particle has the same speed each time but its velocities are in opposite directions.

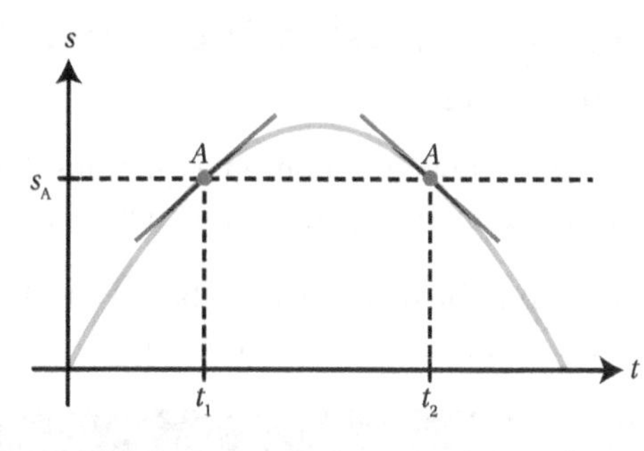

EXERCISE 17A

1. Use the formulae $v = u + at$ and $s = \frac{1}{2}(v+u)t$ to prove that $s = ut + \frac{1}{2}at^2$.

2. The diagram shows a velocity-time graph for a particle moving with constant acceleration, a. Its speed increases from u to v in time t.

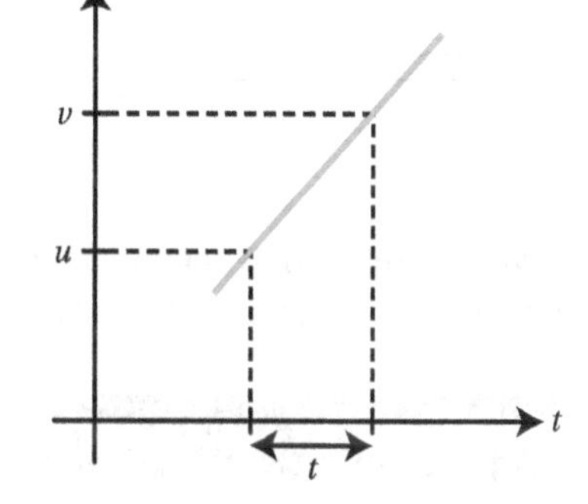

   a Use the graph to explain why $a = \frac{v-u}{t}$.

   b By splitting the area under the graph into a rectangle and a triangle, show that the distance travelled during time t is given by $s = ut + \frac{1}{2}at^2$.

3. a Use the formulae $s = ut + \frac{1}{2}at^2$ and $v = u + at$ to derive the formula $s = vt - \frac{1}{2}at^2$.

   b A particle moves with constant acceleration $3.1\ \text{m s}^{-2}$. It travels 300 m in the first 8 seconds. Find its speed at this time.

4. Use the formulae $v = u + at$ and $s = ut + \frac{1}{2}at^2$ to derive the formula $v^2 = u^2 + 2as$.

5. a If the acceleration is proportional to time, so that $a = kt$, find an expression for v in terms of u, k and t.

   b Hence prove that:

   i $s = ut + \frac{1}{6}kt^3$ ii $s = \frac{1}{3}(2u+v)t$ iii $2(v-u)(2u+v)^2 = 9ks^2$

6. The diagram shows velocity-time graphs for two particles. The initial speed of each particle is u. One particle moves with constant acceleration a.

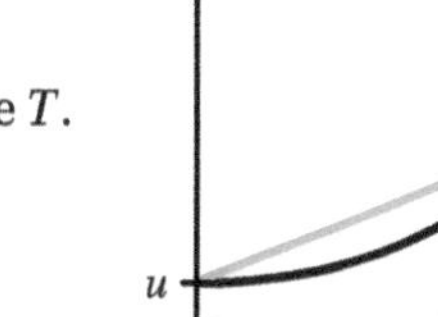

   a For this particle, write down an expression for the velocity at time T.

   The other particle has velocity given by $v = u + at^2$.

   b Find the time when the two particles have the same velocity.

   c Given that the two particles travel the same distance in time T, find the value of T.

Section 2: Using the constant acceleration formulae

You now know five different equations that can be used to solve problems involving motion with constant acceleration.

Selecting which one to use is based on which of the five quantities s, u, v, a and t are involved in the question.

Quantities involved	Equation to use
u, v, a, t	$v = u + at$
s, u, a, t	$s = ut + \frac{1}{2}at^2$
s, u, v, t	$s = \frac{1}{2}(u+v)t$
s, u, v, a	$v^2 = u^2 + 2as$
s, v, t, a	$s = vt - \frac{1}{2}at^2$

If you write down what you are given in the question and what you are trying to find, you can then select the most useful equation.

WORKED EXAMPLE 17.3

A car moves with constant acceleration. When $t = 0$ it passes a junction, where its velocity is $8.2\ \text{m s}^{-1}$. It passes the next junction, 320 m away, 24 seconds later. Find the car's velocity as it passes the second junction.

$s = 320$
$u = 8.2$
$t = 24$
$v = ?$

Write down what you are given and what you are trying to find.

$$s = \frac{1}{2}(u+v)t$$
$$320 = \frac{1}{2}(8.2+v)(24)$$

Looking at the table, s, u, v, t feature in the third equation: use $s = \frac{1}{2}(u+v)t$.

$$8.2 + v = \frac{640}{24}$$
$$8.2 + v = 26.7$$
$$v = 18.5\ \text{m s}^{-1}$$

Now solve the equation to find the value of v.

WORKED EXAMPLE 17.4

A car decelerates uniformly from 26.3 m s^{-1} to 16.2 m s^{-1}, covering a distance of 240 m. Find the deceleration.

$s = 240$
$u = 26.3$
$v = 16.2$
$a = ?$

Write down what you are given and what you are trying to find.

Make sure you get u and v the right way round!

$$v^2 = u^2 + 2as$$
$$16.2^2 = 26.3^2 + 2a(240)$$

s, u, v, a feature in the fourth equation: use $v^2 = u^2 + 2as$.

$$480a = 16.2^2 - 26.3^2$$
$$480a = -429.25$$
$$a = -0.894 \text{ m s}^{-2}$$

So the deceleration is 0.894 m s^{-2}.

You are asked to find the deceleration, which is positive.

WORKED EXAMPLE 17.5

Walking down the street, Imogen sees a bus at the bus stop 25 m away. She starts accelerating uniformly at 0.9 m s^{-2} and reaches the bus stop 4 seconds later. Find her velocity when she arrives at the bus stop.

$s = 25$
$a = 0.9$
$t = 4$
$v = ?$

Write down what you are given and what you are trying to find.

$$s = ut + \frac{1}{2}at^2$$
$$25 = 4u + \frac{1}{2}(0.9)(4^2)$$

This problem could also be solved using the equation in Key point 17.2.

$$4u = 25 - 7.2$$
$$4u = 17.8$$
$$u = 4.45 \text{ m s}^{-1}$$

$$v = u + at$$
$$= 4.45 + (0.9)(4)$$
$$= 8.05 \text{ m s}^{-1}$$

Now use u, a and t to find v (from the first equation).

EXERCISE 17B

1 Choose an appropriate formula to answer each question.

a Find the values of u and s when:

i $a = 2.4 \text{ m s}^{-1}, v = 18 \text{ m s}^{-1}, t = 6 \text{ s}$

ii $a = 0.6 \text{ m s}^{-2}, v = -21 \text{ m s}^{-1}, t = 3.5 \text{ s}$.

b Find the values of t and v when:

i $u = 13 \text{ m s}^{-1}, a = -1.2 \text{ m s}^{-2}, s = 60 \text{ m}$

ii $u = 20 \text{ m s}^{-1}, a = -3 \text{ m s}^{-2}, s = 40 \text{ m}$.

c Find the values of a and t when:

i $u = 12\text{ m s}^{-1}, v = 6\text{ m s}^{-1}, s = 120\text{ m}$ **ii** $u = -3\text{ m s}^{-1}, v = 16\text{ m s}^{-1}, s = 120\text{ m}$.

d Find the values of a and s when:

i $u = 5\text{ m s}^{-1}, v = -6\text{ m s}^{-1}, t = 6\text{ s}$ **ii** $u = 8\text{ m s}^{-1}, v = -5\text{ m s}^{-1}, t = 10\text{ s}$.

2 A car accelerates uniformly from rest to 12.45 m s^{-1} in 6.5 seconds. Calculate:

a the acceleration **b** the distance travelled during this time.

3 A cyclist is travelling at the speed of 4.2 m s^{-1}. She accelerates uniformly at 0.3 m s^{-2} for 7 seconds. Find:

a her final speed **b** the distance she travels in the 7 seconds.

4 A particle reduces its speed from 16.3 m s^{-1} to 7.5 m s^{-1} while travelling 120 m.

a Find the constant acceleration of the particle.

b Find the distance the particle would travel in another 5 seconds.

5 A particle starts from rest and accelerates uniformly at 2.5 m s^{-2}. How long will it take to travel 250 m?

6 A particle reduces its speed from 20 m s^{-1} to 8.2 m s^{-1} while travelling 100 m. Assuming it continues to move with the same constant acceleration, how long will it take to travel another 20 m?

7 A particle moves with constant deceleration of 3.6 m s^{-2}. It travels 350 m while its speed halves. Find the time it takes to do this.

8 A car reduces its speed from 18 m s^{-1} to 9 m s^{-1} while travelling 200 m. Assuming the car continues to move with the same uniform acceleration, how much further will it travel before it stops?

9 **a** A particle moves in a straight line with constant acceleration $a = -3.4\text{ m s}^{-2}$. At $t = 0$ s its velocity is $u = 6\text{ m s}^{-1}$. Find its maximum displacement from the starting point.

b Explain why this is not the maximum distance from the starting point.

Section 3: Vertical motion under gravity

When an object is thrown in the air, the force of gravity acts on it: if the object is moving upwards gravity slows it down; if it is moving downwards gravity speeds it up.

The force of gravity produces an acceleration. This acceleration is the same regardless of the mass of the object, and is called **gravitational acceleration** or **acceleration of freefall**. On Earth, this value is denoted by g and is approximately 9.8 m s^{-2}, although this value varies slightly with geographical position and height above sea level.

The motion of the object may also be affected by **air resistance**: the larger the surface area of the object, the more resistance it would experience.

You will be told the value of g to use in any question – normally 9.81 m s^{-2} (3 s.f.), 9.8 m s^{-2} (2 s.f.) or 10 m s^{-2} (1 s.f.). The degree of accuracy (the number of significant figures) of your answer **must** then reflect the accuracy of g and of other values given in the question. Do **not** give your answer to more significant figures than appear in the question.

However, you can make modelling assumptions.

Key point 17.5

It is assumed that:

- g is constant.
- air resistance can be ignored.

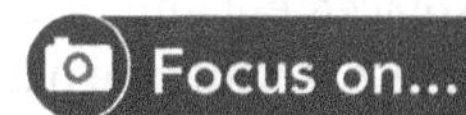

Focus on...

See Focus on ... Modelling 3 for an investigation of these modelling assumptions.

With these assumptions, an object moving vertically under gravity moves in a straight line with constant acceleration. You can therefore use the equations from the previous section to calculate its velocity and displacement.

WORKED EXAMPLE 17.6

A small ball is thrown straight downwards from a window 4.6 m above ground, with an initial velocity of 1.2 m s^{-1}. Air resistance can be ignored. Use $g = 9.8\,\text{m s}^{-2}$, giving your final answer to an appropriate degree of accuracy.

a How long does it take for the ball to reach the ground?

b How would your answer change if air resistance were included?

a

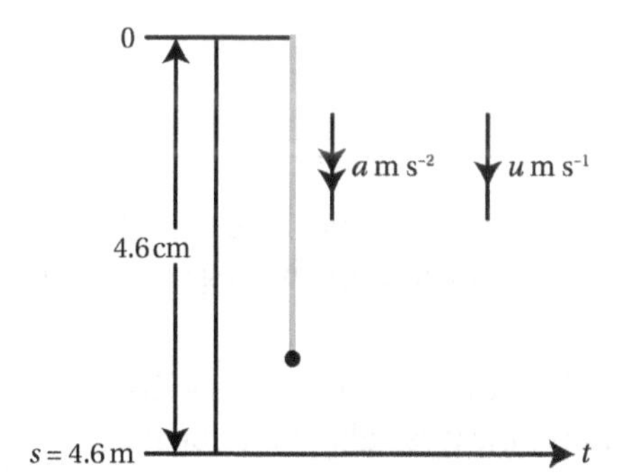

Draw a diagram to show the directions of displacement, velocity and acceleration. Since the motion is only downwards, take this to be the positive direction.

$s = 4.6$
$u = 1.2$
$a = 9.8$
$t = ?$

Write down what you are given and what you are trying to find. Since the displacement is measured downwards, the ground is at $s = +4.6$ m.

$$s = ut + \frac{1}{2}at^2$$
$$4.6 = 1.2t + 4.9t^2$$
$$4.9t^2 + 1.2t - 4.7 = 0$$
$$t = 0.86 \text{ or } t = -1.1$$

Select the equation involving s, u, a, and t: use $s = ut + \frac{1}{2}at^2$.

$$\therefore t = 0.86\text{ s}$$

Since all the values in the question (s, u and g) are given to 2 s.f. this is an appropriate degree of accuracy for the answer.

b It would take longer as air resistance would decrease the acceleration.

In Worked example 17.6 the positive direction was taken to be downward, so the acceleration, velocity and displacement were all positive. The displacement–time graph is a positive parabola with equation $s = 4.9t^2 + 1.2t$. Only the part of the parabola from $t = 0$ to $t = 0.86$ is relevant in this situation. You can see that the velocity increases as the ball approaches the ground.

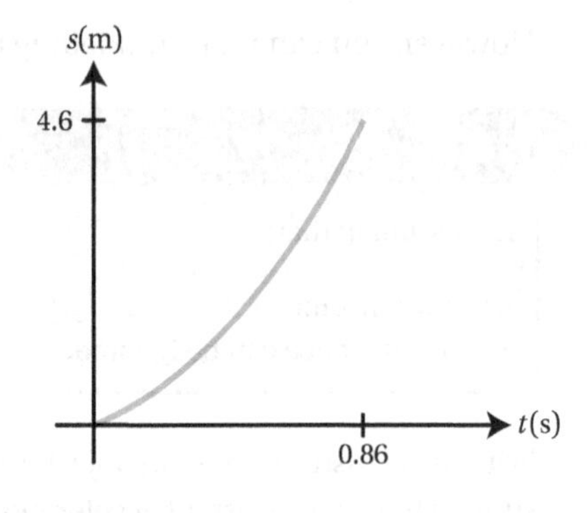

If an object is thrown vertically upwards, it makes more sense to measure displacement in the upward direction. In this case the initial velocity is positive but the acceleration, which always acts downwards, is negative: $a = -9.8\ \text{m s}^{-2}$. The object will eventually stop and start moving downwards; its velocity will become negative. This can be seen on the velocity–time graph, which has equation $v = u - 9.8t$.

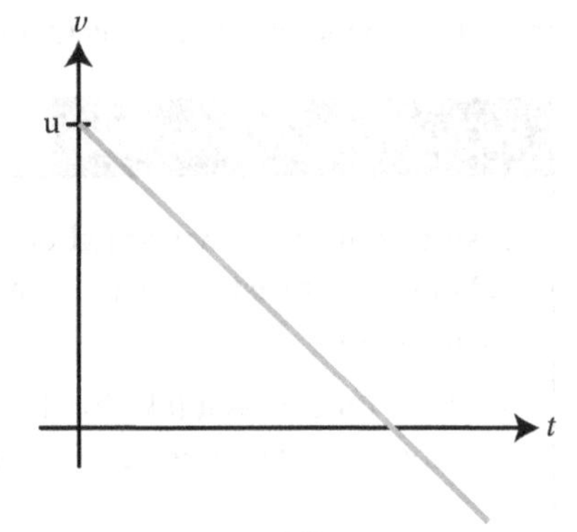

Key point 17.6

When solving problems involving vertical motion:

- take the positive direction for velocity and displacement to be in the direction of initial velocity
- remember that the acceleration is negative if the initial velocity is upwards and positive if it is downwards.

WORKED EXAMPLE 17.7

A stone is thrown upwards, with a velocity of $6\ \text{m s}^{-1}$, from a platform 1.2 m above ground. Find the velocity of the stone when it hits the ground. Use $g = 10\ \text{m s}^{-2}$, giving your final answer to an appropriate degree of accuracy.

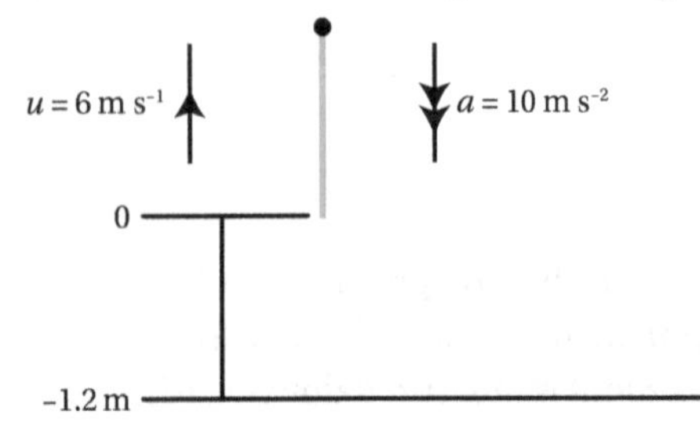

Draw a diagram to show the directions. Since the initial direction of motion is upwards, take this to be the positive direction.

$s = -1.2$
$u = 6$
$a = -10$
$v = ?$

Write down what you are given and what you are trying to find.

The displacement is measured upwards, so the ground is at $s = -1.2$ m.

$v^2 = u^2 + 2as$
$v^2 = 6^2 + 2(-10)(-1.2)$
$v^2 = 60$
$v = \pm 7.75$

Select the equation involving s, u, v and a: use $v^2 = u^2 + 2as$.

The velocity of the stone is $-8\ \text{m s}^{-1}$ (1 s.f.).

Before hitting the ground the stone is moving downwards, so its velocity is negative.

Since both u and g are only given to 1 s.f. you shouldn't give your final answer to any greater degree of accuracy than 1 s.f.

The displacement–time graph for the motion of the stone is a negative parabola with equation $s = -5t^2 + 6t$ (which comes from using $s = ut + \frac{1}{2}at^2$). Its vertex corresponds to the highest point reached by the stone. Note that the value of s at this point is not the maximum height of the stone above ground, because the displacement is measured from the platform.

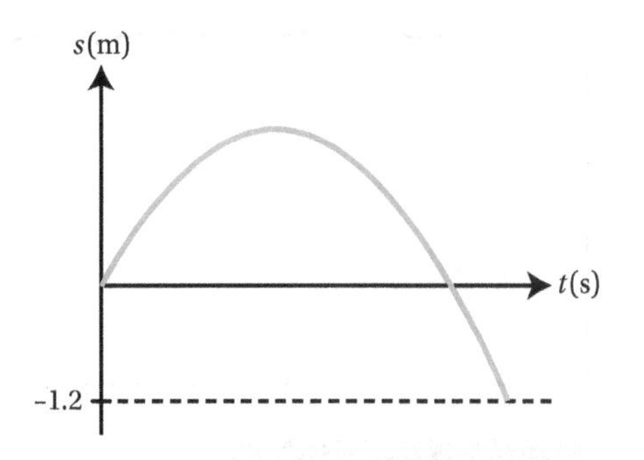

The part of the graph below the horizontal axis represents the motion of the stone below the platform until it hits the ground.

Key point 17.7

The maximum height occurs when $v = 0$.

Elevate

See Support Sheet 17 for another example of vertical motion and further practice questions.

WORKED EXAMPLE 17.8

For the stone from Worked example 17.7, find:

a its greatest height above the ground
b how long it takes the stone to reach the highest point
c for how long it is falling.

a $u = 6$
$v = 0$
$a = -10$
$s = ?$

Write down what you are given and what you are trying to find.

At the greatest height, $v = 0$.

$v^2 = u^2 + 2as$
$0 = 6^2 + 2(-10)s$
$20s = 36$
$s = 1.8$

You can use the same equation as in Worked example 17.7.

$h = s + 1.2$
$= 3\text{ m (1 s.f.)}$

The maximum height of the stone is 3 m.

Take h as the height above the ground. s is the displacement from the starting point, which is itself 1.2 m above the ground.

b $u = 6$
$v = 0$
$a = -10$
$t = ?$

You now want to find t rather than s.

$v = u + at$
$0 = 6 - 10t$
$t = 0.6\text{ s}$

Select the equation involving u, v, a and t: use $v = u + at$.

c $s = 3$
$u = 0$
$a = 10$
$t = ?$

The stone is falling from rest at the maximum point until it hits the ground 3 m below. Since the motion is only downwards now, take that to be the positive direction.

Continues on next page

$s = ut + \frac{1}{2}at^2$ — Select the equation involving s, u, a, t.

$3 = 0 + 5t^2$

$t^2 = \frac{3}{5}$

$\therefore t = 0.8\text{ s (1 s.f.)}$ — Again, 1 s.f. is an appropriate degree of accuracy for the answer.

WORK IT OUT 17.1

A stone is thrown vertically upwards from the top of a cliff with speed 10 m s^{-1}. It hits the sea below with speed 33 m s^{-1}. Find the time taken. Use $g = 9.8\text{ m s}^{-2}$, giving your final answer to an appropriate degree of accuracy.

Which is the correct solution? Identify the errors made in the incorrect solutions.

Solution 1	Solution 2	Solution 3
$u = 10$ $v = 33$ $a = 9.8$ $v = u + at$ $33 = 10 + 9.8t$ $23 = 9.8t$ $t = \frac{23}{9.8} = 2.35$ $t = 2.35\text{ s (3 s.f.)}$	$u = 10$ $v = -33$ $a = -9.8$ $v = u + at$ $-33 = 10 - 9.8t$ $-43 = -9.8t$ $t = \frac{43}{9.8} = 4.4$ $t = 4.4\text{ s (2 s.f.)}$	Highest point when $v = 0$: $u = 10$ $v = 0$ $a = -9.8$ $v = u + at$ $0 = 10 - 9.8t$ $-10 = -9.8t$ $t = \frac{10}{9.8} = 1.0$ Same time to come down: $\therefore t = 1.0 \times 2$ $t = 2.0\text{ s (2 s.f.)}$

WORKED EXAMPLE 17.9

An object is thrown upwards, from ground level, with initial velocity u.

a Find an expression in terms of u and g for the time it takes to:

  i reach the highest point

  ii return to the ground.

b Find the speed of the object when it hits the ground.

a i $u = u$, $v = 0$, $a = -g$, $t = ?$ — Write down what you are given and what you are trying to find. At the greatest height, $v = 0$.

$v = u + at$ — Select the equation involving u, v, a and t: use $v = u + at$.

$0 = u + (-g)t$

$t = \frac{u}{g}$

Continues on next page

ii $s = 0$
$u = u$
$a = -g$
$t = ?$

When the object returns to the ground, $s = 0$.

$s = ut + \frac{1}{2}at^2$

Select the equation involving s, u, a and t: use $s = ut + \frac{1}{2}at^2$.

$0 = ut - \frac{1}{2}gt^2$

$0 = t\left(u - \frac{1}{2}gt\right)$

Factorise.

$t = 0$ or $t = \frac{2u}{g}$

$\therefore t = \frac{2u}{g}$

$t = 0$ is the starting point, so you need the other value of t.

b $v = u + at$

$= u + (-g)\left(\frac{2u}{g}\right)$

You now know the time when the object hits the ground, so you can use it to find the velocity.

$= u - 2u$

$= -u$

$\therefore$ the speed is u.

Speed is the magnitude of velocity.

From these examples, you should notice two results.

Key point 17.8

For an object thrown upwards from ground level

- The time taken to return to the ground is twice the time taken to reach the highest point. Hence the time spent going up is the same as the time spent coming down.
- The speed of the object when it hits the ground equals its initial speed.

This can also be seen from the displacement-time graph and the velocity-time graph, which are both symmetrical.

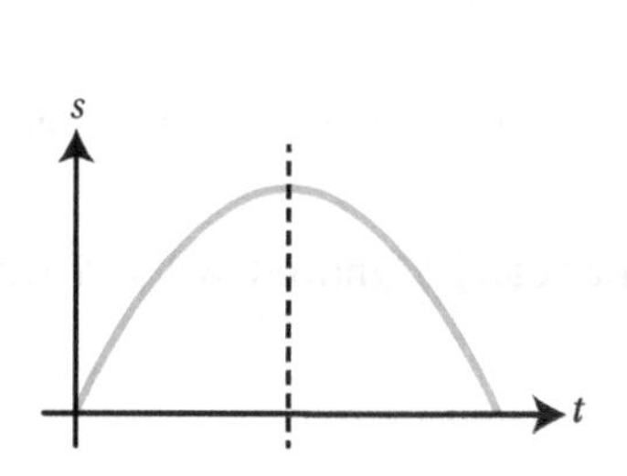

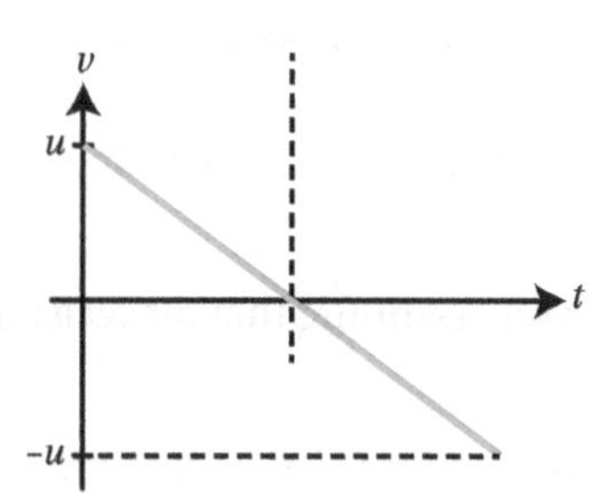

Common error

The speed of a particle when it hits the ground is not zero. It is the speed at the instant it hits the ground that is required.

EXERCISE 17C

In this exercise use $g = 9.8\text{ m s}^{-2}$, unless instructed otherwise, giving your final answers to an appropriate degree of accuracy.

1 A ball is thrown vertically downwards with the given initial speed. Assume the air resistance can be ignored, and that the ball does not reach the ground. Find the speed and the distance travelled by the ball at the specified time.

a **i** Initial speed $= 19\text{ m s}^{-1}$, $t = 0.3$ seconds   **ii** Initial speed $= 23\text{ m s}^{-1}$, $t = 0.8$ seconds

b **i** Initial speed $= 0\text{ m s}^{-1}$, $t = 3.1$ seconds   **ii** Initial speed $= 0\text{ m s}^{-1}$, $t = 6$ seconds

2 A ball is thrown vertically upwards with the given initial speed. Assume that air resistance can be ignored. Find the magnitude and direction of the velocity, and the height relative to the projection point, at the specified time.

a **i** Initial speed $= 9\text{ m s}^{-1}$, $t = 2.5$ seconds   **ii** Initial speed $= 12\text{ m s}^{-1}$, $t = 0.8$ seconds

b **i** Initial speed $= 35\text{ m s}^{-1}$, $t = 3.1$ seconds   **ii** Initial speed $= 20\text{ m s}^{-1}$, $t = 1.1$ seconds

3 In this question use $g = 10\text{ m s}^{-2}$, giving your final answers to an appropriate degree of accuracy.

A stone is dropped from rest from a height of 40 m. Find its velocity when it hits the ground and the length of time for which the stone is in motion.

Tip

Released from rest means that $u = 0$.

4 In this question use $g = 9.81\text{ m s}^{-2}$, giving your final answers to an appropriate degree of accuracy.

A ball is thrown vertically upwards from ground level with a speed of 18.5 m s^{-1}.

a Assuming that air resistance can be ignored, find:

i how long the ball takes to reach the height of 15 m

ii the speed of the ball at this time.

b Explain how your answers to parts **a i** and **ii** would change if air resistance were included.

5 An object is projected vertically upwards with velocity $u\text{ m s}^{-1}$. When it has reached the height of 5.6 m its velocity is 3.5 m s^{-1}. Find the value of u.

6 A ball is projected vertically upwards from the top of a 15 m tall cliff with a velocity of 28 m s^{-1}.

a Find the maximum height of the ball above ground.

b How long does the ball take to reach the ground and how fast is it going when it hits it?

7 A bunch of flowers is projected vertically upwards towards a window 8.3 m above the point from which it is thrown.

a Given that the initial speed is 6.5 m s^{-1}, and assuming that air resistance can be ignored, will it reach the window?

b Find the minimum projection speed required for the flowers to reach the window.

c How would your answer to part **b** change if the air resistance were included?

8 A ball is thrown vertically upwards from a window 12 m above ground level. The initial velocity of the ball is 16 m s^{-1}.

a After what time will the ball reach the highest point?

b How long does the ball take to fall to the ground?

c Sketch the velocity–time graph for the ball's motion.

9 A particle is projected upwards from ground level with initial speed u. Air resistance can be ignored.

a Find an expression, in terms of u, for the time it takes the particle to return to ground level.

b Which of these graphs shows the **distance** travelled by the particle as a function of time? Give reasons for your answer.

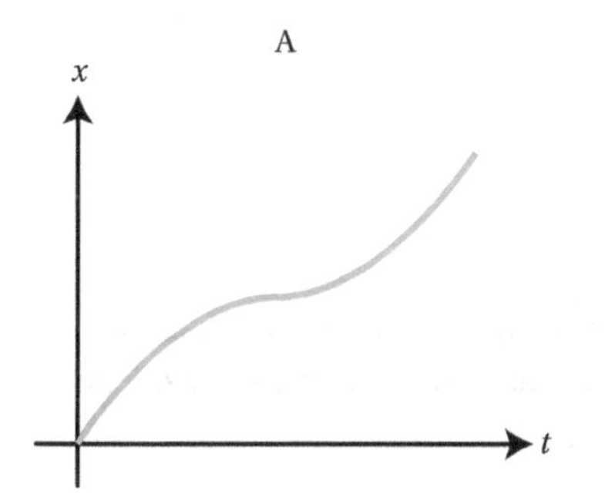

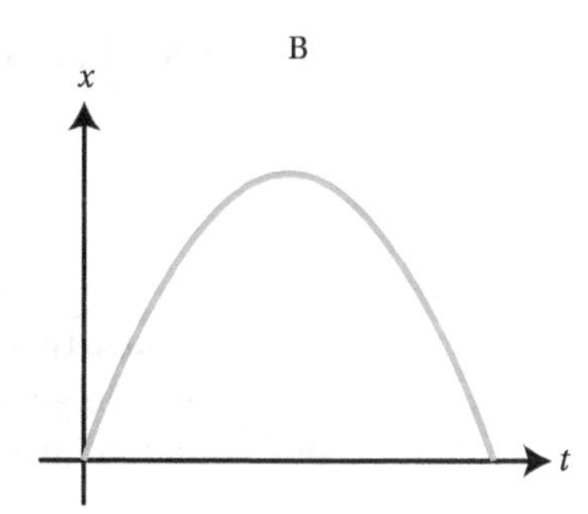

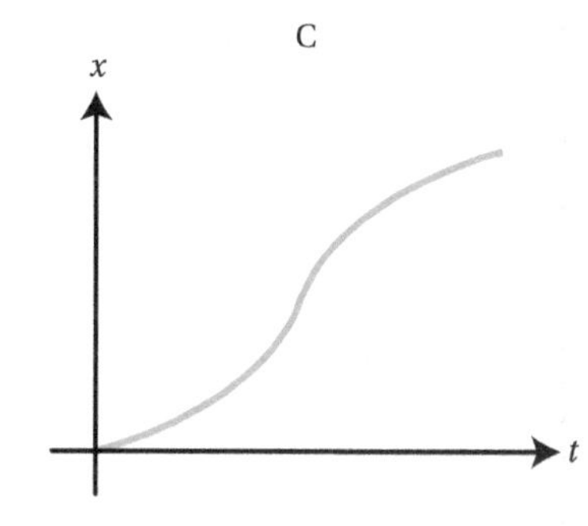

10 In this question use $g = 9.81\text{ m s}^{-2}$, giving your final answers to an appropriate degree of accuracy.

A small stone is projected vertically upwards from the top of a cliff, with speed $u\text{ m s}^{-1}$. It hits the ground with a speed of 18.2 m s^{-1}. A second small stone is also projected vertically upwards with speed $u\text{ m s}^{-1}$ from a nearby cliff which is 4.5 m higher. It takes 2.97 s to hit the ground. Find the value of u and the height of the first cliff.

Section 4: Multi-stage problems

Sometimes you may have information about two separate stages of the motion, which may lead to simultaneous equations.

WORKED EXAMPLE 17.10

Three cameras are positioned on a straight road. The distance between the first and the second camera is 50 m and the distance from the second to the third camera is 30 m.

A car passes the first camera with speed $u\text{ m s}^{-1}$ and immediately starts braking. It passes the second camera 3.1 seconds later and the third camera 2.3 seconds after that. Assuming the deceleration remains constant, find the value of u.

From the first to the second camera:

$s = 50$

$u = u$

$a = a$

$t = 3.1$

Write down the information you have for the first stage of motion.

Choose u and a (rather than v) as these will be the same for the journey from the first to third camera.

Continues on next page

$s = ut + \frac{1}{2}at^2$

$50 = 3.1u + \frac{1}{2}a(3.1^2)$

$50 = 3.1u + 4.805a$

Use $s = ut + \frac{1}{2}at^2$.

There are two unknowns, but you can form a second equation and solve simultaneously for these.

From the first to the third camera:

$s = 80$

$u = u$

$a = a$

$t = 5.4$

Now write down the information for the whole journey.

u is the same as in the first stage; t and s are still measured from the first camera.

$s = ut + \frac{1}{2}at^2$

$80 = 5.4u + \frac{1}{2}a(5.4^2)$

$80 = 5.4u + 14.58a$

Use $s = ut + \frac{1}{2}at^2$ again.

$\therefore \begin{cases} 3.1u + 4.805a = 50 \\ 5.4u + 14.58a = 80 \end{cases}$

Solving simultaneously:

$u = 18\text{ m s}^{-1}$ (2 s.f.)

You now have two simultaneous equations. You can solve them on your calculator, or substitute for a from one equation into the other.

In Worked example 17.10 the acceleration was assumed to remain constant throughout the 5.4 seconds. However, there are many situations in which the acceleration changes part-way through the motion: for example, changing gears while driving a car might change its acceleration. Because the acceleration is not constant, you need to consider the two stages of motion separately.

Tip

Note that in Worked example 17.10 it would have been unhelpful to consider the journey from the second to third camera; although the value of a would have been the same as in the first stage of the motion, the value of u wouldn't so there would have been a third unknown.

WORKED EXAMPLE 17.11

A car starts from rest and accelerates at a constant rate of 2.7 m s^{-2} for 6 seconds. It then changes to a higher gear and accelerates for 10 seconds, reaching a speed of 28.2 m s^{-1}. Find the acceleration for the second stage of the motion.

First stage:

$u = 0$

$a = 2.7$

$t = 6$

The final speed from the first stage is the same as the initial speed for the second stage (which you need).

Continues on next page

$v = u + at$
$= 0 + 2.7 \times 6$
$= 16.2\,\text{m s}^{-1}$

Use $v = u + at$.

Second stage:
$u = 16.2$
$v = 28.2$
$t = 10$
$a = ?$

Using the value of v from the first part as u in the second, you now have enough information to find a.

$v = u + at$
$28.2 = 16.2 + 10a$
$a = 1.2\,\text{m s}^{-2}$

Sometimes it is much easier to use the velocity–time graph rather than forming equations for the two stages.

WORKED EXAMPLE 17.12

A dog accelerates from rest, reaching a speed of V m s^{-1}. It then decelerates back to rest. If the dog ran the total distance of 18 m in 6 seconds, find the value of V.

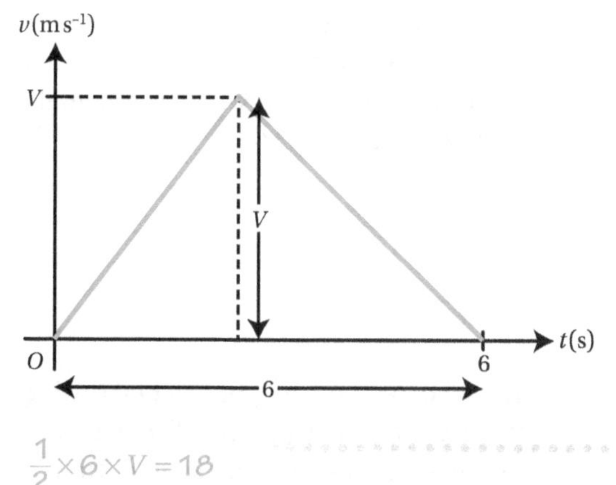

As you only know the total time and total distance, use a velocity–time graph to form a single equation for the entire period of motion.

Note that if you wanted the time for which the dog was accelerating, you would need equations for the two stages.

$\frac{1}{2} \times 6 \times V = 18$
$3V = 18$
$V = 6\,\text{m s}^{-1}$

The area under the graph gives the distance travelled.

Many two-stage problems involve vertical motion. You need to be very careful about the direction of velocity and acceleration.

WORKED EXAMPLE 17.13

A toy rocket with an engine starts from rest at ground level and moves vertically upwards with an acceleration of 3.6 m s^{-2}. After the three seconds the engine is turned off and the rocket moves freely under gravity. Find:

a the greatest height reached by the rocket

b the total time the rocket spends in the air.

Use $g = 9.8$ m s^{-2}, giving your final answers to an appropriate degree of accuracy.

a

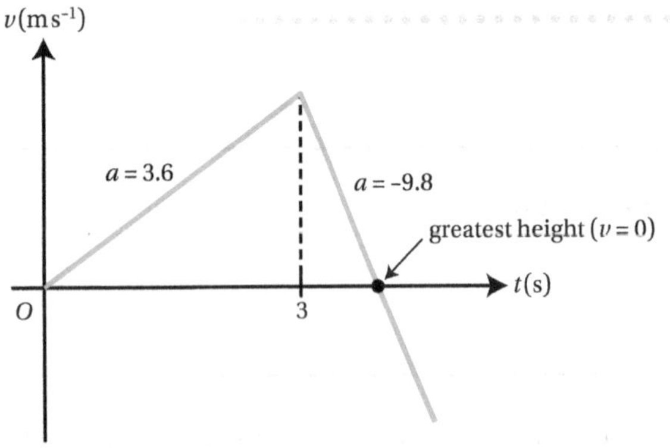

v(m s^{-1})

$a = 3.6$

$a = -9.8$

greatest height ($v = 0$)

t(s)

O

3

There are two stages: first the rocket has upward acceleration of 3.6 m s^{-2} and then it has downward acceleration of 9.8 m s^{-2}.

First stage:

$u = 0$

$a = 3.6$

$t = 3$

$v = ?$

$$v = u + at$$
$$= 0 + 3.6 \times 3$$
$$= 10.8 \text{ m s}^{-1}$$

The greatest height is reached during the second stage. You need to find the initial velocity for that stage, which is the final velocity for the first stage.

$$s = ut + \frac{1}{2}at^2$$
$$= 0 + \frac{1}{2}(3.6)(3^2)$$
$$= 16.2 \text{ m}$$

You also need the height reached during the first stage, because the displacement for the second stage is measured from there.

Second stage:

$u = 10.8$

$v = 0$

$a = -9.8$

$s = ?$

$$v^2 = u^2 + 2as$$
$$0 = 10.8^2 + 2(-9.8)s$$
$$s = 5.95 \text{ m}$$

The initial velocity is upwards but the acceleration is now downwards so it is negative.

u for this stage is v from the first stage.

Continues on next page

The height is $16.2 + 5.95 = 22.2 \approx 22$ m (2 s.f.)

Calculate the total height.

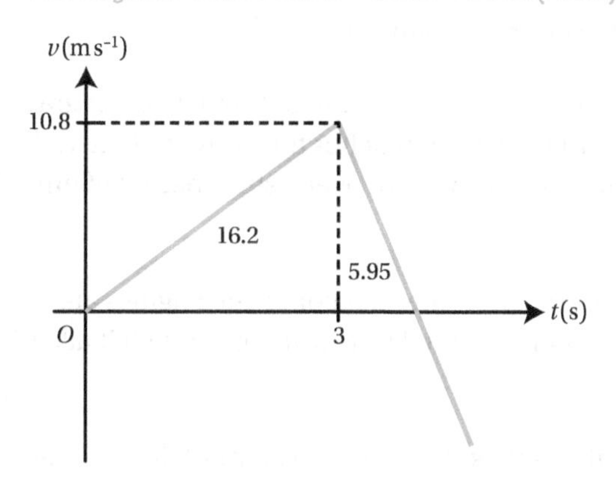

b Second stage:

$u = 10.8$

$a = -9.8$

$s = -16.2$

$t = ?$

The total time spent in the air is 3 seconds from the first stage, plus the time it takes to reach the ground in the second stage.

The ground is 16.2 m below the starting point for the second stage so s is negative.

$s = ut + \frac{1}{2}at^2$

$-16.2 = 10.8t - 4.9t^2$

$4.9t^2 - 10.8t - 16.2 = 0$

$t = -1.02$ or $t = 3.23$

$\therefore t = 3.23$ s

The total time is $3.23 + 3 = 6.23 \approx 6.2$ s (2 s.f.)

Add the time from the first stage.

EXERCISE 17D

In this exercise use $g = 9.8\,\text{m s}^{-2}$, unless instructed otherwise, giving your final answers to an appropriate degree of accuracy.

1. A car accelerates from rest for 8.3 seconds, reaching a speed of $12.8\,\text{m s}^{-1}$. It then travels for another 12 seconds with deceleration of $0.8\,\text{m s}^{-2}$. Find the total distance travelled by the car.

2. A fox is running in a straight line. It passes tree A with speed of $8.3\,\text{m s}^{-1}$ and starts accelerating uniformly. It passes tree B, 120 m away, 13 seconds later. It immediately starts decelerating, coming to rest next to tree C, 250 m from tree B.

   a Find the speed of the fox when it passes tree B.

   b Find the deceleration of the fox.

3. A runner starts with speed u and accelerates uniformly. She covers the first 10 m in 2.1 seconds and the next 10 m in another 1.1 second. Find the value of u.

4. A cyclist starts at the bottom of the hill, moving at a speed of $13.5\ \text{m s}^{-1}$. She moves with constant deceleration of $0.9\ \text{m s}^{-2}$, reaching the top of the hill 9.2 seconds later. She then accelerates down the hill at $1.6\ \text{m s}^{-2}$ for 86 m. Find the speed of the cyclist when she reaches the bottom of the hill.

5. A ball is dropped from the height of 2.6 m above the surface of a water well and falls freely under gravity. After it enters the water, the ball's acceleration decreases to $1.2\ \text{m s}^{-2}$. It reaches the bottom of the well 0.9 seconds later. Assuming the acceleration through the water is constant over a short period of time, find the depth of the water in the well.

6. A rocket is projected vertically upwards with a speed of $260\ \text{km h}^{-1}$. After 6 seconds the engines are switched on and the rocket starts accelerating at $2.8\ \text{m s}^{-2}$. Find the total time from the launch it takes for the rocket to reach the height of 400 m.

7. A car starts from rest at time $t = 0$. It accelerates uniformly until its speed reaches $V\ \text{m s}^{-1}$. It travels at constant speed for 12 seconds and then decelerates uniformly, coming to rest when $t = 26$. The total distance travelled by the car is 840 m. Find the value of V.

8. In this question use $g = 9.81\ \text{m s}^{-2}$, giving your final answers to an appropriate degree of accuracy.

   A model rocket starts from rest. It has an engine that produces upward acceleration of $5.25\ \text{m s}^{-2}$. When the rocket has reached a height of 25 m the engine is switched off. Find the maximum height of the rocket and its speed when it returns to the ground.

9. A ball is dropped from a window 30 m above ground. Half a second later, another ball is projected vertically upwards from the ground, vertically below the window. The balls collide when they are 15 m above ground. Find the initial velocity of the second ball.

10. A car is overtaking a lorry on a straight, horizontal road. The length of the lorry is 15.2 m and the car is modelled as a particle. Initially the car and the lorry are moving at a constant velocity of $18.6\ \text{m s}^{-1}$. The lorry continues to move with constant velocity. The car starts 35 m behind the lorry, accelerates at a constant rate until it reaches speed $V\ \text{m s}^{-1}$, then decelerates at a constant rate back to $18.6\ \text{m s}^{-1}$. It is 40 m in front of the lorry at the point when they again have the same velocity. The overtaking takes 28 seconds.

    a How much further does the car travel than the lorry during the 28 seconds?

    b On the same axes, sketch the velocity–time graphs for the lorry and the car. Hence find the value of V.

Checklist of learning and understanding

- When an object is moving with constant acceleration you can use the equations:
  - $v = u + at$
  - $s = ut + \frac{1}{2}at^2$
  - $s = vt - \frac{1}{2}at^2$
  - $s = \frac{1}{2}(u+v)t$
  - $v^2 = u^2 + 2as$

  where s is the displacement from the starting position, u is the initial velocity, v is the velocity at time t and a is the acceleration.
- You can derive these equations using integration or from a velocity–time graph (which is a straight line).
- If a is negative:
  - maximum displacement from the starting point occurs when $v = 0$
  - the object returns to the starting point when $s = 0$.
- A special case of motion with constant acceleration is vertical motion under gravity where the acceleration is g.
  - You will be given the value of g to use: $10\,\text{m}\,\text{s}^{-2}$ (1 s.f.), $9.8\,\text{m}\,\text{s}^{-2}$ (2 s.f.) or $9.81\,\text{m}\,\text{s}^{-2}$ (3 s.f.).
  - Take the positive direction for velocity and displacement to be in the direction of initial velocity.
  - Acceleration is negative if the initial velocity is upwards and positive if it is downwards.
  - Maximum height occurs when $v = 0$.
- For projection from ground level:
  - time taken going up is the same as time taken coming down
  - the object hits the ground with the same speed with which it was projected.
- Constant acceleration formulae require two modelling assumptions.
  - Air resistance can be ignored (true if the object is modelled as a particle).
  - g is constant (true for small heights).

Mixed practice 17

In this exercise use $g = 9.8\,\text{m s}^{-2}$, unless instructed otherwise, giving your final answers to an appropriate degree of accuracy.

1. A car is travelling at $20\,\text{m s}^{-1}$ when it starts to brake. It decelerates constantly at $4\,\text{m s}^{-2}$.

   Modelling the car as a particle, find how far it has travelled when it is moving at $8\,\text{m s}^{-1}$.

   Choose from these options.

   A 3 m **B** 12 m **C** 42 m **D** 50 m

2. A particle is moving with the speed of $12\,\text{m s}^{-1}$ when it starts to accelerate uniformly at $1.6\,\text{m s}^{-2}$.

   a Find how long it takes for the particle's speed to increase to $26\,\text{m s}^{-1}$.

   b How far does the particle travel in that time?

3. In this question use $g = 10\,\text{m s}^{-2}$, giving your final answers to an appropriate degree of accuracy.

   A stone is projected vertically upwards from ground level with speed of $9\,\text{m s}^{-1}$.

   a Find the maximum height reached by the stone.

   b Find how long it takes before the stone returns to ground level.

4. A cyclist passes point P with a speed of $6.2\,\text{m s}^{-1}$ and starts to decelerate uniformly at $2.1\,\text{m s}^{-2}$. How fast is she moving after she has travelled 8 m?

5. A car is travelling along a horizontal road. It is moving at $14\,\text{m s}^{-1}$ when it starts to accelerate. It accelerates at $0.8\,\text{m s}^{-2}$ for 12 seconds.

   a Find the speed of the car at the end of the 12 seconds.

   b Find the distance travelled during the 12 seconds.

   [© AQA 2014]

6. A particle is projected vertically upwards with speed $24.5\,\text{m s}^{-1}$.

   For how long is it more than 29.4 m above its point of projection?

   Choose from these options.

   A 1 second **B** 2 seconds **C** 3 seconds **D** 4 seconds

7. A golf ball is dropped from the top of a cliff. Two seconds later a tennis ball is thrown vertically downwards from the top of the cliff, with speed u. The tennis ball hits the golf ball before they reach the sea below.

   Initially both balls are modelled as particles and the time from when the tennis ball is thrown until they collide is calculated.

   If the balls are no longer modelled as particles, what would need to happen to the value of u for them to collide after the same time t?

   Choose from these options.

   A No change. **B** It would increase.

   C It would decrease. **D** More information is required.

8 A car travels on a straight horizontal road. It passes point A, travelling at a speed of 32 km h^{-1}, and starts to decelerate uniformly until it reaches speed v km h^{-1}. It then accelerates uniformly. When it reaches point B its speed is 32 km h^{-1} again.

a Draw the velocity–time graph representing the car's journey.

b Given that the distance AB is 550 m and the journey takes 1.2 minutes, find the value of v.

c Find the average speed of the car during its journey from A to B.

9 In this question use $g = 9.81$ m s^{-2}, giving your final answers to an appropriate degree of accuracy.

a Use the formulae $v = u + at$ and $s = ut + \frac{1}{2}at^2$ to derive the formula $v^2 = u^2 + 2as$.

b A ball is projected vertically downwards from the top of a building, with a speed of 8.5 m s^{-1}. It reaches the ground with a speed of 52 m s^{-1}. Find the height of the building.

10 Two particles are projected simultaneously with a speed of 15.4 m s^{-1}. The first particle is projected vertically upwards from ground level. The second particle is projected vertically downwards from a height of 20 m. The two particles move on a same straight line. Find:

a the height above ground at which the particles collide

b the speed of each particle at the moment they collide.

11 A particle moves with constant acceleration a. When $t = 0$ is passes point O, with velocity u. Let s be the displacement from O at time t.

Use integration to show that $s = ut + \frac{1}{2}at^2$.

12 A car travels on a straight horizontal race track. The car decelerates uniformly from a speed of 20 m s^{-1} to a speed of 12 m s^{-1} as it travels a distance of 640 metres. The car then accelerates uniformly, travelling a further 1820 metres in 70 seconds.

a **i** Find the time that it takes the car to travel the first 640 metres.

ii Find the deceleration of the car during the first 640 metres.

b **i** Find the acceleration of the car as it travels the further 1820 metres.

ii Find the speed of the car when it has completed the further 1820 metres.

c Find the average speed of the car as it travels the 2460 metres.

[© AQA 2013]

13 Two cars start from rest, from the same start line, and accelerate uniformly along a racetrack running perpendicular to the start line. After 5 seconds the first car is 30 m in front of the second car. How far in front is it after another 5 seconds?

14 A ball is projected vertically upwards from ground level with speed u_1. At the moment when the first ball is at its maximum height, a second ball is projected vertically upwards from ground level with speed u_2. The two balls fall back to the ground at the same time without colliding in the air. Find the ratio $u_1 : u_2$.

15 A particle travels in a straight line and decelerates uniformly at 2 m s^{-2}. When $t = 0$ its velocity is u m s^{-1} and when $t = 100$ s its velocity is $-v$ m s^{-1} (where $u > v > 0$). The average **speed** of the particle over the 100 seconds is 62.5 m s^{-1}. Find the values of u and v.

Elevate

See Extension Sheet 17 for a selection of more challenging problems.

18 Forces and motion

In this chapter you will learn how to:

- understand what causes motion and the concept of a force (Newton's first law)
- relate force to acceleration (Newton's second law)
- work with situations where several forces act on an object
- work with different types of forces, including gravity
- determine whether a particle is in equilibrium.

Before you start...

Chapter 17	You should be able to use constant acceleration formulae.	1 A particle accelerates from 3 m s^{-1} to 8 m s^{-1} in 12 seconds. Find: a the acceleration b the distance the particle travels in this time.
Chapter 15	You should be able to work with vectors in component form.	2 a Add the vectors $\begin{pmatrix}1\\-3\end{pmatrix}$, $\begin{pmatrix}2\\2\end{pmatrix}$ and $\begin{pmatrix}-5\\7\end{pmatrix}$. b Find vector $\mathbf{v}$ such that $3.5\mathbf{v} = -14\mathbf{i} + 7\mathbf{j}$.
Chapter 15	You should be able to find the magnitude and direction of a vector from its components.	3 Find the magnitude of each vector and the angle it makes with the horizontal direction: a $1.2\mathbf{i} + 2.5\mathbf{j}$ b $\begin{pmatrix}4\\-1\end{pmatrix}$

What causes motion?

In Chapters 16 and 17 you saw how to derive formulae to describe how the displacement, velocity and acceleration of a particle vary with time. So far nothing has been said about the causes of motion: Why should a particle start to move, or change its velocity? You probably already know that motion is caused by forces and you are familiar with some types of force - such as gravitational, electromagnetic or frictional. In this chapter you will investigate the relationship between force and acceleration, for forces acting in one and two dimensions. You will also see how to work out the combined effect of several forces; this requires the application of vectors from Chapter 15.

Section 1: Newton's laws of motion

Imagine a box lying on the table. If you want it to move, you need to push or pull it - you need to act on it with a **force**. If you don't apply a force, the box will remain at rest.

Once the box is moving, a force is required to change its velocity. For example, a frictional force might cause it to slow down, or you may continue to push it to make it accelerate. If there is no force at all, the box will continue to move at a constant speed.

This is one of the most important principles of mechanics: any change in the velocity of an object is caused by a force.

Key point 18.1

Newton's first law

An object continues to move with constant velocity, or remains at rest, unless acted upon by a force.

Fast forward

In Section 2 you will see that when more than one force acts on an object, the force, $\mathbf{F}$, in $\mathbf{F} = m\mathbf{a}$ is the **resultant** or **net force**.

Notice that the law refers to the **velocity**, rather than just the speed of the object. So a force is required to change the direction of motion, as well as the speed.

Once you apply the force, the box will start to accelerate. If you want to produce greater acceleration, you need to push or pull harder. But you also know from experience that the heavier an object is, the more difficult it is to move. So the force required to produce a given acceleration depends on the **mass** of the object; it is, in fact, directly proportional to it.

Did you know?

The unit of force is named after Sir Isaac Newton who developed the laws of motion.

Key point 18.2

Newton's second law

The force, $\mathbf{F}$ newtons, required to make an object of mass m kg move with acceleration $\mathbf{a}$ m s^{-2} is:

$$\mathbf{F} = m\mathbf{a}$$

Force and acceleration are both vectors. If the direction of the force is the same as the direction of motion, the object will continue to move in the same direction but its speed will change. However, if the force acts in a direction different from the direction of motion, it will also change the direction of the velocity vector.

The magnitude of a force is measured in **newtons** (N). The equation in Key point 18.2 tells you how the newton is related to the fundamental units of the SI system: 1 N = 1 kg m s^{-2}.

Did you know?

The force you feel when holding an apple is about 1 newton.

WORKED EXAMPLE 18.1

A truck of a mass 3.2 tonnes is moving in a straight line under the action of a constant driving force. You can ignore friction.

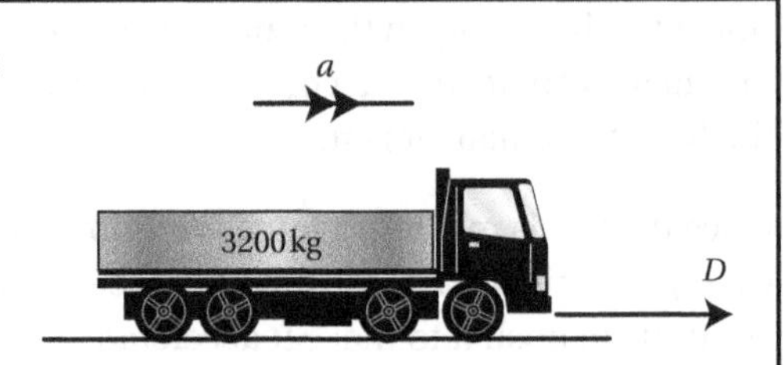

Find the magnitude of this force when the truck is:

a accelerating at a constant rate of 1.6 m s^{-2}

b moving at a constant speed of 58 km h^{-1}.

a 3.2 tonnes = 3200 kg — The mass must be in the SI unit, kilograms.

$F = ma$
$= 3200 \times 1.6$
$= 5120 \text{ N}$ — Use Newton's second law. Since this is action in one direction you do not need to consider F and a as vectors.

b Constant speed $\Rightarrow F = 0 \text{ N}$ — According to Newton's first law, if the velocity is constant, there is no force acing on the object.

Notice that this also fits in with Newton's second law: if v is constant then $a = 0$, so $F = m \times 0 = 0$.

WORKED EXAMPLE 18.2

A box of mass 3.5 kg is being pulled with the force $\mathbf{F} = \begin{pmatrix} 14 \\ -21 \end{pmatrix}$ newtons. Find:

a the vector acceleration of the box

b the magnitude of the acceleration.

a $\underline{F} = m\underline{a}$ — Use Newton's second law with vectors.

$$\begin{pmatrix} 14 \\ -21 \end{pmatrix} = 3.5\underline{a}$$

$$\underline{a} = \frac{1}{3.5}\begin{pmatrix} 14 \\ -21 \end{pmatrix}$$

$$= \begin{pmatrix} 4 \\ -6 \end{pmatrix} \text{m s}^{-2}$$

b $|\underline{a}| = \sqrt{4^2 + (-6)^2}$
$= 7.21 \text{ m s}^{-2}$ — Use Pythagoras' theorem to find the magnitude.

Rewind

You learnt how to find the magnitude and direction of a vector in Chapter 15.

Fast forward

A In Student Book 2, Chapter 16, you will extend what you know about constant acceleration formulae to work with two-dimensional vectors.

Once you know the magnitude of the acceleration you can then use it in the constant acceleration formulae.

WORKED EXAMPLE 18.3

A ball of mass 1.2 kg is rolled across the floor at a speed of 3.6 m s^{-1}. It slows down due to a constant friction force of magnitude 9 N.

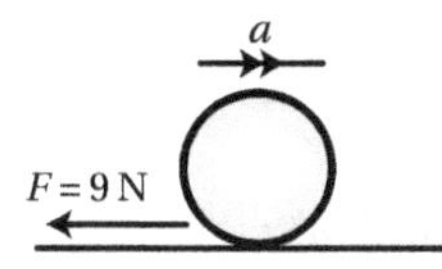

How far does the ball travel before coming to rest?

$F = ma$ — Use $F = ma$ to find a.

$-9 = 1.2a$ — F is negative as it is acting in the opposite direction to the motion.

$a = -7.5\ \text{m s}^{-2}$ — You expect a to be negative as the ball is slowing down.

$u = 3.6$
$v = 0$
$a = -7.5$
$s = ?$ — You can use one of the constant acceleration equations to find the distance.

$v^2 = u^2 + 2as$
$0 = 3.6^2 + 2(-7.5)^2 s$
$15s = 12.96$
$s = 0.864$ — Use $v^2 = u^2 + 2as$.

The ball travels 0.864 m before coming to rest.

Common error

If an object is projected with a given speed, there is no driving force continuing to act throughout the motion, so don't attempt to include it. The force was imparted at the point of projection but does not act thereafter.

Tip

In one dimension (right/left or up/down), always take the positive direction to be the direction in which the object is moving. If a force acts in the opposite direction it will be negative.

EXERCISE 18A

1. Find the magnitude of the force, in newtons, acting on the object in each case.
   - **a** **i** A crate of mass 53 kg moves with constant acceleration of 2.6 m s^{-2}.
     ii A stone of mass 1.5 kg is pushed across ice and decelerates at a constant rate of 0.3 m s^{-2}.
   - **b** **i** A truck of mass 6 tonnes accelerates uniformly at 1.2 m s^{-2}.
     ii A toy car of mass 230 g moves with constant acceleration of 3.6 m s^{-2}.
   - **c** **i** A box of mass 32 kg is dragged across the floor in a straight line, at a constant speed of 5.2 m s^{-1}.
     ii A ball of mass 120 g falls with a constant acceleration of 9.86 m s^{-2}.

2. Find the acceleration of the object in each case.
   - **a** **i** A constant force of magnitude 86 N acts on a box of mass 36 kg.
     ii A toy truck of mass 400 g is pushed with a force of 7.3 N.
   - **b** **i** A car of mass 1.5 tonnes moves under the action of a constant force of magnitude 600 N.
     ii A ball of mass 120 g is slowed down by a force of magnitude 2.6 N.

3 In each part, a particle of mass m moves with constant acceleration **a** under the action of a constant force **F**.

a **i** $m = 3\text{ kg}, \mathbf{a} = \begin{pmatrix} 1.6 \\ -2.5 \end{pmatrix} \text{m s}^{-2}$, find **F**.

ii $m = 5\text{ kg}, \mathbf{a} = \begin{pmatrix} -0.7 \\ 1.3 \end{pmatrix} \text{m s}^{-2}$, find **F**.

b **i** $m = 0.6\text{ kg}, \mathbf{F} = \begin{pmatrix} 3.6 \\ 1.2 \end{pmatrix} \text{N}$, find **a**.

ii $m = 6.3\text{ kg}, \mathbf{F} = \begin{pmatrix} -12.6 \\ 9 \end{pmatrix} \text{N}$, find **a**.

c **i** $\mathbf{a} = \begin{pmatrix} 0.5 \\ -1.5 \end{pmatrix} \text{m s}^{-2}, \mathbf{F} = \begin{pmatrix} 0.7 \\ -2.1 \end{pmatrix} \text{N}$, find m.

ii $\mathbf{a} = \begin{pmatrix} 5 \\ 2 \end{pmatrix} \text{m s}^{-2}, \mathbf{F} = \begin{pmatrix} 2.5 \\ 1 \end{pmatrix} \text{N}$, find m.

4 Discuss these scenarios in class.

a Imagine a car driving at a constant speed around a bend. Is there a force acing on the car? What is its direction?

b Newton's first law states that an object will continue to move with constant velocity if there is no force acting on it. Is it ever the case that there is no force acting on an object?

c In many questions in this chapter and in Chapter 19 it will be stated that forces such as friction or air resistance can be ignored. How realistic are these assumptions?

5 A car of mass 900 kg accelerates from rest to 15 km h^{-1} in 3.5 seconds. Assuming the driving force is constant, find its magnitude.

Tip

If a diagram is not given, always draw one first, indicating forces and the direction of motion.

6 A stone of mass 120 g is pushed across ice with a speed of 3.2 m s^{-1}. It comes to rest 8 seconds later. Find the magnitude of the friction force acting on the stone.

7 A crate of mass 28 kg is pulled across a horizontal floor. The pulling force acting on the crate is 260 N. Assuming that any friction forces can be ignored, how long does it take for the crate to accelerate from rest to 2.5 m s^{-1}?

8 Find, in vector form, the force required to move an object of mass 1.8 kg with acceleration $(0.6\mathbf{i} + 1.1\mathbf{j})$ m s^{-2}.

9 A particle of mass 6.5 kg accelerates under the action of force $\mathbf{F} = \begin{pmatrix} -8.5 \\ 6.5 \end{pmatrix} \text{N}$.

Find:

a the acceleration in vector form

b the magnitude of the acceleration.

10 A van of mass 2.3 tonnes, travelling in a straight line, decelerates under the action of a constant braking force. Its speed decreases from 50 km h^{-1} to 30 km h^{-1} while it covers a distance of 650 m. Find the magnitude of the braking force.

11 A girl pulls a toy truck with a constant horizontal force of 23 N. The truck starts from rest and accelerates uniformly, travelling 16 m in 3 seconds. Find the mass of the truck.

Section 2: Combining forces

In many situations there is more than one force acting on an object. For example, if you are pushing a box across the carpet you are providing a force to accelerate it, but there is also a frictional force slowing it down. A light suspended from a ceiling is being pulled down by **gravity** but is pulled up by the tension in the wire.

When several forces are acting on an object, their combined effect is to produce an acceleration. You can find a single force that would produce the same acceleration; this force is called the **resultant force**.

Key point 18.3

The **resultant force** is a single force that produces the same acceleration as several forces acting together. It is found by adding vectors representing the original forces.

WORKED EXAMPLE 18.4

Two forces act on the particle P, as shown in the diagram.

Find the direction and magnitude of the resultant force.

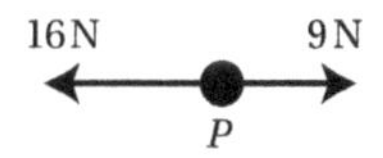

$F = 16 - 9$
$= 7\text{ N}$
The direction is to the left.

You can choose the positive direction to be either left or right. Since the larger force is pointing left it is sensible to make that the positive direction.

When forces are acting in two dimensions, the easiest way to add them is by breaking them into components. You can then use the components to find the magnitude and direction of the force.

Fast forward

A In Student Book 2, Chapter 18, you will learn about geometrical methods of adding forces.

WORKED EXAMPLE 18.5

Three forces are acting on an object, as shown in the diagram. Find:

a the magnitude of the resultant force

b the angle it makes with the horizontal direction.

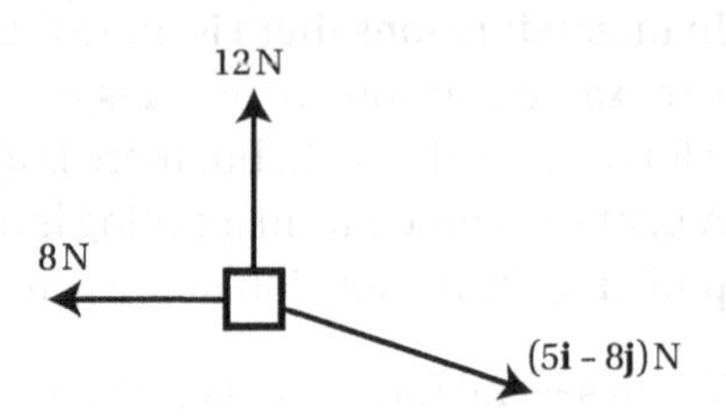

$\underline{F} = (-8\underline{i}) + (12\underline{j}) + (5\underline{i} - 8\underline{j})$

Write all three forces in vector form. Note that the horizontal force is to the left, so it is $-8\mathbf{i}$.

$= (-3\underline{i} + 4\underline{j})$ N

Add the **i** and **j** components separately.

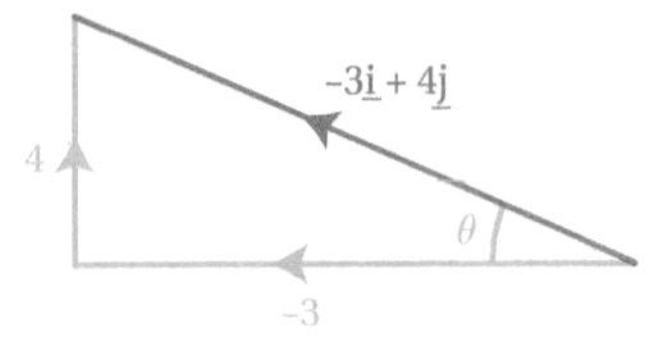

Now use the components to find the magnitude and direction. Draw a diagram to help.

The magnitude is:

$|\underline{F}| = \sqrt{3^2 + 4^2} = 5$ N

$\tan\theta = \frac{4}{3}$

$\theta = 52.1°$

So, the resultant force acts to the left at an angle 52.1° above the horizontal.

Once you have found the resultant force, you can use Newton's second law to find the acceleration.

WORKED EXAMPLE 18.6

Two children are pulling a box of mass 8 kg in opposite directions, as shown in the diagram.

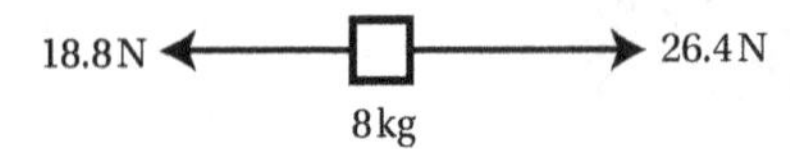

a Find the acceleration of the box.

A third child joins in, pulling with force F_3. The acceleration of the box is now 0.7 m s^{-2} to the right.

b Find the magnitude and direction of F_3.

Continues on next page

Elevate

See Support Sheet 18 for an example of resultant force and acceleration involving the equations of motion.

a $F = ma$

$26.4 - 18.8 = 8a$

$a = 0.95\ \text{m s}^{-2}$

Use $F = ma$ where F is the resultant force. Take the positive direction to be to the right (since that is the direction of the larger force, you expect the box to move in that direction).

b $F = ma$

$26.4 - 18.8 + F_3 = 8 \times 0.7$

Use $F = ma$ again but now with F_3 included. You don't know its direction, so assume that it's to the right (so use $+F_3$).

$F_3 = -2$

The magnitude of F_3 is 2 N and its direction is to the left.

The negative sign means that the force is actually to the left.

Tip

If you do not know which way an unknown force acts, take it to be positive. If you are wrong, the answer for this force will just come out to be negative, which means the force is in the opposite direction to the one you chose.

If the resultant force equals zero, the object will remain at rest or continue to move with constant speed (this is Newton's first law). In case of forces in two dimensions, this means that both components equal zero.

Fast forward

When the resultant force is zero, the object is in equilibrium. You will see more examples of this in Section 5.

WORKED EXAMPLE 18.7

A particle is subject to three forces, as shown in the diagram. Given that the particle moves with constant speed, find $\mathbf{F}_3$ in vector form.

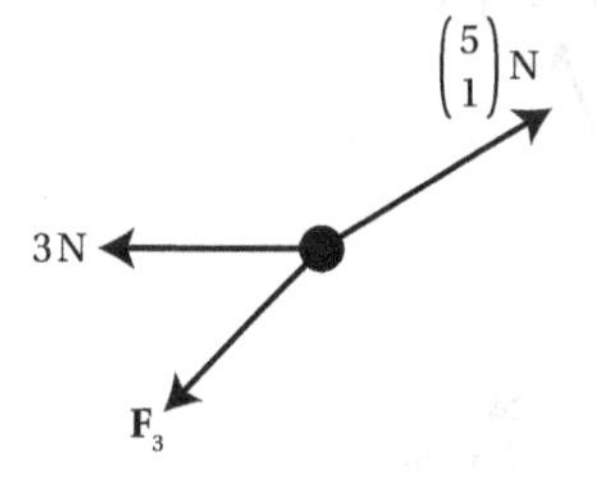

$$\begin{pmatrix}-3\\0\end{pmatrix} + \begin{pmatrix}5\\1\end{pmatrix} + \mathbf{F}_3 = \begin{pmatrix}0\\0\end{pmatrix}$$

Since the particle does not accelerate, the resultant force must be zero.

$$\mathbf{F}_3 = \begin{pmatrix}3\\0\end{pmatrix} + \begin{pmatrix}-5\\-1\end{pmatrix}$$

$$= \begin{pmatrix}-2\\-1\end{pmatrix}\text{N}$$

EXERCISE 18B

1 Find the magnitude and direction of the resultant force in each case.

a **i** 5N ← ● → 3N **ii** 7N ← ● → 12N

b **i**

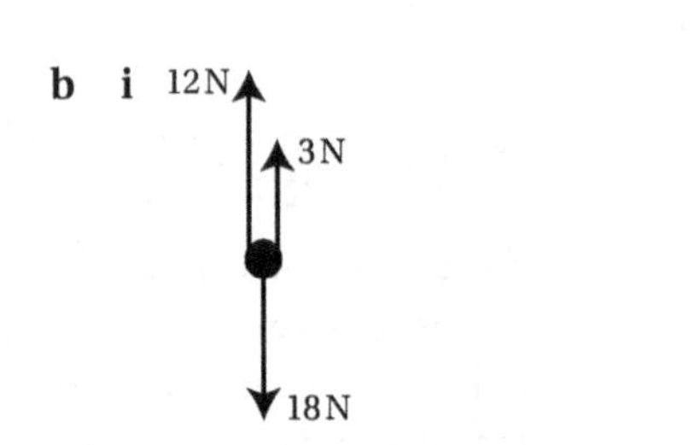

ii

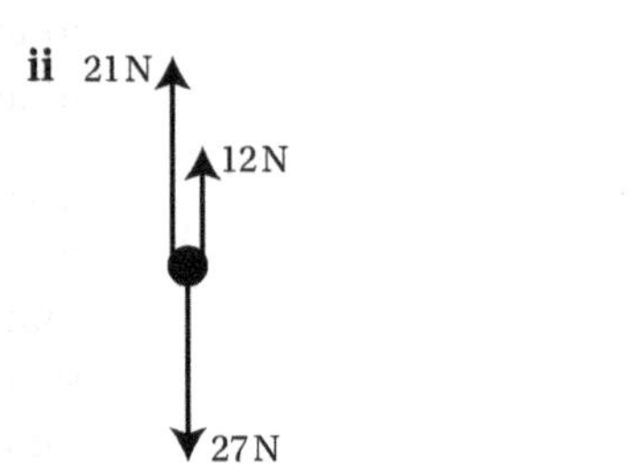

c **i**

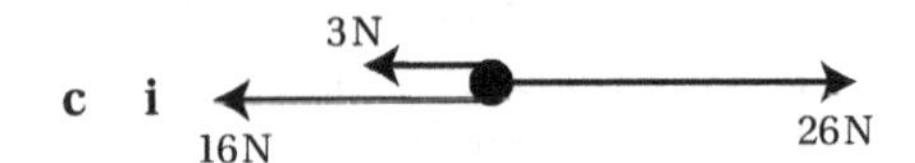

ii

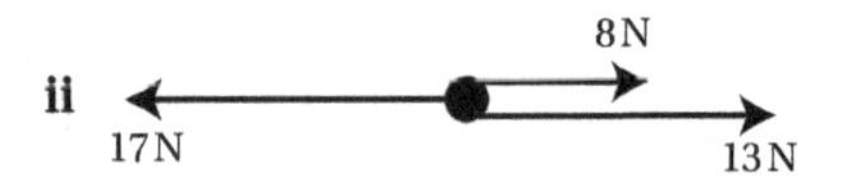

2 In each case, find the resultant force in the form $p\mathbf{i}+q\mathbf{j}$.

a **i** 8N ↑, 13N →

ii

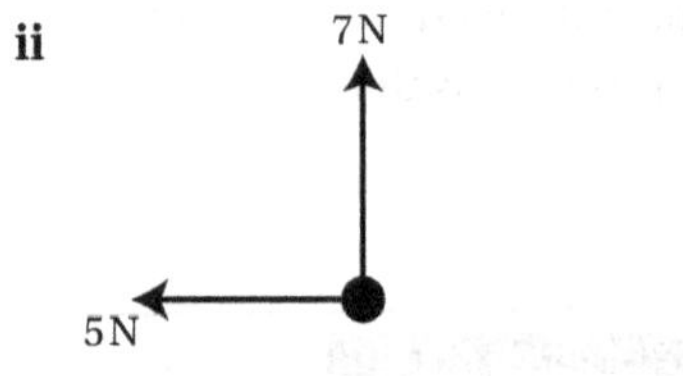

b **i**

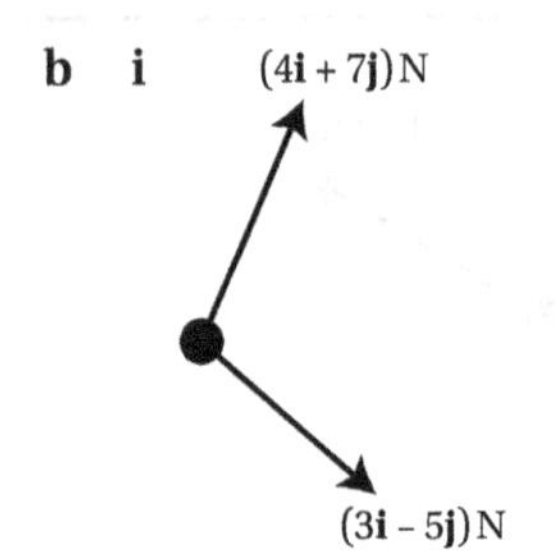

ii

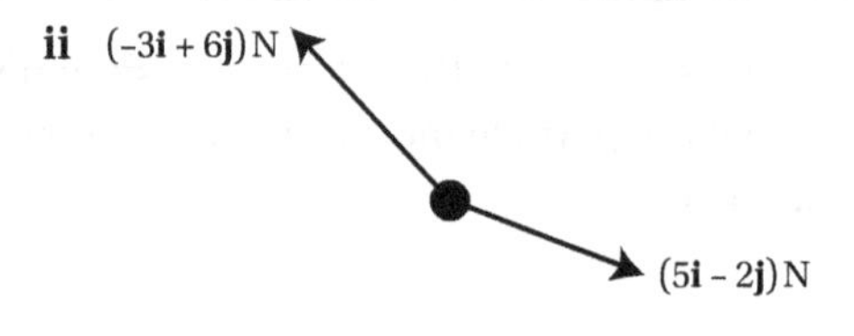

c **i**

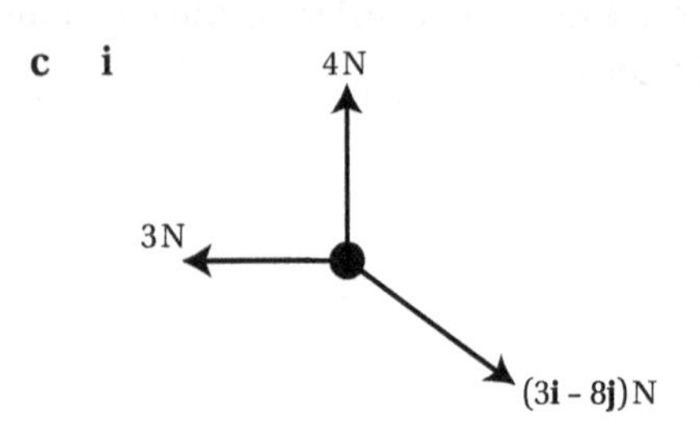

ii

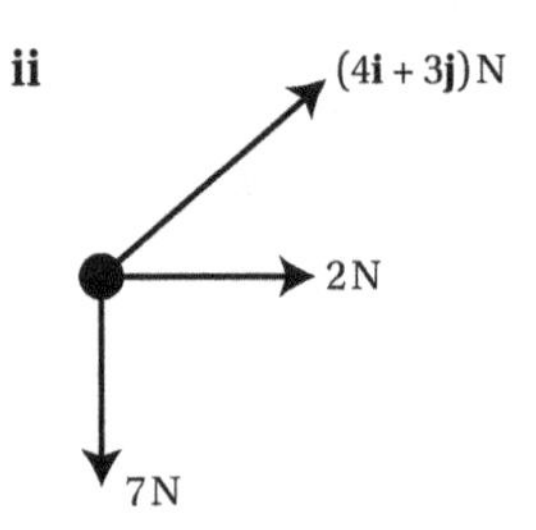

3 Find the magnitude and direction of each resultant force in Question 2.

4 For each object shown in the diagram, find the magnitude and direction of acceleration.

a **i**

6 kg
8 N
26 N

ii

20 N
12 N
$m = 4$ kg

b **i**

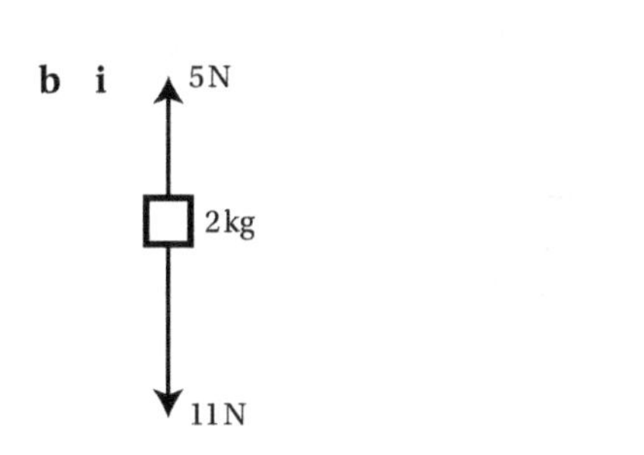

ii

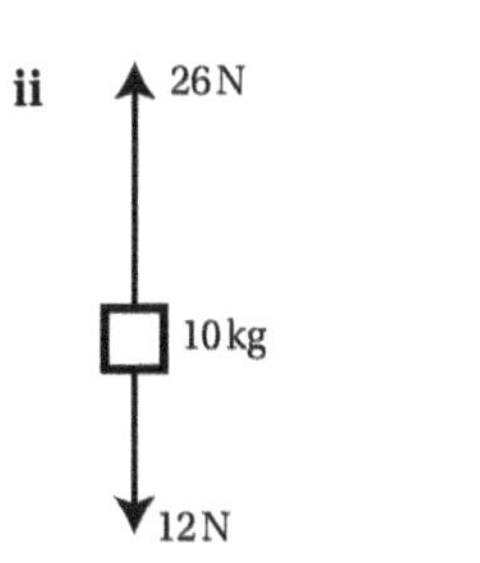

c **i**

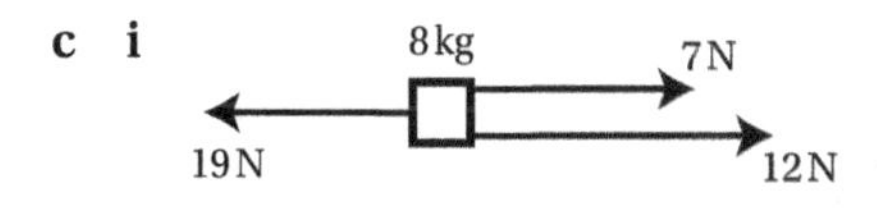

ii

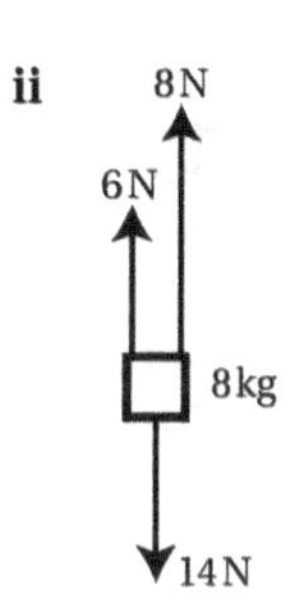

d **i**

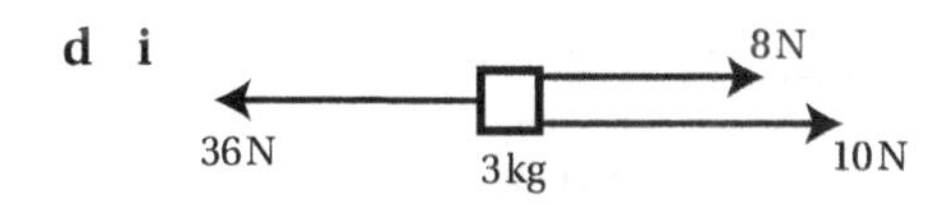

ii

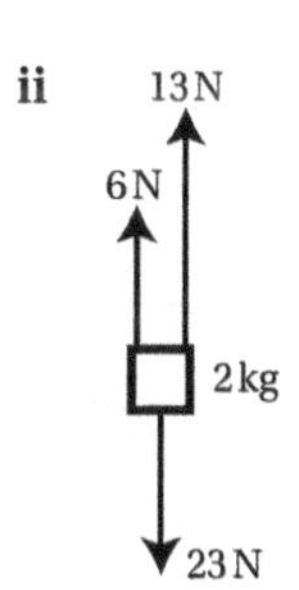

5 Find the acceleration in the form $p\mathbf{i} + q\mathbf{j}$.

a **i**

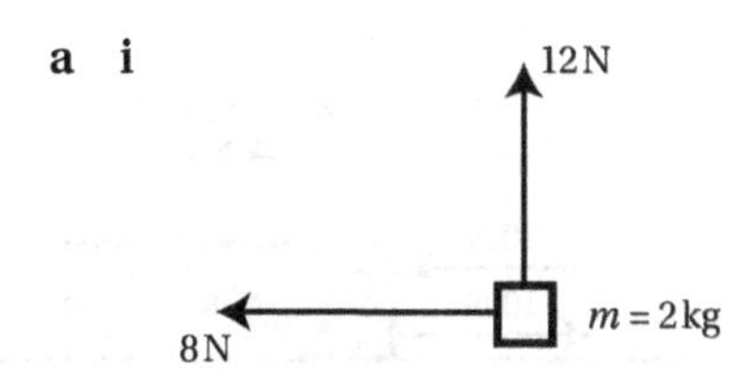

ii

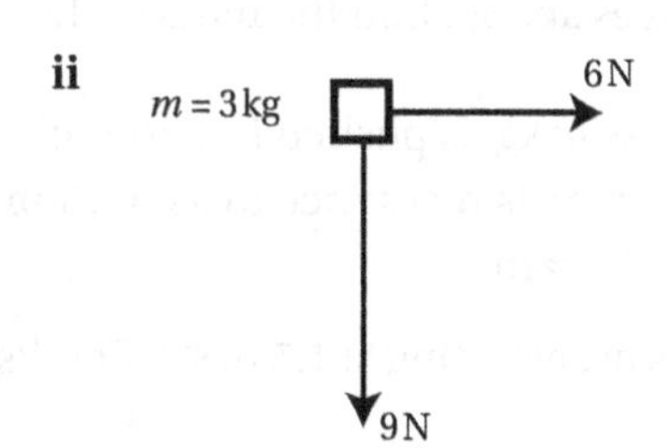

b **i**

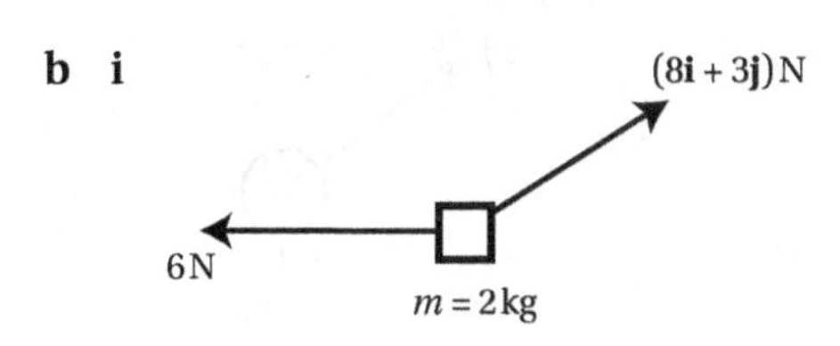

ii

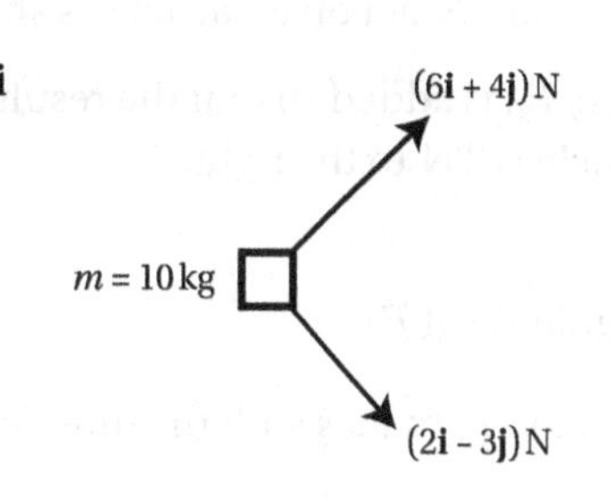

6 In each diagram, the mass of the object and the acceleration are given. Up and right are the positive directions. Find the magnitude of the force marked F.

a i

F 12 N

$m = 10$ kg

$a = -0.3$ m s^{-2}

ii

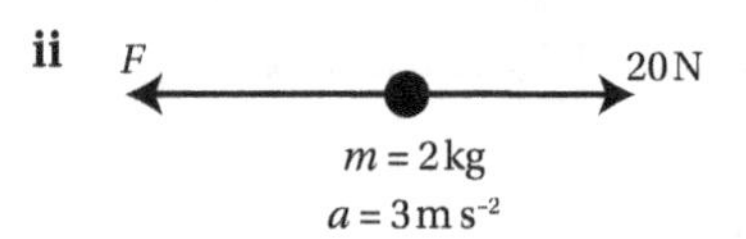

b i

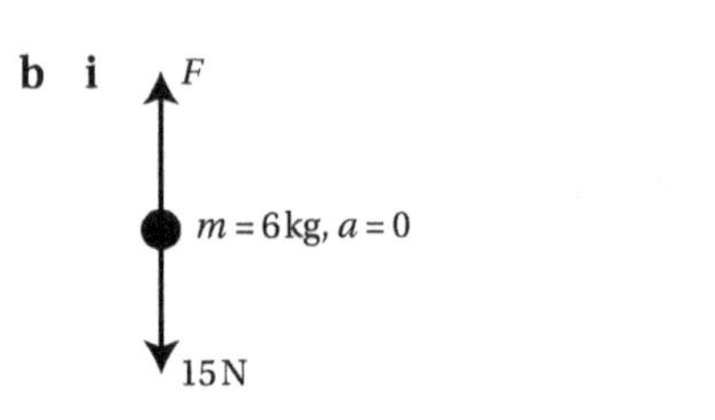

ii

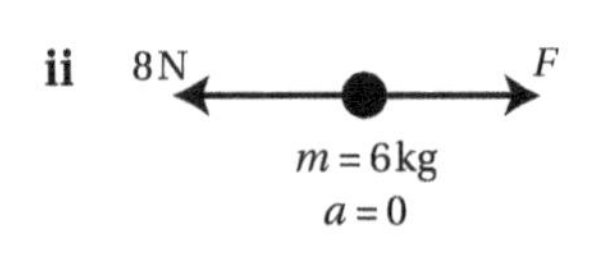

c i

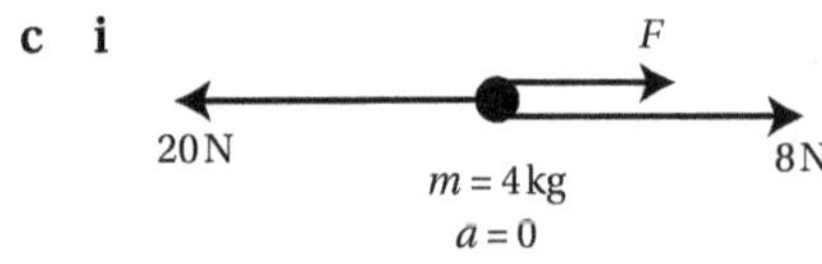

ii

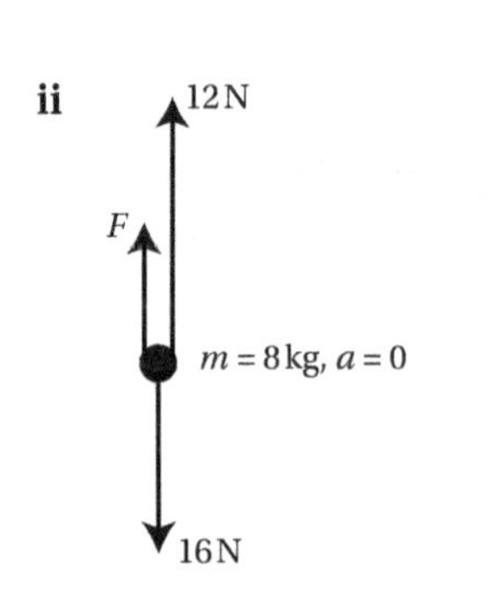

d i

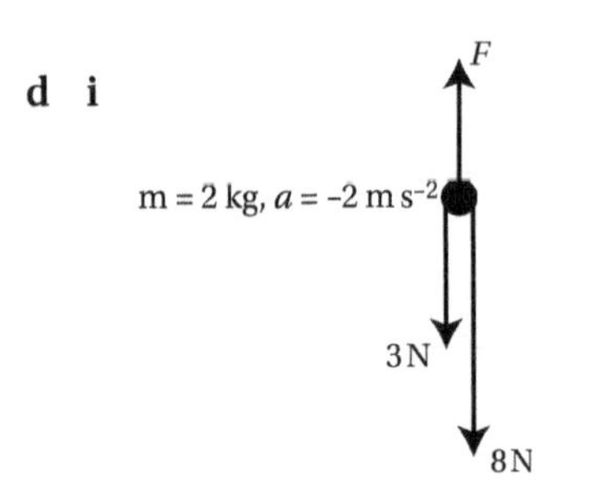

ii

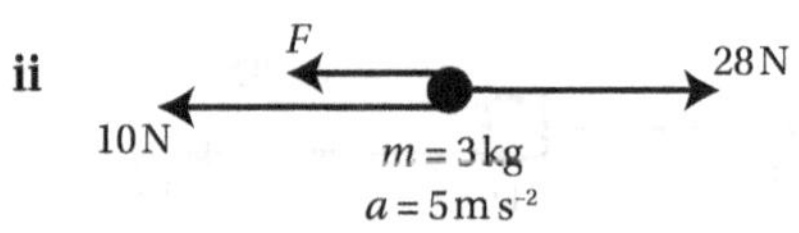

7 Two people attempt to push-start a car on a horizontal road. One person pushes with a force of 100 N; the other with a force of 80 N. The car starts to accelerate constantly at 0.15 m s^{-2}. Assuming these are the only horizontal forces acting, find the mass of the car.

8 A sledge of mass m kg is pushed horizontally through the snow by a force of 40 N. There is resistance to its motion of magnitude 10 N as shown in the diagram.

If the sledge is accelerating at 1.5 m s^{-2}, find its mass.

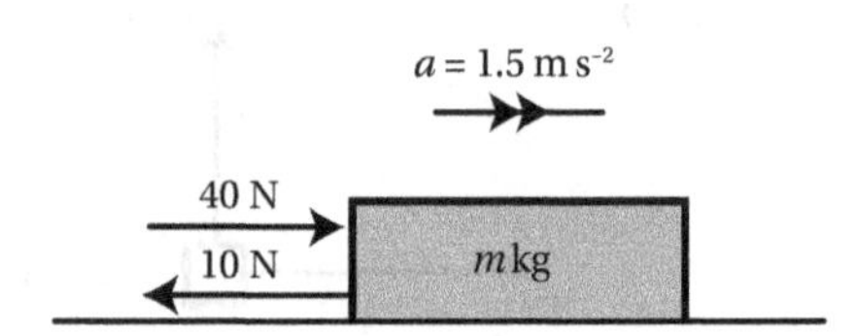

9 Two forces F_1 and F_2 act on a particle as shown.

A third force, F_3, is added so that the resultant force on the particle is 2 N to the right.
Find:

a the magnitude of F_3

b the direction F_3 makes with the direction of motion.

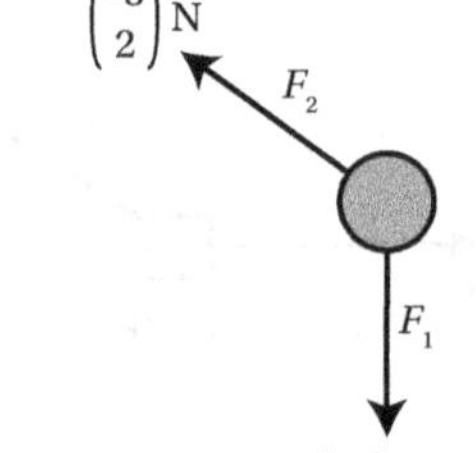

Section 3: Types of force

There are several common examples of forces with which you should be familiar. You need to be able to identify all forces acting on an object in a given situation and draw a force diagram, before using techniques from the previous section to make calculations.

Many examples you will meet involve moving vehicles. To accelerate the vehicle, the engine provides a **driving force**. To slow it down, the brakes provide a **braking force**, acting in the direction opposite to that of the velocity.

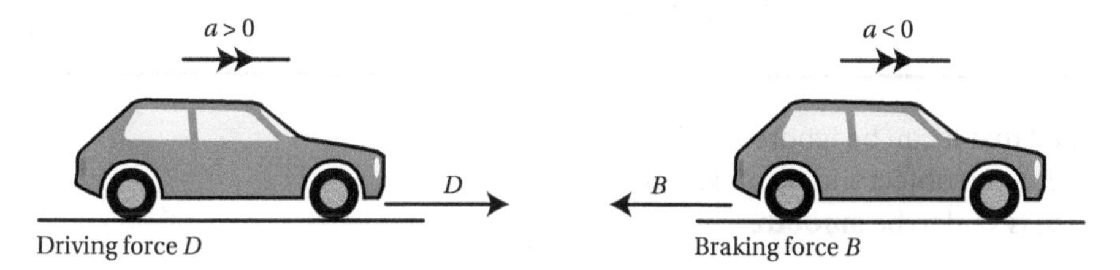

Driving force D

Braking force B

When an object is moving across a surface, there is normally some **friction** resisting the motion. The friction force always acts in the direction opposite to the velocity of the object. There are other types of forces that resist the motion, for example **air resistance**. The magnitude of these resistance forces depends on many different factors, which are beyond the scope of this course. For now, you will usually be told the magnitude of any resistance forces.

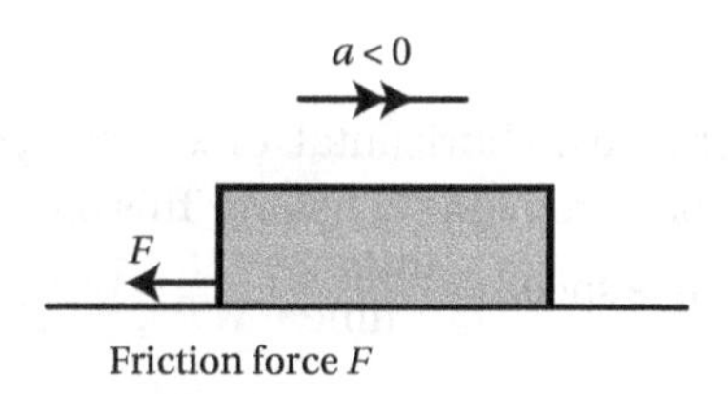

Friction force F

Fast forward

A You can learn more about friction in Student Book 2, Chapter 18.

WORKED EXAMPLE 18.8

a A car moves under the action of a driving force of 1740 N. The total resistance to motion equals 600 N. Given that the acceleration of the car is 1.2 m s^{-2}, find its mass.

b The car starts to brake and decelerates at 2.5 m s^{-2}. Assuming that the total resistance force remains the same, but the driving force has been removed, find the magnitude of the braking force.

Draw a diagram showing all the relevant forces. You do not need to include the car's weight because the question is only about forces and motion in the horizontal direction.

Continues on next page

$F = ma$

$1740 - 600 = m \times 1.2$

$m = 950 \text{ kg}$

Use $F = ma$ with the positive direction to the right.

B ← 600 ←

The braking force is in the opposite direction from the driving force so is negative.

$F = ma$

$-B - 600 = 950 \times (-2.5)$

$B = 1775 \text{ N}$

The acceleration is negative since the car is decelerating.

Sometimes the friction force is so small that it can be ignored. This is the case, for example, when you consider an object sliding on ice. The contact between the object and the ice is said to be **smooth**.

WORKED EXAMPLE 18.9

A small toy of mass 230 g is released along the floor with an initial speed of 8 m s^{-1}.

a The contact between the toy and the floor is modelled as smooth. Predict the time it would take to travel 6 m.

b The toy actually takes 0.82 seconds to travel 6 m. Find the magnitude of the friction force, assuming it is constant.

a No force, so constant speed.

$t = \frac{s}{v}$

$= \frac{6}{8}$

$= 0.75 \text{ s}$

There are no horizontal forces acting on the toy, so its speed remains constant. This means that you can use speed $= \frac{\text{distance}}{\text{time}}$.

b $s = 6$

$u = 8$

$t = 0.82$

$a = ?$

First find the acceleration using the constant acceleration equation involving s, u, a and t, $s = ut + \frac{1}{2}at^2$.

$s = ut + \frac{1}{2}at^2$

$6 = (8 \times 0.82) + \frac{1}{2}a(0.82^2)$

$6 = 6.56 + 0.3362a$

$a = -1.67 \text{ m s}^{-2}$

The acceleration is negative because the friction is slowing the toy down.

Continues on next page

$F = ma$ Now use $F = ma$ to find the force.

$= 0.23 \times 1.67$
$= 0.383$ N

Since you only want its magnitude, you can ignore the minus sign (which tells you that the direction is opposite to the direction of motion).

If you are using a rope to pull a box, you are not acting on the box directly. You are pulling on the rope and the rope pulls on the box; the force exerted by the rope on the box is called **tension**, and is directed away from the box (towards you). If you use a stick, a rod or a tow bar instead of a rope, then you could push the box. The pushing force provided by the tow bar is called **thrust**, and it is directed towards the box.

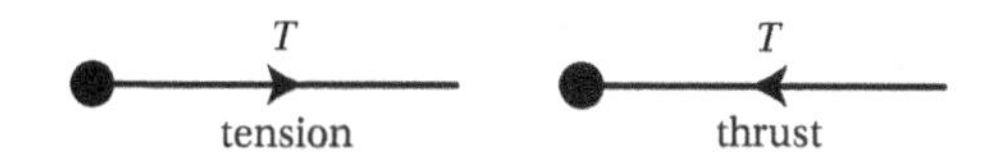

WORKED EXAMPLE 18.10

Two people are using two horizontal ropes to pull a box, as shown in the diagram. The mass of the box is 24 kg. The tension in one rope is 12 N and the acceleration of the box is 0.6 m s^{-2}. Assuming any resistance forces can be ignored, find the tension in the other rope.

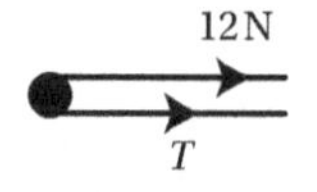

$F = ma$ Use $F = ma$ with positive direction to the right.

$12 + T = ma$
$12 + T = 24 \times 0.6$
$12 + T = 14.4$
$T = 2.4$ N

You may wonder whether you should include the mass of the rope or the rod in your calculations. In practice, its mass is often a lot smaller than the mass of the object you are trying to move, so you can ignore it. In mathematical terms, you are modelling the rope (or rod, tow bar...) as **light**. When you are using a rope or a string, you also need to assume that it does not stretch, otherwise it may be possible to pull the string without moving the object. In mathematical terms, then, the string or rope is **inextensible**.

Fast forward

You will discuss the importance of this assumption further in Chapter 19.

WORKED EXAMPLE 18.11

A boy is using a light stick to pull a toy box of mass 3.2 kg across rough carpet. The tension in the stick is 18 N and the friction force is 7 N. The stick and the carpet are horizontal.

a Find the acceleration of the box, and the time it takes for it to accelerate from rest to 2.1 m s^{-1}.

b Assuming that the friction force remains the same, what tension is required for the box to maintain the constant speed of 2.1 m s^{-1}?

The boy now makes the box slow down and it comes to rest after travelling 0.8 metres. The friction force is still the same.

c Find the magnitude of the force in the stick, and state whether it is a tension or thrust.

7N ← ● → 18N

Draw a diagram showing all the forces acting on the box.

a
$$F = ma$$
$$18 - 7 = 3.2a$$
$$a = 3.44\text{ m s}^{-2}$$

Use $F = ma$. The box is moving to the right, so take that as the positive direction.

$u = 0$
$v = 2.1$
$a = 3.44$
$t = ?$

Use the constant acceleration formula involving u, v, a and t.

$$v = u + at$$
$$2.1 = 0 + 3.44t$$
$$t = 0.611\text{ s}$$

b
$$F = ma$$
$$T - 7 = 0$$
$$T = 7\text{ N}$$

If the box is moving at constant speed, the acceleration is now zero.

c
$s = 08$
$u = 2.1$
$v = 0$
$a = ?$

Find the acceleration using the constant acceleration formula involving s, u, v and a.

$$v^2 = u^2 + 2as$$
$$0 = 2.1^2 + 2 \times 0.8a$$
$$a = -2.76\text{ m s}^{-2}$$

$$F = ma$$
$$T - 7 = 3.2 \times (-2.76)$$
$$T = -1.82$$

Then use $F = ma$. The box is still moving to the right, but now the acceleration is negative. You don't know the direction of the force in the stick; take it as positive.

The force in the stick is a thrust of magnitude 1.82 N.

The negative sign means that the force is in fact to the left. Since this is directed towards the box, it is a thrust rather than tension.

EXERCISE 18C

In each question draw a force diagram first.

1. A child pushes a box of mass 8 kg horizontally with a constant force of 28 N. The friction force between the box and the floor is 12 N. Find the acceleration of the box.

2. A car moves in a straight line on a horizontal road. The engine provides the driving force of 1200 N and the total resistance to the motion is 500 N. Given that the acceleration of the car is 2.6 m s^{-2}, find its mass.

3. A truck of mass 6.2 tonnes moves in a straight line with a constant acceleration of 1.2 m s^{-2}. The driving force of the engine is 8200 N.

   a Find the total resistance to the motion.

   Now assume that the resistance can be ignored.

   b How much less time would it take for the truck to accelerate from rest to 40 km h^{-1}?

4. Two girls are using two horizontal ropes to pull a crate. The mass of the crate is 56 kg. The tension in one of the ropes is 120 N and the friction force is 80 N. Given that the crate moves with constant acceleration of 0.8 m s^{-2} and that the girls are pulling in exactly the same direction, find the tension in the other rope.

5. Two men are pushing a car, each using an equal force of magnitude F N. The resistance to motion has magnitude 420 N. The mass of the car is 850 kg and it is moving at a constant speed of 6 km h^{-1}. Find the value of F.

6. A car of mass 950 kg is moving with a speed of 15.3 m s^{-1} when the driver applies brakes. The total resistance force, excluding the braking force, is 320 N. The car travels 120 m before coming to rest. Find the magnitude of the braking force.

7. A box of mass 45 kg is pulled across a horizontal floor with a light inextensible rope. The rope is horizontal and the tension in the rope is 180 N. The box starts from rest.

   a The contact between the box and the floor is modelled as smooth. According to this model, how long will it take for the box to travel 25 m?

   b The box in fact travels 25 m in 4.2 seconds. Find the magnitude of the friction force between the box and the floor.

8. A car travels along a straight horizontal road. The total resistance to the car's motion is constant and has magnitude 320 N.

   a The car's engine produces a constant driving force of 1200 N. The car passes point A with a speed of 12 m s^{-1} and accelerates to 26 m s^{-1} in 7 seconds. Show that the mass of the car is 440 kg.

   b The driving force is reduced so that the car travels at constant speed of 26 m s^{-1}. State the magnitude of the driving force.

   c The car travels at the constant speed for 12 seconds. Then the driver turns off the engine and the car stops at point B. Find the distance AB.

9 An object of mass 4.6 kg rests on a horizontal surface. It is acted on by two parallel pulling forces of magnitudes 8 N and 6.5 N, as shown in the diagram, and it starts to accelerate uniformly.

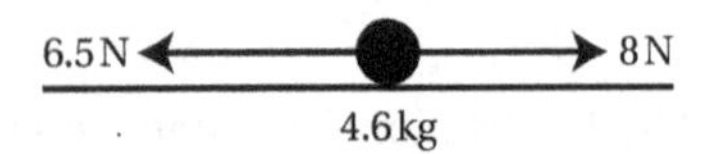

After 4 seconds a third pulling force is added, acting parallel to the other two. When the object has moved a further 2 m in the same direction as before, its speed is 1.24 m s^{-1}. Find the magnitude and direction of the third force.

10 In this question vectors **i** and **j** point east and north, respectively.

Two people are using two horizontal ropes to pull a box with a mass of 145 kg. The ropes are modelled as light and inextensible, and the forces in the ropes are $(p\mathbf{i}+q\mathbf{j})$ N and $(p\mathbf{i}-q\mathbf{j})$ N, as shown in the diagram.

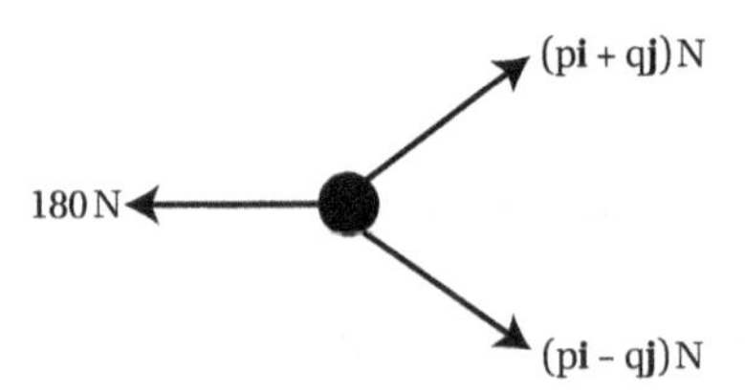

The friction force has magnitude 180 N and is directed west. The box starts from rest.

a Explain why the box moves in a straight line in the east direction.

b Given that the acceleration of the box is $0.003p$ m s^{-2} find the value of p.

c The tension in each rope has magnitude 200 N. Find the value of q.

Section 4: Gravity and weight

If you pick up a ball and let it go, it falls downwards. This is because the Earth exerts gravitational force on it. The force with which the Earth attracts any object is called the **weight** of the object. You know from Chapter 17 that all objects under the influence of gravity alone move with the same acceleration, g. The value of g is approximately 9.8 m s^{-2} but you will be told whether to use $g = 10$ m s^{-2}, 9.8 m s^{-2} or 9.81 m s^{-2}.

Tip

Remember that when you are given a value of g in a question you will need to give your final answers to a degree of accuracy (number of significant figures) that reflects the accuracy of g and of other values given in the question.

Key point 18.4

- The magnitude of the weight force on an object of mass m is $W = mg$.
- Its direction is towards the centre of the Earth (which is normally described as 'downwards').

You should be very careful to use the correct terminology here. In everyday language, you tend to use the word 'weight' to mean 'mass'; so you would say, for example, that the bag of apples weighs 1.2 kg. In Mechanics, 1.2 kg is the mass of the bag; its weight is a **force** with approximate magnitude $1.2 \times 9.8 = 11.76$ N.

The mass is a property of the object itself, independent of where it is. The magnitude of the weight depends both on the object and on the properties of the Earth. On a different planet the weight would be different; for example, the gravitational acceleration near the surface of Mars is about 3.7 m s^{-2}, so the weight of the same bag of apples on Mars would be $1.2 \times 3.7 = 4.44$ N.

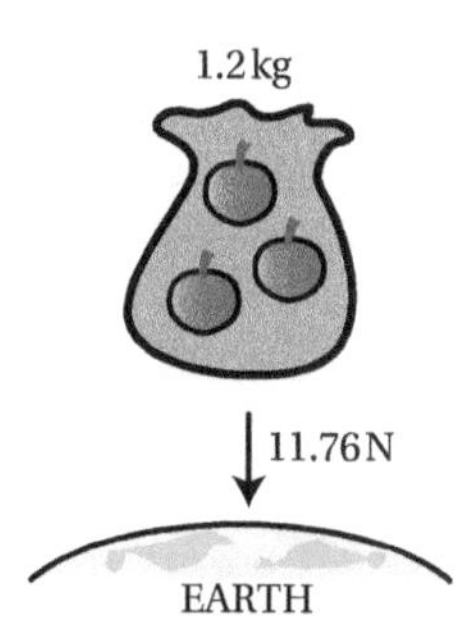

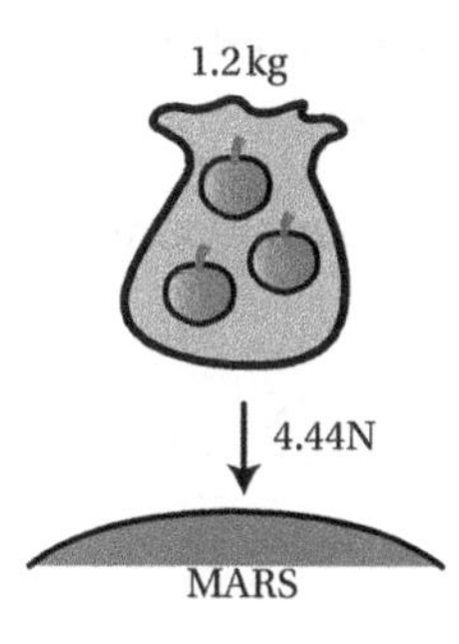

WORKED EXAMPLE 18.12

In his house on Planet X, Zixo has a crystal ball suspended from the ceiling by a light inextensible string. The mass of the ball is 1.6 kg and the tension in the string is 18.7 N. Find the magnitude of gravitational acceleration on Planet X.

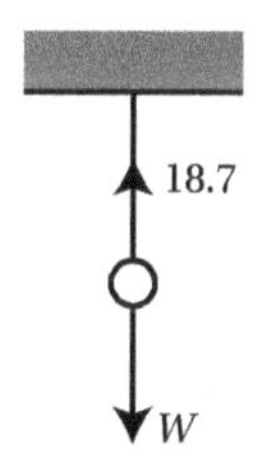

Always draw a diagram showing all the forces. In this case there are two forces acting on the ball: its weight and the tension in the string.

$W - 18.7 = 0$

$W = 18.7$ N

Since the ball is not moving, the net force is zero.

$W = mg$

$18.7 = 1.6g$

$g = 11.7$ m s^{-2}

Now use $W = mg$ to find the value of g.

You should know that the gravitational acceleration is not exactly the same everywhere on Earth. It depends on the latitude (this is because the Earth is not perfectly spherical, and also because of its rotation); it also depends on the altitude: it is lower up a mountain than at sea level. The variation is between around 9.76 m s^{-2} and 9.83 m s^{-2}.

Focus on...

See Focus on ... Modelling 3 for more information about the effect of modelling assumptions.

Whenever there is a possibility of an object moving in the vertical direction, you should include weight on your force diagram. You need to be very careful about the direction of acceleration.

WORKED EXAMPLE 18.13

A crate of mass 87 kg is being lowered by means of a light inextensible rope. Use $g = 9.8\ \text{m s}^{-2}$, giving your final answers to an appropriate degree of accuracy.

a Find the acceleration of the crate when the tension in the rope is 750 N.
b Find the tension in the rope if the crate is being lowered at constant speed.
c Find the tension in the rope required to decelerate the crate from $1.2\ \text{m s}^{-1}$ to rest in 3.5 seconds.

Always start by drawing a diagram.

a $F = ma$

To find acceleration, use $F = ma$. The crate is being lowered, so take the positive direction to be downwards.

$mg - T = ma$
$87 \times 9.8 - 750 = 87a$
$a = 1.2\ \text{m s}^{-2}$ (2 s.f.)

Since all the information in the question is given to 2 s.f. this is an appropriate degree of accuracy for your final answer.

b $F = ma$

Constant speed means that $a = 0$.

$87 \times 9.8 - T = 0$
$T = 850\ \text{N}$ (2 s.f.)

c $u = 1.2$
$v = 0$
$t = 3.5$
$a = ?$

Find acceleration using the formula with u, v, a and t.

$v = u + at$
$0 = 1.2 + 3.5a$
$a = -0.343\ \text{m s}^{-2}$

You expect a to be negative because the crate is decelerating.

$F = ma$

Then use $F = ma$.

$mg - T = ma$
$87 \times 9.8 - T = 87(-0.343)$
$T = 29.8 + 853$
$= 880\ \text{N}$ (2 s.f.)

EXERCISE 18D

In this exercise use $g = 9.8\text{ m s}^{-2}$, unless instructed otherwise, giving your final answers to an appropriate degree of accuracy.

1 The diagram shows an object of mass m kg suspended by a light inextensible string. The tension in the string is T N. Find the direction and magnitude of the acceleration of the object in each case.

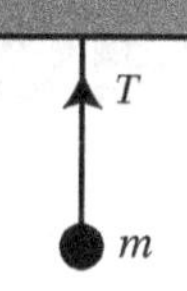

a i $m = 3\text{ kg}, T = 32\text{ N}$ ii $m = 8\text{ kg}, T = 92\text{ N}$

b i $m = 0.6\text{ kg}, T = 4.5\text{ N}$ ii $m = 3.2\text{ kg}, T = 27\text{ N}$

2 An object of mass m is suspended by a vertical rope. The acceleration of the object is $a\text{ m s}^{-2}$ in the stated direction. Find the tension in the rope.

a i $m = 12\text{ kg}, a = 0.6\text{ m s}^{-2}$ downwards ii $m = 3.2\text{ kg}, a = 1.2\text{ m s}^{-2}$ downwards

b i $m = 8\text{ kg}, a = 1.4\text{ m s}^{-2}$ upwards ii $m = 3\text{ kg}, a = 2\text{ m s}^{-2}$ upwards

3 An object is suspended by a string, with tension T N. The acceleration of the object is $a\text{ m s}^{-2}$ in the stated direction. Find the mass of the object.

a i $T = 26\text{ N}, a = 1.2\text{ m s}^{-2}$ downwards ii $T = 18.6\text{ N}, a = 0.7\text{ m s}^{-2}$ downwards

b i $T = 26\text{ N}, a = 1.2\text{ m s}^{-2}$ upwards ii $T = 18.6\text{ N}, a = 0.7\text{ m s}^{-2}$ upwards

4 In this question use $g = 9.81\text{ m s}^{-2}$, giving your final answer to an appropriate degree of accuracy. A crate of mass 105 kg is being lifted using a rope which can be modelled as light and inextensible. The tension in the rope is 1250 N. Find the acceleration of the crate.

5 A ball is at rest, suspended from a ceiling by a light inextensible cable. The tension in the cable is 12 N.

a Find the mass of the ball.

The tension in the cable is increased to 15 N.

b Find the magnitude and direction of the acceleration of the ball.

6 A crate is lowered from a window of a space ship on Mars, using a rope. The tension in the rope is 330 N and the crate is descending at a constant speed. Given that the gravitational acceleration on Mars is 3.7 m s^{-2}, find the mass of the crate, giving your answer to an appropriate degree of accuracy.

7 A crane is lifting a 350 kg load, which is initially at rest on the ground, using a light inextensible cable. It takes 6 seconds to raise it to a height of 84 m. Find the tension in the cable, assuming it is constant.

8 A crate of mass 132 kg is being lowered using a light inextensible rope. The crate decelerates constantly to rest while travelling a distance of $3g$ m in 6 seconds. Find the tension in the rope in terms of g.

9 In this question use $g = 10\text{ m s}^{-2}$, giving your final answers to an appropriate degree of accuracy. A fisherman catches a fish of mass 3.5 kg on the end of his fishing line and lifts it vertically through the water.

a While the fish is moving through the water, the water exerts a net force of magnitude 6 N, acting downwards. Given that the fisherman is raising the fish at a constant speed, find the tension in the fishing line.

b The fish breaks the surface of the water and the tension in the string remains unchanged. Air resistance can be ignored. Find the acceleration of the fish at that moment.

10 A horizontal platform of mass 120 kg is supported by a vertical steel rod, as shown in the diagram. The platform is being lowered and decelerating from 3.2 m s^{-1} to rest in 4.5 seconds. Find the thrust in the rod.

Section 5: Forces in equilibrium

You have already seen examples in which the object is at rest or moving with constant speed; therefore there is no resultant force acting. In such cases you say that the object is in equilibrium.

Key point 18.5

If an object is in **equilibrium** then the resultant force is zero.

WORKED EXAMPLE 18.14

A large box of mass 80 kg hangs in equilibrium, supported by four cables, as shown in the diagram. The tension in each cable has magnitude T.

Find the value of T. Use $g = 10\text{ m s}^{-2}$, giving your final answer to an appropriate degree of accuracy.

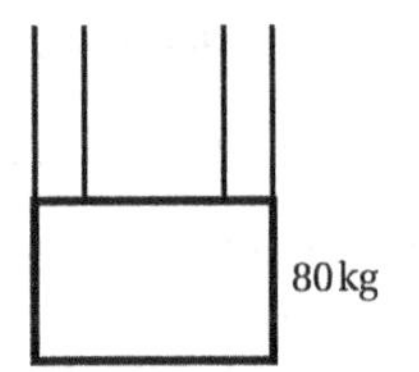

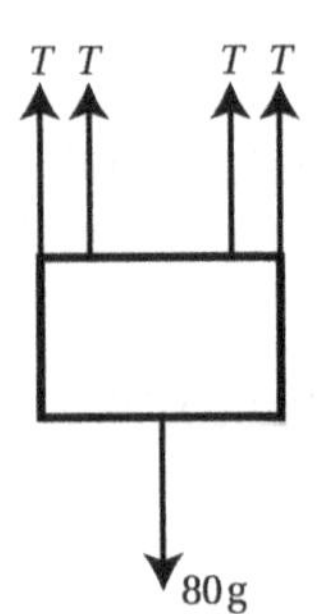

Draw a force diagram. The forces are the four tensions and the weight of the box.

$$80g - 4T = 0$$

$$T = \frac{80g}{4}$$

Since the box is in equilibrium, the resultant force is zero. Take positive direction to be downwards.

$$= 200\text{ N (1 s.f.)}$$

In the question the values are given to 1 s.f. so this is an appropriate degree of accuracy to use for your final answer.

If forces are acting in two dimensions, both horizontal and vertical components need to be zero.

WORKED EXAMPLE 18.15

A particle is in equilibrium under the action of three forces: $\mathbf{F}_1 = \begin{pmatrix} 40 \\ 2x \end{pmatrix}$, $\mathbf{F}_2 = \begin{pmatrix} y \\ -32 \end{pmatrix}$ and $\mathbf{F}_3 = \begin{pmatrix} x \\ y \end{pmatrix}$. Find the values of x and y.

$$\mathbf{F}_1 + \mathbf{F}_2 + \mathbf{F}_3 = 0$$

Since the particle is in equilibrium, the resultant force is zero.

$$\begin{pmatrix} 40 \\ 2x \end{pmatrix} + \begin{pmatrix} y \\ -32 \end{pmatrix} + \begin{pmatrix} x \\ y \end{pmatrix} = 0$$

$$\begin{pmatrix} 40 + y + x \\ 2x - 32 + y \end{pmatrix} = 0$$

$$\Rightarrow \begin{cases} 40 + y + x = 0 \\ 2x - 32 + y = 0 \end{cases}$$

You can write separate equations for the horizontal and vertical components to get two simultaneous equations.

$$\Rightarrow \begin{cases} x + y = -40 \\ 2x + y = 32 \end{cases}$$

$$\Rightarrow x = 72,\ y = -112$$

Solve either algebraically or with your calculator.

Sometimes you don't need to use vector notation; you can simply write separate equations for horizontal and vertical forces.

WORKED EXAMPLE 18.16

A box of mass 65 kg is suspended in equilibrium, supported by five light inextensible cables, as shown in the diagram. The tensions in the two vertical cables are T N and $2T$ N and the tensions in the horizontal cables are 160 N, 50 N and P N.

Find the values of T and P. Use $g = 9.8\ \text{m s}^{-2}$, giving your final answers to an appropriate degree of accuracy.

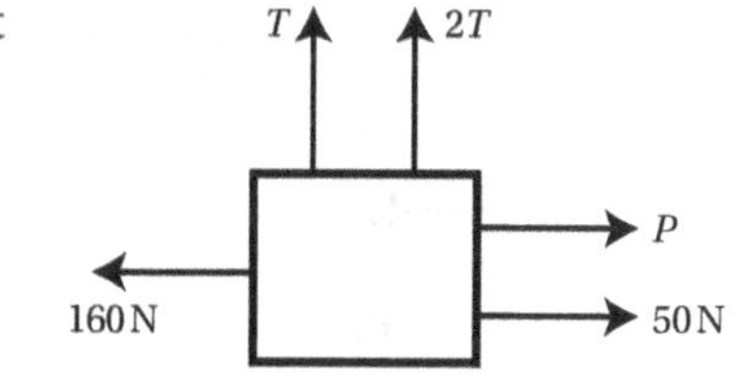

Write separate equilibrium equations for horizontal and vertical directions.

Vertically:

$3T = 65g$

$T = 210\text{ N (2 s.f.)}$

The equilibrium equation is $3T - 65g = 0$. However, sometimes it is easier to think of it as 'forces up = forces down'.

Horizontally:

$P + 50 = 160$

$\Rightarrow P = 110\text{ N}$

And likewise: 'forces to the right = forces to the left'.

EXERCISE 18E

1 Determine which of these particles are in equilibrium.

a

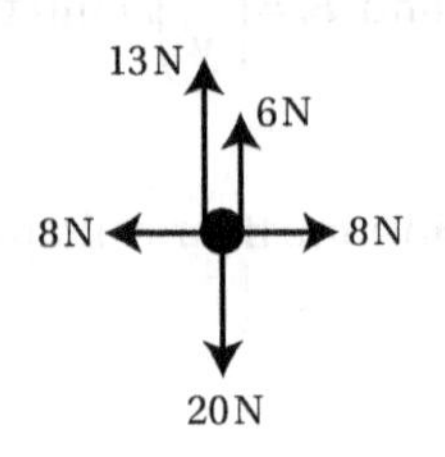

b

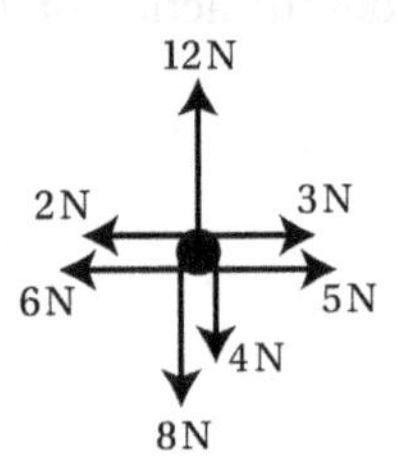

c

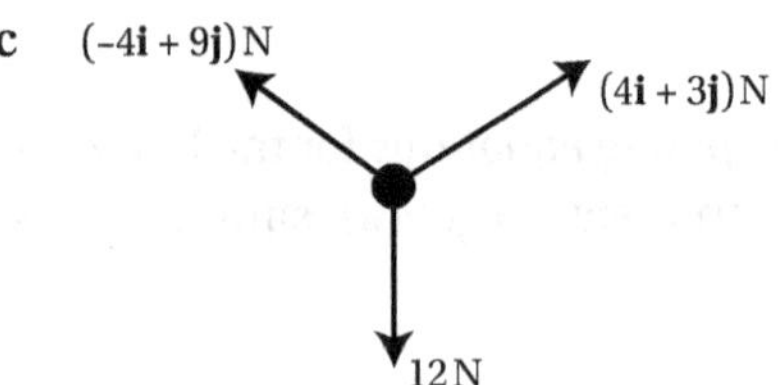

d

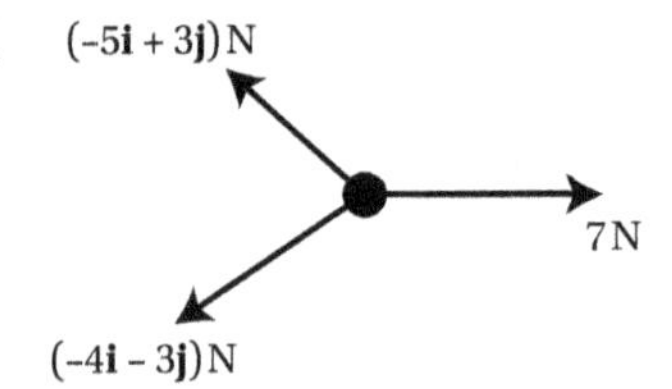

2 Each diagram shows a particle in equilibrium. Find the magnitudes of the forces marked with letters.

a

b

c

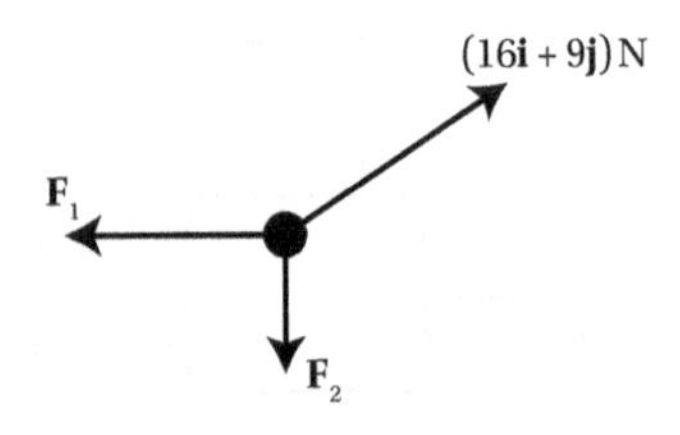

d

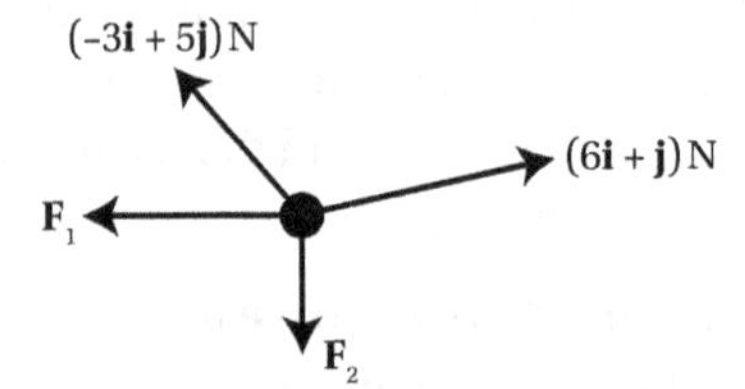

3 A particle is in equilibrium under the action of the three given forces. Find the values of x and y.

a **i** $F_1 = 12\mathbf{i} + 8\mathbf{j}, F_2 = x\mathbf{i} - 15\mathbf{j}, F_3 = 6\mathbf{i} + y\mathbf{j}$

ii $F_1 = 25\mathbf{i} + 18\mathbf{j}, F_2 = x\mathbf{i} + 12\mathbf{j}, F_3 = -31\mathbf{i} + y\mathbf{j}$

b **i** $F_1 = \begin{pmatrix} 3 \\ 8 \end{pmatrix}, F_2 = \begin{pmatrix} -5 \\ 2 \end{pmatrix}, F_3 = \begin{pmatrix} x \\ y \end{pmatrix}$

ii $F_1 = \begin{pmatrix} -2 \\ 1 \end{pmatrix}, F_2 = \begin{pmatrix} 11 \\ 3 \end{pmatrix}, F_3 = \begin{pmatrix} x \\ y \end{pmatrix}$

c **i** $F_1 = x\mathbf{i} + y\mathbf{j}, F_2 = -4\mathbf{j}, F_3 = 8\mathbf{i}$

ii $F_1 = x\mathbf{i} + y\mathbf{j}, F_2 = 15\mathbf{i}, F_3 = -9\mathbf{j}$

d **i** $F_1 = \begin{pmatrix} x \\ -7 \end{pmatrix}, F_2 = \begin{pmatrix} 3y \\ 2x \end{pmatrix}, F_3 = \begin{pmatrix} 11 \\ y \end{pmatrix}$

ii $F_1 = \begin{pmatrix} y \\ 5 \end{pmatrix}, F_2 = \begin{pmatrix} 3x \\ 4y \end{pmatrix}, F_3 = \begin{pmatrix} -2 \\ 25x \end{pmatrix}$

4 A particle is in equilibrium under the action of the forces shown in the diagram. Find the magnitudes of $\mathbf{F}_1$ and $\mathbf{F}_2$.

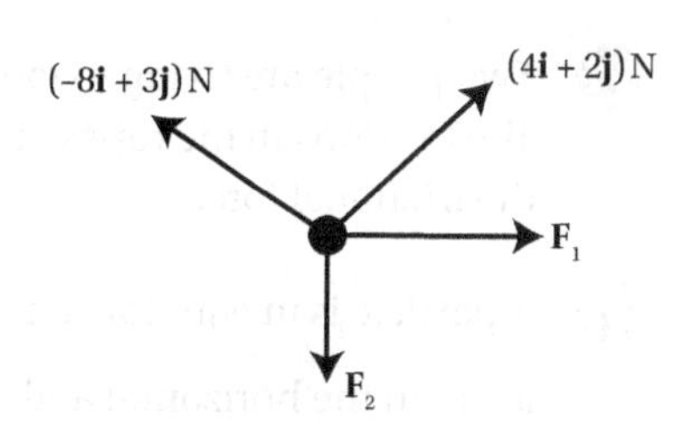

5 A particle is acted on by three forces, $(3\mathbf{i}-6\mathbf{j})$ N, $(-5\mathbf{i}+2\mathbf{j})$ N and $(x\mathbf{i}+y\mathbf{j})$ N. Given that the particle is in equilibrium, find the values of x and y.

6 The diagram shows a particle in equilibrium under the action of four forces. Find the values of a and b.

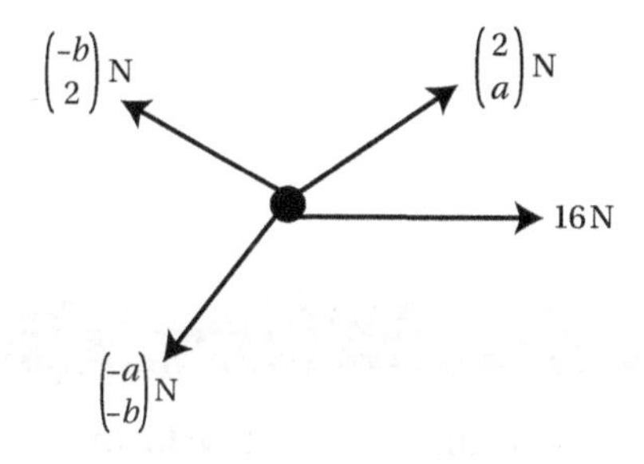

7 In this question use $g = 9.81$ m s^{-2}, giving your final answer to an appropriate degree of accuracy.

A ball of mass 1.4 kg is attached to the floor and the ceiling by two light inextensible strings, as shown in the diagram.

Given that the ball is in equilibrium, find the value of T.

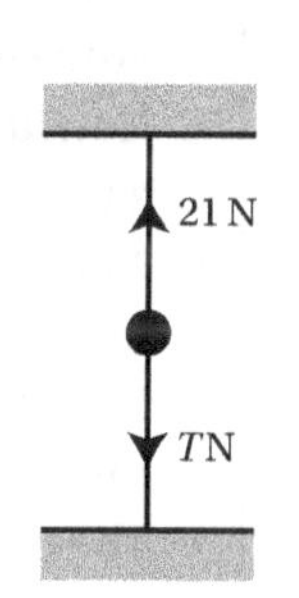

8 Two people are pulling a crate in opposite directions with forces of 16 N and 28 N. Given that the crate is in equilibrium, find the magnitude of the friction force.

9 A particle rests on a smooth floor between two walls. It is attached to the wall on the right by a light inextensible string and the tension in this string is 63 N. It is attached to the wall on the left by a light inextensible string and a light rod. The tension in the string is 82 N.

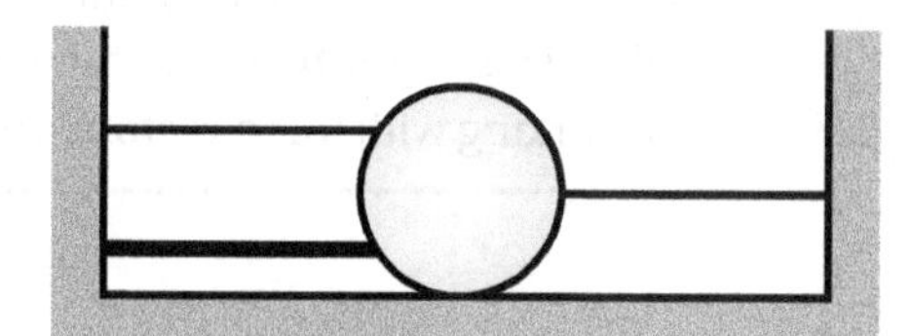

a Find the force in the rod and state whether it is a tension or thrust.

b Find the magnitude and direction of the resultant force on the left wall.

10 In this question use $g = 9.8$ m s^{-2}, giving your final answers to an appropriate degree of accuracy.
A ball of mass m kg is attached to the ceiling by two identical light, inextensible strings. The strings make equal angles with the horizontal. The force on the ceiling from the left string is $\begin{pmatrix} a \\ 7b \end{pmatrix}$ N. The force on the ceiling from the right string is $\begin{pmatrix} a-12 \\ 9b+7 \end{pmatrix}$ N.

Determine the values of a, b and m.

11 Two people are trying to move a heavy box lying at rest on the carpet. They use two light inextensible ropes. The tensions in the ropes are $(32\mathbf{i}+12\mathbf{j})$ N and $(25\mathbf{i}-18\mathbf{j})$ N. The box remains at rest. Find the magnitude of the frictional force.

12 A particle is in equilibrium under the action of three forces, as shown in the diagram.

a Find the horizontal and vertical components of **F**.

b Find the magnitude of **F** and the angle it makes with the horizontal.

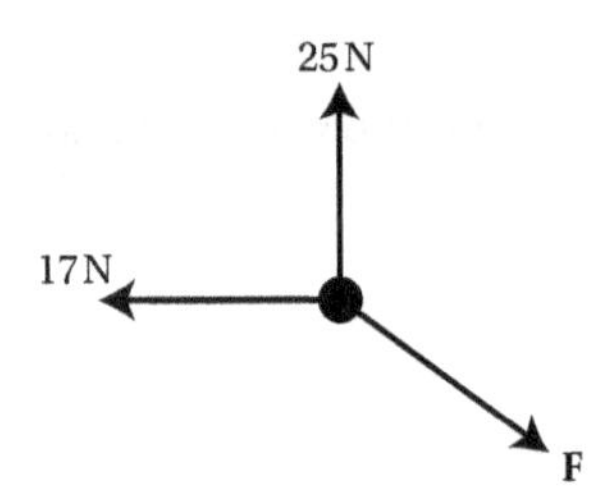

Checklist of learning and understanding

- A **force** can start or stop the motion of an object, change the magnitude or the direction of its velocity. Force is a vector and its magnitude is measured in **newtons** ($1\text{ N} = 1\text{ kg m s}^{-2}$).
- **Newton's first law** states that an object remains at rest or continues to move with a constant velocity unless a force acts on it.
- **Newton's second law** states that the force required to produce a given acceleration is proportional to the mass of the object: $\mathbf{F} = m\mathbf{a}$. The acceleration acts in the same direction as the force.
- If several forces are acting on an object, their combined effect is represented by the **resultant force**. This force is found by adding the vectors corresponding to all the original forces.
- The examples of forces discussed in this chapter are: driving force, braking force, resistance forces (including friction and air resistance), tension, thrust and weight. Whenever you draw a force diagram, you should consider which of these forces need to be included.
- The **weight** of an object is $W = mg$, where m is the object's mass and g is the gravitational acceleration. The mass of an object is fixed but its weight depends on its location in the universe.
- An object is **in equilibrium** if the resultant force is zero.
- When working with force vectors, you can consider horizontal and vertical components separately.

Mixed practice 18

1. A car is travelling with a constant velocity of 20 m s^{-1}, in a straight line, on a horizontal road. A driving force, of magnitude 500 N, acts in the direction of motion and a resistance force opposes the motion of the car. Assume that no other horizontal forces act on the car.

   What is the magnitude of the resistance force acting on the car?

   Choose from these options.

   A 10 000 N **B** 500 N **C** 25 N **D** 0 N

2. A car moves under the action of a constant driving force of magnitude 1360 N. It accelerates from rest to a speed of 12.6 m s^{-1} in 8 seconds.

   a Assuming that any resistance forces can be ignored, find the mass of the car.

   b How would your answer change if a resistance force was included?

3. A box of mass 13 kg slides across a rough floor with an initial speed of 2.6 m s^{-1} and moves in a straight line. It comes to rest after it has travelled 3.7 m. Find the magnitude of the frictional force between the box and the floor.

4. A van of mass 1600 kg travels on a straight horizontal road. The engine produces a constant driving force of magnitude 2170 N. The total resistance force on the van is 865 N.

   a Find the acceleration of the van.

   b Given that the van starts from rest, find the time taken for it to travel 260 m.

5. In this question use $g = 9.8\text{ m s}^{-2}$, giving your final answers to an appropriate degree of accuracy. A box has weight 270 N. The box is being pulled upwards using a light inextensible rope and accelerates with $a = 1.3\text{ m s}^{-2}$.

   a Find its mass.

   b Find the tension in the rope.

   The box is now lowered at a constant speed.

   c Find the new tension in the rope.

   d Explain how you have used the modelling assumption that the rope is:

   i light **ii** inextensible.

6. A particle of mass 2.5 kg is in equilibrium under the action of three forces, $\mathbf{F}_1 = (6.3\mathbf{i} + 1.7\mathbf{j})$ N, $\mathbf{F}_2 = (-3.7\mathbf{i} + 2.1\mathbf{j})$ N and $\mathbf{F}_3 = (p\mathbf{i} + q\mathbf{j})$ N.

   a Find the values of p and q.

   The force $\mathbf{F}_3$ is now removed.

   b Find, in vector form, the acceleration of the particle.

7 Three forces act on a particle. These forces are $(9\mathbf{i}-3\mathbf{j})$ newtons, $(5\mathbf{i}+8\mathbf{j})$ newtons and $(-7\mathbf{i}+3\mathbf{j})$ newtons. The vectors $\mathbf{i}$ and $\mathbf{j}$ are perpendicular unit vectors.

a Find the resultant of these forces.

b Find the magnitude of the resultant force.

c Given that the particle has mass 5 kg, find the magnitude of the acceleration of the particle.

d Find the angle between the resultant force and the unit vector $\mathbf{i}$.

[© AQA 2013]

8 A lift of mass 400 kg is being raised vertically by a cable attached to the top of the lift. It accelerates constantly from rest to a speed of $2\ \mathrm{m\,s^{-1}}$ in 10 seconds.

Modelling the lift as a particle, find the tension in the cable, in terms of g.

Choose from these options.

A $(400g+80)$ N B $(400g-80)$ N C $(400g+2000)$ N D $(400g-2000)$ N

9 A car of mass 750 kg accelerates from $30\ \mathrm{km\,h^{-1}}$ to $40\ \mathrm{km\,h^{-1}}$ while travelling a distance of 200 m. The resistance to the motion of the car has magnitude 380 N.

a Find the driving force of the car.

The car now starts to brake with a braking force of 620 N. The resistance force remains unchanged.

b How long does it take for the car to stop?

10 A box rests in equilibrium on a smooth horizontal floor. Four children pull the box. They use three light inextensible ropes, all in the horizontal plane. The tensions in the ropes are shown in the diagram.

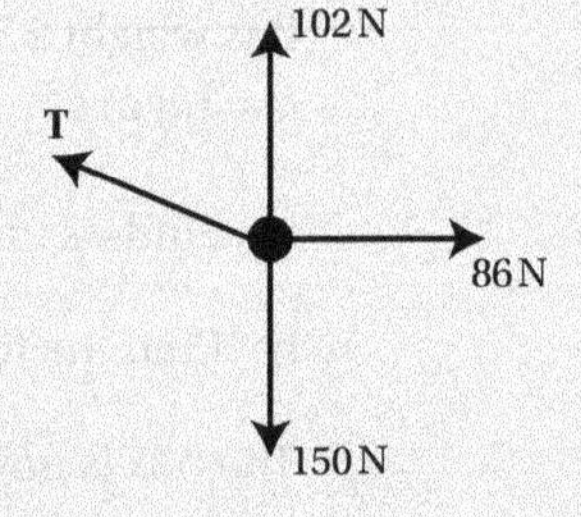

Find the magnitude of the force marked **T** and the angle it makes with the 102 N force.

11 A particle is in equilibrium under the action of the three forces shown in the diagram.

Find the magnitude of **F** and the angle it makes with the horizontal.

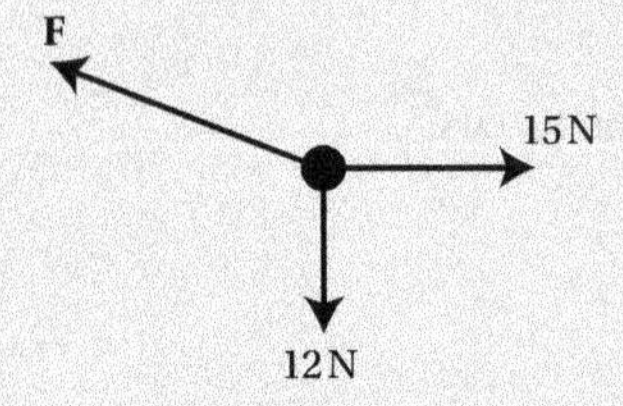

12 A car is travelling at a speed of $20\ \text{m s}^{-1}$ along a straight horizontal road. The driver applies the brakes and a constant braking force acts on the car until it comes to rest.

a **i** After the car has travelled 75 metres, its speed has reduced to $10\ \text{m s}^{-1}$. Find the acceleration of the car.

ii Find the time taken for the speed of the car to reduce from $20\ \text{m s}^{-1}$ to zero.

b The mass of the car is 1400 kg. Given that a constant air resistance force of magnitude 200 N acts on the car during the motion, find the magnitude of the constant breaking force.

[© AQA 2012]

13 A box of mass 24 kg is held in equilibrium, supported by three light rigid rods, as shown in the diagram. All forces act in the vertical plane.

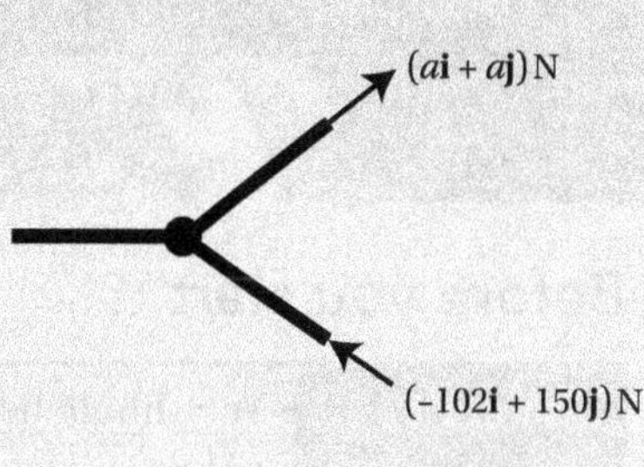

Find the magnitude of the force in the horizontal rod and determine whether it is a tension or a thrust.

14 A particle of mass 4 kg moves under the action of the three forces show in the diagram.

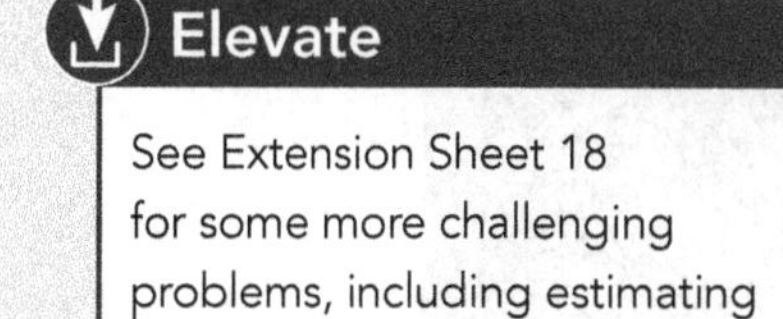

a Find, in vector form, the acceleration of the particle.

b A fourth force is added so that now the particle moves with constant velocity. Find the magnitude of the new force and the angle it makes with the 18 N force.

15 A particle of mass 53 kg is attached to three light inextensible strings, as shown in the diagram. The particle hangs in equilibrium in the vertical plane.

Find the magnitude and direction of **T**.

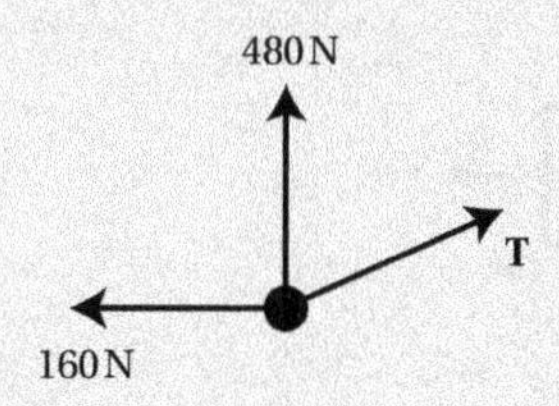

Elevate

See Extension Sheet 18 for some more challenging problems, including estimating forces in different contexts.

19 Objects in contact

In this chapter you will learn how to:

- use Newton's third law: that two objects always exert equal and opposite forces on each other
- calculate the contact force between two objects
- find the tension in a string or rod connecting two objects
- analyse the motion of particles connected by a string passing over a pulley.

Before you start...

Chapter 17	You should be able to use the constant acceleration formulae.	1 A particle accelerates uniformly from 2 m s^{-1} to 8 m s^{-1} while travelling 75 m in a straight line. a Find the acceleration. b How long does the journey take?
Chapter 18	You should be able to find the resultant force and use it in Newton's second law.	2 A particle of mass 2.4 kg is acted on by two horizontal forces, 26 N to the left and 32 N to the right. Find the acceleration of the particle.
Chapter 18	You should be able to calculate and use the weight of an object.	3 Find the weight of a box with mass 34 kg.
Chapter 18	You should know that if a particle is in equilibrium then the resultant force is zero.	4 The particle in the diagram is in equilibrium. Find the values of x and y. y N 5 N 3 N x N 6 N

What happens when particles are connected?

There are many situations in which two objects are in contact or connected in some way; for example, when boxes are stacked on top of each other, or a car pulls a trailer. In such cases, there are contact forces acting between the objects and these need to be taken into account. In this chapter you will examine the normal reaction force and the tension or thrust force in the connecting string, cable or rod. You will also identify modelling assumptions relating to the type of contact or connection between the objects.

Another important force that exists when two surfaces are in contact is friction. You will learn more about it if you study the full A level course.

Section 1: Newton's third law

If you push against a wall, you can feel the wall 'pushing you back.' If you were standing on ice you would probably slide backwards, although you are pushing towards the wall, not away from it. What is the force that you feel? The answer is given by Newton's third law.

Key point 19.1

Newton's third law

If object A exerts a force on object B, then object B exerts a force on object A, with the same magnitude but opposite direction.

This means that whatever force you are exerting on the wall (which is directed towards the wall), the wall exerts a force of equal magnitude back on you. Since this force is away from the wall, you might slip backwards.

force on wall

force on man

Newton's third law is commonly stated as:

'Each action has an equal and opposite reaction.'

An important point to remember is that **the two forces do not act on the same object**, so they do not cancel each other. It is a good idea to draw two separate force diagrams, one for each object.

WORKED EXAMPLE 19.1

Two skaters are standing on ice. They push against each other and start to move away from each other. The first skater, whose mass is 75 kg, moves with acceleration of $3.7\ \text{m s}^{-2}$. The second skater's mass is 63 kg. Assuming that any frictional forces can be ignored, find the acceleration of the second skater.

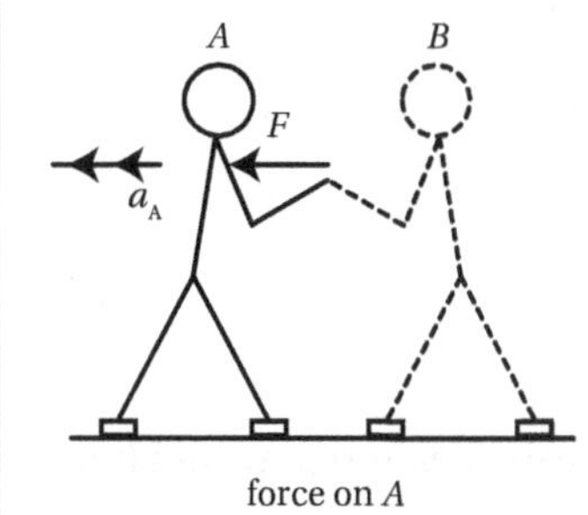

force on A

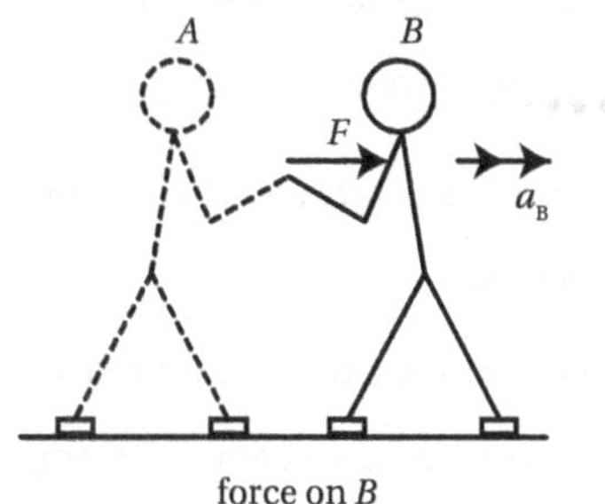

force on B

Draw two separate force diagrams. The force on each skater is away from the other one, so each is pushed backwards.

Continues on next page

Skater A:

$F = m_A a_A$
$= 75 \times 3.7$
$= 227.5\text{ N}$

According to Newton's third law, the two forces have equal magnitude. You can find that force using Newton's second law for the first skater.

Skater B:

$F = m_B a_B$
$277.5 = 63a_B$
$a_B = 4.40\text{ m s}^{-2}$

Now use Newton's second law for the second skater.

EXERCISE 19A

1. Two bumper cars collide. Their masses are 265 kg and 280 kg. While they are in contact, the acceleration of the first car is 5.8 m s^{-2}. Assuming any resistance forces can be ignored, find the acceleration of the second car.

2. Two skaters stand on ice, facing each other. They push off each other and start to accelerate backwards. The mass of the first skater is 78 kg and his acceleration is 3.5 m s^{-2}. The acceleration of the second skater is 4.2 m s^{-2}. Find her mass.

3. Two robots with long extendible arms push against each other with a constant force of 215 N. They start next to each other and slide away from each other in a straight line. The first robot has mass 120 kg and the friction force between its feet and the floor is 96 N. The second robot has mass 90 kg and the friction force between its feet and the floor is 65 N.

   How far are the robots from each other after 2 seconds?

4. Two skaters, of masses 52 kg and 68 kg, stand on ice facing each other and holding hands, with their arms outstretched, 1.2 m apart. They pull towards each other so that the acceleration of the first skater is 0.4 m s^{-2}.

   a Find the acceleration of the second skater.

   The skaters keep holding hands and pulling with the same force.

   b How long does it take for them to come together?

5. Use $g = 9.8\text{ m s}^{-2}$, giving your final answers to an appropriate degree of accuracy.

   An apple of mass 120 g falls from a third floor window, 10 m above the ground.

   a Find the magnitude of the force with which the Earth attracts the apple.

   b How long does the apple take to fall the 10 metres?

   c State the magnitude of the force with which the apple attracts the Earth.

   d The mass of the Earth is $5.972\,18 \times 10^{24}$ kg. If no other forces acted on the Earth, how much would it move in the time it takes the apple to fall 10 m?

Section 2: Normal reaction force

Look at this book resting on your desk. You know that there is weight acting on it, so why is it not accelerating downwards? There must be another, upward force to make the net force zero. This force, exerted by the table on the book, is called the **normal reaction force**.

Key point 19.2

Whenever an object is in contact with a surface, the surface exerts a normal reaction force on it. This force acts in the direction perpendicular to the surface and away from it.

Did you know?

If an object is in contact with a curved surface, you can find the direction of the normal reaction force by calculating the gradient of the normal, as you learnt to do in Chapter 13.

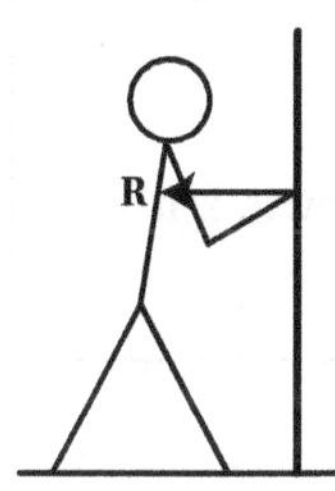

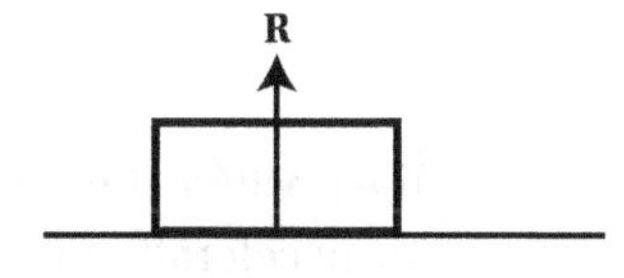

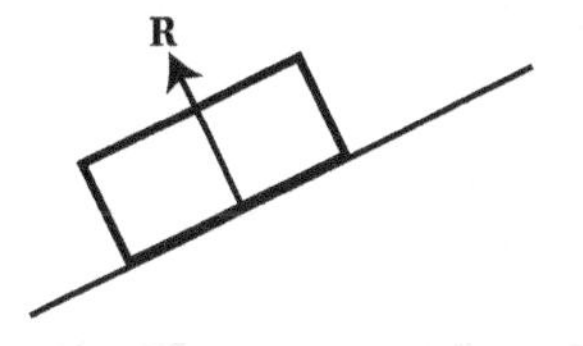

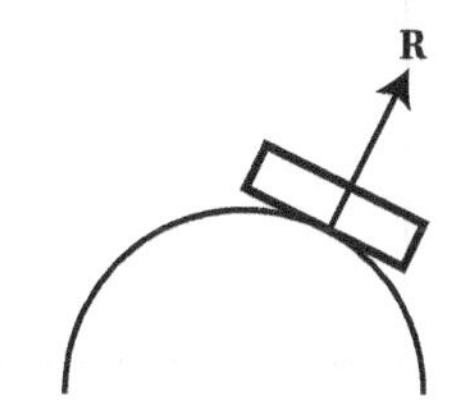

You need to include the normal reaction force, as well as the object's weight, on your force diagram.

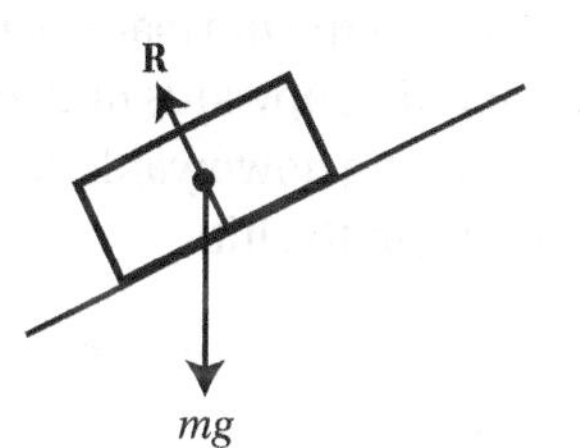

WORKED EXAMPLE 19.2

A person of mass 76 kg is standing in a lift. Use $g = 9.8\,\text{m s}^{-2}$, giving your final answers to an appropriate degree of accuracy. Find the magnitude of the normal reaction force exerted by the floor of the lift on the person when the lift is:

a moving upwards with an acceleration of 2.6 m s^{-2}

b moving downwards with an acceleration of 2.6 m s^{-2}

c moving downwards with a deceleration of 2.6 m s^{-2}.

a

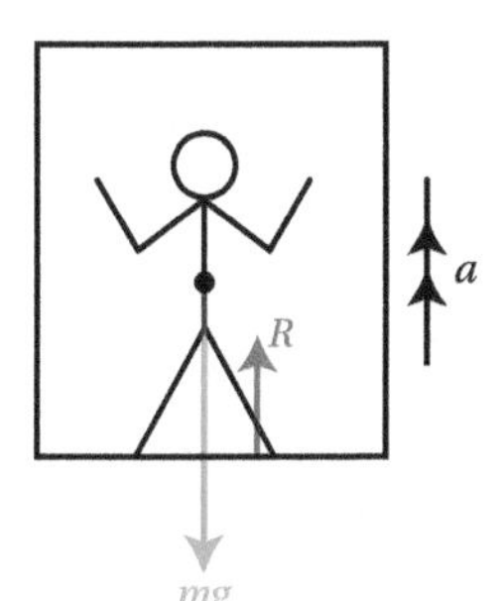

The forces acting on the person are the weight (down) and the normal reaction force (up).

The acceleration of the person is the same as the acceleration of the lift.

$$F = ma$$
$$R - mg = ma$$
$$R - 76 \times 9.8 = 76 \times 2.6$$
$$R = 940\text{ N (2 s.f.)}$$

Since the lift is moving upwards, take the positive direction to be up.

Continues on next page

b

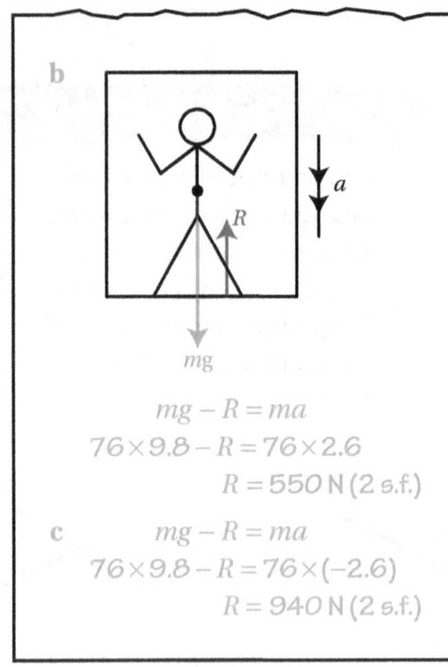

Now it makes sense to take the positive direction to be down as the lift is moving down.

$$mg - R = ma$$
$$76 \times 9.8 - R = 76 \times 2.6$$
$$R = 550 \text{ N (2 s.f.)}$$

c

$$mg - R = ma$$
$$76 \times 9.8 - R = 76 \times (-2.6)$$
$$R = 940 \text{ N (2 s.f.)}$$

The positive direction is still down, but the acceleration is negative.

Notice that the normal reaction force is larger when the lift is accelerating upwards or decelerating downwards, than when it is accelerating downwards. Think about what you would feel if you were standing in the lift:

- if it is accelerating upwards, it feels as if the floor is pushing against your feet
- if it is accelerating downwards it feels as if the floor is moving away from you.

This is because the normal reaction force in the two cases is different.

Common error

Take care not to confuse normal reaction force with Newton's third law; a common misconception is that normal reaction is the 'reaction' to the object's weight. However, the weight and the normal reaction are both acting on the object itself, so they are not a 'Newton's third law' pair. In the example of a book on the table, the reaction to the book's weight is the force with which the book acts on the Earth. The reaction to the normal reaction force is the downward force with which the book pushes the table.

WORKED EXAMPLE 19.3

A book of mass 260 g rests on a horizontal table of mass 65 kg. The table has four legs and the thrust in each leg is the same. Use $g = 9.8 \text{ m s}^{-2}$, giving your final answers to an appropriate degree of accuracy.

a Draw two separate diagrams showing forces acting on the book and the table.
b Find the thrust in each leg of the table.
c State the magnitude and direction of the force exerted on the ground by the table.

a

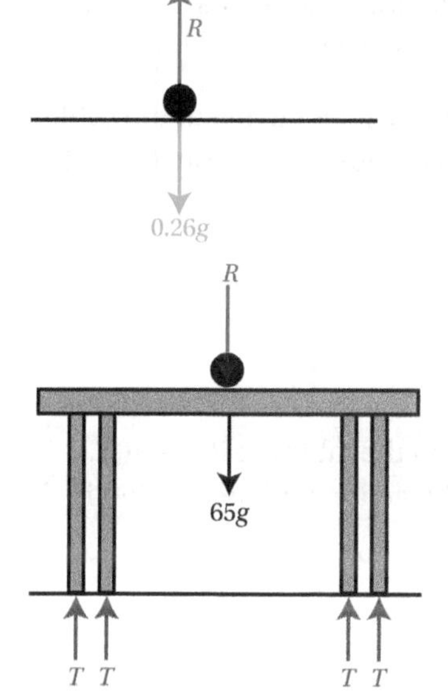

In the diagrams, the 'Newton's third law' pairs of forces are each shown in the same colour.

The forces on the book are its weight and the normal reaction from the table.

The forces on the table are the table's weight, the normal reaction from the book and normal reactions from the ground (the thrust in the legs).

b Forces on the book:

$0.26g = R$

Forces on the table:

$65g + R = 4T$

$\therefore 4T = 65g + 0.26g$

$T = 160 \text{ N}$

The book and the table are in equilibrium, so the net force on each must be zero. This means that 'forces up = forces down.'

c The force on the ground from the table is $4T = 640$ N, directed downwards.

The force exerted on the ground by the table is the same as the normal reaction on the table from the ground.

In this example the normal reaction force exerted by the table on the ground equals the total weight of the book and the table; the ground 'feels' the combined weight of the table and the book. However, this is only the case because all the objects are at rest.

If the table were being lifted by means of a cable attached to it, the normal reaction forces would change depending on the acceleration. You can write Newton's second law equations for both the book and the table, noting that they have the same acceleration.

WORKED EXAMPLE 19.4

A book of mass 260 g rests on a horizontal table of mass 65 kg. The table is being lifted by a cable attached to it and it accelerates upwards. The tension in the cable is 715 N. Find the normal reaction force between the book and the table. Use $g = 9.81\text{ m s}^{-2}$, giving your final answers to an appropriate degree of accuracy.

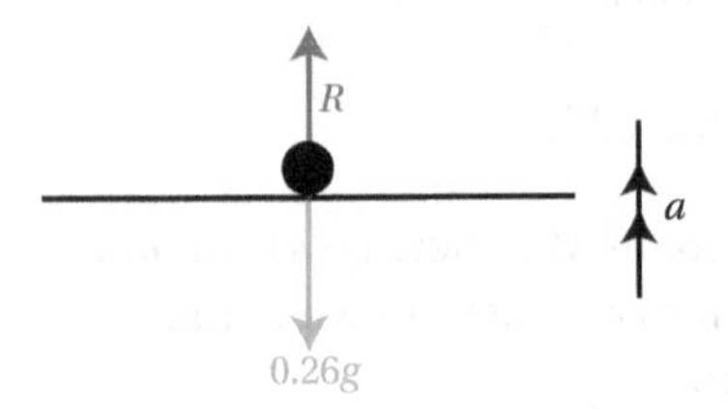

Draw separate force diagrams for the book and the table.

Forces on the book are its weight and the normal reaction.

The book is not attached to the cable, so the tension does not act on it directly.

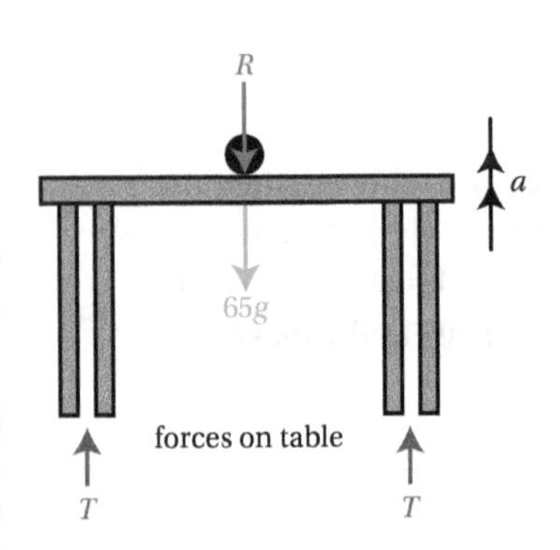

Forces on the table are its weight, normal reaction and the tension in the cable.

Forces on the book:

$$F = ma$$

$$R - 0.26g = 0.26a \quad (1)$$

Forces on the table:

$$F = ma$$

$$715 - 65g - R = 65a \quad (2)$$

Newton's second law for each object: the positive direction is upwards.

(1) + (2):

$$715 - 65g - 0.26g = 65.26a$$

$$74.799 = 65.26a$$

$$a = 1.15\text{ m s}^{-2} \text{ (3 s.f.)}$$

$$R = 0.26a + 0.26g$$

$$= 0.26(1.15) + 0.26g$$

$$= 2.85\text{ N (3 s.f.)}$$

You have two simultaneous equations. You may be able to solve them on a calculator. Otherwise, find a by adding the two equations and then substitute it back to find R.

Notice that the equation for acceleration can be written as $65.26a = 715 - 65.26\,g$. This is the same equation you would get if you considered the table and the book as a single object with mass 65.26 kg (the combined mass of the table and the book) being pulled up, using the tension of 715 N. The normal reaction forces do not appear in this equation.

Key point 19.3

When two objects are in contact and moving the with same acceleration, you can consider them as a single object.

If you want to find the contact force, you need to consider each object separately.

WORKED EXAMPLE 19.5

A person of mass 70 kg stands in a lift of mass 540 kg. The lift is supported by a cable that can be modelled as light and inextensible. Use $g = 10 \text{ m s}^{-2}$, giving your final answers to an appropriate degree of accuracy.

a Draw two diagrams showing all the forces acting on the person and the lift.
The lift is moving downwards and decelerating at 2.1 m s^{-2}.

b Find the tension in the cable.

c Find the magnitude of the normal reaction force exerted on the person by the floor of the lift.

a

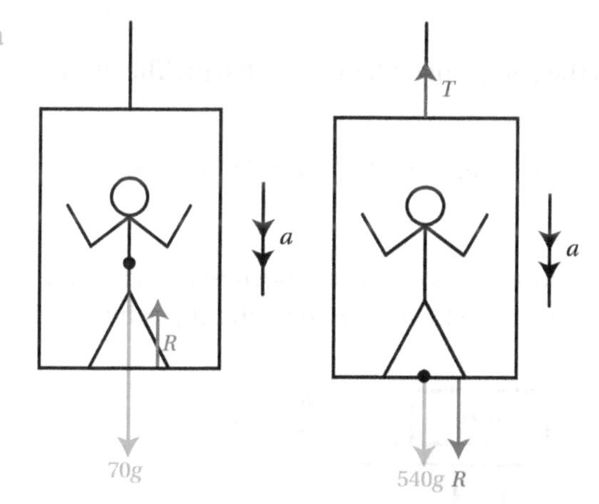

forces on the lift

The forces on the person are weight and normal reaction.

The forces on the lift are its weight, tension (T) and normal reaction.

According to Newton's third law, the two normal reactions have equal magnitudes but opposite directions.

b Newton's second law for the two objects together:

$$F = ma$$

$$610g - T = 610 \times (-2.1)$$

$$T = 7381 \text{ N} = 7000 \text{ N (1 s.f.)}$$

To find the tension, which is an external force on the system, you can treat the lift and the person as a single object of mass $540 + 70 = 610$ kg.

Take the positive direction to be downwards. The acceleration is negative.

c For the person:

$$F = ma$$

$$70g - R = 70 \times (-2.1)$$

$$R = 847 \text{ N} = 800 \text{ N (1 s.f.)}$$

To find the normal reaction force, consider only the forces acting on the person. However, the acceleration is the same as for the whole system.

EXERCISE 19B

In this exercise, unless instructed otherwise, use $g = 9.8\text{ m s}^{-2}$, giving your final answers to an appropriate degree of accuracy.

1 For each situation draw a separate force diagram for each object.

a **i** A book of mass 300 g rests on a table of mass 18 kg. The table is on the floor.

ii A book of mass 120 g rests on the shelf of mass 25 kg. The shelf is on the floor.

b **i** A box of mass 2 kg rests on the table of mass 12 kg. A vertical cable is attached to the table and the tension in the cable is 15 N. The table is not in contact with the floor.

ii A box of mass 5 kg rests on top of a crate of mass 20 kg. The crate is suspended by a vertical cable. The tension in the cable is T.

c **i** A person of mass 65 kg stands in a lift of mass 200 kg. The lift is suspended by a cable and the tension in the cable is T.

ii A person of mass 80 kg stands in a lift of mass 200 kg. The lift is suspended by a cable and the tension in the cable is 2800 N.

d **i** A box of mass 12 kg is suspended by a string from the ceiling of a lift of mass 400 kg. The lift is suspended by a vertical cable.

ii A box of mass 15 kg is suspended by a string from the ceiling of a lift of mass 350 kg. The lift is suspended by a vertical cable.

2 Each diagram shows an object of mass 10 kg resting on a platform. The platform is moving with acceleration shown by a double arrow. Find the magnitude of the normal reaction force exerted on the object by the platform.

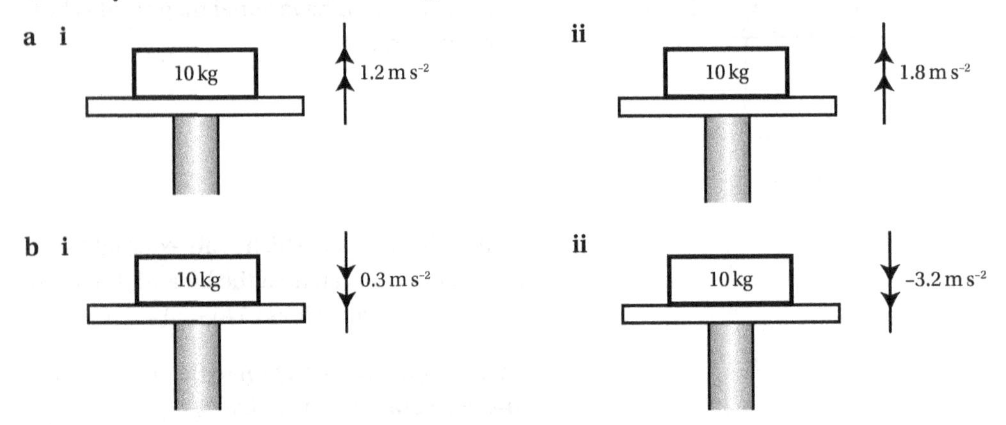

3 In this question use $g = 10\text{ m s}^{-2}$, giving your final answer to an appropriate degree of accuracy.

A book of mass 0.6 kg rests on a horizontal table. A child pushes down on the book with a force of 12 N. Find the magnitude of the normal reaction force between the book and the table.

4 A box of mass 15 kg rests on horizontal ground. The box is attached to a vertical cable that can be modelled as light and inextensible. The magnitude of the normal reaction force between the box and the ground is 68 N. Find the tension in the cable.

5 A crate of mass 120 kg lies on the horizontal floor of a lift. The lift accelerates upwards at 0.4 m s^{-2}. Find the magnitude of the normal reaction force between the crate and the floor of the lift.

6 In this question use $g = 9.81\ \text{m s}^{-2}$, giving your final answer to an appropriate degree of accuracy.

A basket of mass 750 grams is attached to a light inextensible rope and is being lowered at a constant speed. A box of mass 120 grams rests at the bottom of the basket. Find the magnitude of the normal reaction force between the box and the basket.

7 A horizontal plank of mass 27 kg rests on two light vertical supports. A box rests on top of the plank. The thrust in each support is 186 N.

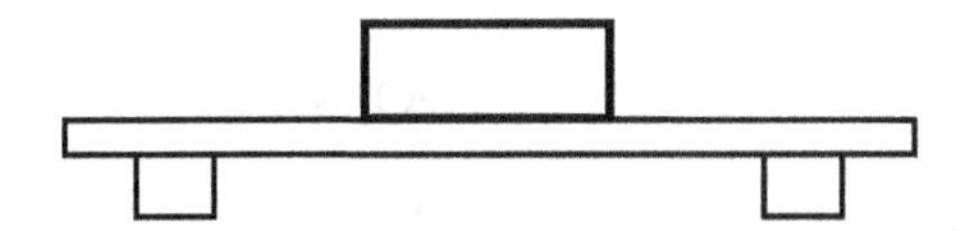

Find the mass of the box.

8 In this question use $g = 9.81\ \text{m s}^{-2}$, giving your final answers to an appropriate degree of accuracy.

A child of mass 40 kg stands in a lift of mass 500 kg. The lift is suspended by a light inextensible cable and accelerates upwards at $0.6\ \text{m s}^{-2}$. Find:

a the tension in the cable

b the magnitude of the normal reaction force between the child's feet and the floor of the lift.

9 A person of mass 85 kg stands in a lift of mass 360 kg. The lift is suspended by a light inextensible cable. Find the magnitude of the normal reaction force between the person's feet and the floor of the lift when the lift is:

a moving downwards and decelerating at $4.2\ \text{m s}^{-2}$

b moving upwards at a constant speed.

10 A cargo container of mass 200 kg is attached to a vertical light inextensible cable. It holds two large boxes, A and B, of masses 450 kg and 350 kg respectively. Box B is on top of box A as shown.

The cargo container is raised with acceleration $a\ \text{m s}^{-2}$. The tension in the cable is 10 000 N.

a Show that $a = 0.2\ \text{m s}^{-2}$

b Find the force exerted by:

i box A on box B. **ii** the container on box A.

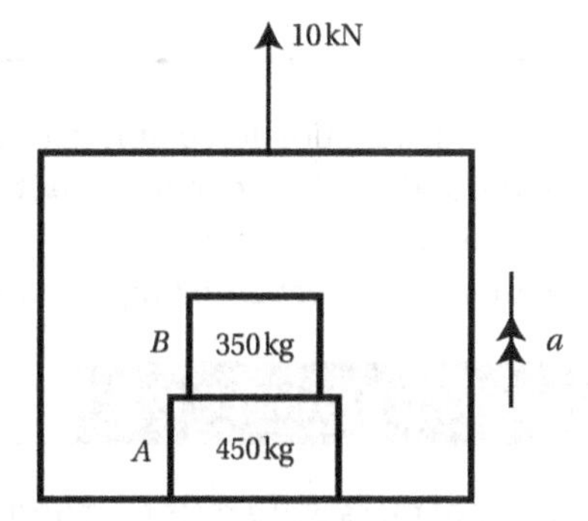

11 A lift of mass 520 kg is supported by a steel rod attached to its base. The rod can withstand the maximum thrust of 15 000 N. The lift can accelerate at $2.5\ \text{m s}^{-2}$ and decelerate at $6.2\ \text{m s}^{-2}$. Find the maximum allowed load in the lift.

12 In this question use $g = 9.81\ \text{m s}^{-2}$, giving your final answers to an appropriate degree of accuracy.

A woman of mass 63 kg stands in a lift of mass 486 kg. The lift is supported by a cable and moves with acceleration of $2.2\ \text{m s}^{-2}$. The magnitude of the normal reaction force between the woman's feet and the floor of the lift is 756 N.

a Is the lift going up or down?

b Find the tension in the cable.

Section 3: Further equilibrium problems

In Chapter 18 you encountered equilibrium, where all the forces on an object balance. When the object is in contact with a surface you need to include a normal reaction force in the calculations.

WORKED EXAMPLE 19.6

A box of mass 16 kg rests on a smooth horizontal table. Four light inextensible strings are attached to the box. The tensions in the string are 12 N, 23 N, 18 N and T N, as shown in the diagram. Use $g = 9.8\text{ m s}^{-2}$, giving your final answers to an appropriate degree of accuracy.

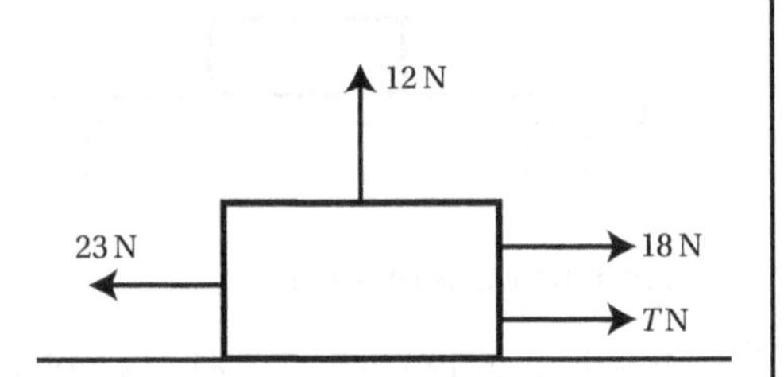

Given that the box is in equilibrium, find:

a the value of T

b the magnitude of the normal reaction force between the box and the table.

a Horizontally:

$18 + T - 23 = 0$

$T = 5\text{ N}$

All the horizontal forces total zero. The components to the right are taken as positive and those to the left as negative.

b Vertically:

$12 + R - 16g = 0$

$R = 16g - 12$

$R = 140\text{ N (2 s.f.)}$

Forces in the vertical direction are tension, weight and normal reaction. All the vertical forces total zero. The components upwards are taken as positive.

The normal reaction force only acts as long as the object is in contact with the surface. When there is another force pulling the object away from the surface, the normal reaction force will decrease. If it reaches zero then the object is no longer in contact with the surface.

WORKED EXAMPLE 19.7

A toy of mass 320 g rests on a horizontal table. A girl pushes vertically down on the toy with a force of 8.2 N. The toy is attached to a light inextensible string, and a boy pulls the string vertically upwards so that the tension in the string is T N. Use $g = 9.81\text{ m s}^{-2}$, giving your final answers to an appropriate degree of accuracy.

a i Express the normal reaction force in terms of T.

ii Hence find the smallest value of T required to lift the toy off the table.

b Find the acceleration of the toy when $T = 12$ N.

Continues on next page

a i

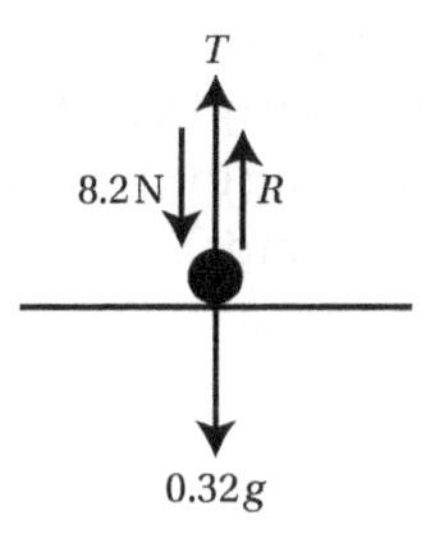

The forces on the toy are its weight, the normal reaction force, the tension and the pushing force of 8.2 N.

$mg + 8.2 - R - T = 0$

$R = 3.136 + 8.2 - T$

$R = 11.34 - T$

When the toy is in equilibrium, the resultant force is zero. Write the equation taking the positive direction to be down.

ii $11.34 - T = 0$

$T = 11.3\text{ N (3 s.f.)}$

The normal reaction force cannot be negative. The toy will leave the table when $R = 0$.

b

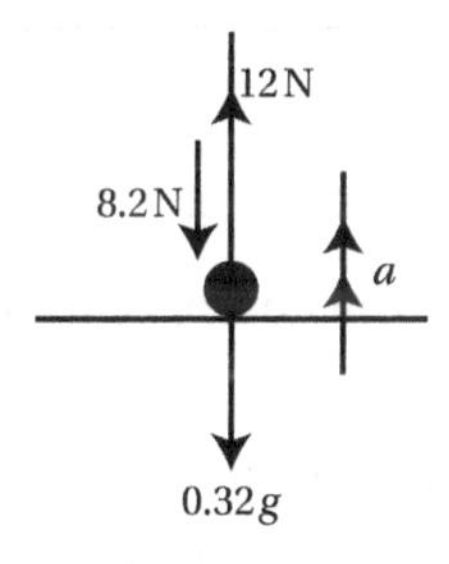

Since $T > 11.3$ N the toy is no longer in contact with the table so $R = 0$. The equilibrium is broken and the toy has upward acceleration, so take up as positive.

$F = ma$

$0.32a = 12 - 0.32g - 8.2$

$0.32a = 0.664$

$a = 2.07\text{ m s}^{-2}\text{ (3 s.f.)}$

Remember that the normal reaction force does not need to act in a vertical direction. It is always perpendicular to the contact surface, and acting away from it.

WORKED EXAMPLE 19.8

A box of mass 1.2 kg is pushed against a rough vertical wall with a force of 40 N and rests in equilibrium. Find:

a the normal reaction force between the box and the wall

b the magnitude and direction of the friction force between the box and the wall.

Use $g = 9.8\text{ m s}^{-2}$, giving your final answers to an appropriate degree of accuracy.

Continues on next page

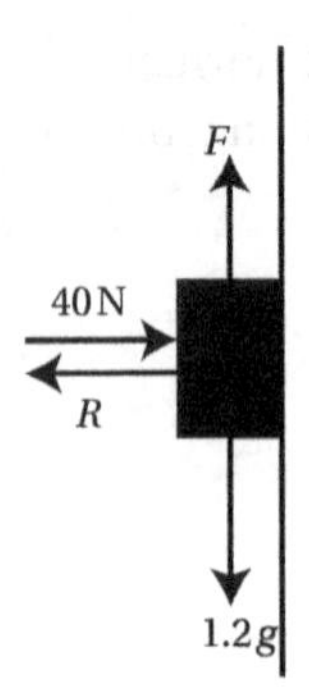

The forces on the box are its weight, the pushing force, the normal reaction force (away from the wall) and the friction force (up along the wall, stopping the box from slipping downwards).

a Horizontally:
$R = 40$ N

Write equilibrium equations for horizontal and vertical directions separately.

b Vertically:
$F = 1.2g$
$F = 12$ N upwards (2 s.f.)

EXERCISE 19C

In this exercise, unless otherwise instructed, use $g = 9.8$ m s^{-2}, giving your final answers to an appropriate degree of accuracy.

1. In this question use $g = 9.81$ m s^{-2}, giving your final answer to an appropriate degree of accuracy.

   A crate of mass 150 kg rests on a horizontal floor. A vertical cable is attached to the crate and the tension in the cable is 820 N. Find the normal reaction force between the crate and the floor.

2. A box of mass 68 kg rests in equilibrium on a horizontal table, as shown in the diagram.

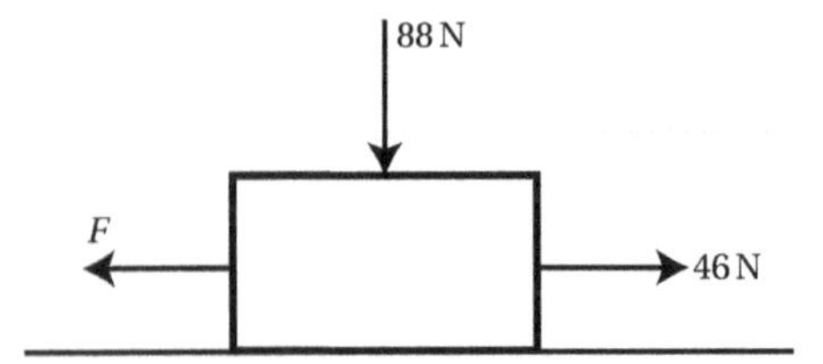

   Find:

   a the value of F

   b the magnitude of the normal reaction force in terms of g.

3. In this question use $g = 10$ m s^{-2}, giving your final answer to an appropriate degree of accuracy.

   A box rests in equilibrium on a horizontal table. The mass of the box is 15 kg and the mass of the table top is 70 kg. The table is supported by four light legs, as shown in the diagram.

   Find the thrust in each leg, assuming they are all the same.

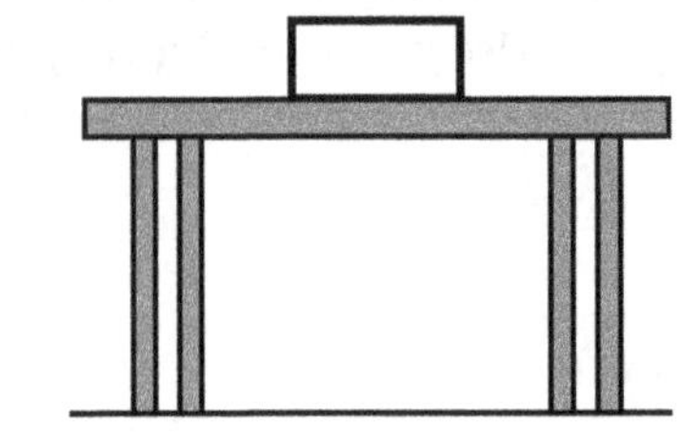

4 In this question use $g = 9.81\ \text{m s}^{-2}$, giving your final answer to an appropriate degree of accuracy.

A book of mass 320 g rests in equilibrium on a horizontal table. Find the magnitude of the normal reaction force between the book and the table in each situation.

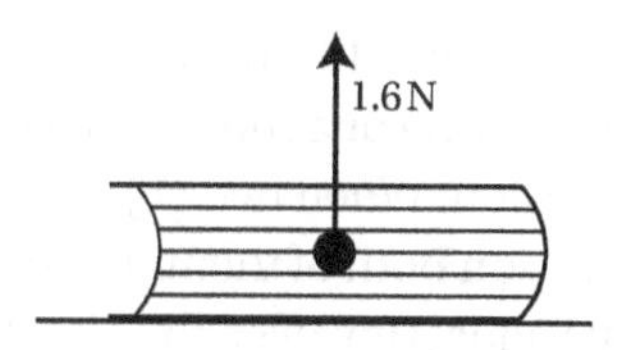

a A light inextensible string is attached to the book, as shown in the diagram. The string is vertical and the tension in the string is 1.6 N.

b The string is now removed and a girl pushes vertically down on the book with a force of 8.2 N.

5 A ball of mass 6.5 kg is suspended by a vertical string and is in contact with a horizontal table. The string can be modelled as light and inextensible. Find, in terms of g:

a the magnitude of the normal reaction force between the ball and the table when the tension in the string is 38 N

b the minimum tension force required to lift the ball off the table.

6 Blocks A, B and C of masses 13 kg, 21 kg and 18 kg, are stacked on top of each other, as shown in the diagram. A light inextensible string is attached to block C and the system is in equilibrium.

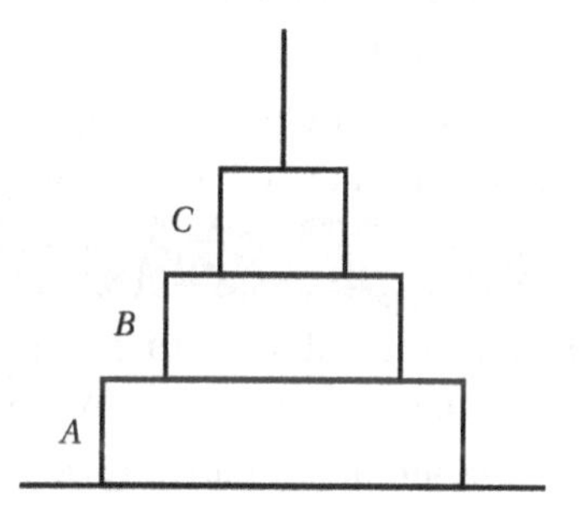

The magnitude of the normal reaction force between blocks A and B is 260 N.

Find the tension in the string and the magnitude of the normal reaction force between blocks B and C.

7 A small box of mass 1.2 kg is held in equilibrium between two rough planks. The planks are vertical and the friction forces between the box and the two planks are equal. Each plank is held in position by a horizontal force of magnitude 140 N.

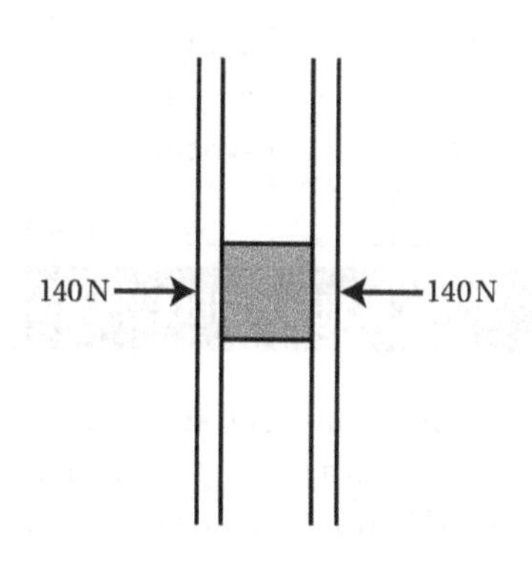

Find:

a the magnitude of the friction force between each plank and the box

b the normal reaction force between each plank and the box.

8 A box of mass 30 kg rests in equilibrium on a rough horizontal floor. A light inextensible string is attached to the box, as shown in the diagram.

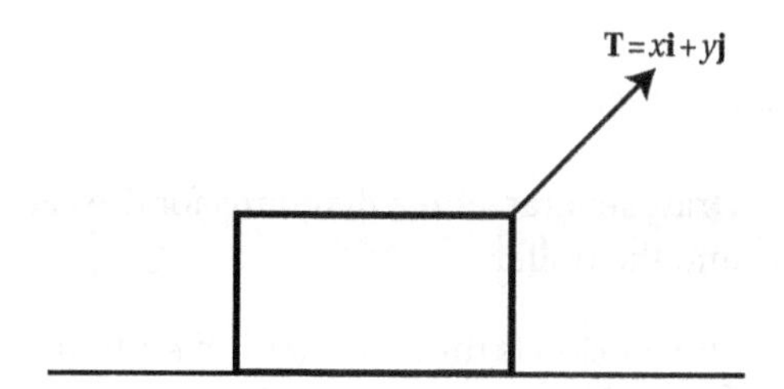

The friction force has magnitude $18g$ N and the magnitude of the normal reaction force is $12g$ N.

Find, in terms of g, the magnitude of the tension in the string.

Section 4: Connected particles

In Chapter 18, you looked at examples of objects being pulled by means of a string or a rod. You saw that one of the forces acting on the object is tension. When two objects are connected by a taut string then the tension acts on both of them. The magnitude of the tension is the same at both of its ends. However, the directions at the two ends are different, because the tension acts away from the object that is attached to its end.

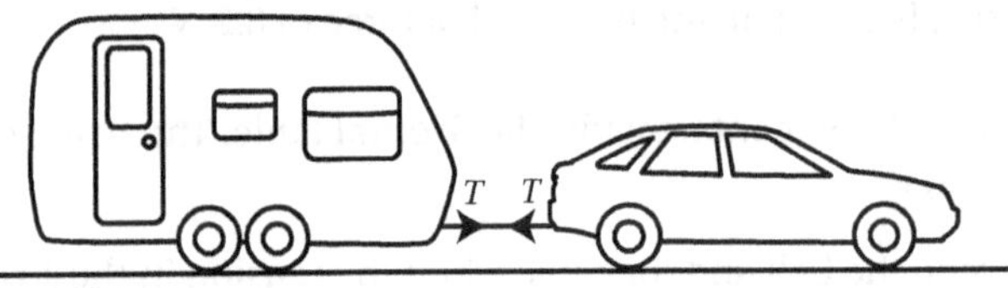

If the string is inextensible, the two connected objects will move with the same speed and same acceleration. Any external forces, such as driving or resistance forces, may be different for each object. Remember that, to find the acceleration, you can consider them as a single object and not include the connecting tension forces.

Two main modelling assumptions have been made here, and it is important that you know what specific effect each has.

Tip

The term string is used for anything that transmits tension but not compression forces, such as a rope, wire, cable, chain. The term rod is used for an object that transmits tension and compression forces, such as a tow bar or plank. A rod cannot stretch or compress itself (it is rigid).

Key point 19.4

It is assumed that a string connecting two particles is:

Term	What it means	How it is used
• Inextensible	The string cannot stretch, e.g. not an elastic band or spring.	The acceleration and velocity of each connected particle is the same (if the string is taut).
• Light	The string has a mass which is negligible in the context of the system.	1. When treating the whole system as a single particle the mass of the string is ignored. 2. In vertical systems the tension of the string is the same throughout. This is because lower parts do not 'pull' on upper parts.

WORKED EXAMPLE 19.9

A car of mass 780 kg is pulling a trailer of mass 560 kg by means of a light, inextensible cable. The engine produces a driving force of magnitude 1800 N. The total resistance forces on the car and the trailer are 800 N and 600 N, respectively.

a Find the acceleration of the car.

b Find the tension in the cable.

c Explain how you used the assumption that the string is:

i light

ii inextensible.

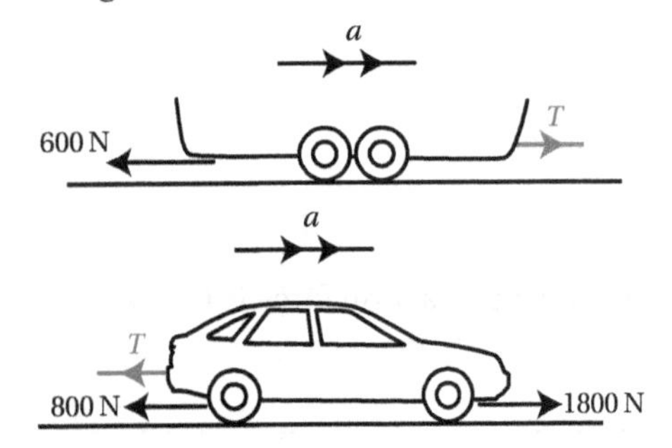

Draw separate force diagrams for the car and the trailer.

The tension is the same at both ends of the cable, and both car and the trailer have the same acceleration.

The driving force acts on the car only.

Continues on next page

a For the car and trailer together:

$$F = ma$$
$$1800 - 800 - 600 = 1340a$$
$$a = 0.299\ \text{m s}^{-2}.$$

Use $F = ma$ for the combined object, taking into account only the external forces.
$m = 780 + 560 = 1340$ kg

b For the trailer:

$$F = ma$$
$$T - 600 = 560 \times 0.299$$
$$T = 767\ \text{N}$$

To find the tension consider each object separately. You can choose which equation to use. In this case use the trailer equation, as there are fewer forces to include.

c i The mass of the cable wasn't included in the total mass of the two objects.

Describing the cable as 'light' means that you can ignore its mass.

ii Both the car and the trailer have the same acceleration.

'Inextensible' means the cable doesn't stretch.

WORK IT OUT 19.1

Particles P and Q, of masses 0.6 kg and 0.4 kg, are connected by a light inextensible string. The particles are initially at rest and the string is taut. A force of magnitude 15 N starts acting on P, in the direction away from Q. The friction force between P and the ground is 2 N and the contact between Q and the ground is smooth. Find the acceleration of the two particles.

Which is the correct solution? (There may be more than one.) Identify the errors made in the incorrect solutions.

Solution 1

Considering P and Q as a single particle:

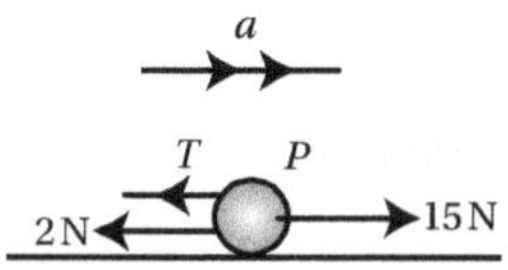

$15 - T - 2 = 1a$
$13 - T = a$

Forces on Q:

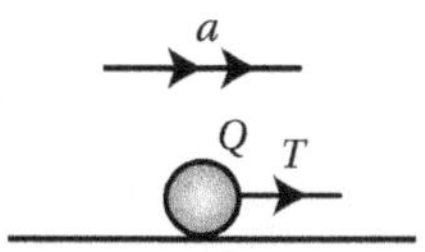

$T = 0.4a$

So:

$13 - 0.4a = a$
$13 = 1.4a$
$a = 9.3\ \text{m s}^{-2}$

Solution 2

Considering P and Q as a single particle: $m = 1$ kg

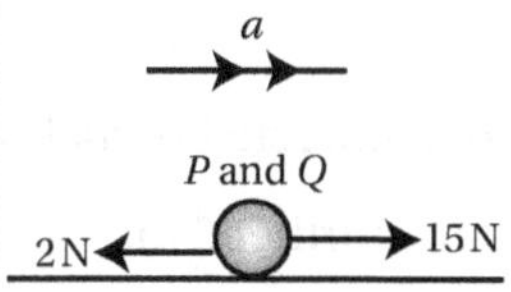

$15 - 2 = 1a$
$a = 13\ \text{m s}^{-2}$

Solution 3

Force on P:

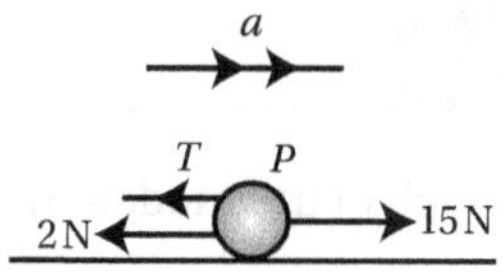

$15 - T - 2 = 0.6a$
$13 - T = 0.6a$

Force on Q:

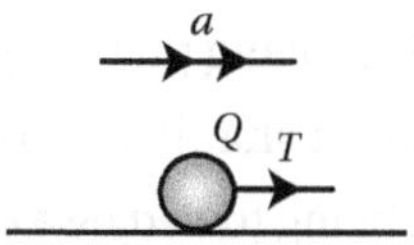

$T = 0.4a$

So:

$13 - 0.4a = 0.6a$
$a = 13\ \text{m s}^{-2}$

What would happen in Worked example 19.9 if the car started to brake? The braking force is acting on the car only, so there is nothing to slow down the trailer. The cable would go slack and the trailer would move closer to the car. This is why, when towing you should always use a rigid object, such as a tow bar, which can exert thrust as well as tension.

WORKED EXAMPLE 19.10

A trailer is attached to a car by a light tow bar. The mass of the trailer is 350 kg and the mass of the car is 680 kg. The car starts to brake and decelerates at 4.6 m s^{-2}. Assuming all other resistance forces can be ignored, find the thrust in the tow bar.

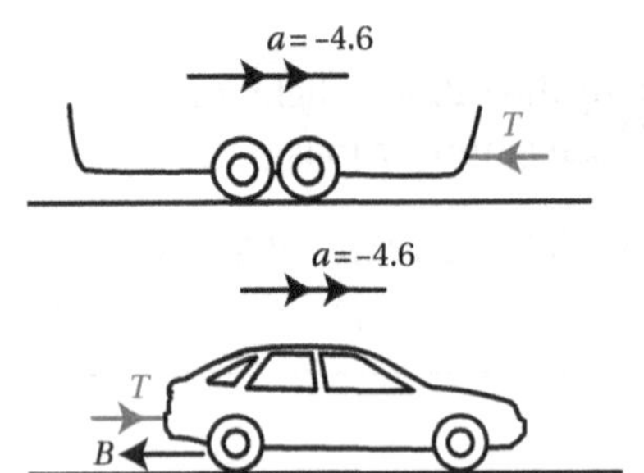

Draw a separate force diagram for each object. The thrust force is directed towards the object. The braking force acts on the car only.

For the car:

$F = ma$

$T - B = 680 \times (-4.6)$

Since you want to find the connecting thrust force between the two objects, write separate $F = ma$ equations for each. The acceleration is negative.

For the trailer:

$F = ma$

$-T = 350 \times (-4.6)$

$T = 1610\text{ N}$

Actually, you only need the second equation to find T.

EXERCISE 19D

1. Discuss which of the modelling assumptions for connected particles hold in each situation.

   a Two cars colliding and one pushing the other a further 5 m.

   b A steel chain being used to lift a chair.

   c A steel chain being used to lift a car.

   d A pendulum hanging on an old clock.

   e A water-skier being pulled by a boat.

   f A glass-blower shaping a bottle between two pliers.

   g A caravan being towed by a car.

   h A steel cable hanging between supports on a bridge.

2. A car of mass 750 kg is towing a trailer of mass 350 kg. The car's engine produces the driving force of 15 kN. The resistance forces on the car and the trailer are 600 N and 400 N, respectively. Find:

   a the acceleration of the car and the trailer

   b the tension in the tow bar.

3 Particle P of mass 4.5 kg is being pulled by a light inextensible string. Another light inextensible string is attached to the other side of P. Particle Q, of mass 6 kg, is attached to the other end of this string. The particles move with acceleration $2.1\ \text{m s}^{-2}$ in a straight line on a smooth horizontal table.

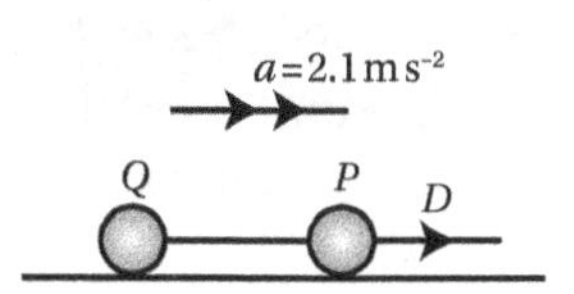

Calculate the tension in each string.

4 Two identical boxes, each of mass 10 kg, are connected by a light inextensible cable. One box is pushed away from the other one with a force of 75 N. The boxes move in a straight line at constant velocity on a rough horizontal table. Find the magnitude of the friction force between each box and the table.

5 A car of mass 1200 kg is towing a trailer of mass 400 kg by means of a light tow bar. The resistance forces acting on the car and the trailer are 500 N and 300 N respectively. The car starts to brake and decelerates at $1.2\ \text{m s}^{-2}$.

Find the magnitude of the braking force.

6 A box of mass 24 kg is pulled across rough horizontal floor with a force of F N. The friction force between this box and the floor is 80 N. Another box, of mass 15 kg, is attached to the first box by a light inextensible string. The friction force between the second box and the floor is 50 N.

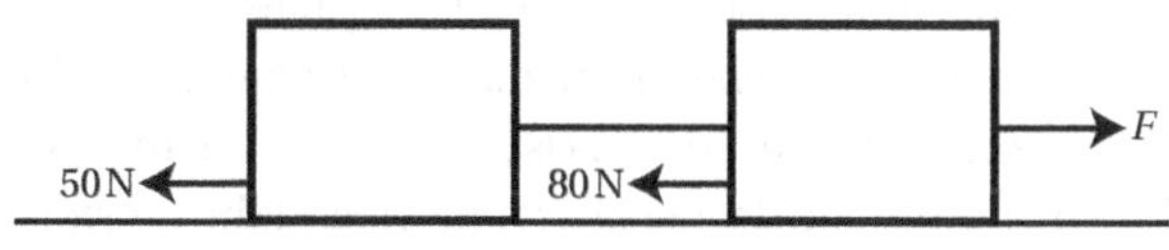

a The string connecting the two boxes will break if the tension exceeds 120 N. Find the largest possible value of F.

The string breaks when the boxes are moving with the speed of $2.4\ \text{m s}^{-1}$.

b Assuming the two boxes do not collide, how long does it take for the second box to stop?

7 A crate of mass 35 kg is suspended by a light inextensible cable. Another crate, of mass 50 kg, is attached to the bottom of the first crate by another light inextensible cable. Find the tensions in the two cables when the crates are:

a being raised with acceleration $0.8\ \text{m s}^{-2}$

b being lowered at constant speed.

8 A train is made up of a locomotive of mass 4500 kg and two carriages each of mass 2500 kg. The train is accelerating at $0.9\ \text{m s}^{-2}$. The resistance force acting on the locomotive is 1200 N and the resistance force acting on each carriage is 500 N. Find:

a the driving force produced by the engine

b the tension in each coupling.

9 A train consists of a locomotive of mass 4000 kg and two carriages each of mass 2000 kg. The train is decelerating at $2\ \text{m s}^{-2}$. The resistance forces are 9000 N on the locomotive, 6000 N on the first carriage and 3000 N on the second carriage.

a Determine whether the locomotive is driving or braking.

b Find the magnitude of the force in each coupling, stating whether it is a tension or a thrust.

Section 5: Pulleys

A string connecting two objects sometimes passes over a pulley, for example, to allow a heavy crate to be raised more easily. If, as before, the string is inextensible then the acceleration of both objects will be the same.

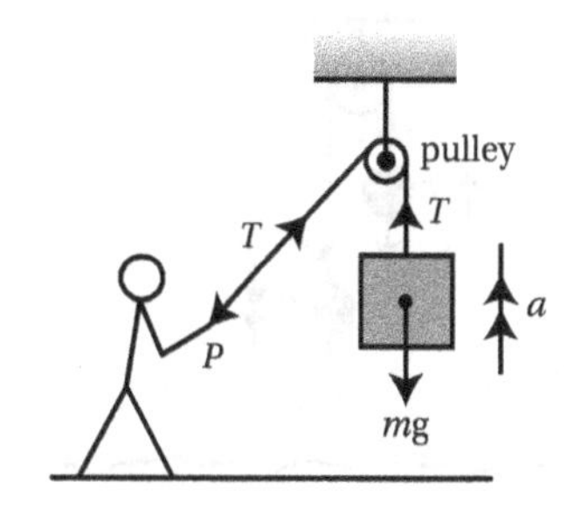

If the pulley has some mass or if there is friction between the pulley and the axis it rotates around, then the magnitude of the tension will not be the same at both ends of the string. However, if you make some further modelling assumptions then the tension is the same throughout the string.

Key point 19.5

It is assumed that a pulley is:

- smooth
- light

The presence of a smooth, light pulley therefore does not alter the method used for connected particles in Section 4.

WORKED EXAMPLE 19.11

A crate of mass 50 kg is attached to a light rope hanging over the edge of a wall and passing over a smooth pulley. A man pulls the other end of the rope, keeping it horizontal. The crate is moving upwards with acceleration of 0.3 m s^{-2}. Use $g = 10$ m s^{-2}, giving your final answer to an appropriate degree of accuracy. Find the tension in the rope.

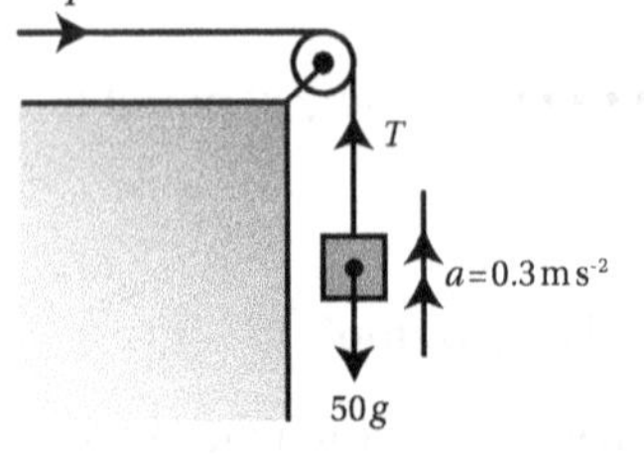

Draw a diagram showing all the forces.

For the crate:

$$F = ma$$

$$T - 50g = 50 \times 0.3$$

$$T = 500 \text{ N (1 s.f.)}$$

Write the force equation for the crate, taking the positive direction to be up.

Tip

When there is a pulley involved, always consider the two particles separately: use $F = ma$ to write an equation of motion for each particle.

Remember that the tension in the rope or string exists only as long as it is taut. Once the string goes slack the tension force becomes zero.

WORKED EXAMPLE 19.12

Particles P and Q are connected by a light inextensible string passing over a smooth pulley. The mass of P is 2.3 kg and the mass of Q is 3.1 kg. Initially Q is held at rest 0.8 m above the floor and P is 0.9 m below the pulley. Use $g = 9.8$ m s^{-2}, giving your final answers to an appropriate degree of accuracy.

The particles are released from rest and the system moves freely under gravity. Find:

a the speed of Q when it reaches the ground

b the time it takes P to reach the pulley from the moment it is first released.

a

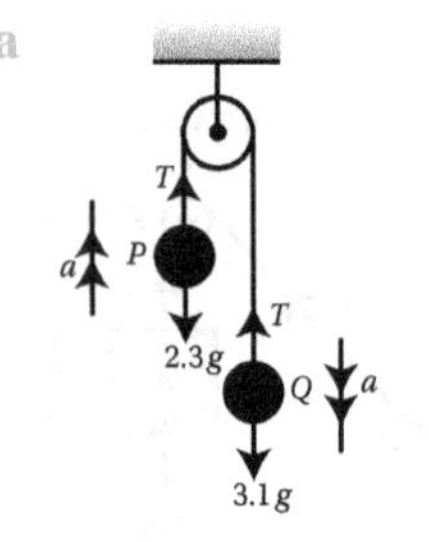

To find the speed of Q you need to know its acceleration.

Draw a force diagram for each particle.

Continues on next page

For particle P:

$F = ma$

$T - 2.3g = 2.3a$

Write an equation for each particle, taking the positive direction to be its direction of motion: up for P...

For particle Q:

$F = ma$

$3.1g - T = 3.1a$

... and down for Q.

$0.8g = 5.4a$

$a = 1.45\ \text{m s}^{-2}$

Eliminate T by adding the two equations.

$s = 0.8$
$u = 0$
$a = 1.45$
$v = ?$

You need the final speed given the initial speed, acceleration and distance.

$v^2 = u^2 + 2as$
$= 0 + 2(1.45)(0.8)$
$= 2.32$

$\therefore v = \sqrt{2.32} = 1.5\ \text{m s}^{-1}$ (2 s.f.)

b

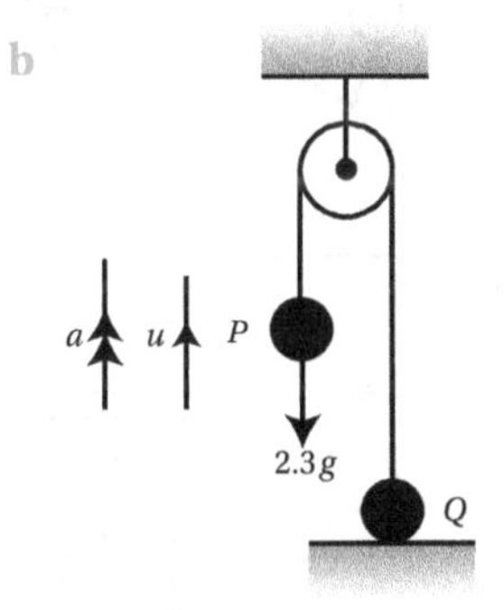

Once Q is on the ground there is no more tension in the string, so the only force acting on P is its weight. This means that its acceleration is g downwards. However, P has an upward speed (equal to the speed Q had when it hit the ground), so it will continue to move upwards for a while; it may or may not reach the pulley before stopping and falling down again.

$s = 0.1$
$u = 1.52$
$a = -9.8$
$t = ?$

P has moved 0.8 m from its initial position, so it is now 0.1 m from the pulley. Its speed is $1.52\ \text{m s}^{-1}$ upwards.

$s = ut + \frac{1}{2}at^2$

$0.1 = 1.52t - 4.9t^2$

$4.9t^2 - 1.52t + 0.1 = 0$

$t = 0.0940$ or $t = 0.217$

$\therefore t = 0.0940\ \text{s}$

P reaches the pulley 0.0940 seconds after Q hits the ground.

P is moving freely under gravity so it will pass the pulley twice, once on the way up and once on the way down. You want the first of the two times.

Continues on next page

Time taken for Q to reach the ground:

$v = 1.52$

$u = 0$

$a = 1.45$

$t = ?$

$v = u + at$

$1.52 = 0 + 1.45t$

$t = 1.05\text{ s}$

Total time taken for P to reach the pulley:

$t = 1.05 + 0.094 = 1.1\text{ s (2 s.f.)}$

You now need to add the time it took for Q to reach the ground. To do this, use the information from part **a**.

EXERCISE 19E

In this exercise, unless instructed otherwise, use $g = 9.8\text{ m s}^{-2}$, giving your final answers to an appropriate degree of accuracy.

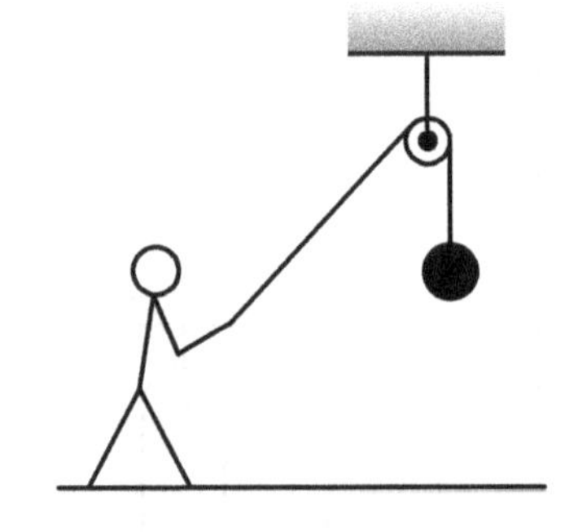

1. In this question use $g = 9.81\text{ m s}^{-2}$, giving your final answer to an appropriate degree of accuracy.

   A man is holding a crate of mass 76 kg by means of a light inextensible rope passing over a smooth pulley.

   Given that the crate is in equilibrium, find the force exerted by the man on the rope.

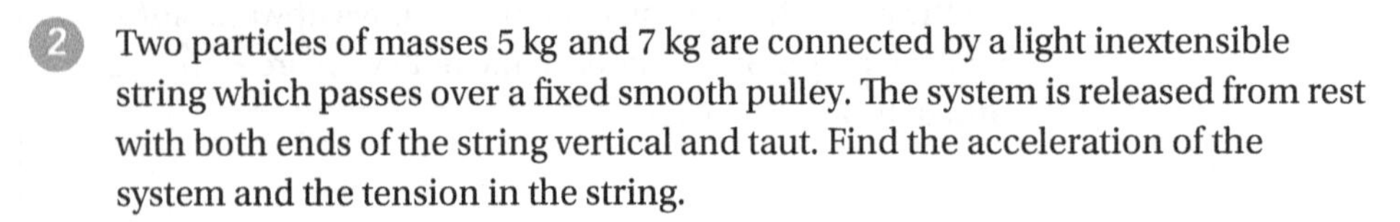

2. Two particles of masses 5 kg and 7 kg are connected by a light inextensible string which passes over a fixed smooth pulley. The system is released from rest with both ends of the string vertical and taut. Find the acceleration of the system and the tension in the string.

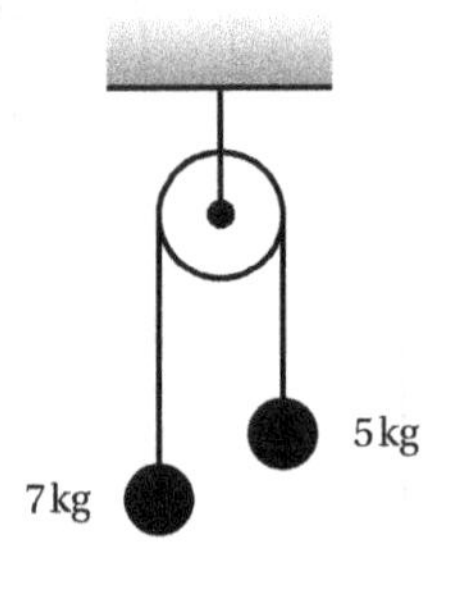

3. Box A, of mass 15 kg, rests on a rough horizontal table. It is connected to one end of a light inextensible string which passes over a smooth pulley fixed to the edge of the table. Box B, of mass 6 kg, is attached to the other end of the string.

   a Given that the system is in equilibrium, find the magnitude of the frictional force between box A and the table.

   b Given instead that the contact between box A and the table is smooth, find the acceleration of the system and the tension in the string.

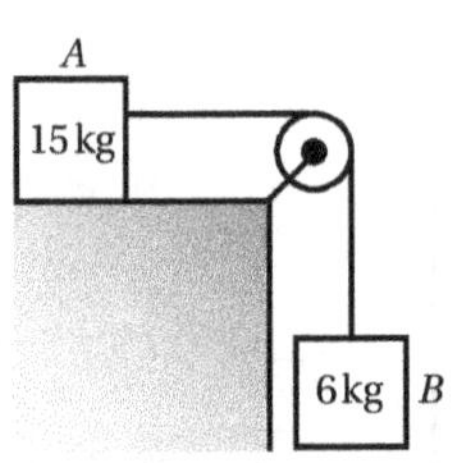

4. In this question use $g = 9.81\text{ m s}^{-2}$, giving your final answer to an appropriate degree of accuracy.

   A box of mass 7 kg lies on a smooth horizontal table. It is attached to one end of a light inextensible string, which passes over a smooth pulley placed at the edge of the table. A ball hangs on the other end of the string. The system is released from rest, with the string taut, and the ball starts to move downwards. The tension in the string is 18.2 N. Find the mass of the ball.

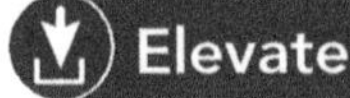

Elevate

See Support Sheet 19 for another example with pulleys and for further practice questions.

5 Two particles have masses m and km, with $k > 1$. The particles are connected by a light inextensible string passing over a smooth pulley. The system is released from rest and moves with acceleration $\frac{2}{3}g$.

Find the value of k.

6 In this question use $g = 9.81\ \text{m s}^{-2}$, giving your final answers to an appropriate degree of accuracy.

A particle A of mass 4 kg is attached by light inextensible strings to two other particles, B and C, of masses 7 kg and 5 kg, respectively. The string connecting A to B passes over a smooth pulley, as shown in the diagram.

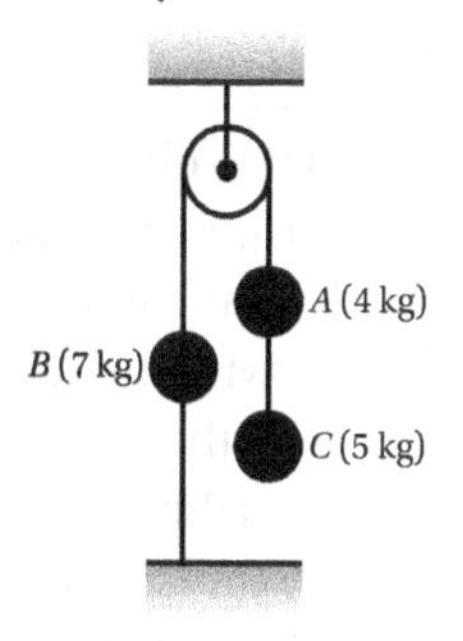

B is attached to the floor by another light inextensible string.

a Given that the system is in equilibrium, find the tension in the string connecting B to the floor.

b This string is now removed. Find the acceleration of the system.

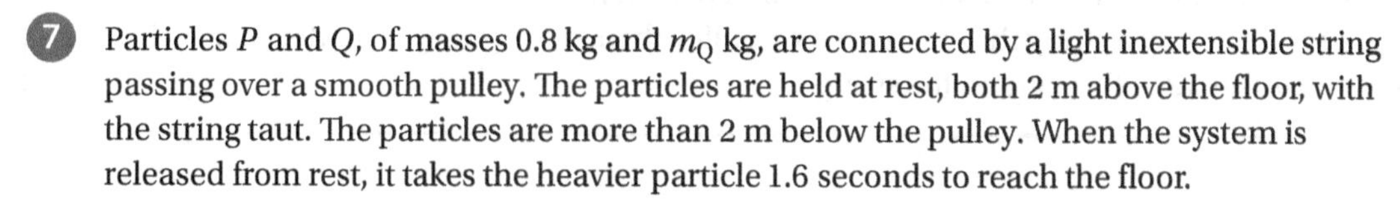

7 Particles P and Q, of masses 0.8 kg and m_Q kg, are connected by a light inextensible string passing over a smooth pulley. The particles are held at rest, both 2 m above the floor, with the string taut. The particles are more than 2 m below the pulley. When the system is released from rest, it takes the heavier particle 1.6 seconds to reach the floor.

Find the two possible values of m_Q.

8 The diagram shows a tape passing over a fixed smooth pulley. One end of the tape is fixed to the ceiling and the other is attached to a box of mass 3 kg. A smooth cylinder of mass 8 kg is placed in a loop formed by the tape.

The system is released from rest and the cylinder starts to accelerate downwards.

a The cylinder moves downwards a distance x. How far upwards does the box move in that time?

b The acceleration of the cylinder is a. Write down an expression, in terms of a, for the acceleration of the box.

c Find the tension in the tape, in terms of g.

9 A box of mass 10 kg rests on a rough horizontal table.
It is attached to two other boxes, of masses 10 kg and m kg, by two light inextensible strings.
Each string passes over a smooth pulley, as shown in the diagram.

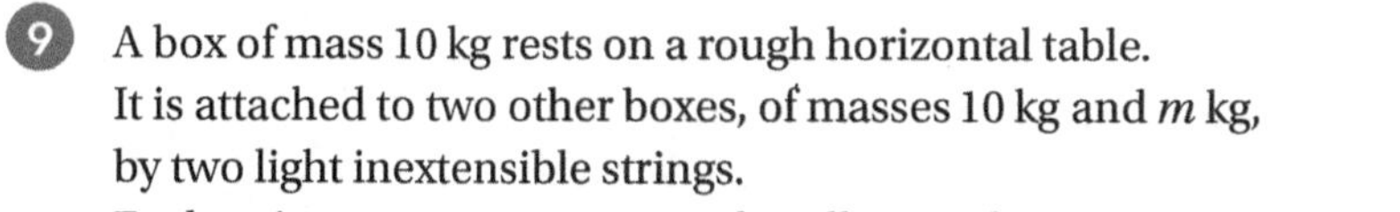

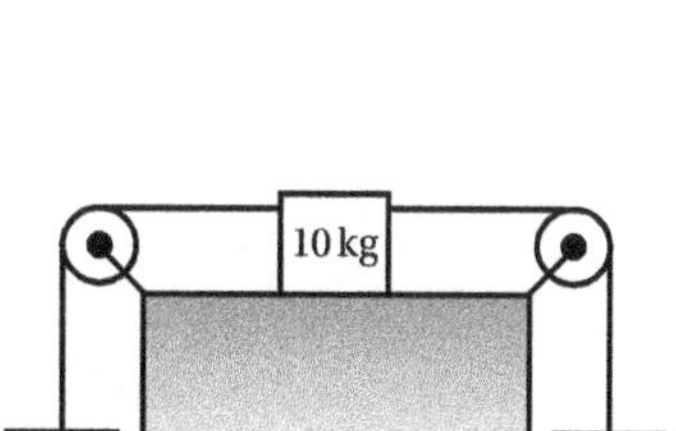

The system is in equilibrium. Given that the maximum possible magnitude of the friction force between the box and the table is 49 N, find the range of possible values of m.

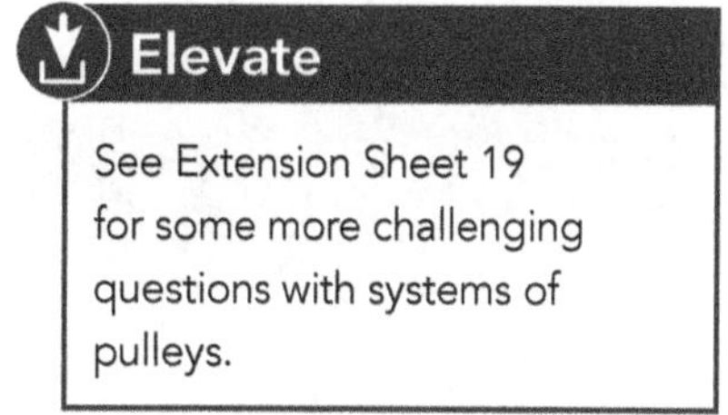

See Extension Sheet 19 for some more challenging questions with systems of pulleys.

Checklist of learning and understanding

- **Newton's third law** states that if object A exerts a force on object B, then object B exerts a force on object A, with the same magnitude but opposite direction.
- Whenever an object is in contact with a surface, the surface exerts a **normal reaction force**. This force is perpendicular to the surface and directed away from it.
- Newton's third law implies that when two objects are in contact, each object exerts a normal reaction force on the other one.
- If two objects are connected by a light inextensible string then the **tension** is the same throughout the string. The tension at the point where the string is connected to an object is directed away from the object.
  - The modelling assumption that the string is **inextensible** is required to ensure that the two objects move with the same acceleration.
  - If the string is replaced by a light rod, then the force can be a **thrust** as well as a tension. The thrust force is directed towards the object.
- To find the acceleration you can treat two connected objects as a single particle, but to find the normal reaction or tension force you need to consider each object separately.
- If two objects are connected by a string that passes over a pulley then, if the pulley is modelled as light and smooth, the tension is the same throughout the string.

Mixed practice 19

In this exercise, unless instructed otherwise, use $g = 9.8\text{ m s}^{-2}$, giving your final answers to an appropriate degree of accuracy.

1. A man of mass 75 kg is standing in a lift moving upwards with constant velocity $v\text{ m s}^{-1}$.

   What will he feel? Choose from these options.

   A Heavier than usual **B** Lighter than usual

   C The same **D** It depends on the value of v.

2. Two skaters, of masses 58 kg and M kg, stand facing each other on ice. They push away from each other and move with initial accelerations of 3.6 m s^{-2} and 4.1 m s^{-2}.

   Find the value of M.

3. A car of mass 850 kg is pulling a trailer of mass 320 kg. The car's engine produces a driving force of 1800 N. The resistance forces acting on the car and the trailer are 450 N and 220 N, respectively. Find:

   a the acceleration of the car **b** the tension in the tow bar.

4. A crate of mass 80 kg lies on a horizontal platform. The platform is being raised and decelerates at 2.6 m s^{-2}.

   Find the magnitude of the normal reaction force between the crate and the platform.

5. In this question use $g = 9.81\text{ m s}^{-2}$, giving your final answer to an appropriate degree of accuracy.

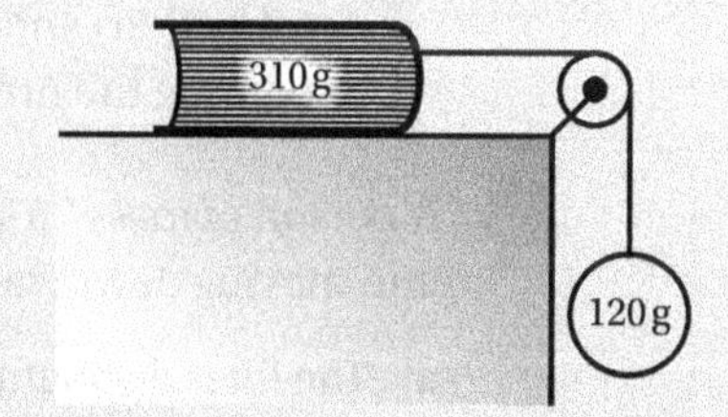

   A book of mass 310 g lies on a rough horizontal table. A light inextensible string is attached to the book. The string passes over a smooth pulley fixed at the edge of the table. A ball of mass 120 g is attached to the other end of the string.

   The system is in equilibrium with the string taut.

   Find the magnitude of the friction force between the book andthe table.

6. In this question use $g = 10\text{ m s}^{-2}$, giving your final answer to an appropriate degree of accuracy.

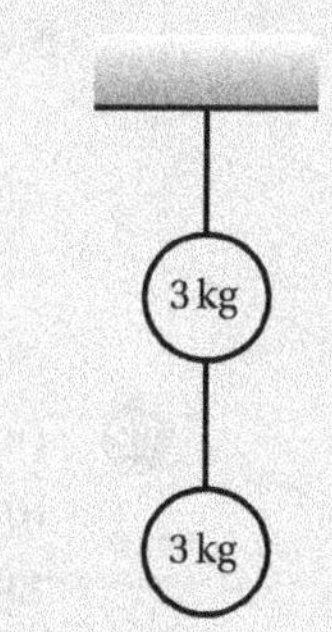

   Two balls are connected and suspended from the ceiling by two light inextensible strings, as shown in the diagram.

   Given that both balls have mass 3 kg, find the tension in each string.

7. A tractor, of mass 3500 kg, is used to tow a trailer, of mass 2400 kg, across a horizontal field. The trailer is connected to the tractor by a horizontal tow bar. As they move, a constant resistance force of 800 newtons acts on the trailer and a constant resistance force of R newtons acts on the tractor. A forward driving force of 2500 newtons acts on the tractor. The trailer and tractor accelerate at 0.2 m s^{-2}.

   a Find R.

   b Find the magnitude of the force that the tow bar exerts on the trailer.

   c State the magnitude of the force that the tow bar exerts on the tractor.

[© AQA 2013]

8 Block A of mass 12 kg lies on a horizontal surface and block B of mass 5 kg lies on block A.

What is the normal reaction force of the ground on block A?

Choose from these options.

A $5g$ **B** $7g$ **C** $12g$ **D** $17g$

9 Two particles, A and B, have masses 3 kg and 7 kg respectively. They are connected by a light inextensible string that passes over a smooth fixed pulley. A second light inextensible string is attached to A. The other end of this string is attached to the ground directly below A. The system remains at rest, as shown in the diagram.

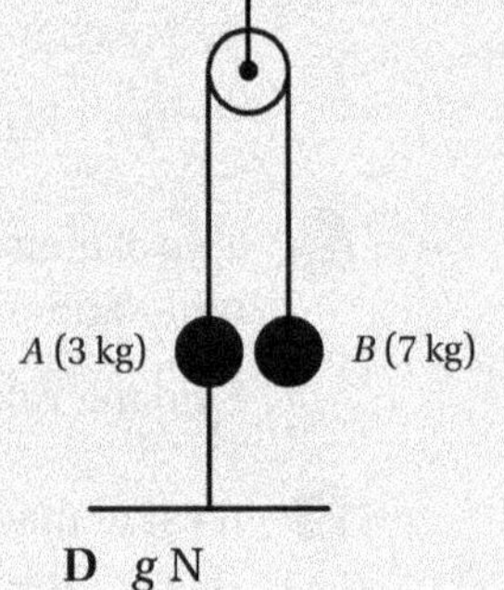

Find the tension in the string connecting A to the ground.

Choose from these options.

A $7g$ N **B** $4g$ N **C** $3g$ N **D** g N

10 A car is pulling a trailer by means of a light rigid tow bar. The mass of the car is 1200 kg and the mass of the trailer is 350 kg. Assume that any resistances to motion can be ignored.

a The car is moving with a speed of 9.2 m s^{-1} when it starts to accelerate at 1.8 m s^{-2}. Find the driving force produced by the car's engine.

b The car continues to accelerate uniformly for 4 seconds. It then starts to brake (with uniform deceleration) and comes to rest after travelling 26 m. Find the magnitude of the thrust in the tow bar during the braking phase.

11 A person of mass 75 kg stands in a lift of mass 450 kg. The lift is suspended by a light inextensible cable and moving downwards.

a The lift is decelerating at 5.2 m s^{-2}. Find the normal reaction force between the person's feet and the floor of the lift.

b Given instead that the normal reaction force between the person's feet and the floor of the lift is 570 N, find:

i the magnitude and direction of the acceleration of the lift

ii the tension in the cable.

12 Two skaters stand on ice 5 m from each other, holding onto the ends of a light inextensible rope. They pull at the rope with a constant force and come together in 1.5 seconds. Any friction can be ignored. Given that the mass of the first skater is 62 kg, and that he moves with acceleration of 1.8 m s^{-2}, find the mass of the second skater.

13 In the diagram, the three strings can be modelled as light and inextensible and the pulley can be modelled as smooth. The masses of the particles A, B and C are 5.2 kg, 3.7 kg and m kg respectively. The system hangs in equilibrium in the vertical plane.

a Find the value of m. **b** Find the tension in each string.

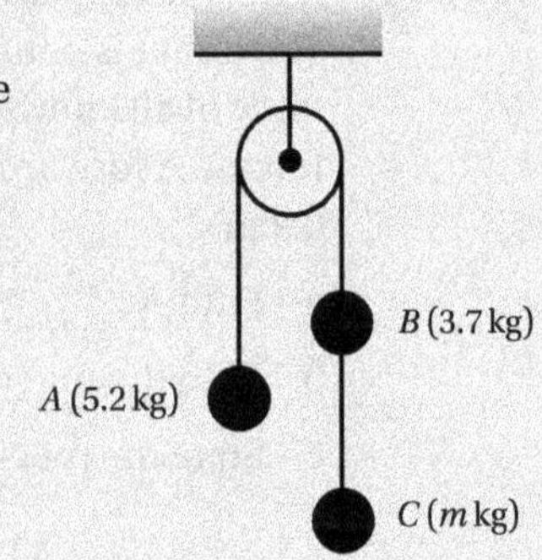

14 A particle of mass 12 kg rests in equilibrium on a rough horizontal table, under the action of two forces, $\mathbf{F}_1 = (16\mathbf{i} + 7\mathbf{j})$ N and $\mathbf{F}_2$, as shown in the diagram.

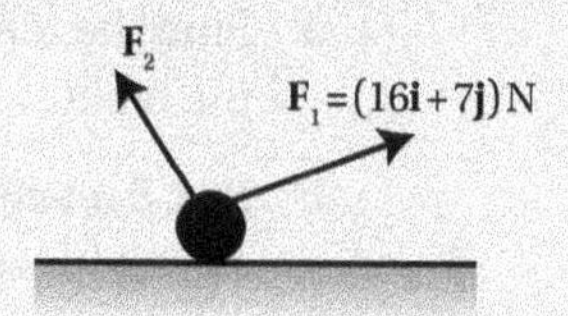

The magnitude of the normal reaction force between the particle and the table is 72 N and the magnitude of the friction force is 9 N.

Find the two possible values for the magnitude of $\mathbf{F}_2$.

15 An engine of mass 25 tonnes pulls two carriages, each of mass 10 tonnes. They are connected by light horizontal rods.

The train moves along a straight horizontal track. A resistance force of magnitude 600 N acts on the engine and a resistance force of magnitude 400 N acts on each carriage. The train is accelerating at $0.5\ \mathrm{m\,s^{-2}}$.

a Find:

i the magnitude of the force that the rod exerts on the second carriage

ii the magnitude of the force that the rod attached to the engine exerts on the first carriage.

b The engine generates a driving force of magnitude D N. Find the value of D.

16 Two particles, A and B, have masses 12 kg and 8 kg respectively. They are connected by a light inextensible string that passes over a smooth fixed peg, as shown in the diagram.

The particles are released from rest and move vertically. Assume that there is no air resistance.

a By forming two equations of motion, show that the magnitude of the acceleration of each particle is $1.96\ \mathrm{m\,s^{-2}}$.

b Find the tension in the string.

c After the particles have been moving for 2 seconds, both particles are at a height of 4 metres above a horizontal surface. When the particles are in this position the string breaks.

i Find the speed of particle A when the string breaks.

ii Find the speed of particle A when it hits the surface.

iii Find the time that it takes for particle B to reach the surface after the string breaks. Assume that particle B does not hit the peg.

[© AQA 2010]

17 In this question use $g = 10\ \mathrm{m\,s^{-2}}$, giving your final answers to an appropriate degree of accuracy.

Box A of mass 6 kg is held at rest at one end of a rough horizontal table. The box is attached to one end of a light inextensible string which passes over a smooth pulley fixed to the other end of the table. The length of that part of the string extending from A to the pulley is 3 m. Box B, of mass 2.5 kg, is attached to the other end of the string and hangs 1.2 m above the ground.

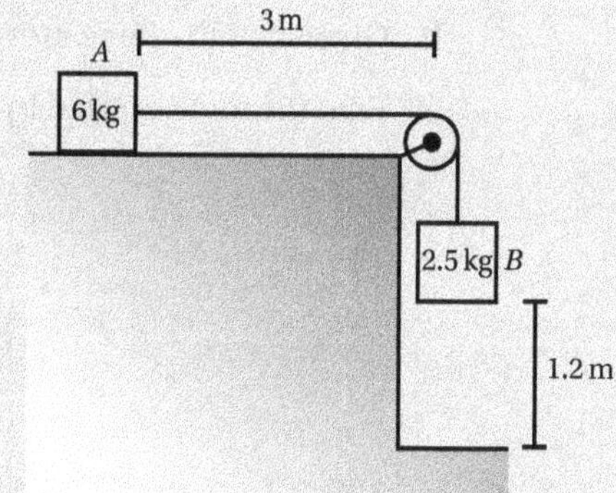

The system is released from rest and moves with acceleration of $0.3\ \mathrm{m\,s^{-2}}$.

a Find the magnitude of the friction force between box A and the table.

b Box B reaches the floor and remains at rest. The magnitude of the friction force between box A and the table remains unchanged. Will box A reach the pulley?

18 Box A, of mass 34 kg, rests on rough horizontal ground. Box B, of mass 49 kg, rests on top of box A. A string is attached to box B and the tension in the string is $(75\mathbf{i}+60\mathbf{j})$ N. The system is in equilibrium.

Find:

a the magnitude of the normal reaction force between box A and the ground

b the magnitude of the friction force between box A and the ground.

The tension in the string is now changed to $(75k\mathbf{i}+60k\mathbf{j})$ N and the value of k is increased from 1. The maximum possible friction force between box A and box B is 120 N and the maximum possible friction force between box A and the ground is 180 N.

c Describe how the equilibrium is broken.

19 Particles P and Q, of masses 3 kg and 5 kg respectively, are connected by a light inextensible string. The string passes over a smooth pulley and the particles hang in the vertical plane with Q 2.5 m above ground.

At time $t = 0$ the system is released from rest with the string taut.

a Find the time required for Q to hit the ground.

Once Q is on the ground, P continues to move. Assume that in subsequent motion, neither particle reaches the pulley.

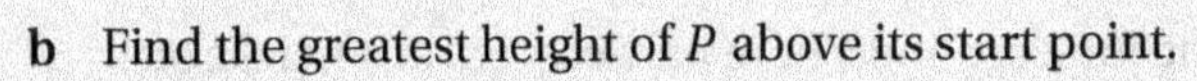

b Find the greatest height of P above its start point.

c Find the time when the string becomes taut again.

20 A light inextensible string passes over a fixed smooth pulley. One end of the string is attached to a scale pan of mass m, and the other end to a weight, R, of mass $10\,m$.

Inside the scale pan are two weights, P and Q, of mass $2\,m$ and $3\,m$ respectively. P is on top of Q.

The $10\,m$ mass is used to raise the scale pan.

a Find the acceleration.

b Given that the force exerted on the pulley is 294 N, find m.

c Find the force exerted by the scale pan on Q.

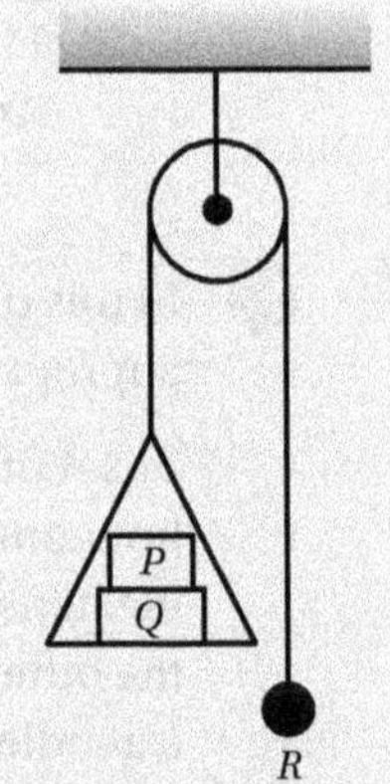

FOCUS ON ... PROOF 3

Using mechanics to derive proofs

You have often used algebra and calculus to derive formulae in Mechanics, for example, in various forms of the constant acceleration formulae. But you don't normally think of using mechanics to prove algebraic results. In this section you will look at two examples of such proofs.

1 Prove this statement about positive real numbers a, b, c and d.

If $\frac{a}{b} < \frac{c}{d}$ then $\frac{a}{b} < \frac{a+c}{b+d} < \frac{c}{d}$.

PROOF 8

Consider a particle moving in a straight line. Suppose it travels at a constant speed v_1 for b seconds, covering the distance of a metres, and then at a different constant speed v_2 for d seconds, covering the distance of c metres.

Then $v_1 = \frac{a}{b}$ and $v_2 = \frac{c}{d}$, so the inequality $\frac{a}{b} < \frac{c}{d}$ means that $v_1 < v_2$.

The expression $\frac{a+c}{b+d}$ represents $\frac{\text{total distance}}{\text{total time}}$, so this is the average speed for the whole journey. This has to be somewhere between the smaller and the larger speeds, so:

$$v_1 < \frac{a+c}{b+d} < v_2,$$

which proves the required inequality.

2 Prove that, for any two positive numbers a and b, $\frac{a+b}{2} \geqslant \sqrt{ab}$.

PROOF 9

Consider two objects moving in a straight line, and let s be the total distance travelled.

- **a** Suppose the first object travels for half the time with speed a and then for the other half of the time with speed b. Find an expression for the total time travelled in terms of a, b and s.
- **b** The second object travels half the distance with speed a and half the distance with speed b. Find an expression for its total time in terms of a, b and s.
- **c** Draw displacement–time graphs to see which object takes longer to travel the distance s. Make sure you consider both possibilities $a < b$ and $a > b$. Can you explain this without referring to the graph?
- **d** Use this to complete the proof that $\frac{a+b}{2} \geqslant \sqrt{ab}$.

Did you know?

The result in the second proof is called the AM–GM inequality. There are several similar inequalities, and they are particularly important in probability and statistics.

FOCUS ON ... PROBLEM SOLVING 3

Alternative representations

It is easy to categorise problems in mathematics by topic, for example, labelling them as 'a mechanics problem' or 'a geometry problem' but sometimes unexpected links provide elegant solutions to otherwise difficult problems.

Consider this problem.

> A farmer wants to build a straight path from his house to the stream and then from the stream to the stables. The positions of the two buildings are shown in the diagram.
>
> What is the shortest possible length of the path?

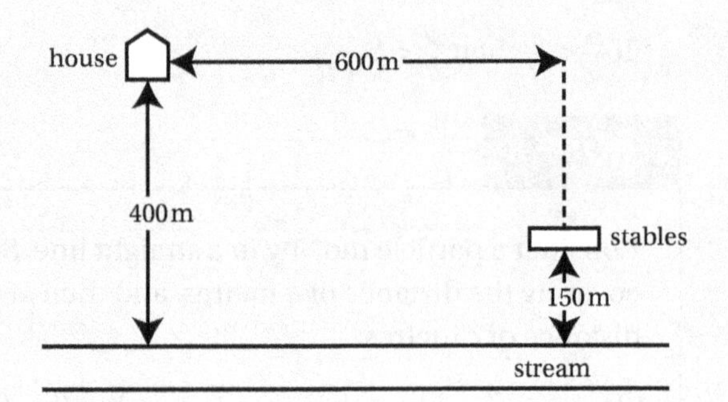

This looks similar to a 'calculus problem' that you encountered in Chapter 13. You could set up a coordinate system, write an expression for the total distance, in terms of x, and use differentiation to find the minimum.

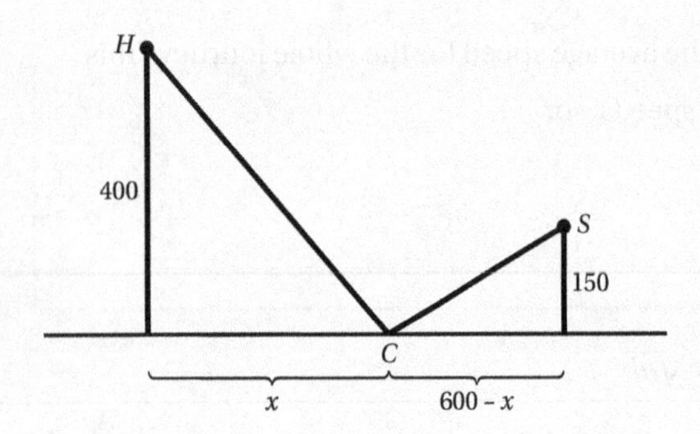

$$d = \sqrt{x^2 + 400^2} + \sqrt{(600 - x)^2 + 150^2}$$

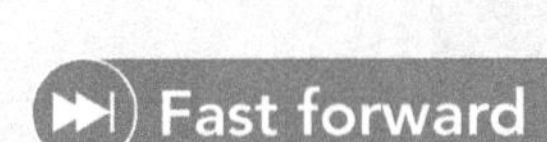

You don't know how to differentiate this expression for d yet, but you will learn how to do so in Student Book 2.

However, there is a much simpler solution if you use an idea from geometry: the shortest distance between two points is along a straight line. Of course, the straight line between the house and the stables doesn't go to the stream. The trick is to find a point on the other side of the stream which is the same distance from the stream as the stables. This is achieved by reflecting the point that represents the stables in the line that represents the stream.

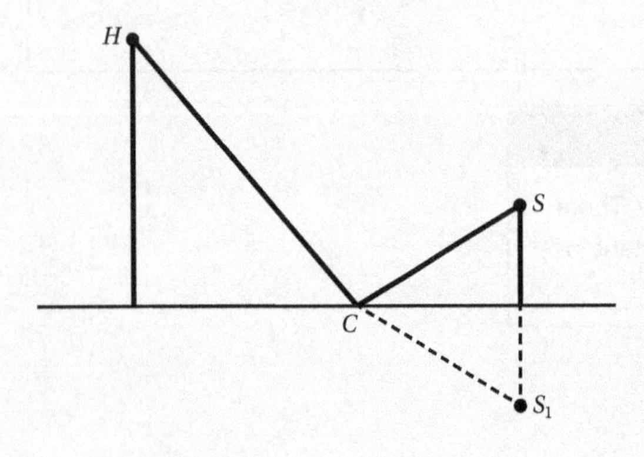

FOCUS ON ... PROBLEM SOLVING 3

Suppose the path goes from the house to the point C, crosses the stream and then goes in a straight line to point S_1. Then the total length of the path is $d = HC + CS = HC + CS_1$. But the shortest distance between H and S_1 is along a straight line, so C should be the point where the line HS_1 crosses the stream.

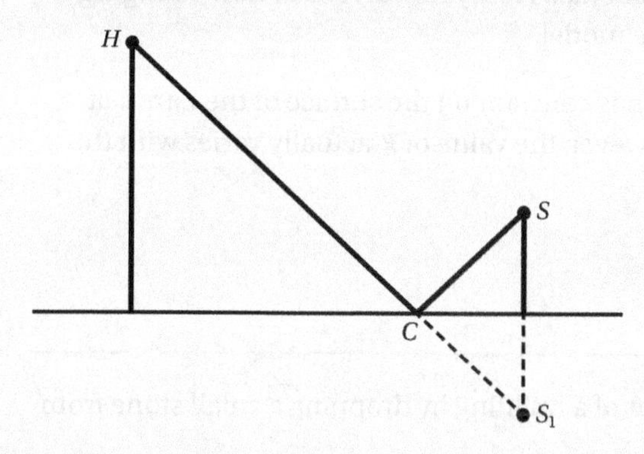

QUESTIONS

1. Find the length of the shortest possible path in the given example.

2. A cube has side of length 1 m. An ant starts at one corner of the cube and crawls to the opposite corner. The ant can only move on the surface of the cube. Find the shortest possible path and calculate its length.

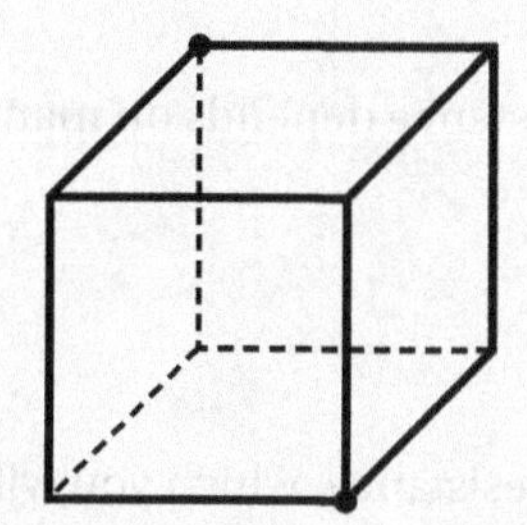

How does this compare to the length of the shortest path if the ant could pass through the cube?

3. Repeat question 2 for an ant moving on the surface of a cuboid with sides of lengths 1 m, 2 m and 3 m.

4. A caterpillar crawls on the surface of a cylinder with radius r and height h. It starts at a point on the edge of the bottom base and crawls to the point on the top base that is diametrically opposite the starting point. Find the length of the shortest possible path it can take.

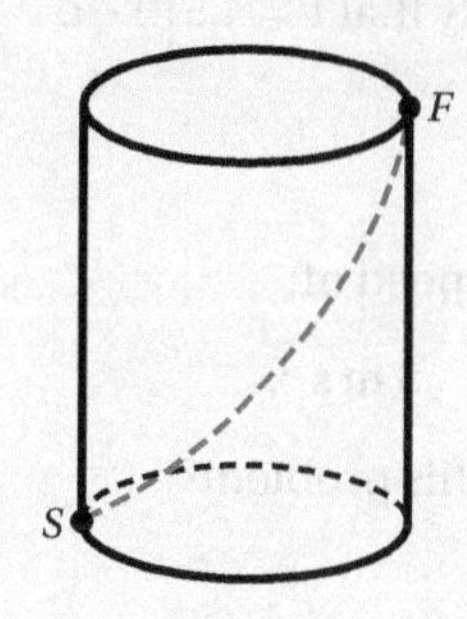

FOCUS ON ... MODELLING 3

Investigating the effect of modelling assumptions

Throughout the mechanics chapters you have made various modelling assumptions that enabled you to write relatively simple equations to describe the motion of an object. Now you will look at how changing some of these assumptions affects the results predicted by the model.

You usually assume that the value of gravitational acceleration is constant on the surface of the Earth, at $9.81\,\text{m s}^{-2}$ (to 3 s.f.), $9.8\,\text{m s}^{-2}$ (to 2 s.f.) or $10\,\text{m s}^{-2}$ (to 1 s.f.). However, the value of g actually varies with the latitude and height above sea level.

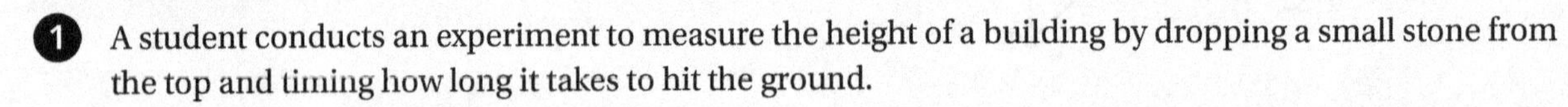

QUESTIONS

1 A student conducts an experiment to measure the height of a building by dropping a small stone from the top and timing how long it takes to hit the ground.

a The stone takes 1.6 seconds to fall.

i Taking $g = 9.8$, estimate the height of the building. Give your answer to 3 s.f.

ii The building is actually located in Greenland, where $g = 9.825\,\text{m s}^{-2}$. What is the percentage error in the estimate from part **i**?

iii If instead the building is located in Denver, USA, where $g = 9.796\,\text{m s}^{-2}$, find the percentage error in the estimate of the height.

b The experiment is repeated with a much taller building, and the stone takes 10.2 seconds to fall. Repeat all the calculations from part **a**.

Another common modelling assumption is that there is no air resistance. Air resistance depends on many different factors, including:

- the size and shape of the object and the material of which it is made
- the density of air (which varies with height and temperature).

It also varies with the speed of the object. There are two common models for air resistance, which you will look at in questions 2 and 3.

2 At low speeds, air resistance can be modelled as being proportional to the speed of the object.

A cyclist starts moving from rest and applies a constant driving force of 180 N. The air resistance has magnitude $40v$, where v is the speed. Assume all other forces can be ignored. The cyclist and her bike have a mass of 120 kg.

a Write the Newton's second law equation for the motion of the cyclist. Verify that $v = 4.5\left(1 - e^{-\frac{t}{3}}\right)$ satisfies this equation.

b Write down an equation for v as a function of t if air resistance is ignored.

c Use each model to predict how long it would take the cyclist to reach the speed of:

i $0.5\,\text{m s}^{-1}$ **ii** $2\,\text{m s}^{-1}$ **iii** $5\,\text{m s}^{-1}$.

d Would you say ignoring air resistance is a good modelling assumption in this problem?

FOCUS ON ... MODELLING 3

3 At higher speeds, a better model is to assume that air resistance is proportional to the square of the speed. In this model, the constant of proportionality depends on the drag coefficient b and the mass of the object. The model leads to this expression for the velocity:

$$v = A\left(\frac{1-e^{-kt}}{1+e^{-kt}}\right)$$

where $A = \sqrt{\frac{mg}{b}}$ and $k = \sqrt{\frac{4bg}{m}}$.

a For a parachutist of mass 80 kg, $b = 0.27$.

i For a model without air resistance, what is the velocity of the parachutist after 1 second? How does this compare to the velocity predicted by this model?

ii How do the velocities after 10 seconds compare in the two models?

b For a 2-pence coin of mass 7.2 g, $b = 2\times10^{-4}$. Answer questions **i** and **ii** in part **a**, based on this revised information.

c Based on a model without air resistance, find how long it takes for an object to reach the ground from a height of:

i 5 m **ii** 500 m.

d Hence investigate whether ignoring air resistance is a suitable modelling assumption for a parachutist and a coin falling from the height of 5 m and 500 m.

4 Models with air resistance predict that the object will reach a **terminal velocity**, as the resistance force increases with speed and eventually balances the weight. How does the weight of an object affect its terminal velocity and the time it takes to reach it?

CROSS-TOPIC REVIEW EXERCISE 3

In this exercise use $g = 9.8 \text{ m s}^{-2}$, unless instructed otherwise, giving your final answers to an appropriate degree of accuracy.

1 In the diagram, $\overrightarrow{AB} = \overrightarrow{DC} = \mathbf{a}$ and $\overrightarrow{BC} = \overrightarrow{AD} = \mathbf{b}$. Q is the midpoint of AD and points M, N, P and Q are such that $AM : MB = 2:1$, $DN : NC = 2:7$ and $BP : PC = 3:1$.

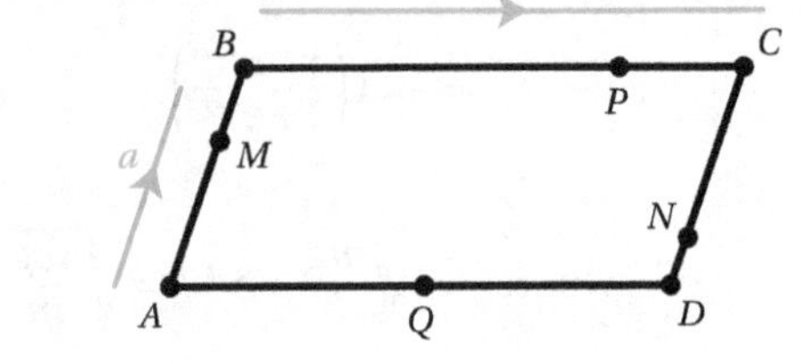

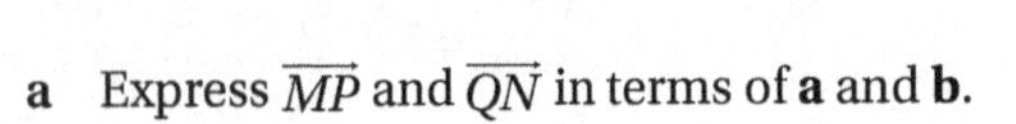

a Express $\overrightarrow{MP}$ and $\overrightarrow{QN}$ in terms of **a** and **b**.

b Hence show that MP and QN are parallel.

2 The diagram shows the velocity–time graph for a particle moving in a straight line.

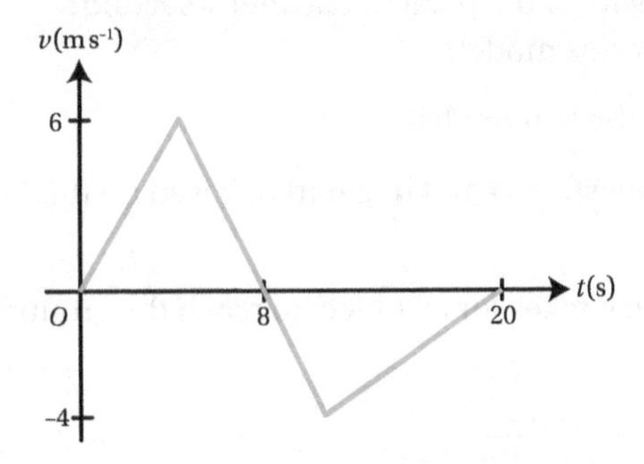

Find the average speed of the particle during the 20 seconds.

3 The diagram shows a velocity–time graph for a train as it moves on a straight horizontal track for 50 seconds.

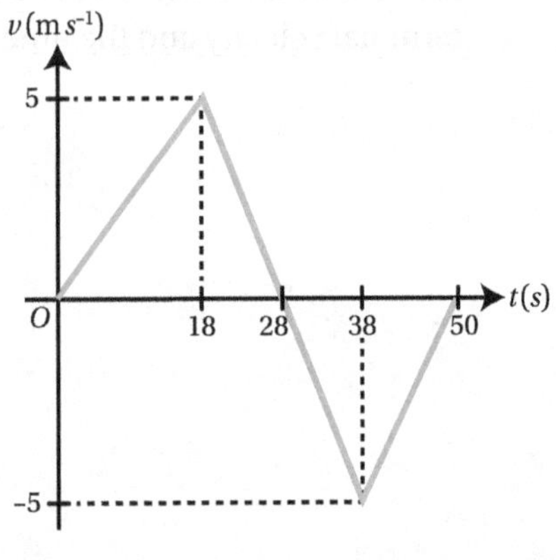

a Find the distance that the train moves in the first 28 seconds.

b Calculate the total distance moved by the train during the 50 seconds.

c Hence calculate the average speed of the train.

d Find the displacement of the train from its initial position when it has been moving for 50 seconds.

e Hence calculate the average velocity of the train.

f Find the acceleration of the train in the first 18 seconds of its motion.

[© AQA 2012]

4 A particle of mass 3 kg moves under the action of two perpendicular forces, as shown in the diagram.

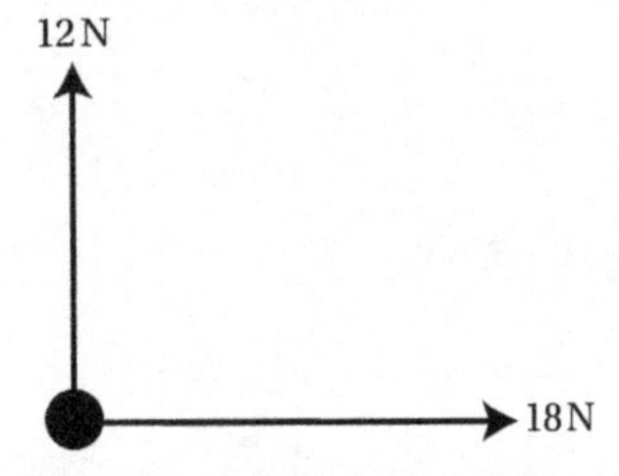

Find, in vector form, the acceleration of the particle.

5 A particle of mass 4.2 kg starts from rest and accelerates uniformly under the action of a constant force of magnitude 13.6 N. Find the speed of the particle after it has travelled a distance of 6.2 m.

6 A particle of mass 3 kg is in equilibrium at the point P under the action of three forces, $\mathbf{F}_1 = (8\mathbf{i} + 3\mathbf{j})$ N, $\mathbf{F}_2 = (-2\mathbf{i} + 6\mathbf{j})$ N and $\mathbf{F}_3 = (a\mathbf{i} + b\mathbf{j})$ N.

a Find the values of a and b.

b The force $\mathbf{F}_3$ is now removed. Find:

i the magnitude of the resultant force

ii the angle the resultant force makes with the vector $\mathbf{i}$.

c Find the displacement of the particle from P 4 seconds after $\mathbf{F}_3$ is removed.

7 A crane is used to lift a load, using a single vertical cable which is attached to the load. The load accelerates uniformly from rest. When it has risen 0.9 metres, its speed is 0.6 m s^{-1}.

a **i** Show that the acceleration of the load is 0.2 m s^{-2}.

ii Find the time taken for the load to rise 0.9 metres.

b Given that the mass of the load is 800 kg, find the tension in the cable while the load is accelerating.

[© AQA 2011]

8 A particle P moves in a straight line, passing the point O with speed 35 m s^{-1}. At time t seconds after leaving O the acceleration $a\text{ m s}^{-2}$ is given by:

$$a = 6t - 22 \quad 0 \leqslant t \leqslant 5$$

a **i** Find an expression for the velocity at time t.

ii Find the times at which P is at rest.

iii Find the maximum speed of the particle in its 5-second journey.

b Find the total distance travelled by P.

9 A particle has velocity given by $v = t^2 + 1$ for $t \geqslant 0$. Velocity is measured in m s^{-1} and time in seconds. The average velocity of the particle from $t = 0$ to $t = T$ is 4 m s^{-1}.

Find the value of T.

10 A car is travelling in a straight line along a horizontal road, with constant acceleration $a\text{ m s}^{-2}$. It passes point A with speed $u\text{ m s}^{-1}$, reaches point B five seconds later and point C two seconds after that.

a Given that the distance $AB = 95$ m and $BC = 80$ m, find the values of u and a.

b The car's engine produces a driving force of 720 N. The resistance to the motion of the car is 250 N. Find the mass of the car.

11 In this question use $g = 9.81\text{ m s}^{-2}$, giving your final answers to an appropriate degree of accuracy.

A person of mass 60 kg stands in a moving lift of mass 420 kg. The normal reaction force between the person's feet and the floor of the lift is 312 N.

a Find the magnitude and direction of the acceleration of the lift.

The lift is suspended by a cable, which can be modelled as light and inextensible.

b Find the tension in the cable.

c Explain how you have used the assumption that the cable is light.

12 A car, of mass 1200 kg, tows a caravan, of mass 1000 kg, along a straight horizontal road. The caravan is attached to the car by a horizontal towbar. A resistance force of magnitude R newtons acts on the car and a resistance force of magnitude $2R$ newtons acts on the caravan. The car and caravan accelerate at a constant 1.6 m s^{-2} when a driving force of magnitude 4720 newtons acts on the car.

a Find R.

b Find the tension in the towbar.

[© AQA 2012]

13 Two particles, P and Q, are connected by a string that passes over a fixed peg, as shown in the diagram. The mass of P is 5 kg and the mass of Q is 3 kg.

$P(5\text{ kg})$ $Q(3\text{ kg})$

The particles are released from rest in the position shown.

a By forming an equation of motion for each particle, show that the magnitude of the acceleration of each particle is 2.45 m s^{-2}.

b Find the tension in the string.

c State **two** modelling assumptions that you have made about the string.

d Particle P hits the floor when it has moved 0.196 metres and Q has not reached the peg.

i Find the time that it takes P to reach the floor.

ii Find the speed of P when it hits the floor.

[© AQA 2011]

14 In this question use $g = 10\text{ m s}^{-2}$, giving your final answers to an appropriate degree of accuracy.

A stone, S, is dropped from the top of a cliff that is 50 m high. One second later another stone, T, is thrown vertically upwards from the foot of the cliff at 20 m s^{-1}.

The stones collide.

a Find the distance from the foot of the cliff at which the stones collide.

b Find the speed and direction of T when the collision happens.

15 A train consists of a locomotive and two carriages. The mass of the locomotive is 5600 kg and the mass of each carriage is 3200 kg. The train is moving at a speed of 40 km h^{-1} when the driver applies the brakes. The train comes to rest after travelling 360 m. The resistance forces throughout this motion are constant, 700 N on the locomotive and 400 N on each carriage.

a Find the force in the coupling between the first carriage and the locomotive.

b Is this force a tension or a thrust?

20 Working with data

In this chapter you will learn how to:

- interpret statistical diagrams including histograms, scatter diagrams, cumulative frequency curves and box-and whisker plots
- calculate mean, median and mode and standard deviation for data
- understand correlation and use a regression line
- clean data to remove outliers.

Before you start...

GCSE	You should be able to interpret basic statistical diagrams such as pie charts and bar charts.	1 Find the percentage decrease in the stock price after the crash in this bar chart. Change in share price Price in pounds: 3, 4, 5, 6 Pre-crash, Post-crash
GCSE	You should be able to calculate the mean, median and mode of a set of data.	2 Find the mean, median and mode of: $1, 1, 4, 5, 9, 10$.
GCSE	You should be able to calculate the range and interquartile range of a set of data.	3 Find the range and interquartile range of: $12, 15, 18, 18, 19, 16, 14, 20, 12$.

Making sense of data

Statistics is an incredibly important part of mathematics in the real world. It provides the tools to collect, organise and analyse large amounts of data. For example, the government uses statistical analysis to plan public services such as building schools and hospitals.

At first, it might seem as if statistics is just about representing data through either diagrams or calculations. However, as you progress through statistics you will see that it is about making inferences about a larger group on the basis of a sample. Medical trials and opinion polls are both examples of using samples to make a prediction about a whole population.

Although many statistical methods were developed more than 200 years ago, technological advances in the last few decades have enabled much larger quantities of data to be analysed, making statistics one of the most common applications of mathematics.

Did you know?

Some of the first users of statistical diagrams (in the 18th and 19th centuries) include Florence Nightingale and the political economist William Playfair. The term histogram was coined by the statistician Karl Pearson.

Gateway to A Level

See Gateway to A Level Section T for a reminder of basic statistical diagrams.

Section 1: Statistical diagrams

Histograms

Once large amounts of data are collected, they are often sorted into groups (or classes). For **continuous data**, these groups need to cover all possible data values in the **range**; this means that there are no gaps between groups. For example, if you had a sample of heights ranging from 120 cm to 200 cm, you might split them into groups as shown in the table.

> **Tip**
>
> Although not strictly necessary, in practice you would round even continuous data. It is therefore important to know where to put a height that is recorded as, say, 180 cm, which is why there is a $\leqslant$ symbol on one end of each interval.

Group	$120 < h \leqslant 160$	$160 < h \leqslant 180$	$180 < h \leqslant 200$
Frequency	60	40	50

You could draw a bar chart for this data.

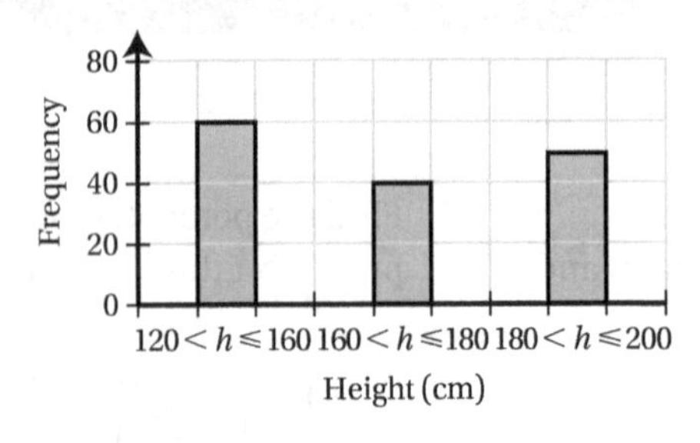

However, this bar chart is misleading. The first group has the greatest frequency, but that might be due to the fact that it covers a wider range of height than the other groups. You can overcome this problem by using a **histogram**. This looks very similar to a bar chart, but the horizontal scale is continuous and the frequency is proportional to area rather than height. The vertical scale is called the **frequency density**.

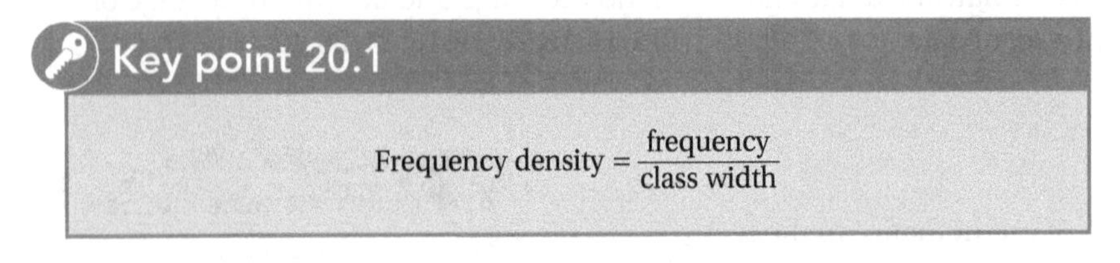

Key point 20.1

$$\text{Frequency density} = \frac{\text{frequency}}{\text{class width}}$$

> **Tip**
>
> The class width is the difference between the largest and smallest possible value in the group.

Plotting the given data on a histogram shows much more clearly the shape of the distribution.

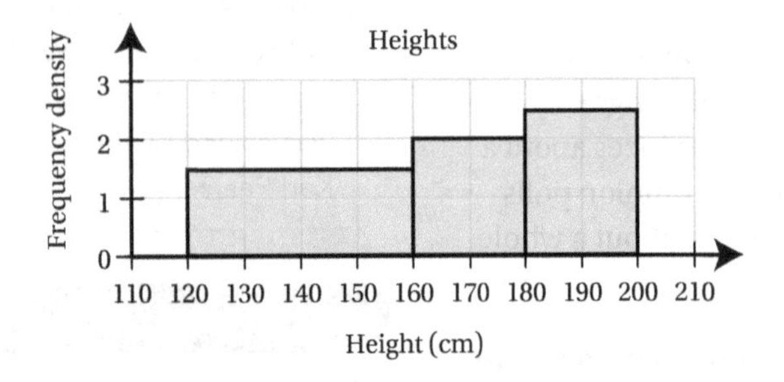

WORKED EXAMPLE 20.1

Use this histogram to estimate the probability of a student living 5 km to 6 km away from school.

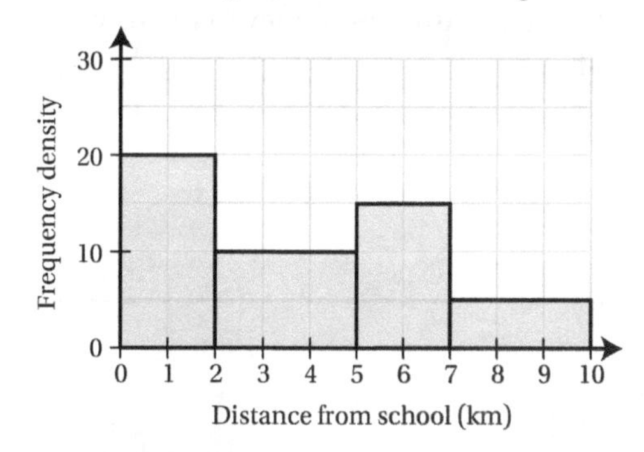

Group	$0<d\leqslant 2$	$2<d\leqslant 5$	$5<d\leqslant 7$	$7<d\leqslant 10$
Width	2	3	2	3
Frequency density	20	10	15	5
Frequency	40	30	30	15

You need to convert frequency frequencies using: frequency density × class width.

So the total frequency is $40+30+30+15=115$

The area from 5 to 6 is $1\times 15=15$ people, so the probability is $\frac{15}{115}\approx 0.13$.

Cumulative frequency graphs

Although these are diagrams, their main purpose is not to provide a visual representation of the data – histograms are much better at that. They are mainly used to estimate the median and quartiles of grouped data.

The **cumulative frequency** is the total number of values that are less than or equal to a particular value.

Tip

Notice that, in Worked example 20.1, you could use the histogram to work out the frequency of part of a group. This is something that you cannot really do from bar charts.

WORKED EXAMPLE 20.2

This table shows the weights of a sample of eggs. Find the cumulative frequency of each group.

Weight, x, of egg (g)	Frequency
$100<x\leqslant 120$	26
$120<x\leqslant 140$	52
$140<x\leqslant 160$	84
$160<x\leqslant 180$	60
$180<x\leqslant 200$	12

Continues on next page

Weight, x, of egg (g)	Cumulative frequency
$x \leqslant 120$	26
$x \leqslant 140$	78
$x \leqslant 160$	162
$x \leqslant 180$	222
$x \leqslant 200$	234

You find the cumulative frequency by adding the frequency in each group to the cumulative frequency in the row above.

Once you have the cumulative frequency you can draw a cumulative frequency diagram. This has the data values along the horizontal axis and the cumulative frequency up the vertical axis. In Worked example 20.2, you can see that 26 eggs weighed 120 g or less, 78 eggs weighed 140 g or less, and so on. Therefore you should plot the cumulative frequency against the upper bound of each group.

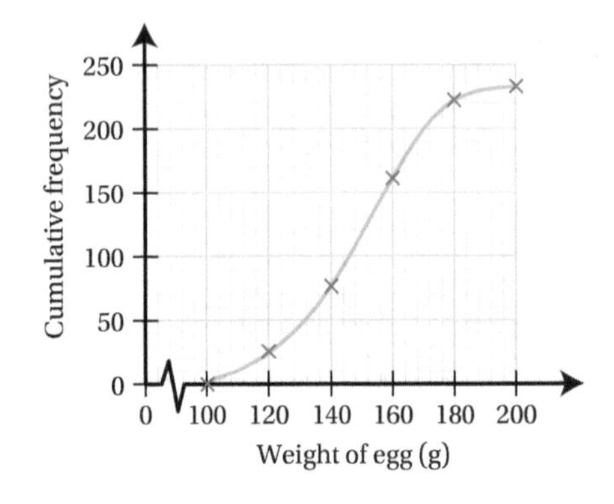

Also, notice that you can plot an additional point at (100, 0) as you know from the data table that there are no eggs weighing less than 100 g. You can then produce a cumulative frequency diagram.

Once you have drawn the cumulative frequency diagram you can use it to estimate the median and the quartiles. The median is the middle value when the data values are arranged in order. On the graph, you can find the median by drawing a horizontal line at half the total frequency to meet the curve, then drawing a vertical line from this point, down to the horizontal axis, to find the median value.

See Gateway to A Level Section U for revision of mean, median, mode and range.

You can find the quartiles by a similar process, drawing horizontal lines at one quarter and three-quarters of the total frequency.

WORKED EXAMPLE 20.3

Estimate the median and interquartile range of the eggs data from Worked example 20.2.

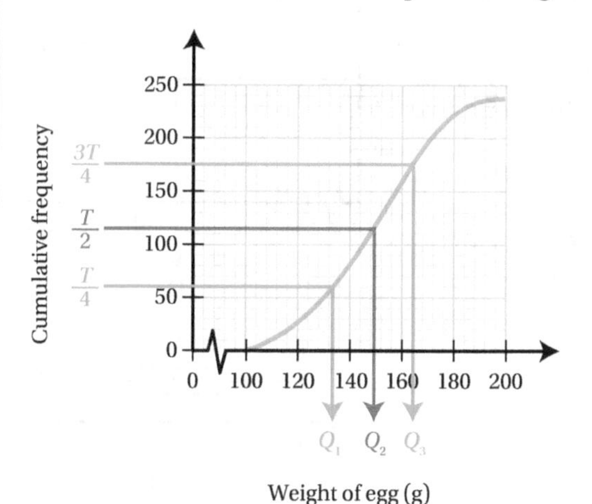

The total frequency is 234, so draw lines across from the frequency axis at $0.5 \times 234 = 117$ for the median, then at $0.75 \times 234 = 175.5$ and $0.25 \times 234 = 58.5$ for the upper and lower quartiles. Where these horizontal lines meet the cumulative frequency curve, draw lines down to the weight axis to find the values of the median and quartiles.

Median $(Q_2) \approx 149$ g
Upper quartile $(Q_3) \approx 164$ g
Lower quartile $(Q_1) \approx 133$ g
$IQR = Q_3 - Q_1 \approx 31$ g

Read off the values on the weight axis shown by the constructed lines. The **interquartile range** (IQR) is the difference between the upper and lower quartiles.

The median and quartiles are specific examples of **percentiles**, which tell you the data value that is a given percentage of the way through the data when it is put in order. The median is therefore the 50th percentile and the lower quartile is the 25th percentile.

WORKED EXAMPLE 20.4

Use the eggs data from Worked examples 20.2 and 20.3.

a Find the 90th percentile of the weight of the eggs.
b The top 10% of eggs are classed as extra large. What range of weights are extra large?

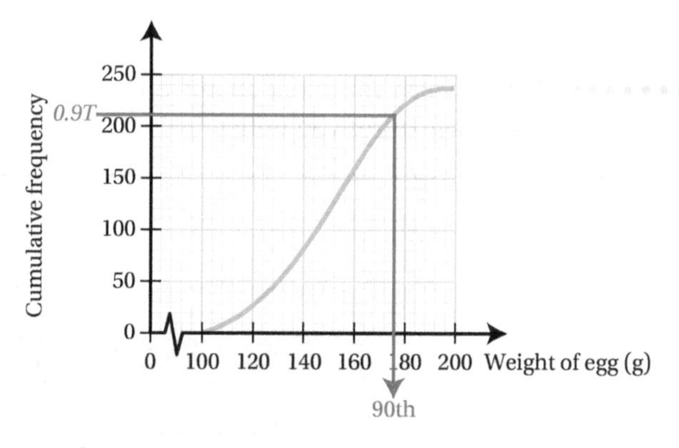

The total frequency is 234, so draw a line across from the frequency axis at $0.9 \times 234 = 211$.

a From the diagram the 90th percentile is approximately 173 g.

Read off the values on the weight axis, as shown by the constructed line.

b Eggs with weights in the range 173 g to 200 g are classified as extra large.

Box-and-whisker plots

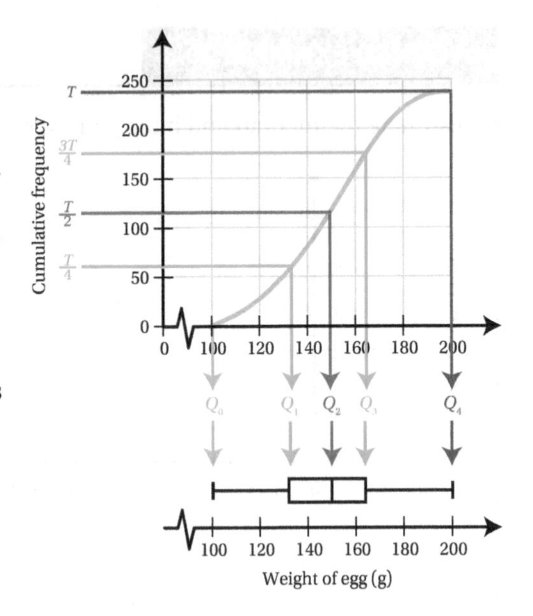

Another useful way of visually representing the information shown in a cumulative frequency diagram is a **box-and-whisker plot**. The ends of the box extend from the lower quartile to the upper quartile. The line in the box indicates the position of the median. Either side of the box, lines extend to the largest and smallest data values in the data set.

You can draw box-and-whisker plots easily from a cumulative frequency diagram. Draw the same lines as you would to find the median and quartiles, but extend them below the horizontal axis, as shown here for the eggs data.

Box-and-whisker plots are useful in comparing two sets of data.

WORKED EXAMPLE 20.5

The box-and-whisker plots show the incomes of a large sample of people in the UK and the USA. Compare the distribution of incomes.

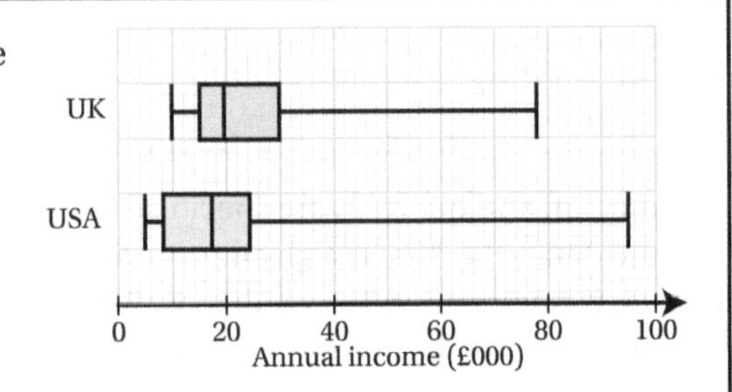

The median income in the UK is slightly higher than in the USA, so people get paid a little more, on average. — You should make one comment on average...

Both the range and the interquartile range for the USA incomes is larger, so they have a larger spread of incomes. — ... and one comment on spread.

The highest incomes in the USA are higher than the highest incomes in the UK. — You can also comment on any other specific features you notice.

EXERCISE 20A

1 For each data set:

i draw a histogram

ii draw a cumulative frequency diagram

iii estimate from your cumulative frequency diagram the median and interquartile range

iv draw a box-and-whisker plot.

Tip

Technology should normally be used to draw statistical diagrams. However, you may find that question 1 helps develop your understanding.

a x is the time taken to complete a puzzle, in seconds (s).

x (s)	Frequency
$0 < x \leqslant 15$	19
$15 < x \leqslant 30$	15
$30 < x \leqslant 45$	7
$45 < x \leqslant 60$	5
$60 < x \leqslant 90$	4

b x is the weight of a plant, in grams.

x (g)	Frequency
$0 < x \leqslant 15$	17
$15 < x \leqslant 30$	23
$30 < x \leqslant 45$	42
$45 < x \leqslant 60$	21
$60 < x \leqslant 90$	5

2 For each histogram, find the probability that $1 < x < 2$.

a i

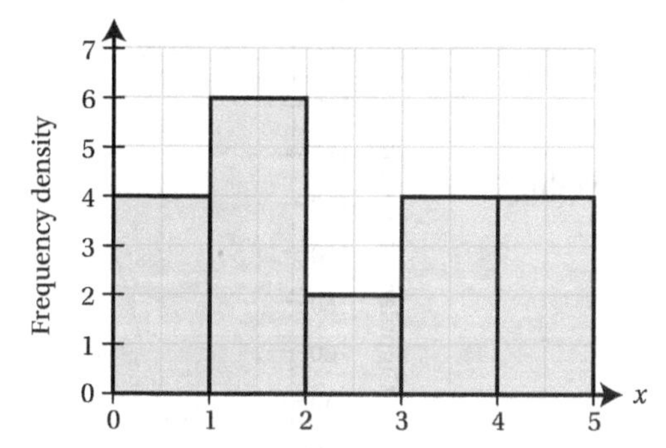

ii

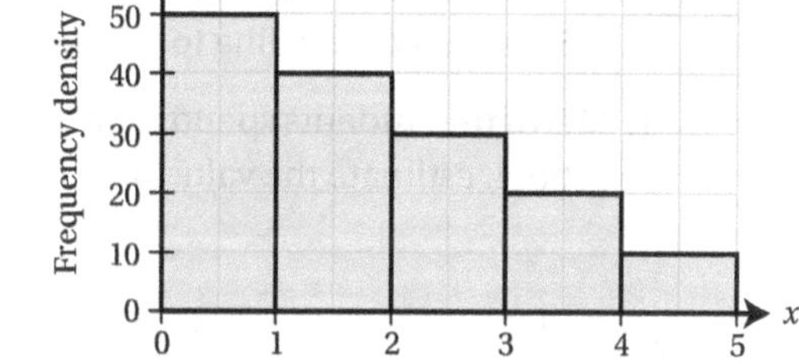

b i

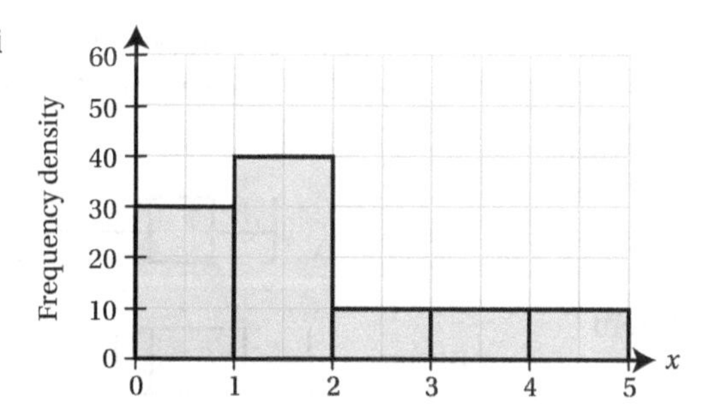

ii

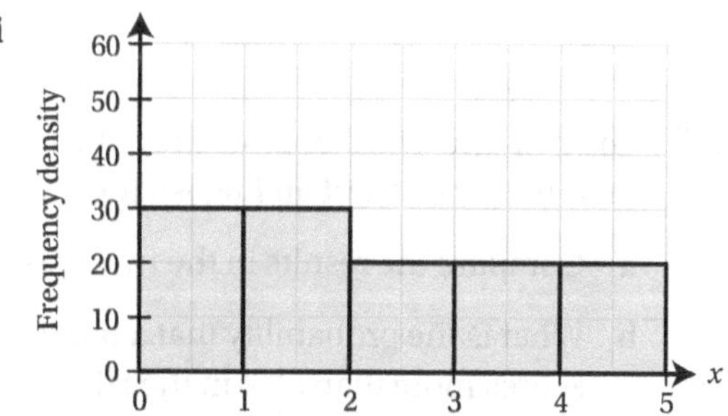

c i

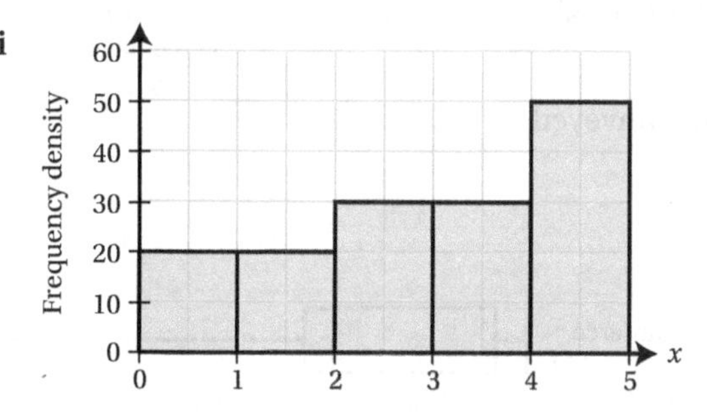

ii

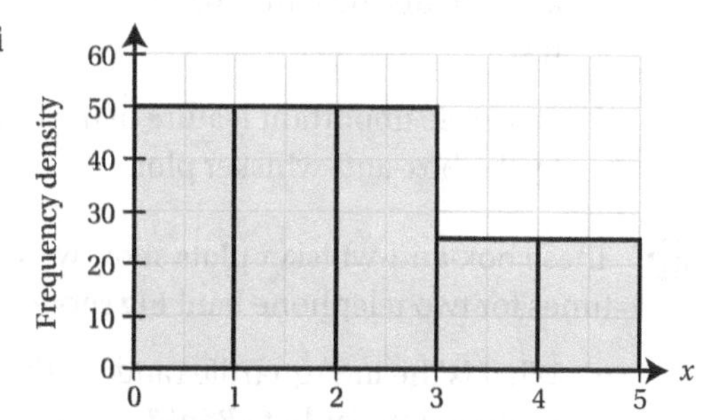

3 From this box-and-whisker plot, state the median and interquartile range.

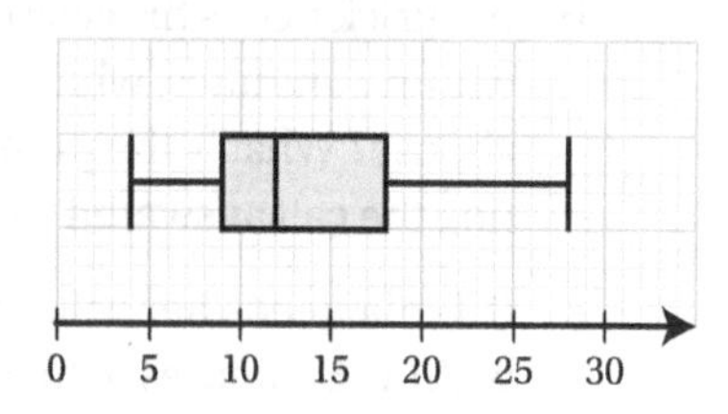

4 80 students were asked to solve a simple word puzzle and their times, in seconds, were recorded. The results are shown on this cumulative frequency diagram.

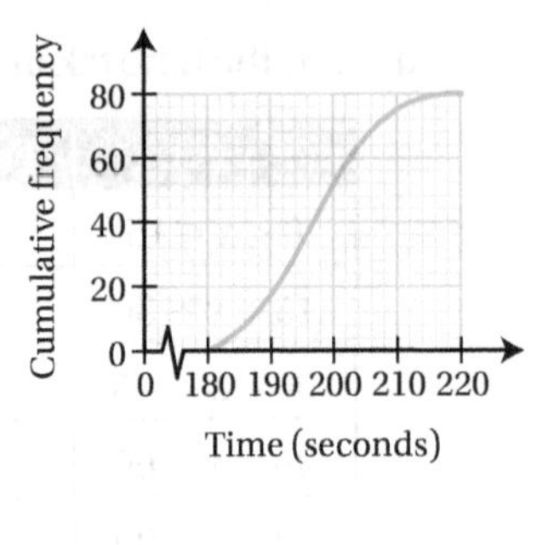

a Estimate the median.

The middle 50% of students took between c and d seconds to solve the puzzle.

b Write down the values of c and d.

c Hence estimate the interquartile range.

5 The cumulative frequency curve indicates the amount of time 200 students spend travelling to school.

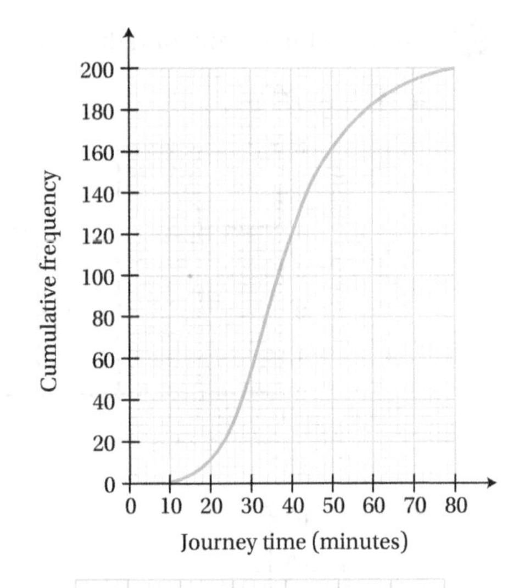

a Estimate the percentage of students who spend between 30 and 50 minutes travelling to school.

b If 80% of the students spend more than x minutes travelling to school, estimate the value of x.

6 The box-and-whisker plots show the results of students in a History test and an English test.

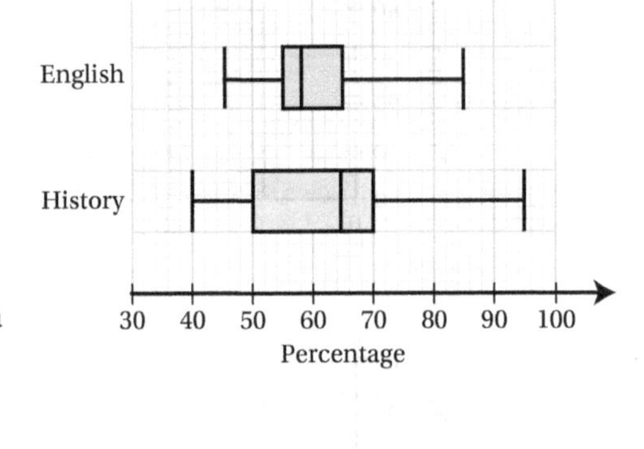

a Compare the results in the two tests.

b What is the probability that a randomly chosen student scores more than 50% in the History test?

c State one further piece of information you would need to know to decide if the History test was easier than the English test.

d State one important feature of the data that is not conveyed by the box-and-whisker plot.

7 These box-and-whisker plots show waiting times for two telephone banking services.

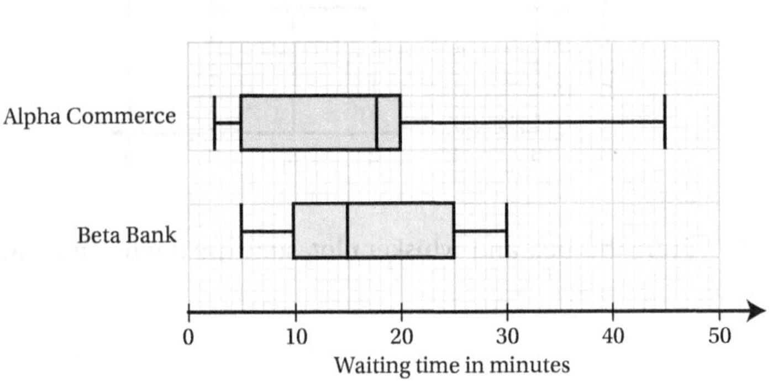

a What is the interquartile range of the waiting time for Beta Bank?

b If Patrick needs his calls to be answered within 5 minutes, which bank should he choose? What is the probability of getting the call answered within 5 minutes?

c If Tania needs her calls to be answered within 15 minutes, which bank should she choose? What is the probability of getting the call answered within 15 minutes?

8 This histogram shows the wages of employees in a company.

a Use the histogram to estimate the probability of a randomly chosen employee earning between £20 000 and £25 000.

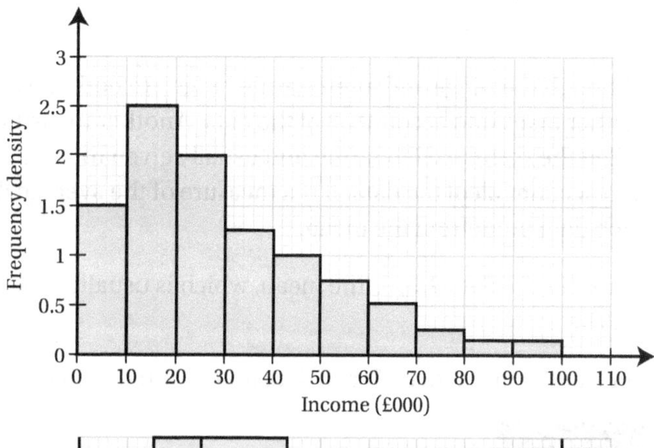

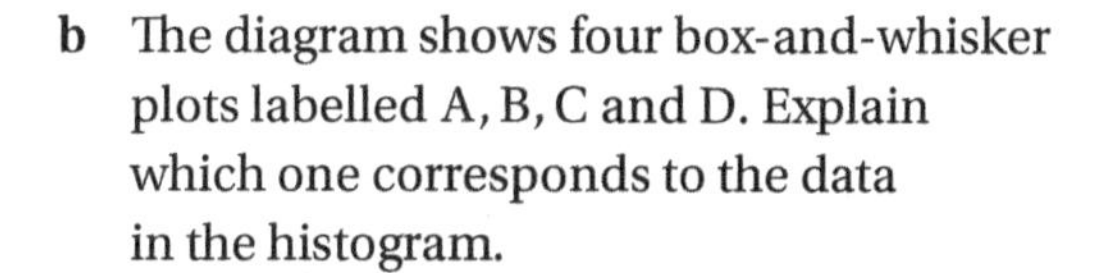

b The diagram shows four box-and-whisker plots labelled A, B, C and D. Explain which one corresponds to the data in the histogram.

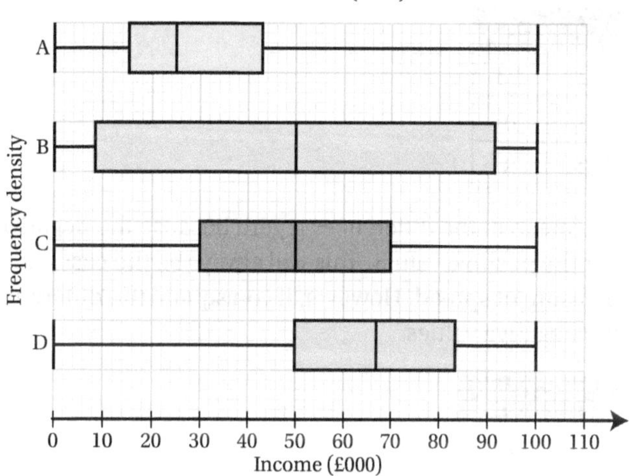

9 Match each histogram with the cumulative frequency diagram coming from the same data:

A

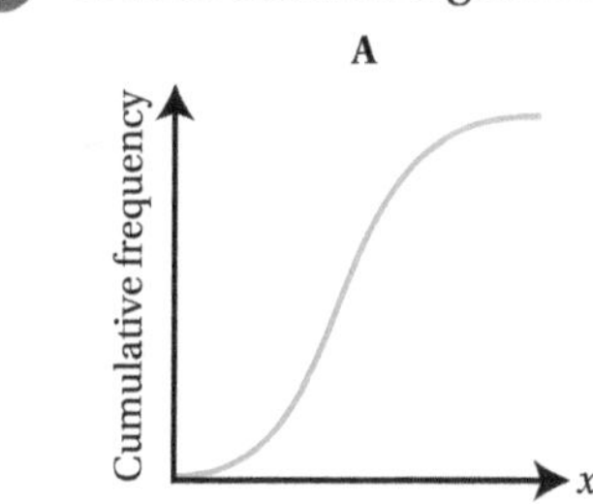

B

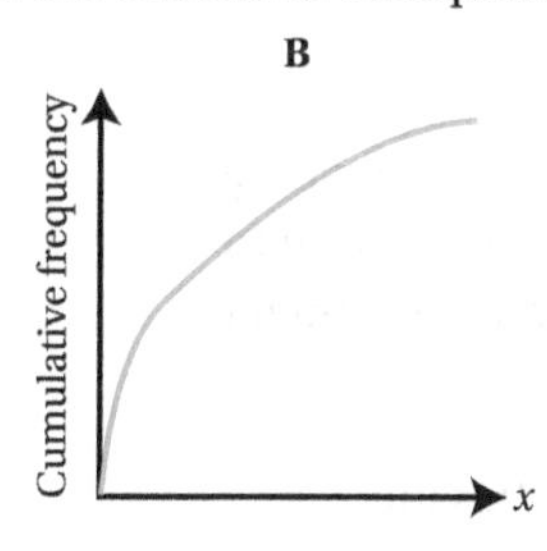

C

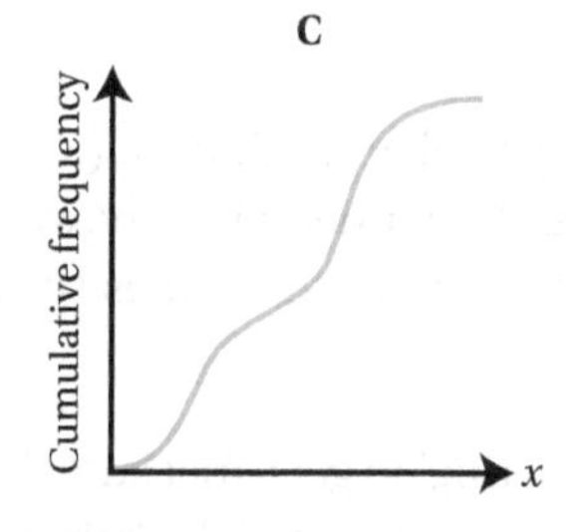

1

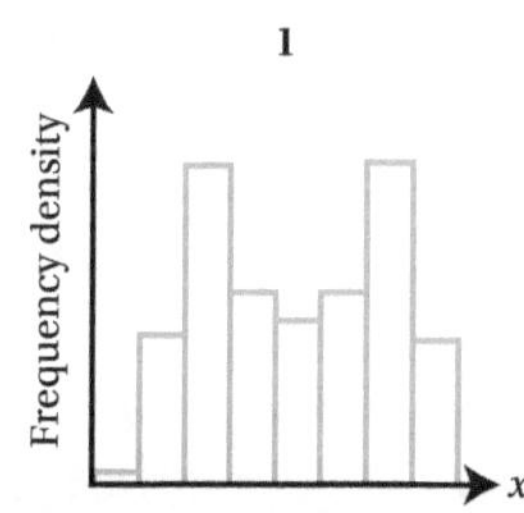

2

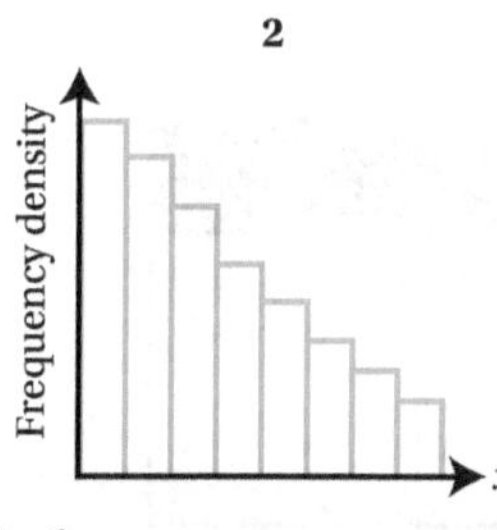

3

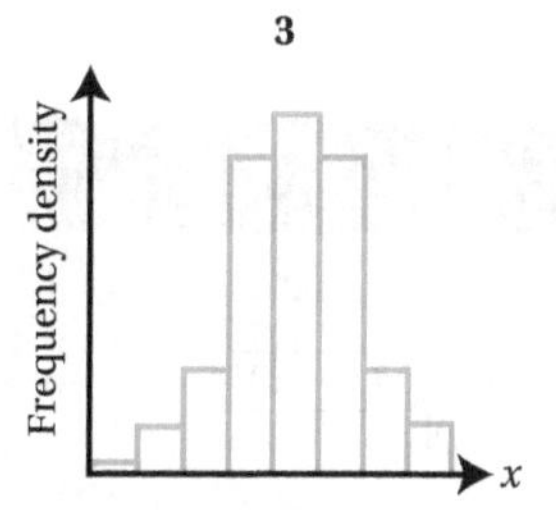

10 A histogram is drawn for this data.

x	$0 < x \leqslant 2$	$2 < x \leqslant 5$
Frequency	a	ab

The height of the bar over the $0 < x \leqslant 2$ group is h. Find, in terms of b and h only, the height of the bar over the $2 < x \leqslant 5$ group.

Section 2: Standard deviation

Although the range and interquartile range are both measures of spread, neither takes into account all of the data. Another measure of spread, called the **standard deviation** and usually given the symbol σ, does take into account all of the data. It is a measure of the average difference of each data point from the mean.

Consider the data 2, 5, 8. The mean, which is usually given the symbol $\bar{x}$, is 5.

Look at the difference of each data point from the mean.

x	$x-\bar{x}$
2	−3
5	0
8	3

The mean of the differences is zero because the negative values cancel out the positive values. This will always be the case so it is not a very good measure of spread. However, if you square the differences you eliminate the negative values.

x	$(x-\bar{x})^2$
2	9
5	0
8	9

The average is given by adding up all the values of $(x-\bar{x})^2$ and dividing by n, the number of data items. In mathematics the symbol Σ is used to mean 'add up all the terms'. In this case, the sum is 18 and the number of data items is 3, and $18 \div 3 = 6$.

You then need to undo the 'squaring' to get a measure that has the same units as x. This means that the standard deviation for this set of data is $\sqrt{6}$.

Key point 20.2

Standard deviation

$$\sigma = \sqrt{\frac{\Sigma(x-\bar{x})^2}{n}}$$

There is an alternative formula for the standard deviation that is sometimes easier to calculate.

Key point 20.3

Standard deviation: alternative formula

$$\sigma = \sqrt{\frac{\Sigma x^2}{n} - \bar{x}^2}$$

This can be written as $\sigma^2 = \overline{x^2} - \bar{x}^2$.

The **variance**, σ^2, is a concept related to the standard deviation. It is the square of the standard deviation.

WORKED EXAMPLE 20.6

Find the range, interquartile range and standard deviation of the numbers: $1, 12, 9, 9, 15, 7, 5$.

The range is $15 - 1 = 14$.

Data in order:

$1, 5, 7, 9, 9, 12, 15$

$LQ = 5, UQ = 12$ so $IQR = 7$

Order the data then split it into two halves. Because there is an odd number of values, discard the middle value.

Find $\bar{x}$ and $\overline{x^2}$.

	x	x^2
	1	1
	5	25
	7	49
	9	81
	9	81
	12	144
	15	225
Mean:	8.29	86.57

$\sigma = \sqrt{86.57 - 8.29^2}$

$= \sqrt{17.92}$

≈ 4.23 (3 s.f.)

In many naturally occurring measurements about two-thirds of the data values will be less than one standard deviation away from the mean. In Worked example 20.6 two data items (1 and 15) out of seven are more than 4.23 away from the mean (8.29). This can serve as a useful quick check. In a large data set, nearly all the data will be within two standard deviations from the mean, and anything more than three standard deviations from the mean is very unusual.

Focus on...

Focus on... Proof 4 shows you how to prove that the two formulae are equivalent.

Common error

Notice that the mean of x^2 ($\overline{x^2}$) is different from the square of the mean or mean squared, ($\bar{x}^2$).

Fast forward

The variance is not itself a good measure of spread, but it has some very useful mathematical properties which you might meet if you study the Statistics option of Further Mathematics.

Fast forward

You will see in Chapter 22 that you can use a different formula if you are using a sample to estimate the variance of the whole population.

Tip

You can use your calculator to find the mean and standard deviation. However, you also need to understand how to use the formula.

Elevate

See Support Sheet 20 for an example of finding the mean and standard deviation when given Σx and Σx^2.

WORK IT OUT 20.1

Find the standard deviation of the data: 2, 4, 5, 9.

Which is the correct solution? Identify the errors made in the incorrect solutions.

Solution 1	Solution 2	Solution 3
$\bar{x} = \frac{2+4+5+9}{4} = 5$ $\Sigma(x-\bar{x})^2 = (-3)^2 + (-1)^2 + 0^2 + 4^2 = 26$ $\sigma = \sqrt{\frac{26}{4}} \approx 2.55$	$\Sigma x = 20$ so $\Sigma x^2 = 400$ $\overline{x^2} = \frac{400}{4} = 100$ $\bar{x} = \frac{20}{4} = 5$ So the standard deviation is $100 - 5^2 = 75$.	From a calculator, the standard deviation is 2.94.

EXERCISE 20B

1 For each set of data, calculate the standard deviation and interquartile range. Use the formula first, then use statistical functions on your calculator to check your answer.

a **i** 19.0, 23.4, 36.2, 18.7, 15.7 **ii** 0.4, −1.3, 7.9, 8.4, −9.4

b **i** 28, 31, 54, 28, 17, 30 **ii** 60, 18, 42, 113, 95, 23

c **i** 1, 2, 1, 3, 5 **ii** 3, −2, 4, −2, 5, 2

2 Sets of data are summarised by the information given. For each set of information find the standard deviation.

a **i** $\Sigma(x-\bar{x})^2 = 42.9, n = 10$ **ii** $\Sigma(x-\bar{x})^2 = 8.9, n = 10$

b **i** $\Sigma x = 49, \Sigma x^2 = 339, n = 8$ **ii** $\Sigma x = 329, \Sigma x^2 = 22\,135, n = 8$

c **i** $\bar{x} = 66.6, \overline{x^2} = 4512.6.$ **ii** $\bar{x} = 24.8, \overline{x^2} = 1072.4.$

3 The interquartile range of the ordered set of data $5, 5, 7, 8, 9, x, 13$ is equal to 7.

a Find the value of x.

b Find the standard deviation of the data set.

4 Ten data items have a sum of 468 and the sum of the squares of the data is 27 172.

a Find the mean of the data.

b Find the variance of the data.

5 The speed, x, in mph, of 10 serves by Tim, a professional tennis player, is summarised as:

$\Sigma x = 1245, \Sigma x^2 = 156\,403.$

a Find the mean speed of the serves.

b Find the standard deviation in the speed of the serves.

c Andy is another professional tennis player. The variance in the speed of Andy's serves is 89.6 (mph)^2. Which player appears to be more consistent in their serving speed?

6 The scores in a Physics test were: 81, 36, 73, 78, 74, 75.

a Find the standard deviation of these scores.

b The standard deviation of the results of the same set of students in a Chemistry test was 5.91. Give two reasons why it would not be appropriate to use a comparison of these two standard deviations to determine whether students were more consistent in Physics or Chemistry.

Tip

In question 6, you can use a calculator to find the standard deviation.

7 Consider the five numbers: 2, 5, 9, x and y. The mean of the numbers is 5 and the variance is 6. Find the value of xy.

8 The mean of a set of 15 data items is 600 and the standard deviation is 12. Another piece of data is discovered and the new mean is 600.25. Find the new standard deviation.

9 The mean IQ of a class of 9 students is 121 and the variance is 226. Another student joins the class and the variance changes to 239.4. What are the possible values of the IQ of the new student?

10 a Explain why, for any piece of data, $x - \bar{x}$ is less than the range.

b By considering the formula:

$$\sigma = \sqrt{\frac{\Sigma(x-\bar{x})^2}{n}}$$

prove that the standard deviation is always strictly less than the range.

Section 3: Calculations from frequency tables

It is very common to summarise large amounts of data in a frequency table. This is a list of all the values that the data takes, together with how often they occur. You could convert this into a list of all the data values and calculate the statistics, as you have previously. However, there is a formula that you can use to find the mean more quickly.

Key point 20.4

$$\text{Mean} = \bar{x} = \frac{\Sigma fx}{n}$$

where f is the frequency of each x value and n is the total frequency.

You can work out $\overline{x^2}$ in a similar way, giving a formula for the variance.

Tip

Worked example 20.7 shows you how to use the formulae, but if your calculator can work with frequency tables, you can use that instead of doing the calculation yourself.

Key point 20.5

$$\text{Variance} = \sigma^2 = \frac{\Sigma fx^2}{n} - \bar{x}^2$$

WORKED EXAMPLE 20.7

The table lists the numbers of passengers (not including drivers) observed in cars passing a school.

Passengers	Frequency
0	32
1	16
2	2
3 or more	0

Find the median, mean and standard deviation in the numbers.

Median = average of 25th and 26th numbers.
Median = 0

There are $32+16+2=50$ data items so the median is the average of the 25th and the 26th data items. Both of these equal 0.

$\Sigma fx = (32\times 0)+(16\times 1)+(2\times 2)$
$= 20$ passengers

The total number of passengers is: none in the first group, 16 in the second group and 4 in the third group.

$\bar{x} = \frac{20}{50} = 0.4$

The total frequency is $32+16+2=50$.
Always check your answer for sense – here 0.4 passengers per car seems reasonable. It does not have to be rounded as means do not have to be achievable numbers.

$\Sigma fx^2 = (32\times 0)+(16\times 1)+(2\times 4) = 24$

To find the standard deviation you need first to calculate the mean of the squares.

$\sigma^2 = \frac{24}{50} - 0.4^2 = 0.32$

Using Key point 20.5.

$\sigma = 0.566$ (3 s.f.)

Remember that standard deviation is the square root of variance.

In Worked example 20.7 you knew the exact data values, but when you are dealing with grouped data, you no longer have this level of precision. In order to assess the mean and standard deviation, the best and simplest assumption is that all the original values in a group were located at the centre of the group, called the **mid-interval value**. To find the centre of the group, take the mean of the largest and the smallest possible values in the group.

WORKED EXAMPLE 20.8

Find the mean and standard deviation of the weight of eggs produced by a chicken farm. Explain why these answers are only estimates.

Weight of egg (g)	Frequency
[100,120)	26
[120,140)	52
[140,160)	84
[160,180)	60
[180,200)	12

x	f
110	26
130	52
150	84
170	60
190	12
Sum:	234

Make a table, assuming each data value lies in the centre of its group.

$$\bar{x} = \frac{\Sigma fx}{n} = \frac{34\,700}{234} = 148.3\text{ g}$$

$$\sigma^2 = \frac{\Sigma fx^2}{n} - \bar{x}^2 = \frac{5\,250\,600}{234} - 148.3^2 = 448.4$$

Your calculator should be able to give you the values of Σfx and Σfx^2. You should write those down as evidence of your method.

Therefore $\sigma = 21.2$ g (3 s.f.)

Always check you answer is sensible. In this case 21.2 g seems like a sensible measure of spread for data that have a range of 100.

These answers are only estimates because you have assumed that all the data in each group is at the centre, rather than using the actual data values.

Sometimes the endpoints of the intervals shown in the table are not the actual smallest and largest possible values in that group. For example, when measuring length in centimetres it is common to round the values to the nearest integer, so 10 – 15 actually means [9.5, 15.5). To find the mid-interval values you must first identify the actual interval boundaries.

Rewind

The bracket notation to describe intervals was covered in Chapter 1.

WORKED EXAMPLE 20.9

Estimate the mean of this set of data.

Age	Frequency
10–12	27
13–15	44
16–19	29

Group	x	f
[10,13)	11.5	27
[13,16)	14.5	44
[16,20)	18	29
Sum:		100

Carefully decide on the upper and lower interval boundaries. There should be no gaps between the groups, because age is continuous. Age is a little bit tricky because you are 16 years old until your 17th birthday.

$$\bar{x} = \frac{\sum fx}{n}$$

$$= \frac{1470.5}{100}$$

$$= 14.7 \text{ (3 s.f.)}$$

Enter the midpoints and frequencies into your calculator.

EXERCISE 20C

1 Calculate the mean, standard deviation and the median for each data set. Use the formula first, then check your answer with a calculator.

a **i**

x	Frequency
0	16
1	22
2	8
3	4
4	0

ii

x	Frequency
–1	10
0	8
1	5
2	1
3	1

b **i**

x	Frequency
10	7
12	19
14	2
16	0
18	2

ii

x	Frequency
0.1	16
0.2	15
0.3	12
0.4	9
0.5	8

2 A group is described as '17–20'. State the upper and lower boundaries of this group if it is measuring:

a age in completed years

b number of pencils

c length of a worm, to the nearest centimetre

d hourly earnings, rounded up to a whole number of dollars.

3 Calculate estimates of the mean and standard deviation of each set of data. Use statistical functions on your calculator.

a i x is the time taken to complete a puzzle, in seconds.

x (s)	Frequency
[0, 15)	19
[15, 30)	15
[30, 45)	7
[45, 60)	5
[60, 90)	4

ii x is the weight of plants, in grams.

x (g)	Frequency
[50, 100)	17
[100, 200)	23
[200, 300)	42
[300, 500)	21
[500, 1000)	5

b i x is the length of fossils found in a geological dig, to the nearest centimetre.

x (cm)	Frequency
0–4	71
5–10	43
11–15	22
16–30	6

ii x is the power consumption of light bulbs, to the nearest watt.

x (W)	Frequency
90–95	17
96–100	23
101–105	42
106–110	21
111–120	5

c i x is the age, in years, of children in a hospital ward.

x (years)	Frequency
0–2	12
3–5	15
6–10	7
11–16	6
17–18	3

ii x is the amount of tips paid in a restaurant, rounded down to the nearest dollar.

x ($)	Frequency
0–5	17
6–10	29
11–20	44
21–30	16
31–50	8

4 The bar chart shows the outcome of a survey into the number of cars owned in each household in a small town called Statham.

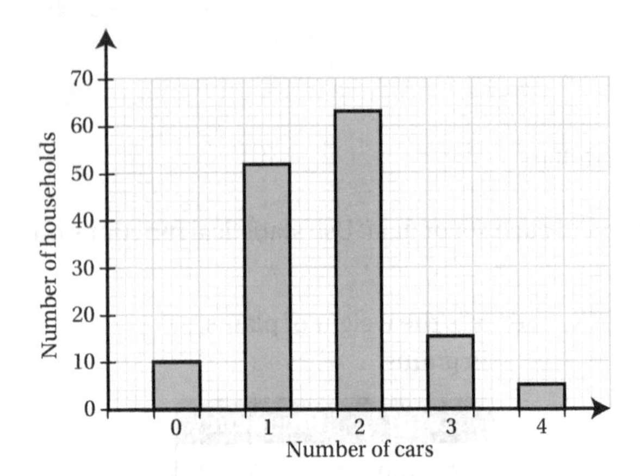

a Use the information from the graph to copy and complete this table.

Number of cars	0	1	2	3	4
Number of households	10	52			

b Hence find the mean and standard deviation of the number of cars in a household.

c The survey is also conducted in a nearby town called Mediton where the mean is found to be 2.17 cars per household with standard deviation 1.12. Make two comparisons, in context, between the two towns.

5 The mass of food eaten by 50 dogs during a week was measured, to the nearest kilogram.

Mass (kg)	1 or 2	3 or 4	5 or 6	7 or 8
Frequency	12	20	16	2

a Estimate the mean and standard deviation of the masses.

b State two ways in which the accuracy of these estimates could be improved, if the observations were to be repeated.

6 The table shows the the numbers of broken eggs per box, in a sample of 50 boxes of 12 eggs.

Number of broken eggs per box	0	1	2	3	4	5	6
Number of boxes	17	8	7	7	6	5	0

a Calculate the median number of broken eggs per box.

b Calculate the mean number of broken eggs per box.

c Calculate the variance of the data.

d The packaging process for the eggs is changed. In a new sample of 50 boxes the mean number of broken eggs was 1.92 and the variance was 2.84. Give one reason to support someone who argues:

i 'The new packaging process is better.'

ii 'The new packaging process is worse.'

iii 'There is no difference between the two packaging processes.'

7 The histogram shows the waiting times for calls to be answered by a helpdesk.

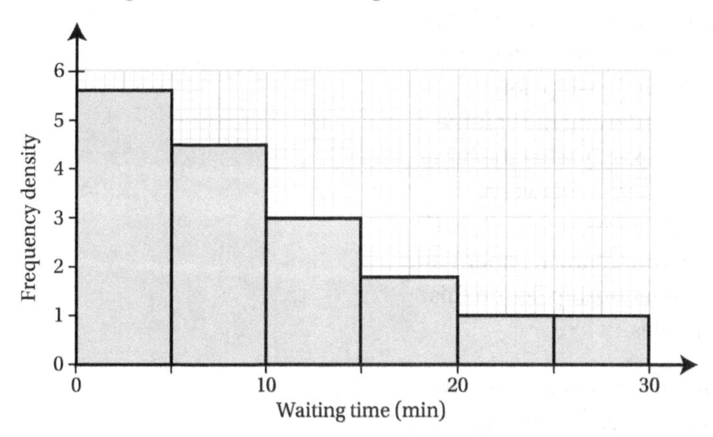

a Use the graph to calculate estimates of the mean and standard deviation of the waiting time.

b What assumptions have you made in your calculations in part **a**?

c Explain, with reference to the graph, why the median will be below the mean for this set of data.

8 The standard deviation of the data in the table is $0.8k$ where k is a positive constant.

x	Frequency
1	8
p	2

a Explain why there will be two possible values for p for each value of k.

b Find expressions for p in terms of k.

c Given that the mean is 2.2, find the value of k.

9 The mean of the data in the table is 32 and the variance is 136. Find the possible values of p and q.

x	Frequency
20	12
40	q
p	8

10 Amy and Bob are both playing a game on their computers. Amy's average score on both level one and level two is higher than Bob's. Show that it is possible for Bob still to have a higher overall average across levels one and two.

Section 4: Scatter diagrams and correlation

So far you have only been considering one variable at a time, such as someone's height or their IQ. Sometimes you will want to find out whether there is a relationship between two variables. By gathering two data values from an individual source, such as a person's age and weight, you can investigate any potential relationship. Data that comes in pairs of values, in this fashion, is said to be **bivariate**.

When you have two sets of data they may have no relationship, they might be independent: for example, the IQ and the house number of a randomly chosen person. In this case, one variable gives you no information about the other one. Alternatively, they may be in a fixed relationship: for example, the length of a side of a randomly created cube and the volume of the cube. When you know one variable you know exactly what the other one will be. However, the situation is usually somewhere in between, whereby if you know one value you can make a good guess at the value of the other variable, but not be certain: for example, your mark in Paper 1 and your mark in Paper 2 of a Maths exam. Where the relationship lies on this spectrum is called the **correlation** of the two variables.

In this course you will focus on linear correlation – the extent to which two variables are related by a relationship of the form $Y = mX + c$. If the gradient of the linear relationship is positive, you describe the correlation as positive, and if the gradient is negative you describe the correlation as negative.

These relationships are often best illustrated in a scatter diagram.

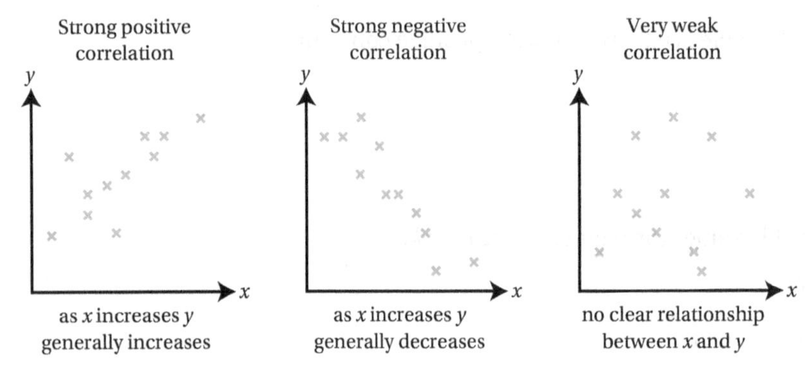

Rather than simply describing the relationship in words you can find a numerical value to represent the linear correlation.

Key point 20.6

A **correlation coefficient**, r, is a measure of the strength of the relationship between two variables. It can take values from -1 to 1.

You must know how to interpret the value of r.

Value of r	Interpretation
$r \approx 1$	Strong positive linear correlation
$r \approx 0$	No linear correlation
$r \approx -1$	Strong negative linear correlation

If $r = \pm 1$ it is perfect correlation – the data lies exactly along a straight line.

Just because $r = 0$ it does not mean that there is no relationship between the two variables – simply that there is no linear relationship.

This graph shows data that has a correlation coefficient of zero, but there is clearly a relationship.

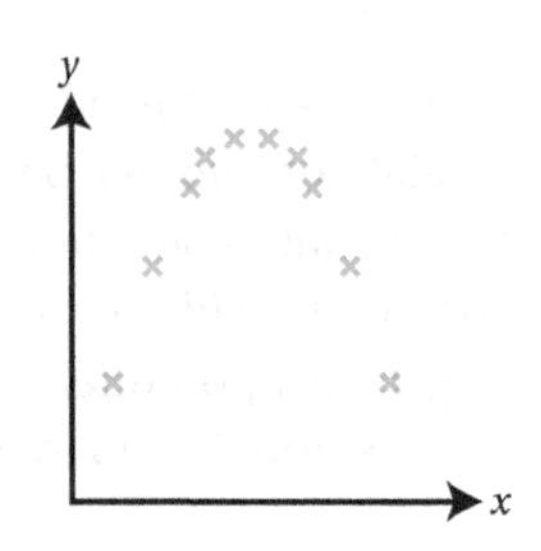

Scatter diagrams can also reveal if there are two separate groups within the data.

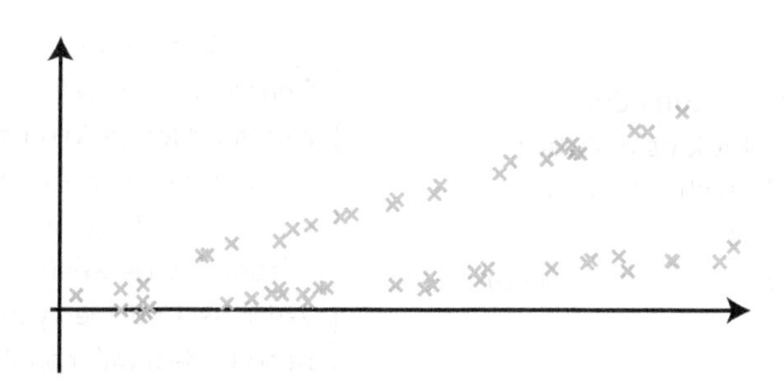

While the product moment correlation coefficient can give a measure of correlation between two variables, it is important to realise that just because r might be close to ± 1, a change in one variable does not necessarily **cause** a change in the other. Such a correlation might be simply coincidence or due to a third, hidden variable.

For example, there might be a strong correlation between ice-cream sales and instances of drowning at beaches in a given location. Clearly, eating more ice-cream does not cause drowning; instead, the hidden variable of temperature could cause both to rise.

WORKED EXAMPLE 20.10

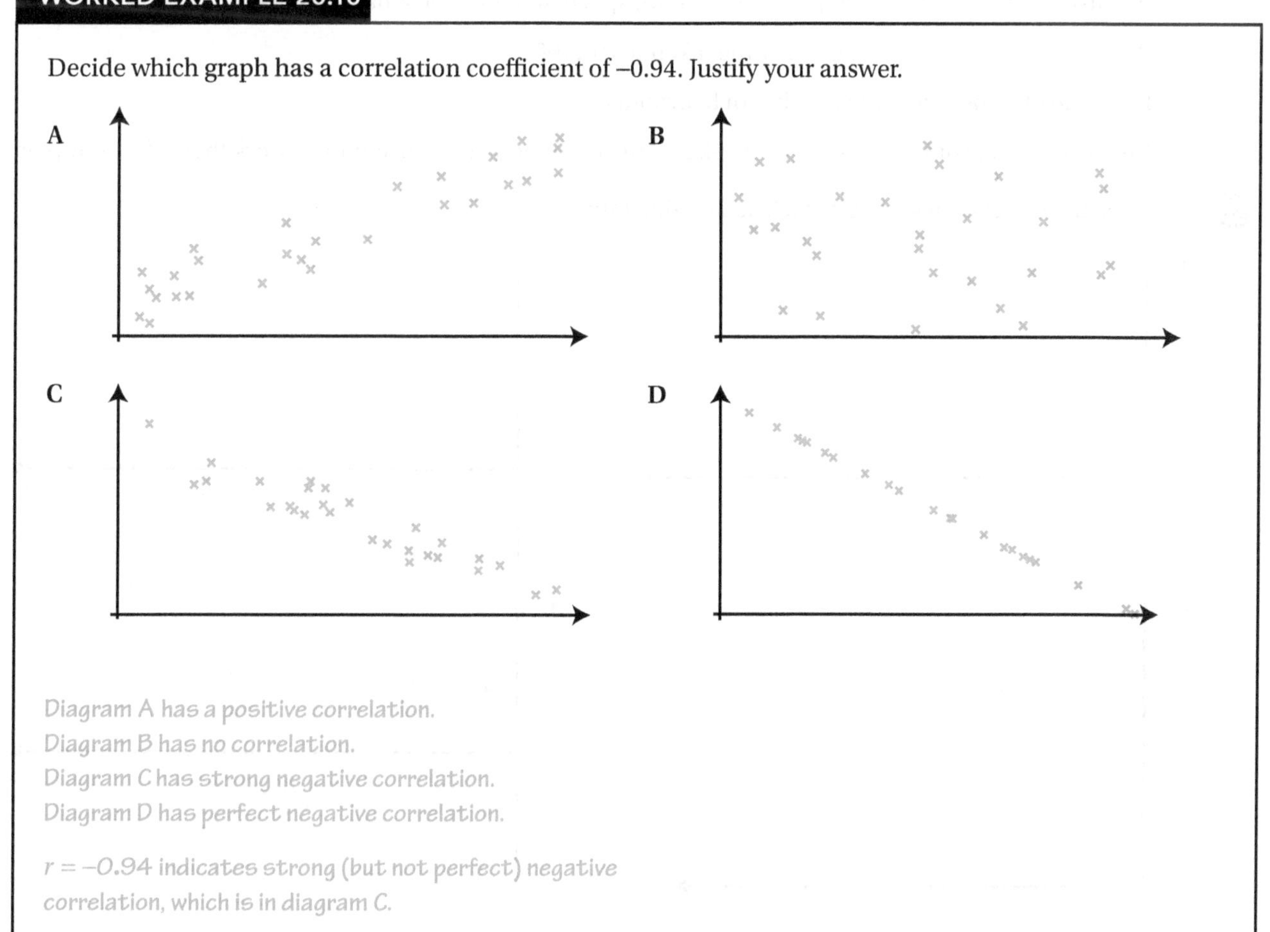

Decide which graph has a correlation coefficient of –0.94. Justify your answer.

Diagram A has a positive correlation.
Diagram B has no correlation.
Diagram C has strong negative correlation.
Diagram D has perfect negative correlation.

$r = -0.94$ indicates strong (but not perfect) negative correlation, which is in diagram C.

Once you have established from the correlation coefficient that there is a linear relationship it is often useful to draw in a line of best fit, sometimes called a **regression line**.

When interpreting regression lines you should always consider whether the data actually follows a linear trend, by looking at either the correlation coefficient or the scatter diagram. You should also be aware that if you are using the regression line to predict values outside the range of the observed data - a process called **extrapolation** - your answer may be less valid.

Fast forward

A In Student Book 2, Chapter 22, you will learn about a method for deciding if a calculated correlation coefficient is evidence of a genuine linear relationship between two variables. Until then you will not need to deal with correlation coefficients where the interpretation is ambiguous.

EXERCISE 20D

1. For each of set of bivariate data describe the correlation you would expect to see.

   a The results in a Physics test and a Mathematics test

   b A person's height and their income

   c The age of a car and the number of miles driven

   d The distance travelled by a car going at constant speed and the time it has travelled

   e The age of a school student and the time taken to run 100 m

   f The value of a house and the number of bedrooms

   g The mass of fruit someone eats each week and the number of days absent from work they take each year

2. Describe the correlation shown in each scatter diagram.

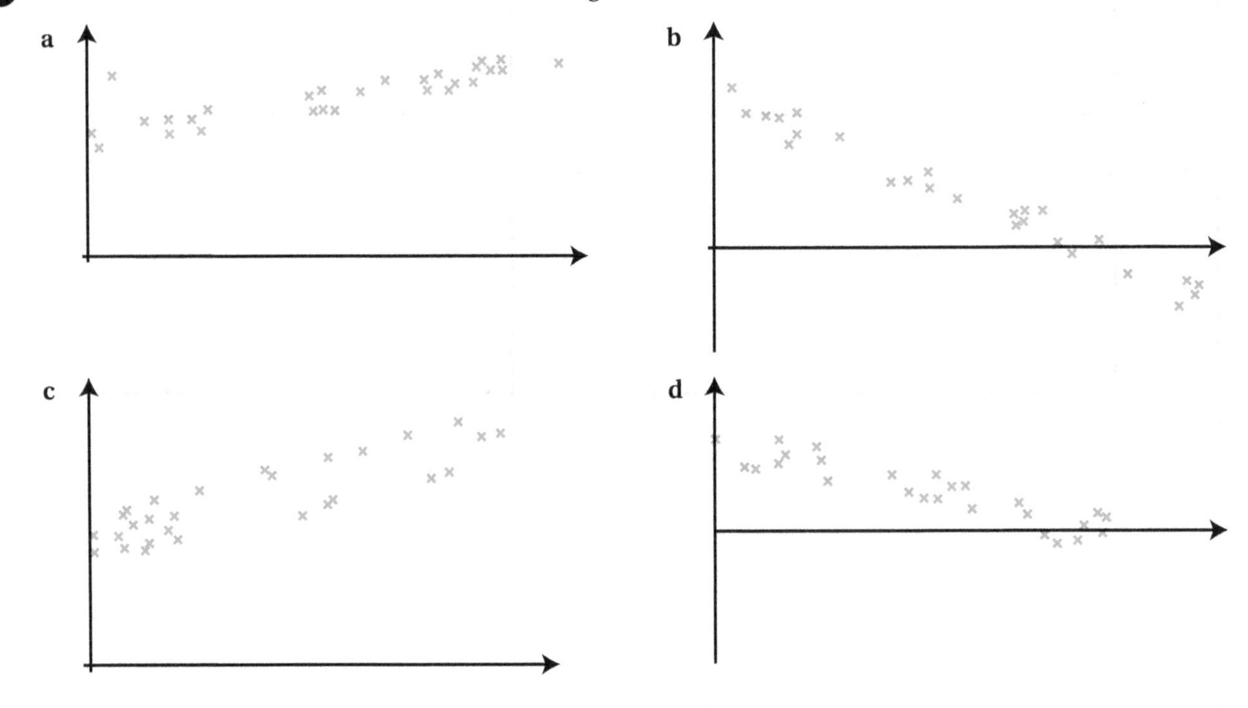

e

f

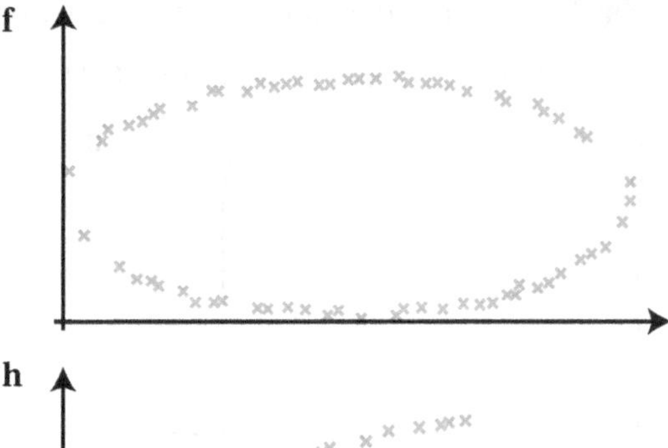

g

h

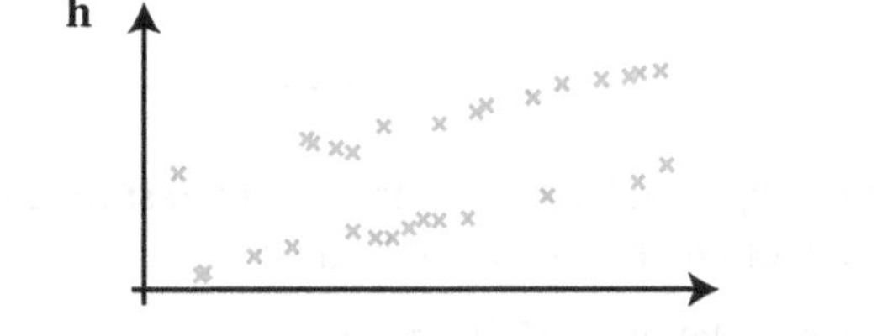

3. The correlation coefficient between the speed of a computer processor and its life expectancy is 0.984, based on a sample of 50 processors.

 a Interpret this correlation coefficient.

 b Does this result imply that processor speed affects the life expectancy? Explain your answer.

4. A road safety group has tested the braking distance of cars of 20 different ages. The correlation coefficient between a car's age and braking distance is 0.92.

 a Interpret the correlation coefficient.

 b Nicole says that this provides evidence that older cars tend to have longer stopping distances. State with a reason whether you agree with her.

5. The weights of babies (Y kg) at age X months are measured for a sample of 100 babies aged between 0 and 18 months.

 a The correlation coefficient is found to be 0.88. Describe what this suggests about the relationship between mass and age for babies.

 b The equation of the line of best fit is found to be $Y = 0.5X + 3.2$. Interpret in context the meaning of:

 i 0.5 **ii** 3.2

 in this equation.

 c Explain why this line of best fit would not be an appropriate model to predict the weight of a 14-year-old boy.

6. The number (x) of years since starting primary school and times (t seconds) to run 100 m of 20 students was measured. The output from a spreadsheet gave this information.

$$t = 24.8 - 0.609x$$
$$r = -0.24$$
$$r^2 = 0.0576$$

 a Interpret, in context, the values:

 i 24.8 **ii** 0.609

 in this output.

 b Give two reasons why it would not be appropriate to use this model to predict the time for a 60-year-old to run 100 m.

7 Match the scatter diagrams with the values of r.

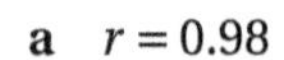
a $r = 0.98$

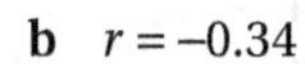
b $r = -0.34$

c $r = -0.93$

d $r = 0.58$

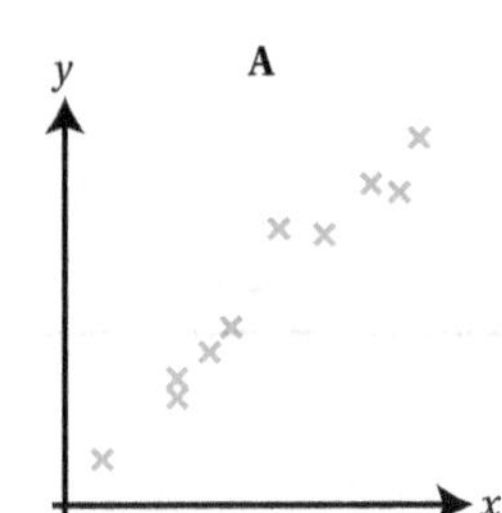

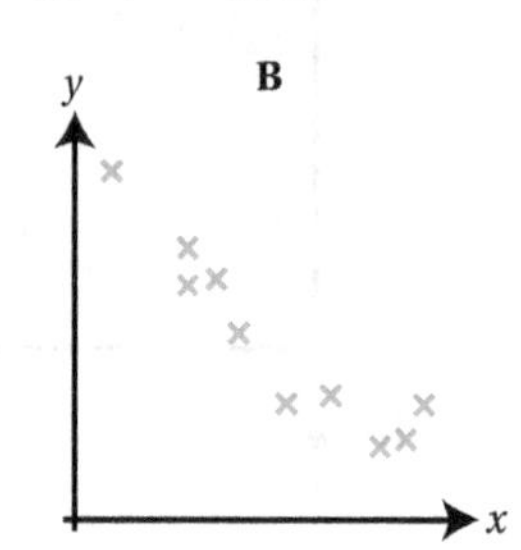

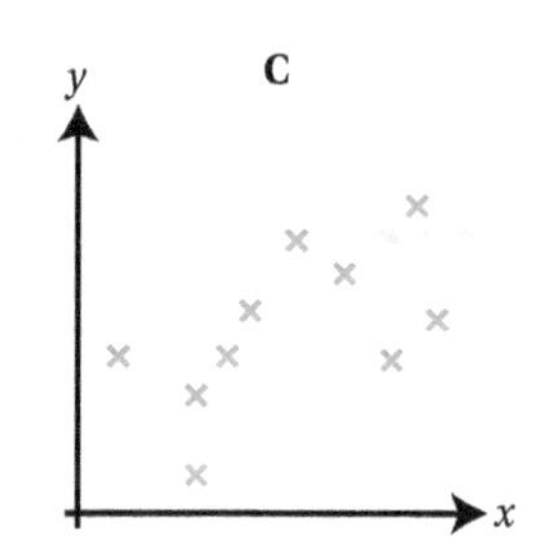

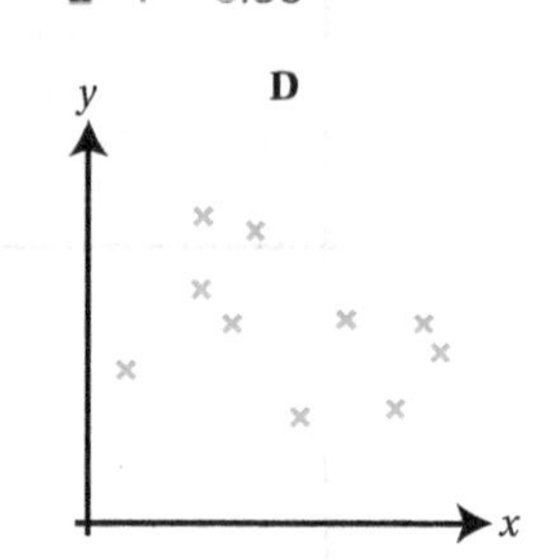

8 The heights and ages of 30 trees in a forest were measured and plotted on this scatter diagram.

The correlation coefficient is 0.96.

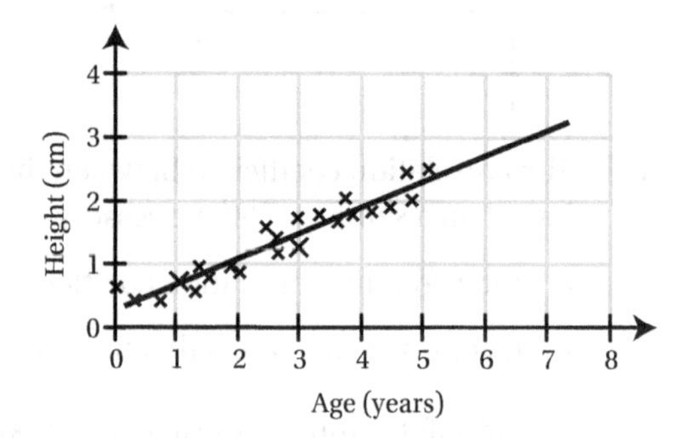

a Use the line of best fit to estimate the height of a tree that is:

i 3 years old **ii** 7 years old.

b Comment on the validity of your answers in part **a**.

9 The graph shows a connected scatter plot of the lowest frequency produced by a sample of 6 speakers against their power.

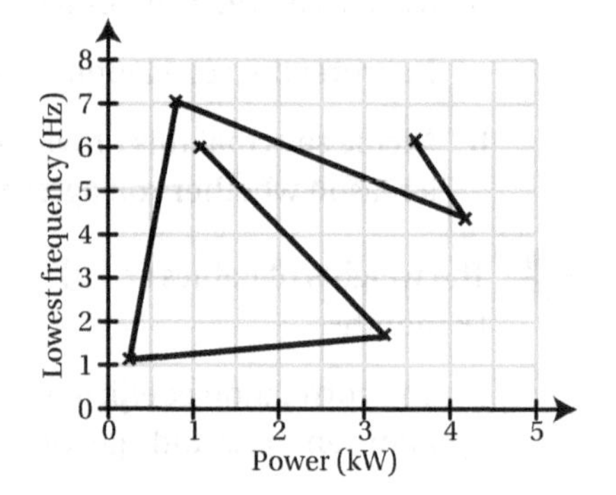

a Use the line provided to estimate, where possible, the lowest frequency, if the power is:

i 3.5 kW **ii** 2kW.

b Explain why it is not appropriate to connect the data with straight lines in this way.

c Describe a situation in which it would be appropriate to connect the points with straight lines in this way.

10 Which statements are true for bivariate data?

a If $r = 0$ there is no relationship between the two variables.

b If $Y = kX$ then $r = 1$.

c If $r < 0$ then the gradient of the line of best fit is negative.

d As r increases then so does the gradient of the line of best fit.

Section 5: Outliers and cleaning data

When you are dealing with real-world data there are sometimes errors, missing data or extreme values that can distort your results. In this section you will look at some standard ways to identify problematic data and how to deal with it.

Often the most useful thing to do is to look at your data graphically. If the underlying pattern is strong, values that are a significant distance from the rest of the data can become obvious. Such values are called **outliers**. For example, on this scatter diagram, the red point does not seem to follow the trend of the other points.

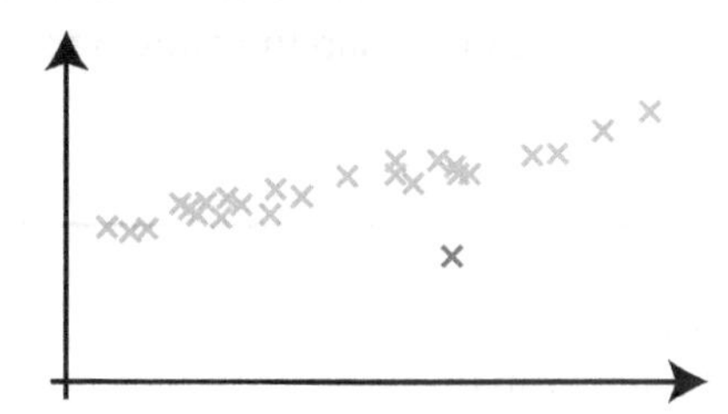

There are also some standard calculations that you can use to check for outliers.

The first is that an outlier is any number more than 1.5 interquartile ranges away from the nearest quartile.

WORKED EXAMPLE 20.11

A group of people were asked to name as many characters from 'Harry Potter' as possible in one minute. The results are illustrated in the box-and-whisker plot. Use the definition of an outlier as being any data value more than 1.5 interquartile ranges away from the nearest quartile to determine if there are any outliers in the data set.

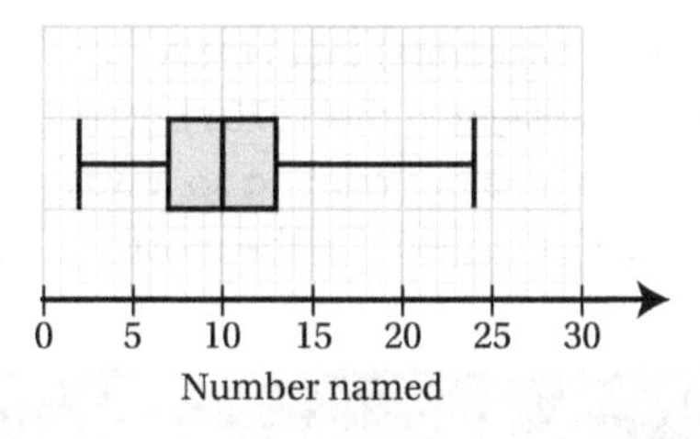

Lower quartile: 7
Upper quartile: 13
Interquartile range: 6

Read the quartiles from the plot to find the IQR.

Upper quartile plus $1.5 \times$ IQR:
$13 + 1.5 \times 6 = 22$
Lower quartile minus $1.5 \times$ IQR:
$7 - 1.5 \times 6 = -2$

Find the largest and smallest values which are not outliers.

The smallest value is 1 which is not an outlier.
The largest value is 24 which is an outlier.

Look at the largest and smallest values.

Another possible method is to classify anything more than two standard deviations from the mean as an outlier.

WORKED EXAMPLE 20.12

The table shows the wages of workers in a factory (in thousands of pounds).

Wages (£000)	10	16	20	145
Number of employees	18	14	4	1

Use the definition of an outlier as anything more than two standard deviations from the mean to determine if the wage of £145 000 is an outlier.

Using a calculator: $\bar{x} = 17$
$= 21.6$

Use a calculator to find the mean and standard deviation.

Two standard deviations from the mean is
$17 + 2 \times 21.6 = 60.2$

Use the full accuracy from a calculator rather than the quoted rounded value.

$145 > 60.2$ so £145 000 is an outlier.

Once you have found that something is an outlier you must then decide whether or not to include it in your calculation. This often requires you to look at the data in context.

If the outlier is clearly an error (for example, the wrong units being used or an impossible value), then it should be excluded from the data.

If there are several outliers it might be a distinctly different group that should be analysed separately. Otherwise, it might simply be that there is an unusual value in your data. This does not mean that it is an error. Unless you have a good reason otherwise, you should keep it in your analysis but report the presence of outliers.

Did you know?

There is a famous story (probably not entirely accurate) that NASA satellites first 'discovered' a hole in the ozone layer in the 1970s but an automatic error checker decided it was an anomalous reading so it was ignored until nearly ten years later. Sometimes outliers are the most interesting part of the data!

EXERCISE 20E

1 Determine, using the definition of an outlier as more than 1.5 IQR from the nearest quartile, if there are outliers in each set of data.

a **i**

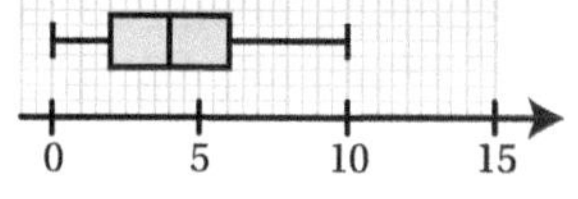

ii

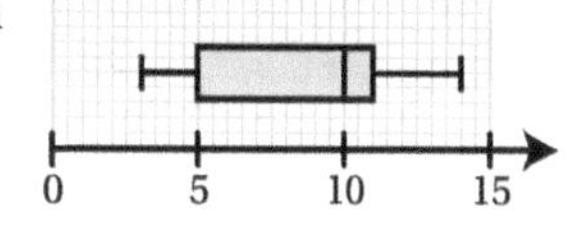

b **i**

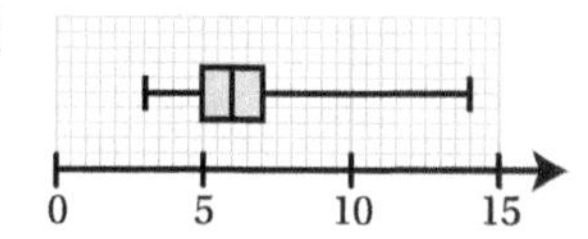

ii

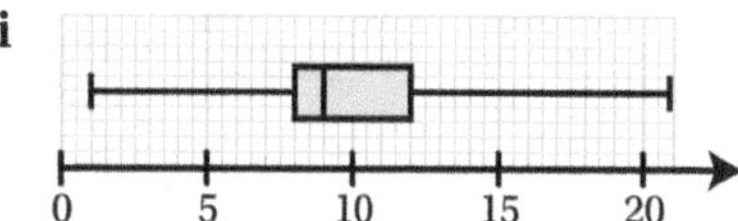

2 Determine if each data set has outliers, defined as data more than 2 standard deviations from the mean.

a **i** $1, 5, 5, 7, 14$ **ii** $80, 85, 90, 90, 125$

b **i** $-3, 8, 8, 10, 14, 15, 16$ **ii** $0, 26, 26, 28, 29, 30, 64$

3 A biologist collects weight and length information on a sample of ants from a forest. He finds the correlation coefficient is 0.22 so concludes that there is not a strong linear relationship between weight and length. A colleague suggests that he should draw a scatter diagram to illustrate the data, which is shown here.

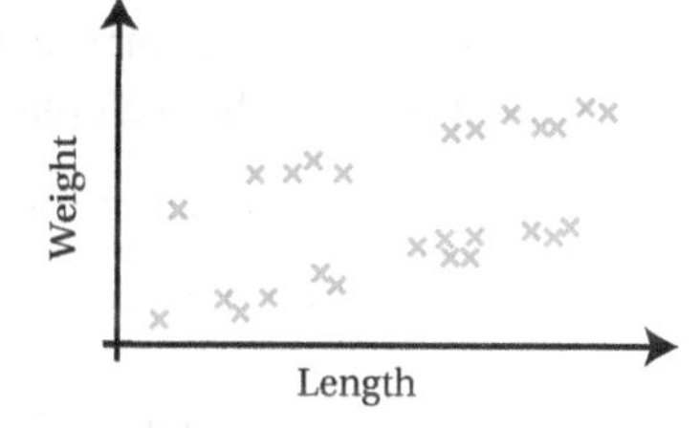

What statistical advice would you give the biologist?

4 30 people attempted to complete a level on a computer game. The box-and-whisker plot shows the numbers of attempts required.

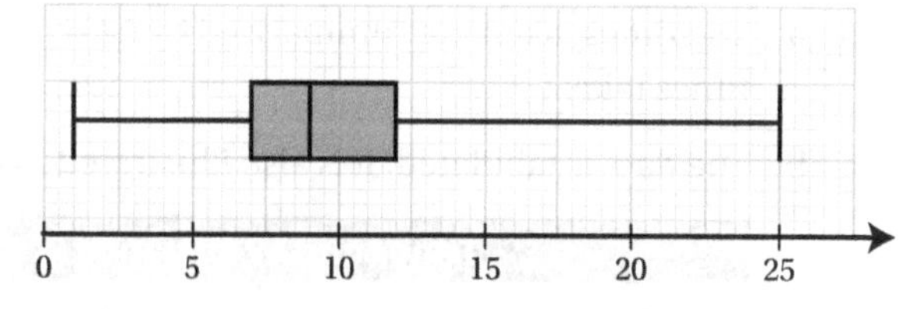

a Find the interquartile range for this set of data.

b Given that outliers are defined as more than 1.5 IQRs from the nearest quartile, show that there must be an outlier in the data.

5 The cumulative frequency graph shows the marks of 100 pupils in a test.

Use the definition that an outlier is more than 1.5 IQR from the nearest quartile to show that there are no outliers in the data.

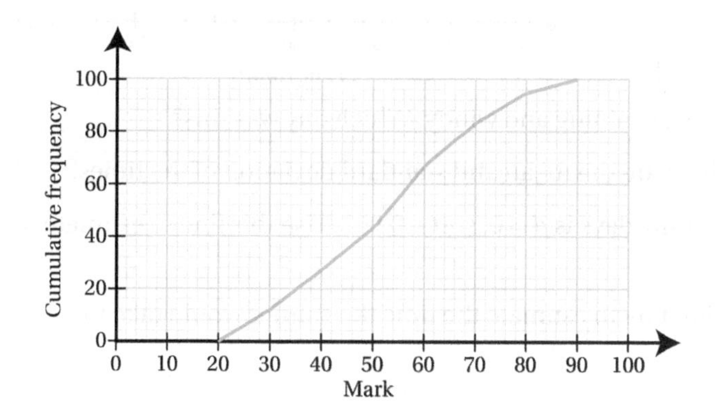

6 The scatter diagram shows the numbers of matches on a tennis court in a week (N) against the average temperature in that week (T).

The mean number of matches per week is 46.4 with standard deviation 7.80.

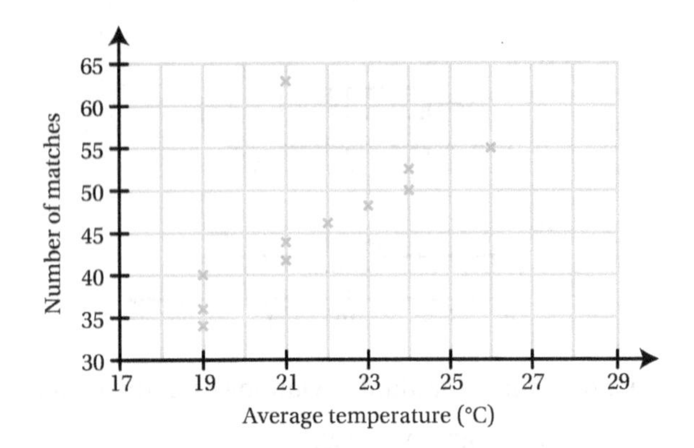

a Use of the tennis court is known to increase noticeably during the first week of the Wimbledon tennis tournament. What was the average temperature during that week?

b The data point corresponding to the first week of Wimbledon is removed. Answer these questions without further calculation.

i Will the mean number of matches per week increase or decrease?

ii Will the standard deviation in the number of matches increase or decrease?

7 A runner uses a smartwatch to track the time (t minutes) taken and distance (d km) covered on her run each day for a 7-day week.

a The times taken are summarised by $\Sigma t = 440$, $\Sigma t^2 = 28\,040$. Find the mean and standard deviation of the times taken.

b The data is tabulated. It is also illustrated in the scatter diagram.

	Time (min)	Distance (km)
Monday	62	18
Tuesday	49	15
Wednesday	75	23
Thursday	62	14
Friday	66	20
Saturday	59	17
Sunday	67	20

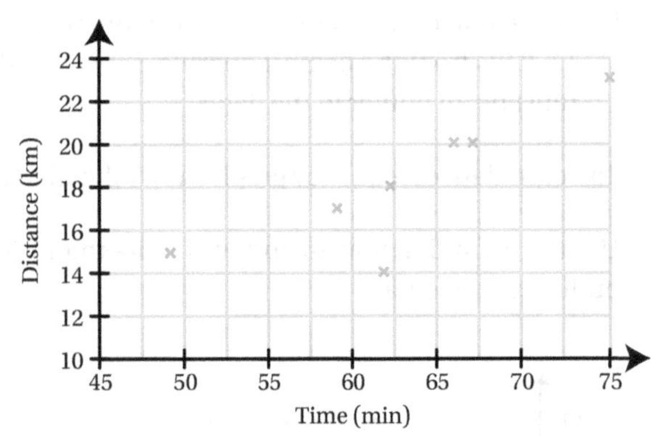

Six of the days were sunny but on one day it was raining and windy. Which day was this?

c Remove the day with bad weather and recalculate the mean and standard deviation of the times.

d The equation of the regression line for the sunny days is $d = 0.314t - 0.957$. Use this line to estimate the speed of the runner.

e Would it be appropriate to use this regression line to estimate the time the runner would take to complete a 40 km race? Justify your answer.

8 **a** A doctor measures the level of infection markers in 6 samples of blood. The mean of the values is 18 with a variance of $\frac{631}{3}$. The largest value is 50. Show that this is an outlier, using the definition that outliers are more than 2 standard deviations away from the mean.

b Find the mean and standard deviation if the value of 50 is removed.

9 In this question, define an outlier as being more than 2 standard deviations from the mean.

a The standard deviation of a set of data is 9 with range 50. Prove that there must be an outlier in the data.

b Consider this data set.

x	Frequency
0	p
30	q

Show that 30 is an outlier if $p > 4q$.

c Hence show that if a data set has a standard deviation of 9 with range 30 it is possible for there to be an outlier.

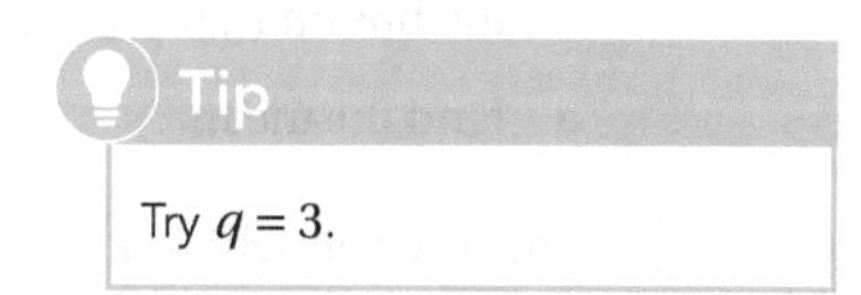

Tip

Try $q = 3$.

d Use a counter example to show that if a data set has standard deviation 9 and range 30 it is **not** certain that there is an outlier.

Tip

Consider the given data set.

x	Frequency
0	p
15	q
30	p

Checklist of learning and understanding

- **Histograms** are a useful visual summary of data, giving an immediate impression of centre and spread.
  - The area represents frequency.
  - The vertical axis represents **frequency density**.
- **Cumulative frequency diagrams** are useful for finding the median and the interquartile range and also facilitate the construction of a **box-and-whisker plot**, another good visual summary of data.
- The centre of the data can be measured using the **mean**, **median** or **mode**. The spread around the centre can be measured using the **range**, **interquartile range** or **standard deviation**.
- The square of the standard deviation is called the **variance**.
- If the data has been grouped you can estimate the mean and standard deviation by assuming that every data value is at the centre of each group.
- The **correlation coefficient** is a value, from –1 to 1, that measures the strength of the relationship between two variables.
- The **regression line** can be drawn on a scatter diagram for a set of **bivariate data**. It should only be used if there is significant linear correlation and there is not too much extrapolation.
- Graphs or calculations can be used to identify **outliers** in data, which then may need to be removed.

Mixed practice 20

1. A random sample of 50 modern cars, of the same make and model, were investigated. The correlation coefficient between the age and the value was calculated. What is a plausible value of r?

   Choose from these options.

   A -1 **B** -0.89 **C** 0.46 **D** 1.16

2. A sample of discrete data drawn from a population is given as $115, 108, 135, 122, 127, 140, 139, 111, 124$.

   Find:

   a the interquartile range **b** the mean **c** the variance.

3. $\Sigma(x-\bar{x})^2 = 12$ and $n = 3$. Find the value of the standard deviation.

4. A student takes the bus to school every morning. She records the length of time, in minutes, she waits for the bus on 12 randomly chosen days. The data set is summarised by $\Sigma x = 49$ and $\Sigma x^2 = 305.7$.

   a Find the mean. **b** Find the variance.

5. The box-and-whisker plot shows the lengths (x) of corn snakes in cm.

   Use the definition that an outlier is more than 1.5 IQR from the closest quartile to find the range of values which would be outliers. Hence show that there are no outliers for this data.

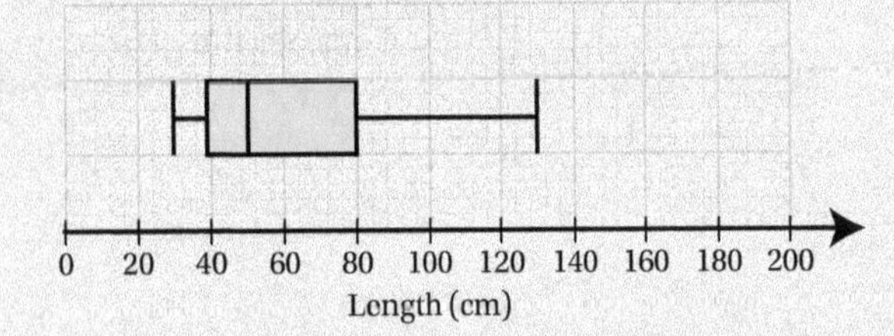

6. **a** Use a counterexample with two data items to show that $\overline{x^2} = \bar{x}^2$ is not always true.

   b If $\overline{x^2} = \bar{x}^2$, find the standard deviation of the data. Hence provide an example of a set of data, containing two items, that has $\overline{x^2} = \bar{x}^2$.

7. The weights, in kilograms, of a random sample of 15 items of cabin luggage on an aeroplane were as follows.

   4.6 3.8 3.9 4.5 4.9 3.6 3.7 5.2 4.0 5.1 4.1 3.3 4.7 5.0 4.8

   For these data:

   a find values for the median and the interquartile range

   b find the value for the range

   c state why the mode is not an appropriate measure of average.

   [© AQA 2014]

8 The table shows the ages of children attending a party.

Age	0–5	6–8	9–12	13–18
Frequency	2	4	12	2

What is the mean age of children at the party? Choose from these options.

A 8.875 **B** 9.375 **C** 9.5 **D** 10

9 **a** From the histogram given here, calculate the mean and the standard deviation of the data.

b The histogram shows the distribution of the length of time, in minutes, Tom has to wait for a bus in the morning. Find the probability that, on a randomly selected day, he has to wait between 15 and 25 minutes.

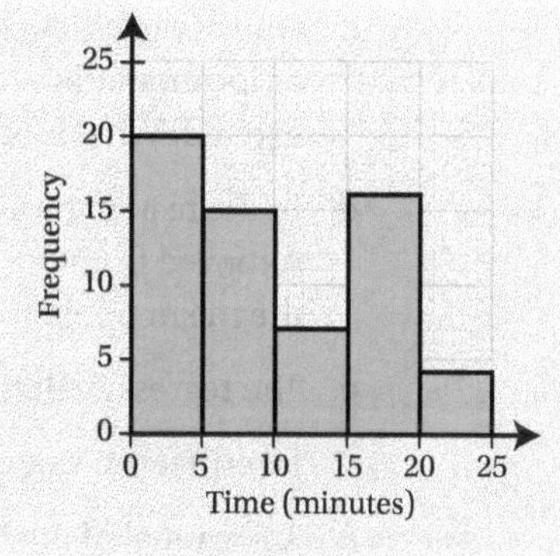

10 Jenny must sit 4 papers for an exam. All papers have an equal weight when their marks are combined. The mean mark of the first 3 papers Jenny has sat is 72% with a standard deviation of 8%.

a If she wants to get a mean of 75% overall what is the lowest percentage she can get in her fourth paper?

b What is the highest possible mean she can get?

c If she does get the highest possible mean, what is her new standard deviation?

11 The cumulative frequency diagram gives the speed, v, of 50 cars in mph as they travel past a motorway checkpoint.

a From the diagram, find the median speed.

b Any car travelling at above 75 mph will be stopped by the police. How many of these cars will be stopped?

c The middle 50% of speeds lie between a and b, where $a < b$. Find the values of a and b.

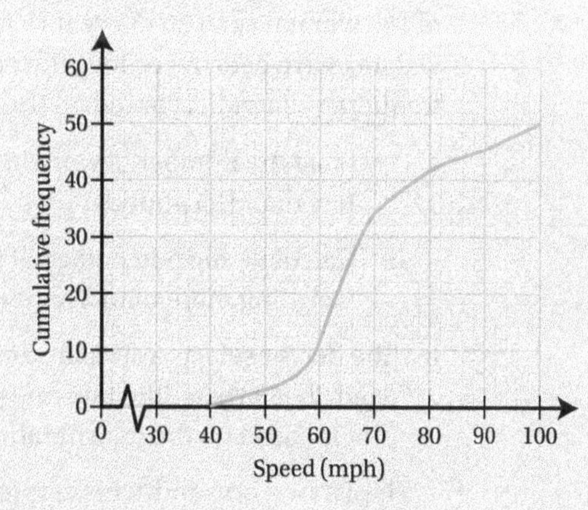

d Copy and complete this frequency table.

v (mph)	Frequency
$40 < v \leqslant 50$	2
$50 < v \leqslant 60$	8
$60 < v \leqslant 70$	
$70 < v \leqslant 80$	
$80 < v \leqslant 90$	
$90 < v \leqslant 100$	

e Hence estimate the mean and the standard deviation of the speeds.

f Use the definition of an outlier as anything more than 2 standard deviations from the mean to show that some of the observed speeds are outliers. Decide, with justification, whether these speeds need to be removed for a valid analysis.

12 A geographer is studying data on the area (A) and population (P) of various cities in a country. He displays his data on a scatter diagram.

The mean population of the cities studied is 604 thousand.

a What is the advantage of displaying the data on a scatter diagram rather than two histograms?

b Describe the correlation of the data.

c One of the cities completely fills an island so that as its population has grown it has not been able to expand. What is the area of the island?

d In the rest of the analysis the city on the island is removed from the data. What effect does this have on the mean population of the cities studied? Explain your answer.

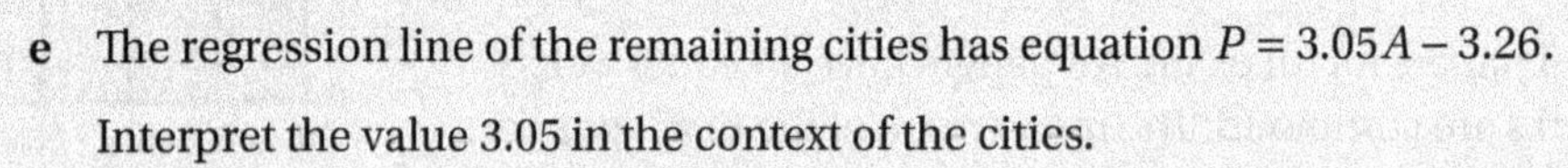

e The regression line of the remaining cities has equation $P = 3.05A - 3.26$.

Interpret the value 3.05 in the context of the cities.

f The capital of the country has an area of $1600\ \text{km}^2$. Why would it not be valid to use the regression line to predict the population of the capital?

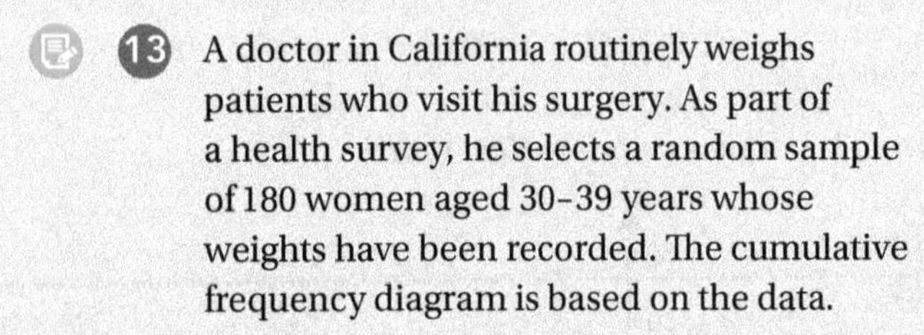

13 A doctor in California routinely weighs patients who visit his surgery. As part of a health survey, he selects a random sample of 180 women aged 30–39 years whose weights have been recorded. The cumulative frequency diagram is based on the data.

a **i** State the number of women weighing less than 125 pounds.

ii Calculate the percentage of women weighing more than 200 pounds.

b The doctor selects a similar sample of men aged 30–39 years. The box-and-whisker plot is based on the data obtained.

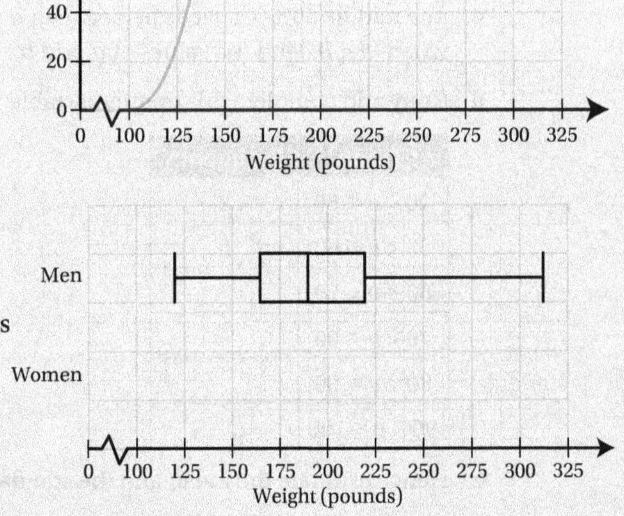

i Draw a box-and-whisker plot to illustrate the data for the women as presented in the cumulative frequency diagram. You may assume that there are no outliers.

ii Make **three** comments on the differences between the way in which the weights of the men are distributed compared with the way in which the weights of the women are distributed, as shown in the box-and-whisker plots.

c Explain why the data obtained from the two samples may not reliably represent the population of Californian people aged 30–39 years.

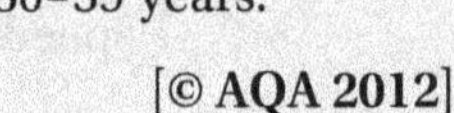

14 The mean of the results of 15 students in a test is 50, with a standard deviation of 10. A new student joins the class and scores 90 in the test. Find the new standard deviation.

Choose from these options.

A 2 **B** 4 **C** 13.7 **D** 187.5

15 The four populations A, B, C and D are the same size and have the same range.

These are histograms for the four populations.

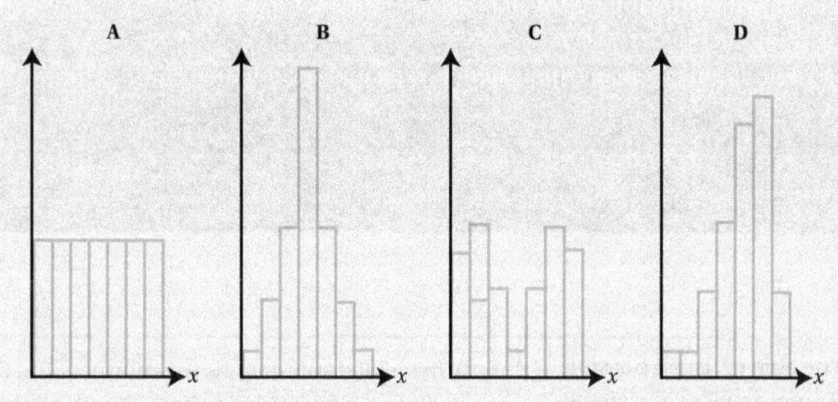

a Each of the three box-and-whisker plots corresponds to one of the four populations. Write the letter of the correct population for each of α, β and γ.

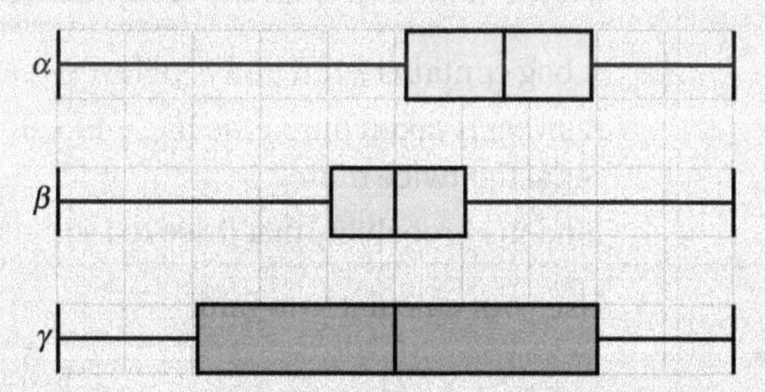

b Each of the three cumulative frequency diagrams corresponds to one of the four populations. Write the letter of the correct population for each of **i**, **ii** and **iii**.

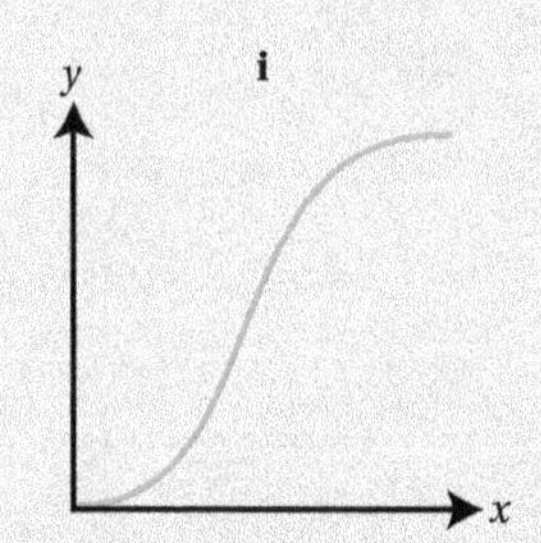

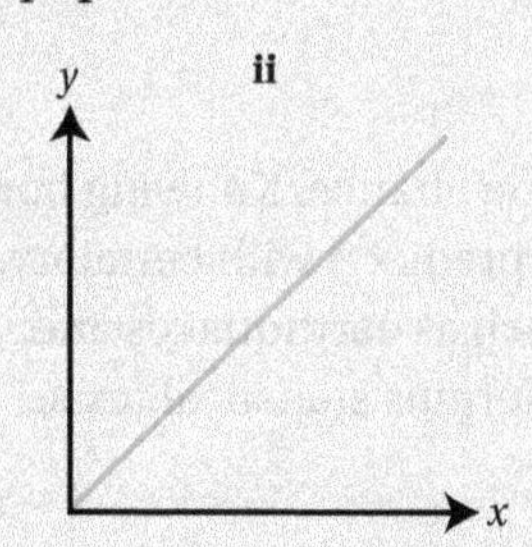

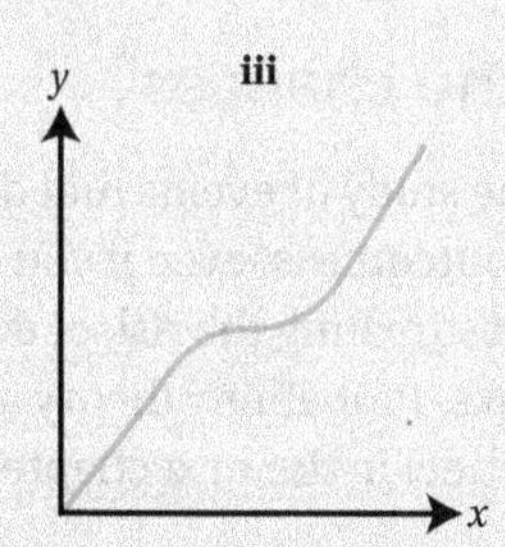

16 The mean of a set of 10 data items is 115 and the variance is 154. Another piece of data is discovered and the new mean is 114. What is the new variance?

17 If the sum of 20 pieces of data is 1542, find the smallest possible value of $\sum x^2$.

See Extension Sheet 20 for questions on using data in different forms to measure risk.

21 Probability

In this chapter you will learn how to:

- work out combined probabilities when you are interested in more than one outcome
- work out the probability of a sequence of events occurring
- construct and use a table showing probabilities of all possible outcomes in a given situation (probability distribution)
- calculate probabilities in a situation when an experiment is repeated several times (binomial distribution).

Before you start…

GCSE	You should be able to list all possible outcomes (sample space) of a single event or a combination of two events.	1 A five-sided spinner has the numbers 1 to 5 written on it, and a four-sided spinner has the letters A to D on it. What is the probability of getting an A and a 3 when the two spinners are spun together?
GCSE	You should be able to use tree diagrams to record probabilities of successive events, and to calculate probabilities of combined events.	2 A bag contains 7 red and 3 yellow sweets. A sweet is taken out of the bag and eaten. This is repeated twice more. Find the probability that three red sweets are picked.
Chapter 9	You should be able to calculate factorials and binomial coefficients.	3 Use your calculator to find: a $7!$ b $\frac{7!}{3!}$ c $^7C_3$ d $\binom{5}{5}$ e $^{10}C_0$.

What are the chances?

Probability is the study of events that depend on chance. Knowing how likely a certain outcome is, even if you cannot predict it with certainty, is important in estimating the risk of events such as earthquakes and disease outbreaks. Probability theory also underpins statistical tests, which you will meet in the next chapter.

In this chapter you will review the concepts of independent and mutually exclusive events, and use them to calculate more complicated probabilities; for example, several events happening at the same time or one after the other.

It is often useful to have a list of all possible outcomes in a given situation, together with the probability of each outcome. This is called a **probability distribution**. A particularly important distribution, called the **binomial distribution**, can be used to model the number of successful outcomes in a series of repeated experiments. It has applications from medical trials to predicting election results from exit polls.

Section 1: Combining probabilities

You will often be interested in probabilities of more than one outcome. For example, your university offer may require you to get an A or a B in Maths. Suppose you know that, last year, 33% of all candidates achieved an A and 18% a B in Maths. You can then work out that the probability of getting an A or a B is $0.33+0.18=0.51$, or 51%. You can write this as $\mathrm{P}(A \text{ or } B)=0.51$.

What if, instead, your offer asks for an A in Maths or Economics? Last year 20% of all candidates got an A in AS Economics; so is the probability of getting an A in at least one of the subjects $0.33+0.20=0.53$? The answer is no, because those who got an A in Maths and those who got an A in Economics are not two separate groups of people – some of them got an A in both. In fact, unless you know how many got two As, it is impossible to find the probability of achieving this combination of grades.

The events 'getting an A in Maths' and 'getting a B in Maths' are **mutually exclusive**. This means that they cannot both happen at the same time; the probability of both happening together is zero. The events, 'getting an A in Maths' and 'getting an A in Economics', are not mutually exclusive because it is possible for both of them to happen at the same time.

Gateway to A Level

See Gateway to A Level Section V for a reminder of how to find combined probabilities when all outcomes are equally likely.

Fast forward

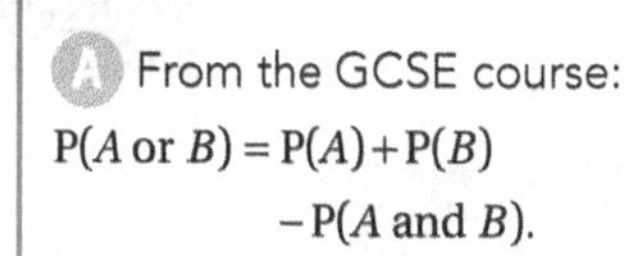
A From the GCSE course:
$\mathrm{P}(A \text{ or } B)=\mathrm{P}(A)+\mathrm{P}(B)-\mathrm{P}(A \text{ and } B)$.

You will meet this again with Venn diagrams in Student Book 2, Chapter 20.

Key point 21.1

Events A and B are **mutually exclusive** if it is impossible for both of them to happen at the same time.

$$\mathrm{P}(A \text{ and } B)=0$$

If events are mutually exclusive, their probabilities can be added.

$$\mathrm{P}(A \text{ or } B)=\mathrm{P}(A)+\mathrm{P}(B)$$

WORKED EXAMPLE 21.1

A fair six-sided dice is rolled once. In each case, state whether the two events are mutually exclusive, and write down the value of $\mathrm{P}(A)$, $\mathrm{P}(B)$ and $\mathrm{P}(A \text{ or } B)$.

a A: rolling a 5; B: rolling a 6

b A: rolling an even number; B: rolling a prime number

a A and B are mutually exclusive. — It is not possible to roll a 5 and a 6 at the same time.

$\mathrm{P}(A)=\frac{1}{6}, \mathrm{P}(B)=\frac{1}{6}$

$\mathrm{P}(A \text{ or } B)=\frac{2}{6}$ — The probabilities can be added.

b A and B are not mutually exclusive. — It is possible to get a number that is both even and prime (2).

$\mathrm{P}(A)=\frac{3}{6}, \mathrm{P}(B)=\frac{3}{6}$ — There are three even numbers (2, 4, 6) and three prime number (2, 3, 5).

$\mathrm{P}(A \text{ or } B)=\frac{5}{6}$ — The probabilities cannot be added; you have to count all possible outcomes (2, 3, 4, 5, 6).

When two events are not mutually exclusive, there is a possibility that they can both happen at the same time. Can the probability of both events happening together be worked out from their individual probabilities?

Consider again the example of Mathematics and Economics grades. Let M be the event 'getting an A in Mathematics' and E the event 'getting an A in Economics.' As before, $P(M) = 0.33$ and $P(E) = 0.20$. What is P(M and E)?

Here are two possible situations that fit in with these numbers. The numbers in cells are percentages.

		Economics grade		Total
		E	not E	
Maths grade	M	15	18	33
	not M	5	62	67
Total		20	80	100

		Economics grade		Total
		E	not E	
Maths grade	M	6.6	26.4	33
	not M	13.4	53.6	67
Total		20	80	100

In both cases $P(M) = 0.33$ and $P(E) = 0.20$. But in the first case P(M and E) $= 0.15$ and in the second P(M and E) $= 0.066$. This suggests that P(M and E) depends on more than just the individual probabilities of M and E. In fact, it depends on how the two events influence each other – how the probability of one event changes when you know the outcome of the other.

Fast forward

A This is called **conditional probability**. You will meet it again in Student Book 2, Chapter 20.

There is one important special case when the probability of two events happening together can be easily calculated: when the two events do not affect each other. Then the events are **independent**.

Key point 21.2

Events A and B are **independent** if knowing the outcome of A does not affect the probability of B. For independent events:

$$P(A \text{ and } B) = P(A) \times P(B)$$

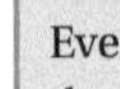

WORKED EXAMPLE 21.2

Two dice are rolled simultaneously. What is the probability that both dice show a prime number?

Let the two events be:

A = 'the first dice shows a prime number'

B = 'the second dice shows a prime number'.

Then:

$$P(A) = \frac{3}{6}, \quad P(B) = \frac{3}{6}$$

There are three prime numbers on a dice: 2, 3, 5.

A and B are independent, so:

$$P(A \text{ and } B) = P(A) \times P(B)$$
$$= \frac{3}{6} \times \frac{3}{6}$$
$$= \frac{1}{4}$$

Knowing the outcome on the first dice does not affect probabilities of the second dice.

WORKED EXAMPLE 21.3

A biased coin has the probability of $\frac{2}{3}$ of showing heads. The coin is tossed three times. Find the probability of throwing either three heads or three tails.

$$\text{P}(3\text{ heads}) = \text{P}(\text{heads}) \times \text{P}(\text{heads}) \times \text{P}(\text{heads})$$
$$= \frac{2}{3} \times \frac{2}{3} \times \frac{2}{3}$$
$$= \frac{8}{27}$$

The three tosses are independent, so multiply the probabilities for each one.

$$\text{P}(3\text{ tails}) = \text{P}(\text{tails}) \times \text{P}(\text{tails}) \times \text{P}(\text{tails})$$
$$= \frac{1}{3} \times \frac{1}{3} \times \frac{1}{3}$$
$$= \frac{1}{27}$$

$$\text{P}(3\text{ heads or }3\text{ tails}) = \frac{8}{27} + \frac{1}{27}$$
$$= \frac{1}{3}$$

The events 'getting three heads' and 'getting three tails' are mutually exclusive, so find the probability of each and add them together.

One important example of mutually exclusive events is an event and its **complement**; the complement occurs when the event does not happen. Since either an event or its complement must occur the total probability is 1.

Key point 21.3

The **complement** of an event A is the event not A or A'.

$$\text{P}(A) + \text{P}(A') = 1$$

This equation is very useful because sometimes the complement is much simpler to work with than the event itself.

WORKED EXAMPLE 21.4

A fair dice is rolled five times. Find the probability of getting at least one six.

$$\text{P}(\text{at least one six}) = 1 - \text{P}(\text{no sixes})$$

There are many possible ways to get at least one six (you could get 1, 2, 3, 4 or 5). But the complement of the event 'at least one six' is 'no sixes', which can happen in only one way: if each dice shows 'not a six'.

$$= 1 - \left(\frac{5}{6}\right)^5$$
$$= 0.598$$

The five rolls are independent, so you need to multiply five probabilities of 'not a six'.

EXERCISE 21A

1 Which events are mutually exclusive? For those events that are mutually exclusive, find P(A or B).

a A fair six-sided dice is rolled.

i A: Rolling a multiple of 3; B: Rolling a multiple of 4

ii A: Rolling an even number; B: Rolling a multiple of 5

b One card is selected from a standard pack of 52 cards.

i A: Selecting a king; B: Selecting a red card

ii A: Selecting an ace; B: Selecting a spade

c Two fair dice are rolled and the scores are added.

i A: The total is a multiple of 6. B: The total is less than 5.

ii A: The total is greater than 7. B: The total is less than 9.

d A bag contains four green and six yellow balls. A ball is taken out of the bag. Its colour is noted and then it is returned to the bag. Another ball is then selected.

i A: Both balls are green. B: Both balls are yellow.

ii A: The first ball is green. B: The second ball is green.

e A bag contains four green and six yellow balls. Two balls are taken out without replacement.

i A: The first ball is green. B: The second ball is green.

ii A: Both balls are green. B: Both balls are yellow.

2 Which of the events in question 1 are independent? Calculate P(A and B) and $\text{P}(A)\times\text{P}(B)$ in each case.

3 Two events, A and B, have probabilities $\text{P}(A) = p$ and $\text{P}(B) = q$.

a Write down an expression for P(A and B) in each situation.

i A and B are independent. **ii** A and B are mutually exclusive.

b If two events are independent, can they also be mutually exclusive?

4 A coin is biased so that the probability of getting tails is $\frac{3}{4}$. The coin is tossed twice. Find the probability that:

a the coin shows heads both times

b the coin shows heads at least once.

5 Two fair six-sided dice are rolled. Find the probability that the product of the scores is 6.

6 Daniel has three blocks with letters C, A and T written on them. He arranges the blocks in a row randomly.

a Write down all possible arrangements of the three letters.

b Find the probability that the blocks make the word 'CAT' or 'ACT'.

7 A fair six-sided dice is rolled once. Define events as:

A: The dice shows an even number

B: The dice shows a prime number.

a Find P(A and B).

b Determine whether events A and B are independent.

8 300 students in years 9, 10 and 11 at a school were asked to say which of Biology, Chemistry and Physics was their favourite science. The results are shown in the table.

		Favourite science			
		Biology	**Chemistry**	**Physics**	**Total**
Year group	**9**	**41**	29	27	97
	10	35	36	34	105
	11	37	30	31	98
	Total	113	95	92	300

a Find the probability that a randomly chosen student:

i prefers Chemistry

ii is in Year 11 and doesn't prefer Biology.

b Determine whether the event 'the student is in Year 9' and the event 'the student's favourite science is Physics' are independent.

9 A fair four-sided spinner, with numbers 1 to 4 written on it, is spun three times. Find the probability of getting either three 1s or three 4s.

10 A fair coin is tossed three times. Work out the probability that it shows:

a three tails

b at least one head.

11 The probability that a student is late for a lesson is 0.15, independently of any other students.

a Find the probability that at least one of the 12 students in a class is late.

b Is the assumption of independence reasonable in this case? Explain your answer.

Section 2: Probability distributions

So far you have only been asked to calculate the probability of a specific event, or a combination of events happening. But sometimes you are interested in probabilities of all possible outcomes in a given situation. For example, if you roll two dice and add up the scores you can get 11 possible outcomes: any integer from 2 to 12. The probabilities of those outcomes are not all equal; a total of 7 is more likely than a total of 12.

The list of all possible outcomes together with their probabilities is called a **probability distribution**. This information is best displayed in a table.

Fast forward

A In the AS course you will only work with **discrete** probability distributions – ones for which all possible outcomes can be listed. In the A Level course you will also meet **continuous** distributions, which can take any value in a given range.

WORKED EXAMPLE 21.5

Two fair dice are rolled and their scores are added. Find the probability distribution of the total.

Possible outcomes for the total:

The numbers in the table show the total score of the two dice.

	1	2	3	4	5	6
1	2	3	4	5	6	7
2	3	4	5	6	7	8
3	4	5	6	7	8	9
4	5	6	7	8	9	10
5	6	7	8	9	10	11
6	7	8	9	10	11	12

The probability distribution of the total:

Each combination of scores is equally likely, so each has a probability of $\frac{1}{36}$.

Total	2	3	4	5	6	7	8	9	10	11	12
Probability	$\frac{1}{36}$	$\frac{2}{36}$	$\frac{3}{36}$	$\frac{4}{36}$	$\frac{5}{36}$	$\frac{6}{36}$	$\frac{5}{36}$	$\frac{4}{36}$	$\frac{3}{36}$	$\frac{2}{36}$	$\frac{1}{36}$

So, for example, $P(\text{total} = 5) = \frac{4}{36}$, because the total of 5 can be obtained in four different ways.

The probabilities in a probability distribution cannot be just any numbers.

Key point 21.4

The total of all the probabilities of a probability distribution must always equal 1.

WORKED EXAMPLE 21.6

In a game at a fair, a ball is thrown at a rectangular target. The dimensions of the target (in metres) are as shown in the diagram. The probability of hitting each region is proportional to its area. The prize for hitting a region is a number of chocolates, equal to the number shown in that region. Find the probability distribution of the number of chocolates won.

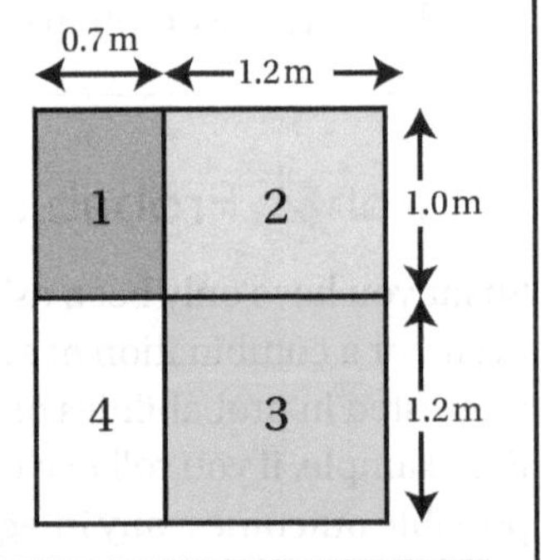

Let X = the number of chocolates won.

The probability is proportional to the area so write $p = k \times \text{area}$.

x	1	2	3	4
$P(X = x)$	$0.7k$	$1.2k$	$1.44k$	$0.84k$

Continues on next page

$0.7k + 1.2k + 1.44k + 0.84k = 1$
$4.18k = 1$
$k = 0.239$

To find k, use the fact that the probabilities add up to 1.

So the distribution is:

x	1	2	3	4
$P(X = x)$	0.167	0.287	0.344	0.201

The notation $P(X = x)$, as used in Worked example 21.6, means 'the probability that the random variable X takes the value x.'

A **random variable** is a variable whose value depends on chance; in this case X is the number of chocolates won – it could take the value 1, 2, 3 or 4, and which value it takes is down to chance. The lower case x is a specific value that the random variable takes. So, for example, $P(X = 2) = 0.287$ means 'the probability that the number of chocolates won was 2 is 0.287.'

Sometimes a probability distribution will be given by a formula.

WORKED EXAMPLE 21.7

A probability distribution is given by

$$P(X = x) = kx, \text{ for } x = 1, 2, 3.$$

Find $P(X = 2)$.

x	1	2	3
$P(X = x)$	k	$2k$	$3k$

Always start by putting the distribution into a table.

To find expressions for each probability, just replace x in the formula with 1, 2, and 3 in turn.

$k + 2k + 3k = 1$
$6k = 1$
$k = \frac{1}{6}$

Use the fact that all the probabilities sum to 1.

So $P(X = 2) = \frac{1}{6} \times 2$
$= \frac{1}{3}$

Now use $k = \frac{1}{6}$ in the table to find $P(X = 2)$.

EXERCISE 21B

1 For each scenario, draw a table to represent the probability distribution.

a A fair coin is thrown three times. T is the number of tails obtained.

b Two fair dice are thrown. D is the difference between the larger and the smaller score, or zero if they are the same.

c A fair dice is thrown once. X is calculated as half the result if the dice shows an even number, or one higher than the result if the dice shows an odd number.

d A bag contains six red and three green counters. Two counters are drawn at random from the bag without replacement. G is the number of green counters remaining in the bag.

e Karl picks a card at random from a standard pack of 52 cards. If he draws a diamond, he stops; otherwise, he replaces the card and continues to draw cards at random, with replacement, until he has either drawn a diamond or has drawn a total of 4 cards. C is the total number of cards drawn.

f Two fair four-sided spinners, each labelled 1, 2, 3 and 4, are spun. X is the product of the two values shown.

2 Find the missing value k for each probability distribution.

a i

x	3	7	9	11
$P(X=x)$	$\frac{1}{2}$	$\frac{1}{4}$	$\frac{1}{8}$	k

ii

x	5	6	7	10
$P(X=x)$	0.2	0.2	k	0.5

b i

x	1	2	3	4
$P(X=x)$	$2k$	$3k$	$2k$	k

ii

x	1	2	3	4
$P(X=x)$	$\frac{k}{2}$	k	$\frac{k}{3}$	$\frac{k}{2}$

c i $P(Y=y)=ky$ for $y=1, 2, 3, 4$

ii $P(X=x)=\frac{k}{x}$ for $x=1, 2, 3, 4$

3 A six-sided dice is biased with the probabilities of each outcome as shown in the table.

Score	1	2	3	4	5	6
Probability	$\frac{1}{12}$	$\frac{1}{6}$	$\frac{1}{12}$	$\frac{1}{4}$	$2k$	$3k$

a Find the value of k.

b The dice is rolled once. Find the probability that the score is more than 4.

4 The number of students absent from a Mathematics lesson on any particular day follows this probability distribution.

Number of absentees	0	1	2	3	4	5
Probability	0.15	0.21	k	0.26	0.12	0.07

a Find the value of k.

b Find the probability that at most 2 students are absent.

5 Ben and Anna both take three shots at a goal. The table shows the probability distribution of the number of goals each of them scores.

Number of goals	0	1	2	3
Probability	0.2	0.3	0.4	0.1

a Find the probability that Anna scores at least one goal.

b Find the probability that both Ben and Anna score three goals.

6 A fair four-sided spinner, with numbers 1 to 4 written on it, is spun twice and the scores are added.

a Find the probability distribution of the total.

b Find the probability that the total is at least 6.

7 A teacher randomly selects how many questions to set for homework after each lesson. The probability distribution of the number of questions is:

Number	2	3	4	5	6
Probability	a	0.1	0.3	0.3	b

The probability that the teacher sets fewer than four questions is 0.2.

Find the values of a and b.

8 A probability distribution of X is given by:

$\mathrm{P}(X = x) = k(x+1)$ for $x = 2, 3, 4, 5, 6$.

a Show that $k = 0.04$.

b Find $\mathrm{P}(X \geqslant 4)$.

Tip

Remember that when a probability distribution is given by a formula, you should start by creating a table.

9 A four-sided dice is biased. The probabilities of the possible scores are listed in the table.

Score	1	2	3	4
Probability	$\frac{1}{3}$	$\frac{1}{4}$	k	$\frac{1}{5}$

a Find the value of k.

b Find the probability that the total score is 4 after two rolls.

10 Ronnie and Jimmy are playing snooker. They both try to pot two balls.

For Ronnie, the probability distribution of the number of successful pots is:

Number of pots	0	1	2
Probability	0.1	0.2	0.7

For Jimmy, the probability distribution is:

Number of pots	0	1	2
Probability	0.3	0.5	0.2

Assuming that their performances are independent, find the probability that, between them,

a they pot exactly one ball

b they pot at least one ball.

11 $\mathrm{P}(X \leqslant x) = kx^2$ for $x = 1, 2, 3$.

a Find the value of k.

b What is the probability distribution of X?

12 $\mathrm{P}(X = x) = \frac{1}{9}$ for integers in the range $[-k, k]$.

a Find the value of k.

b If $Y = X^2$ find the probability distribution of Y.

Section 3: The binomial distribution

Consider these questions.

- A fair six-sided dice is rolled four times. What is the probability of scoring exactly two fives?
- A fair coin is tossed ten times. What is the probability that it shows heads at least six times?
- A multiple-choice test has 20 questions, each with five possible answers, only one of which is right. A student guesses the answer to each question, with an equal probability of guessing any answer. What is the probability that he gets fewer than five correct answers?

All of the questions involve a similar scenario.

- An action is repeated several times (a dice roll, a coin toss, an attempt to answer a question).
- Each time there are two possible outcomes (five or not a five; heads or tails; right or wrong answer).
- You are interested in the probability of one of the outcomes happening a given number of times.

This type of situation is very common, so it is worth asking whether there is a general rule or formula for calculating the probability. The general scenario can be described as:

There is a fixed number of experiments, or 'trials'. Each trial has two possible outcomes, usually labelled 'success' and 'failure'. Then, under certain conditions, the number of successes can be modelled by means of a special probability distribution.

Key point 21.5

The **binomial distribution** models the number of successful outcomes out of n trials, provided these conditions are satisfied.

- Each trial has two possible outcomes.
- The trials are independent of each other.
- The probability of success is the same in each trial.

If n is the number of trials, p is the probability of a success and X denotes the number of successes, you can write $X \sim B(n, p)$.

You can use your calculator to find the probabilities for the binomial distribution. You need to specify the number of trials (n), the probability of success (p) in a single trial and the required number of successes.

WORKED EXAMPLE 21.8

Decide whether each situation can be modelled using the binomial distribution. If it can't, say which of the conditions is not satisfied. If it can, find the required probability.

a A fair dice is rolled until it shows a six. Find the probability of getting two fours.

b Tom and Jerry play eight games of chess. The probability that Tom wins a game is 0.6, independently of any other game. Find the probability that Tom wins exactly four games.

c A student is trying to answer 20 quiz questions. The probability of getting the first question right is 0.9, but the probability halves for each subsequent question. What is the probability that he answers 10 questions correctly?

d In a particular village, 63% of five-year-olds attend the local primary school. What is the probability that, in a group of 15 friends, at least 10 attend that school?

e The probability that it rains on any particular day is 0.3. Assuming the days are independent, find the probability that it rains on more than four days in a week.

a Not binomial; the number of trials is not constant.

The first thing to check is that the situation has a fixed number of trials.

b Binomial, $n = 8$, $p = 0.6$

$P(X = 4) = 0.232$

All the conditions are satisfied. Use the calculator to find the probability.

c Not binomial; the probability of success is not constant.

There is a fixed number of trials, but the probability changes every time.

d Not binomial; the trials are not independent.

If one child attends the school it is more likely that their friends do as well.

e Binomial, $n = 7$, $p = 0.3$

$P(X > 4) = P(X = 5, 6 \text{ or } 7)$

All the conditions are satisfied. The required probability is for three possible outcomes.

$= P(X = 5) + P(X = 6) + P(X = 7)$

$= 0.0288$

The three outcomes are mutually exclusive, so their probabilities can be added.

Cumulative probabilities

In the final part of Worked example 21.8 you were asked to find the probability of more than one outcome. This is straightforward when there are only three probabilities to add up, but what if there were 20 trials and you wanted to find the probability of more than 11 successes? Would you have to add all the probabilities from $P(X = 12)$ to $P(X = 20)$?

Fortunately, most calculators have a function to calculate the probability of totalling up to (and including) a specified number of successes – this is called a **cumulative probability**. For example, if $n = 20$ and $p = 0.3$, you can find that $P(X \leq 7) = 0.772$. You can also find the probability of getting more than a certain number of successes; for example:

$$\begin{aligned} P(X > 11) &= 1 - P(X \leq 11) \\ &= 1 - 0.995 \\ &= 0.005 \end{aligned}$$

If you want to find the probability of getting more than 5 but fewer than 10 successes, this is:

$$P(5 < X < 10) = P(X \leq 9) - P(X \leq 5)$$

You can see this by looking at the number line.

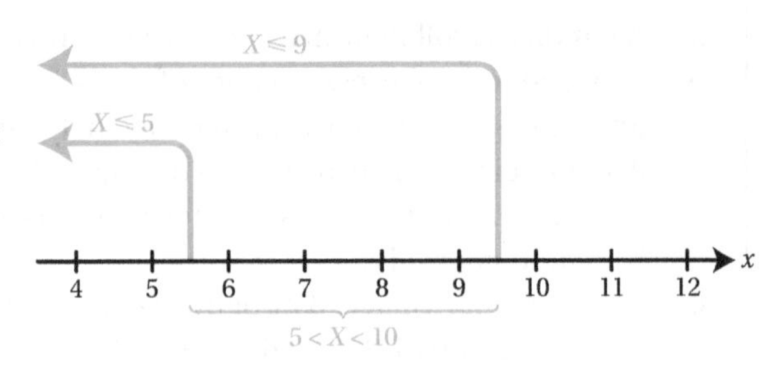

WORKED EXAMPLE 21.9

Anna shoots at a target 15 times. The probability that she hits the target on any shot is 0.6, independently of the other shots. Find the probability that she hits the target more than 5 but at most 10 times.

Let X be the number of times Anna hits the target. Then: $X \sim B(15, 0.6)$

There is a fixed number of trials and you are interested in the number of successes, so this is binomial distribution.

$$\begin{aligned} P(5 < X \leq 10) &= P(X \leq 10) - P(X \leq 5) \\ &= 0.7827 - 0.0338 \\ &= 0.749 \text{ (3 s.f.)} \end{aligned}$$

Express the required probability in terms of cumulative probabilities.

Tip

When you are using a calculator to find probabilities, you should still use correct mathematical notation (not calculator notation) in your answer. You must show what distribution you used and which probabilities you have found from the calculator.

WORK IT OUT 21.1

Four students are trying to answer this question.

Anna shoots at a target 15 times. The probability that she hits the target on any shot is 0.6, independently of the other shots. Find the probability that she hits the target at least 6 times.

Which is the correct solution? Identify the errors made in the incorrect solutions.

Solution 1	Solution 2	Solution 3
$P(X > 6) = 1 - P(X \leq 7)$ $= 1 - 0.213$ $= 0.787$	$P(X \geq 6) = 1 - P(X \leq 6)$ $= 1 - 0.095$ $= 0.905$	$P(X \geq 6) = 1 - P(X \leq 5)$ $= 1 - 0.034$ $= 0.966$

The formula for binomial probabilities

So far you have used a calculator to find probabilities for the binomial distribution. But how are those probabilities calculated?

Consider the first example from the start of this section: a dice is rolled four times; what is the probability of getting exactly two fives?

There are four trials so $n = 4$. If you label a five as a 'success' its probability is $\frac{1}{6}$. The probability of a 'failure' is therefore $\frac{5}{6}$. X stands for the number of fives, so you are interested in $\mathrm{P}(X = 2)$.

One way of getting exactly two fives is if you get a five on each of the first two rolls and something else on the last two rolls. The probability of this happening is $\frac{1}{6} \times \frac{1}{6} \times \frac{5}{6} \times \frac{5}{6} = \left(\frac{1}{6}\right)^2 \left(\frac{5}{6}\right)^2$.

But this is not the only way to score two fives. The two fives may be, for example, on the first and third, or second and fourth rolls. This can be illustrated on a tree diagram.

> **Gateway to A Level**
>
> See Gateway to A Level Section W for revision of using tree diagrams to find probabilities of certain outcomes.

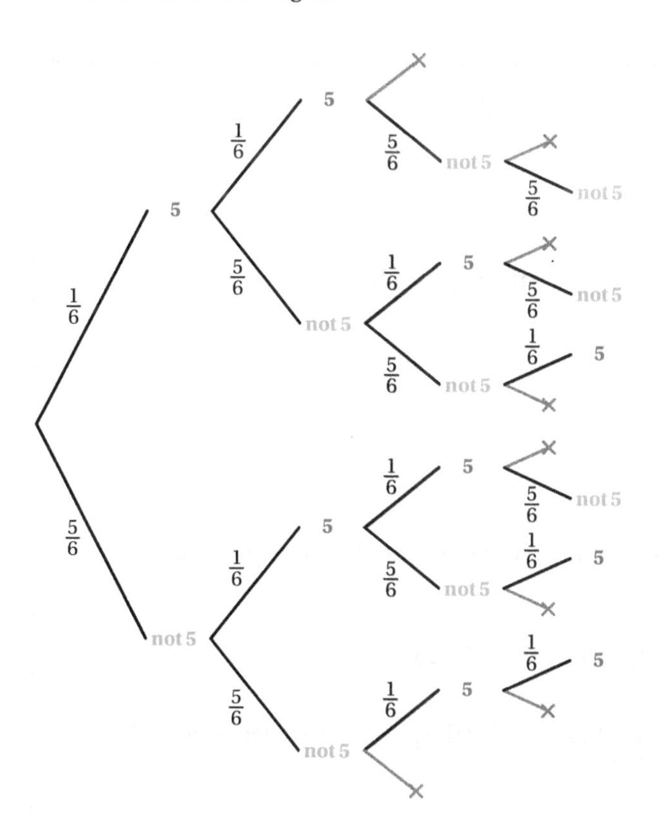

Each of the required paths has the same probability, $\left(\frac{1}{6}\right)^2 \left(\frac{5}{6}\right)^2$, because two of the outcomes are a five and two are something else. The number of paths leading to the outcome 'two fives' is 6, so

$\mathrm{P}(X = 2) = 6\left(\frac{1}{6}\right)^2 \left(\frac{5}{6}\right)^2$.

This reasoning can be generalised. Suppose there are n trials, the probability of a success is p and you are interested in the probability of obtaining r successes. If you can imagine representing this on a tree diagram, each relevant path will have probability $p^r(1-p)^{n-r}$, because r of the outcomes are successes (with probability p) and the remaining $n-r$ outcomes are failures (with probability $1-p$). Then the number of paths that give r successes is given by the binomial coefficient $\binom{n}{r}$.

This leads to the general formula for the probabilities of the binomial distribution.

Rewind

You met binomial coefficients in Chapter 9, when studying the binomial expansion. You can find them either on your calculator or using the formula.

Key point 21.6

If $X \sim B(n, p)$ then:

$$P(X = x) = \binom{n}{x} p^x (1-p)^{n-x} \quad \text{for } x = 0, 1, 2 \ldots n$$

This will be given in your formula book.

Elevate

See Support Sheet 21 for a similar example to Worked example 21.10 and for more practice questions.

WORKED EXAMPLE 21.10

Ten students take a test. They all have a probability p of passing, independent of the results of other students. X is the number of students passing the test.

If $P(X = 6) = 4 \times P(X = 4)$, find the value of p.

$P(X=6) = \binom{10}{6} p^6(1-p)^4$

$= 210p^6(1-p)^4$

$P(X=4) = \binom{10}{4} p^4(1-p)^6$

$= 210p^4(1-p)^6$

Use the formula to write expressions for $X = 6$ and $X = 4$.

So:

Write the equation given in terms of these expressions.

$210p^6(1-p)^4 = 840p^4(1-p)^6$

$p^2 = 4(1-p)^2$

You can divide by $210\,p^4(1-p)^4$ if $p \neq 0, 1$.

$p = 2(1-p)$

$p = 2 - 2p$

$3p = 2$

$p = \frac{2}{3}$

You only need to consider the positive root, since both sides must be positive.

EXERCISE 21C

1 In each example, decide whether the situation can be modelled by the binomial distribution. If it can't, give a reason. If it can, identify the distribution and the required probability (you don't need to calculate the probability).

a A fair coin is tossed 20 times. What is the probability of getting exactly 15 heads?

b Elsa enjoys quizzes. On average she gets 78% of the answers right. What is the probability that in a particular quiz, she gets 9 out of the first 10 questions right?

c A bag contains a large number of balls, with equal proportions of red, blue and green balls. 20 balls are chosen at random. Find the probability that 5 are red and 8 are green.

d A drawer contains 5 black socks and 10 red socks. 6 socks are drawn at random without replacement. Find the probability that at least two black socks are drawn.

e It is known that 2% of a large population carry a gene for diabetes. If 100 people are chosen at random, what is the probability of getting at least 1 person with this gene?

2 Given that $X \sim B(8, 0.2)$, calculate each probability.

a i $P(X=3)$ ii $P(X=4)$ b i $P(X \leqslant 3)$ ii $P(X \leqslant 2)$

c i $P(X>3)$ ii $P(X>4)$ d i $P(X<5)$ ii $P(X<3)$

e i $P(X \geqslant 3)$ ii $P(X \geqslant 1)$ f i $P(3<X \leqslant 6)$ ii $P(1 \leqslant X<4)$

g i $P(1<X<5)$ ii $P(2<X<7)$ h i $P(3 \leqslant X \leqslant 7)$ ii $P(2 \leqslant X \leqslant 5)$

3 In each question identify the distribution, write down the required probability and use your calculator to find it.

a Jake beats Marco at chess in 70% of their games. Assuming that this probability is constant and that the results of games are independent of each other, what is the probability that Jake will beat Marco in at least 16 of their next 20 games?

b On a television channel, the news is shown at the same time each day; the probability that Salia watches the news on a given day is 0.35. Calculate the probability that on five consecutive days she watches the news on exactly three days.

c Sandy is playing a computer game and needs to accomplish a difficult task at least three times in five attempts in order to pass the level. There is a 1 in 2 chance that he accomplishes the task each time he tries, unaffected by how he has done before. What is the probability that he will pass to the next level?

4 Given that $Y \sim B\left(5, \frac{1}{2}\right)$, use the formula to find the exact value of each probability.

a i $P(Y=1)$ ii $P(Y=0)$ b i $P(Y \geqslant 1)$ ii $P(Y \leqslant 1)$

c i $P(Y>4)$ ii $P(Y \leqslant 3)$

5 15% of students at a large school travel by bus. A random sample of 20 students is taken.

a Explain why the number of students in the sample who travel by bus is only approximately a binomial distribution.

b Use the binomial distribution to estimate the probability that exactly five of the students travel by bus.

6 A Biology test consists of eight multiple choice questions. Each question has four answers, only one of which is correct. At least five correct answers are required to pass the test. Sheila does not know the answers to any of the questions, so answers each question at random.

a What is the probability that Sheila answers exactly five questions correctly?

b What is the probability that Sheila manages to pass the test?

7 0.8% of people in a country have a particular cold virus at any time. On a single day, a doctor sees 80 patients.

a What is the probability that exactly two of them have the virus?

b What is the probability that three or more of them have the virus?

c State an assumption you have made in these calculations.

8 On a fair dice, which is more likely: rolling three 6s in four throws or rolling a 5 or a 6 in five out of six throws?

9 This question is intended to help you understand the difference between 'constant probability' and 'independent trials', which are both required for the binomial distribution.

a A bag contains six red and four blue balls. A ball is taken at random and not replaced. This is repeated three times, so that in total three balls are selected.

i Find the probability that the second ball is red.

ii Find the probability that the third ball is red.

iii Find the probability that the first and the second balls are both red. Hence use the formula $\text{P}(A \text{ and } B) = \text{P}(A) \times \text{P}(B)$ to show that the events 'the first ball is red' and 'the second ball is red' are not independent.

iv Find the probability that exactly one of the three balls is red. Compare this to $\text{P}(X = 1)$ when $X \sim \text{B}(3, 0.6)$.

b Repeat parts **i–iv** of part **a**, assuming the three balls are selected with replacement.

Did you know?

Question 8 is the problem which was posed to Pierre de Fermat in 1654 by a professional gambler who could not understand why he was losing. It inspired Fermat (with the assistance of Pascal) to set up probability as a rigorous mathematical discipline.

10 In each situation, discuss whether the trials are independent and whether the probability of success is constant.

a Pulling socks from a drawer; success is pulling a red sock out.

b Consecutive rolls of a fair dice; success is rolling a 6.

c Each member of the class flips a fair coin. If it shows 'heads', they write down the first letter of their first name. If it shows 'tails', they write the last letter; success is writing a vowel.

Tip

This means that the probability of success on the nth trial is not dependent on n.

11 A fair coin is tossed ten times. What is the probability that it shows heads at least six times?

12 A multiple-choice test has 20 questions, each with five possible answers, only one of which is right. A student guesses the answer to each question, with an equal probability of guessing any answer. What is the probability that he gets fewer than five correct answers?

13 $X \sim B(8, p)$. If $P(X = 5) = P(X = 6)$ find the value of p.

14 X is a binomial random variable, where the number of trials is 4 and the probability of success of each trial is p. Find the possible values of p if $P(X = 2) = 0.3456$.

15 Over a one-month period, Ava and Sven play a total of n games of tennis. The probability that Ava wins any game is 0.4. The result of each game played is independent of any other game played. Let X denote the number of games won by Ava over a one-month period.

a Show that $P(X = 2)$ can be given as:

$$P(X = 2) = \frac{2n(n-1)}{9} \times 0.6^n$$

b If the probability that Ava wins two games is 0.121 correct to three decimal places, find the value of n.

16 X is a random variable following $B(n, 0.8)$. If $P(X = n) = 64P(X = 0)$, find n.

17 If $X \sim B(n, p)$ and $P(X = 4) = P(X = 5)$ find n in terms of p.

Checklist of learning and understanding

- **Mutually exclusive** events satisfy $P(A \text{ and } B) = 0$. Their probabilities can be added: $P(A \text{ or } B) = P(A) + P(B)$.
- An event and its **complement** are mutually exclusive, and $P(A) + P(A') = 1$. The probability of the complement is sometimes easier to find.
- For **independent** events, $P(A \text{ and } B) = P(A) \times P(B)$.
- A **probability distribution** is a list of all possible outcomes and their probabilities.
- All probabilities in a probability distribution must add up to 1.
- **Cumulative probability** is the probability of obtaining up to and including a given outcome.
- The **binomial distribution** is a model for the number of successes when an experiment is repeated several times. If there are n trials and the probability of success is p, the distribution is denoted by $B(n, p)$.
- For the binomial distribution to be a good model, these conditions must be satisfied.
  - The number of trials is constant.
  - Each trial has two possible outcomes.
  - The probability of success is the same for each trial.
  - The trials are independent.
- The probabilities for the binomial distribution can be found by using a calculator, or the formula:

$$P(X = x) = \binom{n}{x} p^x (1-p)^{n-x} \quad \text{for } x = 0, 1, 2 \ldots n$$

Mixed practice 21

1 A and B are mutually exclusive events. Which statement is **definitely false?**
Choose from these options.

A $\text{P}(A \text{ or } B) = \text{P}(A) + \text{P}(B)$ B $\text{P}(A \text{ and } B) = 1 - \text{P}(A) - \text{P}(B)$

C $\text{P}(A \text{ and } B) = 0.5$ D $\text{P}(A \text{ or } B) = 1$

2 A factory making bottles knows that on average, 1.5% of its bottles are defective. Find the probability that, in a randomly selected sample of 20 bottles, at least 1 bottle is defective.

3 The mark on a Physics test is an integer between 1 and 5 inclusive. The distribution of test marks is given in the table.

Mark	1	2	3	4	5
Probability	0.05	0.23	0.35	k	0.21

a Find the value of k.

b Find the probability that a randomly selected student scores at least a 3 in the test.

c Write down the most likely mark in this test.

4 A spinner has four equal sections numbered 1, 2, 5 and 7. The spinner is spun twice. Find the probability distribution of the positive difference between the scores (larger minus smaller).

5 When a boy bats in baseball, the probability that he hits the ball is 0.4. In practice he gets pitched 12 balls. Let X denote the total number of balls he hits. Assuming that his attempts are independent, find:

a $\text{P}(X = 3)$ b $\text{P}(X > 5)$.

6 A probability distribution of a variable X is shown in this table.

x	2	3	4	5	6
$\text{P}(X = x)$	p	0.2	0.3	0.3	q

a Write down the value of $p + q$.

b Given that $\text{P}(X \geqslant 4) = 0.75$, find the values of p and q.

7 a Emma visits her local supermarket every Thursday to do her weekly shopping. The event that she buys orange juice is denoted by J, and the event that she buys bottled water is denoted by W. At each visit, Emma may buy neither, or one, or both of these items.

i Copy and complete the table of probabilities for these events, where J' and W' denote the events 'not J' and 'not W' respectively.

	J	J'	Total
W			0.65
W'	0.15		
Total		0.30	1.00

ii Hence, or otherwise, find the probability that, on any given Thursday, Emma buys either orange juice or bottled water but not both.

iii Show that:

A the events J and W are **not** mutually exclusive

B the events J and W are **not** independent.

[© AQA 2011]

8 If $X \sim B(20, p)$ and $P(X > 15) \approx 5\%$ what is the value of p?

Choose from these options.

A 0.5 **B** 0.6 **C** 0.7 **D** 0.8

9 When Robyn shoots an arrow at a target, the probability that she hits the target is 0.6. In a competition she has eight attempts to hit the target. If she gets at least seven hits on target she will qualify for the next round.

a Find the probability that she hits the target exactly four times.

b Find the probability that she fails to qualify for the next round.

10 A test is marked on the scale from 1 to 5. The cumulative distributions of test scores is shown in the table.

Grade (s)	1	2	3	4	5
P(grade $\leqslant$ s)	0.32	0.48	0.65	0.82	1

Find the percentage of candidates whose grade was:

a 2 **b** from 3 to 5 inclusive.

11 A student has the probability 0.7 of answering a question correctly, independently of any other questions.

Find the probability that, in a test containing 15 questions, the student gets more than seven but fewer than 12 correct answers.

12 A biased coin has probability p of showing heads. The coin is tossed six times. The probability that it shows no heads is 0.072. Find the value of p, correct to two significant figures.

13 Asher and Elsa both roll a fair six-sided dice.

a Find the probability distribution of the difference between the scores (Asher's score – Elsa's score).

b Hence find the probability that Elsa gets a higher score than Asher.

14 An amateur tennis club purchases tennis balls that have been used previously in professional tournaments.

The probability that each such ball fails a standard bounce test is 0.15.

The club purchases boxes each containing 10 of these tennis balls. Assume that the 10 balls in any box represent a random sample.

a Determine the probability that the number of balls in a box which fail the bounce test is:

i at most 2

ii at least 2

iii more than 1 but fewer than 5.

b Determine the probability that, in **5 boxes**, the total number of balls which fail the bounce test is:

i more than 5

ii at least 5 but at most 10.

[© AQA 2011]

15 A company producing light bulbs knows that the probability that a new light bulb is defective is 0.5%.

a Find the probability that a pack of six light bulbs contains at least one defective bulb.

b Mario buys 20 packs of six light bulbs. Find the probability that more than four of the boxes contain at least one defective light bulb.

16 A fair coin is tossed repeatedly until it shows tails.

a Find the probability that the first five tosses all show heads.

b Hence find the probability that the first tails appears on the sixth toss.

17 The probability that a student forgets to do homework is 5%, independently of other students. If at least one student forgets to do homework, the whole class has to do a test.

a There are 12 students in a class. Find the probability that the class will have to do a test.

b For a class with n students, write down an expression for the probability that the class will have to do a test.

c Hence find the smallest number of students in the class such that the probability that the class will have to do a test is at least 80%.

18 Two fair dice are rolled and the difference between the two scores is recorded (larger – smaller).

a Find the probability distribution of the recorded number.

b This experiment is repeated ten times. Find the probability that the recorded number is zero on more than three occasions.

19 Four fair six-sided dice are rolled. Let X be the largest number rolled.

a Explain why $\text{P}(X \leqslant k) = \left(\frac{k}{6}\right)^4$, for $k = 1, 2, \ldots 6$.

b Copy and complete this probability distribution table.

x	1	2	3	4	5	6
$\text{P}(X = x)$		$\frac{15}{1296}$	$\frac{65}{1296}$			$\frac{671}{1296}$

20 A fair six-sided dice is rolled until the fourth 6 is obtained.

a Find the probability that there are exactly three 6s in the first seven rolls.

b Hence find the probability that the fourth 6 is obtained on the eighth roll.

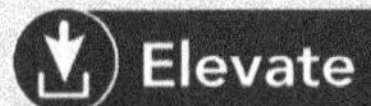

Elevate

See Extension Sheet 21 for a selection of more challenging problems.

22 Statistical hypothesis testing

In this chapter you will learn how to:

- understand the difference between a sample and a population
- understand different types of sampling methods
- understand and use the vocabulary associated with hypothesis tests
- conduct a hypothesis test using the binomial distribution to test if a proportion has changed.

Before you start...

Chapter 21	You should be able to calculate cumulative probabilities for a binomial distribution.	1 Given that $X \sim B(25, 0.6)$ a Find: i $P(X \leqslant 15)$ ii $P(X \geqslant 17)$. b Find the smallest value of k such that $P(X \geqslant k) \leqslant 0.05$.
Chapter 21	You should be able to deduce parameters of a binomial distribution in context.	2 It is known that on average, 35 out of 50 people like coffee. For a random sample of 12 people, let X be the number of those who like coffee. State the distribution of X, including any parameters.

What is hypothesis testing?

The purpose of statistics is to gather information about a population. For example:

- What is the mean height of all 17-year-old boys in the UK?
- What is the range of ages of all professional football players?
- What proportion of the electorate intend to vote for a particular political party in the next election?

It is often impossible to collect all the required data, so statisticians use samples to make inferences about the whole population. A sample value can only give an estimate of the population parameter. How good this estimate is depends on the size and quality of the sample.

Once you have estimated a population parameter it would be useful to know how accurate this estimate is likely to be. This question is, in general, very difficult to answer and requires some advanced probability theory, so you would start by looking at a slightly simpler question, asking whether the parameter value has changed from a previously known, or assumed one. This leads to the procedure known as hypothesis testing, which is one of the most commonly used statistical tools.

Section 1: Populations and samples

If you wanted to know the average height of all adults in the UK, you could simply attempt to measure everyone's height. This is called a **census**. It involves collecting information about the entire **population** – all the individuals of interest. A large organisation may be able to carry out such a survey; indeed, the UK government undertakes a census every ten years in order to plan public services.

For a small organisation, or an individual, a census is generally not an option, because of the time and costs involved. Census data sets also take a very long time to analyse. This means that it is simply not possible to find out the exact average height of all UK adults. Instead you can try to estimate it by taking a **sample** – measuring heights of a part of the population. Most likely, the average of the sample will be different from the population average. However, if the sample is selected well, it may provide a reasonable estimate.

In some situations it is impossible to carry out a census. It may be impossible to identify, or get access to, all members of the population. For example, suppose a zoologist wanted to find out the average weight of an ant. It is impossible even to locate all the ants in the world! What if the process of collecting data destroys the object being measured? For example, if a manufacturer needed to know what maximum load could be placed on a shelf, testing every shelf would mean breaking all of them.

Once you have selected a sample and collected the data you can apply to it any of the techniques you learned in Chapter 20. You then need to decide what this tells you about the whole population.

WORKED EXAMPLE 22.1

A clothing company carries out a survey to find out the average height and the range of heights of 17-year-old girls in the UK. A sample of 200 girls from a large sixth-form college has a mean height of 158 cm and a range of 28 cm.

a Is the true population mean more likely to be larger or smaller than 158 cm?

b State one possible reason why taking a sample from a single college may not result in a good estimate for the mean height.

c Anita says that 28 cm is a good estimate for the range. Priya says that the range is almost certainly larger than 28 cm. Explain who is right.

a The sample mean is equally likely to be larger or smaller than the true population mean.

Assuming the sample has been selected well, the sample mean is a good estimate of the population mean.

Continues on next page

b For example, the college might be located in an area populated by an ethnic group that is, on average, taller or shorter than the whole population.

The part of the population from which the sample was taken may not be typical of the entire country.

c It is unlikely that both the shortest and the tallest 17-year-old girls in the country go to this particular college. So Priya is right, the sample range will be smaller than the population range.

In fact, all measures of spread will tend to be underestimated by a sample. However, in the case of the variance there is a change to the formula you met in Chapter 20, which will provide an estimate that does not systematically underestimate the population variance.

It is called the **unbiased estimate of variance** and it is usually given the symbol s^2.

Key point 22.1

When estimating the variance of a population from a sample, use:

$$s^2 = \frac{\sum (x-\bar{x})^2}{n-1}$$

This will be given in your formula book.

Tip

Check that you can find both types of variance on your calculator.

Common error

When calculating variance, only use $s^2 = \frac{\sum (x-\bar{x})^2}{n-1}$ if you have a sample of data from a population whose variance you are trying to estimate. Otherwise use $\sigma^2 = \frac{\sum (x-\bar{x})^2}{n}$ as normal.

You will often use a sample to find an estimate of a population **parameter** (some numerical characteristic of the population, such as its mean or variance). If you want a sample to provide a good estimate of a population parameter, then the sample needs to be **representative** of the population. This means that the distribution of the values in the sample is roughly the same as in the whole population. This will not be the case if the sampling procedure is **biased**. For example, if you wanted to find out about people's attitudes to music and you selected a sample from those attending a particular concert, then this sample would contain an unusually large number of fans of a particular type of music.

WORKED EXAMPLE 22.2

Comment on possible sources of bias in each of these samples.

a The basketball team is taken as a sample of students at a college used to estimate the average height of all students.

b A sample of people from a particular political party's conference is used to find out about the UK populations' attitudes to taxation.

c A sample is taken from those in a waiting room at a doctor's surgery for a survey to find out how many days sick leave people in the country have had this year.

Continues on next page

a Basketball players are, on average, taller than the general population.

b Attitudes to taxation tend to be related to political affiliation, so members of a particular party will have attitudes not representative of the wider population.

c People at a doctor's surgery are likely to be in poorer health than the general population.

There are several different methods for selecting samples but it is important to remember that even a good (unbiased) sampling method can lead to an unrepresentative sample. This is because the process of sampling is inherently random so there is always a possibility that, for example, only extreme values are selected. The aim of a good sampling method is to minimise the probability of this happening.

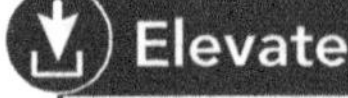

Elevate

See Extension Sheet 22 for questions on ways in which statistics can be used to present a biased view of the data.

Simple random sample

This is the type of sample most people have in mind when they talk about random samples.

Key point 22.2

In **simple random sampling** every possible sample (of a given size) has an equal chance of being selected.

Many common sampling techniques fail to produce a simple random sample. For example, it is common to include equal numbers of each gender when selecting a sample for a social science study. However, this is not a simple random sample because every possible sample doesn't have an equal chance of being selected: a sample consisting of all females, for example, has probability zero of being selected. Another example would be selecting names from a list by choosing the starting point randomly and then taking every tenth name. There are perfectly good reasons for selecting samples in this way in certain situations, you just need to be aware that they are not simple random samples.

Common methods for generating a simple random sample include lottery machines and random-number generators. When using these methods it is a common practice to sample without replacement; this means that you do not include the same individual more than once. If the population is large, it is possible to do this for any reasonably sized sample.

WORKED EXAMPLE 22.3

A student wants to take a sample of students from his college. He has a list of all students, numbered 1 to 478. He uses a random-number generator on his calculator, which can generate three-digit random numbers from 001 to 999, inclusive. The first ten numbers he obtains are:

237, 155, 623, 078, 523, 078, 003, 381, 554, 263

Suggest which could be the first four students in his sample.

237, 155, — Skip 623, as there is no student with this number.

78, — 078 stands for student number 78.

3 — Skip 523 because there is no student with this number, and skip 78 because it has already been included.

Opportunity sampling

It is remarkably difficult to ensure a simple random sample. It may not be possible to acquire the list of all the members of the population, or to obtain measurements from all the individuals you wish to sample.

Opportunity sampling avoids these difficulties.

Key point 22.3

Opportunity sampling means choosing respondants based upon their availability and convenience.

This clearly does not produce a simple random sample, but it may still produce a good estimate of population parameters. However, in some situations it can introduce bias if the group consists of very similar members. It might therefore not be generalisable. For example, if you ask your friends which subject they like most it might not mean the entire school shares that opinion.

WORKED EXAMPLE 22.4

Dina wishes to take a sample of residents from her neighbourhood. She decides to ask some people waiting at the bus stop.

a What sort of sample is this?

b Is the sample likely to be representative if her question is about:

i attitudes to the environment **ii** their favourite football team?

Continues on next page

a Opportunity sample

b i The sample may not be representative because people who use public transport are more likely to have 'green' attitudes.

ii The sample could be representative, as there is no obvious link between use of public transport and football.

You may be able to think of possible sources of bias here, for example, linked to age.

Systematic sampling

A simple random sample might just happen to include only people from London, or only people with the first name 'John'. If these outcomes would be problematic an alternative is **systematic sampling**. This requires a list of all participants that is ordered in some way.

Key point 22.4

Systematic sampling means taking participants at regular intervals from a list of the population.

WORKED EXAMPLE 22.5

A sample is formed by taking a telephone book and calling the person listed at the top of each page.

a State the name given to this type of sampling.

b Explain why this is not simple random sampling.

The calls are made at between 10 am and 2 pm on a Wednesday to enquire about the number of children in the household.

c Suggest a reason why the mean value calculated will be biased.

a Systematic sample

b Not all samples are equally likely, for example, the sample with all the people at the bottom of each page has zero probability of being selected.

Give a reason why every sample isn't equally likely to be chosen.

c Many people may not be home at this time of the day, particularly perhaps people without children who may be working, so they will not be able to answer. This would mean the calculated mean is higher than it should be.

There are many possible reasons biasing the mean in either direction.

Stratified sampling

A simple random sample might not be **representative** of the overall population. There may be more pensioners, or men, or people with Mathematics A Level in your sample than the background population. One way to overcome this is to use a **stratified sample**. You need to decide in advance which factors you think might be important and then separate the population by these factors, before taking a simple random sample within each group.

Key point 22.5

Stratified sampling means splitting the population into groups, based on factors relevant to the research, then random sampling from each group in proportion to the size of that group.

WORKED EXAMPLE 22.6

A school is made up of 250 girls and 150 boys. A sample of size 80 is to be chosen, stratified between boys and girls. How many girls must be included in the sample?

The proportion of the school which is girls is

$$\frac{250}{250+150}=\frac{5}{8}$$

The total school population is $250+150=400$.

$\frac{5}{8}$ of 80 is 50.

Calculate this fraction of the sample of 80.

Quota sampling

Stratified sampling is excellent in principle, but it is often not practical as you need to have access to every member of the population to make a random sample. A common alternative is simply to replace random sampling with opportunity sampling in the stratified sampling method.

Key point 22.6

Quota sampling means splitting the population into groups, based on factors relevant to the research, then opportunity sampling from each group.

WORKED EXAMPLE 22.7

A market researcher is required to sample 100 men and 100 women in a supermarket to find out how much they are spending on that day.

a State the name given to this type of sampling.
b Explain why this method is used rather than stratified sampling.
c State one disadvantage of this method.

Continues on next page

a Quota sampling

b The researcher would have to know in advance who was going to be shopping on that day to create a random sample, and this is not feasible.

c The people who do stop to talk to the researcher might not be representative.

There are several others too!

Cluster sampling

One of the main concerns in real-world sampling is cost. Creating a list of all members of a population, and travelling or contacting the sample, may be very difficult and expensive. One method that is intended to make the process more efficient is **cluster sampling**.

Key point 22.7

Cluster sampling means splitting the population into groups (called clusters), based on convenience, then randomly choosing some clusters to study further.

In stratified sampling all groups were sampled in proportion to their size, but in cluster sampling only some of the clusters are chosen (at random) to be studied. This makes it less accurate than stratified sampling, as choosing an unrepresentative cluster can have a large effect on the outcome.

WORKED EXAMPLE 22.8

Jacob wants to estimate the average life expectancy of all the people in the world. He chooses five countries at random and then finds the life expectancy of each country.

a State the name given to this type of sampling.
b Describe the difference between this sampling and a stratified sample across countries.

a Cluster sampling

b Only some countries are chosen in this sample. For a stratified sample values would be chosen from all countries and combined in proportion to the size of the country.

Comparing sampling methods

Method	Advantage	Disadvantage
Random methods		
Simple random	Produces an unbiased sample.	Hard to do in practice. Requires a list of the entire population and everybody to respond. Is time consuming and expensive.
Systematic	Avoids unwanted clustering of data. Practically easier than using random number generators.	Requires a list of the entire population. Is less random than simple random as no longer independent.
Stratified	Produces a sample representative over the factors identified.	Requires a list of the entire population, with additional information about each member. Is time consuming and expensive. Determining which factors to consider is not always obvious.
Cluster	Cheaper and easier than other random methods.	Less accurate than other random methods - clusters may not be representative.
Non-random methods		
Opportunity	Cheap and convenient.	May not be generalisable.
Quota	Ensures the sample is representative over the factors identified.	May not be generalisable.

Did you know?

The issue of the size of sample to use is of vital importance to statisticians. A sample of size 200 is not twice as good as a sample of size 100 – because of what is called the **law of diminishing returns**. This is not just a rule of thumb – you can prove it using some advanced ideas in statistics which you might meet if you do the Further Mathematics course.

EXERCISE 22A

1. Comment on possible sources of bias in each sample.

   a Determining the attitude towards UK university tuition fees by asking sixth-form students.

   b Finding out about people's perception of the cost of food by asking people in supermarkets to estimate the cost of a pint of milk.

   c Measuring the average height of people in the UK by means of a sample taken from a school.

   d Predicting the 1948 US presidential election result by a telephone survey of US citizens.

2 Name the type of sampling described in each situation.

a A zoologist investigating different flies knows there are roughly equal numbers of males and females. She puts up fly paper that attracts female flies and fly paper that attracts male flies. When each paper has 20 flies on it she takes the paper down.

b A doctor believes that drug resistance may depend on the age of the patient. 20% of patients in the hospital are below 18 and 30% are above 65. She makes a simple random sample to choose four under 18s, ten people aged from 18 to 65 and six over 65s.

c There are about 4000 professional footballers in England, playing for 92 clubs. To estimate their fitness, Gary chooses ten clubs at random and conducts a test on a random sample of players at those clubs.

d The names of all students in a school are put into a hat. A sample is formed from the first ten names taken out.

e To find the average population of countries, George writes a list of all the countries, in alphabetical order, then chooses the 3rd, 13th, 23rd, ... countries and investigates them.

f To determine the outcome of the next election you ask everybody in your class how they would vote.

3 To find the mean and standard deviation of weights of a breed of cat, Ben measures a random sample of 20 cats. The results are summarised as:

$\sum w = 51, \sum w^2 = 138.32$

a Find the mean weight, and show that the standard deviation of this sample is 0.64 to 2 d.p.

b Is the standard deviation of the whole population likely to be larger or smaller than 0.64?

c Ben hopes to obtain a more accurate estimate for the mean by taking a sample of size 100. The mean of this sample is 2.63. Is this necessarily a better estimate of the population mean than the one found in part **b**? Explain your answer.

4 An ecologist wants to establish the proportion of adult fish in the North Sea. She believes that 40% are cod, 40% are haddock and 20% are of other varieties. She catches fish until she has 20 cod, 20 haddock and 10 of other varieties.

a Explain why a random sampling method is not feasible in this situation.

b State the name of the sampling method used.

c Why might this method be better than an opportunity sample of the first 20 fish caught?

5 Work with the data set 1, 2, 5, 8.

a Calculate the variance, using the formula:

$$\sigma^2 = \frac{\sum (x-\bar{x})^2}{n}$$

b Calculate the variance, using the formula:

$$s^2 = \frac{\sum (x-\bar{x})^2}{n-1}$$

c When would it be appropriate to use the formula in **b** rather than the one in **a**?

6 A psychologist, studying the reading age of people in a city, wants to create a stratified sample. He thinks that gender and age are important. According to census data, 40% of the population of the city is female and the median age is 42 for both genders.

a Copy and complete the table to show the numbers of people in each category required in a sample of 80 people.

	Female	**Male**
Under 42		
42 or over		

b State one advantage of stratified sampling over quota sampling.

c Explain why quota sampling might be preferable in this situation.

7 A shop has 40 staff in each of ten branches. The owner wants to interview a sample of 20 staff. These suggestions are made as to how to choose the sample.

A Pick four branches at random and then interview five randomly chosen people at each branch.

B Use a random-number generator to pick 20 staff members from all the shops.

C Ask the manager of each branch to select two staff members and send them for interview.

D Use a 20-sided dice to select two members randomly from each branch to send for interview.

a Name the method of sampling for each suggested method.

b Which method is likely to give the most accurate answer?

c Why might method **A** be used instead of method **B**?

8 A polling firm, investigating the voting intentions of a London borough, makes a numbered list of the registered voters and uses a random-number generator to select 100 participants. They send a questionnaire to each person selected. Explain why this will not necessarily produce a simple random sample.

9 A school is attended by 500 girls and 500 boys.

a A simple random sample of 10 students is obtained by drawing names out of a hat (without replacement). Find the probability that the students are all boys.

b If names are put back in the hat it is possible that the same student gets picked more than once.

i Find the probability of someone being picked more than once.

ii Find the probability of any one student being picked, using this method.

c What is the percentage difference between the probability of all ten students being boys in the situations described in **a** and **b**?

d Instead of simple random sampling an opportunity sample is taken by choosing the first ten students a teacher sees on the playground. Without further calculation, explain whether this will increase or decrease the probability of all ten students being the same gender.

Focus on...

You will look at this problem in more detail in the Focus on ... Problem solving 4.

Section 2: Introduction to hypothesis testing

You will now consider using a sample to make inferences about the population. Think about these questions.

- A particular drug is known to cure a disease in 78% of cases. A new drug is trialled on 100 patients and 68 of them were cured. Does this mean that the new drug is less effective than the old one?
- A sociologist believes that more boys than girls are born during war time. In a sample of 200 babies born in countries at war, 126 were boys. Do these data support the sociologist's theory?
- In the last general election, Party Z won 36% of the vote. An opinion poll surveys 100 people and finds that 45 support Party Z. Does this imply that their share of the vote will change in the next election?

In all these questions you are trying to find out whether the proportion of the population with a certain characteristic is different from a previously known or assumed value by calculating the corresponding proportion from a sample. You would not expect the sample proportion to be exactly the same as the population proportion. If you took a different sample you may well get a different proportion. This means that a sample, however large, cannot provide a definitive answer to a question; it can only suggest what the answer is likely to be.

> **Fast forward**
>
>  In the AS course you will only look at hypothesis tests for the population proportion, using the binomial distribution. In Student Book 2, Chapter 22, you will also meet a hypothesis test for the population mean.

One common procedure for answering this type of question is a **hypothesis test**. It requires the question to be phrased in a specific way.

> **Key point 22.8**
>
> A **hypothesis test** is a procedure for answering a question of the type:
>
> *Does the sample provide sufficient evidence that a population parameter (mean/spread/proportion) has changed from a previously known or assumed value?*

This may be illustrated through the first example at the start of this section. The old drug cured 78% of patients. In a trial of a new drug, 68 out of 100 patients in a sample were cured; this number is called the **test statistic**. The question to ask is:

- *Does this sample provide sufficient evidence that the new population proportion is smaller than 78%?*

The key phrase here is **sufficient evidence**. The sample proportion of $\frac{68}{100} = 68\%$ seems significantly smaller than 78%, but it could happen even if the population proportion is still the same. So how likely would this be?

To calculate this probability you need to assume that the population proportion hasn't changed. This is your default position or **null hypothesis**, which is tested against an **alternative hypothesis**, which represents the idea you have, that there has been a change.

Key point 22.9

- The **null hypothesis**, denoted H_0, specifies the previous or assumed population proportion.
- The **alternative hypothesis**, denoted H_1, specifies how you think the proportion may have changed.

So in this example, the null hypothesis is:

H_0: The proportion of cured patients is 78%

and the alternative hypothesis is:

H_1: The proportion of cured patients is less than 78%.

The question now becomes:

- *Does the sample provide sufficient evidence against the null hypothesis and in favour of the alternative hypothesis?*

If the population proportion of cured patients is still 78%, then the probability of any particular patient being cured is 0.78. There is a sample of 100 patients, so the number of cured patients can be modelled by the binomial distribution $B(100, 0.78)$. The number of 'successes' in the sample is 68. Since you are looking for evidence that the population proportion is less than 0.78, you want to find the probability of 68 or fewer patients being cured.

If $X \sim B(100, 0.78)$ then $P(X \leqslant 68) = 0.0134$.

So if the new drug is as good as the old drug, there is a probability of about 1.3% that only 68 fewer out of a sample of 100 patients get cured.

Tip

It is always the probability of the observed value or more extreme that you need to calculate. So, if you were looking for evidence that the population proportion has increased, you would calculate $P(X \geqslant 68)$.

Is this sufficiently 'unlikely' to conclude that the population proportion for the new drug must be smaller than 78%? This is a matter of judgement, and may be different in different contexts. You need to decide on what probability is 'sufficiently small.' This is called the **significance level** of the test.

Key point 22.10

- The **significance level** of a hypothesis test specifies the probability that is sufficiently small to be accepted as evidence against the null hypothesis.
- Assuming that H_0 is correct, then the probability of the observed, or more extreme, sample value is called the ***p*-value**.
- If the p-value is smaller than the significance level, there is sufficient evidence against H_0 and it can be rejected in favour of H_1. Otherwise, the sample does not provide sufficient evidence against H_0 and you should not reject it.

Common significance levels used in practice are between 1% and 10%. If you conduct the test at the 1% significance level, then the p-value 0.0134 is greater than the significance level, so you don't have sufficient evidence that the population proportion has decreased. However, if you use a 5% significance level then you do have sufficient evidence that the new drug cures less than 78% of the patients.

Key point 22.11

Steps for a hypothesis test for a population proportion

1. State the null and alternative hypotheses, defining any parameters.
2. Decide on the significance level.
3. State the distribution of the test statistic, assuming the null hypothesis is true.
4. Using this distribution calculate the probability of observing the test statistic or more extreme. This is the p-value.
5. Compare the p-value to the significance level. If the p-value is smaller than the significance level, there is sufficient evidence to reject the null hypothesis.
6. Interpret the conclusion in context, remembering to make clear that the conclusion is not a statement of certainty, but of significance.

Notice that there are two possible conclusions you can reach from a hypothesis test:

- the sample provides sufficient evidence to reject H_0 in favour of H_1 or:
- the sample does not provide sufficient evidence to reject H_0.

In the latter case you can't say that H_0 is correct, just that you have not found sufficient evidence to reject it. Once you have reached a conclusion, it is important to interpret it in the context of the question. So in the given example you need to write a statement such as: 'There is sufficient evidence that the new drug is less effective,' rather than just: 'There is sufficient evidence that the proportion is less than 78%.'

Tip

You should always use the word 'evidence' in your conclusion to make it clear that it is not certain.

WORK IT OUT 22.1

Which of these conclusions to hypothesis tests are incorrectly written, and why?

1. Reject H_0. The proportion of red flowers has decreased.
2. There is sufficient evidence to reject H_0, and thus sufficient evidence that the proportion has increased.
3. There is insufficient evidence to reject H_0. The proportion of girls in the club is probably still 45%.
4. Accept H_0, as there is insufficient evidence that the percentage of A grades has increased.
5. There is evidence to accept H_0; the dice is not biased.

WORKED EXAMPLE 22.9

A sociologist believes that more boys than girls are born during war time. In a sample of 200 babies born in countries at war, 116 were boys.

Do these data support the sociologist's theory at the 5% significance level?

Let p be the proportion of boys in the population.
$H_0: p = 0.5$
$H_1: p > 0.5$

You want to find out whether there is evidence that the proportion of boys is more than $\frac{1}{2}$.
So the null hypothesis is that the proportion is $\frac{1}{2}$ and the alternative hypothesis is that it is greater.

Let X be the number of boys in a sample of 200 babies.

If H_0 is correct then $X \sim B(200, 0.5)$.

Test statistic: $X = 116$

The test statistic is the number of boys in the sample. Since you are looking for evidence that the population proportion is greater than 0.5, you need to find the probability of observing this number, or more, if H_0 is correct.

$$P(X \geqslant 116) = 1 - P(X \leqslant 115)$$
$$= 0.0141$$

Use a calculator to find the cumulative probability.

$0.0141 < 0.05$

Compare the probability to the critical value. A small probability means that observing this value or more extreme values is unlikely.

So there is sufficient evidence to reject H_0.

The conclusion is not definite - you must use the phrase 'sufficient/insufficient evidence'.

At the 5% significance level, there is evidence to support the scientist's theory.

The conclusion needs to be interpreted in the context of the question.

In Worked example 22.9 the alternative hypothesis was that the population proportion has increased. This is called a **one-tail test**. Sometimes you might have a reason to believe that the population proportion has changed, but you can't predict in which direction. In this case you have to do a **two-tail test**. The only difference is that you need to compare the p-value to half the significance level, because both very small and very large values of the test statistic would provide evidence against H_0.

Common error

Be sure to make the correct conclusion once you've found the p-value:

- p-value > significance level, then do not reject H_0
- p-value < significance level, then do reject H_0

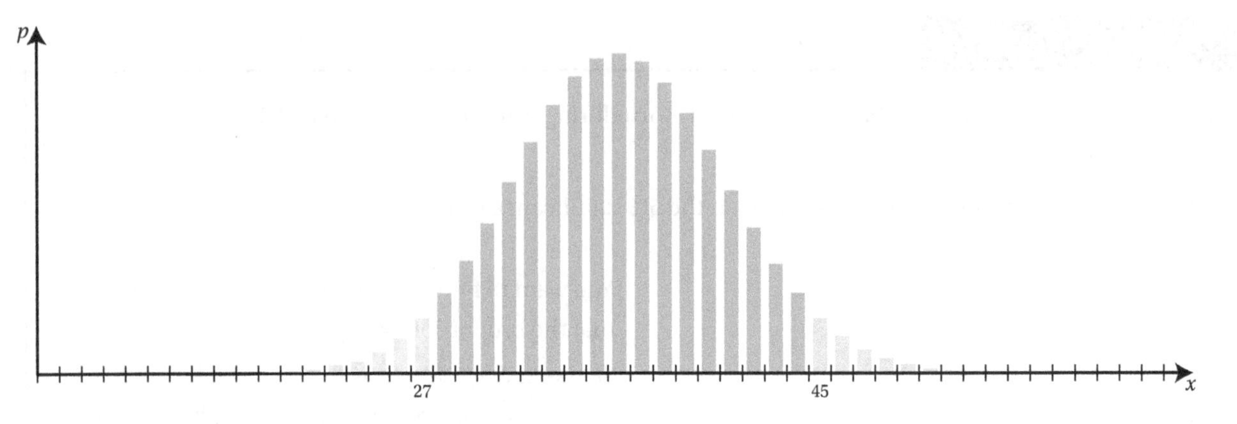

The diagram shows a binomial distribution B(100, p) under H_0: $p = 0.36$.

There is a 5% chance, under H_0, that a sample will take a value in the left (blue) tail and a 5% chance that a sample will take a value in the right (green) tail; in either case, the sample indicates that H_0 should be rejected.

For a 10% two-tailed test, H_0 would be rejected if a sample took a value in either tail, since either end provides sufficient evidence that $p \neq 0.36$.

Elevate

See Support Sheet 22 for a further example like Worked example 22.9 and for more practice questions.

Key point 22.12

- In a **one-tail test** the alternative hypothesis is of the form $p < \alpha$ or $p > \alpha$. You need to compare the p-value to the significance level.
- In a **two-tail test** the alternative hypothesis is of the form $p \neq \alpha$. You need to compare the p-value to half the significance level.

Common error

Be sure to think carefully about whether you need a one-tail or two-tail test. Then set up H_1 accordingly: remember to halve the significance level in each tail with a two-tail test.

The third question at the start of this section was whether the proportion of voters supporting a particular political party has **changed**. Without any further information you cannot anticipate whether the proportion has increased or decreased, so you have to do a two-tail test.

WORKED EXAMPLE 22.10

In the last general election, Party Z won 36% of the vote. An opinion poll company surveys 100 people and finds that 45 support Party Z.

Does this provide sufficient evidence at the 10% significance level that the proportion of voters who support Party Z has changed?

Let p be the proportion of voters supporting Party Z.

H_0: $p = 0.36$

H_1: $p \neq 0.36$

Start by stating the hypotheses. This is a two-tail test because we are looking for evidence of change, rather than just increase or decrease.

Continues on next page

Let X be the number of supporters of party Z in sample of 100 people. If H_0 is true, $X \sim B(100, 0.36)$.	If H_0 is correct, the number of supporters in a sample will follow a binomial distribution with the probability given in H_0.
Test statistic: $X = 45$ $P(X \geqslant 45) = 1 - P(X \leqslant 44) = 0.0397$	Since 45 out of 100 is more than 36%, you need to calculate the probability in the 'right tail.'
$0.0397 < 0.05$ so there is sufficient evidence to reject H_0.	The significance level is 10%, so for a two-tail test you compare the p-value to half the significance level, which is 5%, or 0.05.
There is evidence, at the 10% significance level, that the proportion of voters who support party Z has changed.	Remember to interpret the conclusion in the context of the question, making it clear that it is not certain

WORK IT OUT 22.2

A dice is rolled 5 times and 3 sixes are observed. Test at the 5% significance level if the dice is biased.

Which is the correct solution? Identify the errors made in the incorrect solutions.

Solution A	Solution B	Solution C
$H_0 : p = \frac{1}{6}$	$H_0 : p = \frac{1}{6}$	$H_0 : p = \frac{1}{6}$
$H_1 : p \neq \frac{1}{6}$	$H_1 : p > \frac{1}{6}$	$H_1 : p \neq \frac{1}{6}$
$X \sim B\left(5, \frac{1}{6}\right)$	$X \sim B\left(5, \frac{1}{6}\right)$	$X \sim B\left(5, \frac{1}{6}\right)$
$P(X = 3) = 0.032 < 0.05$	$P(X \geqslant 3) = 0.035 < 0.05$	$P(X \geqslant 3) = 0.035 > 0.025$
So reject H_0.	So reject H_0.	So do not reject H_0.

EXERCISE 22B

1 Write down the null and alternative hypotheses for each test, defining the meaning of any parameters.

a i Daniel wants to test whether the proportion of students in his school who like football is higher than 60%.

ii Elsa wants to find out whether the proportion of households with a pet is higher than 1 in 3.

b i The proportion of faulty components produced by a machine was 6%. The manager wants to check whether this has decreased following a service.

ii Joseph thinks that fewer than half of all children eat five or more pieces of fruit a day and wants to confirm this by using a hypothesis test.

c i Sofia has a coin which she thinks is biased, and wants to use a hypothesis test to check this.

ii Max knows that last year, 26% of entries in AS Level Psychology were graded A and wants to check whether this proportion has changed.

2 In each part you are given null and alternative hypotheses (where p stands for the population proportion), the significance level and the observed data. Decide whether or not there is sufficient evidence to reject the null hypothesis.

a i H_0: $p = 0.3$, H_1: $p > 0.3$, significance level 5%, observed 15 successes out of 40 trials.

ii H_0: $p = 0.6$, H_1: $p > 0.6$, significance level 10%, observed 23 successes out of 45 trials.

b i H_0: $p = 0.5$, H_1: $p < 0.5$, significance level 10%, observed 15 successes out of 40 trials.

ii H_0: $p = 0.45$, H_1: $p < 0.45$, significance level 3%, observed 20 successes out of 60 trials.

c i H_0: $p = 0.4$, H_1: $p \neq 0.4$, significance level 8%, observed 31 successes out of 100 trials.

ii H_0: $p = 0.8$, H_1: $p \neq 0.8$, significance level 5%, observed 19 successes out of 20 trials.

3 It is known that, in the UK, 63% of households own at least one car. David believes that, in his city neighbourhood, car ownership is lower than this. He uses a hypothesis test, based on the binomial distribution, to confirm this.

a State suitable null and alternative hypotheses for his test.

David surveys a random sample of 50 households in his neighbourhood and finds that 29 of them own at least one car.

b Use this data to test David's hypothesis at the 10% significance level. State your conclusion clearly.

4 The 2011 census found that 68% of 16–19-year-olds in a particular town attended a sixth-form college. In 2015 a sample of 60 teenagers in this age range was surveyed and it was found that 46 of them attended a sixth-form college.

Is there evidence, at the 5% significance level, that the proportion of 16–19-year-olds attending a sixth-form college has increased?

5 Rahul believes a six-sided dice is biased so that the probability of rolling a 3 is smaller than $\frac{1}{6}$. He rolls the dice 30 times and gets four 3s.

Does this provide sufficient evidence to support Rahul's belief?

6 The proportion of A grades in AS Mathematics is 33% nationally. In a sample of 38 students from a college, 18 got A grades.

Is there evidence, at the 5% significance level, that the students at this college are getting a higher proportion of A grades than the national average?

7 A treatment for a disease is known to be effective in 82% of cases. A doctor devises a new treatment which she believes is more effective. She uses the new treatment on a random sample of 50 patients and finds that it is effective in 46 cases.

Do these data support the doctor's belief at the 2% significance level?

8 At a teacher's previous school, a third of all sixth-formers had a younger sibling at the school. He wants to find out whether this proportion is different in his new school. He samples 60 sixth-formers and finds that 27 of them have a younger sibling at the school.

Conduct a hypothesis test at the 5% significance level to decide whether there is evidence that the proportion of sixth-formers with a younger sibling at the new school is different from that at the previous school.

9 Angela is playing a board game that involves rolling a dice. She thinks that the probability of rolling a low number (a 1, 2 or 3) is different from the probability of rolling a high number (a 4, 5 or 6). In 40 rolls of the dice she got 14 high numbers.

Test Angela's belief at the 10% significance level.

10 An athletics club had the same running coach for several years. Records show that 28% of his athletes could run 100 metres in under 12 seconds. The club brings in a new coach and, over the following year, 26 out of a sample of 75 athletes recorded 100-metre times under 12 seconds.

Do these data support the hypothesis that the proportion of athletes who can run 100 metres in under 12 seconds has changed? Use the 5% significance level for your test.

11 70% of students in Year 13 are in favour of a new uniform. Rihanna wants to find out whether the proportion is the same in Year 12. She proposes to test the null hypothesis $H_0: p = 0.7$ against two different alternative hypotheses, $H_1: p < 0.7$ and $H_2: p \neq 0.7$, using the 10% significance level.

Rhianna asks a sample of 25 Year 12 students. The data give her sufficient evidence to reject H_0 in favour of H_1, but not in favour of H_2.

How many students in Rhianna's sample were in favour of the new uniform?

12 A student tests the hypothesis $H_0: p = 0.4$ against $H_1: p > 0.4$, where p is the proportion of brown cats in a particular breed.

In a sample of 80 cats of this breed 40 were brown, and this leads him to reject the null hypothesis.

What can you say about the significance level he used for his test?

13 A doctor wants to find out whether the proportion of people suffering from a certain genetic condition has decreased from its previous value, q. She conducts a hypothesis test at the 5% significance level, using a sample of 120 patients. After doing some calculations, she realises that, even if none of her sample had the condition, this would not provide sufficient evidence that the proportion has decreased from q.

Find the maximum possible value of q.

Section 3: Critical region for a hypothesis test

Suppose you toss a coin 200 times and get 86 heads. You now know that you can use a hypothesis test to determine whether this provides evidence that the coin is biased.

It turns out that there is insufficient evidence at the 1% level that the coin is biased, so the next logical question to ask is: how many heads would provide sufficient evidence that the coin is biased?

WORKED EXAMPLE 22.11

In order to find out whether a coin is biased against getting heads, Robert decides to test the hypotheses:

$H_0: p = \frac{1}{2}$ against $H_1: p < \frac{1}{2}$

where p is the probability of the coin showing heads. He tosses the coin 200 times and uses the 1% level of significance.

What is the greatest number of heads he can observe and still have sufficient evidence that the coin is biased?

Let X be the number of heads out of 200 coin tosses.

If H_0 is true, $X \sim B\left(200, \frac{1}{2}\right)$

State the distribution of the test statistic if the null hypothesis is true.

The significance level is 1%, so look for a number k such that $P(X \leqslant k) < 0.01$.

$P(X \leqslant 85) = 0.020 > 1\%$

$P(X \leqslant 84) = 0.014 > 1\%$

$P(X \leqslant 83) = 0.0097 < 1\%$

You know from the start of this section that tossing 86 heads does not provide sufficient evidence, so try values smaller than 86.

In order to have sufficient evidence against H_0 at the 1% significance level, Robert would need to observe 83 or fewer heads out of 200 coin tosses.

The range of values of the test statistic found in Worked example 22.11 ($X \leqslant 83$) is called the **critical region** for the test.

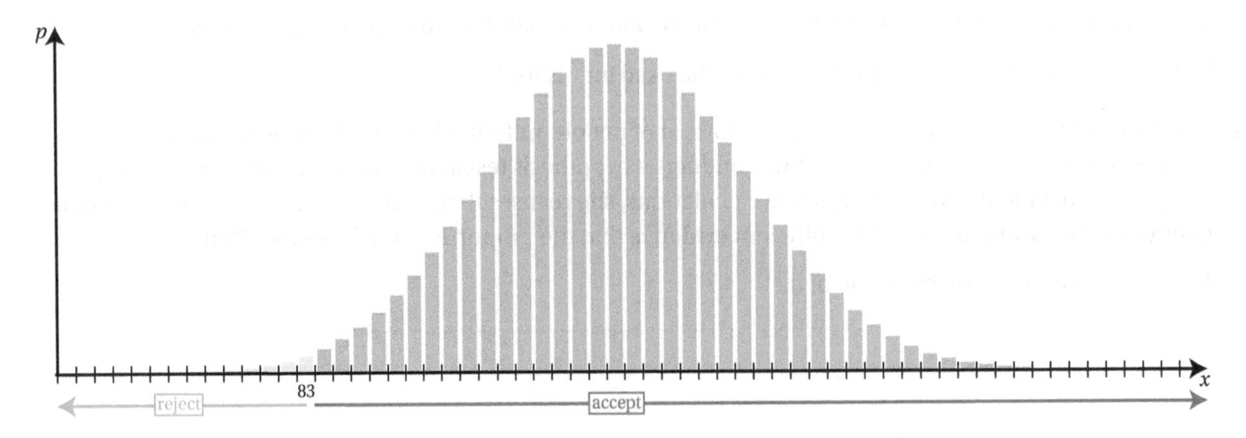

Key point 22.13

- The **critical region** (or **rejection region**) for a test is the set of values of the test statistic that provide sufficient evidence to reject the null hypothesis.
- The value at the edge of the critical region is called the **critical value**.
- The **acceptance region** is the set of values of the test statistic that do not provide sufficient evidence to reject the null hypothesis.

Tip

Remember that the critical value is in the critical region.

In Worked example 22.11 the critical value is 83, the critical region is $X \leqslant 83$ and the acceptance region is $X \geqslant 84$.

Note that the form of the critical region depends on the form of the alternative hypothesis. If you were looking for evidence that the population proportion has increased, the critical region would be of the form $X \geqslant k$.

WORKED EXAMPLE 22.12

The proportion of students getting an A in AS Mathematics is currently 33%. A publisher produces a new textbook which they hope will lead to improved performance. They trial their textbook with a sample of 120 students and want to test their hypotheses at the 5% significance level.

Find the critical region for this test.

Let p be the proportion of A grades.
$H_0: p = 0.33$
$H_1: p > 0.33$

You should always start by stating the hypotheses.

The null hypothesis states the current, known proportion.

Let X be the proportion of A grades. If H_0 is true then $X \sim (120, 0.33)$

Next state the distribution you are going to use.

$P(X \geqslant 47) = 1 - P(X \leqslant 46)$
$= 0.091 \ (> 5\%)$
$P(X \geqslant 48) = 1 - P(X \leqslant 47)$
$= 0.064 \ (> 5\%)$

If H_0 was true you would expect around $120 \times 0.33 \approx 40$ A grades, so you only need to check values of X larger than 40. Use your calculator to find the probabilities.

$P(X \geqslant 49) = 1 - P(X \leqslant 48)$
$= 0.044 \ (< 5\%)$

Look for the first probability smaller than the significance level.

The critical region is $X \geqslant 49$.

Once you have found the critical region you can easily tell whether a value of the test statistic provides sufficient evidence against the null hypothesis.

Key point 22.14

If the value of the test statistic is inside the critical region you have sufficient evidence to reject H_0.

So if, for example, 52 out of the 120 students got A grades, the publisher would have sufficient evidence to claim that their textbook leads to improved performance.

As with any conclusion from a statistical investigation, there is no proof that the improvement was actually caused by the textbook - there could be other factors. Perhaps more importantly, since the results are only based on a sample, there is no guarantee that the proportion of A grades for the whole population has increased. It is possible that the selected sample was unrepresentative, and that most other samples would result in a smaller proportion of A grades.

The meaning of the significance level

To judge how reliable the test is, it is useful to estimate the probability of drawing an incorrect conclusion. It is possible that your sample leads you to reject a correct null hypothesis. In the last example, even if the proportion of A grades is still 33%, a particular sample could suggest that the proportion has increased. That would happen if the test statistic (the number of A grades in the sample) is in the critical region, in this case $X \geqslant 49$. But you have already found the probability of this happening: $P(X \geqslant 49) = 0.044$; so there is a 4.4% chance of rejecting the null hypothesis even though it is correct.

Fast forward

A In Student Book 2, Chapter 22, you will test hypotheses involving continuous distributions; in that case the probability of rejecting a correct null hypothesis will be exactly equal to the significance level.

Key point 22.15

The probability of rejecting a correct null hypothesis is the same as the probability of the test statistic being in the critical region. It is always smaller than (or equal to) the significance level of the test.

WORKED EXAMPLE 22.13

A machine produces smartphone parts. Previous experience suggests that, on average, 7 in every 200 parts are faulty. After the machine was accidentally moved, a technician suspects that the proportion of faulty parts may have increased. She decides to test this hypothesis using a random sample of 85 parts.

a State suitable null and alternative hypotheses.

The technician decides that the critical region for the test should be $X \geqslant 5$. After checking her sample, she finds that four parts are faulty.

b State what conclusion she should draw and justify your answer.

c What is the probability of incorrectly rejecting the null hypothesis?

Continues on next page

a $H_0: p = 0.035$
$H_1: p > 0.035$ (where p is the proportion of faulty parts)

The null hypothesis is that the proportion hasn't changed, so the probability of a part being faulty is $\frac{7}{200} = 0.035$.

b The test statistic $X = 4$ is not in the critical region, so there is not sufficient evidence to reject H_0.

Check whether the test statistic is in the critical region.

There is not sufficient evidence that the proportion of faulty parts has increased.

Interpret the conclusion in the context of the question.

c If $X \sim B(85, 0.035)$ then the probability of rejecting H_0 is:
$P(X \geqslant 5) = 1 - P(X \leqslant 4)$
$= 0.177$

'Incorrectly rejecting the null hypothesis' means that H_0 is correct (so $p = 0.035$) but the test statistic happens to be in the critical region (so $X \geqslant 5$).

EXERCISE 22C

1 Find the critical regions for each hypothesis test. You are given the null and alternative hypotheses, the significance level (SL) and the sample size (n).

a i $H_0: p = 0.6$, $H_1: p < 0.6$, SL = 5%, $n = 50$

ii $H_0: p = 0.2$, $H_1: p < 0.2$, SL = 10%, $n = 40$

b i $H_0: p = 0.5$, $H_1: p < 0.5$, SL = 7%, $n = 120$

ii $H_0: p = 0.8$, $H_1: p < 0.8$, SL = 1%, $n = 75$

c i $H_0: p = 0.6$, $H_1: p > 0.6$, SL = 5%, $n = 50$

ii $H_0: p = 0.2$, $H_1: p > 0.2$, SL = 10%, $n = 40$

d i $H_0: p = 0.5$, $H_1: p > 0.5$, SL = 7%, $n = 120$

ii $H_0: p = 0.8$, $H_1: p > 0.8$, SL = 1%, $n = 75$

2 For each of the tests in question 1, find the probability of incorrectly rejecting the null hypothesis.

3 Ayesha is trying to find out whether it is true that students studying A Level Mathematics are more likely to be boys than girls. She sets up these hypotheses.

$H_0: p = 0.5$, $H_1: p < 0.5$, where p is the proportion of girls studying A Level Mathematics.

She uses a sample of 30 A Level students from her college, and decides to test her hypotheses at the 10% significance level.

Find the critical region for her test.

4 A company is testing a new drug. They want to find out whether the drug cures a certain disease in more than 85% of cases.

a State suitable null and alternative hypotheses, defining any parameters.

The company decide to conduct their test at the 5% significance level, using a sample of 180 patients.

b Let X be the number of patients who are cured after using the drug. Find the critical region for the test.

5 A manufacturer knows that 12% of the population have purchased their products. After a new advertising campaign, they believe that this proportion has increased. The marketing manager wants to test this using a random sample of 200 people.

a Write down suitable null and alternative hypotheses.

The marketing manager decides that the rejection region for the test should be $X \geqslant 32$, where X is the number of people in the sample who have purchased their products.

b After collecting the data, it is found that 30 people have purchased the manufacturer's products. What should the marketing manager conclude?

c Find the probability of incorrectly rejecting the null hypothesis.

6 John has an eight-sided dice and wants to check whether it is biased by looking at the probability, p, of rolling a 4. He sets up these hypotheses.

$H_0: p = \frac{1}{8}, \quad H_1: p \neq \frac{1}{8}.$

To test them he decides to roll the dice 80 times and reject the null hypothesis if the number of 4s is greater than 15 or fewer than 5.

a Let X be the number of 4s observed out of the 80 rolls. State the name given to the region $5 \leqslant X \leqslant 15$.

b What is the probability that John incorrectly rejects the null hypothesis?

7 A hypothesis test is proposed to decide whether the proportion of roses of a certain type that are white is greater than 30%.

a State suitable hypotheses, defining any parameters.

The test is to be carried out at the 5% significance level, using a random sample of 170 roses.

b Find the critical region for the test.

After the data was collected, it was found that 64 of the 170 roses are white.

c State the conclusion of the test.

8 A driving school has records showing that, over a period, 72% of its students passed their test on the first attempt. Under a new management, some procedures change and they want to find out whether the pass rate has changed.

a Defining any parameters, state suitable hypotheses.

The hypotheses are tested using a random sample of 50 students. It is decided that the null hypothesis should be accepted if from 30 to 40 (inclusive) of the students pass the test on the first attempt.

b State the critical region for the test.

c Find the probability of incorrectly rejecting the null hypothesis. What can be said about the significance level of the test?

Checklist of learning and understanding

- A **population** is the set of all individuals or items of interest in a statistical investigation. A **population parameter** is some numerical characteristic of the population (such as the mean or the range).
- A **sample** is a subset of the population. A **statistic** is some numerical characteristic of the sample that is used to estimate the population parameter of interest.
- When estimating the variance of a population from a sample, use:

$$s^2 = \frac{\sum (x - \bar{x})^2}{n-1}$$

- A good sampling procedure avoids **bias**, so the sample is more likely to be **representative** of the population. However, even an unbiased sampling procedure may accidentally produce an unrepresentative sample.
- Common sampling procedures include:
  - simple random sampling
  - opportunity sampling
  - systematic sampling
  - stratified sampling
  - quota sampling
  - cluster sampling.
- A **hypothesis test** determines whether there is evidence that the value of a population parameter has changed from a previously known or assumed one. The conclusion from a hypothesis test is never certain; two possible types of conclusion are:
  - there is insufficient evidence that the value of the population parameter has changed
  - there is sufficient evidence that the value of the population parameter has changed.
- To carry out a hypothesis test for the proportion of the binomial distribution, you must follow these steps.

  1. State the null and alternative hypotheses, defining the meaning of any parameters.
  2. State the binomial distribution which the test statistic would follow if the null hypothesis was correct.
  3. Calculate the probability of observing the given value of the test statistic, or more extreme (the ***p*-value**).
  4. Compare this probability to the **significance level.** Reject the null hypothesis if the p-value is smaller than the significance level.
  5. In a two-tail test, compare the probability to half the significance level.
  6. State the conclusion of the test, interpreting it in context and making it clear that there is some uncertainty (using the word 'evidence').

- The **critical region** (or **rejection region**) for a hypothesis test is the set of values of the test statistic that lead you to reject the null hypothesis. The remaining values form the **acceptance region**.
- For a two-tail test, the critical region is made up of two parts.
- The probability of incorrectly rejecting the null hypothesis is equal to the probability of the critical region. This is always less than or equal to the significance level of the test.

Mixed practice 22

1. A market researcher is asked to conduct a survey outside a library. He is asked to sample 100 male library users, 100 female library users, 50 males who have not been into the library and 50 females who have not been into the library.

   What type of sampling method is this?

   Choose from these options.

   A Quota sampling **B** Systematic sampling

   C Stratified sampling **D** Cluster sampling

2. The organisers of the school concert want to find out how many of the students are planning to attend the concert. The school has 48 different tutor groups, and they decide to select a sample of students, as described.

   - They choose five tutor groups randomly.
   - From each tutor group, they select a random sample of ten students.

   a What name is given to this type of sampling procedure?

   b Explain why this procedure might not give a representative sample in this case.

   The organisers later decide that they should take a simple random sample of 50 students instead.

   c Describe how they might obtain such a sample.

3. Gavin has a six-sided dice which he thinks is biased and shows more 5s than it should. He wants to conduct a hypothesis test to test his belief.

   a State suitable null and alternative hypotheses.

   Gavin rolls the dice 75 times and obtains 18 5s.

   b Conduct the test at the 5% significance level, stating your conclusion clearly.

4. Lisa is late for school on average once in every eight days. She has recently moved closer to the school, and in the last 30 days was late only twice.

   Is there evidence, at the 10% significance level, that the probability of Lisa being late for school has decreased? State your hypotheses and your conclusion clearly.

5. A doctor knows that 20% of people suffer from side effects when treated with a certain drug. He wants to see if the proportion of people suffering from side effects is lower with a new drug. He looks at a random sample of 30 people treated with the new drug.

   What is the largest number of people who could suffer from side effects and still lead to the conclusion at the 5% significance that the new drug has a lower proportion of side effects?

   Choose from these options.

   A 2 **B** 3 **C** 4 **D** 5

6 In the UK, the proportion of families who own their home (as opposed to renting) is 64%. Aneka wants to find out whether this proportion is different in Germany. She surveys a random sample of 180 families in Germany and finds that 98 of them own their homes.

Conduct a hypothesis test at the 5% significance level to test whether the proportion of families in Germany who own their home is different from that in the UK.

7 **a** Define a simple random sample.

Stacey is investigating attitudes to sport among students at her school. She decides to carry out a survey using a sample of 70 students. There are the same number of boys and girls at the school, so Stacey randomly chooses 35 boys and 35 girls.

b **i** State the name for this type of sample.

ii Explain why in this case, this type of sample is better than a simple random sample.

One of Stacey's questions is about participation in school sports teams. She wants to find out whether more than 40% of students play for a school team. She sets up these hypotheses:

$H_0: p = 0.4, \quad H_1: p > 0.4,$

where p is the proportion of students who play for a school sports team.

c Find the critical region for the hypothesis test at the 10% significance level, using a sample of 70 students.

d What is the probability of incorrectly rejecting the null hypothesis?

e In Stacey's sample, 32 students play for a school team. State the conclusion of the test.

8 National records show that 35 per cent of train passengers buy their tickets in advance. A random sample of 25 passengers using a particular railway station is selected, and it is found that 13 of them bought their tickets in advance.

a Investigate, at the 10% level of significance, whether the data support the view that the percentage of passengers from this station who buy their tickets in advance is different from the national figure of 35 per cent.

b It was suggested that, for a follow-up survey, it would be easier to collect all the data from passengers in a particular railway carriage. Explain in context why it would **not** be appropriate to then apply the test that you have used in part **a**.

[© AQA 2014]

9 Adam owns a farm. On the farm, there are 330 sheep, 160 cattle and 10 goats. The animals are each tagged with a number which is recorded in a stock book. The sheep are numbered from 001 to 330, the cattle are numbered from 331 to 490 and the goats are numbered from 491 to 500.

Adam wants to carry out a blood test to check the health of his animals but can only afford to carry out the test on 50 animals.

a **i** Describe how Adam could use random numbers to select a simple random sample of 50 animals for the test.

ii Explain why the small number of goats might cause a problem when using a random sample.

b Adam's farm manager, Ellie, suggests a different sampling method. She suggests randomly choosing a number from 1 to 10, testing the animal corresponding to that number in the stock book, and then testing every tenth animal after that. For example, if 7 were the randomly chosen number, then animals 007, 017, 027 and so on would be tested.

i Name the sampling method suggested by Ellie.

ii State, with a reason, whether a sample chosen in this way is a random sample.

c The animals are kept in many different fields. Adam decides to carry out a quota sample.

i Describe how Adam might carry out a quota sample of size 50.

ii State **one** advantage to Adam of carrying out a quota sample.

iii State **one** disadvantage of carrying out a quota sample.

[© AQA 2012]

10 A test is constructed to see if a coin is biased. It is tossed ten times and if there are 10 heads, 9 heads, 1 head or 0 heads it is declared to be biased.

Which of these is a possible significance level for this test?

Choose from these options.

A 1% **B** 2% **C** 10% **D** 20%

FOCUS ON ... PROOF 4

Using mathematical notation

In Chapter 20 you saw that there are two different formulae for standard deviation:

$$\sqrt{\Sigma\frac{(x-\bar{x})^2}{n}} \quad \text{and} \quad \sqrt{\Sigma\frac{x^2}{n}-\bar{x}^2}$$

Rewind

Recall from Chapter 20 that the symbol σ (lower-case sigma) is used for standard deviation.

In these formulae, x represents the individual data items, n is the number of items, $\bar{x}$ is the mean.

You will learn more about the Σ (upper-case sigma) symbol if you study the A Level course; for now, you just need to know that it means **total** or **sum of.**

You are going to prove that the two formulae are equivalent.

Work with the expression under the square root (this is called the **variance**).

You want to prove:

$$\frac{\Sigma(x-\bar{x})^2}{n}=\Sigma\frac{x^2}{n}-\bar{x}^2$$

PROOF 10

$$\text{LHS}=\frac{\Sigma(x-\bar{x})^2}{n}$$

Start by expanding the brackets on the left and see if you can simplify it to look like the expression on the right.

$$=\frac{\Sigma(x^2-2x\bar{x}+\bar{x}^2)}{n}$$

$$=\frac{\Sigma x^2}{n}-\frac{\Sigma 2x\bar{x}}{n}+\frac{\Sigma\bar{x}^2}{n}$$

The numerator is the sum of three terms. The first one looks like the first term in the expression you are trying to prove, so it seems sensible to split the sum into three parts.

$$=\frac{\Sigma x^2}{n}-2\bar{x}\frac{\Sigma x}{n}+\frac{n\bar{x}^2}{n}$$

Now remember that you are adding up the terms corresponding to different values of x. There are n things to add in each sum.

The mean, $\bar{x}$, is a constant. The sum in the last term just means 'add $\bar{x}^2$ n times', so it equals $n\bar{x}^2$.

In the middle sum, $2\bar{x}$ is a constant multiplying each x, so it can be taken outside the sum.

$$=\frac{\Sigma x^2}{n}-2\bar{x}\bar{x}+\bar{x}^2$$

$$=\frac{\Sigma x^2}{n}-2\bar{x}^2+\bar{x}^2$$

The first term is exactly what you want. The last term equals $\bar{x}^2$, but it has + instead of –.

But look at the middle term: $\frac{\Sigma x}{n}$ means 'add up all the values and divide by n'; this gives the mean, $\bar{x}$.

$$=\frac{\Sigma x^2}{n}-\bar{x}^2$$

= RHS, as required.

You have reached the expression on the right, so the proof is complete.

QUESTIONS

1. **Was the use of Σ notation helpful, or did you find it confusing? Would it be easier to write something like $a+b+c\ldots$ instead? Try re-writing the proof using different notation.**

2. **Somewhat surprisingly, the equivalence of the two standard deviation formulae can be used to prove a pure mathematical result about inequalities.**

   For any real numbers $x_1, x_2, \ldots, x_n$,

   $$\sqrt{\frac{x_1^2+x_2^2+\ldots+x_n^2}{n}} \geqslant \frac{x_1+x_2+\ldots+x_n}{n}.$$

   Try to prove this result.

3. **You know that skewness is a measure of how asymmetrical a distribution is. One measure of skewness is called the Pearson's coefficient of skewness, and has the formula:**

   $$\frac{1}{\sigma^3}\frac{\Sigma(x-\bar{x})^3}{n}$$

   where σ is the standard deviation.

   Prove that this formula is equivalent to $\frac{1}{\sigma^3}\left(\frac{\Sigma x^3}{n}-3\bar{x}\sigma^2-\bar{x}^3\right)$.

FOCUS ON ... PROBLEM SOLVING 4

Experimental design in statistics

Many questions about populations require extremely creative methods to get to an answer. In this section you will look at two examples.

Persuading people to reveal embarrassing traits

Psychologists used to find that when people were asked directly about traits they found embarrassing (such as their drug use, criminal history or infidelity) they would not be truthful. Even anonymous surveys underpredicted the rates compared to a method called **random response**, designed by S.L. Warner in 1965. In this method people were sent to a private booth and asked to flip a coin. If they got a head on the coin flip they were asked to write 'yes', otherwise they were asked to give an honest answer to the question. The researchers could then explain that they could estimate the overall proportion of the population answering 'yes', without knowing the result for any individual who answered 'yes'.

How many adult cod are there in the North Sea?

It is not practical to count all of the fish, so you use a method called **mark and recapture**. A sample of fish is captured and marked, then - some time later - another sample is taken. If it is assumed that the proportion of marked fish in the sample is the same as the proportion in the population, you can estimate the population size.

Did you know?

You could use the internet to research this question – it turns out that the answer varies greatly, depending upon definitions of 'adult', something which caused some media controversy.

QUESTIONS

1 A sample of 1000 people were asked if they had ever stolen goods from a shop. Randomised response was used.

a If 612 people responded 'yes' estimate the proportion of the whole population who had stolen goods from a shop.

b Is this likely to be an underestimate or overestimate?

2 A sample of 10 000 adult cod is caught in the North Sea and marked with microchips. Some time later another sample of 10 000 adult cod is caught and one of the fish is found to have a microchip.

a Use this information to estimate the number of adult cod in the North Sea.

b What assumptions do you have to make in this calculation?

FOCUS ON ... MODELLING 4

Using technology in modelling

Once you have created a probability model you often want to use it to make predictions. This is useful both to check whether the outcomes predicted fit in with known facts and to use the model to influence decisions about future events. Even when the mathematics is very complicated you can use technology to produce a simulation.

For example, the length of a side of a square is chosen at random to be a number between 0 and 10 (not necessarily a whole number), with all values equally likely. This is called a **uniform distribution**. You can use a spreadsheet to do this, using a command such as:

$$=10*\text{RAND}()$$

If this is done over 1000 cells of the spreadsheet you can get a sample of random numbers between 0 and 10. A histogram illustrates one such sample.

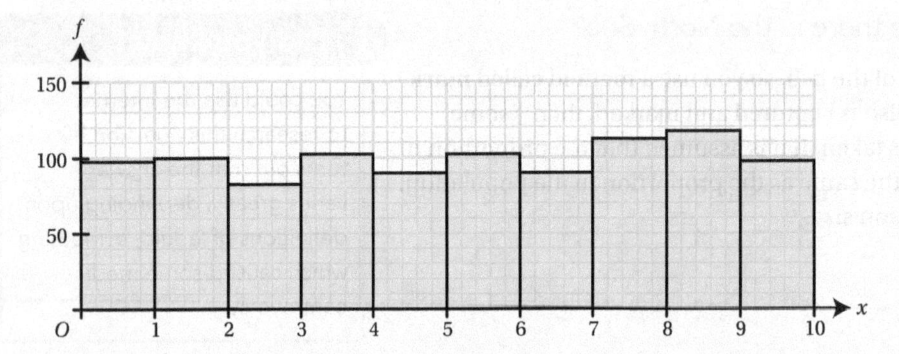

Notice that it is not a perfectly uniform distribution. There are some random variations. If you repeat the experiment you will get a slightly different distribution.

Many people's intuition says that the area should also be a uniform distribution. However, you can square all of your random numbers and form a new histogram:

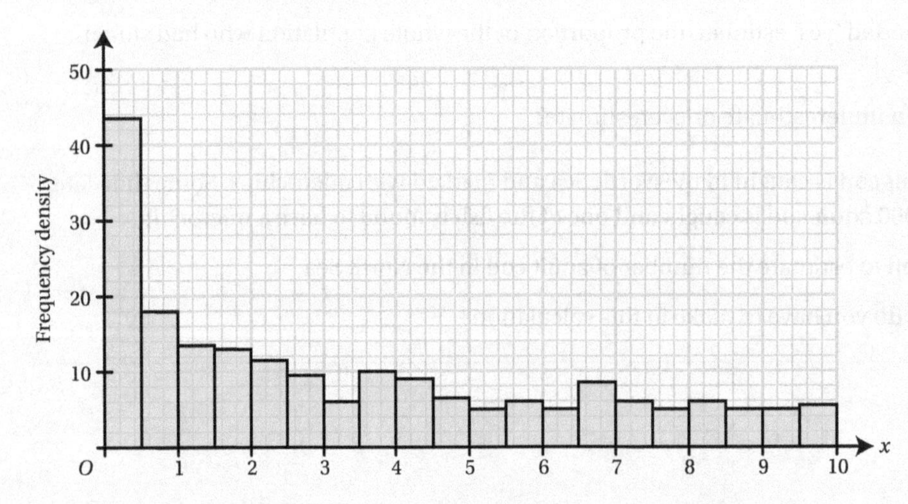

This shows the initially surprising result that the probability of small values of area is significantly higher than that for large values. The probability of the value being less than 50 is about 70.3% and the mean is about 34. It is possible to predict this from theory, but most modern day statistics is done using this type of simulation method – often called **Monte-Carlo simulation** because it uses random numbers.

QUESTIONS

Try to answer these questions by using Monte Carlo simulation.

1. If points are picked at random inside a square of side one unit, what is the probability that they fall within a circle of diameter one unit, centred at the centre of the square? Use the result to estimate a value of π.

2. Every time a man moves he takes a step either left or right, each with probability 0.5. Find his average distance away from his starting point after 100 steps.

3. What is the distribution of a mean of four values taken from the same uniform distribution?

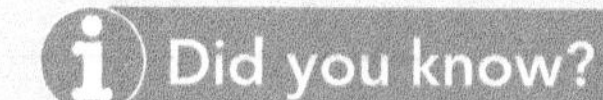

The problem in Question 2 is called a **drunken walk** and it is of huge importance in many areas, from Physics (where it is used to model the movement of atoms in a gas) to finance (where it is used to model the movement of share prices).

CROSS-TOPIC REVIEW EXERCISE 4

1. All students in a class recorded how long, in minutes, it took them to travel to school that morning. The results are summarised in the cumulative frequency table.

Time, t (minutes)	Cumulative frequency
$0 < t \leqslant 6$	0
$0 < t \leqslant 10$	6
$0 < t \leqslant 15$	12
$0 < t \leqslant 20$	22
$0 < t \leqslant 30$	38
$0 < t \leqslant 45$	45

a Copy and complete this frequency table.

Time, t (minutes)	Frequency
$0 < t \leqslant 6$	0
$6 < t \leqslant 10$	6
$10 < t \leqslant 15$	
$15 < t \leqslant 20$	
$20 < t \leqslant 30$	
$30 < t \leqslant 45$	

b Calculate an estimate for the mean and variance of the data.

c Explain why your answer is only an estimate.

2. Rehana wishes to catch a train from her local station to the city centre. She will take a taxi to the station. There are two local taxi companies: Blue Star and Green Star. In the past Rehana has telephoned both companies. The times, in minutes, that she has had to wait between telephoning and the arrival of a taxi are summarised in the box-and-whisker plot.

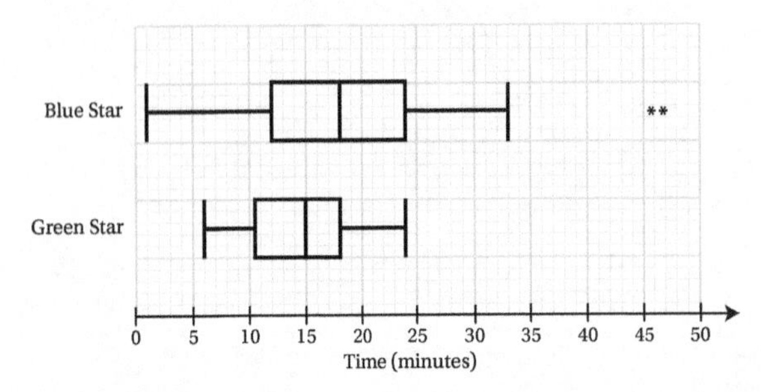

a Compare briefly the waiting times for the two taxi companies.

b Giving a reason for your choice, advise Rehana on which company to telephone if, in order to catch the next train, she needs the taxi to arrive within:

i 5 minutes

ii 25 minutes.

[© AQA 2006]

3 The discrete random variable X has the probability distribution shown in the table.

x	1	2	3	4
$P(X = x)$	$\ln k$	$\ln 2k$	$\ln 3k$	$\ln 4k$

Find the exact value of k.

4 A biased six-sided dice follows this probability distribution.

x	1	2	3	4	5	6
$P(X = x)$	0.12	0.18	p	0.27	0.22	0.08

a Find the probability that the dice lands on either 2 or 3.

b The dice is rolled 12 times. Find the probability that it lands on either 2 or 3 at least five times.

5 What is the probability of getting an average of 3 on two rolls of a fair dice?

6 A newspaper article claimed that fewer than 30 per cent of cyclists stop at zebra crossings when a pedestrian is waiting to use the crossing.

Sabrina collects data at several zebra crossings. She records 20 occasions at random when a cyclist approaches a crossing whilst a pedestrian is waiting to use it. On 4 of these occasions, the cyclist stops.

a Carry out a hypothesis test to investigate the claim made in the newspaper article. Use an exact binomial distribution and the 10% significance level.

b Give one reason why a binomial distribution may not be an appropriate model in this situation.

[© AQA 2011]

7 Three data items are collected: 3, x^2 and x. Find the minimum possible value of the mean.

8 Three data items are collected: 3, 7, x. Find the smallest possible value of the variance.

9 The mean test score of a group of eight students is 34.5 and the variance of the scores is 5.75. Another student with the score of 38 joins the group. Find the new mean and variance of the scores.

10 Aseem records his monthly expenditure over one whole year in order to help him plan his budget. He finds that his average monthly expenditure for the eight months from January to August is £620 and that his average monthly expenditure over the whole year is £586. What was his average monthly expenditure for the four months from September to December?

11 A field study centre is near to a stream of length 1.8 km. The tutors at the centre divide this length into 20 m sections, providing 90 sections where students visiting the centre can collect data about the flow of water in the stream. The sections are numbered from 1 to 90. For the first 840 m, the stream is classed as a 'first order' stream; for the next 360 m, it is classed as a 'second order' stream; and for the remaining sections, it is classed as a 'third order' stream.

A school party visiting the centre is divided into 5 teams of students who are to investigate how the flow of water varies along the stream. Each team will collect data at 3 sections, so 15 different sections must be selected. The collection of data by the teams will be supervised by two teachers from the school and one tutor from the centre.

The students are asked to suggest how the sample of 15 sections should be selected.

a Anders suggests selecting a simple random sample of 15 sections, using random numbers from tables.

i Describe, in detail, how this might be done.

ii State two possible disadvantages, one statistical and one practical, of collecting data from sections selected in this way.

b Barbara suggests rolling a dice to choose the number of the first section to be used, and then selecting every sixth section after that to complete the sample.

i Name this type of sampling.

ii State, with a reason, whether a sample obtained in this way will be **random**.

iii State, with a reason, whether a sample obtained in this way will be **stratified**.

c Caleb suggests choosing a block of 5 consecutive sections along the first order stream and giving one section to each team, then repeating this for the second order and for the third order streams.

i Name this type of sampling.

ii State **two** possible advantages of collecting data from sections selected in this way.

iii State **one** possible disadvantage of collecting data from sections selected in this way.

[© AQA 2011]

12 Greg has a coin with a probability p of showing 'heads' when it is flipped.

a Show that the probability of flipping two or three 'heads' out of three coin-tosses is $f(p) = 3p^2 - 2p^3$.

b Sketch the graph of $y = f(p)$ for $0 \leqslant p \leqslant 1$. Indicate the coordinates of any stationary points.

c The probability of flipping two or three 'heads' out of three coin-tosses is $\frac{5}{32}$. Show that $p = \frac{1}{4}$, and explain whether there are any other solutions.

13 The random variable X has the probability distribution given in the table.

x	1	2	3	4
$P(X = x)$	0.26	e^{-k}	e^{-2k}	0.50

a Find the exact value of k.

b 70 independent observations of X are made. Find the probability that at least 30 of them are $X = 1$.

CROSS-TOPIC REVIEW EXERCISE 4

14 At a building site the probability, $P(A)$, that all materials arrive on time is 0.85. The probability, $P(B)$, that the building will be completed on time is 0.60. The probability that the materials arrive on time and that the building is completed on time is 0.55.

a Show that events A and B are not independent.

b The same company builds 25 buildings.

i Calculate the probability that 20 of them are completed on time.

ii What assumptions did you need to make in your calculation?

c The company made some improvements to their procedures. After this, 20 out of the next 25 buildings were completed on time. Test at the 5% significance level whether the probability of a building being completed on time has increased.

15 The random variable X has the distribution $B(6, p)$ for $0 < p < 1$.

a Given that $P(X = 4) = 3P(X = 3)$, find the value of p.

b Given instead that $P(X = 4) = 3P(X = 2)$, find the value of p.

16 The probability of an event occurring is found to be $\frac{1}{7}(x^2 - 14x + 38)$ where x is known to be an integer parameter. Find all possible values of x.

PRACTICE PAPER 1

Pure mathematics and mechanics

90 minutes, 80 marks

Section A: Pure mathematics

1 The quadratic equation $2kx^2+(2k+3)x+(4k-3)=0$ has a repeated root.

Find the positive value of the constant k.

Choose from these options.

A $\frac{1}{2}$ **B** $\frac{3}{14}$ **C** $\frac{3}{4}$ **D** $\frac{3}{2}$ [1 mark]

2 Given that $\log_b\left(\frac{1}{4}\right)=-4$ find the value of b.

Choose from these options.

A 1 **B** $\sqrt{2}$ **C** -1 **D** 2 [1 mark]

3 Solve the equation $2\cos 2x=1$ for $x\in[0,180°]$. [3 marks]

4 The diagram shows the graph of $y=\mathrm{f}(x)$

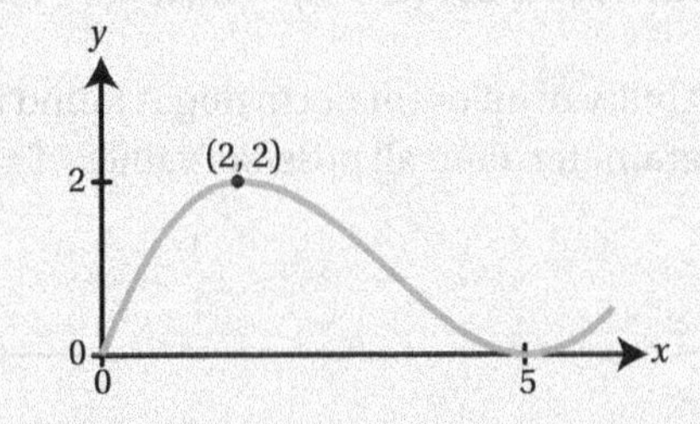

On separate diagrams sketch graphs of:

a $y=\mathrm{f}(2x)$ [2 marks]

b $y=\mathrm{f}'(x)$. [3 marks]

5 Showing all the steps of your method, solve the equation

$3^{2x+1}-28\times3^x+9=0$ [5 marks]

6 A polynomial is defined by $\mathrm{f}(x)=x^3-x^2+ax+12$.

It is given that $(x-2)$ is a factor of $\mathrm{f}(x)$.

a **i** Find the value of a. **ii** Factorise $\mathrm{f}(x)$ completely. [4 marks]

b Hence sketch the graph of $y=\mathrm{f}(x)$, giving coordinates of any axis intercepts. [3 marks]

7 **a** Katya says: 'When you square a prime number, the sum of the digits of the answer is either a prime number or a square number.' (For example, $5^2=25$ and $2+5=7$; $13^2=169, 1+6+9=16$.) Give a counter-example to disprove Katya's statement. [2 marks]

b Prove that the difference of squares of two odd numbers is always a multiple of 4. [4 marks]

8 A scientist models a population of insects using the equation $N=Ae^{kt}$, where N thousand is the number of insects t days after the observations started.

Initially there are 30 000 insects and the population is increasing at a rate of 12 000 insects per day.

a Find the values of A and k. [3 marks]

b Showing all the steps of your method clearly, find how long it will take for the population to reach 1 million. [3 marks]

9 **a** Sketch the graph of $y = \frac{5}{x^2}$, clearly stating the equation of all asymptotes. [3 marks]

b By sketching a second graph on the same axes, show that the curve with equation $y = x^2 - x + \frac{5}{x}$ has only one stationary point. [5 marks]

c Determine whether this is a maximum or minimum point, fully justifying your answer. [3 marks]

10 The tangent to the curve $y = 2x^{\frac{3}{2}} - 36x^{\frac{1}{2}} + 3$ at the point P is parallel to the line $y = 3x + 5$.

Find the x-coordinate of the point P. [6 marks]

Section B: Mechanics

11 A particle moves with velocity $\begin{pmatrix} 3 \\ -4 \end{pmatrix}$ m s^{-1}.

Find the particle's speed. Choose from these options.

A -1 m s^{-1} **B** 1 m s^{-1} **C** $\sqrt{7}$ m s^{-1} **D** 5 m s^{-1} [1 mark]

12 A particle of mass 5 kg moves in a horizontal plane under the action of forces

$\boldsymbol{F}_1 = (23\mathbf{i} - 16\mathbf{j})$ N, $\boldsymbol{F}_2 = (7\mathbf{j})$ N and $\boldsymbol{F}_3 = (11\mathbf{i} + 8\mathbf{j})$ N

Find the acceleration of the particle, giving your answer in the form $(p\mathbf{i} + q\mathbf{j})$ m s^{-2}. [4 marks]

13 In this question use $g = 9.81$ m s^{-2} giving your final answers to an appropriate degree of accuracy.

A ball is projected vertically upwards with a speed of 12.6 m s^{-1}. Assume that air resistance can be ignored.

a Find the speed and the direction of motion of the ball 2.15 seconds after projection. [3 marks]

b How long does the ball take to travel a distance of 15 metres? [5 marks]

c How would your answer to part **b** change if air resistance was included? Give a reason for your answer. [1 mark]

14 In this question use $g = 9.8$ m s^{-2}, giving your final answers to an appropriate degree of accuracy.

A box of mass 12 kg is attached to the roof of a lift by a light inextensible string.

The mass of the lift is 450 kg and it is supported by a cable which is modelled as light and inextensible.

The tension in the cable is 5300 N.

Find the tension in the string. [5 marks]

15 A particle moves in a straight line. Its velocity, v m s^{-1}, at time t seconds is given by the equation

$$v(t) = \begin{cases} 9.6 + 4.2t - 0.3t^2 & \text{for } 0 \leqslant t \leqslant 7 \\ 52.65 - 4.05t & \text{for } t > 7 \end{cases}$$

a Calculate the acceleration of the particle when $t = 9$. [2 marks]

b Find the time at which the particle changes direction. [2 marks]

c Find how long it takes for the particle to travel 190 m. [6 marks]

PRACTICE PAPER 2

Pure mathematics and statistics

90 minutes, 80 marks

Section A: Pure mathematics

1 The function $y = f(x)$ has a turning point at $(6, -3)$.

Find the coordinates of the turning point of $y = f(3x)$.

Choose from these options.

A $(2, -1)$ **B** $(2, -3)$ **C** $(18, -3)$ **D** $(18, -9)$ [1 mark]

2 A curve has equation $y = \dfrac{3}{\sqrt[3]{x}}$.

Find $\dfrac{dy}{dx}$.

Choose from these options.

A $\dfrac{\sqrt[3]{x}}{3}$ **B** $\dfrac{1}{x\sqrt[3]{x}}$ **C** $-\dfrac{1}{x\sqrt[3]{x}}$ **D** $-\dfrac{3}{x\sqrt[3]{x}}$ [1 mark]

3 Find the first three terms in ascending powers of x in the expansion of

$$\left(2-\frac{x}{3}\right)^6$$

Give each term in its simplest form. [4 marks]

4 Find, as a single fraction in terms of a,

$$\int_a^{2a}\left(2-\frac{4}{x^2}\right)dx.$$ [4 marks]

5 Showing all the steps of your method, find the exact solution of the equation

$\log_3(2x+1) = \log_3(x-2)+2$. [5 marks]

6 Points M, N and P have position vectors $\mathbf{m} = 6\mathbf{i} - 3\mathbf{j}$, $\mathbf{n} = 2\mathbf{j}$ and $\mathbf{p} = \mathbf{j} - 5\mathbf{i}$.

a Point Q is such that $MNQP$ is a parallelogram. Find the position vector of Q. [3 marks]

b Find the exact magnitude of the vector $\mathbf{v} = \overrightarrow{MN} + \overrightarrow{MP}$. [3 marks]

7 Prove from first principles that the derivative of $3x^2$ is $6x$. [5 marks]

8 **a** Given that $90° < x < 180°$ and $2\sin^2 x + 2 = 3\cos^2 x$, find the exact value of $\sin x$. [3 marks]

b Hence show that $\tan x = \dfrac{k}{2}$, where k is an integer to be found. [3 marks]

9 In the diagram, ABC is a right angle, $AC = 1$ unit and angle $BAC = \alpha$. Points D and E are one the line AC such that $CD = CE = CB$.

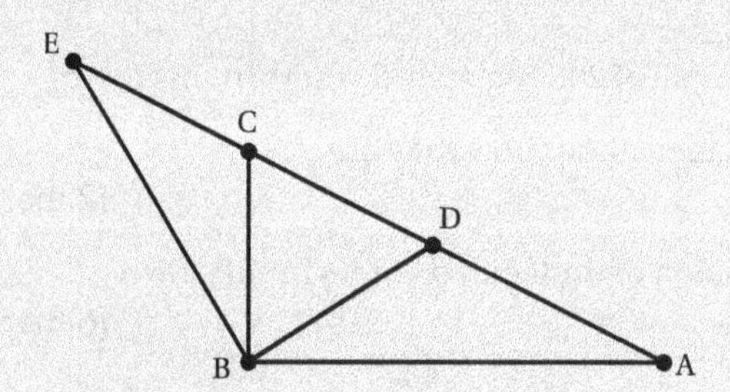

a Express the length AB and BC in terms of α. [2 marks]

b Hence prove that $\frac{1}{AD} + \frac{1}{AE} = \frac{2}{AB^2}$. [5 marks]

10 **a** Find the coordinates of the intersection points of the graphs of $y = 12x - 2x^2$ and $2y = 6 - x$. [4 marks]

b Illustrate the region determined by the inequalities $y \leqslant 12x - 2x^2$ and $x + 2y \geqslant 6$ on a graph. Leave the required region unshaded. [4 marks]

c The inequality $x^2 + px + q \geqslant 0$ is satisfied for $x \in (-\infty, -3] \cup [5, \infty)$. Find the values of p and q. [3 marks]

11 Points A and C have coordinates $(-8, 0)$ and $(0, 20)$ respectively.

Point B lies on the x-axis and CB is perpendicular to AC.

a Find the coordinates of B. [3 marks]

b Hence find the equation of the circle that passes through A, B and C. [4 marks]

Section B: Statistics

12 The correlation coefficient between two variables was found to be -0.8.

Which one of these is a possible equation of the line of best fit?

A $y = 2x$ **B** $y = -2x$ **C** $y = -0.8$ **D** $x = -0.8$ [1 mark]

13 Elena is sometimes late for school, but never more than three times in a week. For any week, the number of days she is late has this probability distribution.

Days late	0	1	2	3
Probability	0.50	0.25	0.15	0.10

Find the probability that, in two randomly chosen weeks, Elena is late for school a total of four times. [4 marks]

14 Pre-election polls suggest that 35% of voters in a certain town are undecided about how they are going to vote. Following a televised debate, a survey is carried out to find out whether the proportion of undecided voters has decreased.

a Describe briefly how to select a simple random sample of 200 registered voters in this town.

In a random sample of 200 voters, 62 are still undecided about how they are going to vote. [2 marks]

b Test, at a 5% significance level, whether the proportion of undecided voters in this town has decreased. [6 marks]

15 The scatter graph shows data about average age and the percentage of people who cycle to work. The information is from a survey done in 2015. Each data point represents one local authority. The data shown is for all 250 local authorities that provided the information.

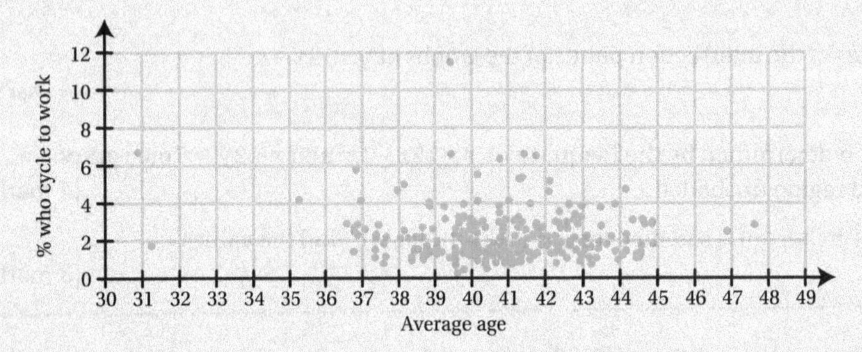

a Describe the correlation between average age and the percentage of people who cycle to work. [1 mark]

The data for the percentage of people who cycle to work, $p\%$, is summarised as:

$\sum p = 561, \sum p^2 = 1699$

b Calculate the mean of the percentage of people who cycle to work, and show that the standard deviation is 1.33. [5 marks]

c Hence identify the outliers in the 'percentage who cycle to work' data. Show your method and circle the outliers on the graph. In this question, outliers are defined as data values that are more than 3 standard deviations from the mean. [3 marks]

d One local authority has a particularly high percentage of people who cycle to work. Is this local authority more likely to be in a built up area or the countryside? Explain your answer. [1 mark]

FORMULAE

Pure mathematics

Binomial series

$$(a+b)^n = a^n + \binom{n}{1}a^{n-1}b + \binom{n}{2}a^{n-2}b^2 + \ldots + \binom{n}{r}a^{n-r}b^r + \ldots + b^n \quad (n \in \mathbb{Z}^+)$$

$$\text{where } \binom{n}{r} = {}^nC_r = \frac{n!}{r!(n-r)!}$$

$$(1+x)^n = 1 + nx + \frac{n(n-1)}{1.2}x^2 + \ldots + \frac{n(n-1)\ldots(n-r+1)}{1.2\ldots r}x^r + \ldots \quad (|x|<1, n \in \mathbb{R})$$

Arithmetic series

$$S_n = \frac{1}{2}n(a+l) = \frac{1}{2}n[2a+(n-1)d]$$

Geometic series

$$S_n = \frac{a(1-r^n)}{1-r}$$

$$S_\infty = \frac{a}{1-r} \quad \text{for} \quad |r|<1$$

Trigonometry: small angles

For small angle θ,

$$\sin\theta \approx \theta$$

$$\cos\theta \approx 1 - \frac{\theta^2}{2}$$

$$\tan\theta \approx \theta$$

Trigonometric identities

$$\sin(A \pm B) = \sin A\cos B \pm \cos A\sin B$$

$$\cos(A \pm B) = \cos A\cos B \mp \sin A\sin B$$

$$\tan(A \pm B) = \frac{\tan A \pm \tan B}{1 \mp \tan A\tan B} \quad \left(A \pm B \neq \left(k+\frac{1}{2}\right)\pi\right)$$

$$\sin A + \sin B = 2\sin\frac{A+B}{2}\cos\frac{A-B}{2}$$

$$\sin A - \sin B = 2\cos\frac{A+B}{2}\sin\frac{A-B}{2}$$

$$\cos A + \cos B = 2\cos\frac{A+B}{2}\cos\frac{A-B}{2}$$

$$\cos A - \cos B = 2\sin\frac{A+B}{2}\sin\frac{A-B}{2}$$

FORMULAE

Differentiation

$\mathbf{f}(x)$	$\mathbf{f}'(x)$
$\tan kx$	$k\sec^2 kx$
$\operatorname{cosec} x$	$-\operatorname{cosec} x \cot x$
$\sec x$	$\sec x \tan x$
$\cot x$	$-\operatorname{cosec}^2 x$
$\dfrac{\mathrm{f}(x)}{\mathrm{g}(x)}$	$\dfrac{\mathrm{f}'(x)\mathrm{g}(x)-\mathrm{f}(x)\mathrm{g}'(x)}{(\mathrm{g}(x))^2}$

Integration

$$\int u\frac{\mathrm{d}v}{\mathrm{d}x}\,\mathrm{d}x = uv - \int v\frac{\mathrm{d}u}{\mathrm{d}x}\,\mathrm{d}x$$

(+ constant; $a > 0$ where relevant)

$\mathbf{f}(x)$	$\int \mathbf{f}(x)\,\mathrm{d}x$
$\tan x$	$\ln\lvert\sec x\rvert$
$\cot x$	$\ln\lvert\sin x\rvert$
$\operatorname{cosec} x$	$-\ln\lvert\operatorname{cosec} x + \cot x\rvert = \ln\lvert\tan \frac{1}{2}x\rvert$
$\sec x$	$\ln\lvert\sec x + \tan x\rvert = \ln\lvert\tan(\frac{1}{2}x + \frac{1}{4}\pi)\rvert$
$\sec^2 kx$	$\frac{1}{k}\tan kx$

Numerical solution of equations

The Newton-Raphson iteration for solving $\mathrm{f}(x) = 0$: $x_{n+1} = x_n - \dfrac{\mathrm{f}(x_n)}{\mathrm{f}'(x_n)}$

Numerical integration

The trapezium rule: $\int_a^b y\,\mathrm{d}x \approx \frac{1}{2}h\{(y_0 + y_n) + 2(y_1 + y_2 + \ldots + y_{n-1})\}$, where $h = \dfrac{b-a}{n}$

Mechanics

Constant acceleration

$s = ut + \frac{1}{2}at^2$	$\mathbf{s} = \mathbf{u}t + \frac{1}{2}\mathbf{a}t^2$
$s = vt - \frac{1}{2}at^2$	$\mathbf{s} = \mathbf{v}t - \frac{1}{2}\mathbf{a}t^2$
$v = u + at$	$\mathbf{v} = \mathbf{u} + \mathbf{a}t$
$s = \frac{1}{2}(u+v)t$	$\mathbf{s} = \frac{1}{2}(\mathbf{u} + \mathbf{v})t$

$v^2 = u^2 + 2as$

FORMULAE

Probability and statistics

Probability

$P(A \cup B) = P(A) + P(B) - P(A \cap B)$

$P(A \cap B) = P(A) \times P(B \mid A)$

Standard discrete distributions

Distribution of X	$P(X = x)$	Mean	Variance
Binomial $B(n, p)$	$\binom{n}{x} p^x (1-p)^{n-x}$	np	$np(1-np)$

Sampling distributions

For a random sample $X_1, X_2, \ldots, X_n$ of n independent observations from a distribution having mean μ and variance σ^2:

S^2 is an unbiased estimator of σ^2, where $S^2 = \frac{\Sigma(X_i - \bar{X})^2}{n-1}$

Answers

1 Proof and mathematical communication

BEFORE YOU START

1 3

2 $(2x+1)(2x-1)$

3 a 180° b 360°

4 π and $\sqrt{2}$

5 15

EXERCISE 1A

1 a i $\Leftarrow$ ii $\Leftarrow$

b i $\Rightarrow$ ii $\Rightarrow$

c i $\Leftrightarrow$ ii $\Leftrightarrow$

d i $\Leftarrow$ ii $\Leftrightarrow$

e i $\Leftarrow$ ii $\Rightarrow$

f i $\Leftrightarrow$ ii $\Leftarrow$

g i No implication. ii No implication.

2 $a=2, b=3$

3 $a=2, b=8$

4 Equals sign is used to match up non-equal values.

5 Not all lines can be connected with $\Leftrightarrow$.

6 a $\sqrt{x^2+9}=3x-7 \quad \Rightarrow x^2+9=9x^2-42x+49$

$\Leftrightarrow 0=8x^2-42x+40$

$\Leftrightarrow 0=4x^2-21x+20$

$\Leftrightarrow 0=(4x-5)(x-4)$

$\Leftrightarrow x=\frac{5}{4}$ or $x=4$

b $x=\frac{5}{4}$ is the solution to $\sqrt{x^2+9}=-(3x-7)$, which gives the same second line of working.

7 a $\Rightarrow, \Leftrightarrow, \Leftrightarrow, \Leftrightarrow$

b First line not reversible.

8 a $\Leftrightarrow, \Leftarrow$

b Dividing by the factor $x+3$ means that the solution $x=-3$ is lost.

9 All is fine until cancelling $(a-b)$. Since $a=b$, this is equivalent to dividing both sides by zero.

10 It depends on exactly how you define the word or. In formal logical proof the terms OR and XOR have two different meanings.

EXERCISE 1B

1 a i $\{x : x > 7\}; (7, \infty)$

ii $\{x : x < 6\}; (-\infty, 6)$

b i $\{x : x \leqslant 10\}; (-\infty, 10]$

ii $\{x : x \geqslant 5\}; [5, \infty)$

c i $\{x : x > 0\} \cap \{x : x \leqslant 1\}; (0, 1]$

ii $\{x : x > 5\} \cap \{x : x < 7\}; (5, 7)$

d i $\{x : x > 5\} \cup \{x : x \leqslant 0\}$; $(-\infty, 0] \cup (5, \infty)$

ii $\{x : x \geqslant 10\} \cup \{x : x < 2\}$; $(-\infty, 2) \cup [10, \infty)$

2 a i $1 \leqslant x < 4$ ii $2 < x \leqslant 8$

b i $1 \leqslant x \leqslant 3$ ii $2 < x < 4$

c i $x < 5$ ii $x \geqslant 12$

d i $8 \leqslant x < 10$ ii $3 \leqslant x < 4$

EXERCISE 1C

1 For example, $x = 4$.

2 For example, $x = 45°$.

3 For example, $x = -1$.

4 For example, 2×3.

5 For example, 4.

6 For example, $-\pi + \pi$.

7 For example, $x = -4$.

8 For example, $n = 41$.

9 For example, skew lines.

EXERCISE 1D

1 Use $n = 2m+1$.

2 Use $a = 2n$ and $b = 2m+1$.

3 Use $n, n+1, n+2$.

4 a Use $5n, 5(n+1)$. b Use $5n, 5(n+1)$.

5 Consider areas.

6 Either use an argument with exterior angles or divide the hexagon into triangles.

7 Use $n = 3m+2$.

8 a x^2+4x+4 b Write as $y=(x+2)^2+6$.

9 Label the exterior angle x and use knowledge of angles on a straight line and interior angles of a triangle.

10 Factorise to $(n+1)(n+2)$.

11 a Position value.

b Consider $n-(a+b+c+d)$.

c Consider $n-(a-b+c-d)$.

12 Consider rationality of $\sqrt{2}^{\sqrt{2}}$.

EXERCISE 1E

1 Proof.

2 Proof.

3 Proof.

4 Proof.

5 Consider all single digit positive integers separately.

6 Consider all cases.

7 Use proof by exhaustion, for even or odd n.

8 Use proof by exhaustion for $n=5m+i$ for $i=0,1,2,3,4$.

9 Factorise and use proof by exhaustion for $n=3m+i$ for $i=0,1,2$.

MIXED PRACTICE 1

1 D

2 Use $2m+1$, $2n+1$.

3 Use $n=2m$.

4 Use a non-reduced fraction.

5 Consider for example $3+(-2)$ and $3-(-2)$.

6 Use definition of rational numbers.

7 Subdivide the n-gon as for Exerise 1D Q6.

8 $a=1, b=-1, c=1$

9 Factorise.

10 Construct further lines to find isosceles triangles; construct a proof using knowledge of angles in triangles.

11 Consider n^3 where $n=3k$ or $n=3k\pm1$.

12 a This does not work in the forward direction. For example, $a=4, b=0.5$.

b This does not work in either direction, for example, $a=\sqrt{2}, b=\sqrt{2}$ so $ab=2$; or $ab=\pi, a=\pi, b=1$.

13 Use $n-1$, n and $n+1$.

14 Use $2m+1$ and $2n+1$.

15 a For example, $n=1601$.

b Factorise.

16 Factorise and use an exhaustive proof considering whether $a+b$ is odd or even.

2 Indices and surds

BEFORE YOU START

1 24

2 3

3 a x^6 b x^7 c x^5

4 a $3x^{-2}$ b $x^{-\frac{1}{3}}$ c 1

5 $2-y+2x-xy$

6 $4a^2-b^2$

WORK IT OUT 2.1

Solution 1 is correct.

EXERCISE 2A

1 a i x^3 ii x^{12}

b i $2x^5$ ii $\frac{1}{2x^4}$

c i $\frac{4}{3x^3}$ ii $\frac{y^{12}}{x^6}$

2 a i $x=\frac{1}{4}$ ii $x=\frac{4}{9}$

b i $x=2$ ii $x=4.5$

3 $\frac{1}{10x^2}$

4 x

5 $2p^2q^{\frac{1}{3}}$

6 $x^{-\frac{1}{2}}-x^{\frac{5}{2}}$

7 $4x^{-\frac{2}{3}}-x^{\frac{4}{3}}$

8 $\frac{3-4b^2}{2b^3}$

9 5×10^{-4}

10 a $\frac{1}{3}$ b 16 cm^2

11 8 cm

12 $a = \frac{3b}{4}$

13 $\frac{3x^3}{2}$

14 $a = 9, b = 1, c = -6, n = \frac{3}{2}$

15 $b = 1, a = \frac{3}{2}$

16 1

17 16

WORK IT OUT 2.2

Solution 2 is correct.

EXERCISE 2B

1 a i $5\sqrt{5}$ ii $2\sqrt{5}$

b i $5\sqrt{5}$ ii $7\sqrt{5}$

c i $2\sqrt{5}$ ii $26\sqrt{5}$

2 a i $\sqrt{32}$ ii $\sqrt{300}$

b i $\sqrt{63}$ ii $\sqrt{80}$

c i $\sqrt{108}$ ii $\sqrt{72}$

3 a i $3+5\sqrt{3}$ ii $2\sqrt{3}$

b i $-4+3\sqrt{3}$ ii $5+3\sqrt{3}$

c i $4+2\sqrt{3}$ ii $7-4\sqrt{3}$

4 a i $\sqrt{7}$ ii $\frac{2\sqrt{5}}{5}$

b i $\frac{\sqrt{6}-2}{2}$ ii $\frac{\sqrt{6}+3\sqrt{2}}{3}$

c i $1+\sqrt{2}$ ii $\frac{\sqrt{7}-\sqrt{5}+\sqrt{35}-1}{6}$

5 $\frac{2}{1-n}$

6 $a = 5, b = 3$

7 $a = 19, b = 13$

8 $(3\sqrt{2})^2 = 18$ and $(2\sqrt{3})^2 = 12$

9 $x = -4$

10 $\frac{2\sqrt{n}+3}{4n-9}$

11 $15n^2 + 5 - 10n\sqrt{3}$

12 a $-ab + (b^2 - a^2)\sqrt{2}$ b $\sqrt{3(a^2+b^2)}$

13 $x = 3, y = 21$

14 a $\sqrt{48}$ b $\sqrt{27} - \sqrt{20} > \sqrt{5} - \sqrt{3}$

15 a Proof. b $\sqrt[3]{9} + \sqrt[3]{6} + \sqrt[3]{4}$

16 No; not if $x < 0$.

MIXED PRACTICE 2

1 B

2 D

3 $n^2 + 5 + 2n\sqrt{5}$

4 $z = 9x^3$

5 $\sqrt{2} + \sqrt{7}$

6 $\frac{1}{16}x^2$

7 $x = \frac{4}{5}$

8 $x^{-\frac{3}{4}} - x^{\frac{5}{4}}$

9 B

10 a i 3 ii −3 iii $\frac{1}{2}$

b $-\frac{7}{2}$

11 $\frac{3}{4x^2}$

12 $\frac{1+n+2\sqrt{n}}{n-1}$

13 A

14 a $a^2 + 2b^2 + 2\sqrt{2}ab$

b Proof.

15 a $a^2 + b^2 + 2ab\sqrt{5}$

b, c Proof.

d i Because $\sqrt{5}$ is closer to 2 than to 3.

ii Because $4 - \sqrt{5} > 1$ so large powers get further away from zero.

3 Quadratic functions

BEFORE YOU START

1 $6x^2 - 7x - 3$

2 a $x = -5, 4$ b $x = -8, \frac{1}{2}$

c $x = 0, \frac{3}{5}$ d $x = \pm\frac{3}{2}$

3 a $2 \pm \sqrt{2}$ b $x = \frac{5 \pm \sqrt{35}}{2}$

4 $x > 2$

EXERCISE 3A

1 a i $2, -\frac{3}{2}$ ii $5, \frac{2}{3}$

b i $\frac{4}{3}$ ii $-\frac{5}{4}$

c i $5, -4$ ii $\frac{5}{2}, -3$

d i $6, -\frac{1}{2}$ ii $\frac{4}{3}, 1$

2 a i $x = \frac{3 \pm \sqrt{5}}{2}$ ii $x = \frac{1 \pm \sqrt{5}}{2}$

b i $x = \frac{-3 \pm \sqrt{11}}{2}$ ii $x = \frac{9 \pm \sqrt{21}}{10}$

c i $x = -2 \pm \sqrt{6}$ ii $x = -1, \frac{4}{3}$

d i $x = \frac{3 \pm \sqrt{7}}{2}$ ii $x = 2 \pm \sqrt{7}$

3 No real solution.

4 $\frac{4}{3}, -\frac{1}{2}$

5 $x = \frac{3 \pm \sqrt{5}}{2}$

6 $2k, 4k$

7 $x = \frac{-b \pm \sqrt{b^2 - 4a(c-y)}}{2a}$

8 $k = \pm 9$

EXERCISE 3B

1 a A i, B iii, C ii

b A ii, B iii, C i

2 a i

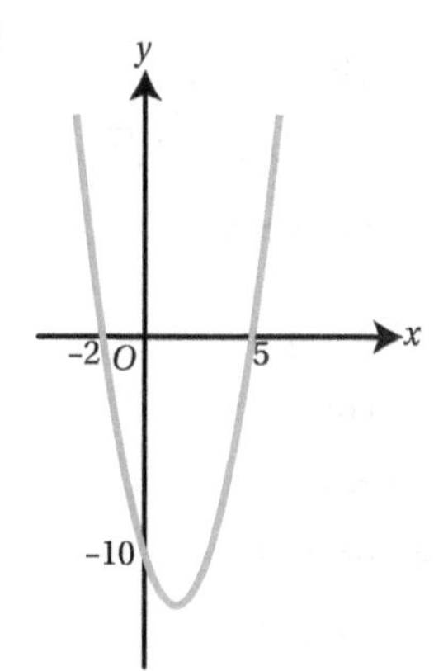

ii

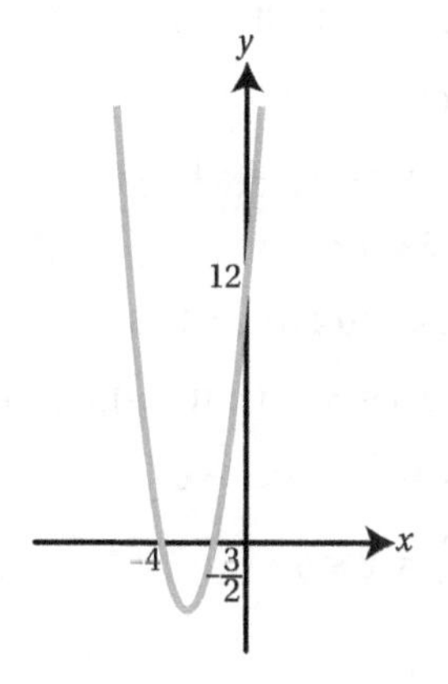

b i

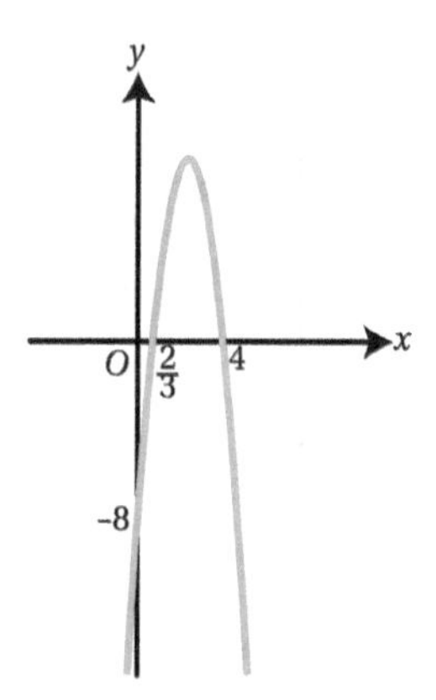

ii

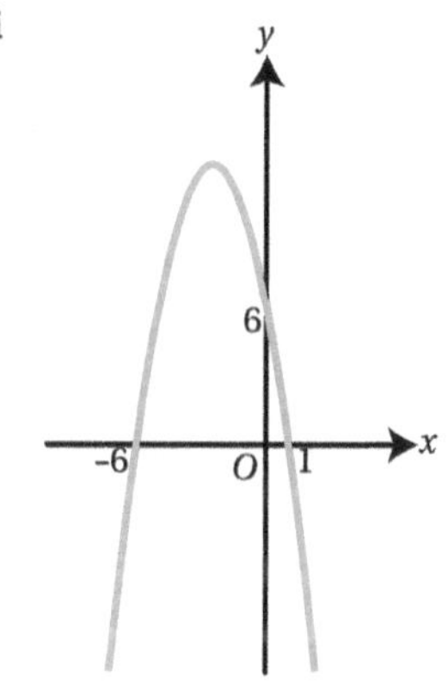

c i

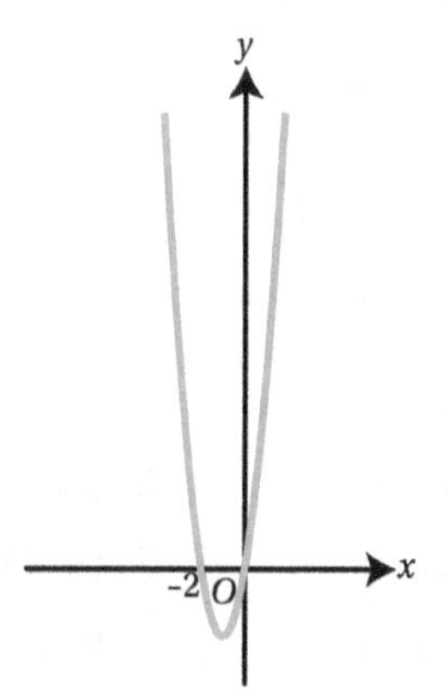

ii

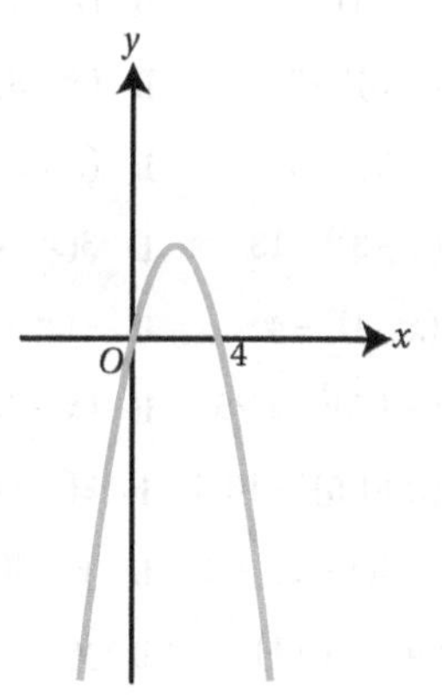

d **i**

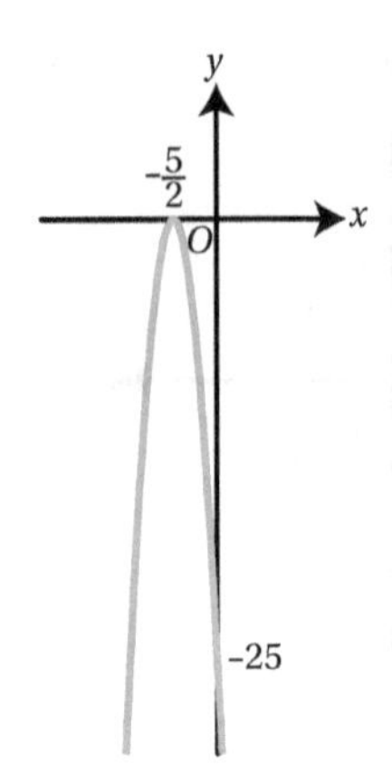

ii

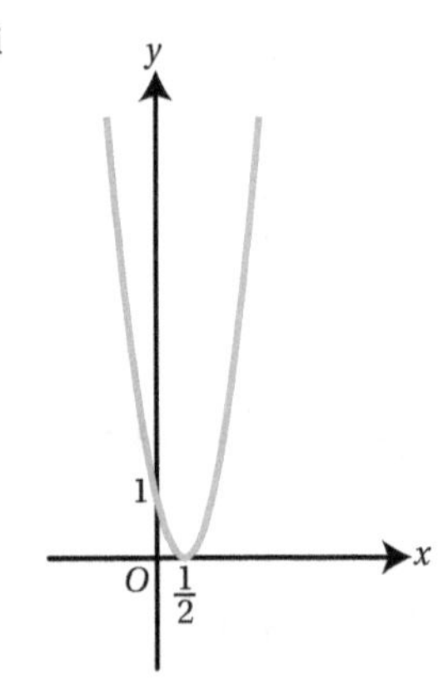

3 **a** **i** $y = 3x^2 - 15x + 12$ **ii** $y = 4x^2 + 4x - 8$

b **i** $y = -2x^2 - 2x + 4$ **ii** $y = -x^2 - 6x - 5$

WORK IT OUT 3.1

Solution 2 is correct.

EXERCISE 3C

1 **a** **i** $(3, 4)$ **ii** $(5, 1)$

b **i** $(2, -1)$ **ii** $(1, -5)$

c **i** $(-1, 3)$ **ii** $(-7, -3)$

d **i** $(-2, -4)$ **ii** $(-1, 5)$

2 **a** **i** $(x-3)^2 - 5$ **ii** $(x-5)^2 - 4$

b **i** $(x+2)^2 - 3$ **ii** $(x+3)^2 - 12$

c **i** $2(x-3)^2 - 13$ **ii** $3(x+1)^2 + 7$

d **i** $-(x-1)^2 - 4$ **ii** $-(x+2)^2 + 5$

e **i** $(x+1.5)^2 - 1.25$ **ii** $(x-2.5)^2 + 3.75$

f **i** $2(x+1.5)^2 + 10.5$ **ii** $2(x-1.25)^2 - 4.125$

3 **a** **i** $y = 2(x-2)^2 + 4$ **ii** $y = 3(x+1)^2 - 5$

b **i** $y = -2(x+1)^2 + 3$ **ii** $y = -3(x-2)^2 + 4$

4 **a** $y = (x-3)^2 + 2$ **b** 2

5 **a** $b = -3, c = 6$ **b** $a = 2$

6 **a** $y = -5 - (x-4)^2$ **b** $(4, -5)$

c Maximum is below x-axis so curve never intercepts x-axis.

7 **a** $2(x+1)^2 - 3$ **b** $x = -1 \pm \sqrt{1.5}$

c

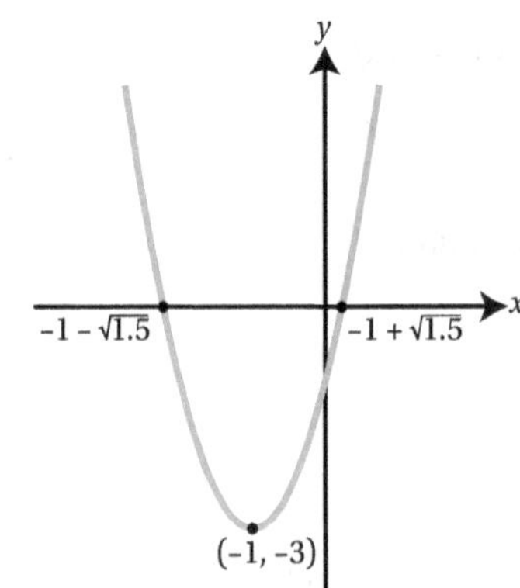

d $x = -1$

8 **a** $(x+3)^2 - 9$ **b** $p \geqslant -9$

EXERCISE 3D

1 **a** **i** $-2\sqrt{2} \leqslant x \leqslant 2\sqrt{2}$

ii $-\sqrt{5} < x < \sqrt{5}$

b **i** $x < -\sqrt{6}$ or $x > \sqrt{6}$

ii $x \leqslant -2\sqrt{3}$ or $x \geqslant 2\sqrt{3}$

c **i** $x < -1$ or $x > 4$

ii $-\frac{2}{3} < x > \frac{5}{2}$

d **i** $x < -1$ or $x > 3$

ii $2 < x < 4$

e **i** $x < 3$ or $x > 12$

ii $-2 < x < 2$

2 **a** **i** $2 < x < 3$ **ii** $-3 < x < 2$

b **i** $x \leqslant -2$ or $x \geqslant 6$

ii $x \leqslant -6$ or $x \geqslant -1$

c **i** $-3 < x < 4$

ii $x < -10.4$ or $x > 1.35$

d **i** $-2.5 \leqslant x \leqslant 1$ **ii** $-1 \leqslant x \leqslant -0.2$

e **i** $x \leqslant -0.5$ or $x \geqslant 6$

ii $\frac{1}{3} \leqslant x \leqslant 5$

3 $x < -2$ or $x > 1.5$

4 $-\frac{5}{3} \leqslant x \leqslant 2$

5 0.904 s

6 a i $x < 2.5$ ii $0.5 < x < 5$

b $0.5 < x < 2.5$

7 $x \leqslant -1$ or $1 \leqslant x < 2$ or $x > 3$

8 $-\frac{2}{5} \leqslant x \leqslant 0$ or $2 \leqslant x \leqslant 3$

9 $5 \leqslant n \leqslant 38$

EXERCISE 3E

1 a i 36 ii 68

b i −47 ii −119

c i 0 ii 0

d i 49 ii 49

2 a i, ii Two b i, ii None

c i, ii One d i, ii Two

3 a i $k < \frac{1}{24}$ ii $k > -\frac{25}{12}$

b i $k = \frac{3}{5}$ ii $k = -\frac{17}{24}$

c i $k \geqslant -\frac{5}{4}$ ii $k \leqslant \frac{1}{16}$

d i $k > \frac{9}{24}$ ii $k < -\frac{25}{12}$

e i $k = \frac{17}{4}$ ii $k = \frac{55}{32}$

f i $k = 1$ ii $k = \frac{1}{32}$

g i $k < 0$ ii $k < 0$

4 $m = \pm\sqrt{2}$

5 $k = \frac{11 \pm 2\sqrt{30}}{2}$

6 $c \geqslant \frac{17}{16}$

7 $0 < k < 6$

8 $-9 < k < -1$

9 $m \leqslant -8$ or $m \geqslant 0$

10 $m < -\frac{9}{16}$

EXERCISE 3F

1 a i $a = \pm\sqrt{3}, \pm\sqrt{7}$ ii $x = \pm 2, \pm\sqrt{3}$

b i $x = -\sqrt[3]{5}, \sqrt[3]{1.5}$ ii $a = 1, -2$

c i $x = \pm\sqrt{2+\sqrt{6}}$ ii $x = \pm\sqrt{6}$

d i $x = 4, 16$ ii $x = 16, 36$

e i $x = 1, 2$ ii $x = 0, 4$

2 $x = 9$

3 $x = \pm 3, \pm 1$

4 $x = 4, 1$

5 a Proof.

b $x = 0, 2$

6 $x = 0, 1$

7 $x = -1, 3$

8 $x = 16$

9 a $x = \pm\sqrt{\frac{b \pm \sqrt{b^2 - 4c}}{2}}$

b $b^2 = 4c, b > 0$

c $c = 0, b \neq 0$

MIXED PRACTICE 3

1 A

2 D

3 $x = 1, 6$

4 $x = \pm 1, \pm 2$

5 a Minimum b $a = 3, b = 7$

6 $a = -3, b = 2, c = 48$

7 a $p = 1, q = 4$ b 1.5

8 a i $25 - (x+3)^2$ ii 25

b i $(x+8)(2-x)$

ii

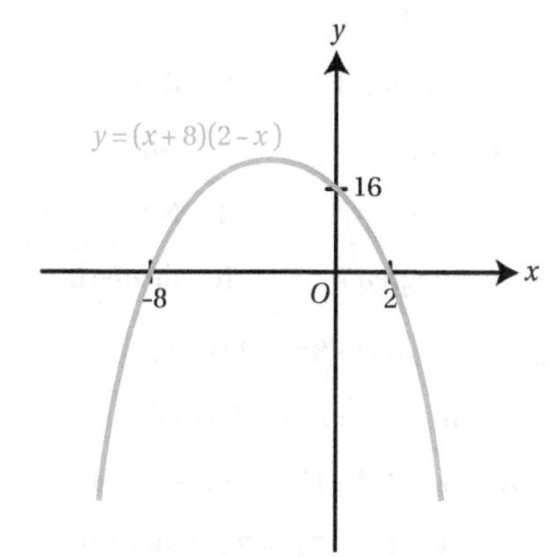

9 A

10 $x \geqslant 3$ or $x < 0$

11

Expression	Positive	Negative	zero
a		✓	
c		✓	
$b^2 - 4ac$			✓
b	✓		

12 a $(x-5)^2+10$ b $\frac{1}{1000}$

13 $3-2\sqrt{2} < k < 3+2\sqrt{2}$

14 1, 16

15 a i $2(x-5)^2+3; p=5, q=3$

ii The minimum is above the x-axis.

b i Proof.

ii $-\frac{1}{7} < k < 1$

16 3, 2

17 $k = \pm 5$

18 a $k-1, 1$ b $-3, 5$

19 Proof.

20 a Proof.

b 6 km

4 Polynomials

BEFORE YOU START

1 $2x^2-5x-3$

2 $(x-3)(x-5)$

3 $x=-2\pm\sqrt{2}$

4 x^6

EXERCISE 4A

1 a degree 3 b degree 5

c No d No

e No f No

g degree 7 h degree 0

2 a i $6x^3+8x^2-29x+14$

ii $3x^3+16x^2+23x+6$

b i $2x^4-15x^3+4x^2+4x-1$

ii $2x^4-7x^3-30x^2+6x+15$

c i $b^4+b^3-3b^2+14b-4$

ii $r^4-11r^3+33r^2-62r+14$

d i $-x^6+2x^5+5x^4-10x^3-x^2+5$

ii $-x^6+2x^4+x^3-x^2-x$

3 Expand: for coefficients and constants.
Factorise: for roots and common factors.

4 a Yes b No

EXERCISE 4B

1 a i x^2+3x+5 ii x^2-2x+9

b i x^2+5x+5 ii x^2-2x-4

c i x^2-2x+9 ii x^2+3x+5

2 a i x^2-8x-3 ii x^2-2x+3

b i x^2+2x+2 ii x^2+3x+4

c i x^2-2 ii x^2+3

EXERCISE 4C

1 a i No ii Yes

b i No ii No

c i Yes ii No

d i Yes ii No

2 a i $(x-2)(x+1)(x+3)$

ii $(x-2)(x-1)(x+1)$

b i $(x-2)(x-1)(x+3)$

ii $(x-2)(x+1)^2$

c i $(x-2)^2(x+2)$ ii $(x-2)^3$

d i $(x-2)(2x+3)(x-4)$

ii $(x-2)(3x-1)(x+6)$

3 a i $(x+1)(x-1)(x+2)$

ii $(x+1)(x-2)(x+2)$

b i $(x-2)^2(x-3)$ ii $(x+2)^3$

c i $(x-1)(x^2-2x+10)$

ii $(x-3)(x^2+x+5)$

d i $(x-1)(2x-1)(3x-1)$

ii $(x+2)(4x+3)(3x-5)$

4 a i $x=-3, 1, 4$

ii $x=-1, -3, 5$

b **i** $x = 2, \frac{3 \pm \sqrt{5}}{2}$

ii $x = 1, \frac{5 \pm \sqrt{17}}{2}$

c **i** $x = -2$ **ii** $x = 2$

5 **a** Proof.

b $p(x) = (x-2)(x-4)(x+3)$; $x = 2, 4, -3$

6 **a, b** Proof.

7 $c = 14, d = 8$

8 **a** $a = -4, b = -17$ **b** $(x+4)$

9 **a** $(x+1)$ **b** $x = -1, 4 \pm \sqrt{2}$

10 0, 4

11 $k = -\frac{1}{2}$

12 $a = 37, b = -30$

EXERCISE 4D

1 **a** **i**

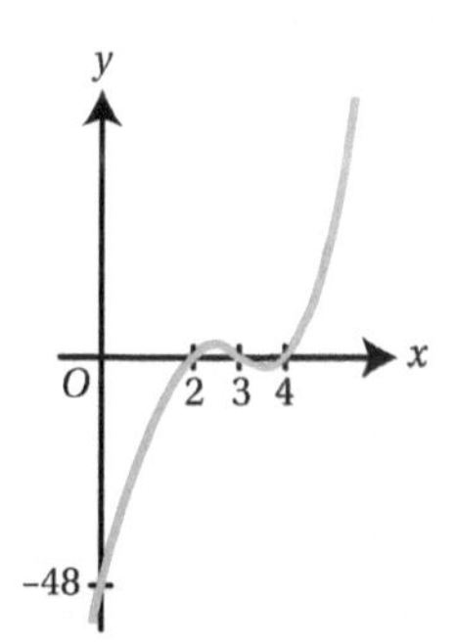

ii

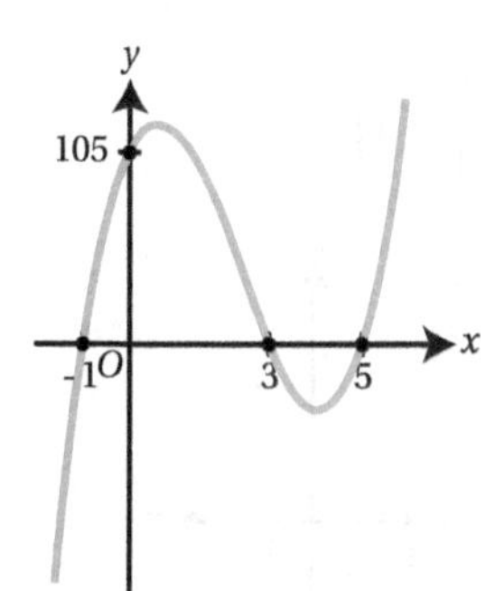

b **i**

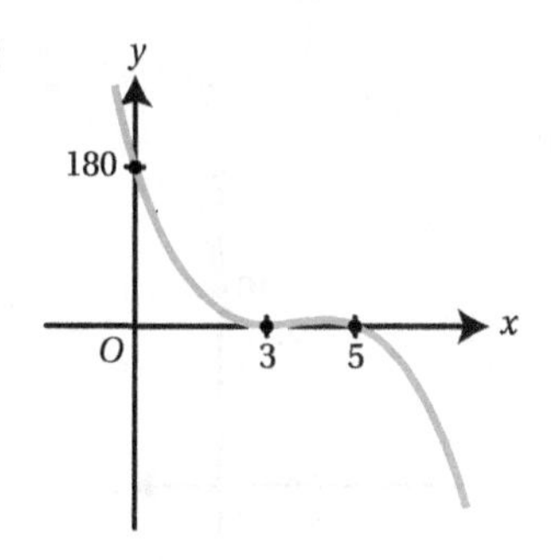

ii

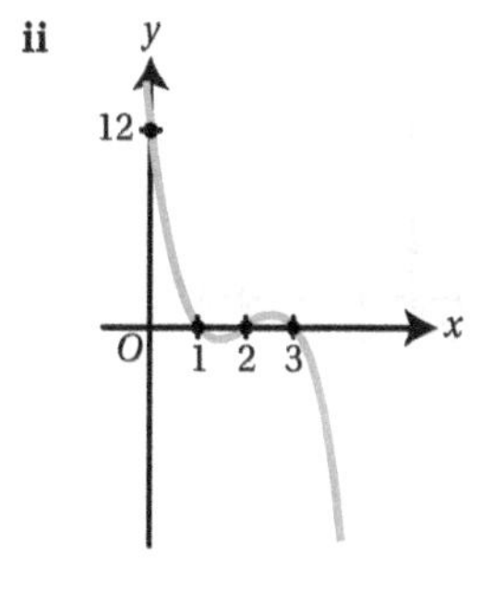

c **i**

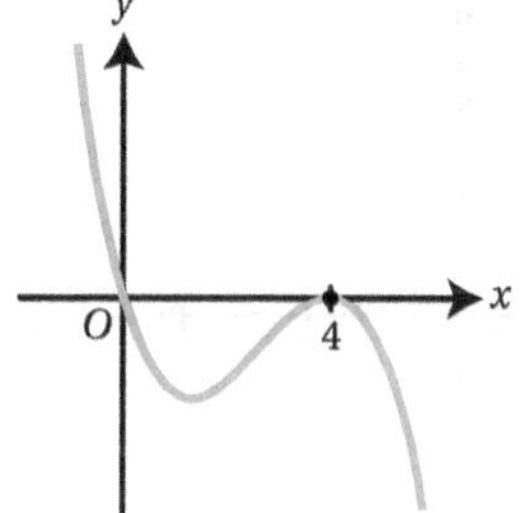

ii

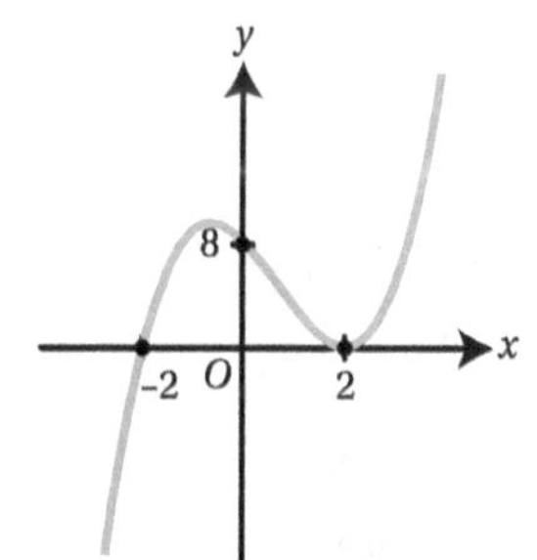

d **i**

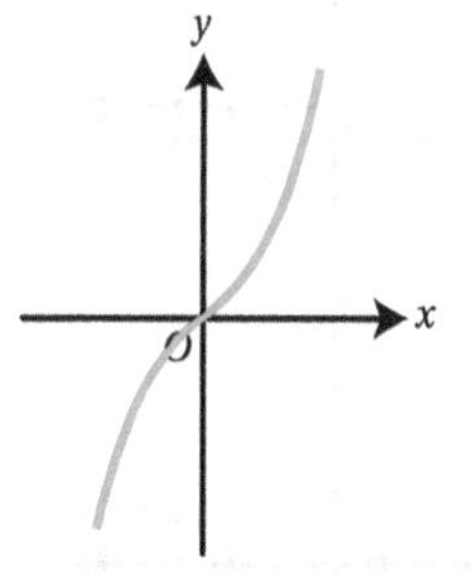

ii

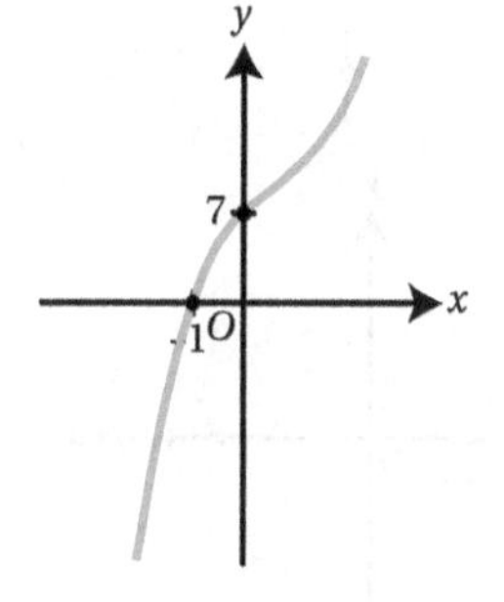

e **i**

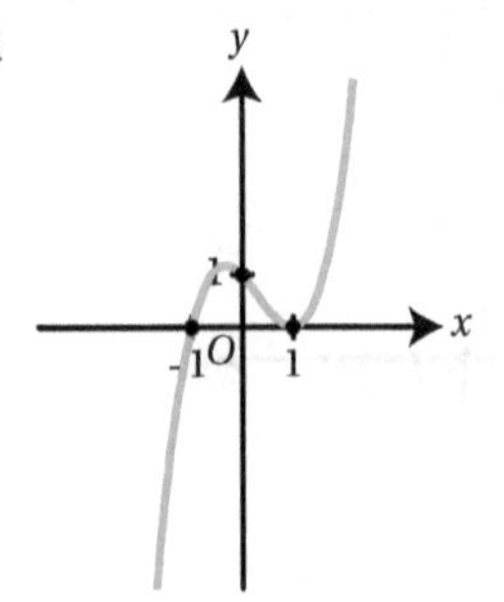

ii

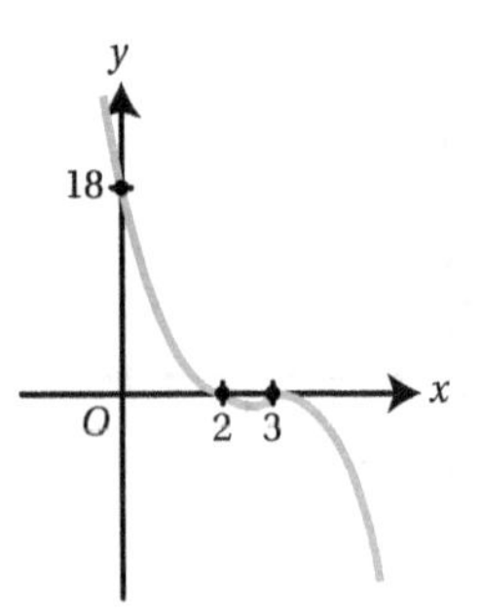

2 **a** **i**

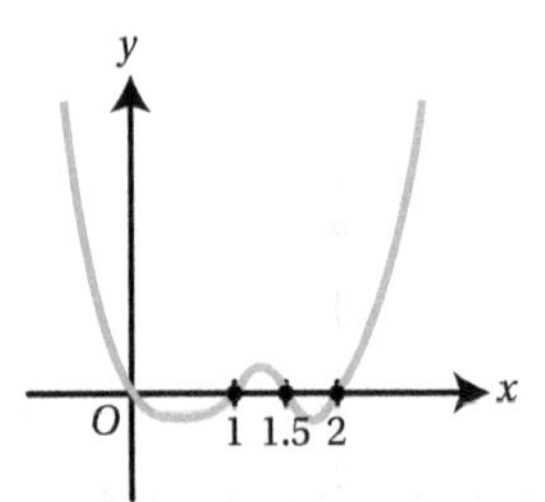

ii

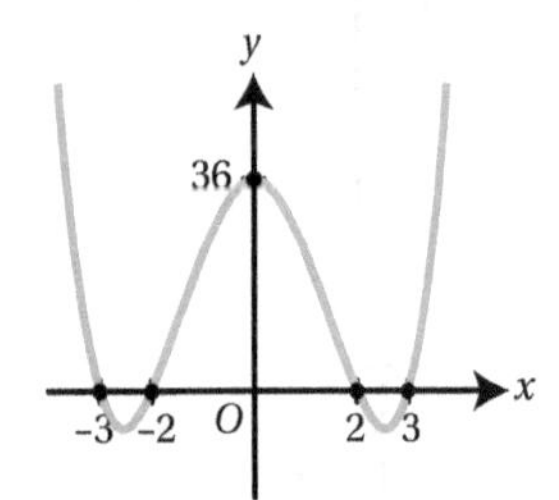

b **i**

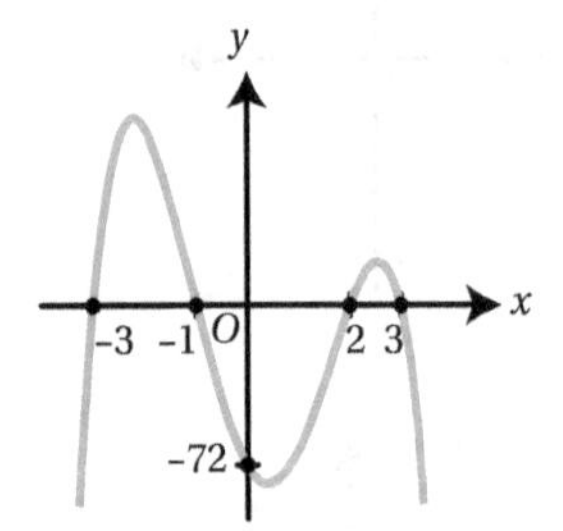

ii

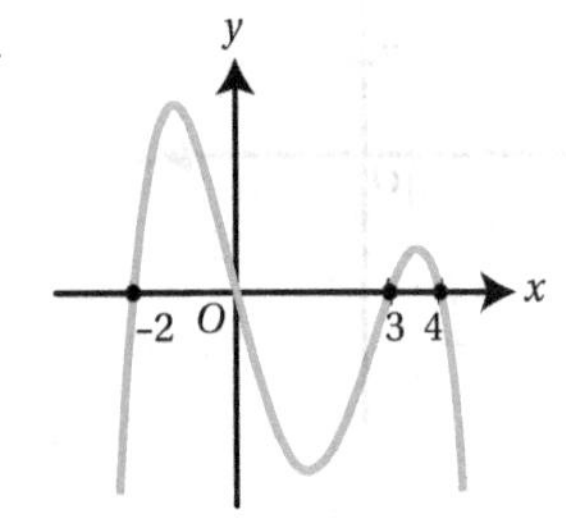

c **i**

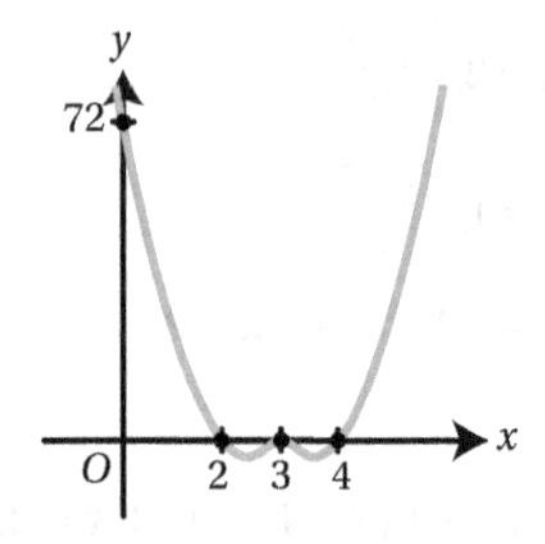

ii

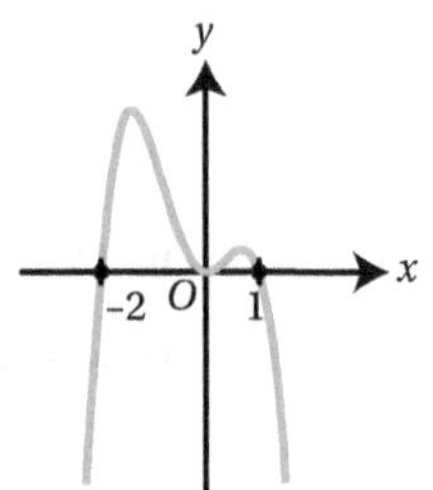

d **i**

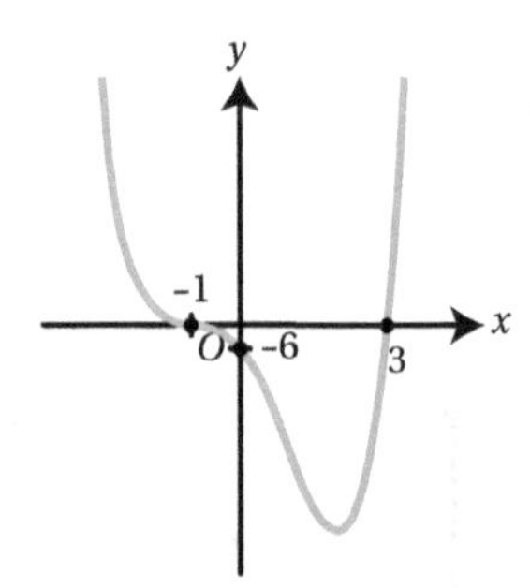

ii

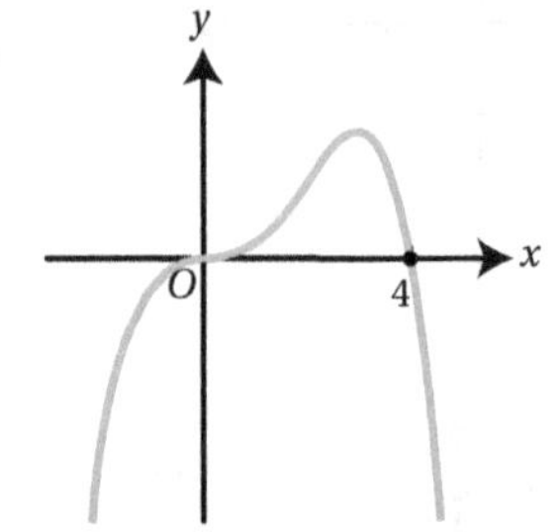

e **i**

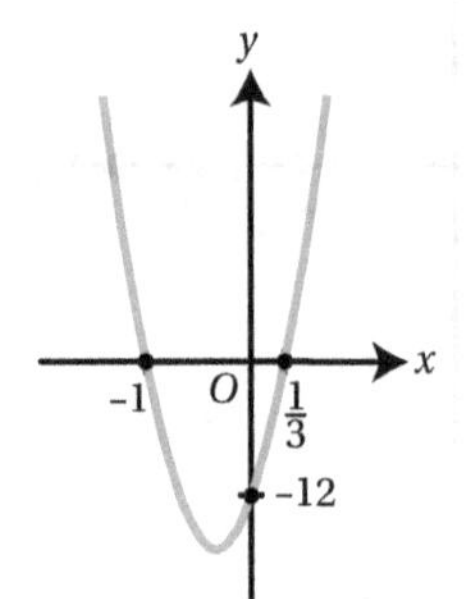

ii

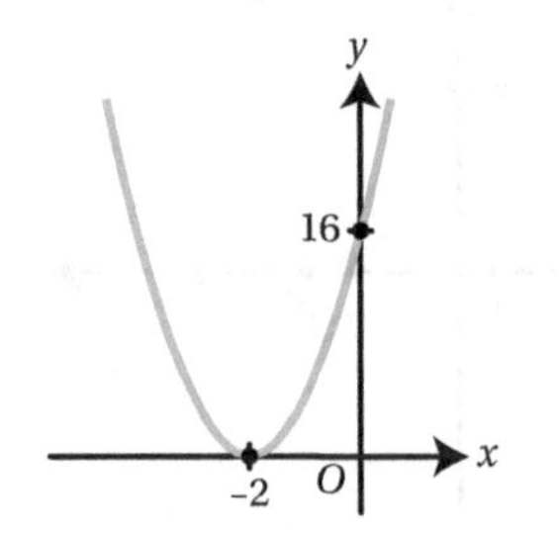

3 **a** **i** $y = 2(x-1)(x-4)(x+2)$

ii $y = 6x(x+2)(x-3)$

b **i** $y = -5x(x-1)(x+2)$

ii $y = -(x-1)(x+2)(x+4)$

c **i** $y = -(x+1)(x-2)^2$

ii $y = (x+1)^2(x-2)$

d **i** $y = x(x+2)(x+3)(2x-1)$

ii $y = -2(x-3)(x-4)(x+1)(x-2)$

e **i** $y = -3x^2(x-1)(x-3)$

ii $y = 5x(x-2)^2(2x+1)$

f **i** $y = 2(x-3)^2(x+1)^2$

ii $y = -(x-3)^2(x-1)^2$

4 **a** Proof.

b $(x-2)(x-1)(2x+1)$

c

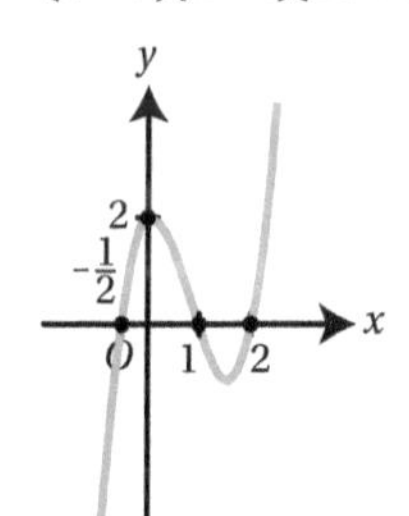

5

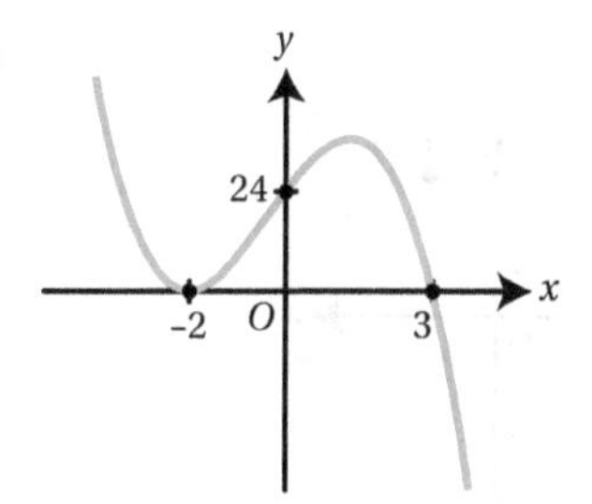

6 **a** 2, −8, −6, 36

b −1, 3, 0, 0

7 **a** $(x-q)(x+q)(x^2+q^2)$

b

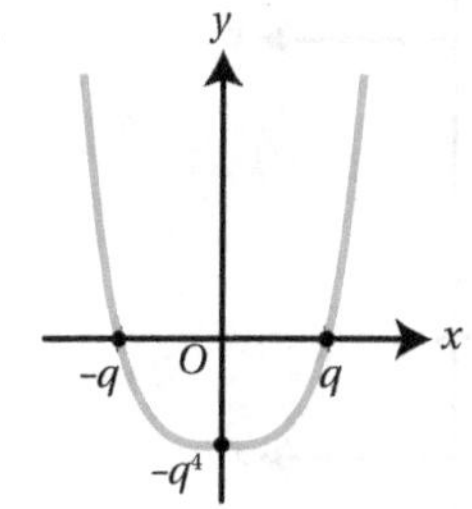

8 **a**

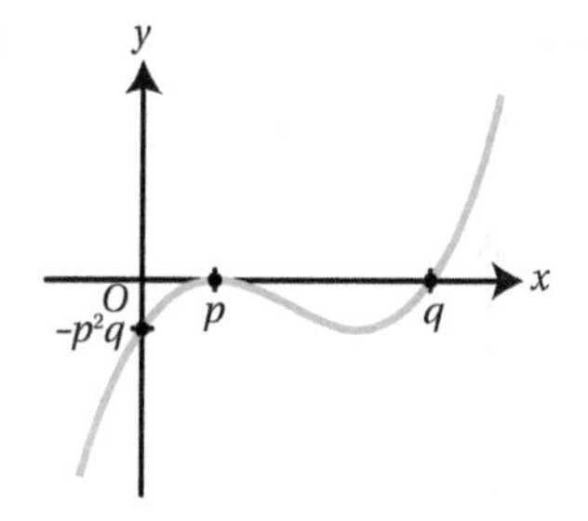

b 1

MIXED PRACTICE 4

1 D

2 A

3 $a = 1, b = 2, c = -12, d = -18, e = 27$

4 $b = 0, c = -3$

5 **a** Proof.

b $f(x) = (x-2)(x+1)(x-3)$

c

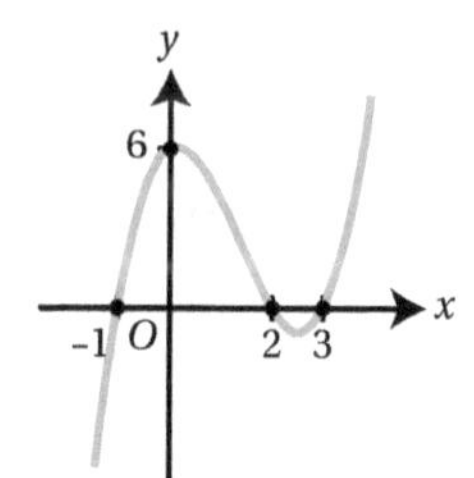

6 **a** **i** Proof.

ii $p(x) = (x+3)(x+5)(x-1)$

7 **a** $a = 1, b = -12$

b

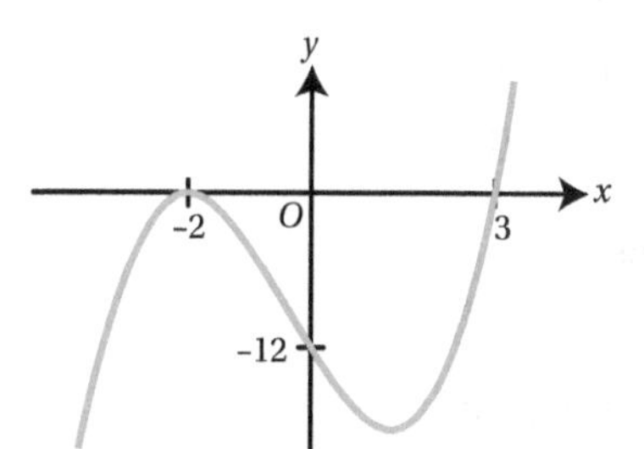

8

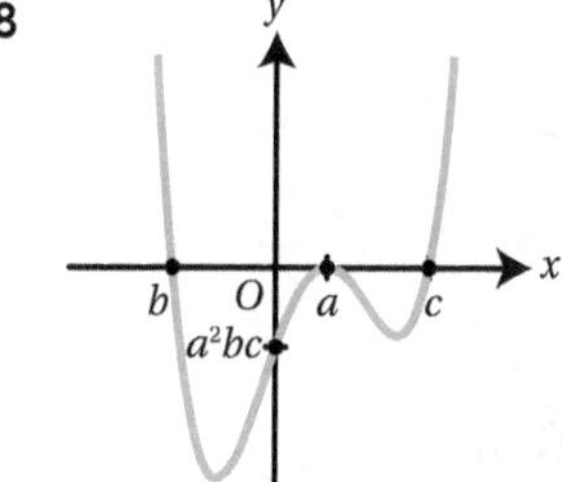

9 **a** $(x+2)$

b $x = -2, -1 \pm \frac{\sqrt{10}}{2}$

10 a i Proof.

ii $p(x) = (x+2)(x-2)(2x-1)$

b

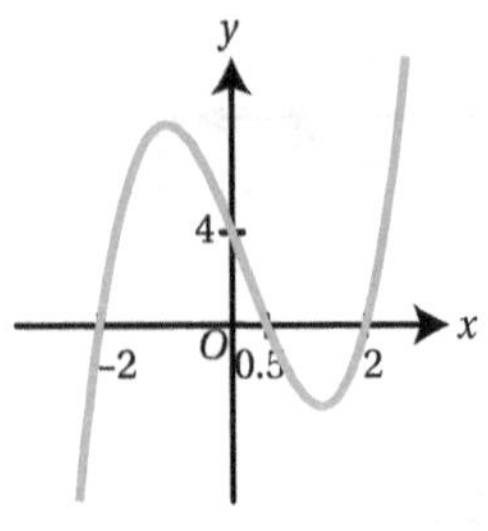

$p(x) > 0$ for $-2 < x < 0.5$ or $x > 2$

11 a i

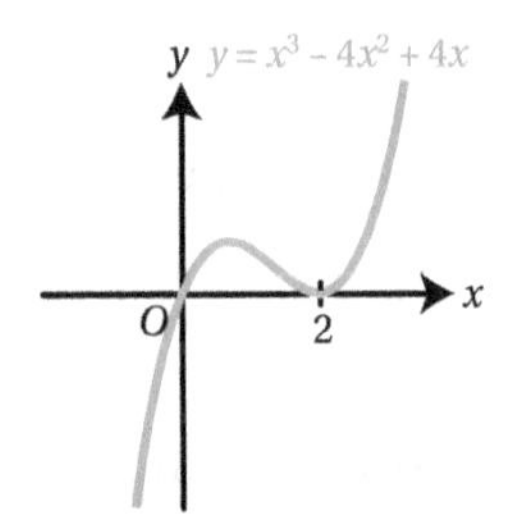

ii Proof.

b i Proof.

ii $p(x) = (x-3)(x^2 - x + 1)$

c $x = 3$

12 $a = -10, b = -18$

5 Using graphs

BEFORE YOU START

1 $x > 4$

2 $x = 1, y = 2$

3 $x = \frac{-1 \pm \sqrt{5}}{2}$

4 One

5 $x = 3$

6 a $m = 2.5n^2$

b $m = \frac{20}{n}$

WORK IT OUT 5.1

Solution 3 is correct.

EXERCISE 5A

1 a i $(-2, -3), (1, 0)$

ii $(3, 0)$

b i $(-3, -9), (4, 5)$

ii No intersection.

2 a i $\left(\frac{-11}{5}, \frac{-8}{5}\right), (3, 1)$

ii $(-3, 3), (5, -1)$

b i $(1, 3), (3, 1)$

ii $(-3, -5), (-5, -3)$

c i $(-1, 6), (2, 3)$

ii $(1, -3), (-1, -5)$

3 $\pm\left(\frac{\sqrt{2}}{2}, \sqrt{2}\right)$

4 $x = 2$

5 $x = 3, y = 8$

6 a $x^2 - 8x + 9.75 = 0$ b 1.5, 6.5

7 $(-\sqrt{3}, -1), (\sqrt{3}, -1), (0, 2), (0, -2)$

8 $m = 0$

EXERCISE 5B

1 Proof.

2 $-1 \pm 2\sqrt{6}$

3 $\pm 6\sqrt{2}$

4 $a < 0$ or $a > 2$

5 Proof.

EXERCISE 5C

1 a i

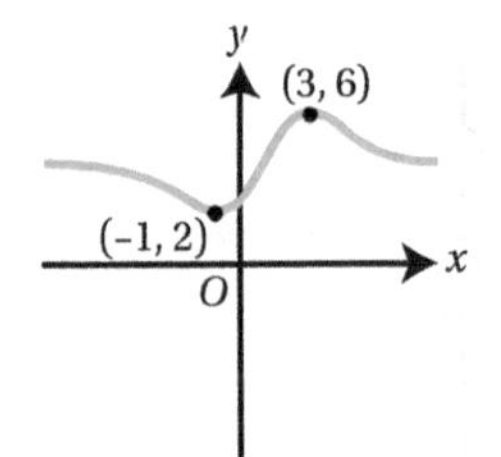

ii

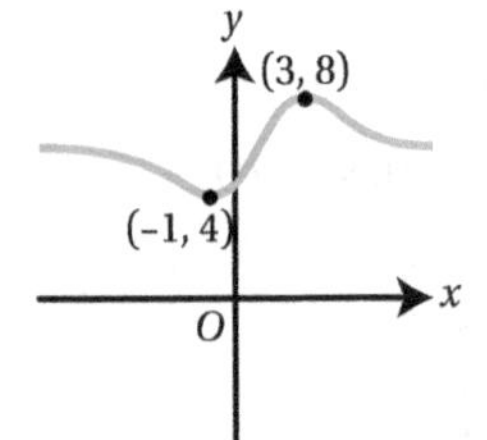

b i

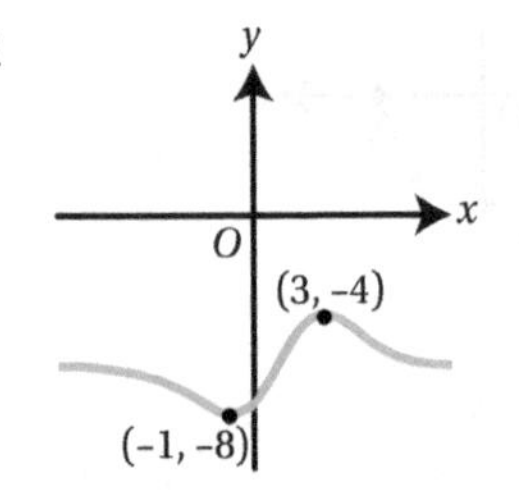

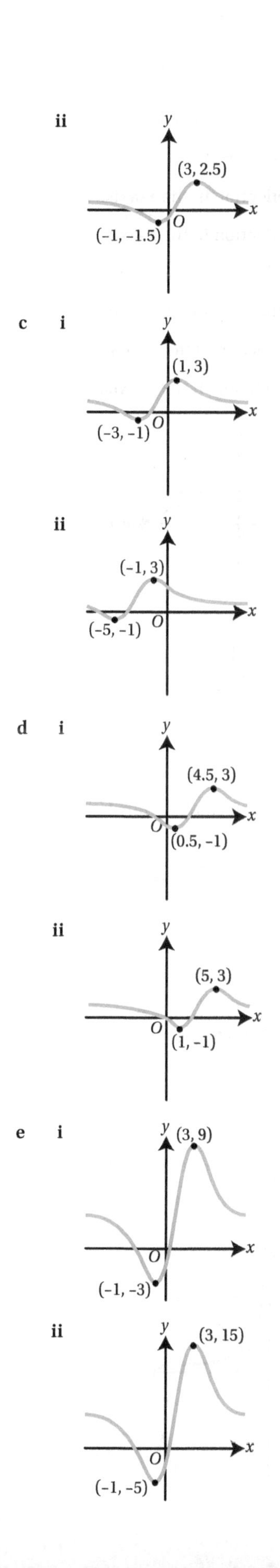
ii
y
(3, 2.5)
x
O
(−1, −1.5)
c i
y
(1, 3)
x
(−3, −1)
O
ii
y
(−1, 3)
x
(−5, −1)
O
d i
y
(4.5, 3)
x
O
(0.5, −1)
ii
y
(5, 3)
x
O
(1, −1)
e i
y
(3, 9)
x
O
(−1, −3)
ii
y
(3, 15)
x
O
(−1, −5)

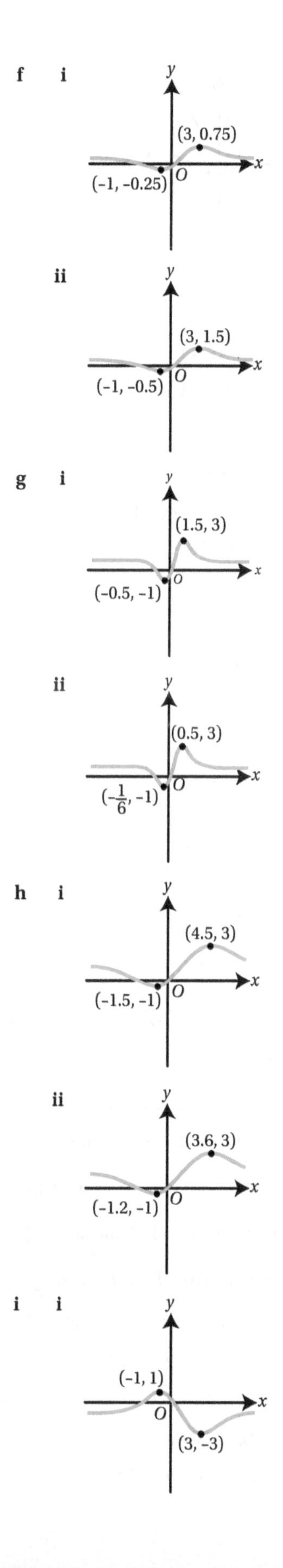
f i
y
(3, 0.75)
x
(−1, −0.25)
O
ii
y
(3, 1.5)
x
(−1, −0.5)
O
g i
y
(1.5, 3)
x
(−0.5, −1)
O
ii
y
(0.5, 3)
x
$(-\frac{1}{6}, -1)$
O
h i
y
(4.5, 3)
x
(−1.5, −1)
O
ii
y
(3.6, 3)
x
(−1.2, −1)
O
i i
y
(−1, 1)
x
O
(3, −3)

ii

2 a i $y=3x^2+3$

ii $y=9x^3-7$

b i $y=7x^3-3x+4$

ii $y=8x^2-7x+6$

c i $y=4(x-5)^2$

ii $y=7(x+3)^2$

d i $y=3(x+4)^3-5(x+4)^2+4$

ii $y=(x-3)^3+6(x-3)+2$

3 a i Vertically down 5 units.

ii Vertically down 4 units.

b i Left 1 unit.

ii Left 5 units.

c i Left 3 units.

ii Right 2 units.

4 a i $y=21x^2$ ii $y=18x^3$

b i $y=\frac{1}{3}(7x^3-3x+6)$

ii $y=\frac{4}{5}(8x^2-7x+1)$

c i $y=x^2$ ii $y=7\left(\frac{x}{5}\right)^2$

d i $y=3(2x)^3-5(2x)^2+4$

ii $y=\left(\frac{3x}{2}\right)^3+6\left(\frac{3x}{2}\right)+2$

5 a i Vertical stretch, scale factor 4.

ii Vertical stretch, scale factor 6.

b i Horizontal stretch, scale factor $\frac{1}{3}$.

ii Horizontal stretch, scale factor $\frac{1}{4}$.

c i Horizontal stretch, scale factor $\frac{1}{3}$.

ii Horizontal stretch, scale factor 2.

6 a i $y=-3x^2$ ii $y=-9x^3$

b i $y=-7x^3+3x-6$

ii $y=-8x^2+7x-1$

c i $y=-4x^2$ ii $y=-7x^3$

d i $y=-3x^3-5x^2+4$

ii $y=-x^3-6x+2$

7 a i Reflection in the x-axis.

ii Reflection in the x-axis.

b i Reflection in the y-axis.

ii Reflection in the y-axis.

c i Reflection in the y-axis.

ii Reflection in the y-axis.

8 a

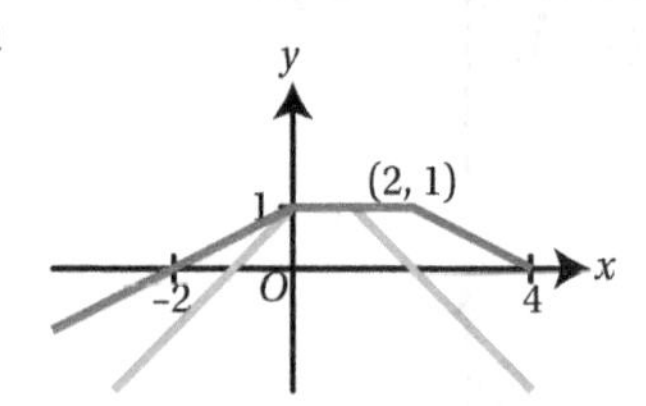

b

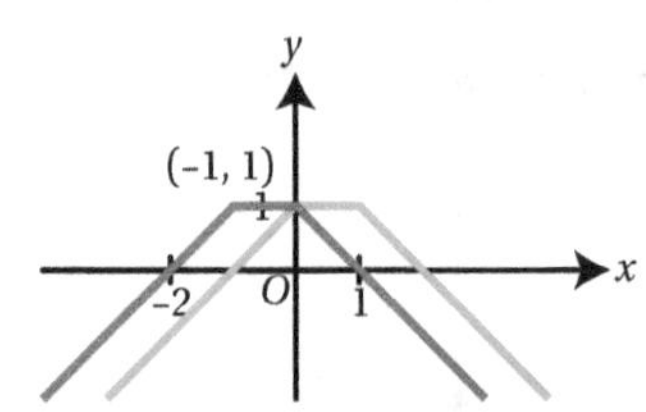

9 Horizontal stretch factor $\frac{1}{2}$.

10 Vertical stretch factor 2 or translation 1 to the left.

11 a k b $\frac{1}{\sqrt{k}}$

12 $k=1+c$

13 $y=x^2-4x+7$

EXERCISE 5D

1 a Translation d to the right.

b $x=d, y=0$

2 $\sqrt{a}$

3 a $\left(\frac{b}{a}, \frac{a^2}{b}\right)$ b Area $=a$

4 $m=-\frac{c^2}{4a}$

5

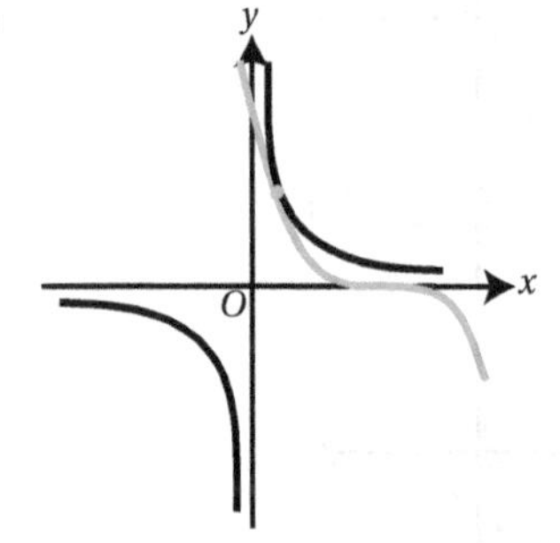

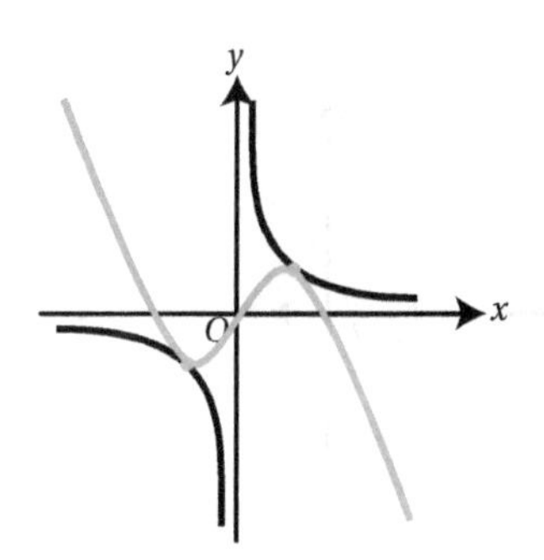

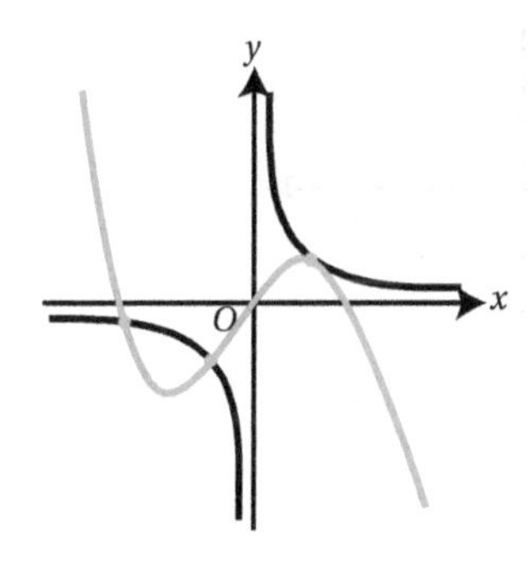

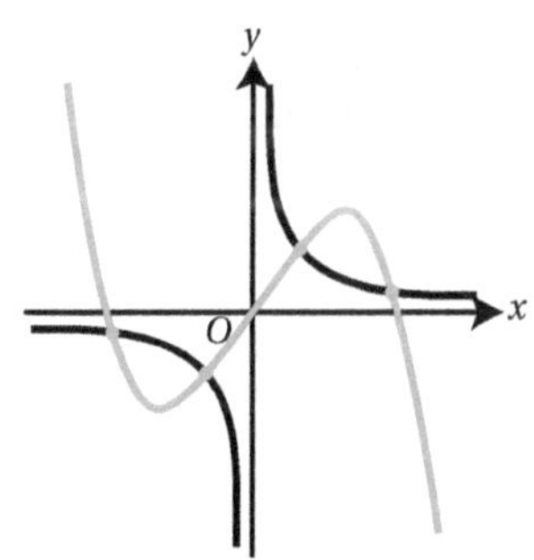

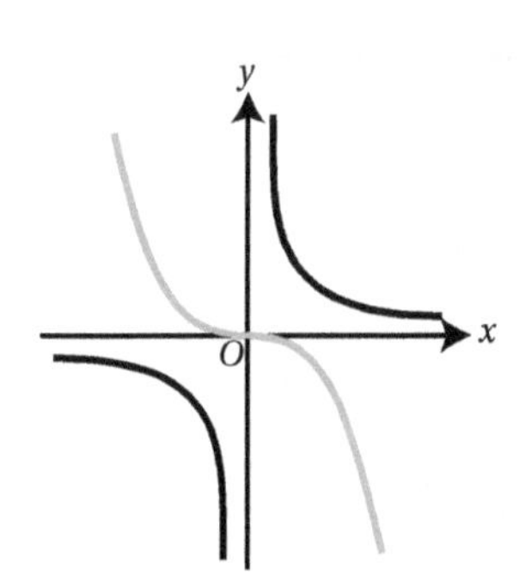

EXERCISE 5E

1 **a** 48

b

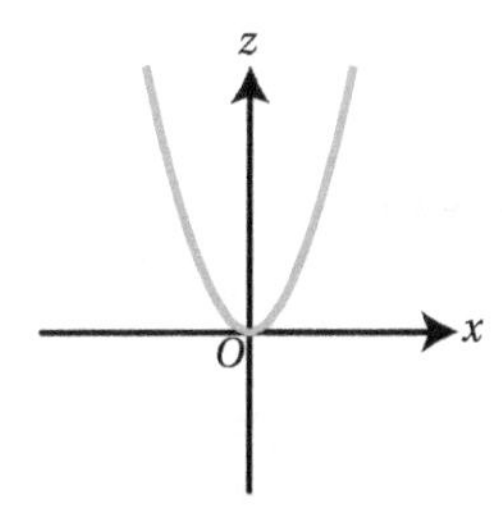

2 **a** 2

b

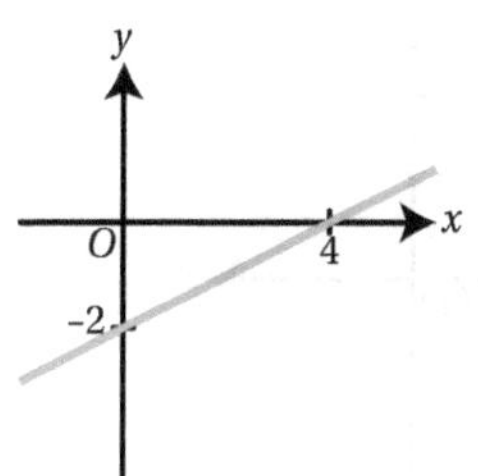

3 **a** 1.25

b

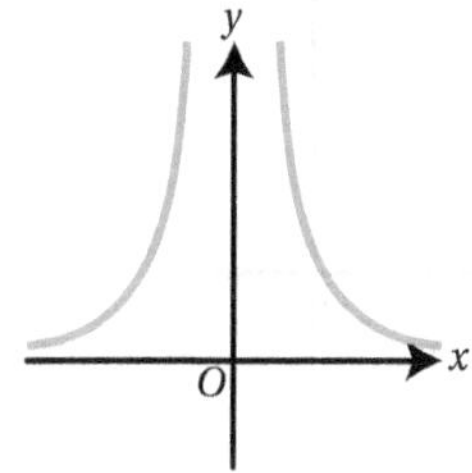

4 **a** 6

b

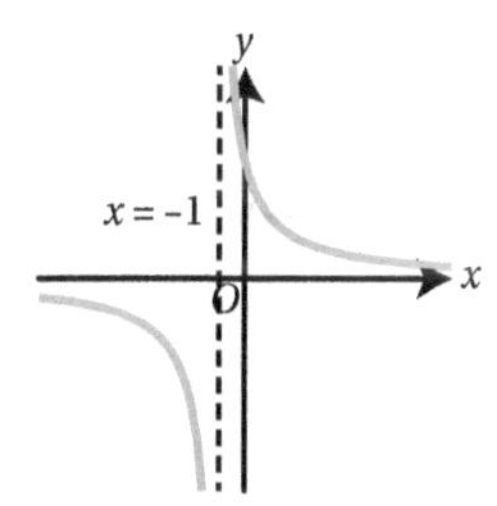

5 **a** $y_S = 8p - 80$

b It is the number of items that would be sold if they were free. It is likely that more would be 'sold' if they were given away for free.

c £39.88

d £25

6 **a** $C_A = 65 + 0.03\,m$, $C_B = 0.05\,m$

b After 3250 minutes

7

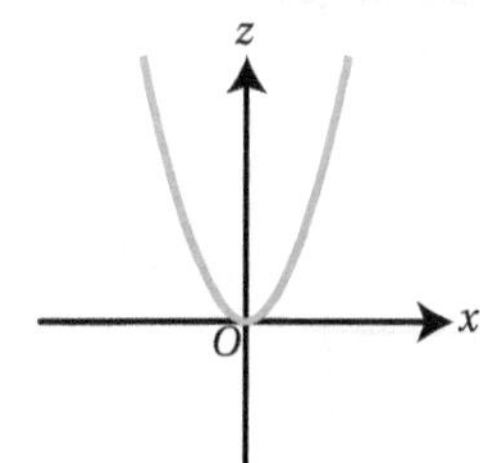

8 17.4%

EXERCISE 5F

1 a i

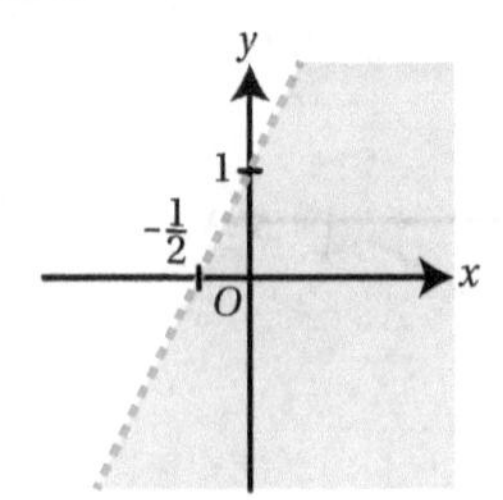

ii

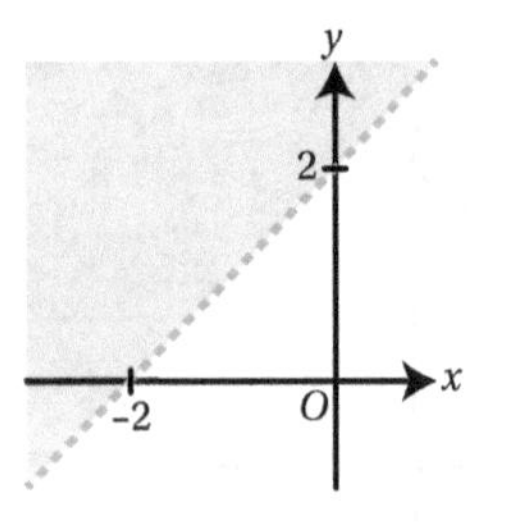

b i

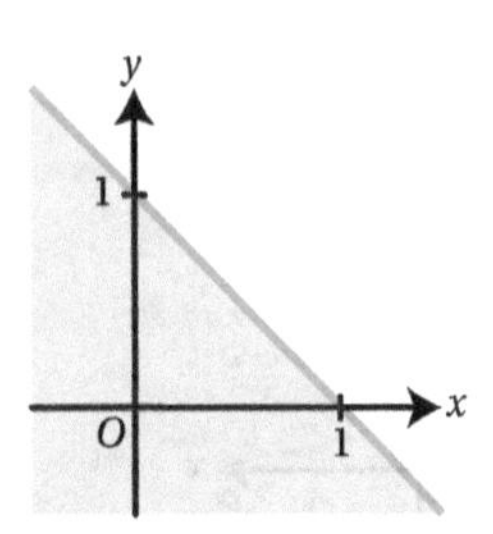

ii

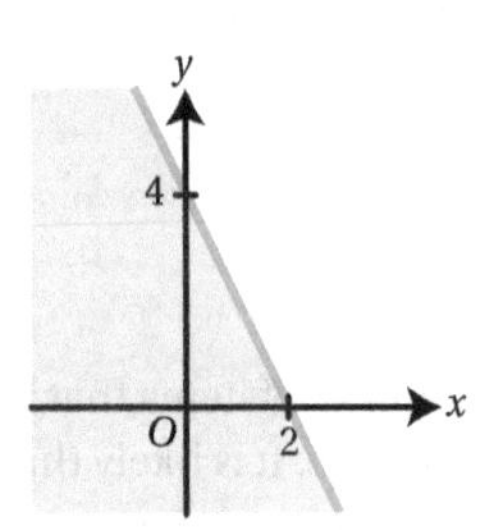

c i

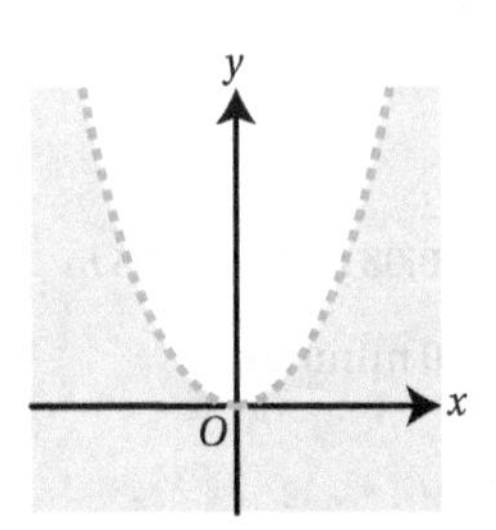

ii

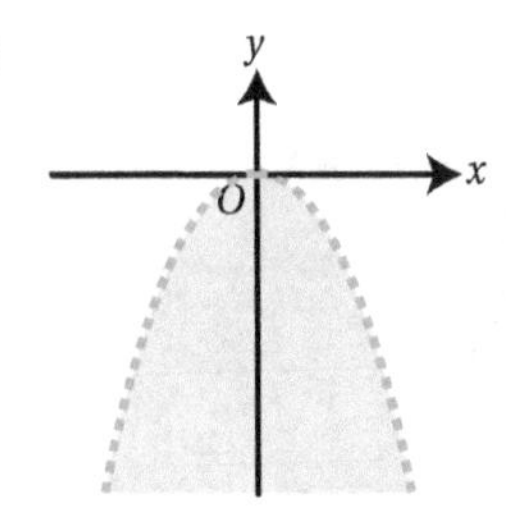

d i

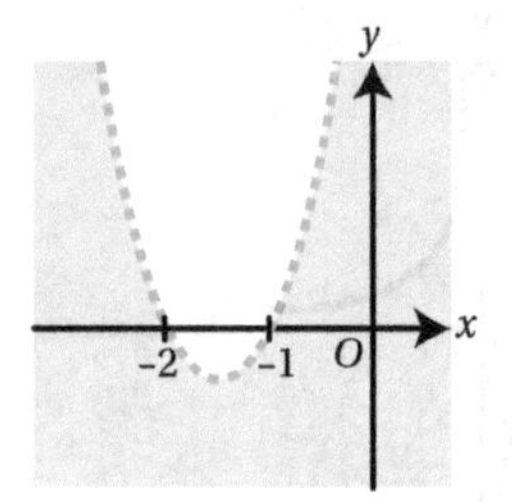

ii

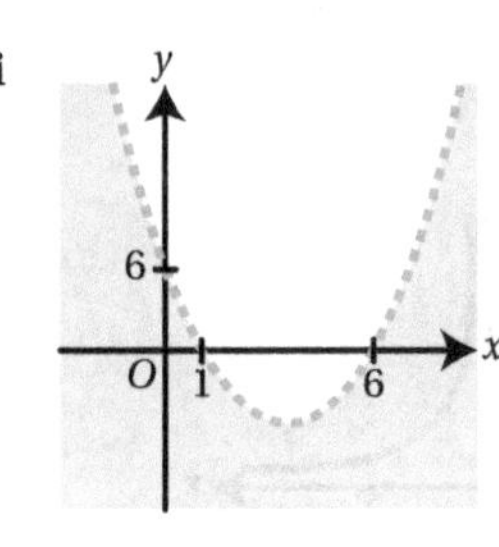

e i

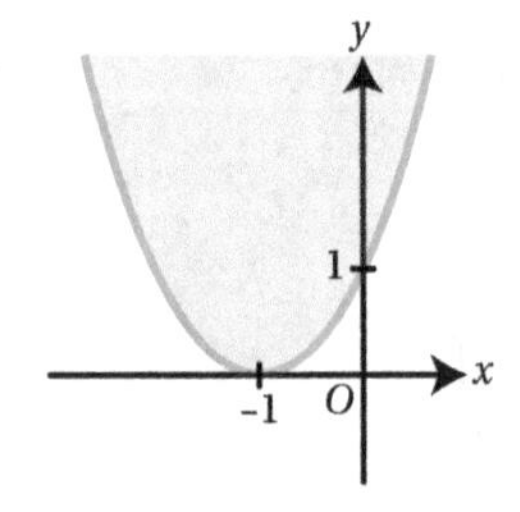

ii

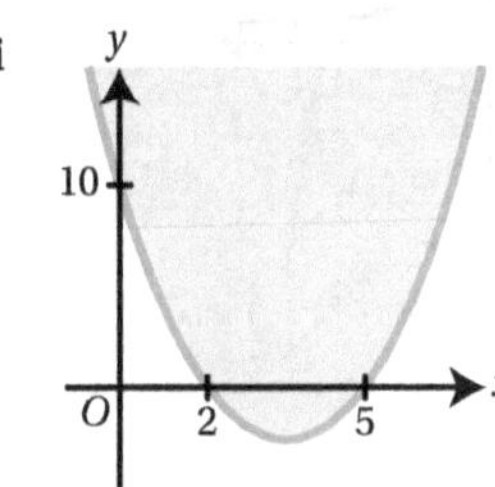

2

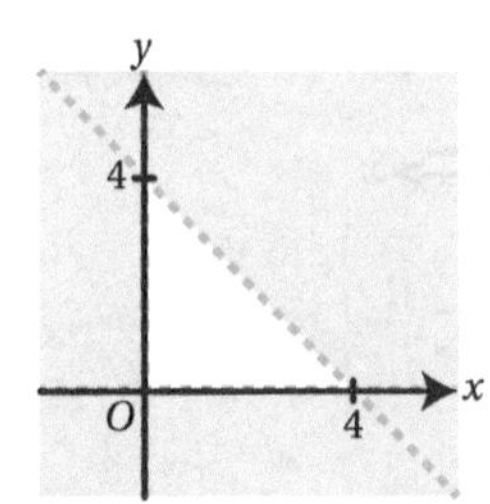

3

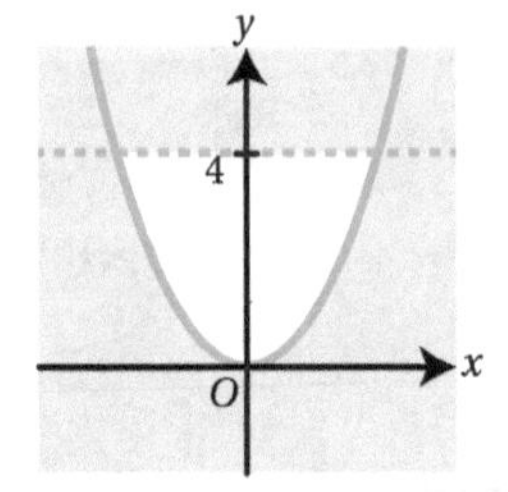

4 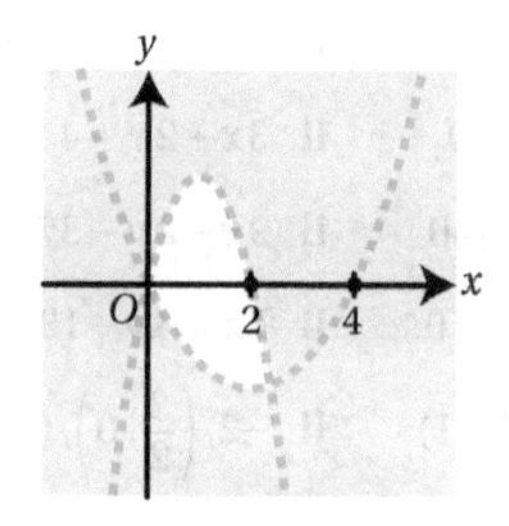

5 $y > x$, $x + y \leqslant 4$

6 $y > (x-1)(x-2)$, $y < 2$

7 54

8 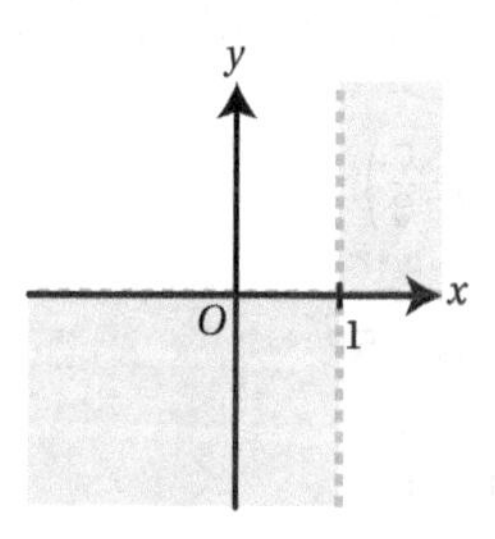

MIXED PRACTICE 5

1 A

2 (3, 4) or (4, 3)

3 a 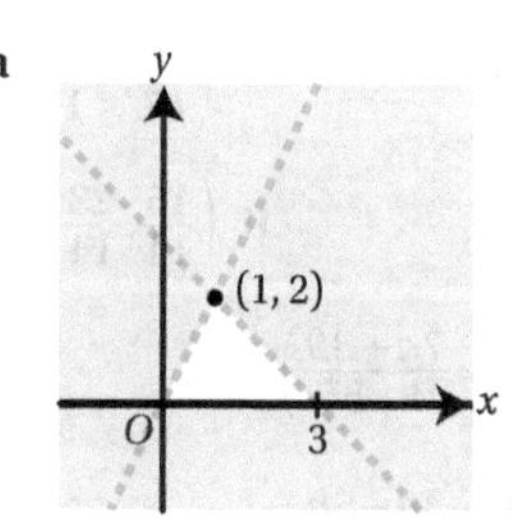

b 2

4 Translation by $\begin{pmatrix} -3 \\ 0 \end{pmatrix}$

5 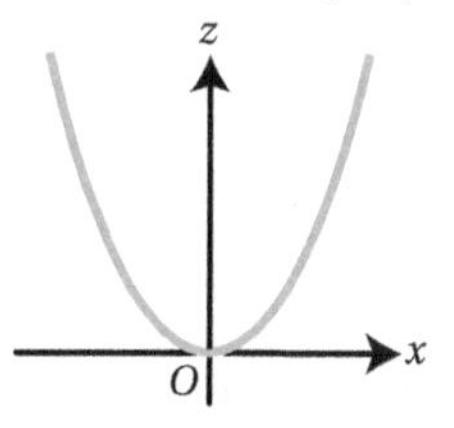

6 3.75 km

7 a 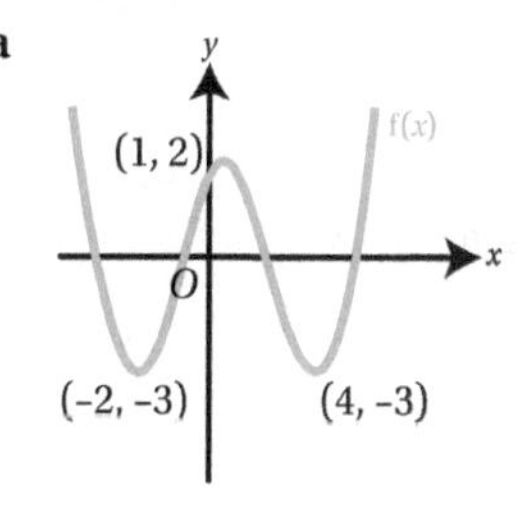

b (1, 2)

8 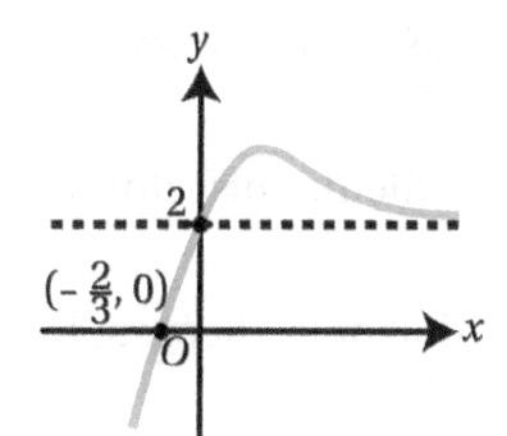

9 a Proof.

b 0.233 is how much weight the boy gains per week, 3.63 is the weight at birth.

c No

10 C

11 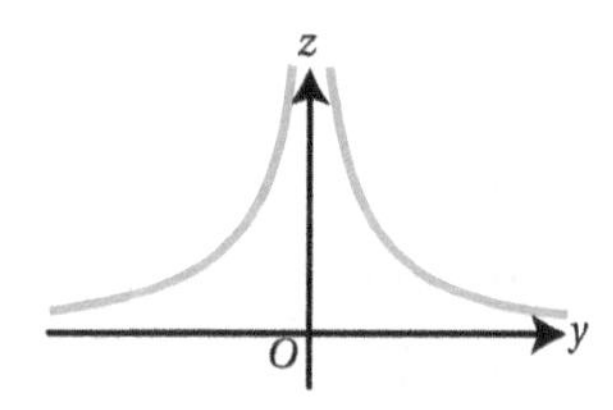

12 a Stretch in x-direction, scale factor 2.

b 2.3

13 a, b Proof.

c $k < -\frac{29}{4}, k > -1$

14 a $x = \pm 2, \pm 3$

b $-3 \leqslant x \leqslant -2$ or $2 \leqslant x \leqslant 3$

6 Coordinate geometry

BEFORE YOU START

1 a $y = 2x - 1$ b $y = -2x + 9$

c $y = 3x$

2 13

3 $x = 3, y = -2$

4 a $w = 70°$ (isosceles triangle and angle ODA $= 90 - 20 = 70°$ [angle in semi-circle])

$x = 5$ (radius)

$y = 90°$ (angle in semi-circle)

$z = 20°$ (isosceles triangle)

b $a = 50°$ (alternate segment theorem)

$b = 90°$ (radius meets tangent)

$c = 130°$ (angle $RPO = 40°$ [angles on straight line], angle $RPQ = 15°$ [angles in a triangle], so angle $QPO = 25°$, then c is apex of isosceles triangle)

c $d = 4$ (OY is perpendicular bisector of chord)

$e = 3$ (applying Pythagoras' theorem in a right-angled triangle)

5 $(x-2)^2 - 7$

6 a $(2, 4), (-3, 9)$

b If $y = 2x+1$ is a tangent to the parabola $y = -x^2$ then $-x^2 = 2x+1$ has only one solution. This means that the discriminant to x^2+2x+1 is 0. $b^2 - 4ac = 2^2 - 4\times1\times1 = 0$, as required.

EXERCISE 6A

1 a i 13 ii 5

b i $\sqrt{5}$ ii $\sqrt{10}$

c i $\sqrt{20}$ ii $\sqrt{8}$

d i $\sqrt{10}$ ii $\sqrt{17}$

2 a i $(2.5, 6)$ ii $(1.5, 2)$

b i $(1.5, 5)$ ii $(2.5, 3.5)$

c i $(1, 3)$ ii $(-2, 2)$

d i $(-2.5, -1.5)$ ii $(-1.5, -3)$

3 $3a\sqrt{5}$

4 $(2-a, 2-b)$

5 a Both are 5.

b It does not lie on the line AC.

6 $1 \pm \frac{2}{\sqrt{5}}$

7 $y = \frac{x^2+1}{2}$

8 Proof.

9 a i $2\sqrt{5}\sqrt{a^2+1}$ ii $(-1+2a, 2+a)$

b It is 2 : 1.

10 $\sqrt{125}$ m

EXERCISE 6B

1 a i $y+1 = 3(x-4)$ ii $y-2 = 5(x+3)$

b i $y-1 = -\frac{1}{2}(x+3)$ ii $y-3 = \frac{4}{3}(x-1)$

c i $y-7 = 4(x-3)$ ii $y-1 = 3(x-4)$

d i $y+1 = -\frac{6}{7}(x-3)$ ii $y+7 = \frac{9}{4}(x+1)$

2 a i $2x+y-1=0$ ii $5x+y-13=0$

b i $x-3y-20=0$ ii $3x+2y+4=0$

c i $5x-3y+11=0$ ii $9x+2y-37=0$

d i $6x+5y-17=0$ ii $7x+4y+13=0$

3 a i $3, \left(-\frac{1}{3}, 0\right), (0, 1)$ ii $-2, \left(\frac{3}{2}, 0\right), (0, 3)$

b i $1.5, \left(-\frac{5}{3}, 0\right), (0, 2.5)$

ii $-0.8, \left(\frac{1}{4}, 0\right), (0, 0.2)$

c i $1, \left(\frac{5}{3}, 0\right), \left(0, -\frac{5}{3}\right)$ ii $4, \left(-\frac{3}{2}, 0\right), (0, 6)$

d i $\frac{4}{3}, \left(\frac{7}{4}, 0\right), \left(0, -\frac{7}{3}\right)$

ii $-2.5, \left(\frac{3}{5}, 0\right), (0, 1.5)$

e i $2, \left(\frac{1}{2}, 0\right), (0, -1)$

ii $3, \left(\frac{16}{3}, 0\right), (0, -16)$

4 a i $(-8, -7)$ ii $(3, 4)$

b i $\left(\frac{7}{3}, -\frac{11}{9}\right)$ ii $\left(\frac{16}{17}, -\frac{3}{17}\right)$

c i $(-1, 1)$ ii $\left(\frac{15}{14}, \frac{23}{14}\right)$

d i $\left(\frac{3a+7b}{a^2+b^2}, \frac{-7a+3b}{a^2+b^2}\right)$

ii $\left(\frac{5a+3b}{a^2-b^2}, \frac{-3a-5b}{a^2-b^2}\right)$

5 a $4x+9y-11=0$

b $4x+9y-2=0$

c $k = \frac{2}{21}$

6 $\frac{361}{12}$

7 a $l_1: x-y=-2, l_2: 3x+2y=-24$

b $(-5.6, -3.6)$

c 10.8

8 a $k = 3$

b $2x-3y+17=0$

9 $\frac{\sqrt{65}}{4}$

10 Proof.

EXERCISE 6C

1 a i $-\frac{3}{5}$ ii -6

b i $-\frac{1}{2}$ ii $-\frac{2}{5}$

c i $\frac{1}{2}$ ii $\frac{1}{3}$

d i $-\frac{7}{3}$ ii -5

2 a i perpendicular ii parallel

b i parallel ii parallel

c i perpendicular ii neither

d i neither ii neither

3 a Proof.

b $2x+5y-22=0$

4 Proof, area = 6.5.

5 a $\left(2, \frac{9}{2}\right)$ b $y=1.2x+2.1$

6 a $2x+3y=-5$ b $\frac{11}{2}$

7 $(6\pm2\sqrt{2}, 0)$

8 $18x-4y-1=0$

9 a Proof.

b $q=\frac{14}{3}$, area $=\frac{53}{3}$

10 a $y=-2x+4$ b $2\sqrt{5}$

11 a Proof. b $k=2$

EXERCISE 6D

1 a i $(x-3)^2+(y-7)^2=16$

ii $(x-5)^2+(y-1)^2=36$

b i $(x-3)^2+(y+1)^2=7$

ii $(x+4)^2+(y-2)^2=5$

2 a i Centre $(-4, 0)$, radius $4\sqrt{2}$

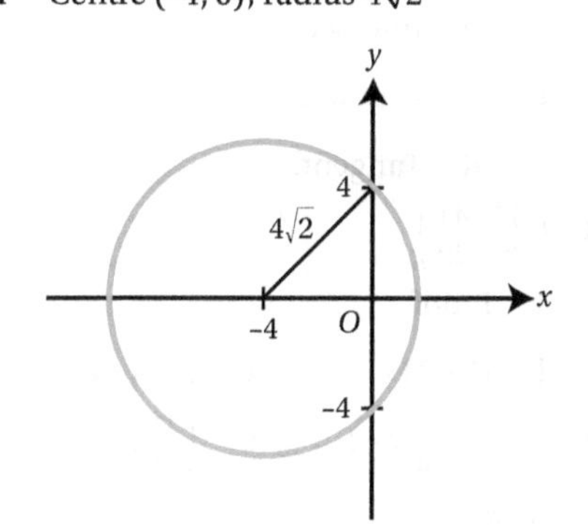

ii Centre $(0, 7)$, radius $2\sqrt{5}$

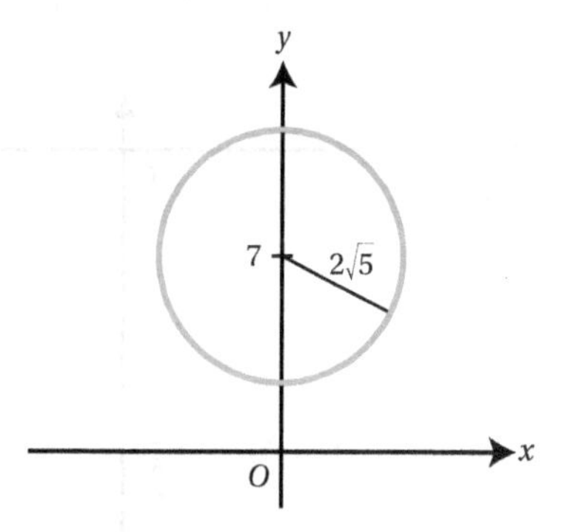

b i Centre $(2, 3)$, radius 2

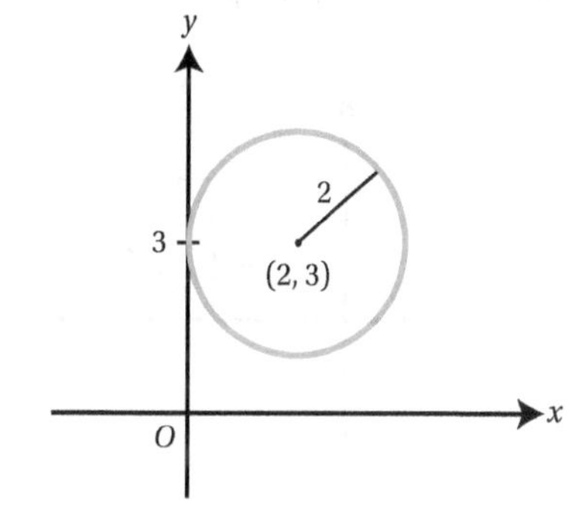

ii Centre $(-1, 4)$, radius 4

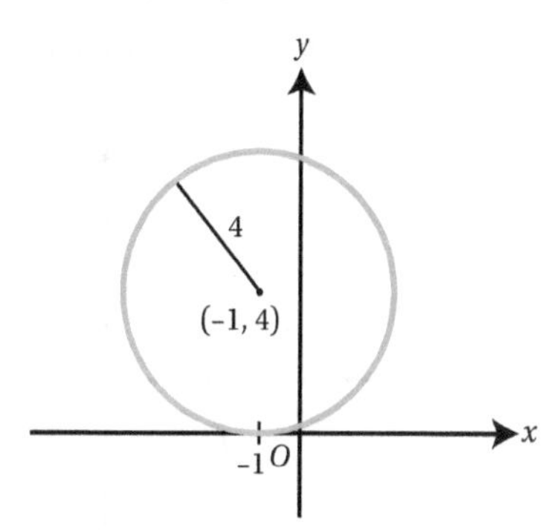

c i Centre $(2, -3)$, radius $\frac{3}{2}$

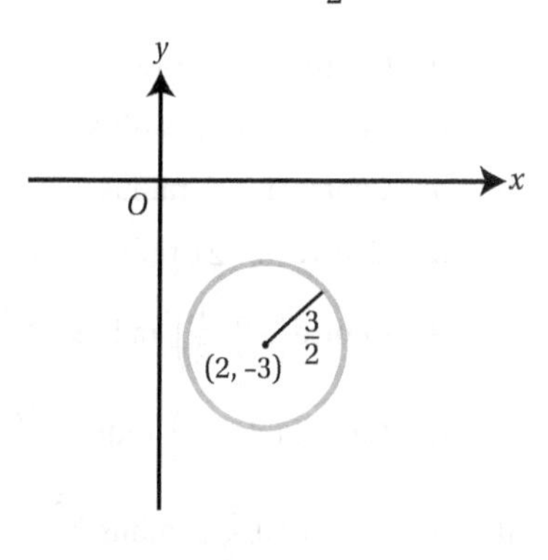

ii Centre $(-1,-5)$, radius $\frac{2}{5}$

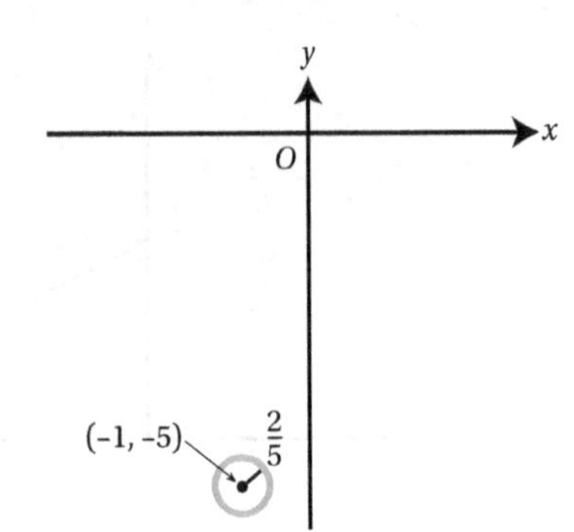

d **i** Centre $\left(3, \frac{1}{2}\right)$, radius $\sqrt{6}$

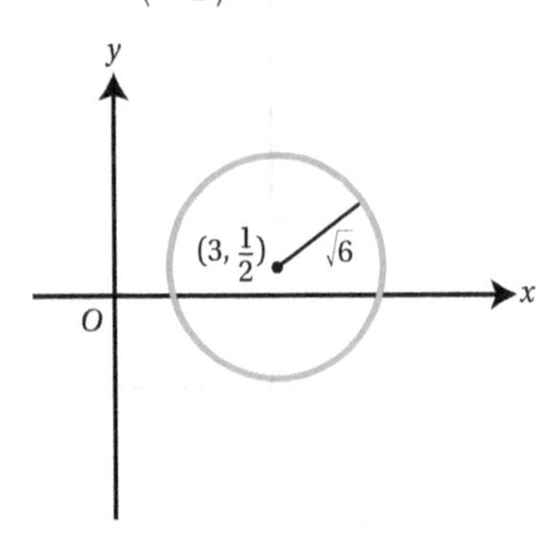

ii Centre $\left(-\frac{3}{4}, \frac{1}{5}\right)$, radius $\sqrt{3}$

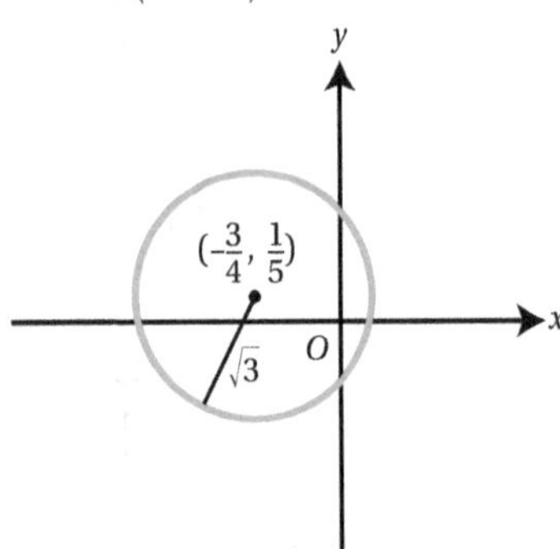

3 **a** **i** Centre $(-2, 3)$, radius 3

ii Centre $(4, -1)$, radius 3

b **i** Centre $(1, -3)$, radius 3

ii Centre $(5, -2)$, radius $\sqrt{30}$

c **i** Centre $\left(-\frac{5}{2}, \frac{1}{2}\right)$, radius $\frac{3\sqrt{2}}{2}$

ii Centre $\left(\frac{3}{2}, -\frac{7}{2}\right)$, radius $\frac{\sqrt{70}}{2}$

d **i** Centre $\left(0, \frac{5}{2}\right)$, radius $\frac{\sqrt{73}}{2}$

ii Centre $\left(-\frac{3}{2}, 0\right)$, radius $\frac{7}{2}$

4 **a** **i** On circumference

ii On circumference

b **i** Outside circle

ii Inside circle

5 **a** $(x+6)^2+(y-3)^2=117$

b $(0,-6), (0,12)$

6 **a** $\left(\frac{5}{2}, -\frac{1}{2}\right), r=\frac{\sqrt{38}}{2}$

b Outside the circle.

7 $4\sqrt{6}$

8 **a** $\sqrt{442}$

b $\left(x-\frac{5}{2}\right)^2+\left(y+\frac{3}{2}\right)^2=\frac{221}{2}$

9 **a** $\sqrt{17}$

b Outside the circle.

10 **a** $90°$

b Proof.

EXERCISE 6E

1 **a** **i** $y=4$ **ii** $5y=3x+19$

b **i** $y=-\frac{2}{3}x+\frac{5}{3}$ **ii** $y=\frac{2}{3}x-\frac{4}{3}$

c **i** $y=-2x+9$ **ii** $y=-x+6$

2 **a** **i** $\left(\frac{-7\pm\sqrt{509}}{10}, \frac{-2\pm\sqrt{509}}{5}\right)$

ii $\left(\frac{7\pm\sqrt{389}}{10}, \frac{2\pm\sqrt{389}}{5}\right)$

b **i** $(2, 7)$ **ii** $(4, 4)$

c **i** No intersection.

ii No intersection.

3 **a** **i** Intersect.

ii Intersect.

b **i** Disjoint.

ii Intersect.

c **i** Disjoint.

ii Tangent.

4 $\left(\frac{57}{29}, \frac{41}{29}\right)$

5 **a** Proof.

b $(1, -2)$ **c** $y=x+5$

6 **a** $y=\frac{1}{2}x+1$ **b** $\left(\frac{18}{5}, \frac{14}{5}\right)$ **c** $\frac{49}{5}$

7 $4\sqrt{5}$

8 **a** $(x-2)^2+(y-5)^2=29$

b Proof.

c 7.43

9 **a** 7 or 21

b $(-2.39, -1.99)$ and $(-2.39, 12.0)$

10 $\frac{200}{3}$

11 1.04, 6.16

12 4.11

13 **a** $(2, -1)$

b Proof.

14 **a, b** Proof.

c $150-25\pi$

15 $a^2+b^2=(1\pm r)^2$

16 **a** $y=8-x, y=-\frac{1}{6}x+\frac{25}{4}$

b $P(2.1, 5.9)$

c $y=\frac{3}{2}x+\frac{11}{4}$

d 3.10

MIXED PRACTICE 6

1 C

2 **a** $-\frac{5}{4}$ **b** $-\frac{7}{12}$

c $18x-12y+29=0$ **d** $\left(-\frac{29}{18}, 0\right)$

3 **a** $(1, 5)$

b Proof.

c $(3.4, 3.8)$

4 **a** $p=-3$ **b** $\frac{3}{4}$

c $k=10$ **d** $(-7, -4)$

5 D

6 **a** Proof.

b $(x-6)^2+y^2=13$

c $3x+2y-5=0$

7 $-6-5\sqrt{5}<k<-6+5\sqrt{5}$

8 **a** $\frac{256}{3}$

b $\frac{104}{3}$

9 $x^2-2x+y^2-6y-22=0$

10 $k=\pm\sqrt{\frac{1}{8}}=\pm\frac{\sqrt{2}}{4}$

11 **a** $(0, -2), (0, 6)$ **b** 5

c **i** $\sqrt{34}$ **ii** 3

12 C

13 $\frac{24\sqrt{13}}{13}$

14 Proof.

15 **a** $p=8; (x-5)^2+(y-7)^2=58$

b $21+5\sqrt{33}$

7 Logarithms

BEFORE YOU START

1 **a** False (18).

b False (20).

c True.

d False $(2^{-1}x^{-3})$.

2 **a** 81 **b** $\frac{1}{27}$

3 **a** $x^{\frac{5}{2}}$ **b** $x^{-\frac{1}{2}}$

4 $x=\frac{7}{3}$

5 **a** $x=-4, 2$ **b** $x=-1, 1$

EXERCISE 7A

1 **a** **i** 3 **ii** 2

b **i** 1 **ii** 1

c **i** 0 **ii** 0

d **i** −1 **ii** −3

e **i** $\frac{1}{2}$ **ii** $\frac{1}{3}$

f **i** $\frac{1}{2}$ **ii** $\frac{1}{2}$

g **i** $\frac{2}{3}$ **ii** $\frac{3}{4}$

h **i** $\frac{3}{2}$ **ii** $\frac{5}{4}$

i **i** $\frac{3}{4}$ **ii** $\frac{9}{4}$

j **i** $-\frac{1}{2}$ **ii** $-\frac{1}{2}$

2 **a** **i** 1.70 **ii** −0.602

b **i** −2.30 **ii** 2.30

c i 1.95 ii −0.317

d i −0.683 ii 2.29

3 a i $5 \ln x$ ii $7 \log x$

b i $\frac{\log a}{\log b}+1$ ii $\log a+1$

4 a i $x=3^y$ ii $x=16^y$

b i $x=a^{y+1}$ ii $x=a^{y^2}$

c i $x=\sqrt[3]{3y}$ ii $x=\sqrt{y}$

d i $x=e^{y-2}$ ii $x=e^2 y$

5 a i $x=32$ ii $x=16$

b i $x=0.4$ ii $x=0.25$

c i $x=6$ ii $x=100$

6 a ae^2 b a^2

7 a 6 b 1.5

8 $x=111$

9 $x=-3$

10 $x=\frac{e^2+1}{3}$

11 $x=10^{1.5}=31.6$

12 $x=9$ or $\frac{1}{9}$

13 $x=81, y=25$

14 $x=e^5$ or e^{-4}

15 It is very close to e.

WORK IT OUT 7.1

Solution 2 is correct.

EXERCISE 7B

1 a i $7y$ ii $2x+y$

b i $x+2y-z$ ii $2x-y-3z$

c i $2-y-5z$ ii $1+y+2z$

d i $x-4y$ ii $2+2x+y+2z$

2 a i $x=7$ ii $x=4$

b i $x=9$ ii $x=\frac{1}{3}$

c i $x=9$ ii $x=4$

3 $x=4$

4 $x=27$

5 $x=8$

6 $x=\frac{5}{3}$

7 a $2x+y$ b $2+x-z$

8 $\ln\left(\frac{x^3 z}{y^2}\right)$

9 a $a+2b$ b $2(a-b)$

10 $A=\frac{1}{3}, B=\frac{3}{2}$

11 $x=2$

12 $x=\frac{2}{3}$ or $\frac{5}{3}$

13 $x=\frac{5}{8}$

14 $x=4$

15 $x=4$ or 36

WORK IT OUT 7.2

Solution 1 is correct.

WORK IT OUT 7.3

Solution 2 is correct.

EXERCISE 7C

1 a i −1.45 ii 1.60

b i 1.36 ii 1.30

c i 2.45 ii 116

d i −0.609 ii 4.62

e i −1.71 ii 0.527

f i 1.11 ii −2.98

2 a i $\frac{\ln 5-\ln 2}{\ln 3}$ ii $\frac{\ln 3-\ln 5}{\ln 7}$

b i $-\ln 4$ ii $-\ln 3$

c i $\frac{\ln 5}{2\ln 2}$ ii $\frac{\ln 7}{3\ln 10}$

d i $\frac{1}{3}\left(\frac{\ln 10}{\ln 2}-1\right)$ ii $\frac{1}{2}\left(\frac{\ln 4}{\ln 5}+3\right)$

e i $\frac{\ln 2}{\ln\left(\frac{5}{2}\right)}$ ii $\frac{\ln 25}{\ln\left(\frac{5}{3}\right)}$

f i $\frac{\ln 3}{\ln(8)-1}$ ii $\frac{\ln 5}{2-\ln 2}$

g i $\frac{\ln \frac{64}{5}}{\ln\left(\frac{3}{8}\right)}$ ii $\frac{\ln\left(\frac{12}{7}\right)}{\ln\left(\frac{2}{9}\right)}$

3 -0.232

4 3.74

5 $x = \ln\left(\sqrt{\frac{e^3}{10}}\right)$

6 $\frac{\log 5}{1-\log 8}$

7 $x = \frac{\ln 50}{\ln 40}$

8 Proof.

EXERCISE 7D

1 a i 0 or $\log_2 5$ ii 1 or $\log_3 4$

b i 0 or $\log_7 8$ ii 1 or $\log_5 4$

c i 1 or $\log_2 3$ ii $\log_3 2$ or $\log_3 4$

d i 2 ii $\log_5 3$

e i $\ln 4$ ii $\ln 4, \ln 5$

f i $1, \frac{1}{\log 5}$ ii $2, \frac{\log 3}{\log 2}$

2 a Proof. b 1.46 or 1.26

3 1 or 3

4 $2\ln 2$ or $-\ln 2$

5 $\log_5 3$

6 $\ln 3$ or $\ln 5$

MIXED PRACTICE 7

1 D

2 $x = 3$

3 $x = \pm 24$

4 1.17

5 $y = \frac{a^4}{7}$

6 a $2a + \frac{b}{2} - c$ b $\frac{a-1}{2}$ c $\frac{b-c}{2}$

7 0.367

8 $x = \frac{\ln 12}{\ln 9}$

9 C

10 $x = e^{\frac{4}{3}} = 3.79, y = e^{\frac{10}{3}} = 28.0$

11 $x = 1 \pm \sqrt{1-e^y}$

12 $x = 1.46$

13 $x = \frac{1}{3}$

14 a $b = a^c$ b $x = -3$

15 a $k = \frac{x^2}{5}$ b 2^{3b+6}

16 A

17 $x = \frac{\ln 3}{\ln 2}$

18 $\frac{\ln\left(\frac{4}{5}\right)}{\ln(36)}$

19 $x = \ln(2 \pm \sqrt{3})$

20 $x = 3$ or 9

8 Exponential model

BEFORE YOU START

1 a $2 + \ln 5$ b $5e$

2 $2 + 3\log x$

3 A horizontal stretch, factor $\frac{1}{2}$.

4 a

b $180°$

EXERCISE 8A

1 a $A = 3, B = 2, C = 1$

b $A = 3, B = 2, C = 1$

c $A = 1, B = 2$

2 a i $3.2e^{3.2x}$ ii $0.6e^{0.6x}$

b i $-1.3e^{-1.3x}$ ii $-e^{-x}$

3 a i 57.6 ii 0.0893

b i -0.0160 ii -0.784

4 a i 25.5 ii 2.4

b i -2.1 ii -0.5

5 a 0.0643 b 18

6 a -1.15 b -0.0363

7 a 3.25 b 0.126

8 -11.3

9 a $\ln 8$ b 0.735

10 a $\ln 0.3$ b -0.0783

EXERCISE 8B

1 a

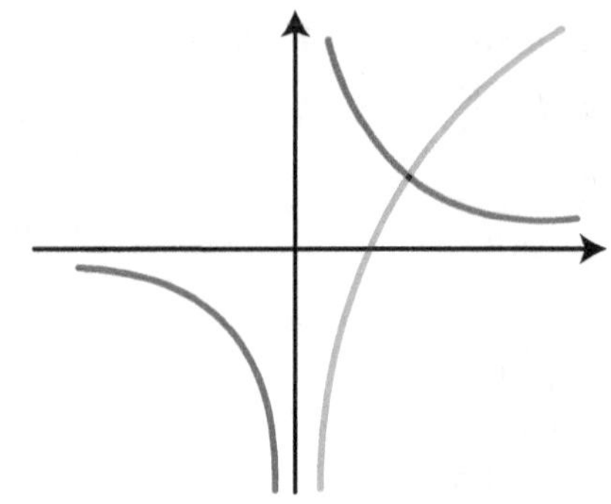

b One

2

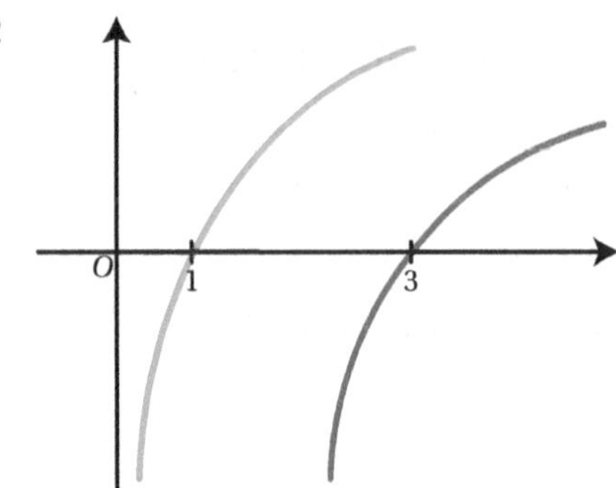

3 Proof.

4 $\begin{pmatrix} 0 \\ \ln 3 \end{pmatrix}, \frac{1}{3}$

5 a

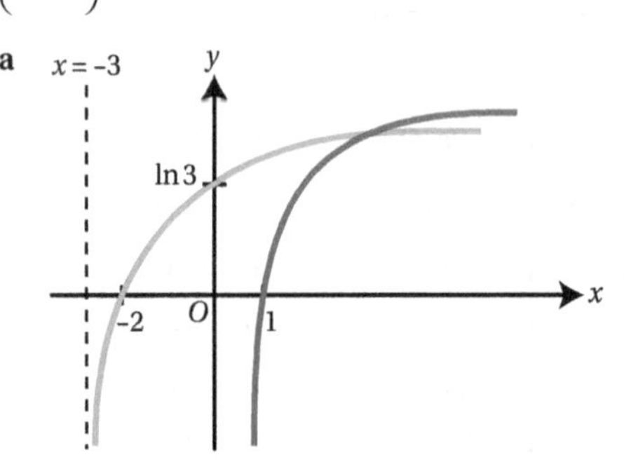

b $x = \frac{1+\sqrt{13}}{2}$

6 a Proof.

b Vertical stretch, scale factor $\frac{1}{\ln(10)}$.

EXERCISE 8C

1 a i $C\left(1+\frac{p}{100}\right)$ ii $C\left(1+\frac{p}{200}\right)^2$

iii $C\left(1+\frac{p}{400}\right)^4$

b Reasoned explanation.

c It tends to (approaches) e.

d $V = Ce^x$

2 a 100 b 48 299

c 2.24 h d 1030 cells per hour

3 a 3 billion b 6.44 billion

4 a 13.31 m^2

b Slowing of growth rate as space and food are used up. Seasonal variation in the growth rate.

5 a

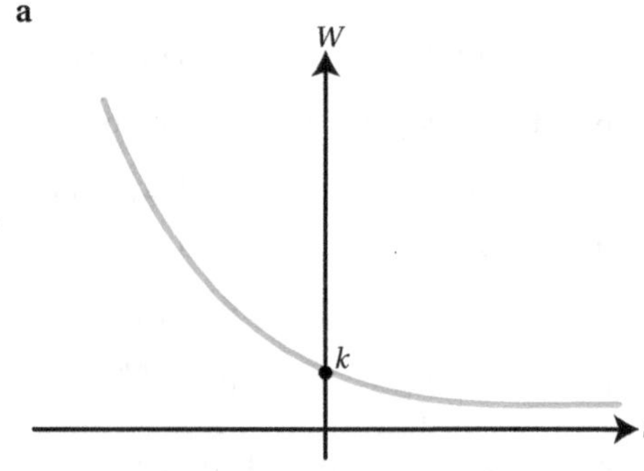

b 2.3 minutes

6 a $N_0 = 450; a = 4.8$ b 3600 per day

c 7.5 hours

7 $N = 500e^{-0.4t}$

8 a $V = 6800e^{-0.223t}$ b £730

c $A = 300, B = 6500$

9 a $P = 7 \times 10^7 \times 1.02^n$

b For example, Population growth rate might change due to changing economic conditions. Immigration is not being taken into account.

10 a 21.2 °C above room temperature.

b i No change.

ii a would get larger, but not go above 1.

11 a 0 m s^{-1} b 40 m s^{-1}

12 a e^{-2} b Proof.

EXERCISE 8D

1 a $A = 2.01, b = 3.32$ b $A = 22.2, b = 0.549$

c $A = 0.0183, b = 9.97$

2 a $C = 3.32, n = 0.7$ b $C = 0.0091, n = 2.1$

c $C = 7.39, n = -0.9$

3 a B b 150 c 0.165

d Not suitable, it predicts indefinite growth.

4 **a**

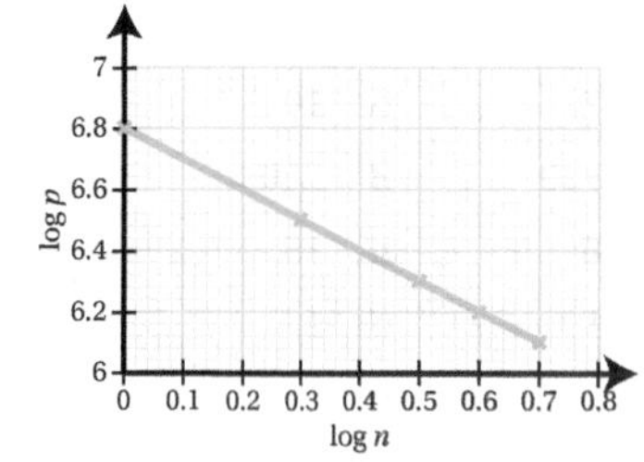

b $p \approx \dfrac{6\,000\,000}{n}$

5 **a** $\ln M = -0.16t + 2$ **b** $K = 7.4, c = 0.85$

c 13 seconds

6 **a** $y = 1.2x + 2$ **b** $C = 7.4, n = 1.2$

c 15.6 years

7 **a** $N = 330\,e^{0.16t}$ **b** 190 mice per week

8 **a** $R = 6e^{-0.2t}$ **b** 4 hours.

c For example: radiation level falls to zero rather than background level.

d $R = 6e^{-0.2t} + 2.7$

9 **a** $\ln\left(\dfrac{1}{N} - 1\right)$ against t will have gradient $-b$ and intercept $\ln J$.

b The logistic function predicts that the population will level off (to one unit) which is more realistic than the exponential model which predicts unlimited population growth.

MIXED PRACTICE 8

1 C

2 C

3 **a**

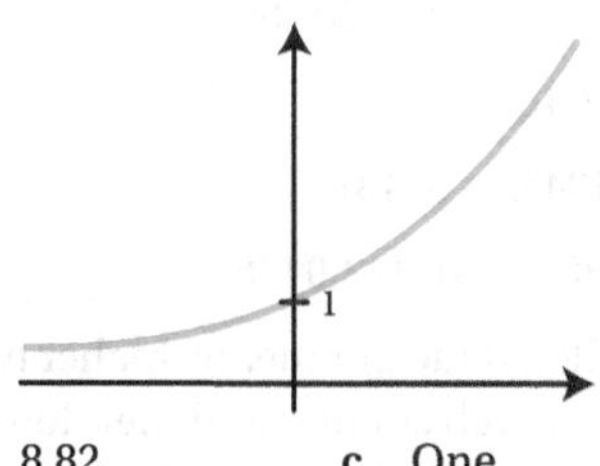

b 8.82 **c** One

4 **a** 32 g **b** -2.9 g s^{-1} **c** 5.0 s

5 **a**

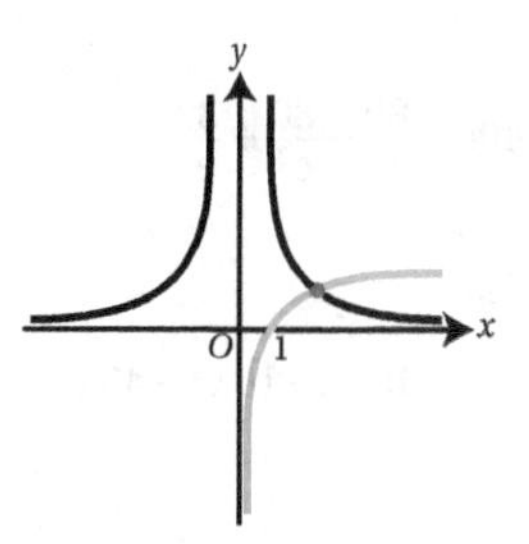

b One

6 **a** 0.4 cm^3 **b** 10 weeks

c Model is for algae in a jar, which limits volume; extrapolation beyond model's validity.

7 **a** $A = 18, k = 0.0141$ **b** 64

c 1.58 p.m.

8 D

9 **a** 12 μU ml^{-1}

b No – long term level is 2 μU ml^{-1}.

10 **a** **i** Proof. **ii** $k = 37\,000, a = 0.949$

b 2750 **c** 2040

11 **a** $\ln N = kt + \ln A$ which will give a straight line.

b 3.74, 4.13

c $\ln N = 0.014t + 3.56$; $N = 35e^{0.014t}$

d 1330

e Size of the lake limits indefinite growth, seasonal variation.

12 $p = -2.5, a = 10^{10}$

13 **a** Explanation.

b $a = 10^4 = 10\,000, n = -\dfrac{2}{3}$

14 0.002 67

15 **a** 0 m s^{-1} **b** 42 m s^{-1} **c** 3.71 s

16 **a** $C = 76, k = 0.027$ **b** 11.3 min

17 **a** 0.5 million **b** 2035 **c** 10 years

d **i** $P = 0.5 \times 2^{0.05t}$

ii $P = 0.5 \times 2^{0.1t}$ for $0 \leqslant t \leqslant 30$, $P = 4$ for $t > 30$.

18 **a** 30 days.

b For example: 30 days is not 4 weeks; does not take into account different rates at different times of year; the fact that number of people is discrete; the fact that the long term number of infections is 500; that another epidemic might occur; the effect of treatment; stochastic effects; geographic effects; that different age groups might recover differently.

c Depends on answer to **b**. For example: $I = 19\,500e^{-0.4t} + 500$.

19 **a** 904 000

b **i** 27.6 minutes **ii** 36.8 minutes

9 Binomial expansion

BEFORE YOU START

1 a 18 b 35

2 a $16x^{12}$ b x^{11} c x^9

3 $4x^2+12x+9$

4 $x=-4, -1$

WORK IT OUT 9.1

Solution 1 is correct.

EXERCISE 9A

1 a i 35 ii 36
b i 1 ii 1
c i 8 ii 45
d i 1 ii 1

2 a i $1+4x+6x^2+4x^3+x^4$
ii $x^5+5x^4+10x^3+10x^2+5x+1$
b i x^3-3x^2+3x-1
ii $1-4x+6x^2-4x^3+x^4$
c i $32-80x+80x^2-40x^3+10x^4-x^5$
ii $729+1458x+1215x^2+540x^3+135x^4+18x^5+x^6$
d i $16-96x+216x^2-216x^3+81x^4$
ii $8x^3-84x+294x^2-343$

3 a i 8 ii 945
b i 14 ii 1792

4 a i 216 ii 20
b i 560 ii −280
c i −5 ii 78 030

5 $27+27x+9x^2+x^3$

6 a $80y^4$ b $-80y^3$

7 −10 500

8 $20\,412x^2$

9 $y^6+18y^7+135y^8+540y^9$

10 a $1+8x+24x^2+32x^3+16x^4$
b $1+\frac{8}{x}+\frac{24}{x^2}+\frac{32}{x^3}+\frac{16}{x^4}$

11 $k=2$

12 56

13 −672

EXERCISE 9B

1 a i 21 ii 10
b i 120 ii 220
c i 8 ii 15
d i 1 ii 1
e i 1 ii 1
f i 7 ii 12

2 a i 10 ii 13
b i 10 ii 12

3 14

4 12

5 a 5 b 80

6 a 10 b 180

7 a 12
b 6

8 $n=2k+1$

EXERCISE 9C

1 a i $2+25x+120x^2$ ii $5+51x+210x^2$
b i $48-224x+264x^2$
ii $972-2025x+270x^2$

2 a $1+10x+40x^2+80x^3+80x^4+32x^5$
b $3+29x+110x^2+200x^3+160x^4+16x^5-32x^6$

3 a $128-2240x+16\,800x^2-70\,000x^3$
b −36 400

4 a $81-540x+1350x^2$
b 80.461

5 a $32\,768+40\,960x+20\,480x^2$
b 33 197.648

6 a $128+1344x+6048x^2$
b i 322.88 ii 142.0448
c ii Smaller value of x means higher order terms much smaller and therefore less important, so the error is less.

7 a $16\,384-28\,672x+21\,504x^2$
b −30 720

8 a $e^5+10e^3+40e+\frac{80}{e}+\frac{80}{e^3}+\frac{32}{e^5}$
b $2e^5+80e+\frac{160}{e^3}$

9 a $(1-x^2)^n$ b $1-10x^2+45x^4$

10 126

11 $m=3, n=15$ or $m=-5, n=-17$

MIXED PRACTICE 9

1 D

2 $4096 - 24\,576x + 67\,584x^2 - 112\,640x^3$

3 $125 + 150x + 60x^2 + 8x^3$

4 720

5 B

6 a $1 - \frac{7x}{10} + \frac{21x^2}{100} - \frac{7x^3}{200}$ b 0.932 065

7 $232 - 164\sqrt{2}$

8 -32

9 $x^8 - 8x^5 + 24x^2 - 32x^{-1} + 16x^{-4}$

10 5733

11 -5

12 $\frac{9}{2}$

13 a $1 + \frac{8}{x} + \frac{16}{x^2}$ b $a = 2, b = \frac{7}{4}, c = \frac{7}{8}$

c 30

14 80

15 $a = 2, n = 5$

16 $k = -1, n = 17$ or $k = 2, n = 5$

17 a

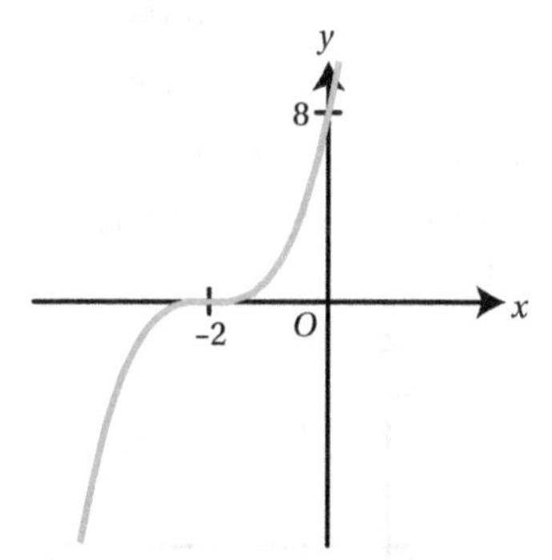

b $x^3 + 6x^2 + 12x + 8$ c 8.012 006 001

d $x = -4$

Focus on ... Proof 1

1 It is important that $q(x)$ is a polynomial to ensure that $q(a)$ is defined. $f(x)$ need not be a polynomial.

2 $r = f\left(\frac{a}{b}\right)$

3 Proof.

4 a Proof.

b No, if $(x - a)$ is a factor then there is no remainder.

Focus on ... Problem solving 1

1 a, b $4\sqrt{6}$

2 $\frac{60}{13}$

3 a, b $k = \pm\frac{\sqrt{2}}{4}$

c No, the y-intercept has to be at the same y-value as the centre of the circle for $\tan\theta$ to be the gradient of the tangent.

4 4

Focus on ... Modelling 1

1 a Graph.

b A is the initial size of the population (at time 0), k affects the rate of growth and decay.

c $k \approx 0.24, A \approx 10$

d $A = 10.29$ (4 s.f.), $k = 0.2448$ (4 s.f.)

e It is not a sensible model as the predicted value of **N** is far larger than the observed value.

2 a A b C

c $k \approx 0.27, A \approx 10, C \approx 460$

3 a N decreases over time. The population is decreasing.

b N tends to 0. The population will die out/become extinct.

c A

d For $D > kA$, the population will grow towards $\frac{D}{k}$. For $D < kA$, the population will decrease towards $\frac{D}{k}$.

e The population is stable.

4 a $A = 7\,430\,000$, $k = 0.005\,59$ to 3 s.f.

b Population decreases by a factor of $e^{-0.005\,59} = 0.9944$ (4 s.f.). Annual death rate is 10.3 people per 1000.

c i About 40 250

ii Assume that the birth and death rates remain constant.

Cross-topic review exercise 1

1 a i $(x-3)^2 - 4$ ii Translation by $\begin{pmatrix} 3 \\ -4 \end{pmatrix}$.

iii

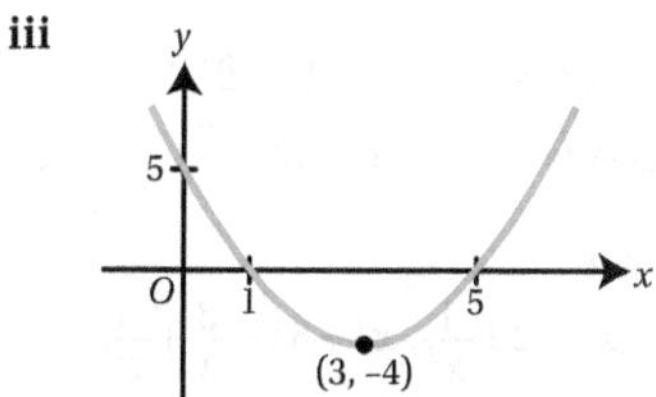

b i

ii 1 or 6

iii

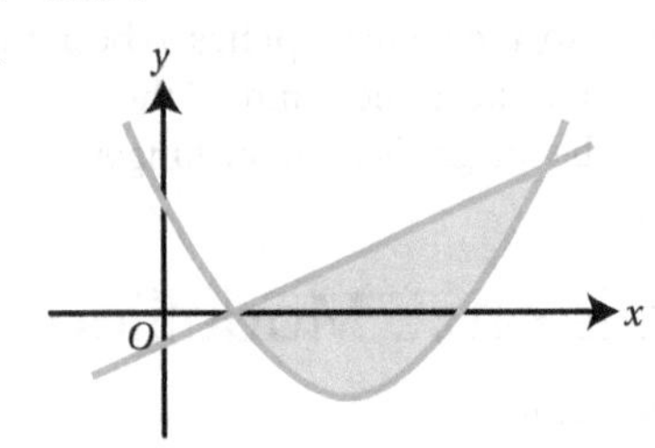

2 **a** Proof.

b $(x-3)^2(x+2)$

c

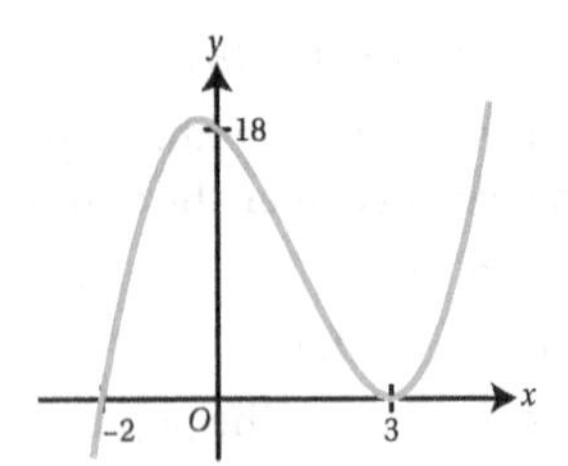

d

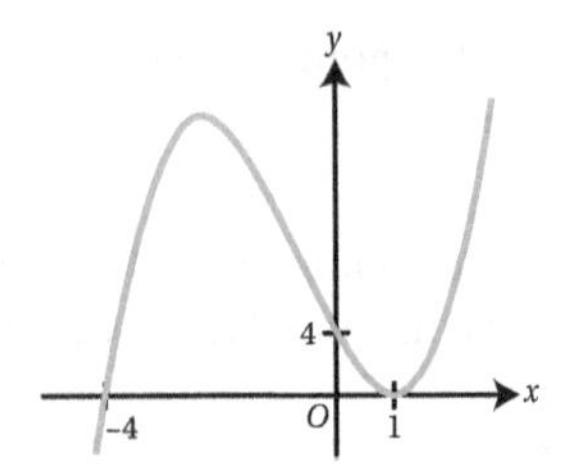

3 84

4 **a** **i** $-\frac{3}{2}$ **ii** $y=-4$

b $(-1, 5)$

c -3 or -11

5 **a** $z=x^2y$ **b** $A=e^2$

6 $-\ln 3$ or $\ln 2$

7 **a** $\frac{1}{k}$ **b** $\ln k$ units up.

8 **a** $2+12x^2+2x^4$ **b** 34

9 $n=\frac{1+\sqrt{1+8k}}{2}$

10 **a** $A=5, C=2$ **b** $\frac{\ln 5}{2}$

11 **a** 3 grams **b** 32 **c** 54

12 **a** Proof.

b **i** $x^2+2+\frac{1}{x^2}$, $x^3+3x+\frac{3}{x}+\frac{1}{x^3}$

ii 7, 18

13 Proof.

14 **a** $x=\ln\left(\frac{y+\sqrt{y^2-4}}{2}\right)$ or $\ln\left(\frac{y-\sqrt{y^2-4}}{2}\right)$

b **i** $y>2$ **ii** Proof.

15 **a** 0.774 **b** $a=\left(\frac{k}{2}\right)^{\frac{3}{2}}$

c **i** $1+6x+12x^2+8x^3$ **ii** $n=-\frac{1}{4}$ or $\frac{3}{2}$

16 **a** 36.8 °C **b** 0 °C

c $T=80e^{-0.1t}+20$ **d** 15.6 minutes

10 Trigonometric functions

BEFORE YOU START

1 3.86 cm

2 21.8°

3 13

4 $x=1$

EXERCISE 10A

1 **a** **i** 0.669 **ii** −0.978

b **i** −0.766 **ii** −0.682

2 **a** **i** 1 **ii** 0

b **i** 1 **ii** −1

c **i** −1 **ii** 0

3 **a** 0.766 **b** 0.766

c −0.766 **d** −0.766

4 **a** 0.766 **b** 0.766

c −0.766 **d** −0.766

5 **a** **i**

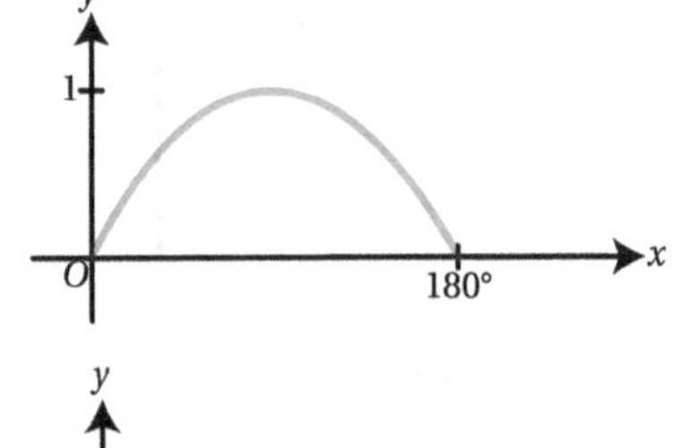

ii

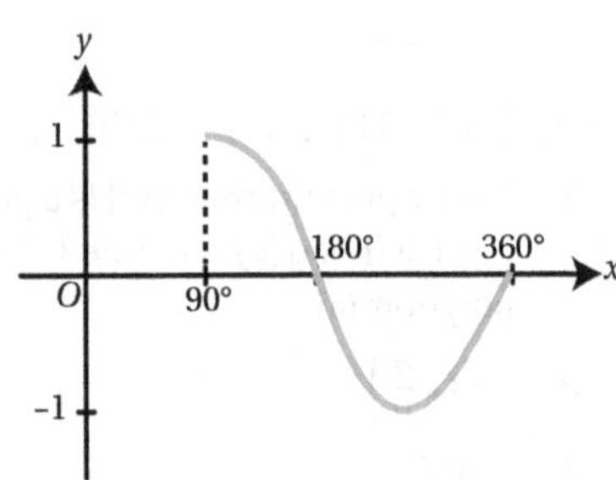

b **i**

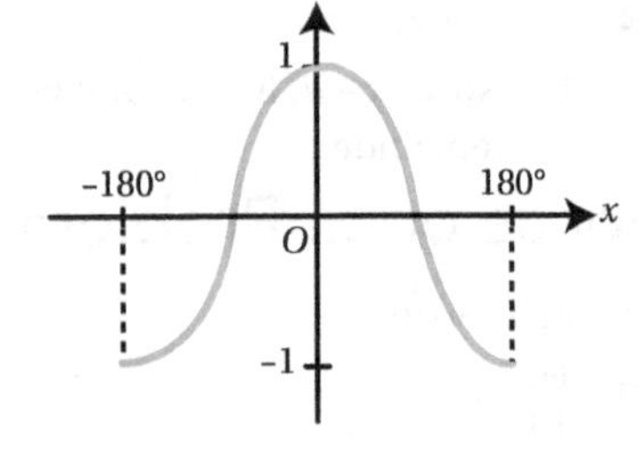

ii

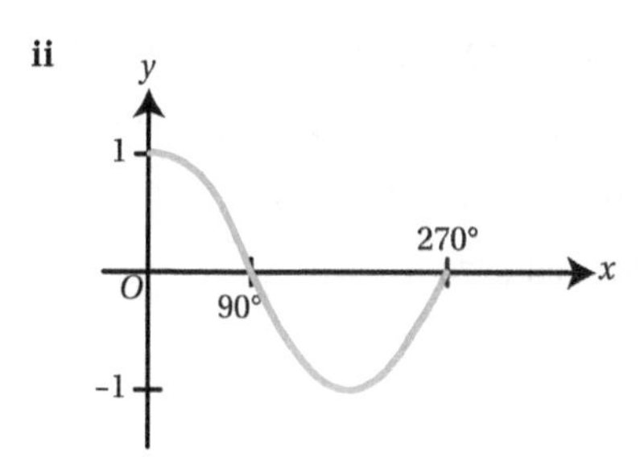

6 0

7 $k = -2$

8 $k = -1$

9 $\sin x$

10 $90 - \ln 3, \ln 3 - 90, -270 - \ln 3, 270 + \ln 3$

EXERCISE 10B

1 a

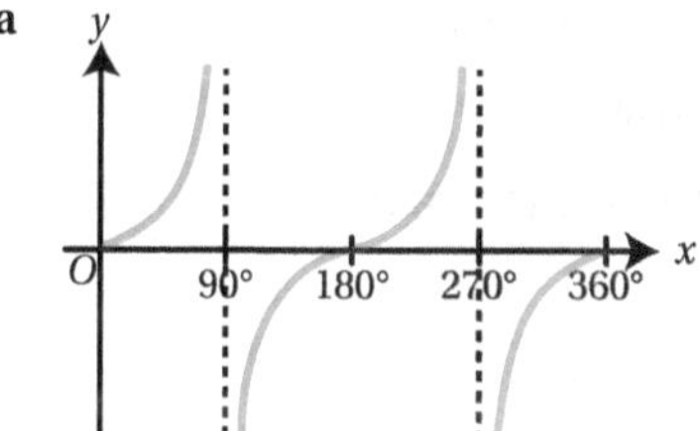

b

2 a i 0.625 ii −0.213

b i 0 ii 1.28

3 a 0.364 b 0.364

c −0.364 d −0.364

4 a $\frac{1}{\tan x}$ b $-\frac{1}{\tan x}$

c $\frac{1}{\tan x}$ d $-\tan x$

5 $2 \tan x$

6 e, e − 180

7 $-t^2$

EXERCISE 10C

1 a $\cos x = \frac{\sqrt{8}}{3}, \tan x = \frac{1}{\sqrt{8}}$

b $\cos x = 0.6, \tan x = \frac{4}{3}$

2 a $\sin x = -\frac{\sqrt{8}}{3}, \tan x = \sqrt{8}$

b $\cos x = -\frac{\sqrt{7}}{4}, \tan x = \frac{3}{\sqrt{7}}$

3 a i $-\frac{2\sqrt{6}}{5}$ ii $\frac{\sqrt{3}}{2}$

b i $-\frac{4}{3}$ ii 0

4 a $\pm\frac{3}{\sqrt{13}}$ b $\pm\frac{1}{\sqrt{5}}$

5 a 3 b 1

c −2 d 1.5

6 a $4 - \sin^2 x$ b $2\cos^2 x - 1$

7 a $-\frac{\sqrt{5}}{3}$ b $-\frac{\sqrt{5}}{2}$

8 $\pm\frac{3}{\sqrt{10}}$

9 $-\sqrt{1 - s^2}$

10 a $5 - \frac{2}{\cos^2 x}$ b $\cos^2 x$

11 1

12 $k = -2$

13 a $\frac{1}{1+t^2}$ b $\frac{t^2}{1+t^2}$

c $\frac{1-t^2}{1+t^2}$ d $\frac{2+3t^2}{t^2}$

EXERCISE 10D

1 a i 44.4° ii 17.5°

b i 128.3° ii 138.6°

c i 81.1° ii −82.0°

2 a i 30°, 150° ii 45°, 135°

b i 60°, 300° ii 30°, 330°

c i 45°, 225° ii 60°, 240°

d i 240°, 300° ii 210°, 330°

e i 135°, 225° ii 180°

f i 150°, 330° ii 135°, 315°

3 a i 26.7°, 153.3° ii 44.4°, 135.6°

b i −138.6°, 138.6° ii −101.5°, 101.5°

c i 18.4°, 198.4°, 378.4°, 558.4°

ii 53.1°, 233.1°, 413.1°, 593.1°

d i −138.2°, −41.8°, 221.8°, 318.2°

ii −165.5°, −14.5°, 194.5°, 345.5°

4 a i 5.7°, 174.3° ii −14.5°, 194.5°

b i 63.6°, 296° ii 57.3°, 303°

c i −121°, 59.0° ii 65.3°, 246°

5 −150°, −30°

6 121°, 301°

7 $\pm30°, \pm330°$

8 $0°, 48.2°, 180°, 312°, 360°$

9 $-336°, -225°, -204°, -45°, 23.6°, 135°, 156°, 315°$

10 For example, $x = 0.5$

11 For example, $x = 180°$

WORK IT OUT 10.1

Solution C is correct.

EXERCISE 10E

1 a i $22.2°, 67.8°, 202°, 248°$

ii $63.8°, 116°, 184°, 236°, 304°, 356°$

b i $37.9°, 82.1°, 158°, 202°, 278°, 322°$

ii $0°, 90°, 180°, 270°, 360°$

c i $14.1°, 59.1°, 104°, 149°, 194°, 239°, 284°, 329°$

ii $58.3°, 148°, 238°, 328°$

2 a i $-25.5°, 126°$ ii $-106°, -23.6°$

b i $-156°, -104°$ ii $-82.5°, 62.5°$

c i $-71.6°, 108°$ ii $-132°, 48.4°$

3 $20°, 80°, 140°$

4 $-120°, -60°, 60°, 120°$

5 $-276°, -83.6°$

6 $205°$

7 $-255°, -225°, -75°, -45°, 105°, 135°, 285°, 315°$

8 $-88.3°, 1.67°, 31.7°$

9 $\pm30°, \pm150°$

10 $\pm\sqrt{30}°, \pm\sqrt{150}° = \pm5.48, \pm12.2$

EXERCISE 10F

1 a i $54.7°, 125°, 235°, 305°$

ii $52.2°, 128°, 232°, 308°$

b i $71.6°, 117°, 252°, 297°$

ii $48.2°, 180°, 312°$

c i $41.4°, 319°$

ii $53.1°, 127°$

d i $0°, 180°, 199°, 341°, 360°$

ii $0°, 129°, 180°, 309°, 360°$

2 a i $\pm180°, \pm66.4°, 0°$

ii $-127°, \pm90°, -53.1°$

b i $\pm90°, 14.5°, 166°$

ii $\pm180°, \pm53.1°, 0°$

3 $-330°, -270°, -210°, -90°, 30°, 90°, 150°, 270°$

4 a -0.5 b $210°, 330°$

5 $-104.0°, -71.6°, 76.0°, 108.4°$

6 $90°, 180°, 270°$

7 $36.5°, 83.5°, 156°$

EXERCISE 10G

1 a i $33.7°$ ii $59.0°$

b i $66.8°$ ii $78.7°$

2 a i $45°, 135°, 225°, 315°$

ii $54.7°, 125.3°, 234.7°, 305.3°$

b i $45°, 135°, 225°, 315°$

ii $0°, 180°, 360°$

3 $67.5°, 158°, 248°, 338°$

4 $-153°, 26.6°$

5 $0°, 135°, 180°, 315°, 360°$

6 $0°, \pm180°$

7 $\pm41.8°, \pm138.2°$

8 $30°, 150°, 270°$

9 $-14.5°, -90°, -166°$

10 a $\frac{1}{3}$ b $70.5°, 289°$

11 a $\frac{2}{3}, -\frac{1}{2}$ b $48.2°, 120°, 240°, 312°$

12 a Proof.

b $-153°, -135°, 26.6°, 45°$

MIXED PRACTICE 10

1 B

2 $-31.8°, 148.2°$

3 $-135°, -45°, 225°, 315°$

4 $\pm131° \pm48.6°$

5 10

6 a Proof.

b $90°, 120°, 240°, 270°$

7 D

8 $570°, 690°$

9 $a = 5, b = 45°$

10 $41.4°, 132°, 228°, 319°$

11 a Proof.

b $25°, 65°, 115°, 155°$, to the nearest degree

12 A

13 $-69.1°, -20.9°, 15°, 75°$

14 a $k = \pm 4$

b Proof.

c i 1 ii $\pm 60°, \pm 300°$

iii $k = 5$ iv 7

11 Triangle geometry

BEFORE YOU START

1 $41.8°$

2 $110°$

3 a $x = -4, 3$ b $2 \pm \sqrt{5}$

4 $8.63°, 171°$

WORK IT OUT 11.1

Solution 1 is correct.

EXERCISE 11A

1 a i 6.04 cm ii 14.4 cm

b i 10.6 cm ii 23.3 cm

2 a i $49.7°$ ii $59.2°$

b i $74.6°$ or $105°$ ii $62.0°$ or $118°$

c i $50.9°$ ii $54.4°$

3 $21.0°, 29.0°$, 8.09 cm

4 $49.9°, 95.1°$, 10.4 cm or $130°, 14.9°$, 2.69 cm

5 9.94 cm

6 23.3 m

7 Proof.

EXERCISE 11B

1 a i 5.37 cm ii 3.44 cm

b i 8.00 cm ii 20.5 cm

2 a i $60.6°$ ii $120°$

b i $81.5°$ ii $99.9°$

3 a $106°$ b $36.2°$

4 6.12 km

5 25.6 km

6 7.95

7 4.4

8 $x = 3$

9 $2\sqrt{2} + \sqrt{41}$ cm

EXERCISE 11C

1 a i 10.7 cm^2 ii 24.3 cm^2

b i 27.6 cm^2 ii 26.2 cm^2

2 a $81.7°, 98.3°$ b $60.9°, 119°$

3 17.7 cm, 29.7 cm^2

4 10 cm

5 5

6 8

7 $4\sqrt{3}$ cm^2

MIXED PRACTICE 11

1 D

2 $58.7°, 121°$

3 6 cm

4 a 8.09 m b 6.58 m

5 a Proof.

b 14.3 cm^2

6 a $123.6°$ b 9.7 cm

7 C

8 $2\sqrt{43}$

9 11.3 cm^2 (3 s.f.)

10 a $\frac{23}{32}$ b $\frac{3}{32}\sqrt{55}$ c $\frac{15}{4}\sqrt{55}$ cm^2

11 a $009°$ or $071°$ b 15.5 km

12 a $\frac{x^2}{4} + 25 - 5x\cos\theta$

b Proof.

c $41.4°$

13 10.8 cm

14 a Proof.

b $\left[-1, -\frac{\sqrt{3}}{2}\right] \cup \left[\frac{\sqrt{3}}{2}, 1\right]$

c $0° < \theta \leqslant 30°$

12 Differentiation

BEFORE YOU START

1 a $x^{\frac{2}{3}} - \frac{5}{2}x^{-1}$ b $2x^{-\frac{1}{2}} + x^{\frac{1}{2}}$

2 a $x \leqslant -3$ b $x \leqslant -2, x \geqslant 6$

3 $-\frac{3}{4}$

4 $x^3 + 6x^2 + 12x + 8$

EXERCISE 12A

1 **a** **i**

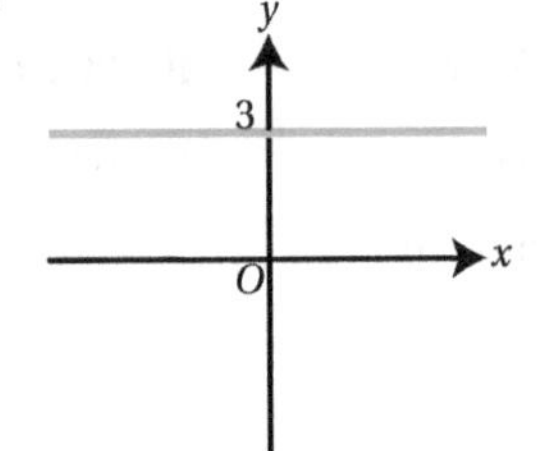

ii

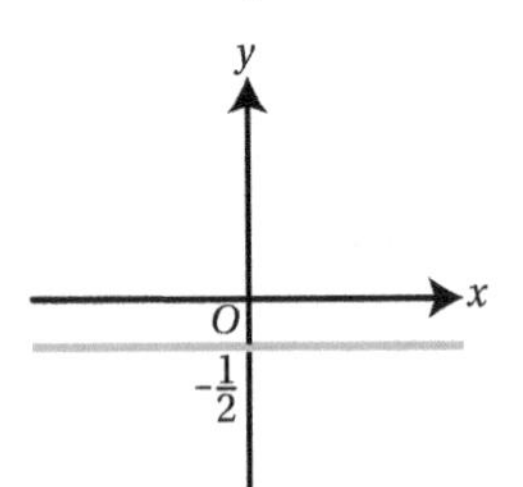

b **i**

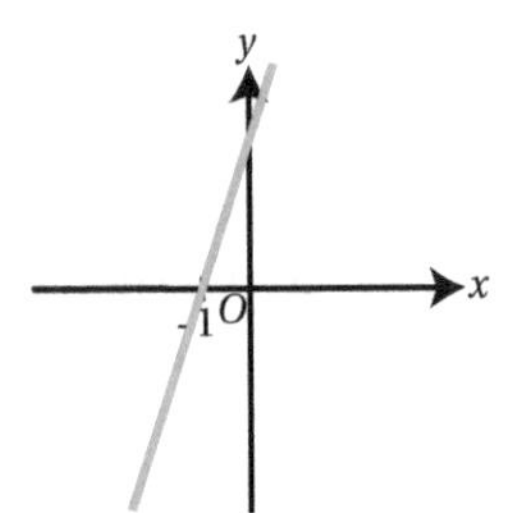

ii

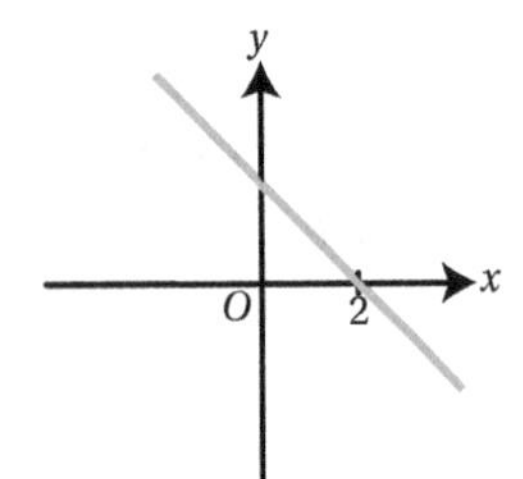

c **i**

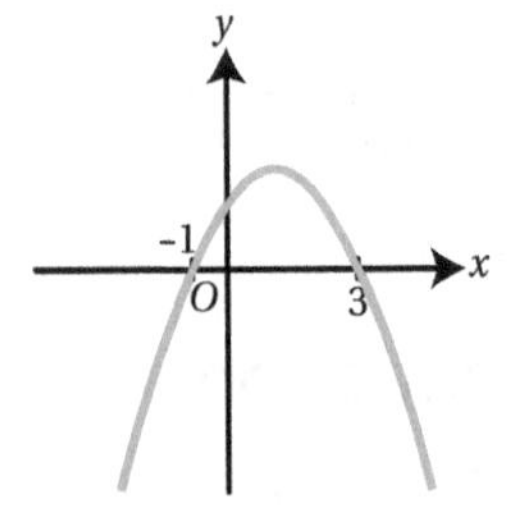

ii

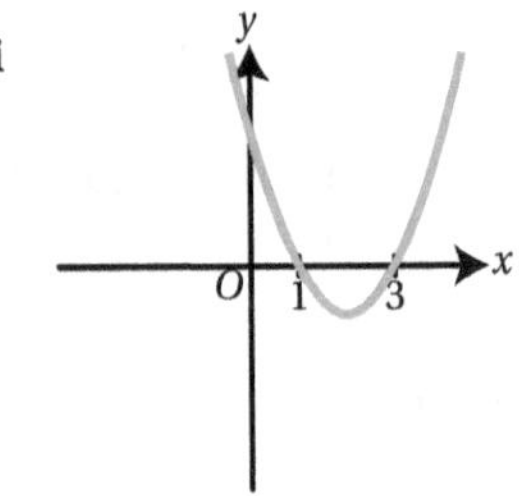

d **i**

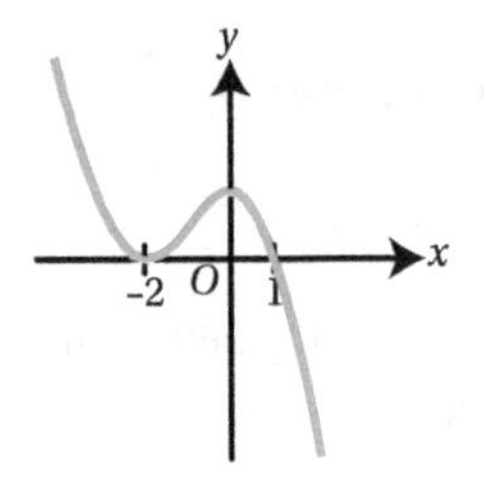

ii

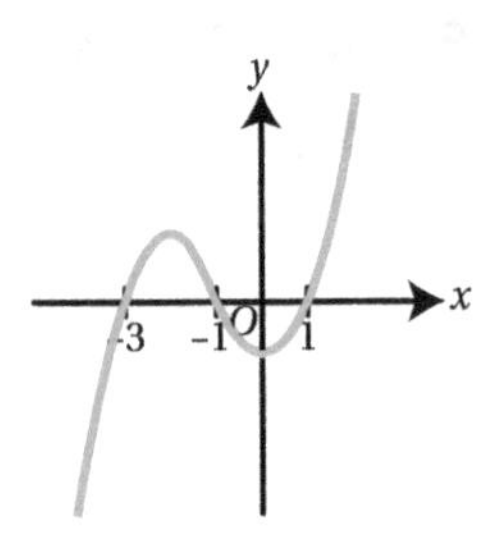

e **i**

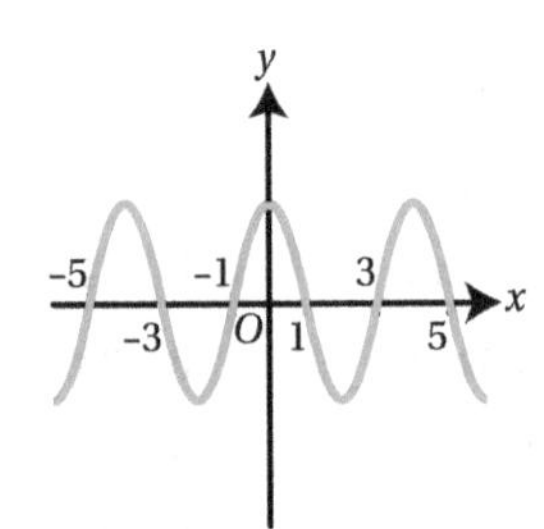

ii

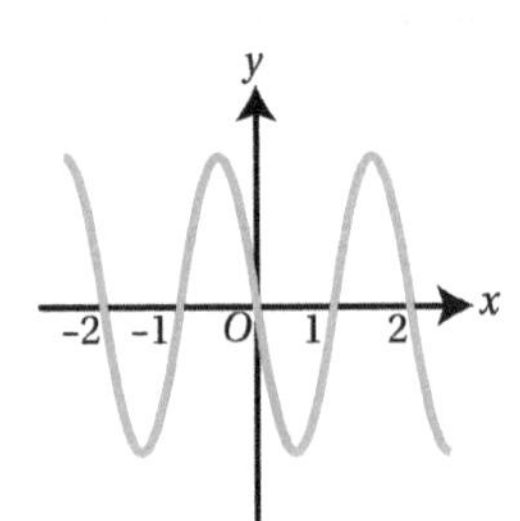

f **i**

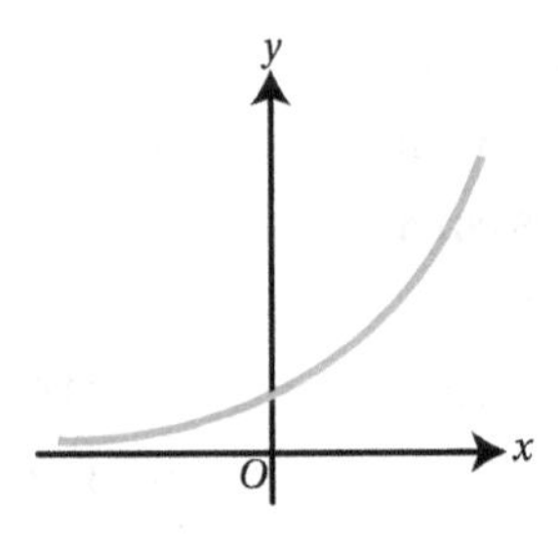

ii

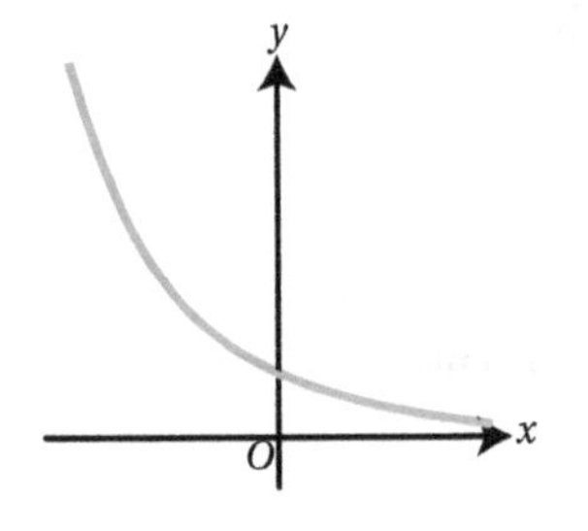

2 **a**

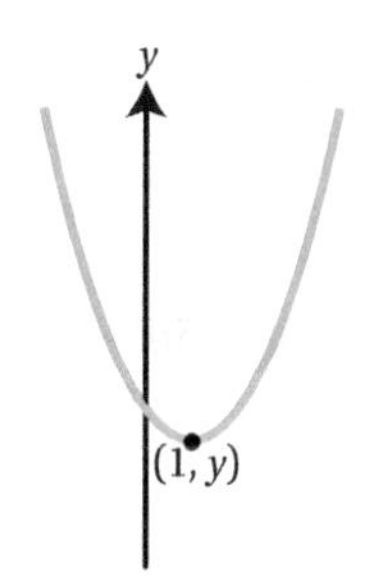

b

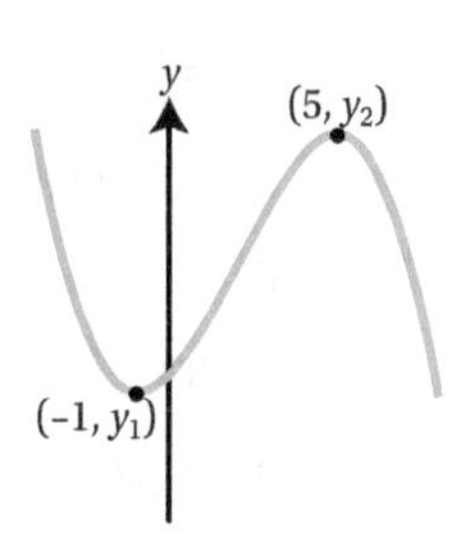

c

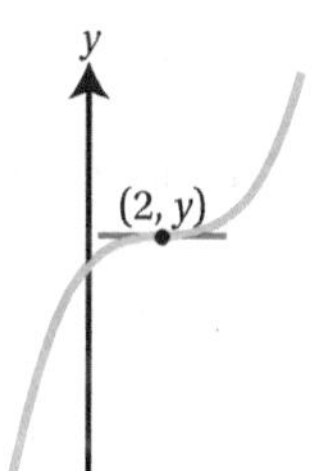

d

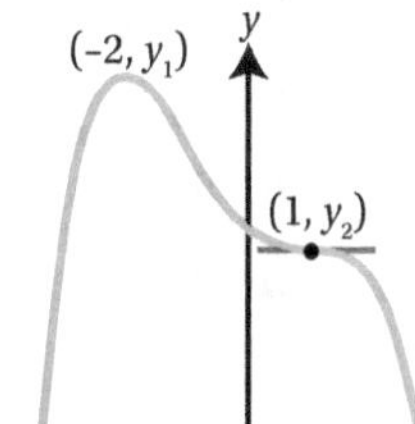

3 **a** Sometimes true. **b** Sometimes true.

c Always true. **d** Sometimes true.

e Sometimes true. **f** Sometimes true.

EXERCISE 12B

1 Proof.

2 -4

3 $6x$

4 Proof.

5 $2x-6$

6 $2x-3$

7 **a** $x^3+3x^2h+3xh^2+h^3$ **b** $3x^2$

8 $9x^2$

9 **a** $x^4+4x^3h+6x^2h^2+4xh^3+h^4$

b Proof.

10 **a** $\frac{f(x+h)-f(x)}{h}$

b Gradient of chord becomes gradient of tangent.

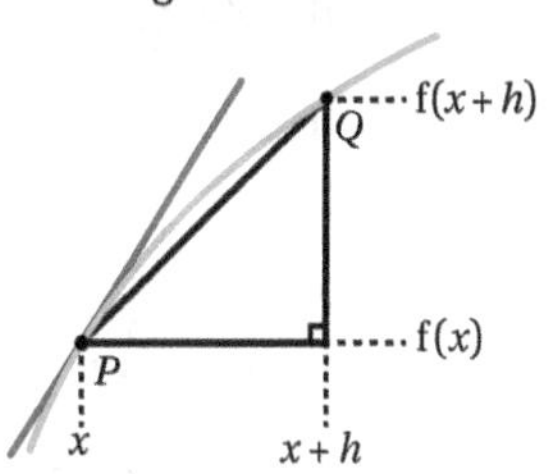

11, 12 Proof.

EXERCISE 12C

1 **a** **i** $y'=4x^3$ **ii** $y'=6x^5$

b **i** $y'=21x^6$ **ii** $y'=-20x^4$

c **i** $y'=-1$ **ii** $y'=3$

d **i** $y'=0$ **ii** $y'=0$

e **i** $y'=2x^5$ **ii** $y'=-\frac{3}{2}x$

f **i** $y'=12x^2-10x+2$

ii $y'=8x^3+9x^2-1$

g **i** $y'=7-\frac{3}{2}x^2$ **ii** $y'=-20x^3+x^4$

2 **a** **i** $f'(x)=\frac{3}{2}x^{\frac{1}{2}}$ **ii** $f'(x)=\frac{2}{3}x^{-\frac{1}{3}}$

b **i** $f'(x)=4x^{-\frac{1}{2}}$ **ii** $f'(x)=8x^{\frac{1}{3}}$

c **i** $f'(x)=\frac{1}{3}x^{-\frac{1}{4}}$ **ii** $f'(x)=\frac{1}{2}x^{-\frac{1}{6}}$

d **i** $f'(x)=12x^3-6x^{-\frac{3}{5}}$

ii $f'(x)=3x^2-x^{\frac{2}{3}}+\frac{2}{3}x^{-\frac{1}{2}}$

e **i** $f'(x)=-x^{-2}$ **ii** $f'(x)=3x^{-4}$

f **i** $f'(x)=-\frac{1}{2}x^{-\frac{3}{2}}$ **ii** $f'(x)=-\frac{2}{3}x^{-\frac{5}{3}}$

g **i** $f'(x)=8x^{-\frac{7}{3}}$ **ii** $f'(x)=6x^{-\frac{7}{4}}$

h **i** $f'(x)=5+\frac{4}{3}x^{-\frac{7}{2}}$

ii $f'(x)=x^{-\frac{10}{7}}-8x^{-7}$

3 $\frac{dy}{dx}=2x-\frac{3}{2}x^{-\frac{1}{2}}$

4 $f'(x)=12x^2-6x-3x^{-\frac{5}{2}}$

5 $f'(x)=-4x^{-\frac{4}{3}}+\frac{1}{3}x^{-\frac{3}{5}}$

WORK IT OUT 12.1

Solution 2 is correct.

EXERCISE 12D

1 a i $y' = \frac{1}{3}x^{-\frac{2}{3}}$ ii $y' = \frac{1}{5}x^{-\frac{4}{5}}$

b i $y' = 2x^{-\frac{3}{4}}$ ii $y' = \frac{1}{6}x^{-\frac{1}{2}}$

c i $y' = x^{-2}$ ii $y' = -4x^{-5}$

d i $y' = -6x^{-3}$ ii $y' = 4x^{-11}$

e i $y' = -\frac{1}{2}x^{-\frac{3}{2}}$ ii $y' = -\frac{1}{3}x^{-\frac{4}{3}}$

f i $y' = 2x^{-\frac{6}{5}}$ ii $y' = -\frac{2}{3}x^{-\frac{5}{4}}$

2 a i $f'(x) = 4x + 5$ ii $f'(x) = 6x - 15$

b i $f'(x) = 6x^{\frac{1}{2}} + \frac{3}{2}x^{-\frac{1}{2}}$

ii $f'(x) = \frac{4}{3}x^{\frac{1}{3}} - \frac{1}{3}x^{-\frac{2}{3}}$

c i $f'(x) = 1 + 8x + 6x^{\frac{1}{2}}$

ii $f'(x) = \frac{1}{2}x^{-\frac{1}{2}} - 2x^{-\frac{3}{4}}$

d i $f'(x) = 2x - 2x^{-3}$

ii $f'(x) = 2x + 8x^{-3}$

3 a i $f'(x) = 2x^{-2}$ ii $f'(x) = -\frac{1}{2}x^{-2} + 2$

b i $f'(x) = -\frac{3}{2}x^{-\frac{5}{2}} + 6x^{-3}$

ii $f'(x) = \frac{3}{2}x^{\frac{1}{2}} - 2x^{-\frac{3}{2}}$

4 $y' = 9x^2 - 8x$

5 $y' = 7x^{\frac{5}{2}} + 4x^{-\frac{1}{2}}$

6 $\frac{dy}{dx} = \frac{4}{5}x^{-\frac{1}{5}}$

7 $f'(x) = -2x^{-\frac{7}{4}}$

8 a $y = 3x^3 - 2x^{-1}$ b $\frac{dy}{dx} = 9x^2 + 2x^{-2}$

9 Proof.

10 $f'(x) = \frac{15}{2}x^{\frac{2}{3}} - \frac{1}{2}x^{-\frac{4}{3}}$

11 $\frac{dy}{dx} = -2x^{-\frac{3}{2}} + 12x^{-2} - \frac{27}{2}x^{-\frac{5}{2}}$

WORK IT OUT 12.2

Solution 2 is correct.

EXERCISE 12E

1 a $\frac{dz}{dt}$ b $\frac{dQ}{dP}$ c $\frac{dR}{dm}$

d $\frac{dV}{dt}$ e $\frac{dy}{dx}$ f $\frac{d^2z}{dy^2}$

g $\frac{d^2H}{dm^2}$

2 a i $\frac{5}{3}x^{-\frac{2}{3}}$ ii $15q^4$

b i $3 - 7t^{-2}$ ii $1 - c^{-2}$

c i $18 + 6x$ ii $6t^{-3}$

3 a i 30 ii $\frac{227}{36}$

b i 7 ii −29 999.8

c i 12 ii −10

d i 24 ii 32

e i 6 ii $\frac{7}{2\sqrt{6}}$

4 a i $2ax + 1 - a$ ii $3x^2$

b i $\frac{1}{2}\sqrt{\frac{b}{a}}$ ii $6a^2v$

5 a i 54 ii 384

b i 8 ii $\frac{1}{108}$

c i 0 ii 42

6 a i ±2 ii $\left(\frac{3}{4}\right)^{\frac{1}{3}}$

b i ±17 ii 6

c i −3 ii 2

7 a i $x > \frac{1}{2}$ ii $x < -1$

b i $x < -1$ or $x > 1$ ii $-\frac{4}{3} < x < 0$

c i $-1 < x < 1$ ii $x < 0$ or $x > \frac{1}{3}$

8 a $-\frac{1}{18}$ b Increasing.

9 0

10 $x < \frac{2}{3}$

11 $x = 5$ or $x = -3$

12 $\pm\sqrt{2}$

13 Proof.

14 (1, 0) and (3, 4)

15 $x < 0$ or $x > 4$

16 $a = -6, b = 2$

17 $x > -\frac{1}{3}$

18 $-1 - \frac{2\sqrt{3}}{3} < x < -1 + \frac{2\sqrt{3}}{3}$

19 $2 < x < 3$

20 $n!$

MIXED PRACTICE 12

1 C

2 $\frac{dy}{dx} = -12x^2 + 24x + 1$

3 $-\frac{1}{2}$

4 a $f'(x) = \frac{3}{2}x^{-\frac{1}{2}} + x^{-\frac{3}{2}}$ b $\frac{7}{8}$

5 $b = 8, c = -7$

6 a $\frac{9}{4}$

b Gradient > 0 so the curve is increasing at this point.

7 $x > \frac{2}{3}$

8 $\frac{109}{54}$

9 a $x^2 + x^{-\frac{1}{2}}$

b i $\frac{dy}{dx} = 2x - \frac{1}{2}x^{-\frac{3}{2}}$

ii $\frac{d^2y}{dx^2} = 2 + \frac{3}{4}x^{-\frac{5}{2}}$

10 a $\frac{dy}{dt} = \frac{1}{2}t^3 - 2t$

b i $-1.5\ \text{m s}^{-1}$

ii $\frac{dy}{dt} < 0$ so the height is decreasing.

c $4\ \text{m s}^{-2}$

11 A

12 D

13 D

14 $3x^2 - 5$

15 a 0 b Increasing.

c

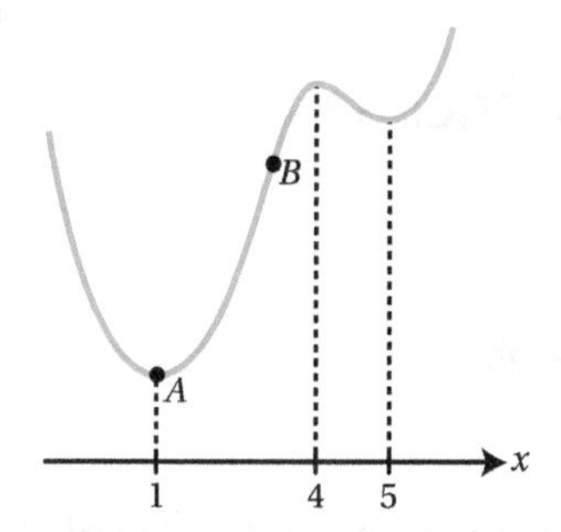

16 $\left(\frac{1}{16}, \frac{7}{16}\right)$

17 -54

18 $a = 4.8, b = -1.8$

19 1 or 25

20 $0 < x < \frac{3}{2}$

13 Applications of differentiation

BEFORE YOU START

1 a $\frac{6x}{5} + \frac{4x^{-3}}{5}$ b $-\frac{3}{4}x^{-\frac{3}{2}}$

2 $\frac{9}{16}$

3 $y = 3x - 5$

4 $y = -\frac{4}{3}x - \frac{1}{3}$

EXERCISE 13A

1 a i tangent $y + 2 = 0$
normal $x - 1 = 0$

ii tangent $9x - y + 11 = 0$
normal $x + 9y - 17 = 0$

b i tangent $x + 3y + 6 = 0$
normal $3x - y + 8 = 0$

ii tangent $3x - 2y - 9 = 0$
normal $4x + 6y + 1 = 0$

c i tangent $x - 4y + 36 = 0$
normal $4x + y - 26 = 0$

ii tangent $x + 6y - 32 = 0$
normal $6x - y - 44 = 0$

2 $11x - 4y - 4 = 0$

3 $32x + 70y - 583 = 0$

4 $(162, 0)$

5 $\frac{9}{4}$

6 $y = 5$ and $y = 6$

7 1.55 or -0.215

8 $a = -6, b = 8$

9 $p = -1, q = -\frac{1}{6}$

10 $\left(\frac{1}{4}, \frac{7}{2}\right)$

11 $(-1, -4)$

12 $(2, 1)$

13 $c = -2$

14, 15 Proof.

EXERCISE 13B

1 a i $(0, 0)$ local maximum; $\left(\frac{10}{3}, -\frac{500}{27}\right)$ local minimum

ii $(0, 1)$ local minimum; $(4, 33)$ local maximum

b i $(0, 0)$ local maximum; $(2, -16)$ local minimum; $(-2, -16)$ local minimum

ii $(-1, 2)$ local minimum; $(0, 3)$ local maximum; $(1, 2)$ local minimum

c i $\left(\frac{1}{2}, 4\right)$ local minimum; $\left(-\frac{1}{2}, -4\right)$ local maximum

ii $\left(\frac{3}{2}, 27\right)$ local minimum

d i $\left(\frac{1}{9}, \frac{1}{3}\right)$ local maximum

ii $\left(\frac{1}{4}, 3\right)$ local minimum

2 $(-4, 92)$ local maximum, $(2, -16)$ local minimum.

3 Proof.

4 a $\pm\frac{1}{3}$

b $\left(\frac{1}{3}, 6\right)$ local minimum, $\left(-\frac{1}{3}, -6\right)$ local maximum.

5 $\left(\frac{1}{4}, -\frac{1}{4}\right)$ local minimum

6 a $(0, 0)$ and $(1, -1)$

b $(0, 0)$ local maximum, $(1, -1)$ local minimum

7 a $\frac{4}{3}$

b Local maximum at $x = -8$
Local minimum at $x = \frac{4}{3}$

8 $a = 3, b = 12$

9 $0 \leqslant \mathrm{f}(x) \leqslant 54$

10 Proof.

11 a $(0, 6), (1, 11), (3, -21)$ b $6 < c < 11$

12 $(0, 0)$ local minimum; $\left(-\frac{4}{k}, \frac{32}{k^2}\right)$ local maximum

EXERCISE 13C

1 a i 2 ii $\frac{49}{12}$

b i $2\sqrt{3}$ ii $4\sqrt{2}$

c i $4\sqrt{2}$ ii $\frac{8}{3}$

2 a $225\ \mathrm{m}^2$ b $60\ \mathrm{m}$

3 a Proof.

b i $r = 10$ ii Proof.

4 a Proof. b $x = 2$

5 Proof.

6 a $V = 225r - \pi r^3$ b $733\ \mathrm{cm}^3$

7 a $V = \frac{(243 - x^2)x}{2}$ b $729\ \mathrm{cm}^3$

8 a Proof.

b $96\sqrt{3}\ \mathrm{cm}^2$

c Proof.

9 a Proof.

b $r = 5.56, h = 7.86$.

10 a $x = y = 3$

b End-point values of 0 and 6 produce the maximum value.

11 Minimum: 2 minutes; maximum: 4 minutes.

12, 13 Proof.

14 $\left(\sqrt{\frac{7}{2}}, \frac{7}{2}\right)$

15 a Proof. b $V = \pi(6 + x)^2\left(\frac{12}{x} + 2\right)$

c 486π

MIXED PRACTICE 13

1 C

2 C

3 $19x - 4y = 28$

4 $x = 2$

5 $x = 2 + \sqrt{2}$ minimum; $x = 2 - \sqrt{2}$ maximum

6 a $(0,-1)(2,3)(4,-1)$

b $(0,-1)$ and $(4,-1)$ local minima; $(2,3)$ local maximum.

7 $\left(\frac{2\sqrt{3}}{3},0\right)$

8 a $y=12x^{-1}+x^{\frac{3}{2}};\left(p=-1,q=\frac{3}{2}\right)$

b i $\frac{dy}{dx}=-12x^{-2}+\frac{3}{2}x^{\frac{1}{2}}$

ii $4x+9y=115$ iii $x=2^{\frac{6}{5}};\left(k=\frac{6}{5}\right)$

9 B

10 D

11 $a=2, b=-13$

12 $(1,-2)$ local maximum; $(4,-3)$ local minimum

13 a $(-\sqrt{3},6\sqrt{3})$ and $(\sqrt{3},-6\sqrt{3})$

b Minimum $-6\sqrt{3}$; maximum 80

14 $\frac{3}{4}$

15 a 4 litres b 41.5 litres c 20 seconds

16 a Proof. b $40\sqrt{2}$

17 a Proof.

b i $72-18x^2$ ii Proof.

c $\frac{d^2V}{dx^2}=-36x$; maximum

18 a Proof. b $3a+b=6$

c $a=1, b=3$ d $(-7,-192)$

19 $a=3^{-\frac{1}{4}}$

20 $a=1.5$

14 Integration

BEFORE YOU START

1 a $x=0$ or $\frac{4}{3}$ b $x=0$ or 4

c $x=0$ or $\pm\sqrt{5}$

2 a $\frac{dy}{dx}=6x-1$ b $\frac{dy}{dx}=\frac{1}{2}x^{-\frac{1}{2}}+2x^{-\frac{5}{3}}$

3 a $\frac{dy}{dx}=-3x^{-3}$ b $\frac{dy}{dx}=\frac{9}{2}x^{\frac{1}{2}}$

EXERCISE 14A

1 a i $y=\frac{1}{6}x^6+c$ ii $y=\frac{1}{7}x^7+c$

b i $y=x^3+c$ ii $y=-x^5+c$

c i $y=\frac{1}{2}x^4+c$ ii $y=-\frac{3}{8}x^8+c$

d i $y=\frac{1}{2}x^2+c$ ii $y=-2x^2+c$

e i $y=c$ ii $y=x+c$

f i $y=-\frac{1}{24}x^4+c$ ii $y=\frac{1}{27}x^9+c$

g i $y=\frac{2x^7}{5}+c$ ii $y=\frac{5x^2}{7}+c$

h i $y=\frac{1}{2}x^6-\frac{1}{2}x^2+4x+c$

ii $y=\frac{2}{5}x^5+x^4-4x^2-x+c$

i i $y=\frac{x^3}{6}+x^2-\frac{3x}{4}+c$ ii $y=\frac{x^3}{3}-\frac{x^6}{16}+c$

2 a i $\frac{2}{5}x^{\frac{5}{2}}+c$ ii $\frac{3}{4}x^{\frac{4}{3}}+c$

b i $3x^{\frac{5}{3}}+c$ ii $8x^{\frac{7}{4}}+c$

c i $x^{\frac{5}{4}}+c$ ii $-x^{\frac{3}{2}}+c$

d i $-\frac{x^{\frac{7}{3}}}{2}+c$ ii $\frac{2x^{\frac{7}{5}}}{3}+c$

e i $\frac{5}{2}x^2-2x^{\frac{7}{2}}+c$ ii $6x+\frac{x^2}{6}-8x^{\frac{3}{2}}+c$

f i $-x^{-1}+c$ ii $-\frac{1}{2}x^{-2}+c$

g i $\frac{3}{2}x^{\frac{2}{3}}+c$ ii $\frac{4}{3}x^{\frac{3}{4}}+c$

h i $-12x^{\frac{1}{3}}+c$ ii $-2x^{\frac{5}{2}}+c$

i i $2x^{\frac{1}{2}}+\frac{2x^{-3}}{5}+c$ ii $-\frac{3x^{-4}}{4}+\frac{42x^{\frac{5}{6}}}{5}+c$

3 a i $3t+c$ ii $7z+c$

b i $\frac{q^6}{6}+c$ ii $\frac{r^{11}}{11}+c$

c i $\frac{15}{2}g^{\frac{8}{5}}+c$ ii $\frac{4}{3}y^{\frac{9}{2}}+c$

d i $20a^{\frac{1}{4}}+c$ ii $-21p^{-\frac{1}{3}}+c$

4 $f(x)=\frac{1}{8}x^4+9x^{-\frac{2}{3}}+2x+c$

5 $\frac{2}{5}x^{\frac{5}{2}}+2x^{-\frac{1}{2}}+c$

6 $\frac{1}{12}x^6+\frac{3}{4}x^{-2}+\frac{1}{4}x^2+c$

WORK IT OUT 14.1

Solution A is correct.

EXERCISE 14B

1 a i $\frac{2}{3}x^{\frac{3}{2}}+c$ ii $\frac{4}{5}x^{\frac{5}{4}}+c$

b i $5x^{\frac{6}{5}}+c$ ii $\frac{1}{8}x^{\frac{4}{3}}+c$

c i $-\frac{1}{2}x^{-2}+c$ ii $\frac{1}{6}x^{-6}+c$

d i $2x^{-4}+c$ ii $-\frac{2}{3}x^{-5}+c$

e i $2x^{\frac{1}{2}}+c$ ii $\frac{3}{2}x^{\frac{2}{3}}+c$

f i $-18x^{\frac{5}{6}}+c$ ii $\frac{3}{2}x^{\frac{3}{4}}+c$

2 a i $\frac{2}{3}x^3+\frac{5}{2}x^2+3x+c$ ii $\frac{3}{4}x^4-2x^3+c$

b i $2x^{\frac{5}{2}}+\frac{8}{3}x^{\frac{3}{2}}+c$

ii $\frac{3}{5}x^{\frac{10}{3}}+\frac{9}{4}x^{\frac{4}{3}}+c$

c i $\frac{2}{3}x^{\frac{3}{2}}+3x^3+\frac{8}{3}x^{\frac{9}{4}}+c$

ii $\frac{1}{2}x^2+\frac{1}{5}x^5-\frac{4}{7}x^{\frac{7}{2}}+c$

d i $\frac{1}{3}x^3-x^{-1}-2x+c$

ii $\frac{4}{3}x^3+\frac{1}{3}x^{-3}+c$

3 a i $-7x^{-1}+x^{-2}+c$ ii $-\frac{1}{4}x^{-2}-\frac{5}{4}x^{-1}+c$

b i $-2x^{-\frac{1}{2}}+3x^{-1}+c$ ii $\frac{2}{5}x^{\frac{5}{2}}-4x^{\frac{3}{2}}+c$

4 $4x^{\frac{5}{2}}+c$

5 $\frac{3}{10}x^{\frac{10}{3}}-4x^3+c$

6 $4x^{\frac{9}{4}}+c$

7 $x^{\frac{1}{3}}+c$

8 $-\frac{2}{3}x^{\frac{3}{2}}+3x-4x^{\frac{1}{2}}+c$

9 $-\frac{1}{6}x^{-2}-\frac{1}{12}x^{-3}+c$

10 a $\mathrm{f}(x)=6x^{-\frac{1}{3}}+x^{\frac{2}{3}}$ b $9x^{\frac{2}{3}}+\frac{3}{5}x^{\frac{5}{3}}+c$

11 a $a=\frac{1}{2}, b=-2$ b $\frac{1}{5}x^{\frac{5}{2}}-4x^{\frac{1}{2}}+c$

12 Proof.

13 $\frac{1}{12}x^{\frac{3}{2}}+2x^{\frac{1}{2}}-4x^{-\frac{1}{2}}+c$

WORK IT OUT 14.2

Solution C is correct.

EXERCISE 14C

1 a i $y=\frac{x^2}{2}+5$ ii $y=2x^3+5$

b i $y=-\frac{1}{2x^2}-\frac{1}{2}$ ii $y=-\frac{1}{x}+4$

c i $y=\frac{3}{2}x^2-5x+10$ ii $y=3x-\frac{1}{2}x^4+\frac{5}{2}$

d i $y=2x^{\frac{3}{2}}-56$ ii $y=2\sqrt{x}+4$

2 a i $y=x^3-x+5$ ii $y=3x^{-1}+2x-2$

b i $\mathrm{f}(x)=\frac{1}{2}x^2-\frac{1}{3}x^3+3x-\frac{19}{3}$

ii $\mathrm{f}(x)=\frac{1}{4}x^4-\frac{1}{6}x^3+\frac{5}{2}x+\frac{89}{12}$

3 $y=\frac{x^2}{2}+\frac{1}{x}+\frac{3}{2}$

4 $\mathrm{f}(x)=\frac{4}{3}x^{\frac{3}{2}}-2x^{\frac{1}{2}}-\frac{14}{3}$

5 $\frac{149}{3}$

6 $y=2x^2-16$

7 a $x=-2$

b Proof.

8 $y=x^2-\frac{x^4}{12}+\frac{10x}{3}-\frac{5}{4}$

9 858

10 $y=\frac{1}{x}+\frac{5}{2}$

EXERCISE 14D

1 a i 320 ii 420.2

b i 0 ii 36

c i 46 ii 48

d i 28.5 ii $\frac{33}{2}$

e i $\frac{76}{3}$ ii $\frac{585}{2}$

f i 36 ii 2

g i $\frac{5}{2}$ ii $-\frac{26}{3}$

2 $-60+28\sqrt{2}$

3 $a=2, b=10$

4 $2k+\frac{1}{k}-3$

5 $8-a^4-a^3$

6 $a=16$

7 $p=-\frac{10}{3}$ or $p=-1$

8 20

9 $\frac{15}{2}$

10 a i 16 ii −16

b i $a^2 - b^2$ ii $b^2 - a^2$

c $\int_b^a f(x)dx = -\int_a^b f(x)dx$

11 a $\frac{a^3}{3} - \frac{1}{3}$ b a^2

c i $f'(a) = a^2$

ii $f'(a) = a^2$, $f'(a)$ is the same in both cases.

d i $4a^3 - 2a$ ii $\frac{3}{a^2}$ iii $1 - 3a^2$

EXERCISE 14E

1 a i $\frac{7}{3}$ ii $\frac{1}{4}$

b i $\frac{2}{3}$ ii $\frac{22}{3}$

2 a i 36 ii $\frac{32}{3}$

b i $\frac{9}{2}$ ii $\frac{4}{3}$

3 a i $\frac{135}{3}$ ii $\frac{17}{2}$

b i $\frac{32}{3}$ ii $\frac{4}{3}$

4 a (3, 0) b $\frac{135}{4}$

5 a $A(0, 12)$, $B(6, 0)$ b 36

6 $\frac{1}{6}$

7 9

8 a $(x-2)(x-5)$ b $\frac{9}{2}$

9 a $(0, 0)$, $(k, 0)$ b 2

10 a i Proof. ii $x = 1$ and $x = 2$

b $\frac{1}{2}$

11 $\frac{38}{4}$

12 108

13 $m = 4$

14 $\frac{1}{6}p^{\frac{3}{2}}$

15 $\frac{a^3}{48}$

MIXED PRACTICE 14

1 A

2 $y = \frac{3}{2}x^2 - \frac{2}{3}x^{\frac{3}{2}} - \frac{59}{3}$

3 $-\frac{1}{x} + 2\sqrt{x} + c$

4 $\frac{2}{3}x^{\frac{3}{2}} - \frac{2}{5}x^{\frac{5}{2}} + 2x - x^2 + \frac{26}{15}$

5 Proof.

6 a $-12 + 8\sqrt{3}$ b $12 - 4\sqrt{3}$

7 $\frac{839}{20}$

8 D

9 B

10 $y = 2x^5 - 2x^3 + 5x - 1$

11 $a = \sqrt{2}$

12 $\frac{4k^3}{3}$

13 a i $-(x-2)^2(x+3)$

ii

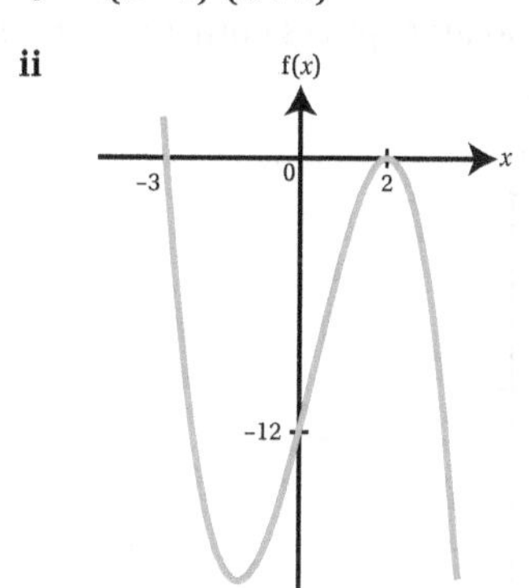

b $\frac{625}{12}$

14 a i a^{n+1} ii $\frac{b^{n+1} - a^{n+1}}{n+1}$ b 3

15 a i $A(-a, 0)$, $B(a, 0)$, $C(0, a^2)$

ii $\frac{1}{3}a^3$ b $\frac{1}{3}a^3$

16 Proof.

17 a (Local) minimum b $x^3 + 3x^2 - 45x + 100$

18 $\frac{27}{2}$

Focus on ... Proof 2

1 Proof.

2 Yes the proof still works.

3 Proof.

Focus on ... Problem solving 2

1 The calculation is similar but is complicated (slightly) by the presence of the square root arising from finding an expression for r in terms of h.

2 1200 cm^2

3 8510 cm^3 (3 s.f.)

4 0.97

5 The x-coordinate is the mean of the x-coordinates of the two points.

6 9.23

Focus on ... Modelling 2

1 a 7:11 a.m. b 4:44 a.m.

2 5 hours 36 minutes after local noon; you would have to consider daylight saving time changes too.

3 a 12 hours 29 minutes

b 11 hours 8 minutes

4 23.44, 13 hours 56 minutes

5 Proof.

6 No sunrise or sunset on those days. ($L > 66.5$ corresponds to places north of the Arctic Circle.)

7 a

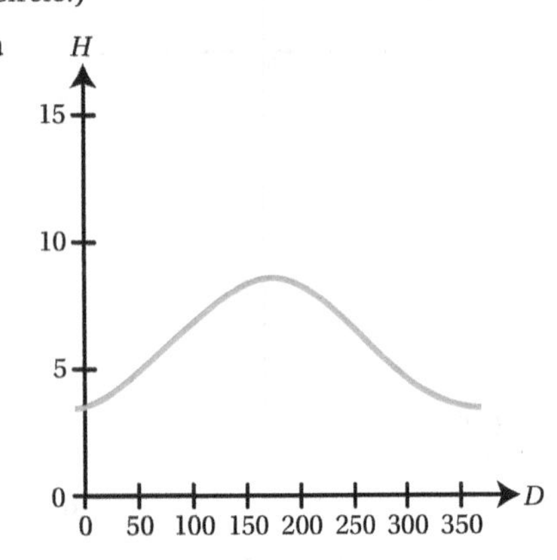

b

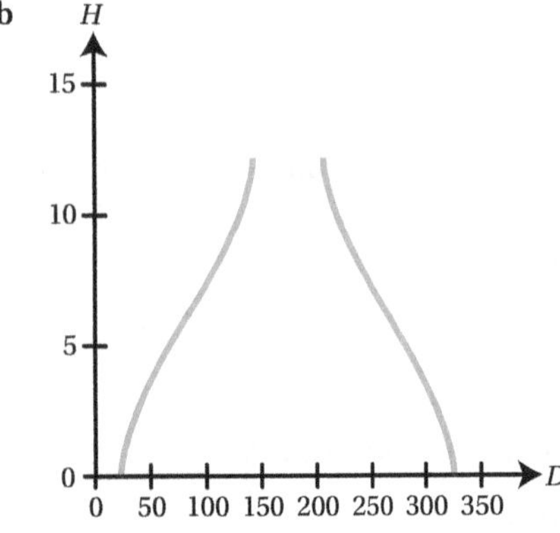

8 a

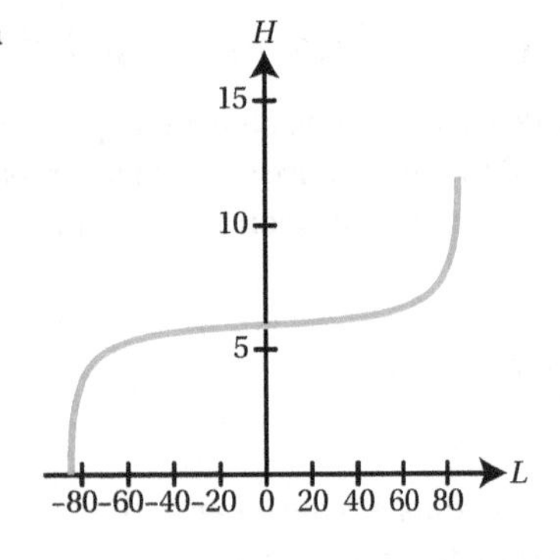

b

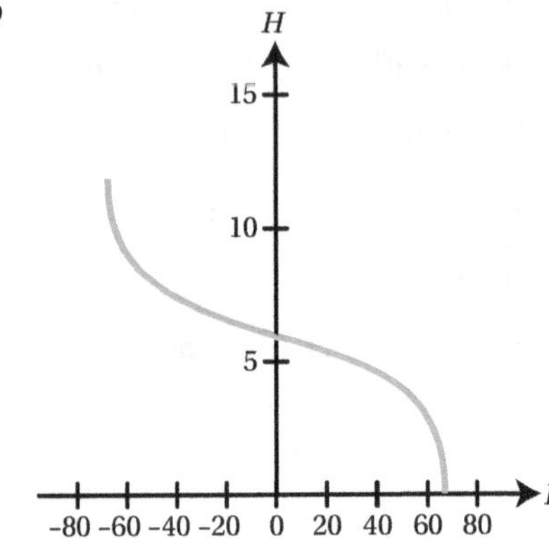

9 $59 \leqslant D \leqslant 60$

10 Proof.

11 Division by 15 rescales the answer from [0, 180] to [0, 12]; 23.44 is the inclination of the Earth's axis; 284 makes 1 January day 1 (it is 284 days after the Spring Equinox); 360 / 365 rescales from [0, 365] to [0, 360] so that the value is in degrees.

12 For example, the Earth is spherical; the Sun is modelled as a point.

Cross-topic review exercise 2

1 $180°$

2 $\theta = 30°, 120°$

3 $a = 2, b = -7$

4 a i Proof. ii $(x+1)(x-1)(x-5)$ b $\frac{128}{3}$

5 a i Proof. ii $(x+3)(x^2-3x+5)$

b i $4x^3-16x+60$

ii, iii Proof.

iv 92

v $\frac{d^2y}{dx^2} > 0 \Rightarrow$ minimum

6 $x = 9.29$

7 a Proof.

b i Proof. ii $x = 225°$

8 a $x = 49°, 229°$

b Translation $\begin{pmatrix} -30° \\ 0 \end{pmatrix}$

c i Proof.

ii $x = 20.7°, 159°, 201°, 339°$

9 a $a = \frac{3}{2}, b = \frac{1}{2}, c = -\frac{1}{2}$

b $y = -4x + 14$

c i $P\left(\frac{7}{2}, 0\right), Q(0, 14)$ ii $\frac{49}{2}$

10 $\frac{20}{3}$

11 a $4x^3h+6x^2h^2+4xh^3+h^4$

b Proof.

12 a $f(x)=x^2-6x+8$ b $y=\frac{x}{2}-1$

c $\left(\frac{9}{2},\frac{5}{4}\right)$

13 a $a=\frac{1}{4}, p=4, q=5$

b $M(0,9)$ and $N\left(\frac{16}{3},\frac{49}{9}\right)$

c i $d=-\frac{3}{4}x^2+4x$ ii $\frac{16}{3}$

14 $\left(\frac{2}{3},\frac{86}{9}\right)$

15 $\frac{\sqrt{8}}{3}$

16 a Proof.

b $r\leqslant\frac{1}{2a}$

17 a i a^4+b^4

ii Proof.

b Proof.

15 Vectors

BEFORE YOU START

1 $\begin{pmatrix}3\\-1\end{pmatrix}$

2 5.39, 21.8°

3 2

WORK IT OUT 15.1

Solution 2 is correct.

EXERCISE 15A

1 a $2\mathbf{i}+\mathbf{j}$ b $-3\mathbf{j}$

c $-2\mathbf{i}+\mathbf{j}$ d $\mathbf{i}-\mathbf{j}$

2

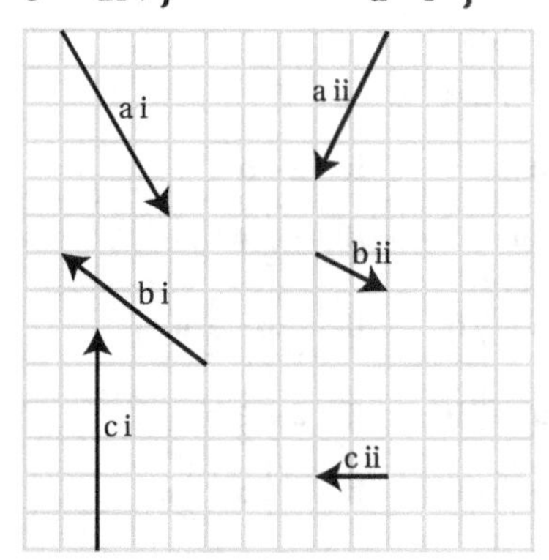

3

	A	B	C
a i	5.83	59.0°	149°
ii	4.47	117°	153°
b i	5	143°	53°
ii	2.24	27°	117°
c i	6	90°	0°
ii	2	180°	90°

4 a i No, magnitude $=\frac{\sqrt{13}}{6}$ ii Yes

b i Yes ii No, magnitude = 3

5 $\pm\frac{1}{\sqrt{29}}$

6 $\pm\frac{1}{\sqrt{13}}$

7 −1 or −3

8 $5.45\mathbf{i}+10.7\mathbf{j}$

9 $\begin{pmatrix}-3\\\sqrt{3}\end{pmatrix}$

EXERCISE 15B

1 a

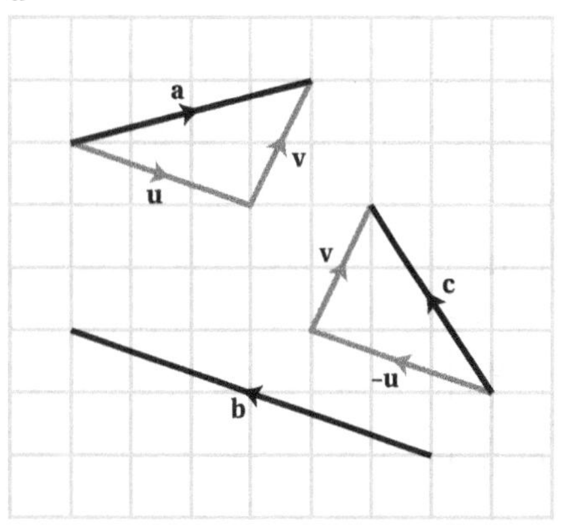

b

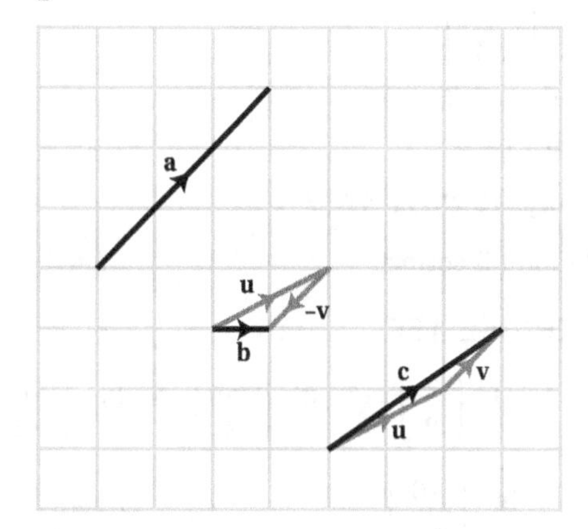

2 **a** **i** $\begin{pmatrix}21\\3\end{pmatrix}$ **ii** $\begin{pmatrix}20\\-8\end{pmatrix}$

b **i** $\begin{pmatrix}2\\3\end{pmatrix}$ **ii** $\begin{pmatrix}6\\-1\end{pmatrix}$

c **i** $\begin{pmatrix}11\\-3\end{pmatrix}$ **ii** $\begin{pmatrix}-3\\5\end{pmatrix}$

d **i** $\begin{pmatrix}10\\-3\end{pmatrix}$ **ii** $\begin{pmatrix}17\\6\end{pmatrix}$

3 **a** **i** $-5\mathbf{i}$ **ii** $4\mathbf{i}+8\mathbf{j}$

b **i** $\mathbf{i}-3\mathbf{j}$ **ii** $2\mathbf{j}$

c **i** $4\mathbf{i}$ **ii** $5\mathbf{i}-4\mathbf{j}$

4 **a** **i** $-4\mathbf{i}+2\mathbf{j}$ **ii** $-\frac{8}{3}\mathbf{i}+\frac{4}{3}\mathbf{j}$

b **i** $4\mathbf{i}-3\mathbf{j}$ **ii** $-2\mathbf{i}+\frac{5}{2}\mathbf{j}$

5 **a** **i** Parallel. **ii** Parallel.

b **i** Not parallel. **ii** Not parallel.

c **i** Parallel. **ii** Not parallel.

6 $\begin{pmatrix}2\\0\end{pmatrix}$

7 -2

8 $-\frac{4}{3}$

9 -2

10 $q=\frac{1}{2}-p$

11 $\frac{11}{7}$

12 $\pm\begin{pmatrix}12\\-16\end{pmatrix}$

13 $\pm\begin{pmatrix}\frac{6}{\sqrt{5}}\\\frac{3}{\sqrt{5}}\end{pmatrix}$

14 $\pm\begin{pmatrix}\frac{-2}{\sqrt{5}}\\\frac{1}{\sqrt{5}}\end{pmatrix}$

15 **a** Proof.

b $t=0$

WORK IT OUT 15.2

Solution 1 is correct.

EXERCISE 15C

1 **a** **i** $\begin{pmatrix}4\\2\end{pmatrix}$ **ii** $\begin{pmatrix}-3\\2\end{pmatrix}$ **iii** $\begin{pmatrix}3\\-2\end{pmatrix}$

iv $\begin{pmatrix}-6\\-5\end{pmatrix}$

b

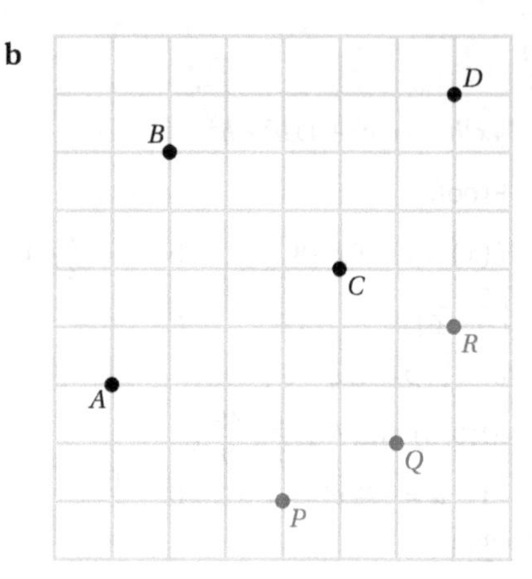

2 **a** **i** $\begin{pmatrix}-1\\3\end{pmatrix}$ **ii** $\begin{pmatrix}3\\-3\end{pmatrix}$

b **i** $\begin{pmatrix}-6\\1\end{pmatrix}$ **ii** $\begin{pmatrix}3\\2\end{pmatrix}$

c **i** $\begin{pmatrix}4\\-6\end{pmatrix}$ **ii** $\begin{pmatrix}2\\5\end{pmatrix}$

3 **a** **i** $(2,1)$ **ii** $(-4,-4)$

b **i** $(1,-2)$ **ii** $(2,0)$

c **i** $(11,-15)$ **ii** $(-9,-1)$

d **i** $(-8,7)$ **ii** $(1,-34)$

4 **a** **i** 6 **ii** $\sqrt{2}$

b **i** $3\sqrt{2}$ **ii** $2\sqrt{5}$

5 **a** $\overrightarrow{PR}=5\mathbf{i}-\mathbf{j}$ **b** $\overrightarrow{SP}=-7\mathbf{i}-4\mathbf{j}$

6 **a** $\overrightarrow{AC}=\begin{pmatrix}-2\\-3\end{pmatrix}, \overrightarrow{DB}=\begin{pmatrix}-4\\-6\end{pmatrix}$ **b** $(0,3)$

7 $D=(3,-19)$

8 **a** $\overrightarrow{AC}=\begin{pmatrix}8\\-4\end{pmatrix}$

b $\overrightarrow{BP}=\begin{pmatrix}p-6\\q+3\end{pmatrix}, \overrightarrow{CP}=\begin{pmatrix}p-8\\q+4\end{pmatrix}$

c $CP=2\sqrt{5}$

EXERCISE 15D

1 **a** **i** $\mathbf{b}$ **ii** $\mathbf{a}+\mathbf{b}$

b **i** $-\mathbf{a}$ **ii** $-\frac{1}{2}\mathbf{a}$

c **i** $\mathbf{a}+\frac{1}{2}\mathbf{b}$ **ii** $\frac{1}{2}\mathbf{b}-\frac{1}{2}\mathbf{a}$

2 **a** **i** $\mathbf{a}+\frac{4}{3}\mathbf{b}$ **ii** $\mathbf{a}+\frac{1}{2}\mathbf{b}$

b **i** $-\frac{3}{2}\mathbf{a}+\mathbf{b}$ **ii** $-\frac{1}{2}\mathbf{b}+\frac{1}{2}\mathbf{a}$

c **i** $\frac{3}{2}\mathbf{a}-\mathbf{b}$ **ii** $-\frac{4}{3}\mathbf{b}+\frac{1}{2}\mathbf{a}$

3 a i Collinear; $AB : BC = 1 : 13$

ii Collinear; $AB : BC = 2 : 5$

b i Not collinear.

ii Not collinear.

c i Not collinear.

ii Collinear; $AB : BC = 1 : 2$

d i Collinear; $AB : BC = 1 : 1 - 2a$

ii Not collinear.

4 $\mathbf{m} = 2\mathbf{i} + 12\mathbf{j}$

5 $AB : BC = 1 : 2$

6 a $\begin{pmatrix} 8 \\ -4 \end{pmatrix}$ b Not a rhombus.

7 a $\mathbf{e} = \begin{pmatrix} 6 \\ 4 \end{pmatrix}$ b $\overrightarrow{AB} = \overrightarrow{DE} = \begin{pmatrix} 2 \\ 1 \end{pmatrix}$

8 Proof.

9 $\begin{pmatrix} 0.8 \\ 1.8 \end{pmatrix}$

10 $\overrightarrow{MN} = \overrightarrow{QP} = \frac{1}{2}(\mathbf{c} - \mathbf{a})$

11 a $C(10 - 8x, 1 + 6x)$ b $x = \frac{6}{25}$

12 a $\frac{1}{2}(\mathbf{p} + \mathbf{q})$

b Proof.

13 a, b Proof.

MIXED PRACTICE 15

1 C

2 $3\sqrt{2}$

3 a $6\mathbf{i} - 5\mathbf{j}$ b $(0, 1)$

4 a $\begin{pmatrix} 4.5 \\ -1 \end{pmatrix}$ b $\frac{3}{2}\sqrt{41}$

5 $6\mathbf{i} - 2\mathbf{j}$

6 $\overrightarrow{MN} = \mathbf{b} - \frac{1}{3}\mathbf{a}$; Proof.

7 D

8 Proof.

9 $\sqrt{9 + 16\cos^2\theta}$

10 $\frac{9}{\sqrt{97}}\mathbf{i} + \frac{4}{\sqrt{97}}\mathbf{j}$ or $-\frac{9}{\sqrt{97}}\mathbf{i} - \frac{4}{\sqrt{97}}\mathbf{j}$

11 a $(4, -6)$

b $|\overrightarrow{OP}| = \sqrt{65}, |\overrightarrow{ON}| = \sqrt{52}, |\overrightarrow{PN}| = \sqrt{13}$; Proof.

12 A

13 a $\overrightarrow{MN} = \frac{1}{2}(\mathbf{c} - \mathbf{a})$; $\overrightarrow{PQ} = \frac{1}{2}(\mathbf{a} - \mathbf{c})$

b Parallelogram

14 a $\mathbf{a} + \mathbf{b}$ b $\mathbf{a} + \frac{1}{2}\mathbf{b}$

c $t\mathbf{a} + (t - 1)\mathbf{b}; t = \frac{2}{3}$

15 a $\begin{pmatrix} 5 - 3t \\ 2 + t \end{pmatrix}; \sqrt{10t^2 - 26t + 29}$

b $\frac{11\sqrt{10}}{10}$

16 a $1 : 1$ b The diagonals bisect each other.

16 Introduction to kinematics

BEFORE YOU START

1 a $-\frac{7}{5}$ b $\frac{2}{3}$

2 $S = 24; T = 42$

3 a 2 m s^{-2} b 40 m

4 a $\frac{dy}{dx} = 6x - 4 - 5x^{-2}$ b -15

5 $(2, 21)$

6 4

7 $y = 5x - 2x^3 - 1$

EXERCISE 16A

1 a Yes b No c No d No

2 These are some suggestions for discussion:

a shape and size of box (which affects) air resistance; change in g at different heights

b size of each ball; where it was hit; friction

c the actual shape of the road; mass of passengers inside the bus; resistance forces (may depend on weather conditions)

d take-off and landing times; changes in height and related changes in g; weather conditions, turbulence; the path is circular, not straight.

EXERCISE 16B

1 a i -410 m ii -290 m

b i 290 m ii 210 m

c i -120 m ii -210 m

2 a i $7.32\text{ m s}^{-1}; 2.93\text{ m s}^{-1}$

ii $8.64\text{ m s}^{-1}; -1.48\text{ m s}^{-1}$

b i $15.3\text{ m s}^{-1}; 0\text{ m s}^{-1}$

ii $9.55\text{ m s}^{-1}; 0\text{ m s}^{-1}$

3 a i 1.00 m s^{-1} ii 17.2 m s^{-1}

b i 18.7 km h^{-1} ii 0.936 km h^{-1}

c **i** $0.009\,26\text{ m s}^{-2}$ **ii** 0.0347 m s^{-2}

d **i** $10\,600\text{ km h}^{-2}$ **ii** $35\,000\text{ km h}^{-2}$

EXERCISE 16C

1 **a** **i** $v = 9t^2 - 4;\ a = 18t$

ii $v = 4t^3 - 6t + 4;\ a = 12t^2 - 6$

b **i** $v = \frac{2t}{5} - 4.2;\ a = \frac{2}{5}$

ii $v = -1.5t^2 + \frac{2t}{3};\ a = -3t + \frac{2}{3}$

2

	$s(0)$	$s(3)$	$v(0)$	$v(3)$	$a(0)$	$a(3)$
a i	2	71	−4	77	0	54
a ii	0	66	4	94	−6	102
b i	0	−10.8	−4.2	−3	$\frac{2}{5}$	$\frac{2}{5}$
b ii	0	−10.5	0	−11.5	$\frac{2}{3}$	−8.33

3 **a** **i** $s = \frac{3}{2}t^2 - 4t + 2$ **ii** $s = t - t^2 + 1$

b **i** $s = 5t - t^3$ **ii** $s = \frac{1}{3}t^3 - t^2$

4 **a** **i** 19.5 **ii** −19

b **i** −100 **ii** $\frac{50}{3}$

5 **a** **i** $v = t - t^2 + 3;\ s = \frac{1}{2}t^2 - \frac{1}{3}t^3 + 3t$

ii $v = 2t^2 + 2t - 2;\ s = \frac{2}{3}t^3 + 2t^2 - 2t$

b **i** $v = -5t + 3;\ s = -\frac{5}{2}t^2 + 3t + 5$

ii $v = 3t - 2;\ s = \frac{3}{2}t^2 - 2t + 7$

6 **a** -3.6 m s^{-2} **b** 0.96 m s^{-1}

c $s = 1.2t^2 - 0.5t^3$

7 **a** $0.1\text{ m s}^{-1};\ 1.2\text{ m s}^{-2}$ **b** 1 m s^{-1}

c 1.92 s; 3.00 m

8 **a** $v = 0.04t^3 - 0.72t^2 + 4.32t;$

$s = 0.01t^4 - 0.24t^3 + 2.16t^2$

b $v = 8.64\text{ m s}^{-1};\ s = 38.9\text{ m}$

9 **a** $v = 7.5 + t - 0.3t^2$ **b** 50 m

c 62.5 m

10 **a** When $t = 0, v = 0$; 8 s and 12 s

b $v = 3.6\text{ m s}^{-1};\ s = 30.6\text{ m}$ **c** 39.5 m

11 -116 m s^{-1}

12 28.9 m s^{-1}

13 1.82 s

14 10.2 m s^{-1}

EXERCISE 16D

1 **a** **i** $0.65\text{ m s}^{-2};\ -0.52\text{ m s}^{-2}$

ii 1950 m

iii 16.25 m s^{-1}

iv 16.25 m s^{-1}

b **i** $1\text{ m s}^{-2};\ -3\text{ m s}^{-2}$ **ii** 888 m

iii 20.2 m s^{-1} **iv** 20.2 m s^{-1}

c **i** $-1.39\text{ m s}^{-2};\ 1.25\text{ m s}^{-2}$

ii 397.5 m

iii 6.31 m s^{-1}

iv 3.69 m s^{-1}

2 **a** 1350 m

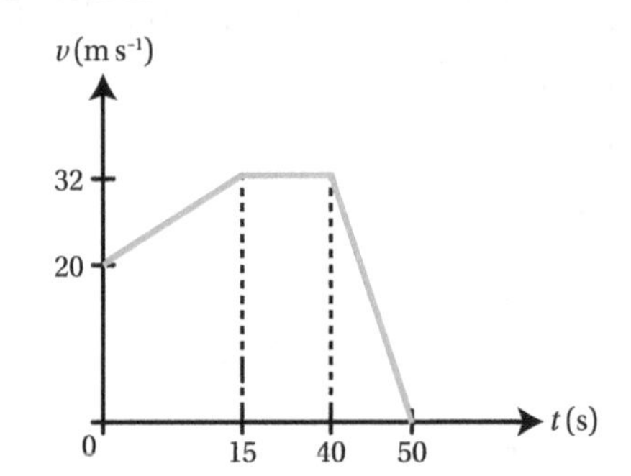

b 495 m

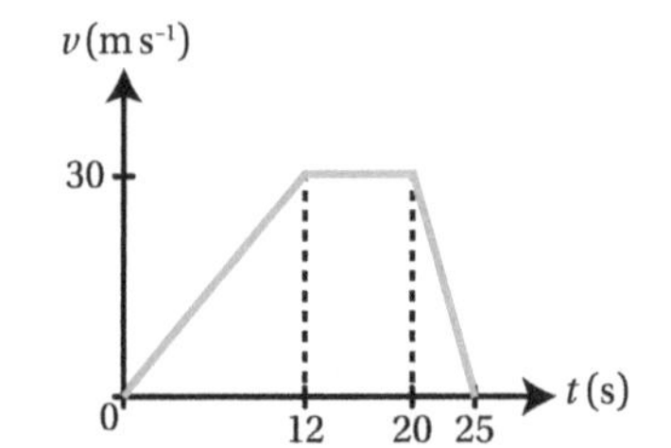

c 862.75 m

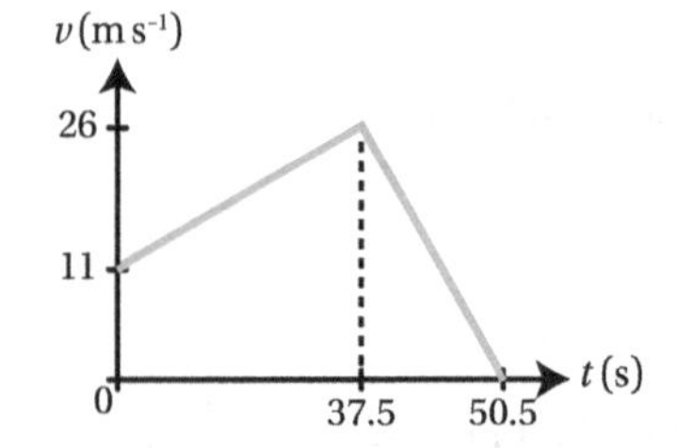

3 **a** $v\ (\text{m s}^{-1})$

1

0

20

40 50

t (s)

−2

b

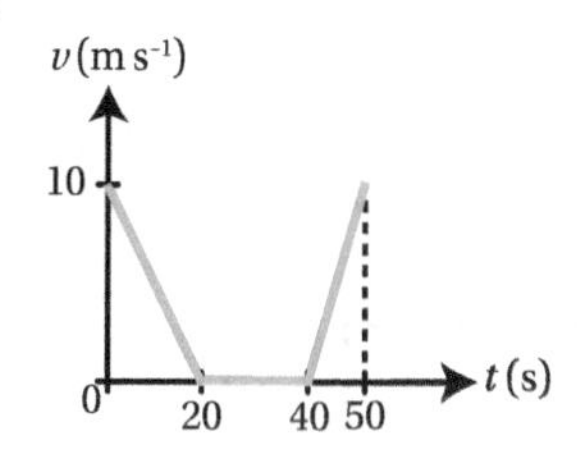

c

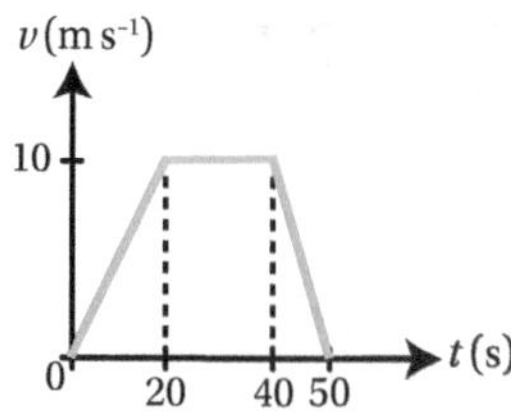

d

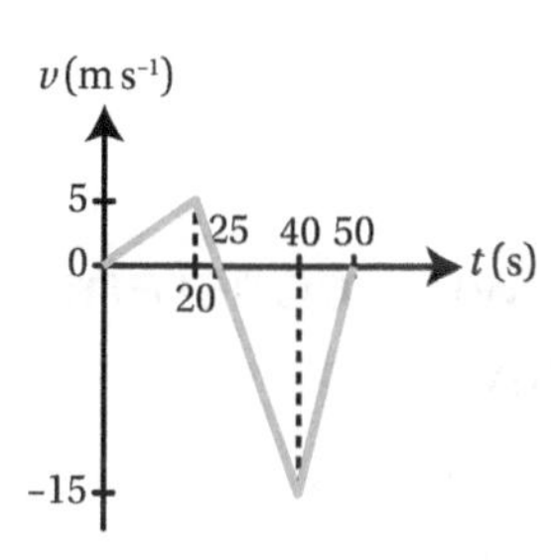

4 **a** 50 m

b Away.

c The particle changes direction.

d The particle stops moving (it is stationary), 30 m from P.

e When $t = 50$

f Decreasing.

g The speed is decreasing, the velocity is increasing.

h 200 m

5 **a**

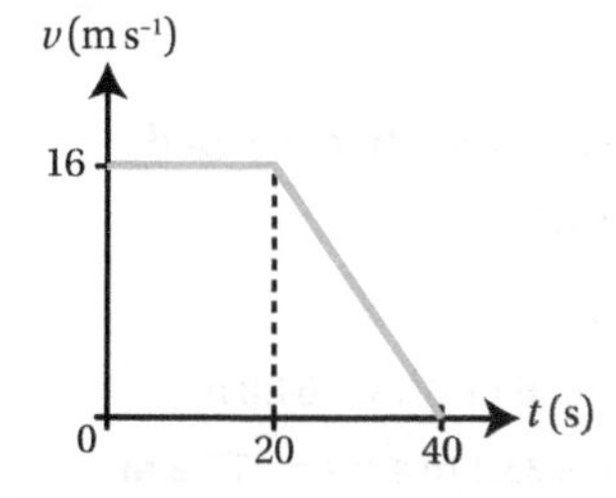

b 480 m

6 **a** $2\ \mathrm{m\,s^{-2}}$ **b** 25 s

c $-7.5\ \mathrm{m\,s^{-2}}$ **d** 175 m (to the right)

7 $2.45\ \mathrm{m\,s^{-1}}$

8 **a**

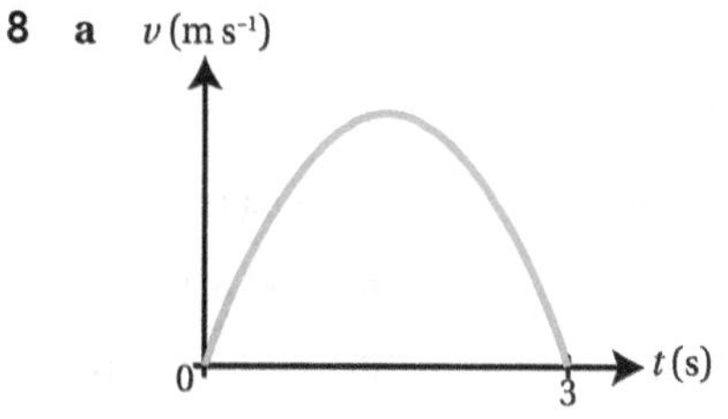

b $9\ \mathrm{m\,s^{-1}}$

9 **a** Stationary. **b** 200 m, $6.67\ \mathrm{m\,s^{-1}}$

c $20\ \mathrm{m\,s^{-1}}$

10 **a**

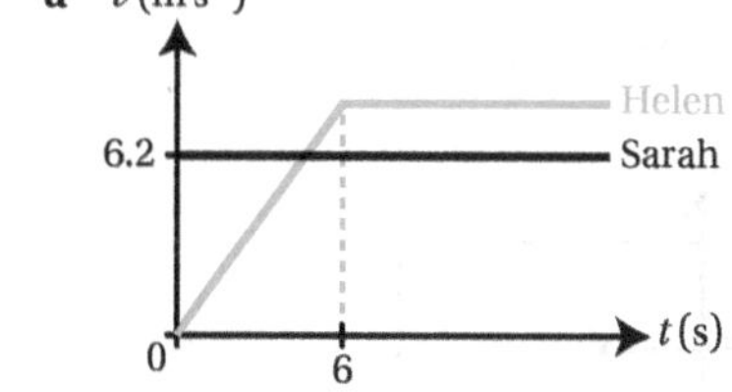

b 14.6 s

11 **a** 8 **b** $0.8\ \mathrm{m\,s^{-2}}$

c C

12 **a** $t = 4$ and $t = 7$ **b** 85.75 m

c $12.25\ \mathrm{m\,s^{-1}}$; $33.5\ \mathrm{m\,s^{-1}}$

13 **a**

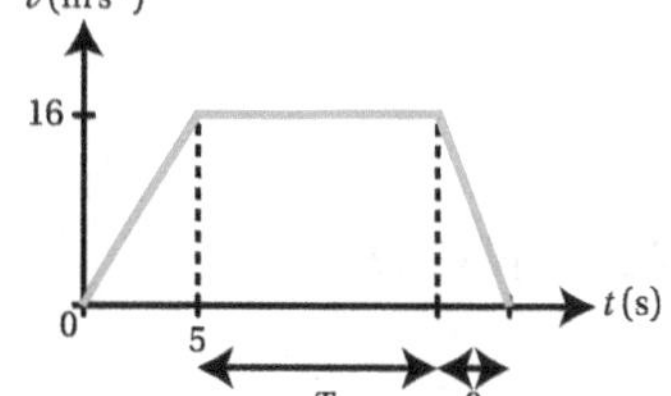

b 13

14 $x = 5$

15 **a** 45 s **b** 115 s

c

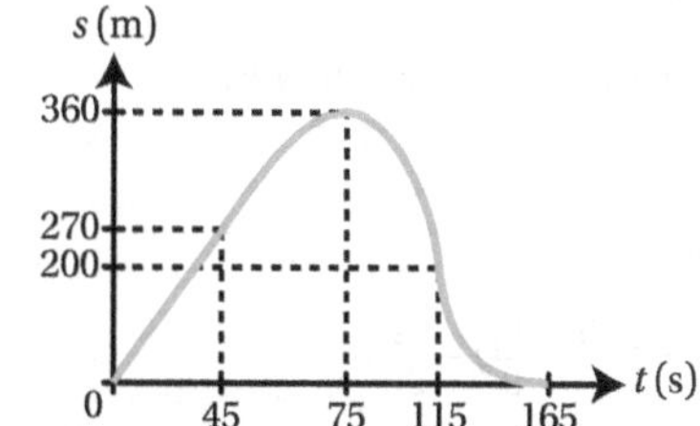

EXERCISE 16E

1 1.5 seconds

2 **a** Proof.

b $5.4\ \mathrm{m\,s^{-1}}$

3 **a** 34.0 m **b** $5.76\ \mathrm{m\,s^{-2}}$

4 5.6 seconds

5 a -2.4 m s^{-2}

b i 13.5 m ii 40.5 m

6 a 3.7 b 1.43 m s^{-1}

c 37.75 m d 7.13 seconds

7 a $v = 6.5t - 0.65t^2$ b $v = 31.85 - 2.6t$

8 a Positive acceleration throughout, starting from rest.

b 3.125 m s^{-1}

c $v(5) = \frac{125}{30} = 4.17 \text{ m s}^{-1}$

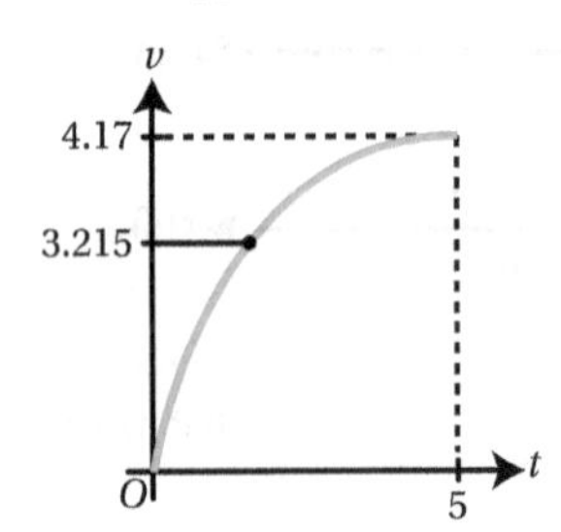

9 a $v(t) = t^3 - 7t^2 + 10t$

b i 4.06 m s^{-1} ii 8.21 m s^{-1}

10 a 15 m b 11.4 seconds

11 13.3

12 34.7 m

MIXED PRACTICE 16

1 B

2 a 4 m s^{-2} b $s = t^3 - 4t^2$

3 a 11.2 m b 2.8 m s^{-1}

4 a 1.4 m s^{-2} b 0 s, 12 s and 15 s

c 3 m s^{-1}

5 a 120 m b 5.71 m s^{-1}

6 D

7 -15 m s^{-2}

8 a $v = \frac{100}{9} + \frac{25}{9}t - \frac{5}{36}t^2$

b $a = \frac{25}{9} - \frac{5}{18}t; t = 10 \text{ s}$

c 253 m

d i No ii Yes

9 4.25 m s^{-1}

10 a 400 m b 64 s

c

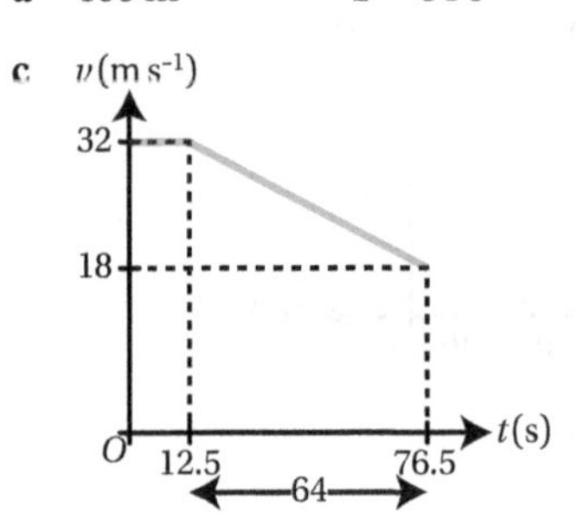

d 26.1 m s^{-1}

11 9.93 s, 30.1 s

12 a 8.73 s

b Proof.

c 0

13 17.3 m s^{-1}

17 Motion with constant acceleration

BEFORE YOU START

1 a $v = 2t - t^3 + 4.2$

b $s = t^2 - \frac{1}{4}t^4 + 4.2t + 14$

2 a $t = 0.249, 2.49$ b $t = 0.563, 1.89$

3 a (0.17, 1.03) b (1.5, 2.275)

EXERCISE 17A

1 Proof.

2 a Proof. b Proof.

3 a Proof.

b 49.9 m s^{-1}

4 Proof.

5 a $v = u + 0.5kt^2$

b i–iii Proof.

6 a $v = u + aT$

b At the start and after 1 second

c $T = 0$ or 1.5 s

EXERCISE 17B

1 a i $u = 3.6 \text{ m s}^{-1}; s = 64.8 \text{ m}$

ii $u = -23.1 \text{ m s}^{-1}; s = -77.2 \text{ m}$

b i $v = 5 \text{ m s}^{-1}; t = 6.67 \text{ s}$

ii $v = 12.6 \text{ m s}^{-1}; t = 2.45 \text{ s}$

c i $a = -0.45 \text{ m s}^{-2}; t = 13.3 \text{ s}$

ii $a = 1.03 \text{ m s}^{-2}; t = 18.5 \text{ s}$

d i $a = -1.83\text{ m s}^{-2}$; $s = -3\text{ m}$

ii $a = -1.3\text{ m s}^{-2}$; $s = 15\text{ m}$

2 a 1.92 m s^{-2} b 40.5 m

3 a 6.3 m s^{-1} b 36.75 m

4 a -0.873 m s^{-2} b 26.6 m

5 14.1 s

6 4.43 s

7 8.05 s

8 66.7 m

9 a 5.29 m

b It will go an unlimited distance in the negative direction.

WORK IT OUT 17.1

Solution 2 is correct.

EXERCISE 17C

1 In this question you should give all final answers to 2 s.f.

a i $v = 22\text{ m s}^{-1}$; $s = 6.1\text{ m}$

ii $v = 31\text{ m s}^{-1}$; $s = 22\text{ m}$

b i $v = 30\text{ m s}^{-1}$; $s = 47\text{ m}$

ii $v = 59\text{ m s}^{-1}$; $s = 180\text{ m}$

2 In this question you should give all final answers to 2 s.f.

a i 16 m s^{-1} downwards; 8.1 m below

ii 4.2 m s^{-1} upwards; 6.5 m above

b i 4.6 m s^{-1} upwards; 61 m above

ii 9.2 m s^{-1} upwards; 16 m above

3 30 m s^{-1}; 3 s (1 s.f.)

4 a i 1.18 s ii 6.92 m s^{-1}

b i Increase. ii Decrease.

5 11 m s^{-1} (2 s.f.)

6 a 55 m

b 6.2 s (2 s.f.); 33 m s^{-1} (2 s.f.)

7 a No b 13 m s^{-1} (2 s.f.)

c Increase.

8 a 1.6 s (2 s.f.) b 3.9 s (2 s.f.)

c

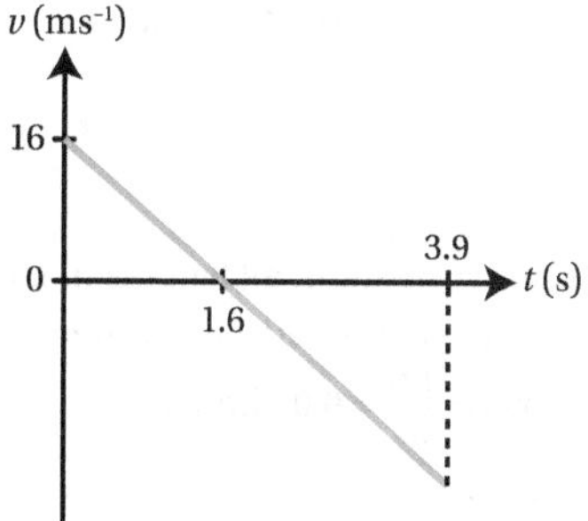

9 a $\frac{u}{4.9}$ s

b Graph A.

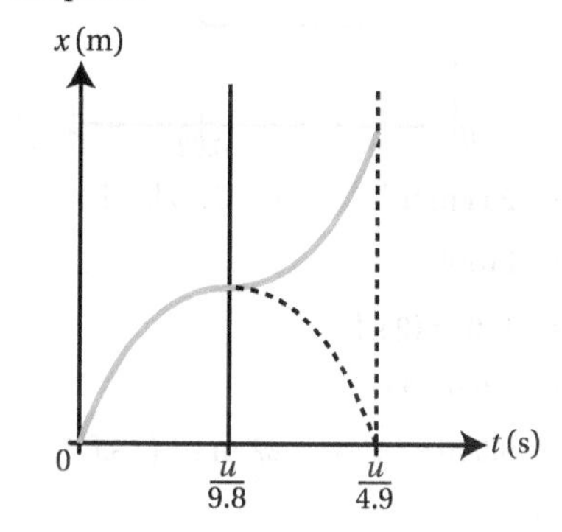

10 13.5 m s^{-1} (3 s.f.); 7.53 m (3 s.f.)

EXERCISE 17D

1 149 m

2 a 10.2 m s^{-1} b 0.207 m s^{-2}

3 1.92 m s^{-1}

4 17.4 m s^{-1}

5 6.9 m (2 s.f.)

6 12 s (2 s.f.)

7 44.2 m s^{-1}

8 Max height is 38.4 m (3 s.f.). Speed as it hits the ground is 27.4 m s^{-1} (3 s.f.).

9 17 m s^{-1} (2 s.f.)

10 a 90.2 m b 25 m s^{-1}

MIXED PRACTICE 17

1 C

2 a 8.75 s b 166.25 m

3 a 4 m (1 s.f.) b 2 s (1 s.f.)

4 2.2 m s^{-1}

5 a 23.6 m s^{-1} b 225.6 m

6 A

7 B

8 Make units consistent: all lengths in kilometres, time in hours and speeds in kilometres per hour.

$1.2 \text{ minutes} = \frac{1.2}{60} = 0.02 \text{ hours}$

a

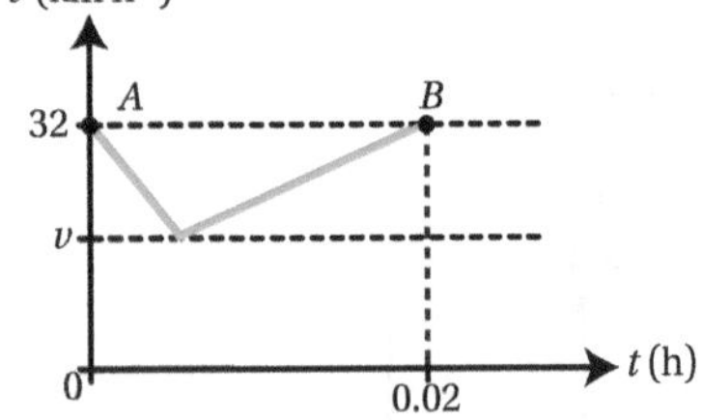

b 23 km h^{-1} c 27.5 km h^{-1}

9 a Proof.

b 130 m (2 s.f.)

10 a 7.9 m (2 s.f.)

b 9.0 m s^{-1} (2 s.f.), 22 m s^{-1} (2 s.f.)

11 Proof.

12 You can use the velocity–time graph and the fact that the area beneath the graph gives the distance travelled.

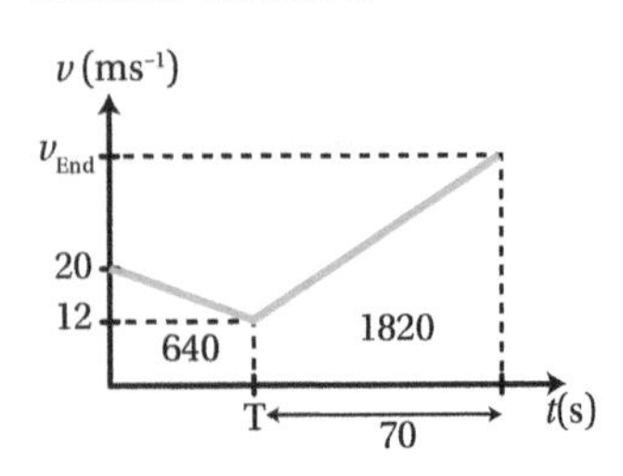

a i 40 s ii 0.2 m s^{-2}

b i 0.4 m s^{-2} ii 40 m s^{-1}

c 22.4 m s^{-1}

13 120 m

14 2 : 1

15 $u = 150 \text{ m s}^{-1}$; $v = 50 \text{ m s}^{-1}$

18 Forces and motion

BEFORE YOU START

1 a 0.417 m s^{-2} b 66 m

2 a $\begin{pmatrix} -2 \\ 6 \end{pmatrix}$ b $-4\mathbf{i} + 2\mathbf{j}$

3 a Magnitude 2.77, 64.4° above horizontal.

b Magnitude 4.13, 14.0° below horizontal.

EXERCISE 18A

1 a i 138 N ii 0.45 N

b i 7200 N ii 0.828 N

c i 0 N ii 1.18 N

2 a i 2.39 m s^{-2} ii 18.3 m s^{-2}

c i 0.4 m s^{-2} ii -21.7 m s^{-2}

3 a i $\begin{pmatrix} 4.8 \\ -7.5 \end{pmatrix}$ N ii $\begin{pmatrix} -3.5 \\ 6.5 \end{pmatrix}$ N

b i $\begin{pmatrix} 6 \\ 2 \end{pmatrix} \text{ m s}^{-2}$ ii $\begin{pmatrix} -2 \\ 1.43 \end{pmatrix} \text{ m s}^{-2}$

c i 1.4 kg ii 0.5 kg

4 a Towards the centre. b Deep space.

c Depends on the object.

5 1071 N (4 s.f.)

6 0.048 N

7 0.27 s

8 $\begin{pmatrix} 1.08 \\ 1.98 \end{pmatrix}$ N

9 a $\begin{pmatrix} -1.31 \\ 1 \end{pmatrix} \text{ m s}^{-2}$ b 1.65 m s^{-2}

10 218 N

11 6.47 kg

EXERCISE 18B

1 a i 2 N to the left. ii 5 N to the right.

b i 3 N down. ii 6 N up.

c i 7 N right. ii 4 N right.

2 a i $13\mathbf{i} + 8\mathbf{j}$ ii $-5\mathbf{i} + 7\mathbf{j}$

b i $7\mathbf{i} + 2\mathbf{j}$ ii $2\mathbf{i} + 4\mathbf{j}$

c i $-4\mathbf{j}$ ii $6\mathbf{i} - 4\mathbf{j}$

3 a i 15.3 N; 31.6° ii 8.60 N; 54.5°

b i 7.28 N; 15.9° ii 4.47 N; 63.4°

c i 4 N; −90° ii 7.21 N; −33.7°

4 a i 3 m s^{-2} right ii 2 m s^{-2} left

b i 3 m s^{-2} down ii 1.4 m s^{-2} up

c i 0 m s^{-2} ii 0 m s^{-2}

d i 6 m s^{-2} left ii 2 m s^{-2} down

5 a i $(-4\mathbf{i} + 6\mathbf{j}) \text{ m s}^{-2}$ ii $(2\mathbf{i} - 3\mathbf{j}) \text{ m s}^{-2}$

b i $(\mathbf{i} + 1.5\mathbf{j}) \text{ m s}^{-2}$ ii $(0.8\mathbf{i} + 0.1\mathbf{j}) \text{ m s}^{-2}$

6 a i 15 N ii 14 N
b i 15 N ii 8 N
c i 12 N ii 4 N
d i 7 N ii 3 N

7 1200 kg

8 20 kg

9 a $\sqrt{65}$ N b 29.7° above the horizontal.

EXERCISE 18C

1 2 m s^{-2}

2 269 kg

3 a 760 N
b 0.86 s less (8.40 compared to 9.26)

4 4.8 N

5 210 N

6 607 N

7 a 3.54 s b 52.4 N

8 a Proof.
b 320 N c 910 m

9 1.69 N against the direction of travel.

10 a Proof.
b 115 c 164

EXERCISE 18D

1 In this question you should give all final answers to 2 s.f.
a i 0.87 m s^{-2} upwards
ii 1.7 m s^{-2} upwards
b i 2.3 m s^{-2} downwards
ii 1.4 m s^{-2} downwards

2 In this question you should give all final answers to 2 s.f.
a i 110 N ii 28 N
b i 90 N ii 35 N

3 In this question you should give all final answers to 2 s.f.
a i 3.0 kg ii 2.0 kg
b i 2.4 kg ii 1.8 kg

4 2.09 m s^{-2} upwards (3 s.f.)

5 a 1.2 kg b 2.5 m s^{-2} upwards

6 89 kg (2 s.f.)

7 5100 N (2 s.f.)

8 $154g$ N

9 a 40 N (1 s.f.) b 2 m s^{-2} (1 s.f.)

10 1300 N (2 s.f.)

EXERCISE 18E

1 a No b Yes
c Yes d No

2 a 14 N b 7 N
c 16 N; 9 N d 3 N; 6 N

3 a i $x = -18; y = 7$ ii $x = 6; y = -30$
b i $x = 2; y = -10$ ii $x = -9; y = -4$
c i $x = -8; y = 4$ ii $x = -15; y = 9$
d i $x = \frac{32}{5}; y = -\frac{29}{5}$ ii $x = -1; y = 5$

4 4 N; 5 N

5 $x = 2; y = 4$

6 $a = 8; b = 10$

7 7.3 N (2 s.f.)

8 12 N

9 a 19 N thrust b 63 N into the room.

10 $a = 6; b = -3.5; m = 5$

11 $F = (-57\mathbf{i} + 6\mathbf{j})$ N

12 a $\mathbf{F} = \begin{pmatrix} 17 \\ -25 \end{pmatrix}$ b 30.2 N; $\theta = 55.8°$

MIXED PRACTICE 18

1 B

2 a 863 kg b Lower mass.

3 11.9 N

4 a 0.816 m s^{-2} b 25.2 s

5 a 28 kg (2 s.f.) b 310 N (2 s.f.) c 270 N
d i Does not contribute to system mass.
ii Constant tension throughout, perfectly transmits tension forces.

6 a $p = -2.6; q = -3.8$ b $\begin{pmatrix} 1.04 \\ 1.52 \end{pmatrix}\text{ m s}^{-2}$

7 a $\begin{pmatrix} 7 \\ 8 \end{pmatrix}$ N b $\sqrt{113}$
c 2.13 m s^{-2} d 48.8°

8 A

9 a 481 N b 8.33 s

10 98.5 N; 60.8°

11 19.2 N; 38.7°

12 a i -2 m s^{-2} ii 10 s iii 2600 N

13 16.8 N, thrust

14 a $(7.5\mathbf{i}-0.75\mathbf{j})\text{ m s}^{-2}$ b 30.1 N; 5.7°

15 165 N; 13.8° above horizontal

19 Objects in contact

BEFORE YOU START

1 a 0.4 m s^{-2} b 15 s

2 2.5 m s^{-2}

3 333.2 N

4 $x = 2$ N to the right; $y = 6$ N upwards

EXERCISE 19A

1 5.49 m s^{-2}

2 65 kg

3 5.32 m

4 a 0.306 m s^{-2} b 1.84 s

5 In this question you should give all final answers to 2 s.f.

a 1.2 N b 1.4 s

c 1.2 N d 2.0×10^{-25} m

EXERCISE 19B

1 a i

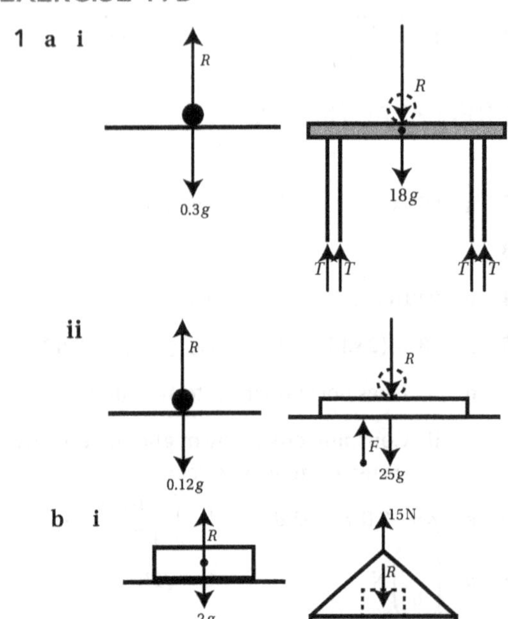

ii

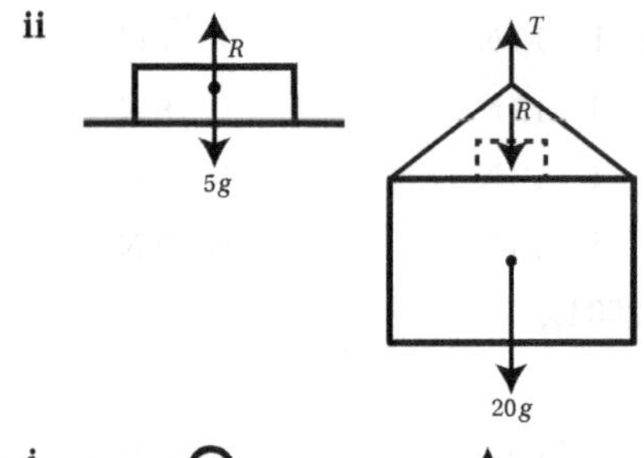

c i

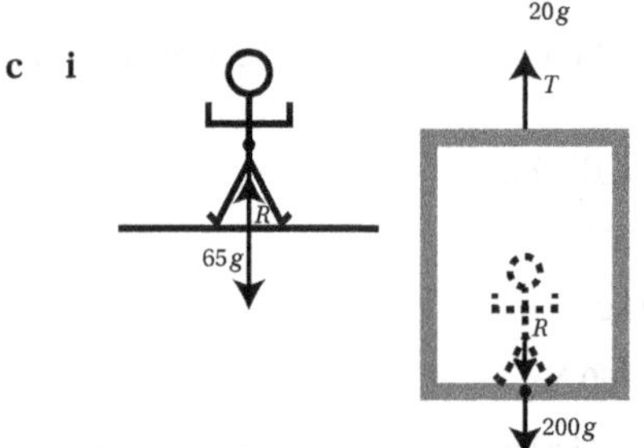

ii

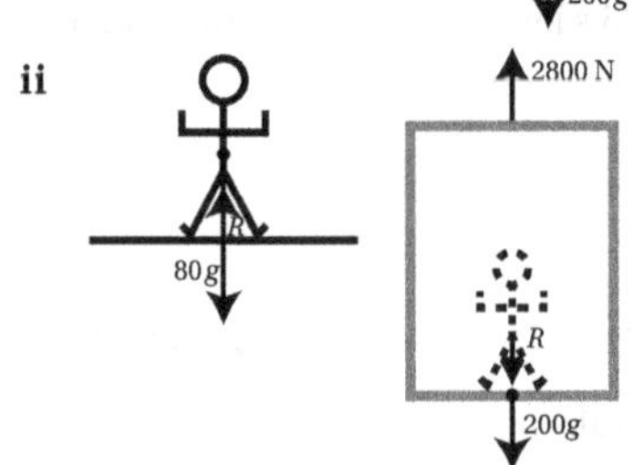

d i

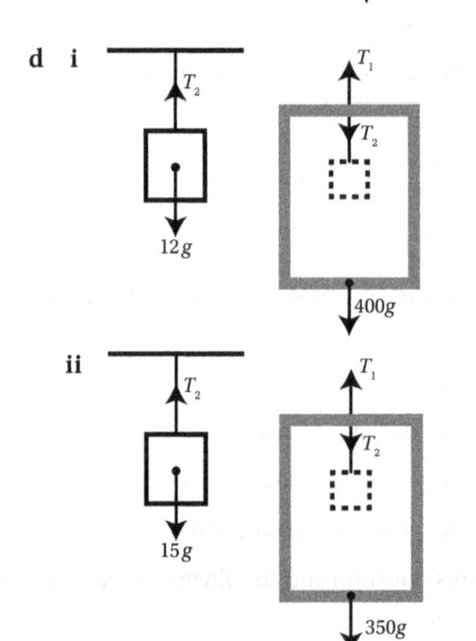

2 In this question you should give all final answers to 2 s.f.

a i 110 N ii 120 N

b i 95 N ii 130 N

3 20 N (1 s.f.)

4 79 N (2 s.f.)

5 1200 N (2 s.f.)

6 1.18 N (3 s.f.)

7 11 kg (2 s.f.)

8 a 6000 N (1 s.f.) b 400 N (1 s.f.)

9 a 1200 N (2 s.f.) b 830 N (2 s.f.)

10 a Proof.

b i 3500 N ii 8000 N

11 420 kg (2 s.f.)

12 a Up b 6590 N (3 s.f.)

EXERCISE 19C

1 652 N (3 s.f.)

2 a $F = 46$ N b $R = (88 + 68g)$ N

3 200 N (1 s.f.)

4 a 1.5 N (2 s.f.) b 11 N (2 s.f.)

5 a $6.5g - 38$ N b $6.5g$ N

6 120 N (2 s.f.); 54 N (2 s.f.)

7 a 5.9 N (2 s.f.) b 140 N

8 $18\sqrt{2}g$ N

WORK IT OUT 19.1

Solutions 2 and 3 are both correct.

EXERCISE 19D

1 These are some suggestions for discussion:

a Connection between cars not rigid.

b Mass of steel chain not negligible compared to mass of chair.

c Mass of steel chain compared to mass of car.

d Mass of rod of pendulum compared to mass of bob on the end.

e Cable needs to be able to stretch.

f Bottle will be drawn out between the pliers.

g Mass of tow bar compared to car and caravan.

h Steel cable needs to stretch when under tension.

2 a 12.7 m s^{-2} b 4855 N

3 12.6 N; 22.05 N

4 37.5 N

5 1120 N

6 a 312 N b 0.72 s

7 a 901 N; 530 N b 833 N; 490 N

8 a 10.75 kN b 5.5 kN; 2.75 kN

9 a Driving force of 2000 N.

b Tension of 1000 N between the locomotive and the first carriage, thrust of 1000 N between the two carriages.

EXERCISE 19E

1 800 N (1 s.f.)

2 1.6 m s^{-2} (2 s.f.); 57 N (2 s.f.)

3 a 59 N (2 s.f.) b 2.8 m s^{-2}; 42 N

4 2.52 kg (3 s.f.)

5 5

6 a 20 N b 1 m s^{-2} (1 s.f.)

7 1.1 kg (2 s.f.) or 0.58 kg (2 s.f.)

8 a $2x$ b $2a$ c $3.6g$ N

9 $5 \leqslant m \leqslant 15$

MIXED PRACTICE 19

1 C

2 50.9 kg

3 a 0.966 m s^{-2} b 529 N

4 580 N (2 s.f.)

5 1.18 N (3 s.f.)

6 30 N; 60 N

7 a 520 N b 1280 N c 1280 N

8 D

9 B

10 a 2790 N b 1810 N

11 a 1100 N (2 s.f.)

b i 2.2 m s^{-2} downwards ii 4000 N (2 s.f.)

12 42.2 kg

13 a 1.5 kg b 51 N (2 s.f.); 15 N (2 s.f.)

14 39.2 N or 46.0 N

15 a i 5400 N ii 10 800 N

b 23.9 kN

16 a Proof. b 94.1 N

c i 3.92 m s^{-1} ii 9.68 m s^{-1} iii 1.39 s

17 a 20 N (1 s.f.) b No (stops after 0.1 m (1 s.f.)).

18 a 750 N (2 s.f.) b 75 N

c Box B moves on top of box A.

19 a 1.43 s b 3.125 m

c 2.14 s

20 a $\frac{g}{4}$ m s^{-2} b $m = 2$ c 122.5 N

Focus on ... Proof 3

2 **a** $\frac{2s}{a+b}$ **b** $\frac{s(a+b)}{2ab}$ **c, d** Proof.

Focus on ... Problem solving 3

1 814 m

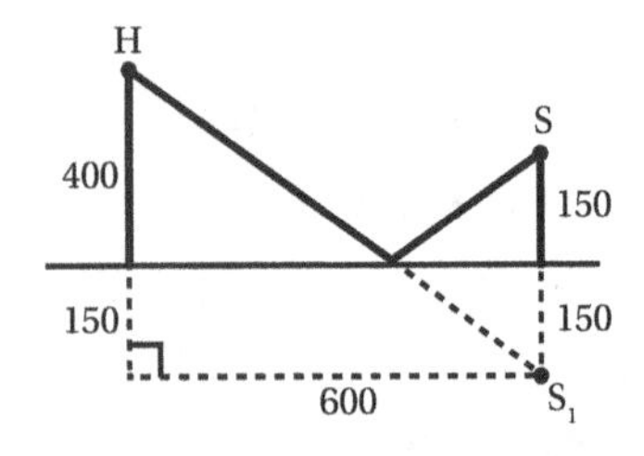

2 $\sqrt{5}$ cm; Shortest distance through cube is $\sqrt{3}$ cm.

3 $3\sqrt{2}$ cm

4 $\sqrt{h^2+\pi^2r^2}$

Focus on ... Modelling 3

1 **a** **i** 12.5 m **ii** 0.25% **iii** 0.04%

b **i** 510 m **ii** 0.25% **iii** 0.04%

2 **a** $120a = 180 - 40v$; remember that $a = \frac{dv}{dt}$.

b $v = 1.5t$

c When there is air resistance, the cyclist can never exceed the speed of 4.5 m s^{-1} (this is called **terminal velocity**).

	i 0.5 m s^{-1}	ii 2 m s^{-1}	iii 5 m s^{-1}
without air resistance	0.33 s	1.3 s	3.3 s
with air resistance	0.35 s	1.8 s	–
% difference	5.7%	28%	–

d Ignoring air resistance is only reasonable in the initial stages of the motion.

3 **a, b**

	speed after 1 second (m s^{-1})		speed after 10 seconds (m s^{-1})	
	air resistance ignored	with air resistance	air resistance ignored	with air resistance
para-chutist	9.8	9.6	98	51
coin	9.8	8.9	98	18

c **i** 1 s **ii** 10 s

d The model is suitable for short distances (a few metres), although there is already a noticeable difference for coin (but not for the parachutist). It is not suitable over distances as long as 10 m; the discrepancy is much larger for the coin.

4 Terminal velocity increases with mass, as does the time taken to reach terminal velocity.

Cross-topic review exercise 3

1 **a** $\overline{MP} = \frac{1}{3}\mathbf{a} + \frac{3}{4}\mathbf{b}$; $\overline{QN} = \frac{2}{9}\mathbf{a} + \frac{1}{2}\mathbf{b}$

b $\overline{MP} = \frac{3}{2}\overline{QN}$

2 2.4 m s^{-1}

3 **a** 70 m **b** 125 m

c 2.5 m s^{-1} **d** 15 m

e 0.3 m s^{-1} **f** 0.278 m s^{-2}

4 $(6\mathbf{i} + 4\mathbf{j})$ m s^{-2}

5 6.34 m s^{-1}

6 **a** $a = -6; b = -9$

b **i** $3\sqrt{13}$ N **ii** 56.3°

c $8\sqrt{13} \approx 28.8$ m

7 **a** **i** Proof. **ii** 3 seconds

b 8000 N

8 **a** **i** $v = 3t^2 - 22t + 35$ **ii** $\frac{7}{3}$ and 5 s

iii 35 m s^{-1}

b 44 m

9 3

10 **a** $a = 6, u = 4$ **b** 78.3 kg

11 **a** 4.6 m s^{-2} (2 s.f.) downwards

b 2500 N (2 s.f.)

c Mass of the cable was not included in the calculation.

12 **a** 400 N **b** 2400 N

13 **a** Proof.

b 37 N (2 s.f.)

c Light and inextensible.

d **i** 0.4 seconds **ii** 0.98 m s^{-1}

14 **a** 20 m (1 s.f.) **b** 5 m s^{-1} downwards.

15 **a** 297 N **b** Thrust

20 Working with data

BEFORE YOU START

1 20%

2 Mean: 5, median: 4.5, mode: 1

3 Range: 8, IQR: 5.5

EXERCISE 20A

1 a i

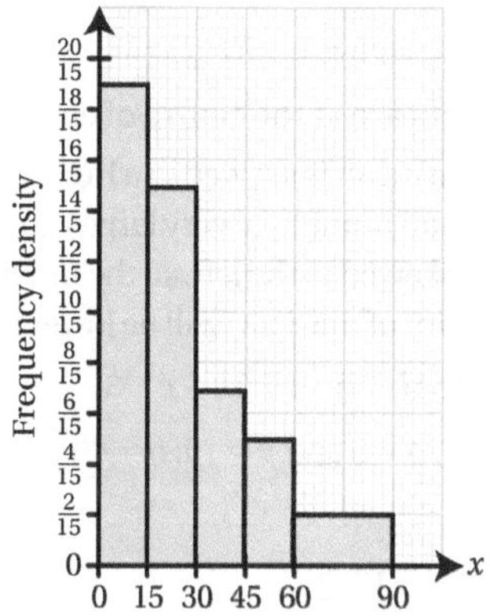

ii

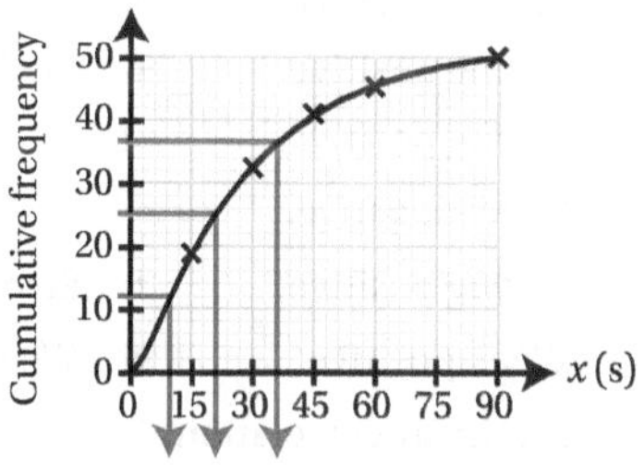

iii median ≈ 21 s, IQR ≈ 26 s

iv

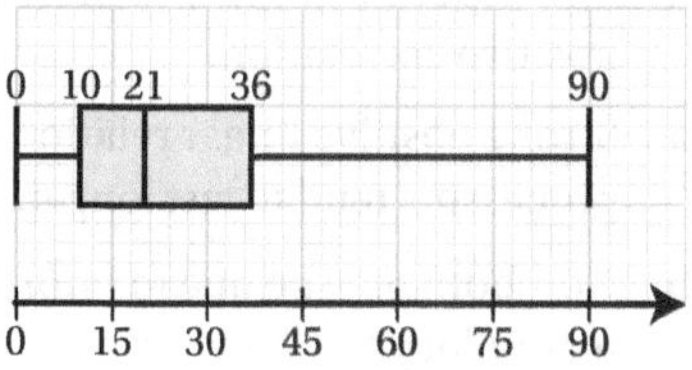

b i

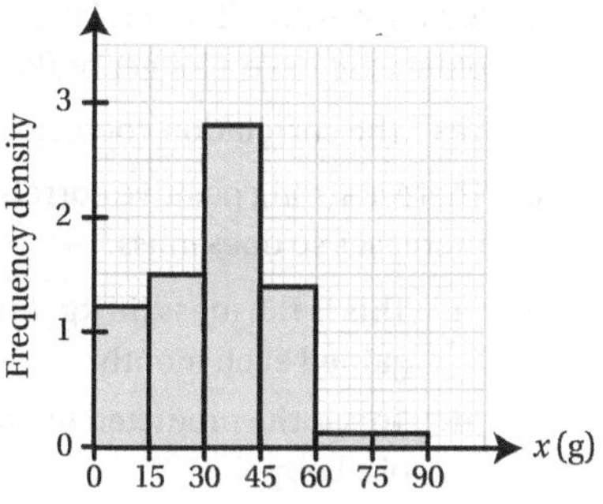

ii

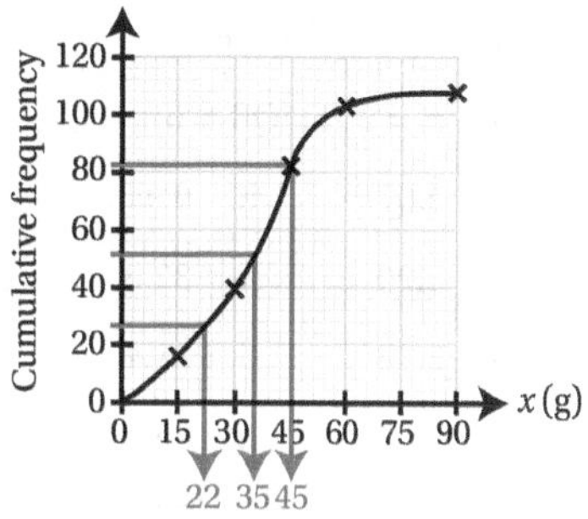

iii median = 35 g, IQR ≈ 23 g

iv

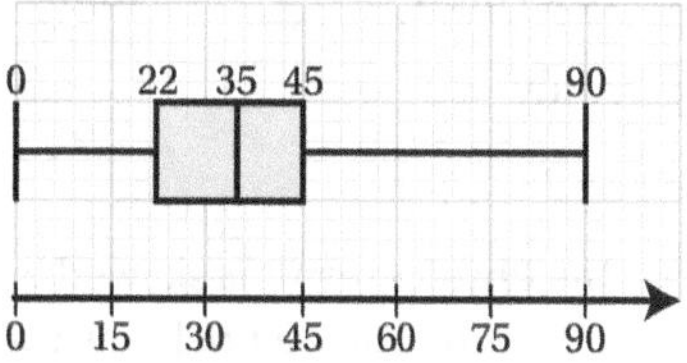

2 a i 0.3 ii 0.27

b i 0.4 ii 0.25

c i $\frac{2}{15}$ ii 0.25

3 Median = 12; IQR = 9.5

4 a 197 b $c = 191; d = 203$ c 12

5 a 53% b 28 minutes

6 a Results in English were generally lower but more consistent.

b 75%

c For example, if the results were for the same group of students.

d The number of students sitting each test.

7 a 15 minutes

b Alpha Commerce; 25%

c Beta Bank; 50%

8 a 0.1 b A

9 A3; B2; C1

10 $\frac{2bh}{3}$

WORK IT OUT 20.1

Solution 1 is correct.

EXERCISE 20B

1 a i $\sigma = 7.23$; IQR = 12.6

ii $\sigma = 6.57$; IQR = 13.5

b i $\sigma = 11.1$; IQR = 3

ii $\sigma = 35.3$; IQR = 72

c i $\sigma = \frac{\sqrt{56}}{5}$; IQR = 3

ii $\sigma = 2.75$; IQR = 6

2 a i 2.07 ii 0.943

b i 2.20 ii 32.8

c i 8.78 ii 21.4

3 a 12 b 2.92

4 a 46.8 b 527

5 a 124.5 mph b 11.8 mph

c Andy

6 a 15.2

b The Physics result is skewed by one very low score. You do not know the total mark available in the two subjects so the values might not be comparable.

7 18

8 11.7

9 101 or 141

10 a, b Proof.

EXERCISE 20C

1 a i $\bar{x} = 1; \sigma = \frac{\sqrt{11}}{4}; q_2 = 1$

ii $\bar{x} = 0; \sigma = \frac{\sqrt{28}}{5}; q_2 = 0$

b i $\bar{x} = 12.1; \sigma = 1.90; q_2 = 12$

ii $\bar{x} = 0.263; \sigma = 0.137; q_2 = 0.2$

2 a 17; one day less than 21 b 17; 20

c 16.5; 20.5 d 16.01; 20

3 a i $\bar{x} = 26.1; \sigma = 20.4$

ii $\bar{x} = 253; \sigma = 151$

b i $\bar{x} = 6.38; \sigma = 5.23$

ii $\bar{x} = 102; \sigma = 5.78$

c i $\bar{x} = 6.58; \sigma = 5.11$

ii $\bar{x} = 15.7; \sigma = 10.8$

4 a

Number of cars	0	1	2	3	4
Number of households	10	52	64	15	5

b $\bar{x} = 1.68; \sigma = 0.875$

c Households in Mediton tend to have more cars but there is greater variation.

5 a $\bar{x} = 3.82; \sigma = 1.67$

b Use more groups; measure more accurately.

6 a 1.5 b 1.84 c 3.01

d i The variance is lower so the process is more consistent.

ii There were more broken eggs, on average.

iii The sample size is too small for small differences to be significant - they may be due to chance.

7 a $\bar{x} = 9.88; \sigma = 7.37$

b Data are at the centre of each group.

c The mean is skewed to a higher value by a small number of very large values. The data is skewed to the left, so the median (halfway point of the data) will be lower than the mean.

8 a Proof. b $1 \pm 2k$ c 3

9 $p = 45; q = 5$ and $p = 10; q = 40$

10 Proof.

EXERCISE 20D

1 a Strong positive. b No correlation.

c Strong positive. d Perfect positive.

e Strong positive. f Strong positive.

g Weak negative.

2 a Strong positive. b Strong negative.

c Weak positive. d Weak negative.

e No linear correlation.

f Non-linear correlation.

g Perfect negative.

h Two separate groups, each with strong positive correlation.

3 a Strong positive linear relationship - faster processors tend to last longer.

b No - both may be due to improvements in technology.

4 a Strong positive linear relationship - older cars have a longer braking distance.

b Yes - the correlation coefficient is positive.

5 a There is strong positive correlation: as age increases so does mass.

b i This is the mass (in kg) typically gained each month.

ii This is the predicted birth mass (in kg) of a baby.

c It does not take into account gender which might be an important factor. 14 years is a large extrapolation from the data.

6 a i 24.8 is the predicted 100 m time of students when they start school.

ii 0.609 is the typical improvement in 100 m time each year.

b The low correlation coefficient suggests that a straight line model is not appropriate. Even if it were appropriate, 60 years is extrapolating from the data.

7 **a** : A, **b** : D, **c** : B, **d** : C

8 a i 1.5 m ii 3 m

b i is valid as the correlation coefficient is large, but

ii is problematic as it is extrapolating from the data.

9 a i 5 Hz ii It is impossible to say.

b The points are not in order, so there is no sense in using a line to interpolate between points.

c If the points were ordered in time, for example, vertical and horizontal displacement of a ball over time.

10 a False b False

c True d False

EXERCISE 20E

1 a i No ii No

b i Yes ii Yes

2 a i No ii No

b i Yes: −3 ii Yes: 64

3 There appears to be two distinct groups – he should find the correlation coefficient for each group separately.

4 a 5

b Proof.

5 Proof.

6 a 21°C

b i Decrease. ii Decrease.

7 a $\bar{t} = 62.9; \sigma = 7.40$ b Thursday

c $\bar{x} = 63; \sigma = 7.98$ d 0.314 km/minute

e No – it is extrapolating from the data.

8 a Proof.

b $x = 11.6; \sigma = 2.58$

9 a, b Proof.

c $q = 3, p = 27$

d Any values $p = 9n$ and $q = 32n$, where n is a positive integer.

MIXED PRACTICE 20

1 B

2 a 24 b 125 c 124

3 2

4 a 4.08 minutes b 8.80 minutes^2

5 $x > 140$ or $x < -20$

6 a For example, 1, 0 b For example, 1, 1

7 a Median = 4.5 kg; IQR = 1.1 kg

b 1.9 kg

c Each value appears exactly once.

8 D

9 a $\bar{x} = 10; \sigma = 6.71$ b 0.323

10 a 84% b 79% c 12.1%

11 a 66 mph b 11

c $a = 61$ mph; $b = 72$ mph

d

v	Frequency
$40 < v \leqslant 50$	2
$50 < v \leqslant 60$	8
$60 < v \leqslant 70$	**26**
$70 < v \leqslant 80$	**8**
$80 < v \leqslant 90$	**4**
$90 < v \leqslant 100$	**2**

e $\bar{x} = 67$ mph; $\sigma = 10.6$ mph

f Do not remove – they are extreme but no reason to believe they are incorrect.

12 a Shows relationships.

b Strong positive.

c 200 km^2

d Decrease.

e Average population density.

f Extrapolating from the data.

13 a i 28 ii 16.7%

b i

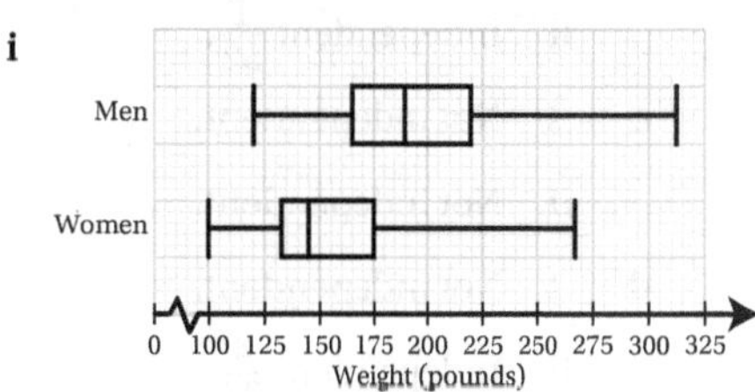

ii For example, women have: smaller range, smaller IQR, greater degree of skew, lower median.

c People visiting the doctor may not have representative weights.

14 C

15 a α: D, β: B, γ: C

b i B ii A iii C

16 150

17 118 888.2

21 Probability

BEFORE YOU START

1 $\frac{1}{20}$

2 $\frac{7}{24}$

3 a 5040 b 840 c 35

d 1 e 1

EXERCISE 21A

1 a i Mutually exclusive $\frac{1}{2}$.

ii Mutually exclusive $\frac{2}{3}$.

b i Not mutually exclusive.

ii Not mutually exclusive.

c i Mutually exclusive $\frac{1}{3}$.

ii Not mutually exclusive.

d i Mutually exclusive $\frac{13}{25}$.

ii Not mutually exclusive.

e i Not mutually exclusive.

ii Mutually exclusive $\frac{7}{15}$.

2 a i Not independent.

ii Not independent.

b i Independent $\frac{1}{26}$.

ii Independent $\frac{1}{52}$.

c i Not independent.

ii Not independent.

d i Not independent.

ii Independent $\frac{4}{25}$.

e i Not independent

ii Not independent

3 a i pq ii 0

b Only if the probability of at least one of them is zero.

4 a $\frac{1}{16}$ b $\frac{7}{16}$

5 $\frac{1}{9}$

6 a ACT, ATC, CAT, CTA, TAC, TCA

b $\frac{1}{3}$

7 a $\frac{1}{6}$

b Not independent.

8 a i $\frac{19}{60}$ ii $\frac{61}{300}$

b i Not independent.

9 $\frac{1}{32}$

10 a $\frac{1}{8}$ b $\frac{7}{8}$

11 a 0.858

b Not reasonable; group behaviour, common experiences and external events.

EXERCISE 21B

1 a

t	0	1	2	3
$P(T=t)$	$\frac{1}{8}$	$\frac{3}{8}$	$\frac{3}{8}$	$\frac{1}{8}$

b

d	0	1	2	3	4	5
$P(D=d)$	$\frac{6}{36}$	$\frac{10}{36}$	$\frac{8}{36}$	$\frac{6}{36}$	$\frac{4}{36}$	$\frac{2}{36}$

c

x	1	2	3	4	6
$P(X=x)$	$\frac{1}{6}$	$\frac{2}{6}$	$\frac{1}{6}$	$\frac{1}{6}$	$\frac{1}{6}$

d

g	1	2	3
$P(G=g)$	$\frac{6}{72}$	$\frac{36}{72}$	$\frac{30}{72}$

e

c	1	2	3	4
$P(C=c)$	$\frac{1}{4}$	$\frac{3}{16}$	$\frac{9}{64}$	$\frac{27}{64}$

f

x	1	2	3	4	6	8	9	12	16
$P(X=x)$	$\frac{1}{16}$	$\frac{2}{16}$	$\frac{2}{16}$	$\frac{3}{16}$	$\frac{2}{16}$	$\frac{2}{16}$	$\frac{1}{16}$	$\frac{2}{16}$	$\frac{1}{16}$

2 a i $\frac{1}{8}$ ii 0.1

b i $\frac{1}{8}$ ii $\frac{3}{7}$

c i $R=$ ii $R=$

3 a $\frac{1}{12}$ b $\frac{5}{12}$

4 a 0.19 b 0.55

5 a 0.8 b 0.01

6 a

X	2	3	4	5	6	7	8
$P(X=x)$	$\frac{1}{16}$	$\frac{2}{16}$	$\frac{3}{16}$	$\frac{4}{16}$	$\frac{3}{16}$	$\frac{2}{16}$	$\frac{1}{16}$

b $\frac{3}{8}$

7 $a=0.1; b=0.2$

8 a Proof.

b 0.72

9 a $\frac{13}{60}$ b 0.207

10 a 0.11 b 0.97

11 a $\frac{1}{9}$

b

x	1	2	3
$P(X \leqslant x)$	$\frac{1}{9}$	$\frac{4}{9}$	$\frac{9}{9}$
$P(X=x)$	$\frac{1}{9}$	$\frac{3}{9}$	$\frac{5}{9}$

12 a 4

b

y	0	1	4	9	16
$P(Y=y)$	$\frac{1}{9}$	$\frac{2}{9}$	$\frac{2}{9}$	$\frac{2}{9}$	$\frac{2}{9}$

WORK IT OUT 21.1

Solution 3 is correct.

EXERCISE 21C

1 a $X \sim B(20, 0.5); P(X=15)$

b Not binomial, the probability may not be the same for each question (e.g. if the questions get harder).

c Not binomial, there are three possible outcomes.

d Not binomial, trials are not independent.

e Approximately binomial.
$X \sim B(100, 0.02); P(X \geqslant 1)$

2 a i 0.147 ii 0.0459

b i 0.944 ii 0.797

c i 0.0563 ii 0.0104

d i 0.990 ii 0.797

e i 0.203 ii 0.832

f i 0.0562 ii 0.776

g i 0.486 ii 0.203

h i 0.203 ii 0.496

3 a $X \sim B(20, 0.7); P(X \geqslant 16) = 0.238$

b $X \sim B(5, 0.35); P(X=3) = 0.181$

c $X \sim B(5, 0.5); P(X \geqslant 3) = 0.5$

4 a i $\frac{5}{32}$ ii $\frac{1}{32}$

b i $\frac{31}{32}$ ii $\frac{6}{32}$

c i $\frac{1}{32}$ ii $\frac{26}{32}$

5 a Drawing from a finite population without replacement so probability is only approximately constant.

b 0.103

6 a 0.0231 b 0.0273

7 a 0.108 b 0.0267

c The probability that a person going to the doctor has the virus is the same as for the whole country.

8 The second one $(0.0165 > 0.0154)$

9 a i 0.6 ii 0.6

iii $\frac{1}{3}$; $P(R_1R_2) \neq P(R_1)P(R_2)$ iv 0.3; 0.288

b i 0.6 ii 0.6

iii 0.36 iv 0.288

10 a Not independent, but constant probability.

b Independent and constant probability.

c Independent, but different probabilities.

11 0.377

12 0.630

13 $\frac{2}{3}$

14 0.6 or 0.4

15 a Proof.

b 10; to do this you will have needed to try some values of n or use tables on your calculator.

16 3

17 $n=\frac{5-p}{p}$

MIXED PRACTICE 21

1 C

2 0.261

3 a 0.16 b 0.72 c 3

4

x	0	1	2	3	4	5	6
$P(X=x)$	$\frac{1}{4}$	$\frac{1}{8}$	$\frac{1}{8}$	$\frac{1}{8}$	$\frac{1}{8}$	$\frac{1}{8}$	$\frac{1}{8}$

5 a 0.142 b 0.335

6 a 0.2 b $p = 0.05; q = 0.15$

7 a i

	J	J'	Total
W	0.55	0.10	0.65
W'	0.15	0.20	0.35
Total	0.70	0.30	1.00

ii 0.25

iii Proof.

8 B

9 a 0.232 b 0.894

10 a 16% b 52%

11 0.653

12 0.36

13 a

x	−5	−4	−3	−2	−1	0	1	2	3	4	5
$P(X=x)$	$\frac{1}{36}$	$\frac{2}{36}$	$\frac{3}{36}$	$\frac{4}{36}$	$\frac{5}{36}$	$\frac{6}{36}$	$\frac{5}{36}$	$\frac{4}{36}$	$\frac{3}{36}$	$\frac{2}{36}$	$\frac{1}{36}$

b $\frac{5}{12}$

14 a i 0.82 ii 0.456

iii 0.446

b i 0.781 ii 0.768

15 a 0.0296 b 2.44×10^{-4}

16 a 0.031 25 b 0.01 5625

17 a 0.460 b $1-(0.95)^n$

c 32

18 a

x	0	1	2	3	4	5
$P(X=x)$	$\frac{6}{36}$	$\frac{10}{36}$	$\frac{8}{36}$	$\frac{6}{36}$	$\frac{4}{36}$	$\frac{2}{36}$

b 0.0697

19 a Proof.

b

1	4	5
$\frac{1}{1296}$	$\frac{175}{1296}$	$\frac{369}{1296}$

20 a 0.0781 b 0.0130

22 Statistical hypothesis testing

BEFORE YOU START

1 a i 0.575 ii 0.274

b 20

2 $X \sim B(12, 0.7)$

EXERCISE 22A

1 a Sixth-form students may be more likely to progress to university and so be more worried about fees than people who are not affected.

b People in supermarkets are more likely to be knowledgeable about food prices.

c Schoolchildren will be shorter than average.

d This is a real-life example. Mostly richer people had telephones, so the poll predicted the wrong winner.

2 a Quota sampling.

b Stratified sampling.

c Cluster sampling.

d Simple random sampling.

e Systematic sampling.

f Opportunity sampling.

3 a 2.55

b Probably larger.

c It is likely to be better, but with random fluctuations it is possible to be worse.

4 a This would require a complete list of all fish in the North Sea and the ability to be able to measure any fish required.

b Quota sampling.

c It will be more representative.

5 a 7.5 b 10

c If the data were a sample from a population and you wanted to estimate the variance of the population.

6 a

	Female	Male	Total
Under 42	16	24	40
42 or over	16	24	40
Total	32	48	80

b It is random therefore less biased or more generalisable.

c It is cheaper. It does not require all selected people to participate. It does not require a list of all people in the city.

7 a A - Cluster, B - Simple random, C - Quota, D - Stratified.

b D

c Less time consuming; no need to travel as far; cheaper.

8 Not everybody will return the questionnaire so not all possible samples are equally likely.

9 a 9.33×10^{-4}

b i 0.0861 ii 0.009 96

c 4.65%

d Increase - genders may cluster together.

WORK IT OUT 22.1

1 Acceptable. You might prefer that the conclusion in context were stated with less certainty; as in: 'Reject H_0. There is sufficient evidence to conclude that the proportion of red flowers has decreased,' rather than stating it as an absolute fact.

2 No comment in context. Proportion of what?

3 Avoid using the word 'probably' in a probability or statistics context. Should read: 'There is insufficient evidence that the proportion of girls in the club has changed (or risen, or fallen - the text gives no indication as to the tail or tails of the test) from 45%.'

4 Do not 'accept' H_0 - either reject it or fail to reject it. The conclusion in context is appropriately stated.

5 Do not 'accept' H_0. 'Cannot reject H_0. There is insufficient evidence to conclude that the dice is biased.'

WORK IT OUT 22.2

Solution C is correct.

EXERCISE 22B

1 a i $H_0: p = 0.6$; $H_1: p > 0.6$; p is the proportion of students who like football.

ii $H_0: p = \frac{1}{3}$; $H_1: p > \frac{1}{3}$; p is the proportion of households with a pet.

b i $H_0: p = 0.06$; $H_1: p < 0.06$; p is the proportion of faulty components produced.

ii $H_0: p = 0.5$; $H_1: p < 0.5$; p is the proportion of children who eat 5 or more pieces of fruit a day.

c i $H_0: p = 0.5$; $H_1: p \neq 0.5$; p is the probability that the coin shows heads.

ii $H_0: p = 0.26$; $H_1: p \neq 0.26$; p is the proportion of entries that were graded A.

2 a i p-value 19.3% (> 5%). Insufficient evidence.

ii p-value 91.4% (> 10%). Insufficient evidence.

b i p-value 7.69% (< 10%). Sufficient evidence.

ii p-value 4.46% (> 3%). Insufficient evidence.

c i p-value 3.98% (< 4%). Sufficient evidence.

ii p-value 6.92% (> 2.5%). Insufficient evidence.

3 a $H_0: p = 0.63$; $H_1: p < 0.63$

b Cannot reject H_0. No evidence of lower car ownership in David's neighbourhood.

4 Insufficient evidence of an increase.

5 Insufficient evidence of a probability less than $\frac{1}{6}$.

6 There is sufficient evidence of a higher proportion of A grades.

7 Insufficient evidence that the new treatment is better.

8 Insufficient evidence that the proportion with a younger sibling at the same school is different.

9 There is sufficient evidence of a difference in probability of getting high and low numbers on the dice.

10 Insufficient evidence that the proportion able to run 100 metres in under 12 seconds has changed.

11 14

12 It was greater than 4.45%.

13 $q < 0.0247$

EXERCISE 22C

1 a i $X \leqslant 23$ ii $X \leqslant 4$

b i $X \leqslant 51$ ii $X \leqslant 51$

c i $X \geqslant 37$ ii $X \geqslant 12$

d i $X \geqslant 69$ ii $X \geqslant 69$

2 a i 0.0314 ii 0.0759

b i 0.0602 ii 0.009 60

c i 0.280 ii 0.0875

d i 0.0602 ii 0.003 88

3 $X \leqslant 10$

4 a $H_0: p = 0.85$; $H_1: p > 0.85$, where p is the proportion of patients the drug cures.

b $X \geqslant 162$

5 a $H_0: p = 0.12$; $H_1: p > 0.12$, where p is the proportion of the population who have purchased the manufacturer's products.

b Cannot reject H_0; no significant rise in product purchase history among the population.

c 0.0556

6 a Acceptance region. b 0.06

7 a $H_0: p = 0.3$; $H_1: p > 0.3$, where p is the proportion of roses which are white.

b $X \geqslant 62$

c Reject H_0: Proportion of roses that are white is significantly greater than 30%.

8 a $H_0: p = 0.72$; $H_1: p \neq 0.72$, where p is the proportion of students passing the test on their first attempt.

b $X \leqslant 29$ or $X \geqslant 41$

c 0.0972; SL $\geqslant$ 9.72%

MIXED PRACTICE 22

1 A

2 a Cluster sampling.

b Students from the same tutor group are likely to go or not go together.

c Obtain a list of all students, number sequentially, use a random number generator to generate 50 numbers, select students with those numbers (ignore repeats and numbers larger than the number of students).

3 a $H_0: p = \frac{1}{6}$; $H_1: p > \frac{1}{6}$, where p is the underlying probability of rolling a 5.

b There is insufficient evidence at the 5% significance level that the probability of rolling a 5 is greater than $\frac{1}{6}$.

4 $H_0: p = \frac{1}{8}$; $H_1: p < \frac{1}{8}$, where p is the underlying proportion of days that Lisa is late. There is insufficient evidence at the 10% significance level that Lisa's probability of being late has decreased from 1 in 8.

5 A

6 $p = 0.005\,23$ (< 0.025), sufficient evidence to reject H_0, there is evidence that the proportion of families who own their homes is different from 64%.

7 a Each possible sample of a given size has an equal probability to be included within the sample.

b i Stratified sampling.

ii The advantage over simple random sampling is that, if boys and girls have a different behavior profile, then a random sample which (by chance) over-represented one group would give a distorted impression of the overall population.

c $X \geqslant 34$

d 9.06%

e There is insufficient evidence that the proportion of students who play in a school sports team is greater than 40%.

8 a Insufficient evidence to conclude that the proportion of passengers buying tickets in advance is different from 35%.

b People travelling in a single party are likely to sit in the same carriage and buy tickets at the same time. A binomial model is not suitable.

9 a i If Adam gets a list of 50 random numbers selected without repeats from 1 to 500, he can select the animals with those number tags to test.

ii Under- or over-representation of goats in the sample is likely.

b i Systematic sampling.

ii Not fully random - no chance of multiple goats being selected.

c i Collect blood samples from 33 sheep, 16 cattle and 1 goat from the first animals of those species he encounters.

ii Speed/simplicity.

iii Not representative if health status is clustered geographically.

10 C

Focus on ... Proof 4

1–3 Proof.

Focus on ... Problem solving 4

1 a 22.4% b Still an underestimate.

2 a 100 million

b Assuming no fish have entered or left the North Sea and that the fish have thoroughly mixed up.

Focus on ... Modelling 4

1–3 Multiple answers are possible.

Cross-topic review exercise 4

1 a 6, 10, 16, 7

b Mean = 21.3; variance = 82.8

c Actual data values are not given.

2 a Blue star more variable. Longer wait on average. Two very long waits (outliers).

b i Blue Star - sometimes arrived within 5 minutes; Green Star never has.

ii Green star - always arrived within 25 minutes; Blue Star sometimes hasn't.

3 $\sqrt[4]{\frac{e}{24}}$

4 a 0.31 b 0.303

5 $\frac{5}{36}$

6 a Cannot reject H_0, since $p = 0.2375 > 0.1$. No significant evidence to support newspapers article's claim.

b p may not be constant - may depend on cyclist/speed/weather. Events may not be independent -2 cyclists may arrive together.

7 $\frac{11}{12}$

8 $\frac{8}{3}$

9 New mean = 34.9 (3 s.f.); new variance = 6.32 (3 s.f.)

10 £518

11 a i
- Use two-digit random numbers.
- Reject repeats and 00 or > 90.
- Continue until 15 numbers obtained.

Use the sites with these numbers.

ii May not cover all 3 orders of stream.

Teams may be spread out all along the stream making supervision hard.

b i Systematic sampling.

ii Not random because not every different group of 15 can be chosen.

iii Stratified. This picks 7 first order, 3 second order and 5 third order, which are correct proportions.

c i Cluster sampling.

ii For example, all the teams are close together, making supervision easier.

Every student/team collects data on each order of stream.

iii For example, the blocks may not be representative of the whole stream.

Five from each order is not the right proportions.

12 **a** Proof.

b

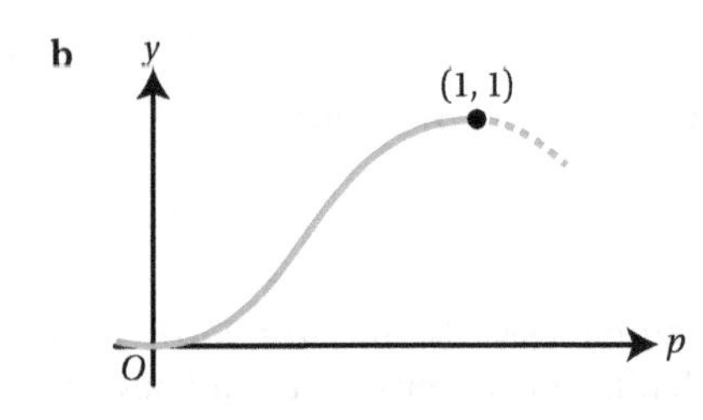

c Proof.

13 **a** ln 5 **b** 0.001 62

14 **a** Proof. **b** **i** 0.0199

ii Same probability for each building; completion of different buildings independent of the others.

c Reject H_0, since $p = 0.0294 < 0.05$. Sufficient evidence of increase in probability of building being completed on time.

15 **a** $\frac{4}{5}$ **b** 0.634

16 3, 11

Practice paper 1

1 D

2 B

3 $30°, 150°$

4 **a**

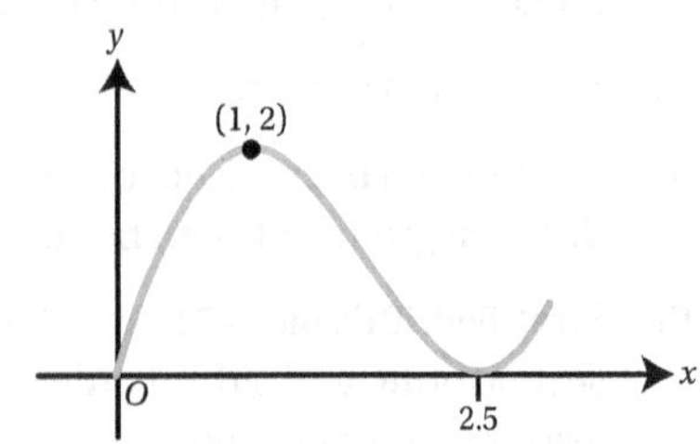

b

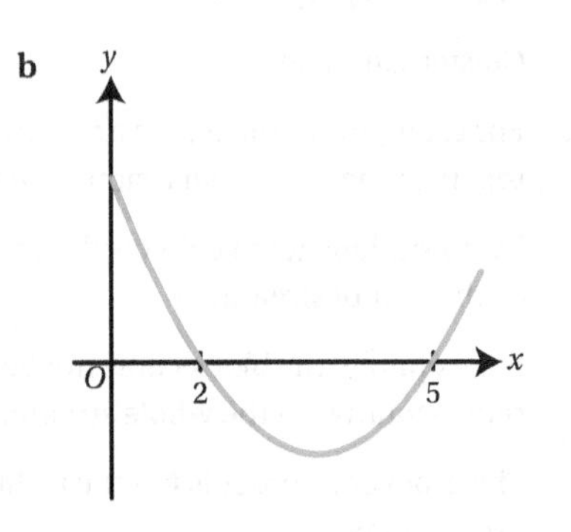

5 $x = -1, 2$

6 **a** **i** $a = -8$ **ii** $f(x) = (x-2)^2(x+3)$

b

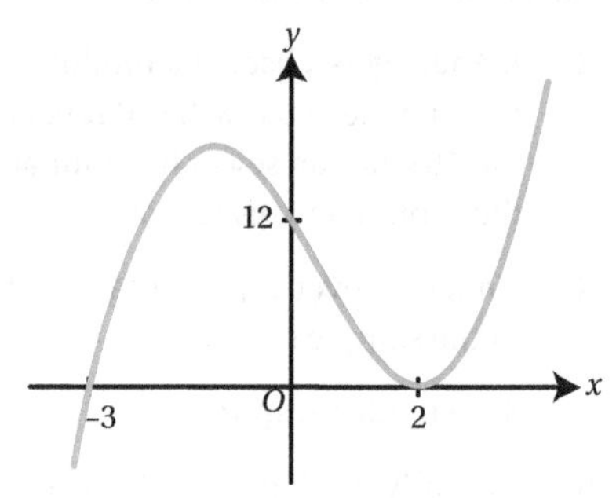

7 **a** For example, 19

b Proof.

8 **a** $A = 30\,000$; $k = 0.4$ **b** 8.77 days

9 **a** Asymptotes: $x = 0, y = 0$

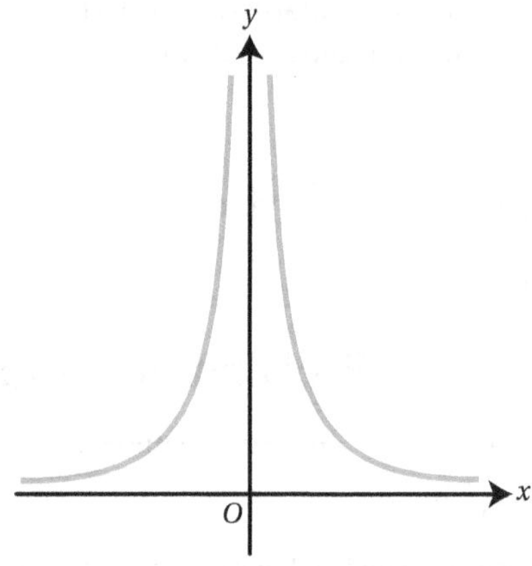

b

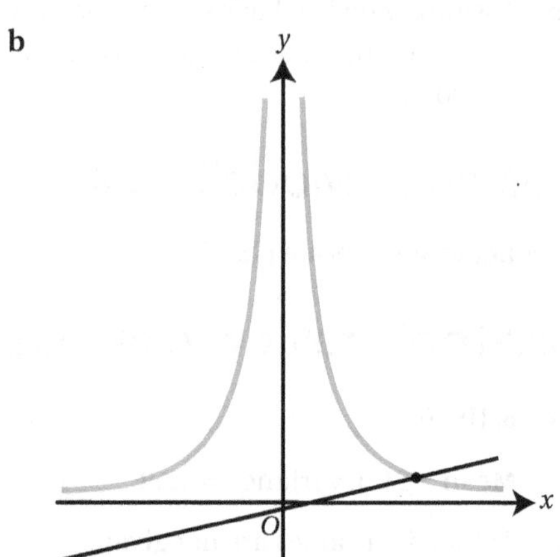

c Minimum.

10 $x = 9$

11 D

12 $(6.8\mathbf{i} - 0.2\mathbf{j})$ m s^{-2}

13 a 8.49 m s^{-1} downwards (3 s.f.)

b 2.5 seconds (2 s.f.)

c t would increase.

14 140 N (2 s.f.)

15 a $a = -4.05$ m s^{-2}

b $t = 13$

c 9.96 seconds

Practice paper 2

1 B

2 C

3 $64 - 64x + \frac{80}{3}x^2$

4 $\frac{2a^2 - 2}{a}$

5 $\frac{19}{7}$

6 a $-11\mathbf{i} + 6\mathbf{j}$ b $\sqrt{370}$

7 Proof.

8 a $\frac{1}{\sqrt{5}}$ b $k = -1$

9 a $AB = \cos\alpha$; $BC = \sin\alpha$

b Proof.

10 a $\left(\frac{1}{4}, \frac{23}{8}\right)$ and $(6, 0)$

b

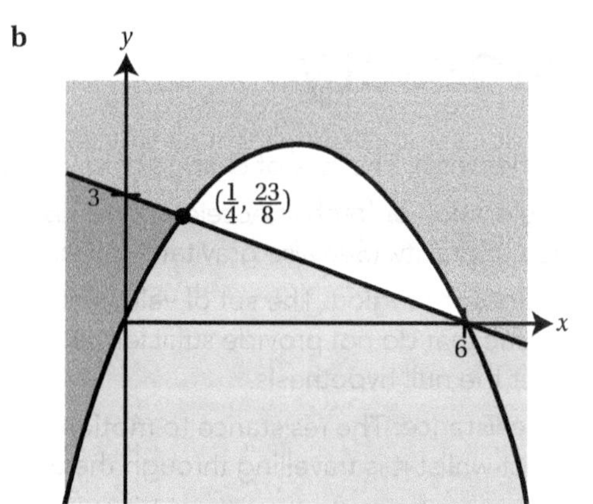

c $p = -2$; $q = -15$

11 a $(50, 0)$ b $(x - 21)^2 + y^2 = 841$

12 B

13 0.0725

14 a Obtain a list of all registered voters, numbered sequentially. Use a random number generator to select 200 numbers from the list, ignoring repeats. The sample is made up of the voters with those numbers.

b $p = 0.133 > 0.05$, do not reject H_0.

15 a No correlation. b 2.24

c 6 outliers above 6.2.

d Urban, as people generally live nearer where they work, making cycling to work more likely.'

Glossary

Acceleration: The rate of change of velocity.

Acceleration of freefall: Acceleration caused by the force of gravity. (*See also* gravitational acceleration.)

Acceptance region: The set of values of the test statistic that do not provide sufficient evidence to reject the null hypothesis.

Air resistance: The resistance to motion of an object whilst it is travelling through the air.

Alternative hypothesis: In hypothesis testing, specifies how you think a population parameter may have changed. It is denoted H_1.

Amplitude: The maximum 'height' of a periodic function, which is half of the distance from the minimum value of y to the maximum value of y.

Asymptote: A line that a function gets increasingly close to (but never meets).

Average speed: Total distance divided by the total time taken.

Biased: A sample that is not a good representation of a population.

Binomial coefficients: The constants of the terms in the expansions of expressions of the form $(a+b)^n$.

Binomial distribution: Probability distribution of the number of 'successes' out of a fixed number of independent trials.

Binomial expression: An expression that contains two terms.

Binomial theorem: A 'pattern' for quickly expanding expressions of the form $(a+b)^n$.

Bivariate: Data in pairs coming from the same source (for example, a person's height and weight).

Box-and-whisker plot: A statistical diagram showing the range, quartiles and median of a set of data.

Braking force: A force that causes an object to slow down.

Census: Collection of data from the whole of a population.

Chord: The line segment between two points on a curve.

Cluster sampling: The population is split into groups (clusters) based on convenience; a simple random sample of clusters is then chosen to study further.

Coefficient: A constant in front of (multiplying) a variable.

Collinear: Points that lie along the same straight line.

Complement: Everything except a given event. The sum of the probabilities of an event and its complement is one.

Components (of a vector): The forces acting in the x and y directions.

Conditional probability: A probability that depends on the result of a preceding event.

Congruent (expressions): Expressions connected by the identity symbol; they are equal for all values of the variable(s).

Constant of integration: The constant introduced whenever indefinite integration takes place. Its value can be found if other information is provided.

Constraint: A condition that allows a variable to be eliminated in situations where a function would normally depend on two variables.

Continuous (data): Data that can take any value within a range, such as height.

Correlation: A relationship between two variables whereby as one variable increases, the other either increases or decreases.

Correlation coefficient: A number measuring the 'strength' of correlation between two variables.

Counter example: An example used to prove that some proposition is not always true.

Critical region: The set of values of the test statistic that provide sufficient evidence to reject the null hypothesis. (*See also* rejection region.)

Critical value: The value at the edge of the critical region.

Cumulative frequency: The total number of data values less than or equal to a particular value.

Cumulative probability: The sum of the probabilities up to and including a specified value of the variable.

Deceleration: Negative acceleration. The rate of change of velocity as an object slows down.

Definite integration: The evaluation of an integral between two points (this eliminates the need for the constant of integration and gives a numerical answer).

Degree (of a polynomial): The highest power of x occurring in a polynomial function.

Derivative: A function that gives the gradient of $y = f(x)$ at any point. (Also sometimes called the **gradient function**.)

Differentiation: The process of finding a derivative or gradient function of a given function.

Direction (of a vector): The angle the vector makes with a fixed line (usually the horizontal).

Discrete (data): Data that can only take certain specific values (for example, the scores when rolling a dice).

Discriminant: In the quadratic formula, the expression $b^2 - 4ac$ (within the square root) which determines how many roots a quadratic equation has.

Displacement: The vector representing the translation from one point to another.

Driving force: A force provided by, for example, the engine of a car, that causes an object to accelerate in the direction of motion.

Equal (vectors): Two or more vectors that have each of their components equal.

Equation: A mathematical statement involving an = sign.

Equilibrium: A state in which the resultant of the forces on an object is zero.

Exponential decay: A relationship of the form $y = a^x$, where $0 < a < 1$. As x increases, y decreases.

Exponential equation: An equation with the variable in the power.

Exponential function: A function with the variable in the power.

Exponential growth: A relationship of the form $y = a^x$, where $a > 1$. As x increases, so does y.

Extrapolation: Predicting results beyond the limits of the collected data.

Factor theorem: A theorem relating to polynomials, $p(x)$: if $p(a) = 0$ then $(x - a)$ is a factor of $p(x)$.

Factorial: The product of consecutive integers from 1 to n. Denoted $n!$

Force: An interaction that acts to change the velocity of an object.

Friction: A force (between surfaces) that acts to resist motion. Always acts parallel to the surface.

Function: A rule that assigns to each input value a unique output value.

Gradient: A number measuring the steepness of a line or curve.

Gravitational acceleration: Acceleration caused by the force of gravity. (*See also* acceleration of freefall.)

Gravity: The force of attraction of a massive object (for example, the Earth) on all objects around it.

Histogram: A chart that resembles a bar chart but in which the area is used to represent the frequency. The vertical scale is the **frequency density**.

Hypothesis test: A procedure for determining whether a given sample provides significant evidence that a population parameter (mean/spread/proportion) has changed from a previously known or assumed value.

Identity: An equation that is true **for all values** of the unknown. It is shown by the symbol ≡.

Indefinite integration: Integration without limits. A constant of integration is always required.

Independent (events): Two (or more) events for which the probability of the second (subsequent) event does not depend on the outcome of the first.

Inextensible: *(In Mechanics, relating to string, rope, etc.)* Does not stretch.

Instantaneous acceleration: Acceleration at a specific point in time; it equals the derivative of the velocity function evaluated at that point (and the gradient of the velocity–time graph).

Instantaneous velocity: Velocity at a specific point in time; it equals the derivative of the displacement function evaluated at that point (and the gradient of the displacement–time graph).

Instantaneously at rest: An object is instantaneously at rest when $v = 0$. The object may change direction at this point.

Integer: Whole number.

Integration: The process of reversing differentiation.

Interquartile range: A measure of spread. The difference between the lower and upper quartile of a data set.

Intersection: The overlap between two or more sets or intervals.

Interval notation: A form of notation to represent all numbers in a range. This uses the end points of the range in ( ) brackets if the end points are not included, and [ ] brackets if the end points are included.

Light: *In Mechanics*, when an object's weight is small enough to be negligible and so can be ignored in calculations.

Limits of integration: The lower and upper values used for a definite integral.

Local maximum: A turning point at which the value of the function is higher than the values to either side of it. (This may not be the largest value of the function overall.)

Local minimum: A turning point at which the value of the function is lower than the values to either side of it. (This may not be the smallest value of the function overall.)

Logarithm: The power to which a base needs to be raised to produce a given value.

Magnitude: The 'size' or 'length' of a vector. (*See also* modulus.)

Mathematical model: A simplification of a real-world situation into mathematical equations.

Mid-interval value: For grouped data, the centre value of a group.

Modulus: The magnitude of a vector, usually denoted $|\mathbf{a}|$.

Mutually exclusive: Events than cannot occur at the same time.

Natural logarithm: The logarithm to base e.

Net force: The resultant force when more than one force acts on an object. (*See also* resultant force.)

Newton's third law: If object A exerts a force on object B, then object B exerts a force on object A, with the same magnitude but opposite direction.

Newton: The unit of magnitude of force.

Normal: A line perpendicular to a tangent at its point of contact with a curve.

Normal reaction force: The force exerted by a surface on any object in contact with that surface. It acts in the direction perpendicular to the surface and away from it.

Null hypothesis: In hypothesis testing, specifies the previous or assumed population parameter. Denoted by H_0.

One-tail test: A hypothesis test where the alternative hypothesis takes the form $p < \alpha$ or $p > \alpha$.

Opportunity sampling: Choosing respondents based upon their availability and convenience.

Outlier: A value that does not appear to fit the 'trend' of a data set.

Parabola: A curve that is the shape of a quadratic graph – $\cup$ or $\cap$ shaped.

Parallel (vector): A vector that is a scalar multiple of another vector.

Parameter (of a statistical distribution): A numerical characteristic of the population, such as its mean or variance.

Particle: An object that has mass but no size: it occupies a single point in space.

Particle model: A mathematical model which assumes that an object occupies a single point in space.

Percentiles: The data value a given percentage of the way through the data when it is put in order.

Period (of a function): A value p such that $\mathrm{f}(x+p) = \mathrm{f}(p)$ for all x.

Periodic: A function that repeats after a certain period.

Perpendicular bisector: A line through the mid-point of a line segment between two points and at right angles to it.

Polynomial: A function that is a sum of terms containing non-negative (positive or zero) integer powers of x.

Population: All the individuals of interest for collection of data.

Position vector: The vector connecting a point to a fixed origin.

Probability distribution: The list of all possible outcomes together with their probabilities.

Proof: A series of correct mathematical statements that follow one from the other to form a valid argument.

Proof by exhaustion: A method of proof where all possibilities are checked and shown to be true.

p-value: The probability of the observed or a more extreme sample value occurring if H_0 is correct.

Quota sampling: The proportion of members of the sample with certain characteristics is fixed to be the same as in the whole population; the individuals for each group are then selected by opportunity sampling.

Random variable: A variable whose value is subject to variations due to chance.

Range: A measure of spread. The difference between the largest and smallest values in a data set.

Rate of change: The rate at which one variable changes in relation to another variable.

Rationalising the denominator: The process of removing surds from a denominator.

Regression line: A line of best fit to bivariate data.

Rejection region: The set of values of the test statistic that provide sufficient evidence to reject the null hypothesis. (*See also* critical region.)

Representative: A sample that is a good representation of a population.

Resultant force: A single force producing the same effect as several forces acting together. (*See also* net force.)

Resultant vector: The result of the sum of two (or more) vectors.

Sample: A part of a population.

Scalar: A quantity that has magnitude but no direction (for example, mass).

Set notation: Notation used to describe sets.

Significance level: In a hypothesis test, the probability that is sufficiently small to be accepted as evidence against the null hypothesis.

Simple random sampling: A method of sampling where every sample (of a given size) has an equal chance of being selected.

Smooth: A surface for which the frictional force is so small it can be ignored.

Standard deviation: A measure of spread. This uses all the data in the data set.

Stationary point: A point on a curve where the tangent is horizontal and so the gradient is zero. (*See also* turning point.)

Stratified sampling: The proportion of members of the sample with certain characteristics is fixed to be the same as in the whole population; the individuals for each group are then selected randomly.

Substitution: The replacement of every occurrence of one variable in one equation by its expression from another equation.

Surd: A root that is an irrational number.

Systematic sampling: Participants are taken at regular intervals from a list of the population.

Tangent: A straight line that touches the curve at a point but does not intersect it again (near the given point).
The trigonometric function, $\tan(x)$.

Tension: The force exerted by a rope (rod, stick, etc.) when it is pulling an object.

Test statistic: In hypothesis testing, a value determined from the sample, which is then compared to what is expected under the null hypothesis.

Thrust: A pushing force.

Trigonometric functions: Sine, cosine and tangent.

Turning point: A local maximum or minimum point on a curve. (*See also* stationary point and vertex.)

Two-tail test: A hypothesis test where the alternative hypothesis takes the form $p \neq \alpha$.

Unit vector: A vector with magnitude one.

Union: The union of two or more sets or intervals is the collection of the contents of all those sets or intervals.

Variance: The square of the standard deviation.

Vector: A quantity that has both size (magnitude) and direction, for example, velocity.

Velocity: Speed in a certain direction.

Vertex: For a parabola, the maximum or minimum point. (*See also* turning point.)

Weight: The force with which the Earth attracts an object.

y-intercept: The point(s) where a line or curve crosses the y-axis.

Zero vector: A vector with zero magnitude.

Index

Acknowledgements

The authors and publishers acknowledge the following sources of copyright material and are grateful for the permissions granted. While every effort has been made, it has not always been possible to identify the sources of all the material used, or to trace all copyright holders. If any omissions are brought to our notice, we will be happy to include the appropriate acknowledgements on reprinting.

Thanks to the following for permission to reproduce images:

Cover image: Peter Medlicott Sola/ Getty Images

Back cover: Fabian Oefner
www.fabianoefner.com

thomas lieser /Getty Images; Huw Jones/Getty Images; Anindo Dey Photography/Getty Images; Sebastian Almes/EyeEm/Getty Images; monsitj/Getty Images; Brian T. Evans/Getty Images; Manuela Schewe-Behnisch/EyeEm/Getty Images; Damir Zorcic/EyeEm/Getty Images; Michael Betts/Getty Images; Tom Hoenig/Getty Images; Mina De La O/Getty Images; SCIEPRO/Getty Images; David Wells/EyeEm/Getty Images; Henrik Sorensen/Getty Images; Pavleta Miteva/EyeEm/Getty Images; srgktk/Getty Images; Rene Keller/Getty Images; Photo Antonio Larghi (Fotografia)/Getty Images; DEA PICTURE LIBRARY/Getty Images; Michelle Shinners/Getty Images; mehmettorlak/Getty Images; Diana David-Wells/EyeEm/Getty Images; Thai Yuan Lim/EyeEm/Getty Images; Leo Mason/Popperfoto/Getty Images.

AQA material is reproduced by permission of AQA.